AF520695

Free Radicals in Human Health and Disease

Vibha Rani • Umesh Chand Singh Yadav
Editors

Free Radicals in Human Health and Disease

Editors
Vibha Rani
Department of Biotechnology
Jaypee Institute of Information Technology
Noida, Uttar Pradesh, India

Umesh Chand Singh Yadav
School of Life Sciences
Central University of Gujarat
Gandhinagar, Gujarat, India

ISBN 978-81-322-2034-3 ISBN 978-81-322-2035-0 (eBook)
DOI 10.1007/978-81-322-2035-0
Springer New Delhi Heidelberg New York Dordrecht London

Library of Congress Control Number: 2014952303

Printed on acid-free paper

Springer is part of Springer Science+Business Media (www.springer.com)

Foreword

I am pleased to write a foreword to the book entitled *Free Radicals in Human Health and Diseases*, edited by Dr. Vibha Rani and Dr. Umesh C. S. Yadav and published by Springer. Both Dr. Rani and Dr. Yadav have been among the brightest students during their doctoral studies at the School of Life Sciences, Jawaharlal Nehru University, New Delhi, where I had the opportunity to teach, guide, and interact with them. The authors have an expertise in oxidative stress and inflammatory pathologies that is evident from their reputed publications. The book discusses comprehensively and succinctly about the generation of myriads of free radicals, oxidative stress-induced redox signaling pathways, various oxidative stress pathologies, and strategies to modulate a cell's antioxidant system to negate the effects of oxidative stress.

This book also has contributions from eminent scientists working in the field of oxidative stress biology who have provided a comprehensive and updated review of their respective topics. The expert contributors have elucidated the mechanisms of ROS production and their regulations, balance between oxidative stress and antioxidant system, oxidative stress-induced pathologies, and strategies to ameliorate them. The elaborated description in chapters will enhance the understanding of oxidative stress biology which will help the readers gain an in-depth insight and latest development in the field of oxidative research. The better understanding of regulation of oxidative stress can be utilized for devising strategies for development of novel therapeutic strategies for clinical intervention in oxidative stress-induced diseases. I am certain that the readers including students, researchers, and faculties will find this book extremely informative, interesting, and stimulating.

With best wishes

School of Life Sciences, JNU
New Delhi, India

R.K. Kale

Preface

This book entitled *Free Radicals in Human Health and Diseases* is our collaborative effort to combine approaches and ideas from different experts working on oxidative stress. The advent of oxidative hypothesis of pathogenesis led to a paradigm shift in how we understand the disease mechanism today. The oxidative stress caused by free radical generation in cells and its environment is implicated in a plethora of human diseases as well as in health and has become an area of intense research globally. Free radicals and reactive oxygen species (ROS), synthesized as metabolic by-products, are essential for a number of biochemical and physiological processes. Environmental stress drastically increases the levels of free radicals inside and around cells, thereby disturbing the equilibrium between free radical production and inbuilt antioxidant capabilities. When cellular production of ROS overwhelms its antioxidant capacity, it causes damage to cellular macromolecules such as lipids, protein, and DNA. ROS that are generated in all aerobic cells are in balance with biochemical antioxidants, and an imbalance due to excessive ROS may lead to degradation of antioxidant capacity of cells. Increased ROS can result in age-related diseases like Alzheimer and Parkinson disease, cancer and metastasis, cardiovascular diseases like atherosclerosis and cardiomyopathy, diabetic complications, and other inflammatory disorders like rheumatoid arthritis as well as reproductive deficiencies like polycystic ovary syndrome and loss of sperm motility. For scientists working in the field of oxidative stress, it is imperative to understand several aspects involved in the regulation and maintenance of oxidative balance. The defined methodology and techniques are required to estimate the dysregulation of this imbalance. Additionally, it is essential to understand the upstream as well as downstream regulators of this imbalance in a specific disorder. Therefore, an in-depth knowledge and understanding would help in designing new strategies to address specific questions of interest in one's research area.

This book is a consolidated effort of authors working in the field of oxidative stress to provide comprehensive and up-to-date information on it. The book contains a total of 26 chapters divided into 4 distinct parts for better understanding and interpretation. The first part covers the players of oxidative stress which include ROS, free radicals, and other submicroscopic particles. It also talks about the tools and techniques that can be used to measure oxidative stress and free radicals. The second part gives an in-depth mechanistic outlook of oxidative stress at both molecular and genetic levels. The third part focuses on various disease conditions that emanate from ensuing oxidative stress and due to loss of cellular oxidative balance. The last part discusses the strategies to ameliorate oxidative stress-induced diseases focusing on antioxidative therapies and current approaches in this field.

Although several books on the topic of oxidative stress are available, which is justified from the fact that ROS are implicated in the mediation of a number of pathogenesis, this book presents a comprehensive account of the redox biology including the topics like generation of free radicals, modification of biomolecules by ROS, ROS-induced signaling pathways and their regulations, role in disease development, role in molecular and genetic pathways leading to gene regulation and expression, and suggestions of strategies to prevent and treat oxidative stress pathologies based upon the current research and their future perspectives. This broad coverage of topics enhances the appeal of this book to its readers and thus is of extreme interest to the target audience such as researchers, scientists, pathologists, students, faculties, and the broad scientific community. To enhance and update the understanding of oxidative stress-related research, the authors have presented diverse theories to cover the advancement on this important subject. We hope that by combining these topics and presenting them as a book, we will help a wider section of scientific readers in understanding recent developments in oxidative stress biology, addressing their queries, enhancing inquisitiveness, and generating fruitful research ideas and tools to help direct and progress their own research.

We, as editors of the book, would like to extend our immense gratitude to all the contributors. We would like to especially acknowledge Prof. R. K. Kale, Founding Vice-Chancellor, Central University of Gujarat, and Professor, School of Life Sciences, JNU, New Delhi, for writing a foreword for our book. We would also like to acknowledge and thank our mentors Prof. Shyamal Goswami and Prof. Najma Z. Baquer, School of Life Sciences, JNU, New Delhi, India, and Prof. Satish K. Srivastava, Department of Biochemistry and Molecular Biology, University of Texas Medical Branch (UTMB), Galveston, Texas, USA, for their incessant encouragement and valuable training. In addition, we would like to acknowledge Neha Atale, Research Scholar at Jaypee Institute of Information Technology, Noida, and Shrey Kohli, Research Scholar at Otto von Guericke University (Medical Faculty) Magdeburg, Germany, for their sincere efforts at different stages in preparing the book. We also acknowledge the efforts, time, and patience of our colleagues, students, and family who helped in the various stages during the preparation of the book. We are grateful to the Department of Biotechnology (DBT) and Department of Science and Technology (DST),

Govt. of India. We are highly grateful to the publisher, Springer India (Pvt.) Ltd., for agreeing to publish our book, and to Dr. Mamta Kapila (Editor, Life Sciences, Springer India) for her patience and understanding during the preparation of this book.

Noida, UP, India Vibha Rani

Gandhinagar, Gujarat, India Umesh Chand Singh Yadav

Contents

Part III Oxidative Stress and Diseases

Part IV Strategies to Ameliorate Oxidative Stress Induced Diseases

About the Editors

Dr. Vibha Rani is working as an Assistant Professor in the Department of Biotechnology, Jaypee Institute of Information Technology, Noida, India. After completing her doctoral studies from School of Life Sciences, Jawaharlal Nehru University, New Delhi, she has continued her academic career where she teaches as well as guides doctorate students in the area of oxidative stress-induced cardiomyopathy and other development-related heart diseases. She is an academician as well as an accomplished young scientist and has been in research for 14 years. She has received extramural funding for her research work from the Department of Science and Technology (DST) and Department of Biotechnology (DBT), Govt. of India. Her research focus is to understand the mechanism of action of phytomolecules and also to develop an ideal drug molecule for the prevention of cardiac diseases. She has been successfully able to communicate her research findings to various international journals.

Dr. Umesh Chand Singh Yadav is currently working as an Assistant Professor at the School of Life Sciences, Central University of Gujarat, Gandhinagar. Dr. Yadav has a Ph.D. degree from School of Life Sciences, Jawaharlal Nehru University, New Delhi. He has over 9 years of research experience, 6 years as postdoctoral fellow, and more than 3 years as faculty, from the University of Texas Medical Branch (UTMB), Galveston, Texas, USA, where he performed excellent research in oxidative stress-induced inflammatory pathologies such as diabetic and cardiovascular complications, lung inflammatory diseases including asthma and chronic obstructive pulmonary disease (COPD), and ocular inflammatory diseases. He has published more than 30 research articles in high-impact journals including reviews and book chapters which indicate his expertise in the area of oxidative stress and inflammatory diseases and associated signaling pathways. Dr. Yadav has been awarded a prestigious Ramanujan Fellowship from the Department of Science and Technology (DST), New Delhi, and has recently returned to his native country, India, where he has established his laboratory.

Part I

Free Radical Generation

Production of Reactive Oxygen Species and Its Implication in Human Diseases

Shalini Mani

Abstract

Reactive oxygen species (ROS) are chemical molecules with one unpaired electron and mostly derived from molecular oxygen. It is produced in all the mammalian system by various exogenous and endogenous sources. Mitochondria are major sources of ROS production and they are produced as a respiratory by-product. The main sites of superoxide radical production in the respiratory chain are complexes III and I; however, other mitochondrial enzymes are also involved in the production of ROS. Because of the presence of one unpaired electron, ROS is highly reactive, and it may cause oxidative damage to the biomolecules and cell organelles and hence may affect the cellular physiology and their survivability.

A variety of diseases have been associated with excessive ROS production leading to mitochondrial damage, apoptosis, and necrosis. The interrelationship between ROS and mitochondria suggests shared pathogenic mechanisms in mitochondrial and ROS-related diseases. Some common diseases, known to be caused by ROS and mitochondrial damages, are several mitochondrial diseases, neurodegenerative diseases, and aging. In the present chapter, we have summarized the molecular mechanisms of ROS production, its damaging effect on cellular physiology, as well as the existing evidence of mitochondrial ROS involvement in human diseases.

Keywords

Reactive oxygen species • Superoxides • Mitochondria • Electron transport chain • Mitochondrial DNA • Permeability transition pore

S. Mani (✉)
Department of Biotechnology, Jaypee Institute of Information Technology,
A-10, Sector-62, Noida 201307, Uttar Pradesh, India
e-mail: shalini.mani@jiit.ac.in;
mani.shalini@gmail.com

V. Rani and U.C.S. Yadav (eds.), *Free Radicals in Human Health and Disease*,
DOI 10.1007/978-81-322-2035-0_1, © Springer India 2015

1 Introduction

During the course of evolution, organisms developed the capacity to utilize molecular oxygen as the terminal oxidant in respiration in order to gain energy as an advantage over the anaerobic pathway by a factor of 18. By utilizing the molecular oxygen as a terminal electron acceptor, these aerobic organisms were now enabling metabolism of their organic carbon for energy production. In this respiratory process, oxygen is consumed along with metabolic substrates, while ATP, water, and carbon dioxide are generated. However, the presence of intracellular oxygen also allowed inadvertent redox reactions by oxygen radicals to damage critical biomolecules. Based upon the work done by Priestley, Scheele, and Lavoisier, the dual role of oxygen as sustainer and destroyer of life was realized [1], but the exact mechanism was not known. During 1959, for the first time, Gerschman showed that oxygen toxicity is due to the generation of reactive oxygen species (ROS) or free radicals [1].

2 Production of ROS

ROS is a phase used to describe a variety of molecules and free radicals (chemical species with one unpaired electron) derived from molecular oxygen. The ground state oxygen may be converted to the much more reactive forms either by energy transfer or by electron transfer reactions. The former leads to the formation of singlet oxygen, whereas the latter results in the sequential reduction to superoxide, hydrogen peroxide, hydroxyl radical, etc. ROS is a collective term that includes not only oxygen radicals (superoxide and hydroxyl) but also some non-radical derivatives of molecular oxygen (O_2) such as hydrogen peroxide (H_2O_2). Thus, it is a broader expression and includes hydrogen peroxide and lipid peroxide with no unpaired electron, superoxide ($^{\bullet}O_2^-$), hydroxyl ($^{\bullet}OH^-$), peroxyl ($ROO^{\bullet -}$), alkoxy ($RO^{\bullet}$) radicals, radicals of nitric oxide ($^{\bullet}NO$), nitrogen dioxide ($^{\bullet}NO_2$), peroxynitrite ($^{\bullet}ONOO^-$), ozone (O_3), and possibly singlet oxygen. Though hydrogen peroxide and lipid peroxide are not free radicals, they act as reservoirs for the highly reactive $^{\bullet}OH^-$, $ROO^{\bullet}$, and $RO^{\bullet}$ radicals. The production of these reactive species occurs continuously in the organism, and depending upon the sources, this production may be endogenous or exogenous.

2.1 Exogenous Sources of ROS

Environmental agents including non-genotoxic carcinogens can directly generate or indirectly induce ROS in cells. Exposure of the cell to gamma irradiation results in the production of a whole range of radical and non-radical species from ionization of intracellular water (e.g., aqueous electron, $^{\bullet}OH^-$, H_2O_2). Even exposure to nonionizing irradiation such as UV-C (<290 nm), UV-B (290–320 nm), and UV-A (320–400 nm) can indirectly produce a variety of ROS including O_2, H_2O_2, and $^{\bullet}O_2^-$ radicals [2]. Air pollutants such as car exhaust, cigarette smoke, and industrial contaminants encompassing many types of NO derivatives constitute major sources of ROS that attack and damage the organism either by direct interaction with the skin or following inhalation into the lung. Many of the drugs, such as bleomycin and Adriamycin, whose mechanism of action is mediated via ROS production, are also a major source of ROS. Narcotic drugs and anesthetizing gases are further considered major contributors to the production of ROS [3]. A large variety of xenobiotics (e.g., toxins, pesticides, and herbicides such as paraquat) and chemicals (e.g., mustard gas, alcohol) produce ROS as a by-product of their metabolism in vivo [3]. One of the major exogenous sources of oxidants is food. A large portion of the food consumed is oxidized to a higher degree and contains different kinds of oxidants such as peroxides, aldehydes, oxidized fatty acids, and transition metals. Food debris that reaches the intestinal tract places an enormous oxidative pressure on the intestinal tract mucosa [3].

2.2 Endogenous ROS Production in Cell Organelles

Although the exposure of the organism to ROS is extremely high from exogenous sources, the exposure to endogenous sources is much more important and extensive, because it is a continuous process during the life-span of every cell in the organism. Among the very varied endogenous sources, mitochondria, endoplasmic reticulum (ER), and peroxisomes are important cellular organelles which are involved in the ROS production.

2.2.1 ROS Production in Mitochondria

The reduction of oxygen to water in the mitochondria for ATP production occurs through the donation of four electrons to oxygen to produce water. Mitochondrial electron transport chain (ETC) reduces 95 % of O_2 by tetravalent reduction to H_2O without any free radical intermediates [4, 5]. However, the remaining 5 % of oxygen is reduced via the univalent pathway in which free radicals are produced. Mitochondria from different tissues may vary conspicuously in their capacity to produce ROS using different substrates, and this capacity may be related to membrane composition, animal species, and age.

During the process of ROS production, several major oxygen derivatives are formed, and considerable quantities of superoxide and hydrogen peroxide (H_2O_2) are formed. The mitochondrial electron transport chain is a multicomponent system involved in a series of oxidation-reduction reactions between redox couples and pairs, transfer of electrons from a suitable donor (reductant) to a suitable electron acceptor (oxidant) (Fig. 1) [6]. These oxidation-reduction reactions involve either the transport

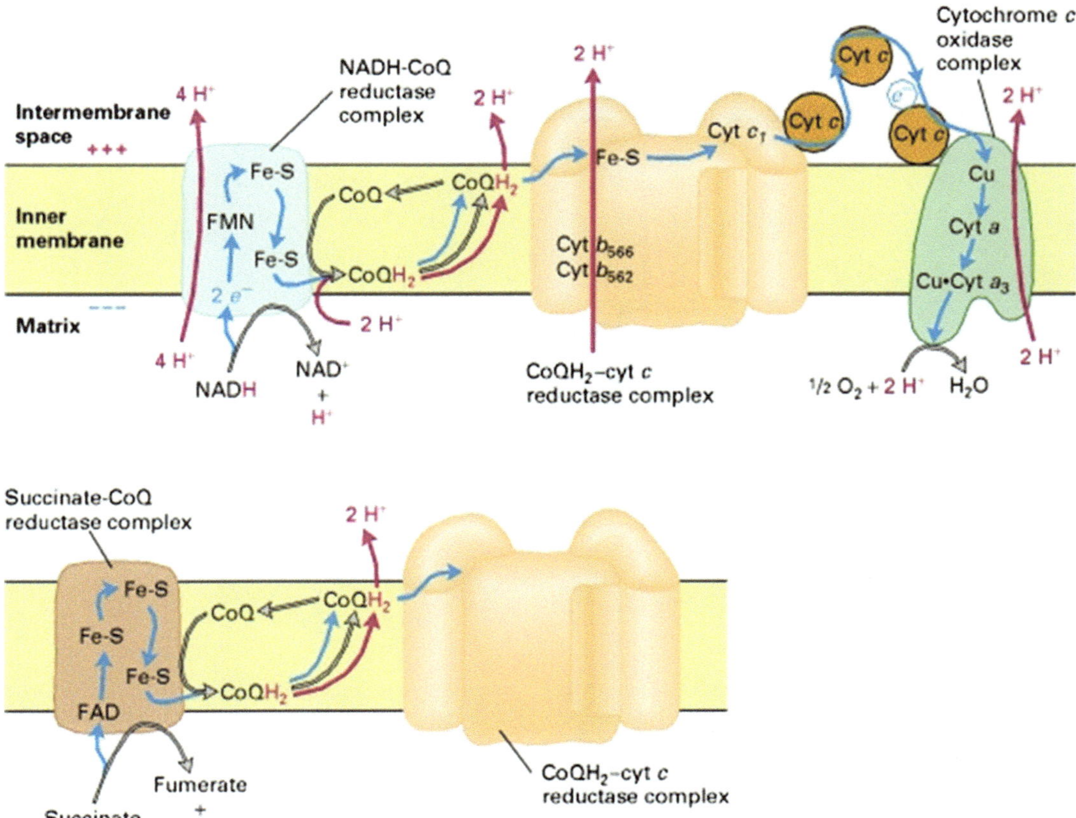

Fig. 1 Stepwise flow of electrons through the electron transport chain from NADH, succinate, and $FADH_2$ to O_2 [6]

of electrons only, as in the case of the cytochromes, or electrons and protons together, as occurs between NADH and FAD. The part of the ETC that actually uses O_2 is the terminal oxidase enzyme, cytochrome oxidase. Cytochrome oxidase releases no detectable oxygen radicals into free solution. However, during the transfer of electrons through earlier components of the transport chain, a few electrons do leak out directly on to O_2, resulting in the generation of $^{\bullet}O_2^-$ [6–8].

The major sites of ROS formation in the respiratory chain lay within respiratory complexes I and III, with a general consensus that production at complex I is about half of that at complex III. The roles of the various complexes that constitute the ETC along with their contribution to the generation of ROS are discussed as follows (Fig. 2) [9]:

2.2.1.1 Complex I

Complex I also known as the NADH-coenzyme Q reductase or NADH dehydrogenase. It is huge, 850,000 kD, and is composed of more than 30 subunits. It contains an FMN prosthetic group and seven or more Fe-S clusters. This complex binds NADH and transfers two electrons in the form of a hydride to FMN to produce NAD^+ and FMNH2, which involves the transfer of electrons one at a time to a series of iron-sulfur complexes. It is the major site for ROS production. It has been proposed that the participation of the reduced flavin mononucleotide in the active site for NADH oxidation and this mechanism are supported in mitochondria by correlations between the NAD(P) + potential and O_2 reduction. Two possible sites of oxygen reduction in complex I are either the flavin moiety or the quinone-binding site [10].

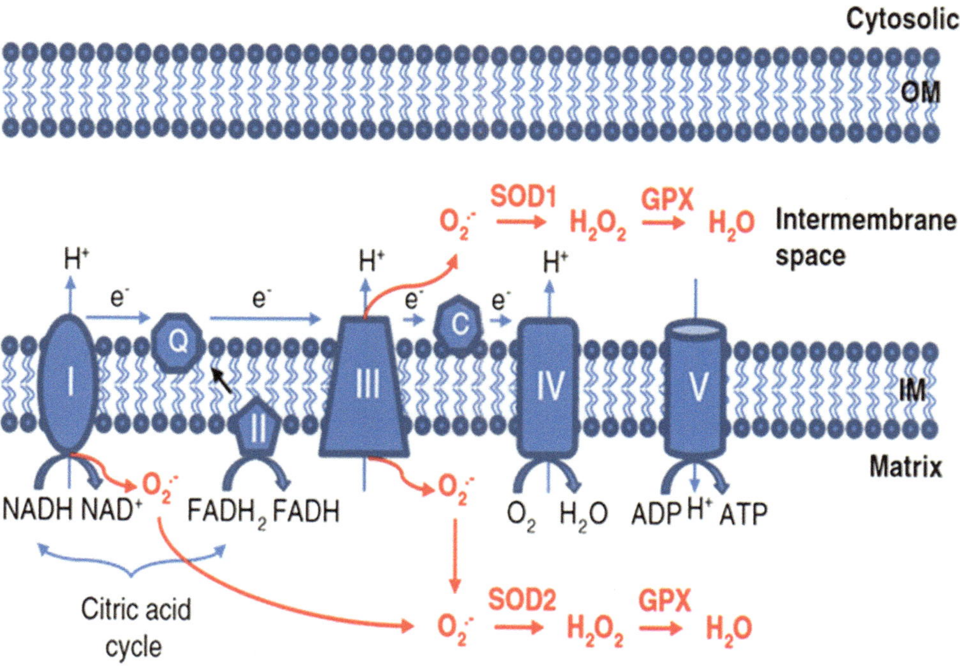

Fig. 2 Production and disposal of mtROS. Electrons (e^-) donated from NADH and $FADH_2$ pass through the electron transport chain and ultimately reduce O_2 to form H_2O at complex IV. MtROS are produced from the leakage of e^- to form superoxide (O_2^-) at complex I and complex III. O_2^- is produced within matrix at complex I, whereas at complex III O_2^- is released toward both the matrix and the intermembrane space. Once generated O_2^- is dismutated to H_2O_2 by superoxide dismutase 1 (SOD1) in the intermembrane space and by SOD2 in the matrix. Afterward H_2O_2 is fully reduced to water by glutathione peroxidase (GPX). Both O_2^- and H_2O_2 produced in the process are considered as empty ROS. *OM* outer membrane, *IM* inner membrane (Reproduced with permission from [9])

Fast rates of H_2O_2 generation have been observed in succinate-supported respiration, the mechanism involving reverse electron flow from succinate to NAD^+, providing reducing equivalents to a redox carrier of complex I that serves as a site of ROS formation [11–15]. The process is inhibited by rotenone which blocks the electron flow between the ubiquinone and the iron-sulfur center N-2 [4, 16, 17].

In addition, studies of intact mitochondria or submitochondrial particles have suggested the role of quinone-binding site in ROS production [10]. This site reduces several quinones to unstable semiquinone forms, which further reduced the oxygen to superoxide, whereas water-soluble CoQ homologues used as electron acceptors from isolated complex I stimulate H_2O_2 generation [18]. The evidence that inhibitors binding to the three quinone-binding sites of the complex stimulate superoxide production proposes that the oxygen reduction site lies upstream of the quinone-binding sites [18, 19].

2.2.1.2 Complex II

It is also known as succinate dehydrogenase. It is composed of four subunits. Two of which are iron-sulfur proteins, and the other two subunits together bind FAD through a covalent link to a histidine residue. These two subunits are called flavoprotein 2 or FP2. Complex II contains three Fe-S centers. In the first step of this complex, succinate is bound and a hydride is transferred to FAD to generate FADH2 and fumarate. FADH2 then transfers its electrons one at a time to the Fe-S centers. FAD functions as a two-electron acceptor and a one-electron donor. The final step of this complex is the transfer of two electrons one at a time to coenzyme Q to produce CoQH2.

ROS is generated in the reverse reaction, with electrons being supplied from the reduced ubiquinone pool. The contribution of complex II in generation of ROS is relatively lower than complex I [20].

Complex II as a source of ROS was demonstrated when electrons are channeled through complex II rather than complex I, while in both cases, reaching complex III yielded higher ROS levels [21]. Carboxin inhibits ROS production stimulated by antimycin in complex II and cause destabilization of stable semiquinone SQs of complex II [18]. A related inhibition of ROS generation was observed by carboxin in COS-7 cells respiring on glucose [22] showing that complex II may be a primary source of ROS in intact cells [18]. Auto-oxidation of flavin is the source of superoxide generation stimulated by one-electron acceptor cytochrome c, and ferricyanide [23] was shown to exhibit the direct role of complex II in ROS generation in purified succinate dehydrogenase (SDH) [18].

2.2.1.3 Complex III

This complex is also known as coenzyme Q-cytochrome c reductase because it passes the electrons from CoQH2 to cyt c through a very unique electron transport pathway called the Q cycle. In complex III, there are two b-type cytochromes and one c-type cytochrome, and along with complex I, it is also a major site of ROS generation [24].

The first evidence that complex III generates superoxide was shown by antimycin insensitive reduction of cytochrome c mediated by superoxide radicals [25]. The source of superoxide in the enzyme has been either assumed to be cytochrome b566 or ubisemiquinone (SQ) or Rieske iron-sulfur center [26]. The coenzyme Q is fully reduced in the inner side of the mitochondrial membrane (ubiquinol, QH2) and then migrates to the outer side of the inner membrane carrying 2 H^+ that become part of the pool needed to sustain ADP phosphorylation. Once on the outer side of the membrane, one electron is transferred to cytochrome c1 (via the Rieske Fe-S protein), resulting in the formation of $Q^{\bullet}$. The second electron is needed to reduce cytochrome b, but eventually some electrons leak to oxygen, producing $O_2^{-\bullet}$ [27]. SQ at the center o is assumed to be a major candidate in univalent oxygen reduction: CoQ may be transformed by a safe electron carrier to a superoxide generator when protons are allowed to penetrate the inner membrane of mitochondria [11].

2.2.2 ROS Production in Endoplasmic Reticulum

Similar to mitochondria, ER is another membrane-bound intracellular organelle, but unlike mitochondria, it is primarily involved in lipid and protein biosynthesis. ER when under stress produces ROS mainly by two mechanisms during disulfide bond formation [28]. First, ROS are produced as a by-product during transfer of electrons from protein thiol to molecular oxygen by endoplasmic reticulum oxidoreductin-1 (ERO-1) and protein disulfide-isomerase (PDI) [28]. Alternatively, ROS can be created during misfolding of protein due to depletion of GSH [29, 30], since after GSH is consumed, thiols are repaired enabling them to interact with ERO-1/PDI and to be re-oxidized [28]. These steps generate consecutive cycles of disulfide bond formation and breakage, with each cycle producing more ROS as a by-product [31]. The second mechanism presumes ROS are generated by unfolded proteins, independent of the formation of disulfide bonds [28]. Accordingly, accumulation of unfolded proteins in the ER elicits Ca^{2+} leakage into the cytosol, increasing ROS production in the mitochondria [32]. The mechanism of ROS production in ER is explained in Fig. 3 [28].

2.2.3 ROS Production in Peroxisomes

Peroxisomes participate in fatty acid oxidation and contain peroxide-producing enzymes. Peroxisomes are an important source of total cellular H_2O_2 production. Peroxisomes in mammals play an important role in a variety of metabolic pathways such as fatty acid α- and β-oxidation, ether phospholipid biosynthesis, glyoxylate metabolism, amino acid catabolism, polyamine oxidation, and oxidative part of the pentose phosphate pathway [33]. Peroxisomes contain a variety of enzymes that generate H_2O_2 as part of their normal catalytic cycle. These enzymes, which are essentially flavoproteins, include acyl-CoA oxidases, urate oxidase, D-amino acid oxidase, D-aspartate oxidase, L-pipecolic acid oxidase,

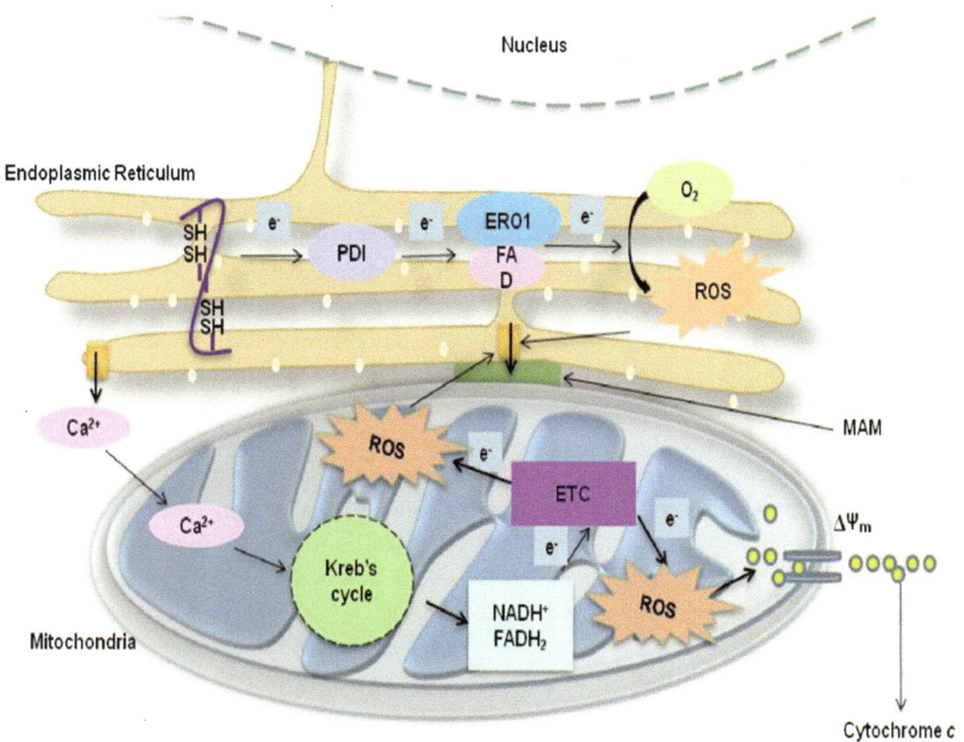

Fig. 3 ER and mitochondrial associated reactive oxygen species (ROS) production under ER stress (Reproduced with permission from [28])

L-α-hydroxy acid oxidase, polyamine oxidase, and xanthine oxidase [34]. As peroxisomes contain a large number of ROS-producing enzymes, hence using all the abovementioned metabolic pathways, different types of ROS such as hydrogen peroxide, superoxide, nitric oxide radicals, hydroxyl radical, and peroxynitrites are produced. Catalase is also a peroxisomal enzyme which metabolizes the hydrogen peroxide formed in these organelles. Peroxisomal catalase utilizes H_2O_2 produced by these oxidases to oxidize a variety of other substrates by "peroxidative" reactions. These types of oxidative reactions are particularly important in liver and kidney cells in which peroxisomes detoxify a variety of toxic molecules (including ethanol) that enter the circulation.

3 ROS-Induced Cellular Damage

These free radicals and other activated oxygen species are continuously formed in our body and important for cellular physiology at its low concentration (will be discussed in the next chapter). However, on top of their physiological function, they may also be damaging to the cellular integrity due to its high reactivity, at its high concentration. Due to the presence of an unpaired electron ($^{\bullet}O_2^-$ and $^{\bullet}OH^-$) or the ability to extract electrons from other molecules (H_2O_2, $HOCl^-$), ROS can readily react with all types of biomolecules and damage the cellular structure. Among the biological targets, most vulnerable to oxidative damage, are proteinaceous enzymes, lipidic membranes, and DNA. By targeting these biomolecules, ROS may directly harm its main site of production (mitochondria) and influence the cellular viability, via mitochondria dependent or independent pathways.

3.1 Effect on Mitochondria

The mitochondrion is the major site for ROS generation. Thus, the first site of damage by the ROS is the various biomolecules (like proteins, lipid, and nucleic acids) present within the mitochondrion itself. It has been demonstrated that after exposing cells to oxidative stress, mitochondrial DNA (mtDNA) damage is more extensive and persists longer than damage in nuclear DNA (nDNA). Several reasons may contribute to this selective vulnerability: (i) mtDNA lacks histones which are protective against free radical damage; (ii) mtDNA lacks an adequate repair system, rendering it unable to cope with the extensive damage, especially strand breaks; (iii) mtDNA has very few noncoding sequences, therefore increasing the likelihood of a DNA alteration to affect a gene; and (iv) mtDNA is located near the inner mitochondrial membrane, a major site of oxygen radical production [35]. The highly reactive ROS induces damage to mtDNA which includes single- and double-strand breaks, abasic sites, and purine and pyrimidine base damage. The mtDNA codes primarily for the proteins that form the complexes of the electron transport chain. Accumulation of mutations in the mtDNA as a result of oxidative damage results in the production of proteins that are less efficient. This dysfunction of the OXPHOS system may eventually lead to a hindrance in the ATP generation capacity and decreased ATP synthesis of the cell as well as increased leakage of ROS [36].

Excessive ROS production in the mitochondria also results in the damage to the lipids. ROS formation triggers lipid peroxidation which adversely affects the OXPHOS system as well as the mitochondrial membrane potential. The products of lipid peroxidation will interact with the lipids present in the mitochondrial bilayer, impairing its function and resulting in the opening of the mitochondrial permeability transition pore (MPTP). The opening of the MPTP leads to dysregulation of calcium homeostasis and may affect the entire cellular metabolism [37]. The susceptibility of mitochondrial function to oxidative damage may further lead to the induction of various reactions affecting the cellular viability in the affected cell, as explained in the next sections.

3.2 Effect on Cellular Viability

When a cell is exposed to oxidative stress, it may result in the activation of various signaling pathways that lead to the elimination of the damaged cell. These mechanisms include apoptosis, necrosis, and autophagy among others [38]. The entire mechanism is summarized in Fig. 4.

3.2.1 Apoptosis

Apoptosis is a form of programmed cell death characterized by several morphological changes in the dying cell like nuclear condensation and fragmentation, blebbing of the cell membrane, and swelling. It is controlled by extrinsic (receptor-mediated) as well as intrinsic (mitochondria-mediated) signaling pathways. Mitochondrion, which is the site of ROS generation, is an important regulator of cell survival as well as cell death. ROS plays an important role in the activation of the signaling pathways that ultimately lead to apoptosis [39]. The MPTP is a major player in apoptosis as well as necrosis as its induction under various conditions triggers cell death via either of the mechanisms. Increase in the intracellular levels of ROS triggers the opening of the MPTP and a subsequent decrease in the mitochondrial membrane potential. The opening of the MPTP leads to the release of mitochondrial intermembrane proteins like cytochrome c from the mitochondria and into the cytosol [39]. The release of the cytochrome c into the cytosol results in the

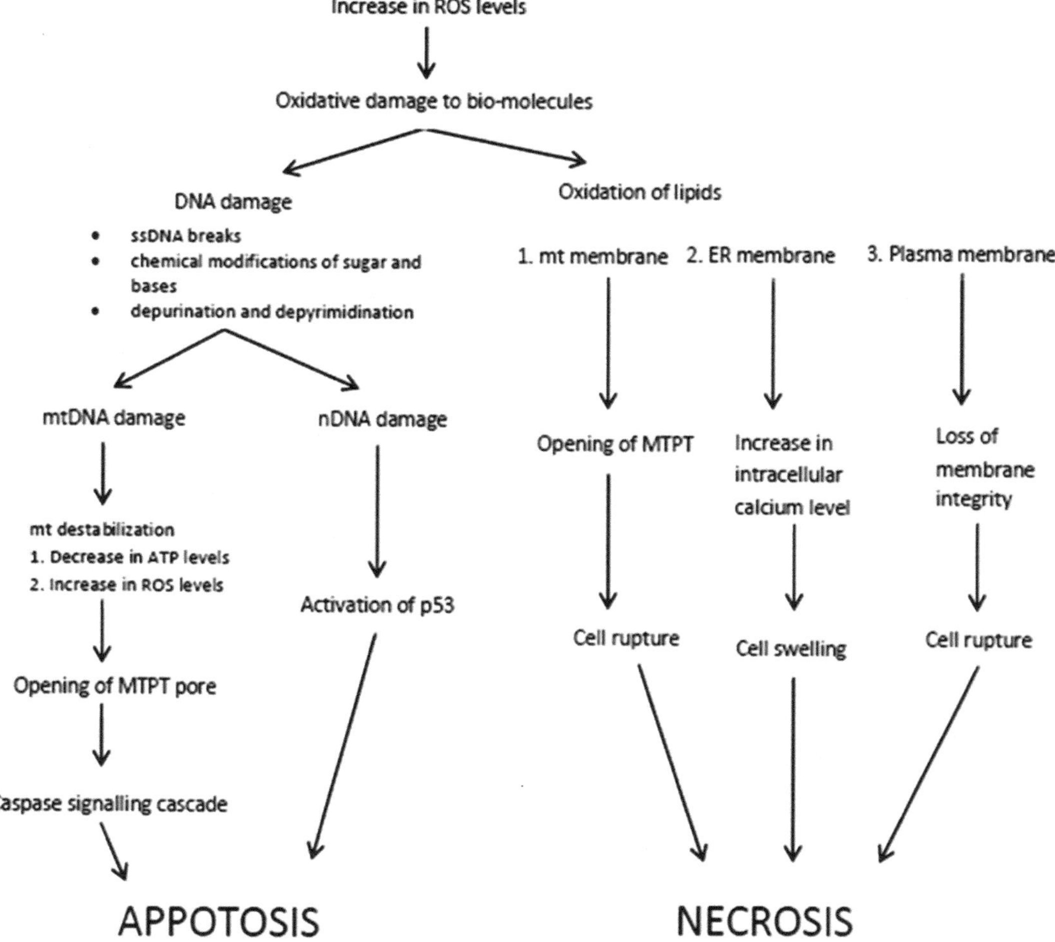

Fig. 4 Detailed mechanism of ROS-mediated apoptosis and necrosis

formation of the apoptosome complex by its interaction with apoptotic protease-activating factor 1 (Apaf-1). In this caspase-dependent signaling pathway, the apoptosome complex recruits procaspase 9 which induces the activation of the downstream effectors, caspase 3 and caspase 7, leading to apoptosis [40]. Oxidative stress also affects the mitochondrial regulation of calcium homeostasis. Rise in calcium levels also results in the opening of the MPTP, thus triggering apoptosis [32]. The intrinsic pathway is mediated by death receptors that stimulate the activation of initiator caspases which lead to apoptosis (41). The majority of the death receptors belong to the tumor necrosis factor-α (TNF-α) family of death receptors (like Fas, TRAIL, TNFR1, etc.). ROS serve as an important signaling molecule in the caspase-signaling pathway initiated by TNF-α. The ligation of TNF-α to its receptor, tumor necrosis factor receptor 1(TNF-α-TNFR1 pathway), is of particular interest as it activates the c-Jun N-terminal kinase (JNK) pathway. The JNK pathway has diverse functions including cellular differentiation, proliferation, and apoptosis [41]. The levels of the ROS in the cell as well as the duration of JNK activation will determine whether the pathway would lead to cell survival or cell apoptosis. The sustained JNK activation by high ROS levels leads to receptor-mediated cell death [40, 41]. Increased in the level of mitochondrial ROS may also cause damage to nuclear DNA, and p53 is capable of sensing the DNA damage and may further direct the cell toward apoptosis [42].

3.2.2 Necrosis

Necrosis, unlike apoptosis, is an unregulated process of cell death and is morphologically characterized by a gain in cell volume, swelling of organelles, and plasma membrane rupture, which results in the loss of intracellular contents [38].

Necrosis can occur in response to varied physiological conditions like ischemia, hypoxia, irradiation, pathogen attack, oxidative stress, etc. ROS is an important mediator of cellular necrosis. Excessive production of ROS can contribute to necrotic cell death by causing degradation of biomolecules and thus cause damage to intracellular organelles [43]. ROS initiates damage to lipids by targeting the double bonds present in the polyunsaturated fatty acids which have increased propensity for oxidative damage. The lipid oxidation of these fatty acids which make up the plasma membrane as well as organelle membranes results in the loss of the integrity of these membranes. Plasma membrane degradation can lead to the loss and failure of ion channels (such as ATP-dependent sodium-potassium pumps, calcium pumps, etc.) which are important for maintaining critical ion balance of the cell. The loss of the membrane potential across the mitochondrial inner membrane leads to the rupture of plasma membrane and causes the opening of the MPTP. It further causes the loss of the proton gradient and a shutdown of the OXPHOS [44]. Damage to the membranes of ER, which is a store of calcium ions, leads to an increase in intracellular calcium concentration as a result of ER leakage which acts as a signal for cell necrosis. Calcium ions may activate various calcium-dependent proteases like calpains and cathepsin thus triggering necrosis because of damage to intracellular proteins [44].

3.2.3 Autophagic Cell Death

Autophagy (self-eating) is a multistep process that is characterized by the vesicular sequestration and degradation of long-lived cytoplasmic proteins and organelles, for example, mitochondria [45]. The resulting double-membrane vesicle is termed an autophagosome. Autophagy is typically observed in cells that are exposed to a variety of metabolic and therapeutic stresses, including growth factor deprivation, inhibition of the receptor tyrosine kinase/Akt/mammalian target of rapamycin (mTOR) signaling, shortage of nutrients, ischemia/reperfusion, inhibition of proteasomal degradation, accumulation of intracellular calcium, and ER stress [46, 47]. ROS may provide a common link between cellular stress signals and the initiation of autophagy, as ROS accumulation has been reported to result in inactivation of the cysteine protease ATG4, which in turn causes accumulation of the ATG8-phosphoethanolamine precursor that is required for the initiation of autophagosome formation [48].

4 Mitochondrial Dysfunction and ROS in Human Diseases

Mitochondrial diseases are the diseases which are caused by inherited or spontaneous mutations in mtDNA or nDNA which lead to altered functions of the proteins or RNA molecules of mitochondria. As mitochondria perform so many different functions in different tissues, there are different mitochondrial diseases due to dysfunction of the mitochondria, which can be broadly categorized into mitochondrial diseases, neurodegenerative disorders, and aging. The detail of ROS and human diseases will be discussed in the later section of this book. However, a brief overview is mentioned here.

4.1 Mitochondrial Diseases

Several clinical syndromes are associated with mtDNA mutations, the most common being NARP (neurogenic muscle weakness, ataxia, and retinitis pigmentosa), MELAS (mitochondrial encephalomyopathy, lactic acidosis, and stroke-like episodes), MERRF (myoclonic epilepsy and ragged-red fibers), LHON (Leber hereditary optic neuropathy), KSS (Kearns-Sayre syndrome), ophthalmoplegia, ataxia, retinitis pigmentosa, cardiac conduction defect, and elevated cerebrospinal fluid protein. mtDNA encodes few proteins which are involved in the electron transport chain, the main source of ROS in cells. This special situation highlights the theory of the "vicious cycle," a theory attractive within the realm of degenerative processes [49]. In which, random primary mitochondrial mutations initially induce a defect in the respiratory chain that leads to the leakage of ROS from the ETC. Later, ROS may trigger accumulation of secondary mtDNA mutations which intensify the mitochondrial respiratory defects and increasing production of ROS from mitochondria [49].

Several studies have demonstrated that mtDNA mutations associated with human disease lead to defects like ETC complex dysfunction, increased ROS production, and oxidative damage, as is the situation in MELAS, where hydroxyl radical damage to mtDNA can be accelerated by a specific mitochondrial genotype associated with the disease [50]. Many clinical syndromes including fatal infantile lactic acidosis, adult onset exercise intolerance, focal dystonia, LHON, cardiomyopathy with cataracts, hepatopathy with tubulopathy, Leigh's disease, cataracts and developmental delay, and lactic academia have been associated with isolated complex I deficiencies [51].

4.2 Mitochondrial Dysfunction in Neurodegenerative Diseases

Increased level of ROS in neurodegenerative processes may affect normal mitochondrial parameters like ATP production, mitochondrial membrane potential, MPTP activation, and calcium uptake. Overproduction of free radicals leads to neurodegeneration. Neural cells suffer functional or sensory loss in neurodegenerative diseases. Apart from several other environmental or genetic factors, oxidative stress leading to free radical attack on neural cells contributes grievous role to neurodegeneration. These changes can lead to neuronal death, mainly through excitotoxic pathways, involving oxidation of macromolecules and apoptosis, thus affecting pathogenesis of common neurodegenerative diseases such as Parkinson disease (PD), Alzheimer's disease (AD), amyotrophic lateral sclerosis (ALS), Huntington's disease (HD), and Friedreich's ataxia (FA) [52]. Subsequently, enzymatic deficiencies in the electron transport chain were identified in additional neurodegenerative diseases: complex IV deficiency in AD and ALS [53–55] and complex II and III in HD and FA. In HD and FA, the genetic defects appear to be in nuclear genes that encode nonrespiratory proteins (huntingtin for HD and frataxin for FA) [52]. This fact suggests that the observed respiratory deficiencies are secondary to these primary pathogenic factors.

4.3 Mitochondrial Dysfunction in Aging

Mitochondrial ROS involvement in aging is strongly suggested by experimental data [56]. As mt DNA is naked and hence it is constantly exposed to ROS generated by mitochondrial electron transport chain. As a result, large number of mutations in mtDNA may exponentially get accumulated with age. The simultaneous increase in lipid peroxidation and oxidation of mitochondrial proteins adds to the oxidative stress effects, initiating the vicious cycle of molecular degeneration. This putative vicious cycle can operate at different rates in various tissues, leading to differential accumulation of oxidative damage, which could explain the differences in functional impairment and deterioration of different tissues in the aging process. There is a substantial evidence that damage to mtDNA accumulates with age. Finally, several mtDNA point mutations also increase with normal aging [57]. The point mutation of mitochondrial DNA characteristic for MERRF disease is found also in healthy people of different ages [58]. But still it is not clear whether such mutations are generated by ROS-mediated damage. Apart from these common clinical features, ROS is also known to participate in many pathological conditions including cardiovascular diseases, malignancies, autoimmune diseases, and neurological degenerative diseases.

5 Conclusion

ROS is produced in all the cells as default process, and mitochondria plays a central role in ROS generation. The excess amounts of ROS are known to trigger the oxidative damage to various biomolecules of the cell like proteins, lipids, and nucleic acids that result in the degradation and damage of cellular organelles. The excessive accumulation of ROS may trigger the process of cell death via apoptosis, necrosis, and autophagy. The side effect of ROS on cellular physiology, morphology, and viability may further get affected in the form of several pathological conditions.

More than 200 disorders have been described in literature in which ROS were important for the initiation state of a disease or produced during its course.

Our understanding of the intricate relationship between mitochondrial function, ROS production, ROS damage, and the development of a clinical phenotype is still very limited. Mitochondria are cellular organelles that perform vital functions essential for ATP production, homeostasis, and metabolism. They are integral to different cell death and survival pathways. These roles identify mitochondria as a potential target for drugs to treat metabolic and hyperproliferative diseases.

Therapies that target the generation of ROS in the mitochondria can help in the treatment of various diseases associated with oxidative damage. Hence, there is a need to do extensive research in this area, in such a way that we can decrease the ROS level in the mitochondria and thus improving or managing the health of the diseased person.

Acknowledgment Author wishes to acknowledge Nancy Taneja, Arshpreet Kalsi, and Alok Tandon for their contribution in editing the manuscript.

References

1. Gerschman R (1959) Oxygen effects in biological systems. Sym Spec Lect XXI Int Congr Physiol Soc 21:222–226
2. Zhang X, Rosenstein BS, Wang Y et al (1997) Identification of possible reactive oxygen species involved in ultraviolet radiation-induced oxidative DNA damage. Free Radic Biol Med 23:980–985
3. Halliwell B (1991) Reactive oxygen species in living systems: source, biochemistry, and role in human disease. Am J Med 91:14S–22S
4. Chance B, Sies H, Boveris A (1979) Hydroperoxide metabolism in mammalian organs. Physiol Rev 59:527–605
5. Shigenaga MK, Hagen TM, Ames BN (1994) Oxidative damage and mitochondrial decay in aging. Proc Natl Acad Sci U S A 91:10771–10778
6. Lodish H, Berk A, Zipursky SL et al (2000) Molecular cell biology, 4th edn. W. H. Freeman, New York, Section 16.2, Electron transport and oxidative phosphorylation
7. Nohl H, Hegner D (1978) Do mitochondria produce oxygen radicals in vivo? Eur J Biochem 82:563–567

8. Turrens JF, Boveris A (1980) Generation of superoxide anion by the NADH dehydrogenase of bovine heart mitochondria. Biochem J 191:421–427
9. Li X, Fang P, Mai J et al (2013) Targeting mitochondrial reactive oxygen species as novel therapy for inflammatory diseases and cancers. J Hematol Oncol 6:1–19
10. Hirst J, King M, Pryde K (2008) The production of reactive oxygen species by complex I. Biochem Soc Trans 36:976–980
11. Cino M, Del Maestro RF (1989) Generation of hydrogen peroxide by brain mitochondria: the effect of reoxygenation following post decapitative ischemia. Arch Biochem Biophys 269:623–638
12. Hansford RG, Hogue BA, Mildaziene V (1997) Dependence of H_2O_2 formation by rat heart mitochondria on substrate availability and donor age. J Bioenerg Biomembr 29:89–95
13. Korshunov SS, Skulachev VP, Starkov AA (1997) High protonic potential actuates a mechanism of production of reactive oxygen species in mitochondria. FEBS Lett 416:15–18
14. Korshunov SS, Korkina OV, Ruuge EK et al (1998) Fatty acids as natural uncouplers preventing generation of O2d– and H_2O_2 by mitochondria in the resting state. FEBS Lett 435:215–218
15. Kwong LK, Sohal RS (1998) Substrate and site specificity of hydrogen peroxide generation in mouse mitochondria. Arch Biochem Biophys 350:118–126
16. Turrens JF, Boveris A (1980) Generation of superoxide anion by the NADH dehydrogenase of bovine heart mitochondria. Biochem J 19:421–427
17. Liu Y, Fiskum G, Schubert D (2002) Generation of reactive oxygen species by the mitochondrial electron transport chain. J Neurochem 80:780–787
18. Lenaz G (2001) The mitochondrial production of reactive oxygen species: mechanisms and implications in human pathology. IUBMB Life 52:159–164
19. Hinkle P, Butow RA, Racker E et al (1967) Partial resolution of the enzymes catalyzing oxidative phosphorylation. XV. Reverse electron transfer in the flavin-cytochrome beta region of the respiratory chain of beef heart submitochondrial particles. J Biol Chem 242:5169–5173
20. Quinlan C, Orr A, Perevoshchikova I (2012) Mitochondrial complex II can generate reactive oxygen species at high rates in both the forward and reverse reactions. J Biol Chem 287:27255–27264
21. Cortopassi G, Wang E (1995) Modelling the effects of age-related mtDNA mutation accumulation: complex I deficiency, superoxide and cell death. Biochim Biophys Acta 1271:171–176
22. McLennan HR, Degli Esposti M (2000) The contribution of mitochondrial respiratory complexes to the production of reactive oxygen species. J Bioenerg Biomembr 32:153–162
23. Zhang L, Yu L, Yu CA (1998) Generation of superoxide anion by succinate cytochrome c reductase from bovine heart mitochondria. J Biol Chem 273:33972–33976
24. Chen Q, Vazquez E, Moghaddas S et al (2003) The reverse reaction, with electrons supplied from the reduced ubiquinone pool. J Biol Chem 278:36027–36031
25. Loschen G, Azzi A, Flohe L (1973) Mitochondrial H_2O_2 formation: relationship with energy conservation. FEBS Lett 33:84–87
26. Nohl H, Gille L, Scho¨nheit K et al (1996) Conditions allowing redox-cycling ubisemiquinone in mitochondria to establish a direct redox couple with molecular oxygen. Free Radic Biol Med 20:207–213
27. Turrens JF (2003) Mitochondrial production of reactive oxygen species. J Physiol 2:335–344
28. Bhandary B, Marahatta A, Kim HR et al (2003) An involvement of oxidative stress in endoplasmic reticulum stress and its associated disease. Int J Mol Sci 14:434–456
29. Santos CX, Tanaka LY, Wosniak J et al (2009) Mechanisms and implication of reactive oxygen species generation during the unfolded protein response: roles of endoplasmic reticulum oxidoreductase, mitochondrial electron transport, and NADPH oxidase. Antioxid Redox Signal 11:2409–2427
30. Tu BP, Weismann JS (2002) The FAD and O(2) dependent reaction cycle of Erol-mediated oxidative protein folding in the endoplasmic reticulum. Mol Cell 10:983–994
31. Higa A, Chevet E (2012) Redox signalling loops in the unfolded protein response. Cell Signal 24:1548–1555
32. Malhotra JD, Kaufman RJ (2007) Endoplasmic reticulum stress and oxidative stress: a vicious cycle or a double edged sword? Antioxid Redox Signal 9:2277–2293
33. Angermüller S, Bruder G, Völkl A et al (1987) Localization of xanthine oxidase in crystalline cores of peroxisomes. A cytochemical and biochemical study. Eur J Cell Biol 45:137–144
34. Fransen M, Nordgren M, Wang B et al (2012) Role of peroxisomes in ROS/RNS- metabolism: implications for human disease. Biochim Biophys Acta 1822:1363–1373
35. Zeviani M, Antozzi C (1997) Mitochondrial disorders. Mol Hum Reprod 3:133–148
36. Gutierrez J, Ballinger S, Darley-Usmar V et al (2006) Free radicals, mitochondria, and oxidized lipids: the emerging role in signal transduction in vascular cells. Circ Res 99:924–932
37. Ott M, Gogvadze V, Orrenius S et al (2007) Mitochondria, oxidative stress and cell death. Apoptosis 12:913–922
38. Fulda S, Gorman A, Hori O et al (2010) Cellular stress responses: cell survival and cell death. Int J Cell Biol 2010:214074
39. Datta K, Sinha S, Chattopadhyay P (2000) Reactive oxygen species in health and disease. Natl Med J India 13:304–310
40. Circu M, Yee A (2010) Reactive oxygen species, cellular redox systems and apoptosis. Free Radic Biol Med 48:749–762

41. Han-Ming S, Pervaiz S (2006) TNF receptor superfamily-induced cell death: redox dependent execution. FASEB J 20:1589–1598
42. Liu Y, Kulesz-Martin M (2001) p53 protein at the hub of cellular DNA damage response pathways through sequence-specific and non-sequence-specific DNA binding. Carcinogenesis 22:851–860
43. Nele Vanlangenakker N, Berghe T, Krysko D et al (2008) Molecular mechanisms and pathophysiology of necrotic cell death. Curr Mol Med 8:207–220
44. Zong XW, Thompson C (2006) Necrotic death as a cell fate. Genes Dev 20:1–15
45. Eskelinen EL (2008) New insights into the mechanisms of macroautophagy in mammalian cells. Int Rev Cell Mol Biol 266:207–247
46. Lum JJ, Bauer DE, Kong M et al (2005) Growth factor regulation of autophagy and cell survival in the absence of apoptosis. Cell 120:237–248
47. Ogata M, Hino SI, Saito A et al (2006) Autophagy is activated for cell survival after endoplasmic reticulum stress. Mol Cell Biol 26:9220–9231
48. Scherz-Shouval R, Shvets E, Fass E et al (2007) Reactive oxygen species are essential for autophagy and specifically regulate the activity of Atg4. EMBO J 26:1749–1760
49. Kirkinezosa IG, Moraesa CT (2001) Reactive oxygen species and mitochondrial diseases. Cell Dev Biol 12:449–457
50. Yakes FM, Van Houten B (1997) Mitochondrial DNA damage is more extensive and persists longer than nuclear DNA damage in human cells following oxidative stress. Proc Natl Acad Sci U S A 94:514–519
51. Tanaka M, Kovalenko SA, Gong JS et al (1996) Accumulation of deletions and point mutations in mitochondrial genome in degenerative diseases. Ann N Y Acad Sci 786:102–111
52. Parker WD, Boyson SJ, Parks JK (1989) Abnormalities of the electron transport chain in idiopathic Parkinson's disease. Ann Neurol 26:719–723
53. Mutisya EM, Bowling AC, Beal MF (1994) Cortical cytochrome oxidase activity is reduced in Alzheimer's disease. J Neurochem 63:2179–2184
54. Borthwick GM, Johnson MA, Ince PG et al (1999) Mitochondrial enzyme activity in amyotrophic lateral sclerosis: implications for the role of mitochondria in neuronal cell death. Ann Neurol 46:787–790
55. Xu GP, Dave KR, Moraes CT et al (2001) Dysfunctional mitochondrial respiration in the wobbler mouse brain. Neurosci Lett 300:141–144
56. Brookes PS, Land JM, Clark JB et al (1998) Peroxynitrite and brain mitochondria: evidence for increased proton leak. J Neurochem 70:2195–2202
57. Munscher C, Rieger T, Muller-Hocker J et al (1993) The point mutation of mitochondrial DNA characteristic for MERRF disease is found also in healthy people of different ages. FEBS Lett 317:27–30
58. Zhang C, Linnane AW, Nagley P (1993) Occurrence of a particular base substitution (3243 A to G) in mitochondrial DNA of tissues of ageing humans. Biochem Biophys Res Commun 195:1104–1110

Reactive Oxygen Species and Cellular Defense System

Susinjan Bhattacharya

Abstract

Reactive oxygen species (ROS) is a collective term used for oxygen-derived free radicals (superoxide, hydroxyl radical, nitric oxide) and non-radical oxygen derivatives of high reactivity (singlet oxygen, hydrogen peroxide, peroxynitrite, hypochlorite). ROS can be either harmful or beneficial to the body. An imbalance between formation and removal of free radicals can lead to a pathological condition called as oxidative stress. However, the human body employs molecules known as antioxidants to counteract these free radicals. But late several studies have indicated that antioxidants can also have deleterious effects on human health depending on dosage and bioavailability. This makes it essential to analyze the extent of utility of antioxidants in the improvement of human health. It is noteworthy that if the generation of free radicals exceeds the protective effects of antioxidants, this can cause oxidative damage which accumulates during the life cycle, and this has been implicated in aging and age-dependent diseases such as cardiovascular disease, cancer, neurodegenerative disorders, and other chronic conditions. This chapter highlights the main themes from studies on free radicals, antioxidants, and oxidative stress and effect of oxidative stress in diseases.

Keywords

Free radicals • Reactive oxygen species • Antioxidants • Oxidative stress • Diseases

S. Bhattacharya (✉)
Department of Biotechnology, Jaypee Institute of Information Technology, A-10, Sector-62, Noida 201307, Uttar Pradesh, India
e-mail: s.bhattacharya@jiit.ac.in

V. Rani and U.C.S. Yadav (eds.), *Free Radicals in Human Health and Disease*,
DOI 10.1007/978-81-322-2035-0_2, © Springer India 2015

1 Introduction

In recent years, the role of free radical reactions in human disease, biology, toxicology, and food deterioration has become an area of intense interest. The free radicals have a special affinity for lipids, proteins, and DNA [1]. A free radical is an atom, molecule, or compound that is highly unstable because of its atomic or molecular structure (i.e., the distribution of electrons within the molecule). This instability makes free radicals very reactive, and they attempt to pair up with other molecules, atoms, or even individual electrons to create a stable compound. To achieve a more stable state, free radicals can "steal" a hydrogen atom from another molecule, bind to another molecule, or interact in various ways with other free radicals.

Free radicals can be defined as reactive chemical species having a single unpaired electron in an outer orbit and are continuously produced by the organism's normal use of oxygen [2]. This unstable configuration creates energy that is released upon reaction with adjacent molecules, such as proteins, lipids, carbohydrates, and nucleic acids. The majority of free radicals that damage biological systems are derived from oxygen and more generally referred to as "reactive oxygen species."

Oxygen (O_2) is an essential element for cell function and life. It plays an important role in a series of biochemical reactions occurring in the respiratory chain, which is responsible for most of the production of adenosine triphosphate (ATP), which provides the energy required for a multitude of cellular reactions and functions. This process of respiratory chain takes place in membrane-enclosed cell structures called mitochondria.

Molecular oxygen can accept a total of four electrons, one at a time, and the corresponding number of protons to generate two molecules of water. This process leads to the generation of different by-products that are generally ROS and reactive nitrogen species (RNS) [3]. Intracellular generation of ROS mainly comprises oxygen radicals, like superoxide ($O_2^{\cdot-}$); peroxide ($O_2^{\cdot=}$), which normally exists in cells as hydrogen peroxide (H_2O_2); and the hydroxyl radical ($^{\cdot}OH$) [4]. Superoxide, peroxide, and the hydroxyl radical are considered the primary ROS and have sparked major research on the role of free radicals in biology and medicine. Though only about 2–3% of the O_2 consumed by the respiratory chain is converted to ROS [5], the toxic effects of oxygen in biological systems—such as the breakdown (i.e., oxidation) of lipids, inactivation of enzymes, introduction of changes (i.e., mutations) in the DNA, and destruction of cell membranes and, ultimately, cells—are attributable to the reduction of O_2 to ROS [6–8].

The term ROS is often used to include not only the radicals $^{\cdot}OH$, $RO_2^{\cdot}$, $NO^{\cdot}$ and $O_2^{\cdot}-$ but also the non-radicals HOCl, $^1O_2$, $ONOO^-$, O_3, and H_2O_2 [9]. However, the biological system undergoes damage mainly by radicals generated from oxygen. Oxygen has two unpaired electrons in separate orbitals in its outer shell. This electronic structure makes oxygen especially susceptible to radical formation. Sequential reduction of molecular oxygen (equivalent to sequential addition of electrons) leads to formation of a group of reactive oxygen species: superoxide anion, peroxide (hydrogen peroxide), and hydroxyl radical (Fig. 1).

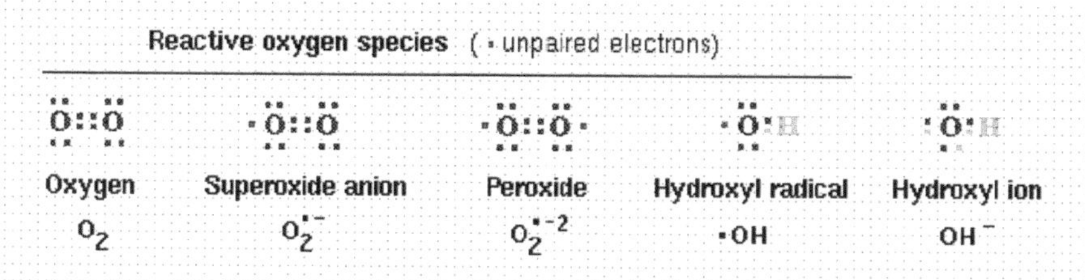

Fig. 1 Structure of reactive oxygen produced in biological systems derived from oxygen. Note the notation used to denote them and the difference between hydroxyl radical and hydroxyl ion, which is not a radical

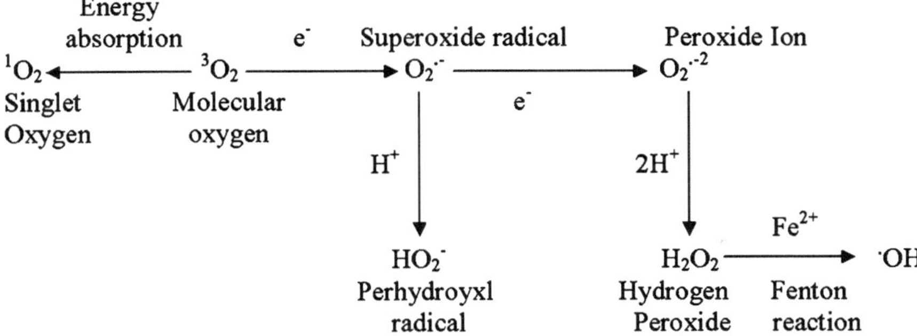

Fig. 2 Schematic representation of ROS generation. The single-electron reduction of O_2 results in the generation of the $O_2^{\cdot -}$. At low pH, dismutation of $O_2^{\cdot -}$ is unavoidable, with one $O_2^{\cdot -}$ giving up its added electron to another $O_2^{\cdot -}$ and then with protonation resulting in the generation of H_2O_2. Again, $O_2^{\cdot -}$ can be protonated to form the $HO_2^{\cdot}$. Additionally, in the presence of transition metals such as copper and iron, further reactions take place, e.g., through the Haber–Weiss mechanism or the Fenton reaction to give up ·OH. O_2^{-} can also react with another very influential signaling-free radical species, NO^{-}, to give up peroxynitrite ($OONO^{-}$)

Another radical derived from oxygen is singlet oxygen, designated as $^1O_2$ (Fig. 2). This is an excited form of oxygen in which one of the electrons jumps to a superior orbital following absorption of energy, thereby freeing oxygen from its spin-restricted state. Further, the singlet state can release a modest amount of energy and get transformed to a triplet state which involves change of electron spin.

2 Sources of ROS Generations

The amount of free radical production is determined by the balance of many factors, and ROS are produced both endogenously and exogenously. The endogenous sources of ROS include mitochondria, cytochrome P-450 metabolism, peroxisomes, and inflammatory cell activation [5]. In general, ROS can be (i) generated during UV light irradiation and by X-rays and gamma rays; (ii) produced during metal-catalyzed reactions; (iii) present in the atmosphere as pollutants; (iv) produced by neutrophils, eosinophils, and macrophages during inflammation; and (iv) by-products of mitochondrial-catalyzed electron transport reactions and various other mechanisms [3, 4].

There are numerous cellular systems that can produce ROS. The major source of ROS production in the cell is the mitochondrial respiratory chain that utilizes approximately 80–90 % of the O_2 a person consumes and generates most of the ROS produced in the body.

Another major source of ROS, especially in the liver, is a group of enzymes called the cytochrome P-450 mixed function oxidases. There are many variants of these iron-containing enzymes, some of which are responsible for removing or detoxifying a variety of compounds present in our environment and ingested (e.g., foods or drugs), including alcohol [10]. Some cytochrome P-450 enzymes also are important for metabolizing substances that naturally occur in the body, such as fatty acids, cholesterol, steroids, or bile acids [11]. The cytochrome P-450 molecules in their biochemical reactions catalyzed use O_2, and during these reactions small amounts of ROS are generated. The extent of ROS generated varies considerably depending on the compound to be degraded and on the cytochrome P-450 molecule involved. One type of cytochrome molecule that is especially active in producing ROS is known as CYP2E1, whose activity increases after heavy alcohol exposure [12].

ROS also are produced by a variety of oxidative enzymes present in cells, such as xanthine oxidase. Xanthine oxidase under normal physiological conditions acts as a dehydrogenase, wherein it removes hydrogen from xanthine or hypoxanthine and attaches it to NAD, thereby

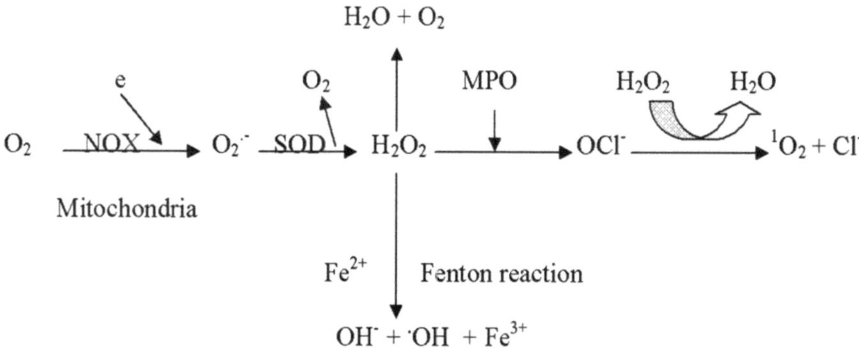

Fig. 3 Generation of ROS in neutrophils. The generation of superoxide ($O_2^{\cdot-}$) from oxygen (O_2) is mediated either by the NADPH oxidase complex (NOX) or in mitochondria by cytochrome c peroxidase or xanthine oxidase. Superoxide will be converted to hydrogen peroxide (H_2O_2) either spontaneously or mediated by superoxide dismutase (SOD). Hydrogen peroxide can be converted to H_2O + O_2 in the presence of catalase and glutathione (GSH) peroxidase, or hydrogen peroxide can act as a source of hydroxyl radical (·OH) via the Fenton reaction. Myeloperoxidase (MOP) uses hydrogen peroxide as substrate for the formation of halogenated ROS such as hypochlorite (OCl^-). Reaction of hypochlorite with hydrogen peroxide results in the formation of singlet oxygen ($^1O_2$)

generating NADH. However, under certain conditions, such as the disruption of blood flow to a tissue, xanthine dehydrogenase is converted to a ROS-producing oxidase form [13].

Other sources of ROS in the body are two types of immune cells called macrophages and neutrophils, which defend the body against invading microorganisms (Fig. 3). ROS production here is beneficial and even essential to the organism as it functions to destroy foreign pathogens [14]. Macrophages and neutrophils contain a group of enzymes called the NADPH oxidase complex, which, upon activation generates superoxide radicals and hydrogen peroxide. Hydrogen peroxide then interacts with chloride ions present in the cells to produce hypochlorite, which in turn destroys the pathogen. The NADPH oxidase complex and the resulting ROS production are critical to the body's defense against all kinds of diseases, as is evident in patients with a condition called chronic granulomatous disease, in which ROS production by the NADPH oxidase complex is drastically reduced. Patients with this condition are highly sensitive to infections and usually die at an early age [15].

Another peroxidase enzyme that is abundantly expressed in neutrophils is myeloperoxidase (MPO) (Fig. 3). MPO in the presence of heme as a cofactor produces hypochlorous acid (HOCl) from hydrogen peroxide and chloride anion (or the equivalent from a non-chlorine halide) [16]. It also oxidizes tyrosine to tyrosine radical in the presence of hydrogen peroxide. Both HOCl and tyrosine radical are cytotoxic and are used by the neutrophil to kill pathogenic organisms [17].

Humans are constantly exposed to environmental free radicals, including ROS, in the form of radiation, UV light, smog, tobacco smoke, and certain compounds called as redox-cycling agents, which include some pesticides as well as certain medications used for cancer treatment. The toxicity of these medications against tumor cells (as well as normal body cells) results because of their modification by cellular enzymes to an unstable intermediate that then reacts with molecular oxygen to produce the original product plus a superoxide radical. Thus, a vicious cycle of chemical reactions involving these compounds continually produces ROS [11].

3 Role of Metals

Earlier studies suggested the possibility that the superoxide radicals or hydrogen peroxide radicals in ROS production process interact with each other to produce the most reactive ROS, the hydroxyl radical ($^{\bullet}OH$). Direct interaction between these two radicals under normal physiological conditions does not play a significant role in generating hydroxyl radicals. However, in the presence of certain metals, particularly free iron or copper ions, a sequence of two reaction steps can occur that results in hydroxyl radical generation. In the first step, hydrogen peroxide can produce the hydroxyl radical as the primary oxidant by removing an electron from the participating metal ion [18]. In the second step, the original metal ions are regenerated involving the superoxide radical ($O_2^{\bullet-}$), so that they are again available for reaction with the hydrogen peroxide. This combination of two chemical reactions appears to account for most of the hydroxyl radical production in biological systems and supports the role of metals such as iron and copper to produce oxidative stress and ROS-induced injury in cells. Because of iron's critical contribution to hydroxyl radical formation, anything that increases the levels of free iron in the cells promotes ROS generation and oxidative stress [19, 20].

4 Oxidative Stress

The formation and removal of free radicals are balanced in a normal cell. However, with more formation of free radicals or when levels of antioxidants are diminished, the cell enters a state called as "oxidative stress." This state if prolonged can cause cell damage and death. Oxidative stress plays a major role in the development of chronic and degenerative diseases such as cancer, arthritis, aging, autoimmune disorders, and cardiovascular and neurodegenerative diseases [21]. In mammalian cells, oxidative stress leads to increase in intracellular levels of free Ca^{2+} and iron, and excessive rises in intracellular free Ca^{2+} may cause DNA fragmentation by activating endonucleases [22]. Endonuclease activation results in single- and double-strand DNA breaks that subsequently lead to increased levels of nuclear proteins which are essential in DNA repair, p53, and poly-ADP ribosylation. Intranuclear Ca^{2+} fluctuations can also affect chromatin organization, induce gene expression, and influence protease and protein kinase activities. The induced kinases like MAP kinase and calmodulin kinases are involved in the activation of transcription factors that initiate transcription of early response genes, for example, c-fos and c-jun, and also activation of phospholipase A_2 which results in permeabilization of the plasma membrane and of intracellular membranes such as the inner membrane of mitochondria leading to apoptosis [23].

5 Antioxidants and Prevention of Oxidative Damage

Uncontrolled generation of ROS can lead to their accumulation causing oxidative stress in the cells. The human body has several mechanisms to counteract oxidative stress by producing antioxidants which are either naturally produced in the body or externally supplied through foods and/or supplements. Antioxidants are those molecules that are present at low concentrations and significantly delays or prevent oxidation of the oxidizable substrate [24]. Endogenous and exogenous antioxidants are effective as free radical scavengers by donating their own electrons to ROS and thereby neutralize adverse effects of the latter [4]. Thus, they can enhance the immune defense and lower the risk of cancer and degenerative diseases [25]. In general, an antioxidant in the body may work at three different levels: (a) **prevention**, maintains formation of reactive species to a minimum level, e.g., desferrioxamine; (b) **interception**, scavenges reactive species either by using catalytic and non-catalytic molecules, e.g., ascorbic acid and alpha-tocopherol; and (c) **repair**, repairs damaged target molecules, e.g., glutathione [23].

The antioxidant systems are classified into two major groups, protective or enzymatic antioxidants and nonenzymatic antioxidants.

6 Enzymatic Antioxidants

Enzymatic antioxidants involved in the elimination of ROS include superoxide dismutase (SOD), catalase, and glutathione peroxidase (GPx) (Figs. 4 and 5) [26]. There are different types of SODs that are differentially located in the mammalian cells and requirement of metal ions as cofactors for their function. For example, a copper–zinc SOD is present in the cytosol and in the space between two membranes surrounding mitochondria. Again, manganese-containing SOD is present in the matrix of mitochondria [27].

The enzymes, catalase, and glutathione peroxidase help to remove hydrogen peroxide. Catalase is an iron-containing enzyme found primarily in the small membrane-enclosed cell components called peroxisomes and detoxifies hydrogen peroxide by catalyzing a reaction between two hydrogen peroxide molecules, resulting in the formation of water and O_2. In addition, catalase can promote the interaction of hydrogen peroxide with compounds that can

$$2\,O_2^{\cdot -} + 2\,H^+ \xrightarrow{\text{SOD}} H_2O_2 + O_2$$

$$2\,H_2O_2 \xrightarrow{\text{CAT}} 2\,H_2O + O_2$$

$$H_2O_2 + 2\,GSH \xrightarrow{\text{GPx}} 2\,H_2O + GSSG$$

Fig. 4 The antioxidant enzymes and the reactions they catalyze. Note the notations used: *SOD* for superoxide dismutase, *CAT* for catalase, *GPx* for glutathione peroxidase, *GSH* for reduced glutathione (monomeric glutathione), *GSSG* for oxidized glutathione (glutathione disulfide)

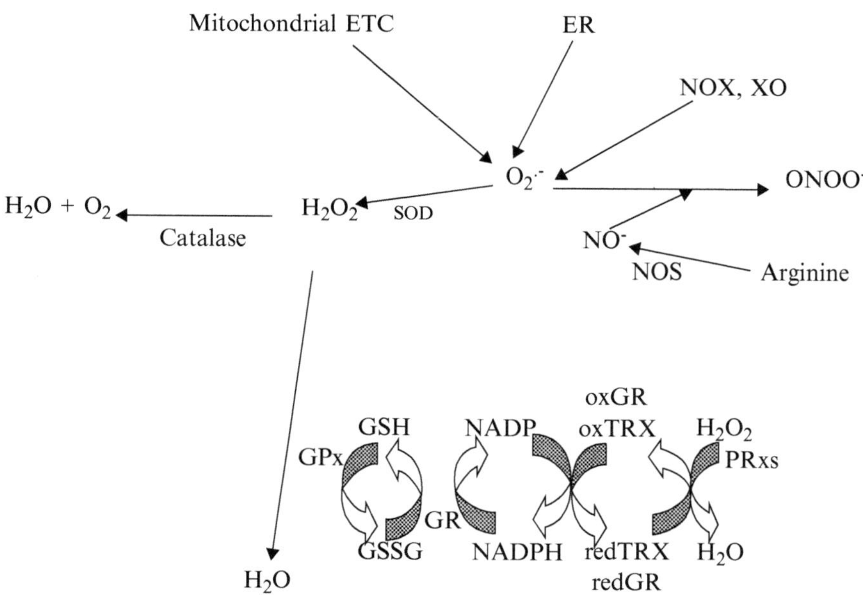

Fig. 5 Reactions of enzymatic antioxidants. Major sites of cellular ROS generation include mitochondrial electron transport chain (ETC), endoplasmic reticulum system (ER system), and NAD(P)H oxidase (NOX) complex. Nitric oxide synthases (NOS) are key enzymes for production of NO. GSH and NADPH play roles in maintaining reduced cellular state. Symbols used: *GPx*, glutathione peroxidase; *GR*, glutathione reductase; *oxTRX*, oxidized thioredoxin; *redTRX*, reduced thioredoxin; *oxGR*, oxidized glutaredoxin; *redGR*, reduced glutaredoxin; H_2O_2, hydrogen peroxide; NO^-, nitric oxide; $ONOO^-$, peroxynitrite; *SOD*, superoxide dismutase; *GSH*, reduced glutathione; *GSSG*, oxidized glutathione; *NADPH*, reduced nicotinamide adenine dinucleotide phosphate; *XO*, xanthine oxidase; *PRxs*, peroxiredoxins

serve as hydrogen donors so that the hydrogen peroxide can be converted to one molecule of water, and the reduced donor becomes oxidized (a process sometimes called the peroxidatic activity of catalase) [28].

The glutathione peroxidase system consists of several components, including the enzymes glutathione peroxidase and glutathione reductase and the cofactors monomeric glutathione (GSH) and reduced nicotinamide adenosine dinucleotide phosphate (NADPH) [29]. Glutathione peroxidase contains an amino acid that is modified by addition of a molecule of the metal selenium; therefore, low amounts of selenium are critical for the body's antioxidant defense. Together, these molecules effectively remove hydrogen peroxide. GSH serves as a cofactor for the enzyme glutathione transferase, which helps to remove certain drugs and chemicals as well as other reactive molecules from the cells. GSH can also interact directly with certain ROS (e.g., the hydroxyl radical) to detoxify them, as well as performing other critical activities in the cell [4].

7 Nonenzymatic Antioxidants

GSH due to its functions is probably the most important antioxidant present in cells. Therefore, enzymes that generate GSH are critical for the body's ability to protect itself against oxidative stress. NADPH is involved in a much more diverse range of reactions in the cell than GSH, and due to its role played in the glutathione peroxidase system, NADPH or the enzymes that generate this compound are sometimes considered antioxidants [30].

In addition to GSH and NADPH, numerous other nonenzymatic antioxidants are present in the cells, most prominently vitamin E (α-tocopherol) and vitamin C (ascorbic acid). Vitamin E is a major antioxidant found in the lipid phase of membranes and, like other chemically related molecules, acts as a powerful terminator of lipid peroxidation. During the reaction between vitamin E and a lipid radical, the vitamin E radical is formed, from which vitamin E can be regenerated in a reaction involving GSH and ascorbate (Fig. 6) [31].

Ascorbic acid can exert its antioxidant activity both directly and also in concert with vitamin E and glutathione. The hydrogen atoms produced in the oxidation of ascorbic acid first to ascorbyl radical and then to dehydroascorbate react with a polyunsaturated fatty acid and/or phospholipid peroxyl radical (PUFA-OO·) to form the nonradical product, PUFA-OOH. Trapping of PUFA-OO· prevents this radical from attacking polyunsaturated fatty acid side chains and other lipoproteins in the plasma membrane and thus breaks the chain reaction of lipid peroxidation. Dehydroascorbic acid is reduced back to ascorbic acid by glutathione and enzymatic reduction. Ascorbic acid also helps alpha-tocopherol in trapping aqueous radicals. Alpha-tocopherol also traps peroxyl radicals before they propagate radical chain reactions leading to lipid peroxidation and regenerate alpha-tocopherol by reducing the alpha-tocopheroxyl radical produced in this trapping reaction (Fig. 6) [32].

Under certain conditions, ascorbic acid may work as a prooxidant rather than as an antioxidant by reducing transition metal ions, which in turn drives the Fenton reaction potentially resulting in oxidative stress. A few ions of transition metal elements have redox transitions with potentials of a magnitude that allows the catalytic decomposition of hydroperoxides. The redox couples of most importance to biological systems are Cu^{+}/Cu^{2+} and Fe^{2+}/Fe^{3+}. The one electron redox cycle results in the formation of peroxyl and alkoxyl radicals, the latter of which can rearrange and react with oxygen to form peroxyl radicals [33].

8 ROS Toxicity

ROS may be toxic to cells because they can react with most cellular macromolecules, including proteins, lipids, and DNA and generate different types of secondary radicals like lipid radicals, sugar- and base-derived radicals, amino acid radicals, and thiol radicals. These radicals in the presence of oxygen are converted to peroxyl radicals, which can induce chain reactions [34]. The

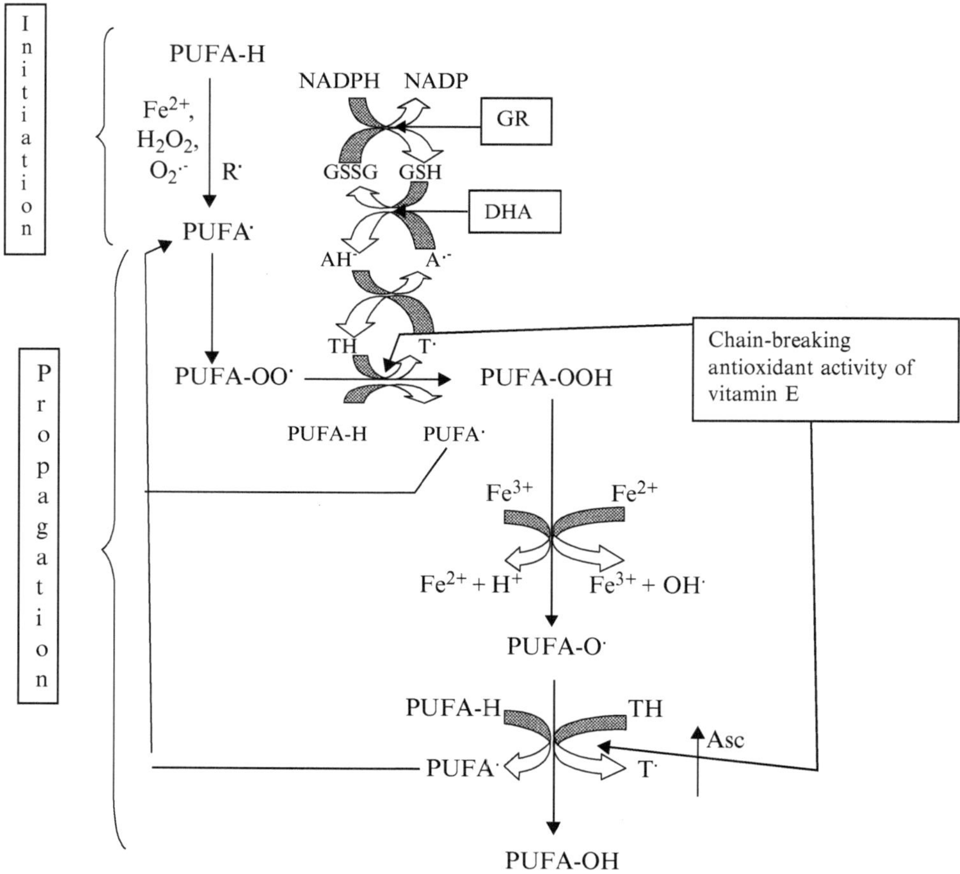

Fig. 6 Main peroxidative reaction on PDFA. α-Tocopherol (TH, Vitamin E) efficiently transfers a H-atom to a lipid free radical peroxyl (PUFA-OO·), alkoxyl (PUFA-O·), and carbon-centered (PUFA·) radicals giving the corresponding nonradical product of PUFA· (PUFA-OO·H, PUFA-OH, and PUFA-H) and an α-Tocopheroxyl radical (T). T once formed reacts with the second free radical (PUFA-OO·, PUFA-O·, and PUFA·), or each other to make a nonradical product (T-OOL, T-OL, or T-T). *Symbols used*: *R·*, a radical; $O_2^{\cdot-}$, superoxide anion; *PUFA-H*, polyunsaturated fatty acid; *PUFA·* carbon-centered PUFA· radical (lipid alkyl radical); *PUFA-OO·* hydroperoxyl radical; *PUFA-OO·*, hydroperoxide; *PUFA-O·*, alkoxyl radical; *PUFA-OH*, hydroxylated PUFA; *AH·*, Vitamin C or ascorbic acid; *ZA·⁻*, ascorbyl radical/Ascorbate; *DHA*, dehyrdoascorbate; *GR*, glutathione reductase; *GSH*, reduced glutathione; *GSSG*, oxidized glutathione

biological implication of such reactions primarily depends on several factors like site of generation, nature of the substrate, reactivation mechanisms, and redox status [25, 31].

The ROS-induced oxidation of proteins can lead to changes in the proteins' three-dimensional structure as well as fragmentation, aggregation, or cross-linking of the proteins [35]. The side chains of all amino acid residues of proteins, in particular tryptophan, cysteine, and methionine residues, are susceptible to ROS oxidation. Protein oxidation products are usually carbonyls such as aldehydes and ketones. Finally, protein oxidation often will make the marked protein more susceptible to degradation by cellular systems responsible for eliminating damaged proteins from the cell [36].

Membranes that surround the cells as well as other cellular structures, such as the nucleus and mitochondria, contain high concentration of unsaturated fatty acids in their lipid components. The complete degradation (i.e., peroxidation) of membrane lipids is a hallmark of oxidative damage [35] and can result in the formation of lipid hydroperoxide (PUFA-OOH) which can further breakdown to aldehydes such as malonaldehyde and 4-hydroxynonenal (4-HNE) or form cyclic endoperoxide, isoprotans, and hydrocarbons. The lipid peroxidation can lead to cross-linking of membrane proteins, change in membrane fluidity, and formation of lipid–protein and lipid–DNA adducts which may be detrimental to cell functions [31].

DNA is the cell's genetic material, and any permanent damage to the DNA can result in malfunctions of the proteins or complete inactivation of the affected proteins. Thus it is essential for the viability of individual cells or even the entire organism that the DNA remains intact. ROS are a major source of DNA damage, such as modification of DNA bases, single- and double-strand DNA breaks, loss of purines (apurinic sites), damage to the deoxyribose sugar, DNA–protein cross-linkage, and damage to the DNA repair system [37]. Of the reactive oxygen species, hydroxyl (•OH) radical is one of the potential inducers of DNA damage. A variety of adducts are formed on reaction of •OH radical with DNA. The •OH radical can attack purine and pyrimidine bases to form •OH radical adducts, which are both oxidizing and reducing in nature which in turn can induce base modifications and sometimes release of bases. Some of the important base modifications include 8-hydroxydeoxyguanosine (8-OHdG), 8 (or 4-, 5-)-hydroxyadenine, thymine peroxide, thymine glycols, and 5-(hydroxymethyl) uracil [38]. Free radicals can also attack the sugar moiety to produce sugar peroxyl radicals leading to strand breakage. Although cells have developed repair mechanisms to correct naturally occurring changes in the DNA, additional or excessive changes caused by ROS or other agents can lead to permanent changes or damage to the DNA with potentially detrimental effects like cell death, mutagenesis, carcinogenesis, and aging of the cell.

9 ROS in Normal Physiology

ROS act as stimulating agents for biochemical processes within the cell. It plays a role in low concentration for normal physiological functions like gene expression, cellular growth, and defense against infection by induction of transcription factors such as nuclear factor-kappa B (NF-κB) and activator protein-1 (AP-1) and activation of signal transduction pathways [39, 40]. A particular example of signal transduction molecules activated by ROS is the mitogen-activated protein kinases (MAPKs).

Research studies show that ROS molecules modulate multiple redox-sensitive intracellular signaling pathways in mammalian cells generated from NADPH oxidase (Nox), which include inhibition of protein tyrosine phosphatases, activation of certain redox-sensitive transcription factors, and modulation of the functions of some ion channels [41]. A prominent feature of Nox/ROS-mediated signaling is the heterogeneity of its activating stimuli, and at the cellular level, Nox can be activated by a large collection of chemical, physical, environmental, and biological factors [42]. For example, growth factor-mediated responses in endothelial cells (EC) leading to angiogenesis are signaled by ROS such as superoxide and H_2O_2. The Nox acts as major source of ROS in endothelial cells, which consists of Nox2 (gp91phox) and the homologues Nox1 and Nox4 with p22phox, p47phox, p67phox, and the small G protein Rac1 as subunits {"phox" stands for phagocytic oxidase and "Rac" stands for Rho-related C3 botulinum toxin substrate}. Vascular endothelial growth factor (VEGF), a key angiogenic growth factor along with angiopoietin-1, activates the endothelial Nox. Consequently, ROS derived from this Nox stimulate diverse redox-signaling pathways leading to angiogenesis-related gene induction as well as EC migration and proliferation that may contribute to postnatal angiogenesis in vivo (Fig. 7) [43].

ROS also appear to serve as secondary messengers in many developmental stages; for example, in sea urchins ROS levels are elevated during fertilization. Also, prenatal and embryonic development in mammals has also been suggested to be regulated by ROS [44]. Apart from these, ROS also participate in the biosynthesis of molecules such as thyroxin and prostaglandin that accelerate developmental processes [3]. In thyroid cells, regulation of H_2O_2 concentration is critical for thyroxine synthesis, as it is needed to catalyze the binding of iodine atoms to thyroglobulin [45]. It is noteworthy that ROS are also used by the immune system, like ROS have been shown to

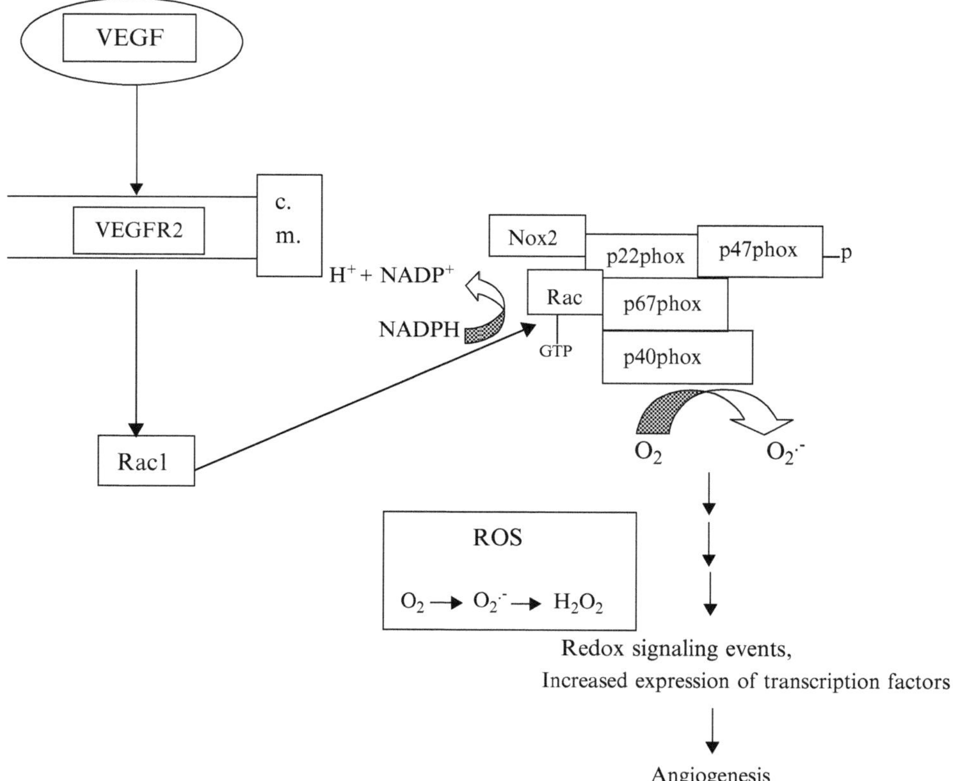

Fig. 7 Schematic role of ROS derived from NADPH oxidase in angiogenesis binding of vascular endothelial growth factor (VEGF) to VEGF receptor VEGFR2 stimulates ROS ($O_2^{\cdot-}$, H_2O_2) production via Rac1-mediated NADPH oxidase activation. NADPH oxidase enzyme complex in endothelial cells consists of a flavocytochrome b558 reductase composed of a "Nox2" catalytic subunit bound to a smaller subunit, p22phox (maturation and stabilization partner); regulatory subunits including the cytosolic organizer proteins, p47phox and p40phox; cytosolic activator proteins, p67phox; and the small molecular weight G protein Rac1. The enzyme catalyzes transfer of electrons from NADPH to molecular oxygen to form $O_2^{\cdot-}$ across the membrane, which in turn can be spontaneously or catalytically converted to H_2O_2. The ROS generated induces downstream signaling events which converge and integrate to induce angiogenesis

trigger proliferation of T cells through NF-κB activation. Macrophages and neutrophils generate ROS in order to kill the bacteria that they engulf by phagocytosis. Furthermore, tumor necrosis factor-alpha (TNF-α) mediates the cytotoxicity of tumor and virus-infected cells through ROS generation and induction of apoptosis [14].

10 Diseases Involving Excessive ROS Levels

Overproduction of ROS generates oxidative stress which reflects an imbalance between the systemic manifestation of reactive oxygen species and a biological system's ability to readily detoxify the reactive intermediates or to repair the resulting damage. Disturbances in the normal redox state of cells can cause toxic effects through the production of peroxides and free radicals [45]. Excess of free radicals damages essential macromolecules of the cell, leading to abnormal gene expression, disturbance in receptor activity, proliferation or cell dynamics, immunity perturbation, mutagenesis, and protein deposition and damages all components of the cell, including proteins, lipids, and DNA [46]. ROS has been implicated in many major diseases that plague humans [44]. Some of these noteworthy diseases due to oxidative stress-induced disruptions in normal mechanisms of cellular signaling include cancer [47, 48], inflammatory diseases [49–51], cardiovascular diseases [52, 53], respiratory diseases [54], diabetes [55, 56], male infertility [57], aging process [58, 59], neurological diseases [59–61], etc.

11 Antioxidants and Their Supplementation

Antioxidants are the substances that may protect cells from the damage caused by free radicals. Antioxidants interact with and stabilize free radicals and may prevent some of the damage that free radicals might otherwise cause. However, these antioxidants whenever are consumed in large doses can act as prooxidants [28]. The antioxidants depending on their source of availability can be endogenous or exogenous in nature. The endogenous antioxidants can either be enzymatic, like superoxide dismutase (SOD), catalase (CAT), glutathione peroxidase (GPx), and glutathione reductase (GRx) [26], or nonenzymatic in nature. The nonenzymatic antioxidants can be further grouped to metabolic antioxidants, such as lipoic acid, glutathione, L-arginine, uric acid, bilirubin [62], and nutrient antioxidants. Some of the nutrient antioxidant can be exogenous in nature as they cannot be produced in the body and must be provided through foods such as vitamin E, vitamin C, carotenoids, and trace elements (Se, Cu, Zn, Mn) [63–65].

12 Conclusion

Free radical formation is a continuous process in humans and can be involved in development of diseases. Although cells are rich in antioxidant enzymes as well as small antioxidant molecules, these agents may not be sufficient enough to normalize the redox status during oxidative stress, and there is a need of antioxidant supplementation. The health benefits of administering antioxidants such as vitamins E and C or other compounds are the subject of current research, and clinical trials employing antioxidants in the treatment of various conditions are under way. However, there are many unsolved questions about antioxidant supplementation in disease prevention. This throws the future direction for research if this supplementation can be recommended as an adjuvant therapy.

References

1. Sivanandham V (2011) Free radicals in health and diseases. Pharmacol Online 11:1062–1077
2. Tiwari AK (2004) Antioxidants: new- generation therapeutic base for treatment of polygenic disorders. Curr Sci 86:1092–1102
3. Shinde A, Ganu J, Naik P (2012) Effect of free radicals & antioxidants on oxidative stress. J Dent Allied Sci 1:63–66
4. Kunwar A, Priyadarsini KI (2011) Free radicals, oxidative stress and importance of antioxidants in human health. J Med Allied Sci 1:53–60
5. Chance B, Sies H, Boveris A (1979) Hydroperoxide metabolism in mammalian organisms. Physiol Rev 59:527–605
6. De Groot H (1994) Reactive oxygen species in tissue injury. Hepatogastroenterology 41:328–332
7. Nakazawa J, Genka C, Fujishima M (1996) Pathological aspects of active oxygens/free radicals. Hepatogastroenterology 46:15–32
8. Toykuni S (1999) Reactive oxygen species–induced molecular damage and its application in pathology. Pathol Int 49:91–102
9. Aruoma OI (1994) Nutrition and health aspects of free radicals and antioxidants. Food Chem Toxicol 32:671–683
10. Proulx M, du Souich P (1995) Inflammation-induced decrease in hepatic cytochrome P450 in conscious rabbits is accompanied by an increase in hepatic oxidative stress. Res Commun Mol Pathol Pharmacol 87:221–236
11. Cederbaum AI (2001) Introduction—serial review: alcohol, oxidative stress, and cell injury. Free Radic Biol Med 31:1524–1526
12. Lieber CS (1997) Cytochrome P450 2E1: its physiological and pathological role. Physiol Rev 77:517–544
13. Sultatos LG (1988) Effects of acute ethanol administration on the hepatic xanthine dehydrogenase/xanthine oxidase system in the rat. J Pharmacol Exp Ther 246:946–949
14. Rosen GM, Pou S, Ramos CL et al (1995) Free radicals and phagocytic cells. FASEB J 9:200–209
15. Kohchi C, Inagawa H, Nishizawa T, Soma G (2009) ROS and innate immunity. Anticancer Res 29:817–821
16. Klebanoff SJ (2005) Myeloperoxidase: friend and foe. J Leukoc Biol 77(5):598–625
17. Heinecke JW, Li W, Francis GA, Goldstein JA (1993) Tyrosyl radical generated by myeloperoxidase

catalyzes the oxidative cross-linking of proteins. J Clin Invest 91(6):2866–2872
18. McCord JM (1988) Iron, free radicals, and oxidative injury. Semin Hematol 35:5–12
19. Tsukamoto H, Lu SC (2001) Current concepts in the pathogenesis of alcoholic liver injury. FASEB J 15:1335–1349
20. Sadrzadeh SM, Nanji AA, Price PL (1994) The oral iron chelator, 1,2– dimethyl–3–hydroxypyrid–4–one reduces hepatic free iron, lipid peroxidation and fat accumulation in chronically ethanol–fed rats. J Pharmacol Exp Ther 269:632–636
21. Lian AP, Hua H, Chuong PH (2008) Free radicals, antioxidants in disease and health. Int J Biol Sci 4:89–96
22. Orrenius S, McConkey DJ, Bellomo G, Nicotera P (1989) Role of Ca2+ in toxic cell killing. Trends Pharmacol Sci 10:281–285
23. Zhivotovsky B, Orrenius S (2011) Calcium and cell death mechanisms: a perspective from the cell death community. Cell Calcium 50:211–221
24. Kohen R, Nyska A (2002) Oxidation of biological systems: oxidative stress and antioxidants. Toxicol Pathol 30:620–630
25. Sevanian A, Ursini F (2000) Lipid peroxidation in membranes and low-density lipoproteins: similarities and differences. Free Radic Biol Med 29:306–311
26. Mates JM (2000) Effects of antioxidant enzymes in the molecular control of reactive oxygen species toxicology. Toxicology 153:83–104
27. Fridovich I (1997) Superoxide anion radical, superoxide dismutases, and related matters. J Biol Chem 272:18515–18517
28. Seifried HE, Anderson DE, Fisher EI, Milner JA (2007) A review of the interaction among the dietary antioxidants and reactive oxygen species. J Nutr Biochem 18:567–579
29. Schafer FQ, Buettner GR (2001) Redox environment of the cell as viewed through the redox state of the glutathione disulfide/glutathione couple. Free Radic Biol Med 30:1191–1212
30. Block G, Pattersen B, Subar A (1992) Fruit, vegetables and cancer prevention: a review of the epidemiological evidence. Nutr Cancer 18:1–29
31. Auroma OI (1998) Free radicals, oxidative stress, and antioxidants in human health and disease. JAOCS 75:199–212
32. Peake JM (2003) Vitamin C: effects of exercise and requirements with training. Int J Sports Nutr Exerc Metab 13:125–151
33. Hogg N, Kalyanaraman B (1999) Nitric oxide and lipid peroxidation. Biochim Biophys Acta 1411:378–384
34. Winterbourn CC (2008) Reconciling the chemistry and biology of reactive oxygen species. Nat Chem Biol 4:278–286
35. Beckman KB, Ames BN (1997) Oxidative decay of DNA. J Biol Chem 272:19633–19636
36. Wu D, Cederbaum AI (2004) Alcohol, oxidative stress and free radical damage. Alcohol Res Health 27(4):277–284
37. Spencer JPE, Jenner A, Aruoma OI, Cross CE et al (1996) Oxidative DNA damage in human respiratory tract epithelial cells. Time course in relation to DNA strand breakage. Biochem Biophys Res Commun 224:17–22
38. Breen AP, Murphy JA (1995) Reactions of oxyl radicals with DNA. Free Radic Biol Med 18:1033–1077
39. Lander HM (1997) An essential role for free radicals and derived species in signal transduction. FASEB J 11:118–124
40. Droge W (2002) Free radicals in the physiological control of cell function. Physiol Rev 82:47–95
41. Bedard K, Krause KH (2007) The Nox family of ROS-generating NADPH oxidases: physiology and pathophysiology. Physiol Rev 87:245–313
42. Jiang F, Zhang Y, Dusting GJ (2011) NADPH oxidase-mediated redox signaling: roles in cellular stress response, stress tolerance, and tissue repair. Pharmacol Rev 63:218–242
43. Fukai M (2006) Redox signaling in angiogenesis: role of NADPH oxidase. Cardiovasc Res 71:226–235
44. Knight JA (1998) Free radicals: their history and current status in aging and disease. Ann Clin Lab Sci 28:331–346
45. Schreck R, Baeuerle PA (1991) A role for oxygen radicals as second messengers. Trends Cell Biol 1:39–42
46. Kehrer JP (1993) Free radicals as mediators of tissue injury and disease. Crit Rev Toxicol 23:21–48
47. Diplock AT, Rice-Evans AC, Burton RY et al (1994) Is there a significant role of lipid peroxidation in the causation of malignancy and for antioxidants in cancer prevention? Cancer Res 54:19525–19565
48. Waris G, Ahsan H (2006) Reactive oxygen species: role in the development of cancer and various chronic conditions. J Carcinog 5:14
49. Chapple IL (1997) Reactive oxygen species and antioxidants in inflammatory diseases. J Clin Periodontol 24:287–296
50. Li X, Fang P, Mai J, Choi ET et al (2013) Targeting mitochondrial reactive oxygen species as novel therapy for inflammatory diseases and cancers. J Hematol Oncol 25:6–19
51. Geronikaki AA, Gavalas AM (2006) Antioxidants and inflammatory disease: synthetic and natural antioxidants with anti-inflammatory activity. Comb Chem High Throughput Screen 9(6):425–442
52. Dhalla NS, Temsah RM, Netticadan T (2000) Role of oxidative stress in cardiovascular diseases. J Hypertens 18:655–673
53. Zhang N, Bradley TA, Zhang C (2010) Inflammation and reactive oxygen species in cardiovascular Disease. World J Cardiol 2:408–410
54. Suzy A, Comhair A, Erzurum SC (2002) Antioxidant responses to oxidant- mediated lung diseases. Am J Physiol Lung Cell Mol Physiol 283:L246–L255

55. Johansen JS, Harris AK, Rychly DJ et al (2005) Oxidative stress and the use of antioxidants in Diabetes. Cardiovasc Diabetol 4:1–11
56. Kaneto H, Katakami N, Matsuhisa M et al (2010) Role of reactive oxygen species in the progression of type 2 diabetes and atherosclerosis mediators of inflammation. Mediat Inflamm 2010:453892
57. Makker K, Agarwal A, Sharma R (2009) Oxidative stress & male infertility. Indian J Med Res 129:357–367
58. Miquel J, Economos AC, Fleming J et al (1980) Mitochondrial role in cell aging. Exp Gerontol 15:575–591
59. Afanas'ev I (2010) Signaling and damaging functions of free radicals in aging-free radical theory, hormesis, and TOR. Aging Dis 1:75–88
60. Pimentel C, Batista-Nascimento L, Rodrigues-Pousada C, Menezes RA (2012) Oxidative stress in Alzheimer's and Parkinson's diseases: insights from the yeast *Saccharomyces cerevisiae*. Oxidative Med Cell Longev 2012:1–9
61. Gilgun-Sherki Y, Melamed E, Offen D (2001) Oxidative stress induced- neurodegenerative diseases: the need for antioxidants that penetrate the blood brain barrier. Neuropharmacology 40:959–975
62. Kohen R, Nyska A (2002) Oxidation of biological systems: oxidative stress phenomena, antioxidants, redox reactions, and methods for their quantification. Toxicol Pathol 30:620–630
63. Willett WC, Macmahon B (1984) Diet and cancer-an overview. N Engl J Med 310:697–703
64. Willcox JK, Ash SL, Catignani GL (2004) Antioxidants and prevention of chronic disease. Crit Rev Food Sci Nutr 44:275–295
65. Radimer K, Bindewald B, Hughes J et al (2004) Dietary supplement use by US adults: data from the national health and nutrition examination survey, 1999–2000. Am J Epidemiol 160:339–349

The Noxious Nanoparticles

Sudha Srivastava, Kushagr Punyani, and Shuchi Arora

Abstract

With the advent of nanotechnology and the phenomenal escalation in the industrial and commercial applications of nanoparticles, their exposure to the human population has also increased. This calls for a need to meticulously study their intracellular as well as immunological effects, which have been covered under the umbrella of nanotoxicity. This chapter addresses the sources and routes of nanoparticle exposure, followed by the in vivo translocation and their elimination. We have also discussed the physicochemical properties of nanoparticles modulating nanotoxicity and their intracellular biomolecular interactions. A major stress is laid on nanoparticle-mediated oxidative stress, the role of NF-κB signalling, receptor-mediated nanotoxicity and the complement system-mediated immune response. The role of complement system as clearance mechanism has also been briefly discussed.

Keywords

ROS • Nanoparticles • Nanotoxicity • Oxidative stress • NF-κB signalling

1 Introduction

Nanotechnology is defined as the design, synthesis and application of materials and devices engineered at the nanoscale. Providing the prospects of miniaturization and fabrication of novel materials with unique traits and unparalleled development, nanotechnology has also contributed to the roll of iotas taking a toll on human health. Nanomaterials are finding their way into industry and household as cosmetics, fillers, catalysts, semiconductors and drug carriers and thereby into human lives, at a pace which demands a restraint owing to the associated exposure, their safety and the short- and long-term risks [1, 2]. These petite particles may provide exposure and internalization via inhalation, ingestion, skin uptake, implants and injection.

S. Srivastava (✉) • K. Punyani • S. Arora
Department of Biotechnology, Jaypee Institute of Information Technology, A-10, Sector-62, Noida 201307, Uttar Pradesh, India
e-mail: sudha.srivastava@jiit.ac.in

V. Rani and U.C.S. Yadav (eds.), *Free Radicals in Human Health and Disease*, DOI 10.1007/978-81-322-2035-0_3,

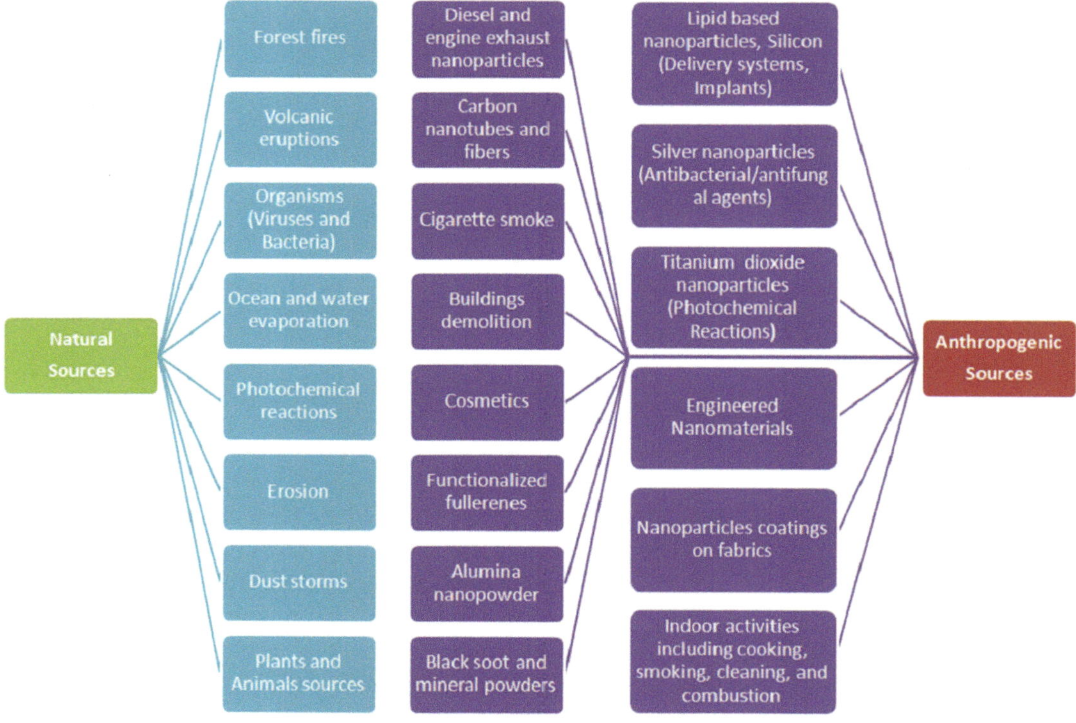

Fig. 1 Natural and anthropogenic sources of nanoparticles

Humans have been exposed to nano-sized particles present in the atmosphere throughout the evolutionary stages. However, owing to anthropogenic sources, the grade of exposure and the detrimental impact have increased tremendously over the last century [3]. Hence, the concern pertaining to the increase in the titres of naturally occurring nanoparticles in the atmosphere and introduction of engineered nanoparticles is relevant (Fig. 1). Human skin, respiratory tract and the gastrointestinal tract are in a dynamic but steady contact with the environment. While the former provides a physical barricade to foreign substances, the latter are more susceptible to damage. It is of merit to deem all of these ways to be potential routes of entry for natural or anthropogenic nanoparticles, besides injections and implant-mediated exposure to engineered materials [4]. Post-ingress, these nanoparticles can undergo systemic navigation to dock to various tissues and organs in the body, affecting cellular damage via oxidative stress and organelle injury. Figure 2 outlines the various routes of exposure, in vivo translocation and elimination from the body. A failure in clearance from the body leading to accumulation leads to various pathological conditions as a consequence of their toxicological effects.

The toxicity of the nanoparticles is dictated by their chemical composition, material properties, surface modifications, size and dosage or exposure. Further, it is crucial to appreciate that toxicity or biochemical interferences are a function of the individual's response and hence the genome and metabolome. The accelerated exposure to nanoparticles has called the need for nanotoxicology, the science of the impact of engineered nanodevices and nanostructures on living organisms.

Owing to at least one dimension less than 100 nm, nanoparticles resemble in aspect to cellular components and proteins, thereby providing them with elevated odds of interference with intracellular pathways, causing and exacerbating cellular damage and evasion of natural defences. Besides, their interaction with biomolecules,

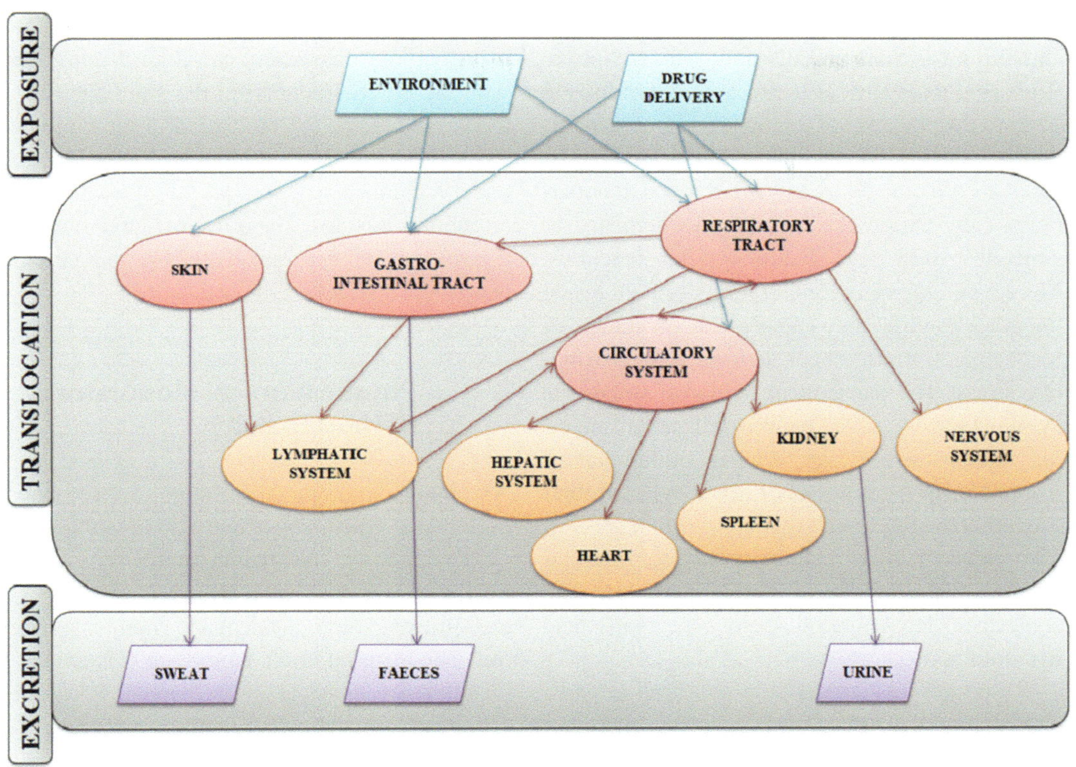

Fig. 2 Entity relationship diagram of nanoparticle exposure, translocation and excretion

organelles and cells is a departure from the bulk material or larger particles. This is attributed to the surface effects and the quantum effects, hence determining their chemical kinetics and dynamics, and the mechanical, optical, electric and magnetic properties. The consequent biological interaction of the nanoparticles can be advantageous or detrimental. For instance, they may demonstrate antioxidant activity or may be exploited for enhanced carrier capacity and barrier penetration for drugs and therapeutics. However, toxicity, oxidative stress and cellular dysfunction remain serious adversities. The evident outcome has been the prospect of utilizing nanotechnology for the potential medical and environmental exploits, shadowed by the toxicological aspects and the fear of the unknown.

Comparable to cellular components in size, having highly reactive large surface area, nanoparticles readily interact with cells and molecules inside the human system. Cellular interactions with nanoparticles cause oxidative stress, inflammation and DNA damage. Nanoparticles due to their small sizes pass across the cell membranes and interact with intracellular organelles such as mitochondria, cytoplasm, lipid vesicles or the structural elements based on factors such as size, shape and charge. Entry of nanoparticles inside the cell can be through passive uptake or adhesive interaction. It has been suggested that nanoparticles create reactive oxygen species (ROS) and thereby modulate intracellular calcium concentrations, activate transcription factors and induce cytokine production. The cellular interactions exhibited by nanoparticles are capable of causing abnormalities or even cell death.

Attributed to their larger surface area to volume ratio than their larger-sized counterparts, nanoparticles have the ability to generate more free radicals and ROS in vitro and in vivo. ROS have damaging effects on cells including lipid peroxidation, structural and functional attenuation of proteins, DNA damage, meddling with the signalling cascades and disrupting transcription. Primarily, nanoparticle-induced ROS production owes to the presence of oxidants and free radicals on the surface of the particles. Alternatively, catalysis effectuated by transition metal nanoparticles can intrude the cellular biochemistry to generate ROS by Fenton type reactions. Moreover, nanoparticles, if translocated inside the mitochondria, may cause physical damage or biochemical interferences and precipitate oxidative stress. The consequent oxidative stress, via cytokine release, or the nanoparticles themselves, via opsonisation, can trigger inflammatory responses by recruitment of phagocytes, further exacerbating ROS and reactive nitrogen species. The elevated ROS can lead to biomolecular damage including that of the nuclear DNA. The consequent DNA modifications may finally lead to cytotoxicity.

Respiratory internalization of ultrafine particles of quartz, asbestos fibres, silicates and other mineral dust particles can lead to pulmonary inflammation, oxidative damage, fibrosis and cytotoxicity. Bronchiolar inflammation can further exacerbate chronic obstructive pulmonary disorders, including asthma, emphysema, bronchiolitis and bronchitis, thereby oppressing alveolar function and gaseous exchange. Inhalation of nanoparticles has also been associated with lung cancer and neurodegenerative diseases. In the gastrointestinal tract, nanoparticle exposure has been associated with Crohn's disease and colon cancer. Once they invade the systemic circulation, nanoparticles may lead to arteriosclerosis, thrombosis, arrhythmia and cardiac failures. Moreover, exposure to organs including liver and spleen may cause pathological manifestations at respective sites. Autoimmunity has also been associated with exposure to certain nanoparticles, including rheumatoid arthritis, scleroderma and lupus. Figure 3 summarizes the pathologies in various organ systems resulting from systemic nanoparticle exposure.

The current pool of knowledge in the field of nanotoxicology is sparse; however, the mobility of these particles and the chemical reactivity necessitate concern and further research. The consideration needs to be a three-tiered investigation of interactions of the nanoparticles with the environment from the site of access to the target organs, determination of the biological responses at the said sites and analysis of the secondary immune reactions at the ultimate site of retention. Besides, the physical and chemical aspects pertaining to toxicity of the nanoparticles need to be studied.

2 Physiochemical Modulators of Nanotoxicity

An obvious and crucial question that rises upon highlighting the consequent oxidative stress and cytotoxicity of the nanoparticles is *'are all nanoparticles toxic and to what extent?'* The answer lies in their physical and chemical properties. The material of the nanoparticle is crucial in dictating its physical and chemical properties. However, it is interesting to notice the degrees to which these properties can be modified or even overridden by the size dependence and surface modifications. This section attempts to encompass most of the relevant aspects pertaining to the determiners of toxicity of nanoparticles.

When particles of two different diameters are compared, for a given material mass, smaller particles show a higher particle density and greater surface area in contrast to the particles with larger sizes. Similarly, nanoparticles have higher particle density and greater surface area compared to bulk particles. The direct consequence of availability of a larger surface area is the enhanced chemical reactivity and hence a higher ability to achieve elevated ROS levels by biochemical meddling as well as increased toxicity. Another analogy that can be drawn from the effect of their size is the effect of agglomeration. Based on their surface energy and chemistry, nanoparticles tend to agglomerate, which is further accelerated by a high particle density or concentration. As the agglomerate size increases beyond the nano range, the toxic effects fade. This is attributed to the increased clearance of the larger-sized agglomerates by the macrophages, thereby reducing the toxicity.

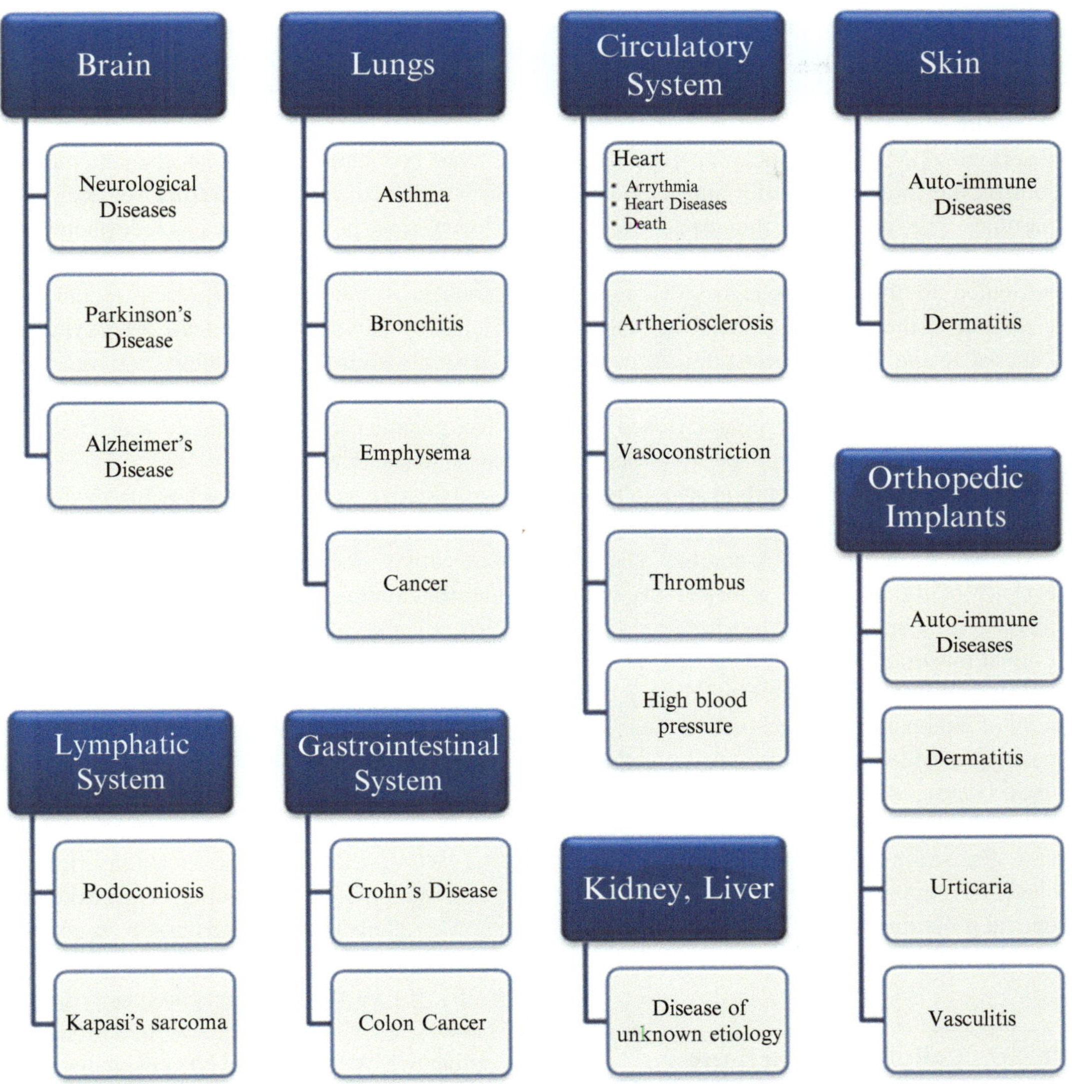

Fig. 3 Pathologies in various organ systems owing to systemic exposure of nanoparticles

Besides ROS generation, the size of the nanoparticles falls in the range of subcellular components, which provides them with higher odds of biological interaction and interference. Nanotoxicity is shown to follow dependence on the particle surface area rather than the administered dose [5].

Further, the material properties of the nanoparticles dictate their chemistry, which in turn governs cellular endocytosis, subcellular localization and ROS catalysis. The toxic effects of a nanoparticle are also dependent on the particle crystallinity. Jiang et al. [5] have demonstrated the ROS generation capacity using crystal variants of TiO_2 nanoparticles – amorphous, anatase, rutile and anatase/rutile mixtures [5]. It was observed that amongst particles with diameter less than 10 nm, at 50 μg/ml, 3 nm amorphous TiO_2 nanoparticles exhibited higher ROS activity versus 4 nm anatase TiO_2. For particles with diameter exceeding 30 nm, amorphous TiO_2 particles had greater ROS activity than anatase TiO_2. Further, pure anatase TiO_2 demonstrated higher ROS activities than anatase/rutile mixtures, and in the anatase/rutile mixtures, rutile TiO_2 nanoparticles showed lower activities than amorphous as well as anatase forms.

Besides the particle size, aspect ratio also determines the nanoparticle toxicity. Aspect ratio is the ratio of diameter and length of the particle. The greater is the aspect ratio, the higher toxicity the nanoparticle shows. This is attributed to the facile phagocytosis-mediated clearance of the particles with shorter aspects in contrast to the larger aspect ratio fibres by the macrophages. Owing to the same reasons, long asbestos fibres fail to be phagocytosed by the macrophages leading to respiratory disorders. Similarly, single-walled carbon nanotubes have been demonstrated to be considerably more toxic than the multiwalled nanotubes. The higher toxicity of the former is achieved via reduced phagocytosis and altered mitochondrial function even at low doses.

Another interesting factor modulating the toxicity of nanoparticles is the halo of their surface coating and functionalization. An effective non-toxic coating on toxic nanoparticles can render them non-toxic and vice versa. Surfactants too can drastically modify their physicochemical characteristics, including magnetic, electric and optical properties, as well as their chemical reactivity, thereby modifying their toxic effects.

3 Cellular Interactions of Nanoparticles

3.1 Oxidative Stress

Oxidative damage has been established as a major mode of nanoparticle-mediated cytotoxicity. The resultant oxidative stress affects the gene expression and the cellular transduction, including Ca2+ and cytokine signalling. The effect has been confirmed for a number of particles, and generation of reactive oxygen species has been demonstrated in vitro and in vivo mediated by nanoparticles ranging from fullerenes to single-walled carbon nanotubes and quantum dots. Further, UV exposure and transition metals have been proven to augment the impact. A plausible theory to explain generation of ROS is the preferential localization of the nanoparticles to mitochondria – the redox hubs – thereby interfering in ROS production and antioxidant fortifications.

The peculiarities of this mechanism mediated by various nanoparticles are not yet deciphered. However, the possible means include photoexcitation of the nanoparticles, redox active metabolic products of the nanoparticles and reticuloendothelial system-mediated release of oxyradicals. These possibilities are certainly not comprehensive and can be further attributed to their small size, chemical composition, surface defects and oxidation state vacancies.

Oxidative stress refers to a redox dysfunction wherein ROS production overpowers the antioxidant activity, thereby manifesting adverse molecular functioning and biological outcome. To map the oxidative stress, a common device is the intracellular ratio of glutathione (GSH) to glutathione disulphide (GSSG). The GSH/GSSG redox couple plays a regulatory role in the maintenance of the redox poise as well as the cellular responses. The harbinger of phase II response is the downstream protective regulator, transcription factor Nrf2, which activates transcription of >200 antioxidants including superoxide dismutase, glutathione S-transferase isoenzymes, haem oxygenase 1, catalase and glutathione peroxidise [6]. Failure of these antioxidant responses is associated with attenuation of ROS production, leading to MAP kinase and NF-κB-mediated proinflammatory responses, and mitochondria-mediated cytotoxicity. While the proinflammatory effects lead to expression of cytokines, chemokines and adhesion molecules, the proapoptotic factors released by the mitochondria lead to cytotoxicity.

Nanoparticle-induced oxidative stress has been studied in vitro as well as in vivo. Using primary mouse embryo fibroblasts, it has been shown that intracellular GSH levels are significantly reduced after exposure to a variety of nanoparticles (carbon black, carbon nanotubes, silica and zinc oxide) [7]. Superoxide dismutase titres have also been correlated linearly with GSH, and lipid peroxidation was demonstrated in the cells in a dose-dependent manner [7].

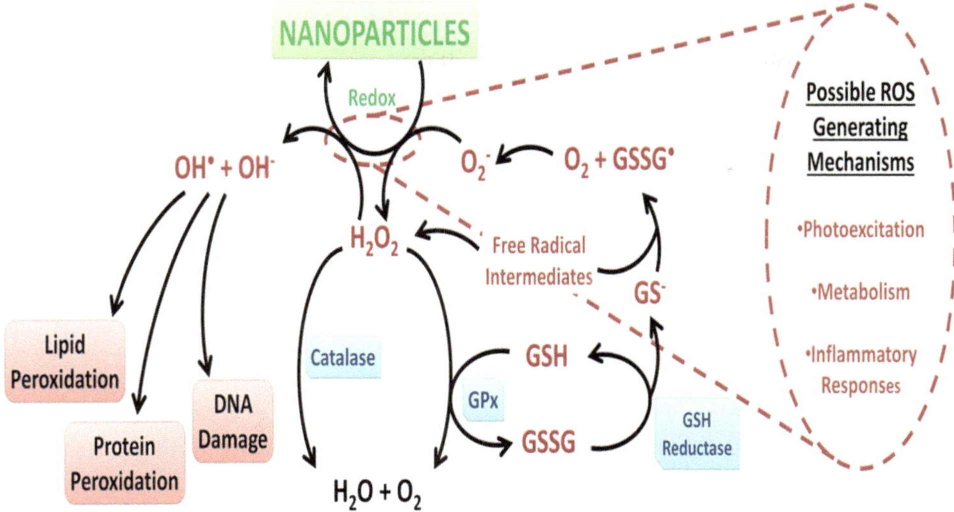

Fig. 4 Schematic representation of ROS generation by nanoparticles with release of oxyradicals

Research in pathophysiologies owing to inhalation of ultrafine particles hailing from air pollution and mineral dust particles has indicated that the ambient ultrafine particles lead to respiratory and cardiovascular distress, with oxidative stress as the underlying mechanisms. ROS and oxidative stress have been demonstrated to generate inflammatory responses and cytotoxicity. Further, it has been observed that the manufactured nanoparticles lead to ROS elevation; fullerenes induce ROS, lipid peroxidation and cytotoxicity; and metal oxide nanoparticles and carbon nanotubes have been shown to be the cause of ROS, lipid peroxidation and inflammatory responses in vitro and in vivo [8].

A schematic representation of plausible mechanism of nanoparticle-mediated ROS generation perturbing antioxidant defence system is shown in Fig. 4. Till date, the biochemical mechanism of ROS generation by different nanoparticles (organic, inorganic NPs like fullerene, CNTs, SWCNTs, MWCNTs, Au NPs, Ag NPs, etc.) has not been deciphered completely. However, the following factors play a major role in oxyradical release – (a) photoexcitation generating free electrons, (b) cytochrome P450-mediated redox active intermediates of nanoparticles and (c) oxyradical released from macrophages following inflammatory response [8].

The following sections discuss the putative mechanisms involved in nanoparticle-mediated cellular damage and nanotoxicity, in general.

3.1.1 Oxidative Stress and NF-κB

Oxidative stress remains the foremost candidate amongst the grounds for nanoparticle-induced cellular damage. Particles including quartz are responsible for manifestation of oxidative stress to yield their proinflammatory effects. The primary factor responsible in generation of oxidative stress is the chemical reactivity at the large surface area of the nanoparticles, for instance, the free radicals generated at the surface of the quartz nanoparticles. These radicals can bring about their effect by further damaging the plasma membranes or interfering in the redox poise and rendering a misbalance in the oxidative stress-mediated signalling.

It is important to highlight the role of NF-κB in nanoparticle-mediated inflammatory responses. Nuclear factor kappa-light-chain-enhancer of activated B cells, abbreviated as NF-κB, is a ubiquitous animal cell transcription factor participating in transcriptional responses to a variety of stimuli essentially responsible for stress, both extracellular and endogenous, thereby effectuating cytokine production and cell survival. The transcriptional regulation of NF-κB varies based on the cell lineage and may be altered/co-regulated by other transcription factors, however, in general, and strongly applies to several genes expressing proinflammatory molecules like IL-2, IL-6, IL-8, GM-CSF and TNF, cell adhesion molecules such as ICAM-1 and E-selectin, inducible nitric oxide synthase (iNOS) and cyclooxygenase-2 (COX-2) [9].

NF-κB, upon cellular stimulation, dissociates from the inhibitory IκB bound to its DNA-binding domain (DBD), undergoes nuclear translocation and accomplishes target gene expression [10]. Both kinase- and redox-signalling pathways, independently involved in the NF-κB activation, are potentially significant for the effects of nanoparticles. NF-κB has been the focus of nanotoxicology research serving as the bridge between the nanoparticle-mediated oxidative stress and the consequent inflammatory signalling.

Schins et al. [11] have reported the significance of NF-κB signalling in response to quartz particles [11]. Quartz participates by depleting IκB, thereby activating NF-κB which subsequently contributes to the production of IL-8. In a recent study by Sharma et al. [12], gold nanoparticles have been demonstrated to activate NF-κB in B lymphocytes, by interacting with NF-κB signalling transduction proteins IκB kinases – IκBα and IκBβ – and also leading to an altered B cell response [12]. The importance of NF-κB in nanoparticle-mediated oxidative stress and inflammation calls for the need for further investigation into particle-specific characteristics of the NF-κB response.

3.2 Receptor-Mediated Effects

Internalization of the nanoparticles can be attributed to two major mechanisms – passive endocytosis and active endocytosis (Fig. 5). Endocytosis has been established as a common way of internalization of viruses and other foreign particles, wherein the particle is encapsulated in a vesicle or endosome derived from the cell membrane, followed by introduction of the encapsulated particle into the cytosol. Endocytosis, mostly, when associated with typical life processes of the cell, is ATP driven, i.e. active. However, ligands targeting the cell surface receptors can also be responsible for passive endocytosis.

One of the most probable receptors involved in the later mechanisms is the epidermal growth factor receptor, EGFR. EGFR is a receptor tyrosine kinase present on the cell surface of the epithelial cells and fibroblasts and activated by the epidermal growth factor (EGF). Post EGF binding, activation cascades of the receptor involve conformational including receptor dimerization, following transphosphorylation of the tyrosine residues. This further leads to the activation of the phospholipase C gamma (PLC-γ) associated to the receptor and the downstream hydrolysis of the phospholipid phosphatidylinositol 4,5-bisphosphate (PIP2) in the second messengers inositol 1,4,5-triphosphate (IP3) and 1,2-diacylglycerol (DAG). IP3 is further responsible for increase in cytosolic concentrations of Ca^{2+} from intracellular reserves, while DAG activates protein kinase C (PKC) to accomplish cellular responses. Also, the associated MAPK/ERK cascade finally leads to increased DNA synthesis and cell proliferation along with elevated turnover of extracellular matrix components (fibronectin, collagen, laminin and glycosaminoglycan).

Metals including arsenic, copper, zinc and vanadium have been known to activate EGFR [13], which has brought EGFR to the centre stage of nanotoxicology since inorganic nanoparticles might affect this receptor by non-specific receptor activation.

3.3 Role of Complement System

Complement is a component of the innate immunity system comprising of proteinaceous biomolecules, circulating as inactive precursor pro-proteins, which, when recruited and activated, complement cell-mediated and humoral immunity. The complement system acts upon bacterial stimulation in a 4-tiered sequence upon recruitment – opsonization to enhance bacterial phagocytosis, chemotaxis to attract neutrophils

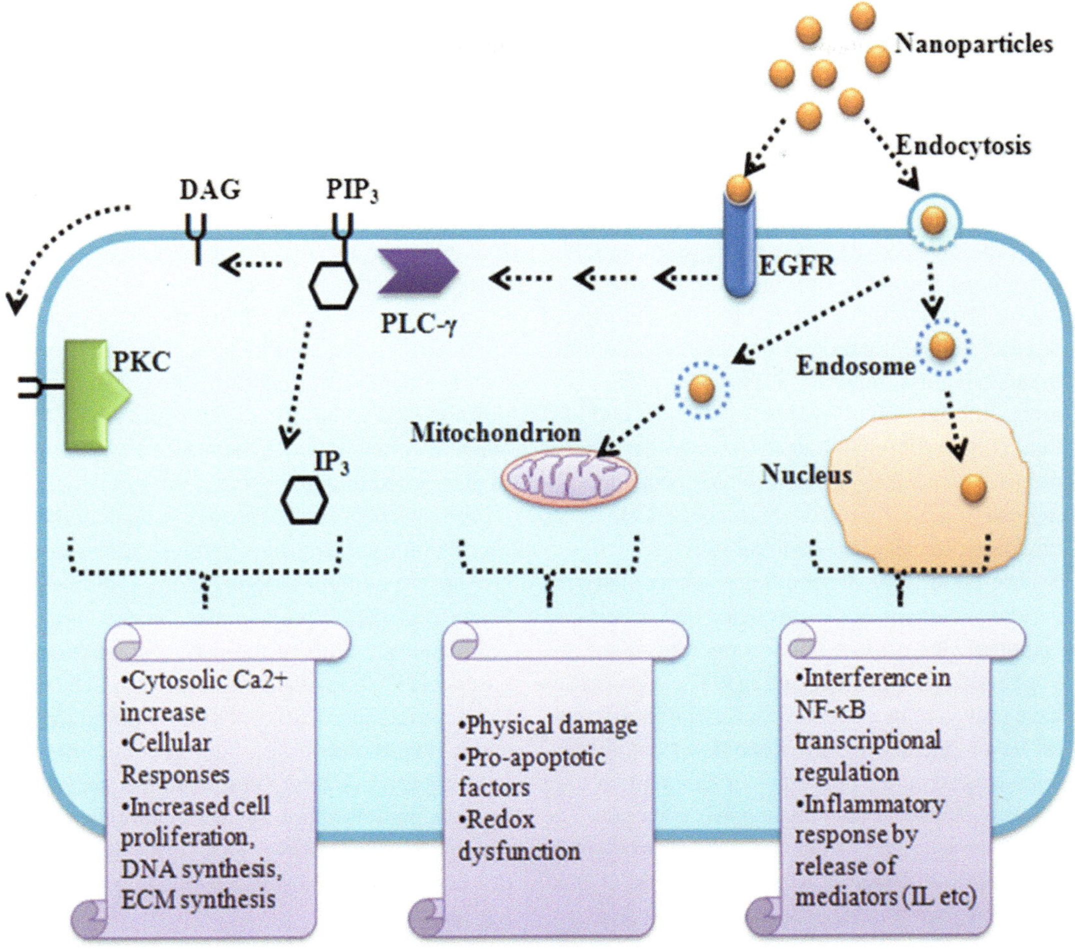

Fig. 5 Schematic showing cellular internalization and biochemical disruption by nanoparticles

and macrophages, activation of membrane attack complex to achieve bacterial lysis and clearance of the clumped residues. However, the complement system is also involved in modulating cellular interactions with nanoparticles. In this case, chemotaxin C5a is generated which is responsible for activation and recruitment of the polymorphonuclear leukocyte (neutrophils), followed by the activity of the opsonin C3b to enhance phagocytosis [14].

For nanoparticles, the complement system can skew the balance on either side, i.e. following opsonization, the nanoparticles may undergo rapid clearance from the body via complement receptor-mediated phagocytosis or on the other hand, elevate the immune response to lead to an allergic reaction or anaphylaxis. This necessitates research in penchant to complement activation by nanoparticles intended to be administered or probable to undergo exposure systemically.

When an elevated antigen presentation is of merit, for instance, vaccines, nanoparticle-mediated complement activation manifested locally upon subcutaneous or intradermal administration is beneficial. However, if the need be, surface properties of the nanoparticles can be altered to tune the described interactions. In 2003, Chanan-Khan et al. investigated hypersensitivity reactions in patients administered with Doxil – a nanoliposome formulation of doxorubicin – and demonstrated complement activation as the mechanism of the dose-rate-dependent hypersensitivity towards Doxil [15]. Further, the surface charge of the nanoparticles has been recognized as a crucial parameter governing the activation and recruitment of the complement system. In general, nanoparticles bearing surface charges have a better ability to activate the complement system. Additionally, immobilization of polyethylene glycol (PEG) or poloxamine molecules have shown to diminish complement activation [16]. On the contrary, surface coatings of dextran augment nanoparticle-mediated complement activation; however, it depends on the polymer conformation [17]. Hence, it can be deduced that nanoparticle-mediated complement activation is governed by a plethora of aspects, including the charge, surface coating, as well as its thickness, density and configuration.

The toxicity of nanoparticles is modulated by various factors, both physical and chemical. Nanoparticles possess novel properties, quite distinguished from the bulk particles. These properties have proved to be extremely beneficial for successful developments in various fields of application. Furthermore, a few properties of nanoparticles are the causal factors behind nanoparticle toxicity. The toxic characteristic of nanoparticles can be attributed to the smaller size, large surface area, crystalline form, aspect ratio, concentration and dose.

4 Nanotoxicology and Challenges

With the recent increase in the use of nanoparticles commercially, the exposure and the associated concerns have also grown. Consequently, the field of nanotoxicology has been gaining importance. However, relatively young, there exist lack and lacunae in the research of nanotoxicology. With increasing evidences of nanoparticle-mediated cytotoxicity, there is a need of thorough investigation in the underlying pathological mechanisms of various nanoparticles and the extent of toxicity. There exists no comprehensive repertoire of the relevant information on nanoparticles. Further, there is no regulatory agency that looks into the toxicological information generated out of the research in the field. The potential adverse effects of the nanoparticles included in commercial products are limited to the on-label specifications, which more than often are completely omitted.

Further, due to absence of any golden standards in the field, the research lacks in merit. Most of the data has been generated from cell-free assays or in vitro studies, which may not be correlated to the in vivo toxicities. Moreover, due to interaction with the body fluids, the surface chemistry of nanoparticles may undergo modifications, thereby altering its toxicity. The nanoparticles may also be deposited in the body if they evade clearance, which increases the possibility of attenuated toxicities.

When introduced in the environment or the biological fluids, many nanoparticles undergo massive agglomeration or aggregation. Agglomeration leads to modified particle size, surface area and sedimentation traits of the particles and hence its tissue localization as well as its toxicity. The toxicity of the nanoparticles is obviously tissue specific, and nanoparticles have a high systemic mobility; hence, there is a need to estimate the toxicities of the nanoparticles upon systemic administration in suitable animal models. This further necessitates the development of suitable imaging techniques for nanoparticles.

References

1. Sioutas C, Delfino RJ, Singh M (2005) Exposure assessment for atmospheric ultrafine particles (UFPs) and implications in epidemiologic research. Environ Health Perspect 113:947–955
2. Afshari A, Matson U, Ekberg LE (2005) Characterization of indoor sources of fine and ultrafine particles: a study conducted in a full-scale chamber. Indoor Air 15:141–150
3. Buzea C, Pacheco II, Robbie K (2007) Nanomaterials and nanoparticles: sources and toxicity. Biointerphases 2:MR17–MR71
4. Choi HS, Ashitate Y, Lee JH et al (2010) Rapid translocation of nanoparticles from the lung airspaces to the body. Nat Biotechnol 28:1300–1303
5. Jiang J, Oberdörster G, Elder A et al (2008) Does nanoparticle activity depend upon size and crystal phase? Nanotoxicology 2:33–42
6. Cho HY, Reddy SP, Kleeberger SR (2006) Nrf2 defends the lung from oxidative stress. Antioxid Redox Signal 8:76–87

7. Kim YJ, Yu M, Park HO (2010) Comparative study of cytotoxicity, oxidative stress and genotoxicity induced by silica nanomaterials in human neuronal cell line. Mol Cell Toxicol 6:336–343
8. Oberdörster G, Oberdörster E, Oberdörster J (2005) Nanotoxicology: an emerging discipline evolving from studies of ultrafine particles. Environ Health Perspect 113:823–839
9. Hallenbeck JM (2002) The many faces of tumor necrosis factor in stroke. Nat Med 8:1363–1368
10. Perkins ND (2007) Integrating cell-signalling pathways with NF-κB and IKK function. Nat Rev Mol Cell Biol 8:49–62
11. Schins RP, McAlinden A, MacNee W et al (2000) Persistent depletion of I kappa B alpha and interleukin-8 expression in human pulmonary epithelial cells exposed to quartz particles. Toxicol Appl Pharmacol 167:107–117
12. Sharma M, Salisbury RL, Maurer EI et al (2013) Gold nanoparticles induce transcriptional activity of NF-kappaB in a B-lymphocyte cell line. Nanoscale 5:3747–3756
13. Wu W, Graves LM, Jaspers I et al (1999) Activation of the EGF receptor signaling pathway in human airway epithelial cells exposed to metals. Am J Physiol 277:L924–L931
14. Donaldson K, Tran CL (2002) Inflammation caused by particles and fibers. Inhal Toxicol 14:5–27
15. Chanan-Khan A, Szebeni J, Savay S et al (2003) Complement activation following first exposure to pegylated liposomal doxorubicin (Doxil): possible role in hypersensitivity reactions. Ann Oncol 14:1430–1437
16. Owens DE 3rd, Peppas NA (2006) Opsonization, biodistribution, and pharmacokinetics of polymeric nanoparticles. Int J Pharm 307:93–102
17. Bertholon I, Vauthier C, Labarre D (2006) Complement activation by core–shell poly (isobutylcyanoacrylate)–polysaccharide nanoparticles: influences of surface morphology, length, and type of polysaccharide. Pharm Res 23:1313–1323

Tools and Techniques to Measure Oxidative Stress

Vibha Rani, Suryma Asthana, Mohit Vadhera, Umesh Chand Singh Yadav, and Neha Atale

Abstract

Oxidative stress overwhelms the natural antioxidant defense system by creating imbalance in production and consumption of reactive oxygen species (ROS). A number of reactive molecules and free radicals exemplify the ROS. Accurate measurement of ROS may help in the diagnosis of various diseases such as diabetes, cancer, and cardiovascular diseases. The robust and sensitive assays are required for its detection and quantification. In this chapter, we describe various techniques to measure the oxidative stress by formation of oxidative by-products of lipids, proteins, and nucleic acids as well as the probing with various compounds. Methods including trapping, spectrofluorimetry, flow cytometry, ELISA, and antibody-based assays have been discussed. Understanding the tools and techniques to measure oxidative stress will help researchers to overcome various complications due to overproduction of reactive species (RS).

Keywords

Reactive species (RS) • Electron spin resonance (ESR) • Malondialdehyde (MDA) • Advanced glycation end products (AGEs)

V. Rani (✉) • S. Asthana • M. Vadhera • N. Atale
Department of Biotechnology, Jaypee Institute of Information Technology, A-10, Sector-62, Noida 201307, Uttar Pradesh, India
e-mail: vibha.rani@jiit.ac.in

U.C.S. Yadav
School of Life Sciences, Central University of Gujarat, Sector 30, Gandhinagar, Gujarat 382030, India

V. Rani and U.C.S. Yadav (eds.), *Free Radicals in Human Health and Disease*, DOI 10.1007/978-81-322-2035-0_4, © Springer India 2015

1 Introduction

Oxidative stress is found to be associated with the pathophysiology of various diseases. The overproduction of reactive oxygen species (ROS) is a principal cause of disease [1, 2]. Estimation of ROS in living organisms conveys a noteworthy analytical challenge. ROS processed or detoxified in cellular compartments requires methods employed directly to the particular cell confinement [3]. The protocols for measuring cellular ROS levels may be an important tool to contribute to a better understanding of its metabolism in cell-based systems. Knowledge of these methods and the actions that decrease it would prove to be therapeutically beneficial. Oxidative damage to the cells due to ROS can be prevented by antioxidant treatment. Such a treatment should put off the onset of that disease or at least delay it [4, 5].

Most reactive species (RS) are highly reactive and have a brief life. Thus, they are hard to estimate in intricate cellular environment [6]. RS processed or detoxified in cellular compartments requires methods employed directly to the particular cell confinement. RS might be measured either specifically or after the establishment of oxidative by-products of lipids, proteins, or nucleic acids (fingerprinting) [7]. Procedures to accurately measure these intermediates have been widely investigated. While correlating oxidative stress with the development of various diseases, the free radicals (or the oxidative damage it causes) should be observed at the site of injury. RS formation and the observed tissue damage should have a parallel time course. Application of the RS *in vitro* over the same time course should recreate most tissue harm caused *in vivo*. Eliminating the RS or repressing their formation should reduce the tissue damage proportional to the amount of reduction [8, 9].

It has become critical that methods be adapted, modified, and standardized to estimate the levels of these molecules in cells and tissues. The chapter describes about various techniques to measure ROS and oxidative damage with those applicable to human studies. Following are some techniques described to measure oxidative stress (Fig. 1).

2 Electron Spin Resonance

Electron spin resonance (ESR) is a technique which detects the vicinity of unpaired electrons. Since reactive radicals do not aggregate to sufficiently high levels to be measured, poorly reactive radicals, with less sensitivity, are detected by this technique. Spin trapping is used to overcome this sensitivity issue. In spin trapping, ROS are allowed to react with specially selected trap molecules to produce less reactive and more stable species that can be readily detected by ESR [10].

A wide range of ESR traps include:

- N-tert butyl-p-phenylnitrone (PBN)
- 5,5-Dimethyl-1-pyrroline N-oxide (DMPO)
- 1,1,3-Trimethyl-isoindole N-oxide (TMINO)
- 5,5-Diethylcarbonyl-1-pyrroline N-oxide (DECPO)
- N-2-(2-ethoxycarbonyl-propyl)-a-phenylnitrone (EPPN)
- 5-Diethoxyphosphoryl-5-methyl-1-pyrroline N-oxide (DEPMPO)
- 5-Tert-butoxycarbonyl-5-methyl-1-pyrroline N-oxide (BMPO)

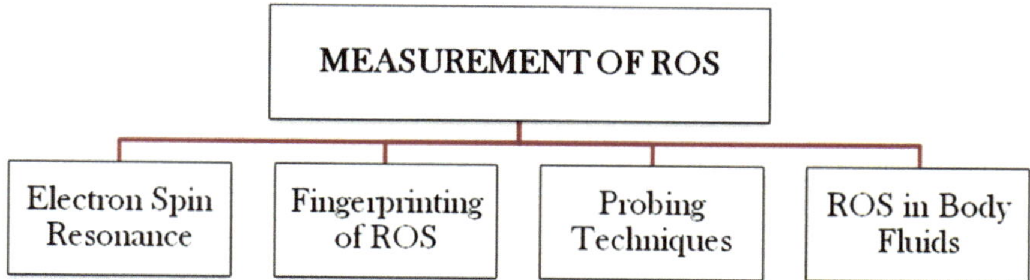

Fig. 1 Methods for the measurement of ROS generation

2.1 Trapping of RS

There are various traps which are used in ESR to scavenge the radicals.

2.1.1 Ex Vivo Trapping

Spin traps cannot be used in humans because of unknown toxicity at the high levels. However, these traps might be utilized on extracted body fluids and tissue samples [11]. Since highly reactive radicals and non-radical species do not survive to be detected in ex vivo material, ESR of *ex vivo* samples with traps detects secondary radicals resulting from reaction of RS with biomolecules. These secondary radicals include lipid-derived and protein radicals.

2.1.2 Ascorbate as a Trap

Ascorbic acid (vitamin C) reacts with many free radicals. Semidehydroascorbate radical, an oxidation product of ascorbic acid, can easily be detected by ESR. Its measurement can be used as a biomarker for RS production in organs and blood plasma. This is a semiquantitative method because:

- Ascorbate radicals react quickly with each other and generate ESR-silent molecules.
- This radical can also be reduced by enzymes *in vivo* [12].

2.1.3 Aromatic Traps

These are known to be acceptable for human consumption. Traps like salicylate and phenylalanine are widely used to measure ex vivo radical formation.

- Salicylate is hydroxylated by OH radical to form 2-3dihydroxybenzoate which is not produced enzymically in vivo [13].
- Both L- and D-phenylalanine are hydroxylated by OH radical to yield ortho- and meta-tyrosines, which are not enzymically produced in humans [14]

The success of these traps depends on their concentration at sites of free radical generation. Aromatic hydroxylation techniques are not quantitative measures of radical generation, especially if the end products are measured in the plasma, where their origin is unclear.

2.1.4 Urate as a Trap

An endogenous compound like urate, which is readily oxidized by a range of RS including peroxynitrite, can be used as a trap. Several groups have used urate as a "selective" scavenger of ONOO in animal studies. Allantoin is an oxidation product of urate. Its measurement may be a promising method for human application [15].

3 Fingerprinting of ROS

When a free radical reacts with a biological molecule to leave a unique "finger print," then the vicinity of that unique fingerprint might be utilized to gather that the RS has been generated [16]. Such "biomarkers" can then be utilized to research impacts of dietary supplements or engineered antioxidants on oxidative harm caused. In fact, since strategies as of now accessible for the immediate estimation of RS are of restricted application to humans, most clinical research concentrates on the estimation of oxidative harm. The injury caused by these RS is more important than the aggregate amount of such species produced.

3.1 Biomarkers of Superoxide Anion

3.1.1 GSH/GSSG Ratio

Indeed, reduced glutathione (GSH) is recognized as one of the most important nonenzymatic oxidant defenses within the body. It exists in very large amounts (mM levels) within cells where it acts to detoxify peroxides as well as maintain other physiologically important antioxidants in their reduced forms.

Reduced glutathione is continuously regenerated from its oxidized form GSSG by the action of an NADPH-dependent reductase. Because the rates of synthesis of GSH, export of GSH and GSSG from the intracellular compartment to the extracellular space, and formation of protein-bound GSH mixed disulfides are slow relative to the rates of reduction of GSSG and oxidation of GSH, "the balance of GSH and GSSG provides a dynamic indicator of oxidative stress" in vivo [17].

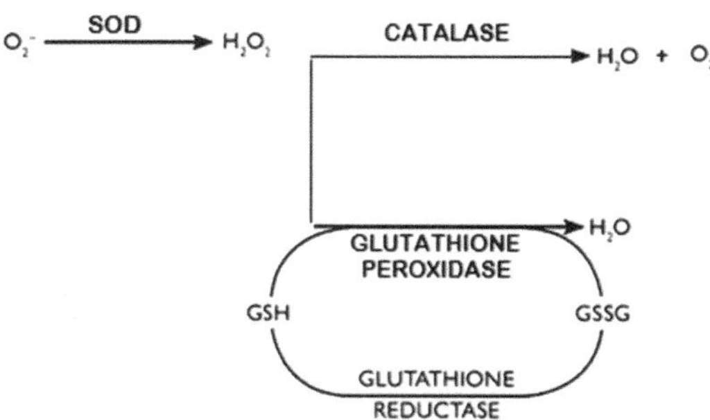

Fig. 2 Mechanism of action of antioxidative enzymes

A few distinctive tests have been developed to measure glutathione in samples. By utilizing a luciferin-derived compound as a part of conjunction with glutathione S-transferase, the measured level of GSH might be proportional to the luminescent indicator produced when luciferase is added subsequently. DTNB (Ellman's reagent) on reacting with GSH generates glutathione. The level of glutathione can be resolved colorimetrically with in the vicinity of glutathione reductase at 412 nm (Fig. 2) [18].

3.1.2 Cytochrome C Reduction

Ferricytochrome c is reduced to ferrocytochrome c on reacting with O_2^-. The absorption spectrum of ferrocytochrome c is different compared to ferricytochrome and can easily be estimated spectrophotometrically. However, sensitivity of this technique is limited [19].

3.2 Biomarkers for Hydrogen Peroxide

3.2.1 Horseradish Peroxidase-Linked Assays

Numerous assays for checking the presence of H_2O_2 are based on the oxidation of another compound. Hydrogen donor compounds are oxidized in the presence of H_2O_2 by HRP.

Hydrogen peroxide (H2O2) is the most important ROS in regard to signaling or cell cycle regulation [20]. Many substrates serve as hydrogen donors and have been used along with horseradish peroxidase (HRP) enzyme to produce fluorescence. Commonly used substrates include:

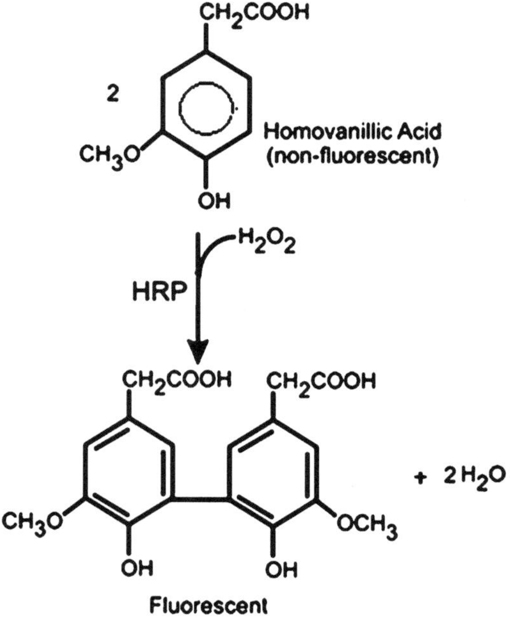

Fig. 3 Dimerization of homovanillic acid by the action of HRP and hydrogen peroxide

- Homovanillic acid – It forms a dimer when oxidized by hydrogen peroxide through the catalysis of horseradish peroxidase. As a dimer, its peak excitation wavelength is 315 nm, with an emission wavelength of 425 nm (Fig. 3) [21].
- Amplex® Red – It is converted to resorufin by hydrogen peroxide in the presence of HRP, which is a highly colored compound and can be detected colorimetrically at 570 nm or by fluorescence using excitation of 570 nm and emission of 585 nm (Fig. 4) [22].

3.3 Measuring Lipid By-products

Lipids can undergo oxidation, chlorination and nitration by a wide range of RS. Lipid peroxidation is a complex phenomenon, and many products are formed in varying amounts. Lipid oxidation can be measured in many ways but commonly used methods are the following.

3.3.1 TBARS Assay

Thiobarbituric acid-reactive substances – TBARS – are generated as by-products of lipid

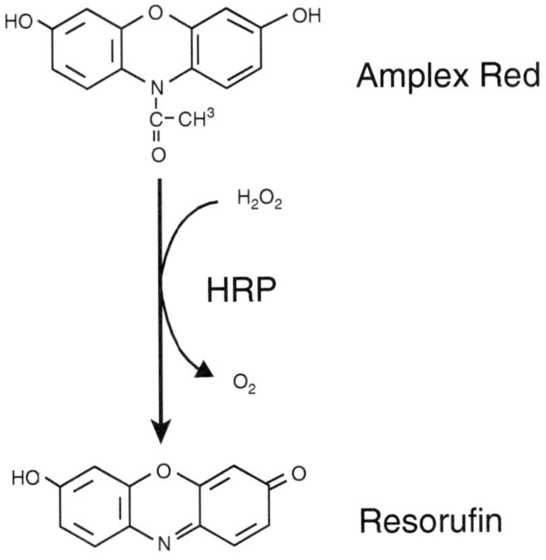

Fig. 4 Conversion of Amplex Red to resorufin by HRP using H_2O_2

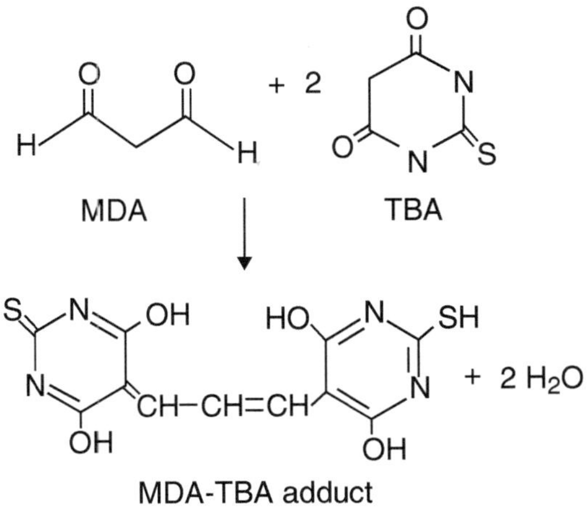

Fig. 5 Illustration of lipid peroxidation. Detection of MDA by formation of MDA-TBA adducts. Colorimetry or fluorescence-based measurement can be performed

peroxidation (i.e., on degradation of fats) which can be detected by the TBARS assay using thiobarbituric acid as a reagent.

Since RS have extremely short half-lives, it is difficult to measure them directly. Hence, several products of the damage caused by oxidative stress can be measured, such as TBARS [23].

The TBARS assay measures malondialdehyde (MDA) present in the sample, as well as MDA formed by the hydrolytic reaction of lipid hydroperoxides. MDA is among the several low-molecular-weight end products formed due to the breakdown of primary and secondary lipid peroxidation products (Fig. 5).

However, only certain lipid peroxidation products generate MDA, and MDA is neither the sole end product of fatty peroxide formation and decomposition nor a substance generated exclusively through lipid peroxidation [24].

Since MDA can also arise from free radical attack on sialic acid and deoxyribose, direct measurement of MDA is not preferred.

3.3.2 The Isoprostanes

Isoprostanes are specific end products of the peroxidation of polyunsaturated fatty acids. The most extensively used, F2-isoprostanes, are derived from arachidonic acid, but some others are also available. These are neuroprostanes or F4-isoprostanes from eicosapentaenoic and docosahexaenoic acids [25]. Isoprostanes are best measured by mass spectrometry. Although isoprostanes can be detected in foods, they do not appear to pass through the gut in sufficient quantities to affect plasma or urinary levels [26].

Isoprostanes are rapidly metabolized and undergo quick turnover. Thus, slower metabolism could also be a possible reason for the rise in plasma isoprostane levels. Apart from whole-body measurement, they can also be measured in fluids drawn from specific sites, such as synovial fluid, pericardial fluid, wound exudate, and breath condensate.

3.3.3 Exhaled Hydrocarbons

Exhaled air contains a range of hydrocarbons, including ethane and pentane. Evidences suggest that ethane may be a biomarker of lipid peroxidation, as well as pentane. However, they are minor end products of lipid peroxidation, and their formation also depends on the environment of the reaction. They are also difficult to estimate routinely in human studies and require complex equipment.

3.4 Biomarkers for DNA Damage

Oxidative DNA damage seems to relate to an increased risk of disease development later in life. DNA subjected to attack by hydroxyl radical generates a huge range of base and sugar modification products. Such products can be measured by HPLC, gas chromatography (GC)-MS, liquid chromatography (LC)-MS, and antibody-based techniques [27]. Initial products of RS attack upon purines, pyrimidines, and deoxyribose form stable end products. The relative amount of these end products depends on the reaction conditions. Thus, it is not reliable to measure any single reaction product as an indication of oxidative DNA damage.

The HO• attacks DNA strands when it is produced adjacent to cellular and mitochondrial DNA, which lead to the formation of many oxidation products. The interaction of HO• with the nucleobases of the DNA strand, such as guanine, leads to the formation of C8-hydroxyguanine (8-OHGua) or its nucleoside form deoxyguanosine (8-hydroxy-2-deoxyguanosine) [28]. Initially, the addition of HO• leads to the generation of radical adducts; subsequently 8-hydroxy-2-deoxyguanosine (8-OH-dG) is formed by one-electron abstraction.

3.4.1 Antibody-Based Techniques

The problems of artifactual DNA oxidation during DNA isolation and analysis can be avoided by measuring oxidative DNA damage in the intact cell. Antibody-based methods have been developed for 8OHdG and are useful for visualization of damage but seem likely to be semiquantitative.

3.4.2 Comet Assay

The comet assay can be applied directly to cells and measures DNA strand breaks. The number of DNA strand breaks can be used to estimate the level of oxidized DNA bases in the cell. Strand breaks arise not only during oxidative DNA damage but also during repair of that damage.

The comet assay (single-cell gel electrophoresis) is a simple method for measuring deoxyribonucleic acid (DNA) strand breaks in eukaryotic cells [29]. Cells are embedded in agarose on a microscope slide and lysed with detergent and high salt. Nucleoids are formed containing supercoiled loops of DNA which are linked to the nuclear matrix. Electrophoresis at high pH generates structures resembling comets, which can be observed by microscopy. The intensity of the comet tail relative to the head gives an estimate of the number of DNA breaks. The loops containing a break lose their supercoiling and are free to extend towards the anode. Visual analysis with staining of DNA and calculation of fluorescence is done to determine the extent of DNA damage. This can be done by manual scoring or automatically by imaging software.

3.4.3 Ultraviolet Light

Ultraviolet light produces a wide variety of lesions in the DNA, for example, thymine dimers. The levels are increased after exposure to sunlight and in psoriasis. Sunlight-induced DNA damage can be estimated by using thymine dimers as biomarker.

3.5 Biomarkers for Proteins

Oxidative damage to proteins leads to their by-products and affects the function of receptors, enzymes, transport proteins, and antigen-antibody-based immune responses. The damage to other biomolecules such as inactivation of DNA repair enzymes and loss of fidelity of damaged DNA polymerases in replicating DNA can be observed by different labeling techniques. The analysis of proteins is more complex than that of DNA because 20 amino acids are found in literature and the different pathways of mechanism are found for reactive species functioning. Free radical attack on proteins can generate amino-acid radicals, which may crosslink or react with O_2 to give peroxyl radicals. Abstraction of H triggers the formation of more free radicals forming protein peroxides, which decompose in complex ways, accelerated by transition metal ions, to generate yet more radicals [30]. Oxidation of proteins during cooking of food can confound the measurement of these products in body fluids as potential biomarkers, since they could be absorbed through diet.

3.5.1 The Carbonyl Assay

It is the most commonly used biomarker of protein damage. Carbonyl assay is the estimation of protein carbonyl groups. Carbonyls can be formed as a result of:

- Glycation of proteins by sugars
- By the binding of aldehydes
- By the direct oxidation of amino-acid side chains by RS

Products like glutamate and aminoadipic semialdehydes are measured spectrophotometrically through ELISA. Immunochemical tests for protein-bound aldehydes such acrolein and 4-HNE are generally utilized. The carbonyl test as connected to tissues and body liquids measures the average amount of protein alteration.

3.5.2 Advanced Glycation End Products (AGEs)

Although, the formation of AGEs is not dependent entirely on oxidative stress, they can be used as potential biomarkers. AGEs result from nonenzymatic glycation reactions to proteins and DNA [31]. The reaction is slow and begins with the formation of a Schiff base which is converted to a more stable Amadori product. Nε-(caroboxymethyl) lysine (CML) and pentosidine are some of the common biomarkers.

Depending on conditions, AGEs can increase or decrease the oxidative stress by increased RS production or activation of antioxidant enzymes respectively.

AGEs have known to be associated with several diseases including cardiovascular disease, Alzheimer's, cancer, and diabetes. Hence, AGEs are useful biomarkers of metabolic diseases, but because they are not necessarily RS-mediated, their use for measuring oxidative stress is generally avoided. AGEs can be measured using fluorescence-based assays such as spectrofluorometry, FACS, etc.

4 Probing Methods for RS

The fluorescence methodology, associated with the use of suitable probes, is an appropriate approach to measure RS in biological and nonbiological environments because of its high sensitivity, ease in data collection, and high spatial resolution in microscopic imaging techniques. Some common fluorescent probes used for the detection of RS includes:

(a) DCFH
(b) Dihydrorhodamine 123 (DHR)
(c) Dihydroethidium (dihydroethidine) (DHE)
(d) Luminol and lucigenin (LC)
(e) 2-[6-(4′-Hydroxy) phenoxy-3H-xanthen-3-on-9-yl] benzoic acid (HPF) and 2- [6-(4′-amino) phenoxy-3H-xanthen-3-on-9-yl] benzoic acid (APF),
(f) Fluorescein
(g) 2-Methyl-6-(4-Methoxyphenyl)-3,7-Dihydroimidazo [1,2-A] pyrazin-3-One, Hydrochloride (MCLA)
(h) *Cis*-parinaric acid,
(i) 4,4-difluoro-5-(4-phenyl-1,3,-butadienyl)-4-bora-3a,4a-diaza-s-indacene-3-undecanoic acid (C11-BODIPY $^{581/591}$)
(j) Coelenterazine
(k) 8-Amino-5-chloro-7-phenylpyridazo [3,4-*d*] pyridazine-1,4-(2H, 3H) dione (L-O12)
(l) Dipyridamole

4.1 Dichlorofluorescin Diacetate (DCFDA)

DCFDA is one of the most frequently used probes to detect "cellular peroxides" DCFDA enters cells and accumulates mostly in the cytosol. DCFDA is deacetylated by esterases to dichlorofluorescein (DCFH). This nonfluorescent product is converted by RS into DCF, which can easily be visualized by strong fluorescence at 525 nm when excited at 488 nm. Washing cells after loading allows unreacted DCFDA to diffuse out. DCFH and DCF can also diffuse out and undergo extracellular reactions [32, 33]. The major RS that can oxidize DCFH are alkoxyl, $NO_2^•$, carbonate ($CO_3^{•-}$), peroxyl, peroxynitrite, and $OH^•$ radicals. Both H_2O_2 and $O_2^{•-}$ cannot react with DCFH.

DCF fluorescence is a generalized assay for oxidative stress. It does not directly measure H_2O_2, $NO^•$, lipid peroxides, and singlet O_2 or $O_2^{•-}$. Cytochrome *c* is a powerful catalyst of

DCFH oxidation. Thus, the use of DCFDA as a probe for oxidative stress during apoptosis should be done carefully, since a rise in cytoplasmic cytochrome *c* levels could give a bigger "signal" without any change in cellular peroxide levels. One-electron oxidation of DCFH by different radicals and heme proteins produces intermediate radicals, including phenoxyl radicals that can interact with such cellular antioxidants as GSH and ascorbate and with NADH to create more RS [34].

Cells should be loaded with low concentrations of DCFDA to avoid any cytotoxicity. Serum-free media must be used since serum contains endogenous esterase activity and the permeability of de-esterified dichlorofluorescein (DCF) is less. Fluorescence micrographs of DCFH DA stained cells indicate the ROS overproduction and confer the red signal on TRITC filter. Care must be taken so as to avoid measuring light emission from the medium as well as from the cells in case DCF and DCFH leaks out of cells (Fig. 6).

4.2 Dihydrorhodamine 123 (DHR)

DHR is a probe widely used to detect several RS ($OH^•$, $ONOO^-$, $NO_2^•$, peroxidase-derived species) but is less reactive with $O_2^{•-}$, H_2O_2, or $NO^•$. DHR is oxidized to rhodamine 123, which is highly fluorescent at 536 nm when excited at about 500 nm. Rhodamine 123 is lipophilic and positively charged which accumulates in the mitochondria due to the membrane potential. Thus after formation, minimal amount of rhodamine 123 leaks out of cells. DHR is also more sensitive at detecting HOCl than DCFDA. Rhodamine 123 and ethidium can be ejected from cells by drug conjugate efflux pumps, so the presence and activity of the membrane transport systems in the cells being studied is a factor that needs to be considered [35].

At high levels, rhodamine 123 can sensitize singlet O_2 formation in mitochondria and cause NAD (P) H oxidation. Exposure to the synthetic hydroperoxide *tert*-butylhydroperoxide caused

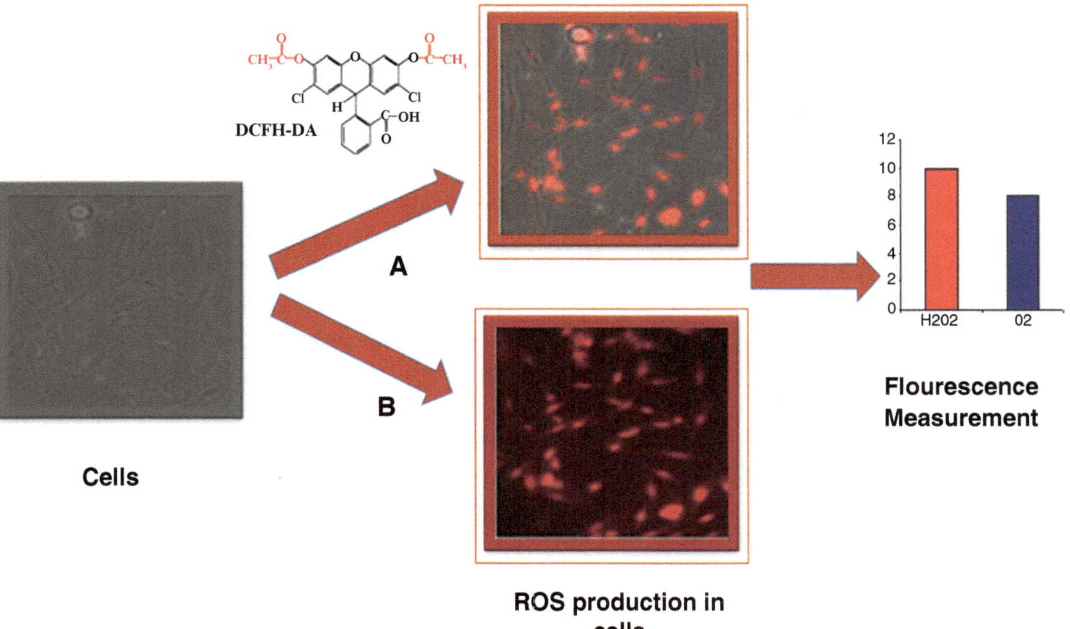

Fig. 6 ROS overproduction. Detection of ROS by DCFH-DA fluorescence and its quantification (**A**) superimposed image of bright field and DCFH-DA fluorescence showing ROS generation (**B**) fluorescence micrograph of DCFH-DA fluorescence on TRITC filter. Images were taken at 40× magnification under fluorescence microscope

light emission from dihydrorhodamine. This appeared to be due to increased formation of $NO^{\bullet}$ leading to $ONOO^-$ formation, which in turn activated the enzyme phospholipase A_2 to raise cellular arachidonic acid levels. The source of DHR response was thought to be arachidonic acid metabolism to RS [36].

4.3 Dihydroethidium (Dihydroethidine) (DHE)

This is frequently used as a probe for $O_2^{\bullet-}$, being oxidized to a fluorescent product. This is usually thought to be ethidium, which tends to intercalate into nuclear DNA.

Ethidium fluoresces strongly at around 600 nm when excited at 500–530 nm. Recent work, however, suggests that the product is not ethidium. DHE is readily oxidized and often gives a high background. It can also undergo a direct redox reaction with ferricytochrome *c* [37].

4.4 Luminol and Lucigenin (LC)

These two compounds are often used to detect the production of RS by activated scavenger cells. A luminol analogue L-012 was reported to be more effective than luminol for sensing $O_2^{\bullet-}$ and $ONOO^-$ and better than dihydroethidine for $O_2^{\bullet-}$ detection. The use of luminol to detect $O_2^{\bullet-}$ is a problematic orbit. It does *not* react directly with $O_2^{\bullet-}$ but must first be oxidized in a one-electron step (e.g., by $OH^{\bullet}$, $ONOO^-$, or peroxidase plus H_2O_2). The resulting luminol radical reacts with $O_2^{\bullet-}$ to generate a light-emitting product. The luminol radical can also reduce O_2 to generate $O_2^{\bullet-}$, that is, the presence of luminol plus an oxidizing agent can lead to artifactual $O_2^{\bullet-}$generation. Subsequently luminol is an inconsistent test; any oxidizing executor that can oxidize luminol by one electron will result in light discharge inhibitable by SOD and the luminol is both the source and the finder of the $O_2^{\bullet-}$.

LC is accepted to be more specific for the identification of $O_2^{\bullet-}$than luminol, yet again it doesn't react straightforwardly with $O_2^{\bullet-}$. It should first be lessened to LC cation radical ($Lc^{\bullet+}$), which then responds with $O_2^{\bullet-}$ to give the fluorescent product. Transformation of LC to $Lc^{\bullet+}$ can't be accomplished quickly by $O_2^{\bullet-}$, and requires other cellular reducing frameworks (e.g., xanthine oxidase, the mitochondrial electron transport chain, or the phagocyte NADPH oxidase), presenting a clear intricacy in deciphering outcomes. $Lc^{\bullet+}$ can additionally diminish O_2 to $O_2^{\bullet-}$, that is, addition of LC can artifactually produce more $O_2^{\bullet-}$ [38–40]. The degree to which this can meddle with correct estimation of $O_2^{\bullet-}$is still being wrangled about; however, it seems, by all accounts, to be critical in some cell frameworks. Alternative tests that could be more particular might be coelenterazine and the luciferin analogue 2-methyl-6-(p-methoxyphenyl)-3,7-dihydroimidazo [1,2-] pyrazin-3-one (MCLA) .

4.5 Probes of Lipid Peroxidation/ Membrane RS

A few membrane-partitioning tests have been acquainted to measure lipid peroxidation or RS inside membranes. One is cis-parinaric acid. At the point of encountering lipids experiencing peroxidation, it is quickly oxidized, losing its fluorescence. Parinaric acid has a polyunsaturated structure which makes it profoundly prone to nonspecific oxidation, thus it ought to be stored in the dark with a nitrogen environment. It might be included as the free acid or consolidated into particular phospholipids (parinaroyl lipids) and used to study their relative susceptibilities to oxidative harm inside the membrane [41].

Another probe is diphenyl-1-pyrenylphosphine, reported to react with peroxides to generate a product that fluoresces at 380 nm when excited at 351 nm. It is light and air sensitive.

Once more, the vicinity of these different tests could change the rate, and maybe, the complex pathway, of lipid peroxidation. Serum proteins, particularly albumin, devotedly bind fatty acids and can inhibit entry of probes into cells and may additionally cause moderate filtering out when cells are incubated in the media with included serum. Yet, serum-free cells undergo oxidative stress, which can increase the background generation of RS.

5 Output Measurement Techniques

1. The least complex strategy is the fluorescence microplate reader, where information is exhibited as increments or abatements in relative fluorescence. Plate readers measure the total fluorescence. They don't recognize intracellular and extracellular fluorescence from reactions in the medium.
2. Flow cytometry has an advantage of having the capacity to measure the intracellular fluorescence of cells. Quantitative information on the amounts of cells emanating fluorescence can be acquired instead of simply relative fluorescence units. An extra drawback of flow cytometry is that the cytometer is at room temperature as opposed to 37 °C, which can give variable data with some cell types.
3. Confocal microscopy is another effective instrument. Cells are stacked with fluorescent colors and observed in real time in situ at 37 °C. The intracellular localization of RS can be seen, and the contribution of mitochondrial, endoplasmic reticulum, or lysosomal reactions in oxidative stress may be pictured utilizing counter stains such as ER Tracker or Lysotracker.

6 Measuring ROS in Body Fluids

ROS can be measured in body fluids like blood, plasma, semen, and urine. Many biomarkers are found in body fluids, and they can be used to assess rates of "whole-body" DNA damage. Problem with measuring free radicals in body fluids is that direct (or semi direct) detection of free radical species is possible in very special cases, and generally only secondary free radical products are detectable in body fluids, for example, biomarkers in urine gives no information about its tissue origin, although in some cases, sample from specific sites such as cerebrospinal fluid can be taken for direct detection.

Urine has the potential to measure simultaneously range of oxidative stress products and other inflammatory markers. We will now discuss different biomarkers present in body fluids due to oxidative modification of lipids, proteins, and DNA.

6.1 Oxidative Modifications of Lipids in Body Fluids

6.1.1 F_2-Isoprostanes

F_2-isoprostanes are generated throughout nonenzymatic oxidation of arachidonic acid by distinctive sorts of free radicals, including ROS. Contingent upon the position where oxygen is added to arachidonic acid, four regioisomers are formed, giving each of the four F2-isoprostane series [42, 43].

The levels of urinary F2-isoprotanes are moderately stable in people, yet are broadly variable in human populations. Thus, they are very functional as biomarkers for human studies. Urinary F2-isoprotanes levels have been accepted as sensitive biomarkers of oxidative stress, making them an important device for surveying oxidative stress. F2-isoprostanes are proposed for correlations of oxidative stress between people.

6.1.2 Aldehydes Formed in Lipid Peroxidation

Like F2-isoprostanes, a few profoundly reactive and subsequently harmful aldehydes are formed as a consequence of peroxidation of polyunsaturated unsaturated fatty acids. Some of these aldehydes have been utilized as biomarkers of oxidative stress including:

Malondialdehyde (MDA). MDA is a frequently used biomarker that is measured in plasma, urine, or tissue as thiobarbituric acid-reactive (TBAR) material but can't be used as a systemic biomarker of oxidative status [44].

4-Hydroxy-2-nonenal (4-HNE) and 4-oxo-2-nonenal (4-ONE). The free levels of 4-HNE and 4-ONE are very unstable since they are exceedingly reactive aldehydes and effortlessly form covalent bonds with protein thiol and amino groups and with other molecules.

Urinary levels of 1,4-dihydroxynonane-mercapturic acid (DHN-MA), an end product of 4-HNE metabolism, have been measured frequently [45].

Acrolein. Another reactive aldehyde produced from lipid peroxidation is Acrolein. Lysine adducts of acrolein are measured in the urine [46].

6.2 Oxidative Modifications of Proteins in Body Fluids

6.2.1 Protein Adducts Produced by Lipid Peroxidation Products

Each of the products formed by lipid peroxidation can form adducts with the amino-acid residues of protein. Cysteine, histidine, and lysine residues are the common targets.

4-HNE-amino-acid adducts were detected in hepatocytes oxidized with tert-butylhydroperoxide or metal ions and in rat plasma in response to iron overload, a known promoter of ROS [47].

Acrolein-lysine adducts were formed in vitro in LDL, as a result of copper-catalyzed oxidation. Acrolein-lysine is the main product caused by the reaction of acrolein to protein.

6.2.2 Dityrosine

Tyrosine radical is generated by many RS like hydroxyl radicals and peroxynitrite. It results in the formation of dityrosine on reaction with proteins [48]. The amount of dityrosine produced is proportional to the oxidative stress and radical formation rate. It cannot differentiate between oxidative and nitrosative stress because it is produced from a lot of RS.

Dityrosine imprints damaged proteins for degradation and is discharged in urine. Urinary dityrosine could be measured through chromatographic techniques with mass spectrometry recognition. The range of urinary dityrosine between individuals is no less than fourfold. Intraindividual levels don't fluctuate altogether throughout the day. In this way, dityrosine seems, by all accounts, to be a guaranteeing biomarker of oxidative status.

6.3 Oxidative Modifications of DNA

An extensive number of purine- and pyrimidine-derived products are formed due to RS reaction with DNA. 8-Hydroxy-2′-deoxyguanosine (8-oxodG) is the most considered biomarker produced by nonenzymatic DNA oxidation [49, 50]. 8-OxodG levels can be measured in the urine and have been approved in animal models with the help of CCl_4 but not in humans yet.

An important source of interindividual variability of 8-oxodG levels in fluid is the variety in DNA repair capacity among different individuals. This may not be linked with RS-induced attack and damage. Healthy individuals and diseased patients can have the same levels of 8-oxodG. But the former is due to less DNA repair capacity, while the latter may be due to oxidative DNA damage.

8-OxodG levels are a valid biomarker of oxidative stress. However, interindividual variation due to DNA repair is a potential problem and requires further studying.

Three fundamental systems used to examine F2-isoprotanes in urine samples are gas chromatography with mass spectrometry detection (GC-MS), liquid chromatography with coupled mass spectrometry identification (LC-MS/MS), and enzyme-linked immunosorbent assay (ELISA). The ELISA-based test is naturally less effective against the chromatography-based methods because of cross-reactivity.

Estimations of urinary 8-oxodG include chromatography-based systems and ELISA. Chromatography-based methods indicate low variability among different assays.

7 Conclusion

Different methods were used to measure the oxidative stress but their mechanisms need to be understand and also to study that what is likely to confound and how quantitative the methods can be (how, what and how much). With careful attention to understanding these points, errors can be minimized and the field of free radical

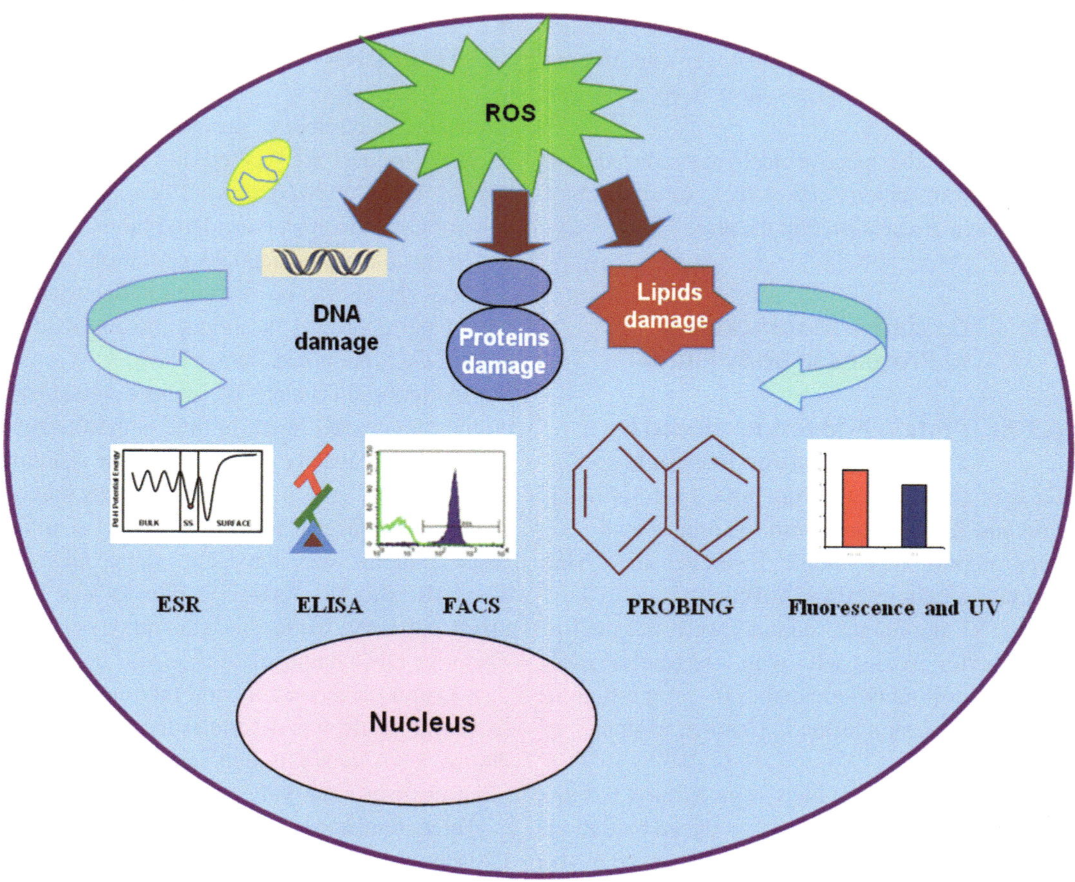

Fig. 7 Diagrammatic representation of oxidative stress-associated damages and their assessment by various techniques

research can continue to move forward. Various methods of ROS detection in cells have been summarized in Fig. 7 and stated the quantitation of free radicals.

The biggest difficulty reported with much of the cellular ROS research has been with the lack of reporter agents specific for discrete molecules. ROS moieties by their nature are reactive with a number of different molecules; as such designing reporter agents has been difficult. Protein adducts are more stable than the aldehydes from which they were derived; thus, they are believed to be better biomarkers of oxidative status. These adducts reflect the extent of oxidative damage to proteins that cause protein dysfunction. These adducts can be measured using ELISA. Further study of these biomarkers is necessary to understand the role of these biomarkers.

To further sharpen the tools for measuring ROS, it calls for continued efforts in designing better techniques like small-molecule fluorescent ROS probes with improved selectivity, reversible kinetics, and compartment-targeting property. Meanwhile biologically inspired novel ROS indicators could be developed on the basis of their sensitivity and specificity.

By searching the enormous chemical space of small molecules and of proteins, the current repertoire of ROS indicators can be extended and ultimately a level of reliability can be achieved. This knowledge can help us in verifying the effects of natural compounds on subsequently arising diseases. In synergy with the exponentially increasing numbers of indicator-expressing organisms and disease models, the booming technology for super-resolution and single-molecule imaging, and the trend for using miniature, plant-in devices to obtain images in conscious, free-moving animals, imaging ROS in vivo will serve as the most powerful force to transform the landscape and push forward the frontiers in ROS signaling.

Acknowledgment We acknowledge the Department of Biotechnology, DBT (Govt. of India) for financial support. This work was supported by a research grant awarded to Dr Vibha Rani by the Department of Biotechnology (DBT), Government of India (BT/PR3978/17/766/2011).

References

1. Valkoa M, Leibfritzb D, Moncol J (2007) Free radicals and antioxidants in normal physiological functions and human disease. Int J Biochem Cell Biol 39:44–84
2. Li X, Fang P, Mai J et al (2013) Targeting mitochondrial reactive oxygen species as novel therapy for inflammatory diseases and cancers. J Hematol Oncol 6:19
3. Shulaev V, Oliver DJ (2006) Metabolic and proteomic markers for oxidative stress. New tools for reactive oxygen species research. Plant Physiol 141:367–372
4. Lobo V, Patil A, Phatak A et al (2010) Free radicals, antioxidants and functional foods: impact on human health. Pharmacogn Rev 4:118–126
5. Osakwe ON, Siegel A (2013) A novel standardized oxygen radical absorbance assay for evaluating antioxidant natural products. J AOAC Int 96:1365–1371
6. Dreive C, Rice-Evans C (2001) The mechanisms for nitration and nitrotyrosine formation in vitro and in vivo: impact of diet. Free Radic Res 35:215–231
7. Tarpey MM, Wink DA, Grisham MB (2004) Methods for detection of reactive metabolites of oxygen and nitrogen: in vitro and in vivo considerations. Am J Physiol Regul Integr Comp Physiol 286:431–444
8. Rahman K (2007) Studies on free radicals, antioxidants, and co-factors. Clin Interv Ageing 2:219–236
9. Turrens JF (2003) Mitochondrial formation of reactive oxygen species. J Physiol 552:335–344
10. Halliwell B, Whiteman M (2004) Measuring reactive species and oxidative damage in vivo and in cell culture: how should you do it and what do the results mean? Br J Pharmacol 142:231–255
11. Kaur H, Edmonds SE, Blake DR et al (1996) Hydroxyl radical generation by rheumatoid blood and knee joint synovial fluid. Ann Rheum Dis 55:915–920
12. Ingelman-Sundburg M, Kaur H, Terelius Y et al (1991) Hydroxylation of salicylate by microsomal fractions and cytochrome P-450. Lack of production of 2,3-dihydroxybenzoate unless hydroxyl radical formation is permitted. Biochem J 276:753–757
13. Li M, Carlson S, Kinzer JA et al (2003) HPLC and LC–MS studies of hydroxylation of phenylalanine as an assay for hydroxyl radicals generated from Udenfriend's reagent. Biochem Biophys Res Commun 312:316–322
14. Kaur H, Halliwell B (1990) Action of biologically-relevant oxidizing species upon uric acid. Identification of uric acid oxidation products. Chem Biol Interact 73:235–247
15. Halliwel B, Gutteridge JM (1999) Free radicals in biology and medicine. Oxford University Press, Oxford
16. Jones DP (2002) Redox potential of GSH/GSSG couple: assay and biological significance. Methods Enzymol 348:93–112
17. Rahman I, Kode A, Biswas SK (2006) Assay for quantitative determination of glutathione and glutathione disulfide levels using enzymatic recycling method. Nat Protoc 1:3159–3165
18. Bolt HM, Hengstler JG, Stewart J (2009) Analysis of reactive oxygen species. EXCLI J 8:241–245
19. Wang X, Fang H, Huang Z et al (2013) Imaging ROS signaling in cells and animals. J Mol Med 91:917–927
20. Ruch W, Cooper PH, Baggiolini M et al (1983) Assay of H_2O_2 production by macrophages and neutrophils with Homovanillic acid and horseradish peroxidase. J Immunol Methods 63:347–357
21. Reszka KJ, Wagner BA et al (2005) Effects of peroxidase substrates on the Amplex red/peroxidase assay: antioxidant properties of anthracyclines. Anal Biochem 342:327–337
22. Wasowicz W, Neve J et al (1993) Optimized steps in fluorometric determination of thiobarbituric acid-reactive substances in serum: importance of extraction pH and influence of sample preservation and storage. Clin Chem 39:2522–2526
23. Janero DR (1990) Malondialdehyde and thiobarbituric acid-reactivity as diagnostic indices of lipid peroxidation and peroxidative tissue injury. Free Radic Biol Med 9:515–540
24. Morrow JD, Awad JA, Kato T et al (1992) Formation of novel non-cyclooxygenase-derived prostanoids (F2-Isoprostanes) in carbon tetrachloride hepatotoxicity: an animal model of lipid peroxidation. J Clin Invest 90:2502–2507
25. Halliwell B (1999) Oxygen and nitrogen are pro-carcinogens. Damage to DNA by reactive oxygen, chlorine and nitrogen species: measurement, mechanism and the effects of nutrition. Mutat Res 443:37–52
26. Sung CC, Hsu YC (2013) Oxidative stress and nucleic acid oxidation in patients with chronic kidney disease. Oxid Med Cell Longev 301982:1–56
27. Collins AR (2004) The comet assay for DNA damage and repair: principles, applications, and limitations. Mol Biotechnol 26:249–261

28. Collins AR, Dušinská M et al (2001) Inter-individual differences in repair of base oxidation, measured in vitro with the comet assay. Mutagenesis 16:297–301
29. Adlam HA, Davies MJ (2003) Cell-mediated reduction of protein and peptide hydroperoxides to reactive free radicals. Free Radic Biol Med 34:44–55
30. Petrat F, Pinduir S, Kirsch M et al (2003) NAD(P)H, a primary target of 1O2 in mitochondria of intact cells. J Biol Chem 278:3298–3307
31. Ohashi T, Mizutani A, Murakami A et al (2002) Rapid oxidation of dichlorodihydrofluorescin with heme and hemoproteins: formation of the fluorescein is independent of the generation of reactive oxygen species. FEBS Lett 511:21–27
32. Rota C, Fann YC, Mason RP (1999) Phenoxyl free radical formation during the oxidation of the fluorescent dye 20,70-dichlorofluorescein by horseradish peroxidase. Possible consequences for oxidative stress measurements. J Biol Chem 274:28161–28168
33. Eruslanov E, Kusmartsev S (2010) Identification of ROS using oxidized DCFDA and flow-cytometry. Methods Mol Biol 594:57–72
34. Buxser SE, Sawada G, Raub TJ (1999) Analytical and numerical techniques for evaluation of free radical damage in cultured cells using imaging cytometry and fluorescent indicators. Methods Enzymol 300:256–275
35. Zhao H, Kalivendi S, Zhang H et al (2003) Superoxide reacts with hydroethidine but forms a fluorescent product that is distinctly different from ethidium: potential implications in intracellular fluorescence detection of superoxide. Free Radic Biol Med 34:1359–1368
36. Benov L, Sztejnberg L, Fridovich I (1998) Critical evaluation of the use of hydroethidine as a measure of superoxide anion radical. Free Radic Biol Med 25:826–831
37. Daiber A, Oelzeb M, August M et al (2004) Detection of superoxide and peroxynitrite in model systems and mitochondria by the luminol analogue L-012. Free Radic Res 38:259–269
38. Faulkner K, Fridovich I (1993) Luminol and lucigenin as detectors for O_2. Free Radic Biol Med 15:447–451
39. Tarpey MM, White CR, Suarez E et al (1999) Chemiluminescent detection of oxidants in vascular tissue. Lucigenin but not coelenterazine enhances superoxide formation. Circ Res 84:1203–1211
40. Liochev SI, Fridovich I (1997) Lucigenin luminescence as a measure of intracellular superoxide dismutase activity in Escherichia coli. Proc Natl Acad Sci U S A 94:2891–2896
41. Ritov VB, Banni S, Yalowich JC et al (1996) Non-random peroxidation of different classes of membrane phospholipids in live cells detected by metabolically integrated cis-parinaric acid. Biochim Biophys Acta 1283:127–140
42. Milne GL, Musiek ES, Morrow JD (2005) F2-isoprostanes as markers of oxidative stress in vivo: an overview. Biomarkers 10:S10–S23
43. Milne GL, Sanchez SC, Musiek ES et al (2007) Quantification of F2-isoprostanes as a biomarker of oxidative stress. Nat Protoc 2:221–226
44. Nielsen F, Mikkelsen BB, Nielsen JB et al (1997) Plasma malondialdehyde as biomarker for oxidative stress: reference interval and effects of life-style factors. Clin Chem 43:1209–1214
45. Spickett CM (2013) The lipid peroxidation product 4-hydroxy-2-nonenal: advances in chemistry and analysis. Redox Biol 1:145–152
46. Kawai Y, Furuhata A, Toyokuni S (2003) Formation of Acrolein-derived 2′-deoxyadenosine adduct in an iron-induced carcinogenesis model. J Biol Chem 278(50):50346–50354
47. Uchida K (2003) 4-Hydroxy-2-nonenal: a product and mediator of oxidative stress. Prog Lipid Res 42:318–343
48. Pacher P, Beckman JS, Liaudet L (2007) Nitric oxide and peroxynitrite in health and disease. Physiol Rev 87:315–424
49. Valavanidis A, Vlachogianni T, Fiotakis CJ (2009) 8-hydroxy-2′ -deoxyguanosine (8-OHdG): a critical biomarker of oxidative stress and carcinogenesis. Env Sci Health Part C Env Carcinog Ecotoxicol Rev 27:120–139
50. Martinet W, Knaapen MVW, Guido RY et al (2002) Elevated levels of oxidative DNA damage and DNA repair enzymes in human atherosclerotic plaques. Circulation 106:927–932

Part II

Mechanism of Oxidative Stress

Impact of Oxidative Stress on Human Health

M.M. Towhidul Islam and Hossain Uddin Shekhar

Abstract

The evolution of aerobic metabolic processes such as respiration unavoidably led to the production of reactive oxygen and nitrogen species (ROS and RNI) in mitochondria and peroxisomes of human. A common feature among the different ROS types is their capacity to cause oxidative damage to proteins, DNA, and lipids. Depending on the nature of the ROS species, some are highly toxic and rapidly detoxified by various cellular enzymatic and nonenzymatic mechanisms. In other circumstances, human purposefully generates ROS as signaling molecules to control various processes including defence against pathogen and programmed cell death. Information are accumulating steadily which shows oxidative damage of tissue, and cellular components may act as a primary or secondary causative factor in many different human diseases and aging processes. This chapter describes the roles of ROS in different human diseases and ultimate human mechanisms which are applied to control these circumstances.

Keywords

Oxidative stress • Reactive oxygen species • Antioxidant • Disease

1 Introduction

The imbalance between ROS production and antioxidant defense leads to "oxidative stress." "Oxidative stress" was first introduced by Helmut Sies in 1985 to describe the disturbance in pro-oxidant-antioxidant balance. It is a situation when steady-state ROS concentration is transiently or chronically enhanced; as a result it disturbs cellular metabolism and its regulation and finally damages cellular constituents [1]. In the simplest case, pathology originates from the perturbations in either reactive species formation, their elimination, or in both simultaneously. But cell produces ROS not only under stress; recently it was found that many of the cellular mechanisms are also influenced by

M.M.T. Islam • H.U. Shekhar (✉)
Department of Biochemistry and Molecular Biology, Faculty of Biological Sciences, University of Dhaka, Dhaka 1000, Bangladesh
e-mail: hossainshekhar@yahoo.com; shekhardu@hotmail.com

V. Rani and U.C.S. Yadav (eds.), *Free Radicals in Human Health and Disease*,
DOI 10.1007/978-81-322-2035-0_5,

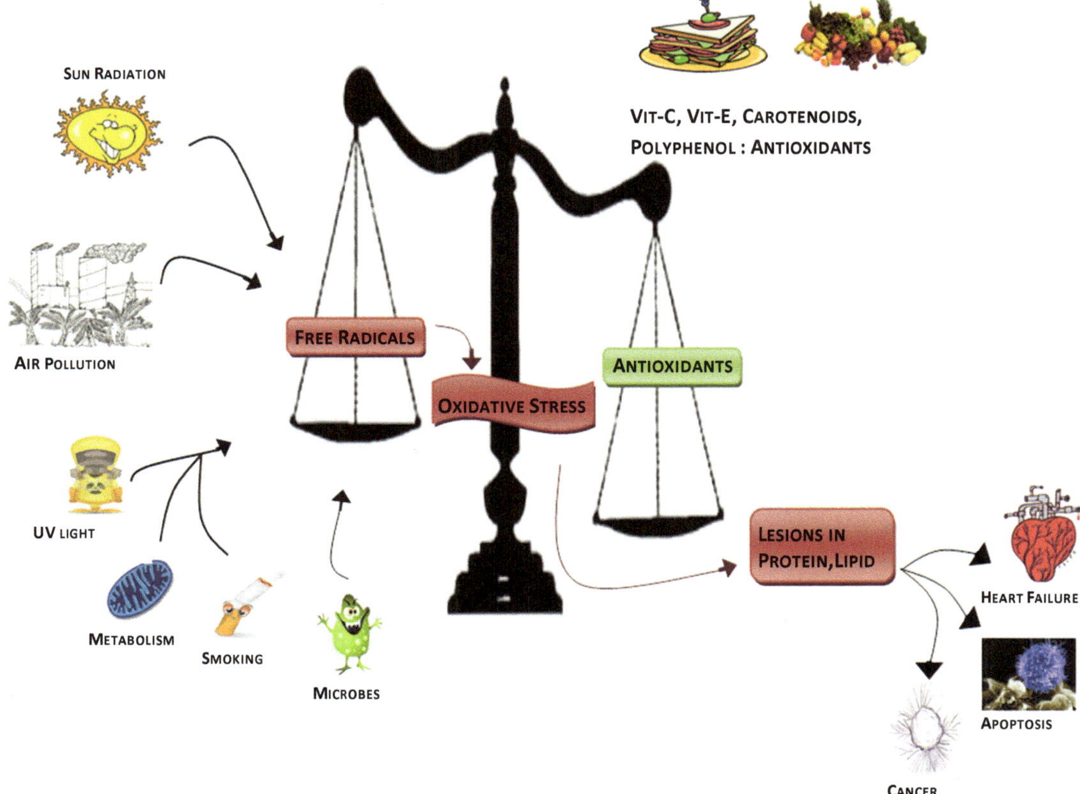

Fig. 1 Oxidative modification of biomolecules. In normal physiological conditions, superoxide, hydrogen peroxide, or other radicals are generated. Generation of intracellular ROS may activate the intracellular signaling pathways. Overproduction of the hydroxyl radical causes oxidative modification of biomolecules, such as lipids, proteins, and nucleic acids, and this ultimately causes various diseases (adapted from http://robinthomas.biz/so-what-causes-oxidative-stress-anyway/)

these compounds as it was found that excess intake of antioxidants has some adverse effects on the body [2]. Thus lowering ROS levels below the homeostatic set point may interrupt the physiological role of oxidants in cellular proliferation and host defense. Similarly, increased ROS may also be detrimental to cell death or to acceleration in aging and age-related diseases (Fig. 1).

Free radicals or reactive species can be defined as any chemical species that contains unpaired electrons. Unpaired electrons increase the chemical reactivity of an atom or molecule. Among the reactive oxygen species (ROS) such as hydroxyl or superoxide radicals, and hydrogen peroxide, singlet oxygen are most important ones. Among the reactive nitrogen species (RNI) such as nitric oxide (NO), peroxynitrite and, others, among them, *S*-nitrosothiols are considered as the major oxidative stress-inducing compounds (Table 1). These free radicals have one or more unpaired electrons in their outer orbital, which make these species very unstable and tending to react with other molecules to have pair electrons and thereby generate more stable species.

Free radicals can be produced by several different biochemical processes within the body for example: reduction of molecular oxygen during aerobic respiration yielding superoxide and hydroxyl radicals; by-products of catecholamines' oxidation and activation of the arachidonic acid cascade produce electrons, which can reduce molecular oxygen to superoxide; production of superoxide and hypochlorous

Table 1 Classification of different reactive species

Reactive species	Radicals	Non-radicals
Reactive oxygen species (ROS)	Superoxide: O_2^-	Hydrogen peroxide: H_2O_2
	Hydroxyl: OH^-	Hypochlorous acid: HOCl
	Peroxyl: RO_2^-	Hypobromous acid: HOBr
	Alkoxyl: RO^-	Ozone: O_3
	Hydroperoxyl: HO_2^-	Singlet oxygen: Δ g
Reactive nitrogen species (RNS)	Nitric oxide: NO^-	Nitrogen dioxide: NO_2
	Nitrous acid: HNO_2	Nitrosyl cation: NO
		Nitrosyl anion: NO^-
		Dinitrogen tetroxide: N_2O_4
		Dinitrogen trioxide : N_2O_3
		Peroxynitrite: $ONOO^-$
		Peroxynitrous acid: ONOOH
		Alkylperoxynitrites: ROONO

acid (HOCl), a powerful oxidant, by activated phagocytes; and nitric oxide production by vascular endothelium and other cells. In addition, free radicals can be produced in response to external electromagnetic radiation, such as gamma rays, which can split water to produce hydroxyl radicals (Fig. 2). In the cell they interact with proteins, carbohydrates, and lipids, with a consequence of alteration of both in the intracellular and intercellular homeostasis, leading to possible cell death and regeneration (Fig. 1) [3].

Under physiological steady-state conditions, these molecules are scavenged by different antioxidative defense components that are often confined to particular compartments like mitochondria, peroxisomes, or cytosol. This system consists of different antioxidative enzymes like superoxide dismutase (SOD), catalase (CAT), glutathione peroxidase (GPx), and glutathione reductase to neutralize these radicals, together with a number of low molecular weight antioxidants such as ascorbate, α- tocopherol and glutathione, cysteine, thioredoxin, vitamins, histidine peptides, the iron-binding proteins transferrin and ferritin, dihydrolipoic acid, melatonin, urate, and plasma protein thiols to protect against free radicals [4]. If the equilibrium between production and scavenging of ROS is perturbed by a number of adverse pathological and environmental factors, then intracellular levels of ROS may rise rapidly. These suggest that antioxidant defense system may play a vital role in the pathogenesis of many human diseases. Many studies have also found strong association between dietary intakes of antioxidant-rich nutrients and blood levels of antioxidant with reduced level of oxidative stress (Fig. 3) [5].

2 Cellular Mechanisms to Control Reactive Species and Radicals

The body has evolved major antioxidant defense mechanisms to protect it from free radical attacks. These defenses can be conveniently considered as cellular, membrane, and extracellular mechanisms.

Cellular antioxidant defenses include the superoxide dismutase (SOD), glutathione peroxidase (GPx), glutathione reductase, and catalase enzymes. Superoxide dismutases (containing copper or zinc at the active site) in cytosol and in mitochondria (containing manganese) catalyze the dismutation of superoxide to hydrogen peroxide and oxygen. The product of this reaction, hydrogen peroxide, is a weak oxidant and is relatively stable. However, unlike superoxide, hydrogen peroxide can rapidly diffuse across cell membranes, and then in the presence of transition metal ions, it can be converted to hydroxyl radicals via Fenton chemistry. Two enzyme systems

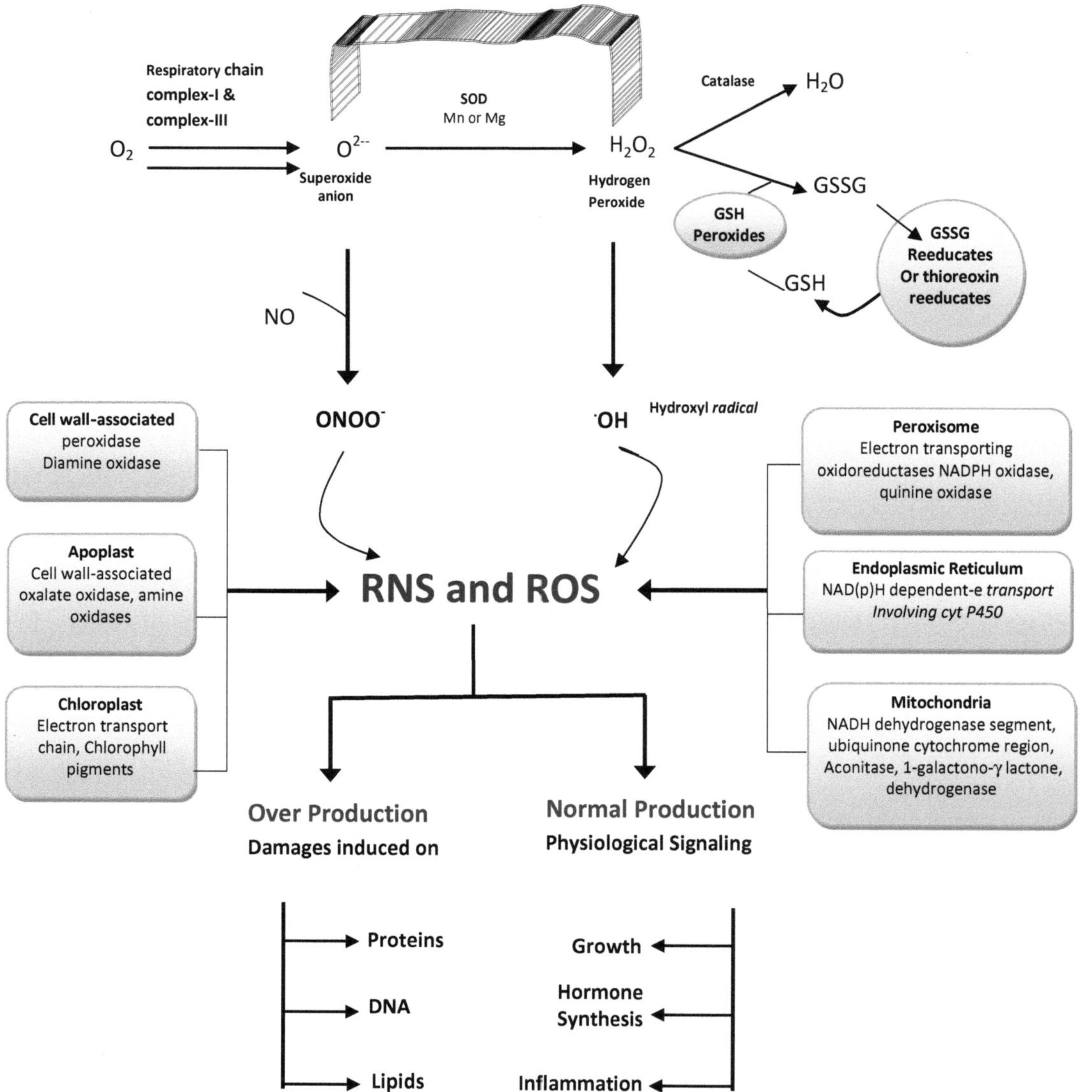

Fig. 2 Production and removal of reactive oxygen species (*ROS*). Superoxide radical ($^{\bullet}O_2^-$) is generated in low levels under physiological states, but its production is greatly enhanced under pathological situations via enzymes such as NADPH oxidase, xanthine oxidase, and via a dysfunctional mitochondrial respiratory chain. $^{\bullet}O_2^-$ is neutralized to water via a two-step process involving superoxide dismutase (*SOD*) in the first step and glutathione peroxidase (*GPx*) or catalase in the second step. Increased production of $^{\bullet}O_2^-$ and/or impairment of antioxidant defense systems leads to a buildup of the intermediate hydrogen peroxide (H_2O_2). H_2O_2 forms the toxic oxygen species hydroxyl anion ($^{\bullet}OH$) via Fenton biochemistry, which is highly reactive and causes lipid peroxidation forming lipid hydroperoxides (*LOOH*). The functional importance of GPx resides in its ability to remove H_2O_2 and LOOH and neutralize these to water and lipid alcohol, respectively. Additionally, the increase in $^{\bullet}O_2^-$ also favors the formation of peroxynitrite ($ONOO^-$) which reduces the bioavailability of nitric oxide ($^{\bullet}NO$). GPx also functions to neutralize $ONOO^-$. If these radicals are not neutralized, they can damage biomolecules (adapted from Frontiers in Bioscience 14, 4015–4034, January 1, 2009)

can break down hydrogen peroxide, one is that glutathione peroxidases present in cytosol and mitochondria have a major role in removing hydrogen peroxide generated by superoxide dismutase with the oxidation of glutathione (GSH) (Fig. 2). The second one is Catalase which are present in peroxisomes in many tissues remove hydrogen peroxide when present in high concentrations [5, 6].

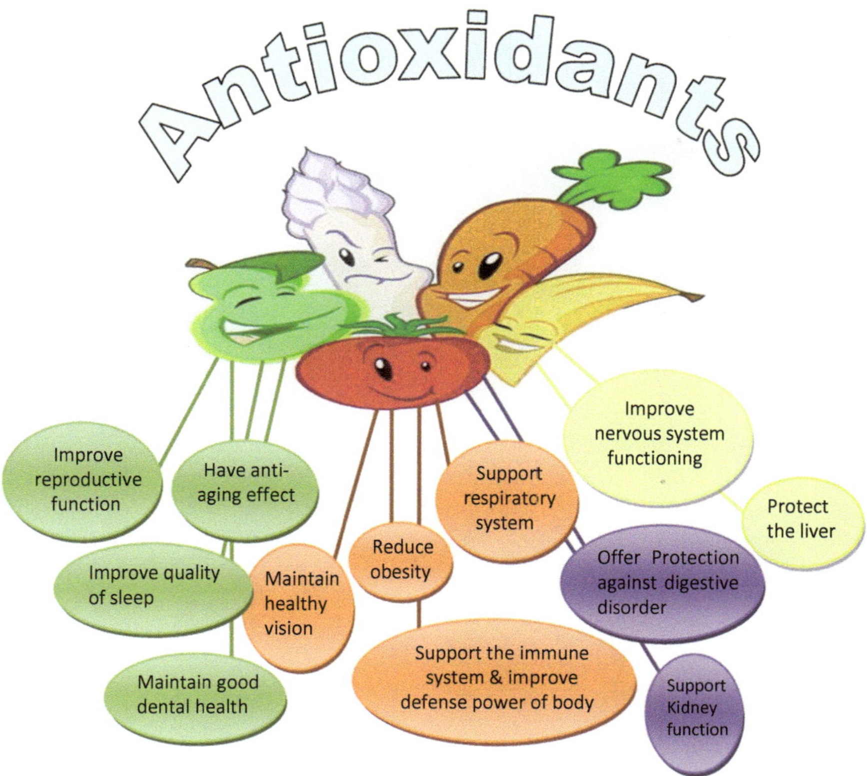

Fig. 3 Beneficial roles of antioxidants in human health (adapted from www.thevemmasolution.com/resources-center/all-about-antioxidant/)

3 Oxidative Stress and Various Human Diseases

Oxidative stress has been involved in the pathogenesis of several diseases including cancer, atherosclerosis, rheumatoid arthritis, and neurodegenerative diseases such as Parkinson's disease, Alzheimer's disease, and Huntington's disease (Fig. 4). The reactive species produced in oxidative stress can cause direct damage to the cellular DNA and are therefore mutagenic, and it may also cause apoptosis and promote proliferation, invasiveness, and metastasis of different cell types. For example, it was known that infection by *Helicobacter pylori,* which is one of the important mediators of gastric cancer, causes an increase in the production of reactive oxygen and nitrogen species in human stomach, which ultimately accelerates the cancer progression [7]. The following sections will discuss in detail the roles of ROS in different human diseases.

3.1 Neuronal Diseases

The brain is highly susceptible to oxidative stress due to its high polyunsaturated fatty acid content, high metabolic rate, and limited regeneration capability [8]. On the other hand, the brain is poor in catalytic activity and has moderate amounts of glutathione peroxidase and superoxide dismutase. That's why neuronal cells may be among all the cell types of the body most vulnerable to oxidative stress. This oxidative stress has been implicated in a variety of neurodegenerative disease, including Alzheimer's disease, Parkinson's disease (PD), and amyotrophic lateral sclerosis. These diseases are associated with protein aggregation and defined by the progressive loss of specific neuronal cell populations. A common feature of these diseases is oxidative damage of neurons, which might be responsible for the dysfunction or death of neuronal cells that contributes to ultimate disease pathogenesis.

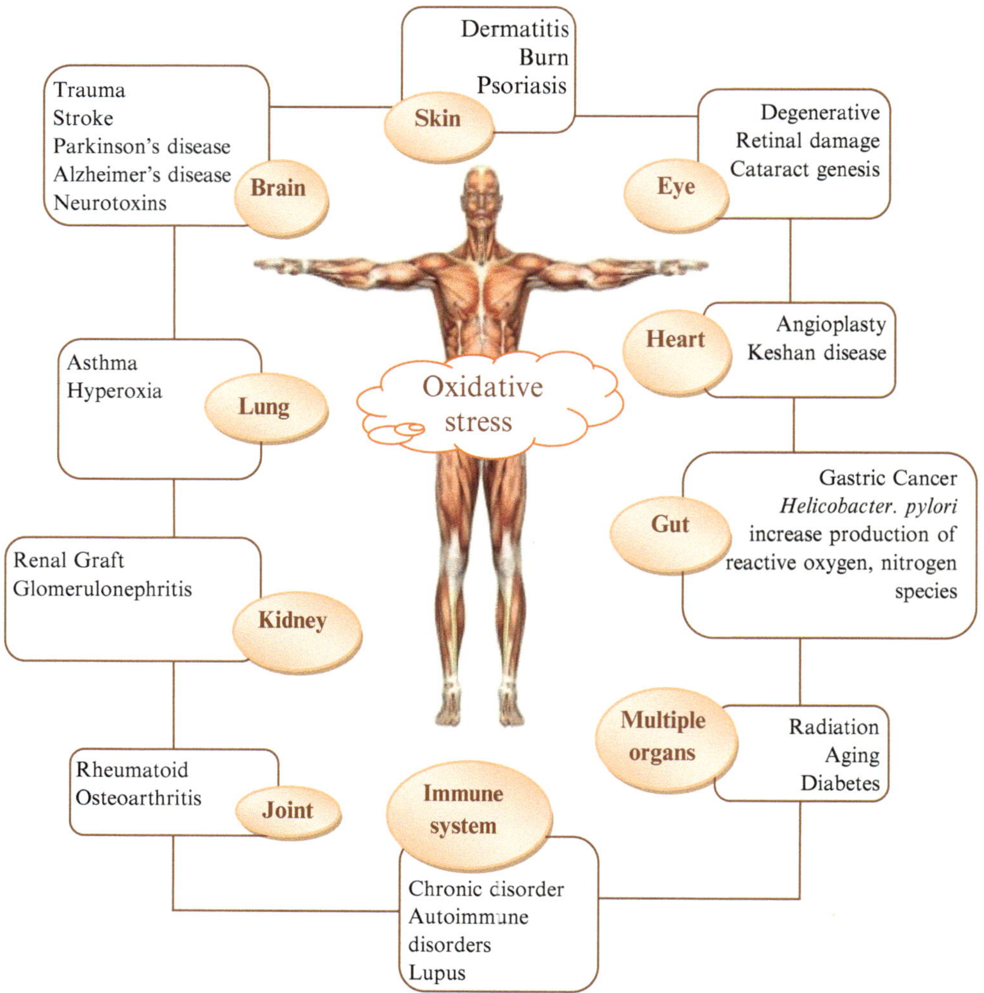

Fig. 4 Different human organs are affected by oxidative stress; if persists, then it ultimately causes deleterious diseases (adapted from www.wisechoiceliving.com/its-radical-part-1/)

In Alzheimer's disease (AD), DNA bases are modified by oxidative stress with hydroxylation, for example, conversion of cytosine to 5-hydroxymethylcytosine [8]. To handle this situation, activity of the antioxidant proteins catalase, superoxide dismutase (SOD), glutathione peroxidase, and glutathione reductase is increased in the hippocampus and amygdala which tries to minimize the oxidative damage of the brain [9].

Parkinson's disease (PD) is characterized by the loss of dopaminergic neurons of the substantia nigra and the deposition of intracellular inclusion bodies. A diverse array of evidence suggests that α-synuclein has a role in modulating the activity of dopamine. The A53T mutation of α-synuclein is associated with familial PD and impairs vesicular storage of dopamine [10], which leads to the accumulation of dopamine in the cytoplasm, and subsequently ROS is generated through its interaction with iron, a process that increases with age. This ROS then cause damage to other neurons.

Amyotrophic lateral sclerosis (ALS) is considered by the deposition of a misfolded protein in neural tissue. In ALS it is mostly copper/

zinc SOD proteins [11]. Mutations of SOD can convert the protein from an antioxidant to a pro-oxidant capable of causing oxidative damage to the cells. There are more than 100 mutations of SOD associated with the familial forms of the disease. Through transgenic mouse studies it has been shown that these mutations lead to a toxic gain of function by SOD [12]. The nature of this gain of function is widely debated, and there are two main theories: one suggests that the toxicity is due to misfolded forms of SOD, whereas the other proposes that SOD becomes a pro-oxidant protein which generates ROS. It should be noted that the two theories might not be mutually exclusive.

There is a potential mechanism depicting association of oxidative stress in *autism* with membrane lipid abnormalities, immune dysregulation, inflammatory response, impaired energy metabolism, and cytotoxicity, leading to clinical symptoms and pathophysiology of the disease (Fig. 4). Oxidative stress is known to be associated with premature aging of cells and can lead to tissue inflammation, cell membrane damage, autoimmunity, and ultimately to the cell death [13].

3.2 Eye Diseases

The retina is a highly specialized sensory organ that transduces light energy into neural signal. It also has high energy requirement and an extensive vascular network. Reactive oxygen species (ROS) generated via light exposure, normal energy production, phagocytosis of spent photoreceptor membranes by retinal pigment epithelium (RPE) cells, and circulating toxins render retina to an increased risk for oxidative stress. This stress has been implicated in the pathogenesis of several eye disorders such as cataract, macular degeneration, diabetic retinopathy, retinitis pigmentosa, and corneal disease [14].

Oxidative stress is an important factor in *senile cataract*. UV-induced oxidative damage seems to play a major role in cataract and retinal degeneration. As the crystalline lens is constantly subjected to oxidative stress from radiation and other environmental sources, the crystalline proteins, lipids, polysaccharides, and nucleic acids can be easily damaged [15]. Eyes have several mechanisms to protect its components from oxidative stress and to maintain its redox state. However, with aging, accumulation of oxidized lens components and decreased efficiency of repair mechanisms can contribute to the development of cataract.

Age-related macular degeneration is a leading cause of blindness in many countries worldwide. The retina is highly susceptible to photochemical damage from the continuous exposure to UV and visible lights, its high oxygen consumption, and due to its high proportion of polyunsaturated fatty acids. But the relationship between UV light exposure and AMD is unclear. Although short wavelength radiation and blue light induce significant oxidative stress to the retinal pigment epithelium and may cause disease progression. There are several reports that relates oxidative stress with the cellular damage caused by ROS in the pathogenesis of AMD [16]. The macular pigment is formed by two dihydroxycarotenoids, lutein and zeaxanthin, which act as a natural barrier protecting the macula against this oxidative stress [17].

Diabetes has been implicated in the increased oxidative stress which is thought to play an important role in the pathogenesis of *diabetic retinopathy* [18]. However, the mechanism of the hyperglycemia-induced oxidative stress is not clear. In diabetes, cells are unable to utilize all of its blood glucose, and ultimately hyperglycemia results with higher level of intracellular glucose. This high level of glucose increases NADH and FADH2 levels of the cell, increasing the proton gradient beyond a particular threshold at which the complex III prevents further increase by stopping the electron transport chain. This whole situation then continuously produces ROS. It was found that these ROS generation and resulting lipid peroxidation play a vital role in the development of diabetic complications in the eyes. The administration of vitamin E may prevent

ROS-induced lipid peroxidation and thereby limit the development of diabetic complications in the eyes [19].

3.3 Heart Diseases

The pathophysiology of cardiovascular morbidity is complex and multifactorial. Oxidative stress is an important contributory factor to the etiology of many cardiovascular diseases, including atherosclerosis, coronary heart disease (heart attack), cerebrovascular disease (stroke), cardiomyopathies, peripheral vascular disease, ischemic heart disease, heart failure, and hypertension [20].

It was reported that moderate exercise poses an acute oxidative stress, and regular endurance exercise is associated with improved cardiovascular function and a reduction in traditional *coronary heart disease (CHD)* risk factors. These findings are consistent with the hypothesis that adaptations induced by acute exposures to exercise-induced oxidative stress lead to long-term vascular protection. This occurs through activation of signaling pathways that lead to increased synthesis of intracellular antioxidants and antioxidant enzymes and decreased ROS production during exercise [21].

In *atherosclerosis*, there is a deposition of plasma lipoproteins that occur in the artery wall, which ultimately causes atherosclerotic plaque formation, provides a barrier to arterial blood flow and may contribute to clinical events. It was suspected that free radical-mediated oxidative processes play a key role in atherogenesis. At the center of this hypothesis are low-density lipoproteins (LDL), which, as part of their normal circulation, occasionally leave the antioxidant-rich plasma, entering the subendothelial space of arteries. Then LDL lipids are oxidized and initiating the process of atherosclerotic lesions. oxLDL is taken up by macrophages and induces the release of factors that recruit other cells and stimulate smooth muscle cell proliferation. oxLDL may also upregulate expression of cellular adhesion molecules that facilitate leukocyte binding. All of these events speed up the formation of plaque, which may result to heart attack and stroke in many patients [22]. Interestingly, this oxidization can be inhibited by nutritional antioxidants. Several epidemiological evidences and interventional studies correlate higher level of antioxidant-rich food uptake with lower incidence of atherosclerosis.

Ischemic stroke is the consequence of the interruption or severe reduction of blood flow in arteries followed by physiological and metabolic changes. When anoxia is followed promptly by reperfusion, tissue can be saved, but reperfusion might potentially have negative consequences: upon reoxygenation, ROS are rapidly built up, and numerous nonenzymatic oxidation reactions take place both in the cytosol and/or in other cellular organelles. Consequently, a large excess of O_2-derived free radicals appears during the first minute of reperfusion and peaks some 4–7 min after the onset of reperfusion. This high amount of ROS leads to cell death in the heart [23].

Associations between *obesity* and markers of oxidative stress have been observed in humans. Several hypotheses have been proposed for these observed associations. For example, it has been suggested that oxidative stress in obesity may result, in part, from the accumulation of intracellular triglycerides. Intracellular triglycerides are proposed to elevate superoxide radical production within the electron transport chain by inhibiting the mitochondrial adenosine nucleotide transporter. Inhibition of this transporter leads to a decrease in intramitochondrial adenosine diphosphate (ADP) that, in turn, reduces the flux of protons through the adenosine triphosphate-synthase reaction (i.e., the adenosine triphosphate-synthase reaction requires ADP as substrate). As a result, electrons build up within the electron transport chain, which then causes reduction of O_2 to form $O_2^{\bullet -}$. Another hypothetical source of increased oxidative stress may be the presence of excess adipose tissue as because adipocytes and preadipocytes have been identified as sources of inflammatory cytokines [24]. Cytokines are potent stimuli for the production of ROS/RNS by macrophages and monocytes, which upregulates the activity of oxidant-generating enzymes, including NAD(P)H oxidase, inducible NOS, and myeloperoxidase. As the accumulation of intracellular triglycerides or tissue adipocytes promotes increased oxidative stress, reduction of total body fat through diet and/or exercise may be an effective means of reducing systemic inflammation and oxidative stress. Consistent with this prediction, it was found that reductions in plasma markers of oxidative stress and ROS production by isolated leukocytes were reduced after 4 weeks of diet restriction and weight loss [25].

3.4 Blood Disorders

Beta-thalassemia major is an inherited disease resulting from reduction or total loss of beta globin chains. But oxidative stress and decreased antioxidant defense mechanism play an important role in the pathogenesis of beta-thalassemia major. In patients with beta-thalassemia major, frequent blood transfusions are required due to severe anemia, which cause oxidative stress as a result of increased blood levels of iron, lipid peroxides, and free-radical intermediates, as well as the decrease in total antioxidant level. One study showed a significant increase in the levels of lipid peroxide and iron but decrease in levels of vitamin E and total antioxidant capacity. Use of iron chelator agents in combination with antioxidants can be helpful to treat the patients with beta-thalassemia major [26].

Another study revealed that the levels of glutathione (GSH) and vitamin C were significantly decreased in *acute lymphoblastic leukemia (ALL)* patients and in the treatment group compared to controls. This finding indicates a possible link between decreased GSH and increased levels of oxidative damage to cells, supporting the idea that there is a persistent oxidative stress in acute lymphoblastic leukemia [27].

3.5 Liver and Pancreatic Diseases

The liver is an important organ which has a central role in metabolic homeostasis. It is responsible for the synthesis, storage, and redistribution of nutrients, like proteins, carbohydrates, fats, and vitamins. The liver is also responsible for xenobiotic metabolism [28]. That's why it is very much vulnerable to free radicals. The situation is even worst during the diseased conditions. It was found that chronic hepatitis B and C patients had significantly higher concentration of malondialdehyde (MDA) and nitric oxide (NO) level. Also they had higher activities of myeloperoxidase (MPO), arylesterase (AE), paraoxonase-1 (PON1), and other oxidative stress parameters but unfortunately lower level of ascorbic acid and other antioxidants in blood than normal patients. Oxidative stress is also closely related to the pathological damage during hepatic fibrosis [28]. Several studies suggested that markers of oxidative stress increase in patients with alcoholic hepatitis, such as CYP2E, one of the cytochromes which increased 15–20-fold, leading to excessive electron leakage and release of ROS, causing oxidative stress to those patients. Redox reaction followed by oxidative damage and imbalance of antioxidants often leads to subclinical hepatitis without jaundice, inflammatory necrotic hepatitis, liver cirrhosis, and ultimately cancer [29].

Pancreas contains mostly α-and β-cells. But among them β-cells have low level of antioxidant enzymes such as SOD, CAT, and GPx and thereby have increased susceptibility to oxidative stress [30]. In the β-cells increased mitochondrial ROS production results from enhanced glucose or fatty acid flux through glycolysis and the TCA cycle. The insufficiency of antioxidant enzymes to scavenge these ROS and RNS leads to oxidative stress. Ultimately, this high level of ROS and RNS causes destruction of β-cells, which impairs insulin production and diabetics results. Evidence shows that overexpression of antioxidant enzymes and/or antioxidants such as N-acetyl-L-cysteine (NAC) in islets or transgenic mice protects the pancreas against ROS-induced β-cell toxicity [31].

3.6 Lung Diseases

Many observations suggest that oxidative stress plays an important role in the pathogenesis of *asthma*. In the airway, high amount of oxidants initiates Th2-dominant immunity instead of inducing immune tolerance in the initial phase of development of airway allergic inflammation [32]. Furthermore, enhanced oxidative stress may contribute to the progression or perpetuation of existing airway inflammation through enhanced airway hyperresponsiveness, stimulation of mucin secretion, and induction of various proinflammatory chemical mediators all of which are believed to be related to severe asthma [33]. High incidences of bronchial asthma have been reported in areas with air pollution, which is a representative stimulus among exogenous oxidants [34]. Despite these findings, it is still unclear whether increased oxidative stress in the asthmatic airway is simply a consequence of chronic airway inflammation or a principal contributor to the development of allergic inflammation.

Pulmonary fibrosis (PF) is the end result of a diverse group of lung disorders. Although there are multiple initiating agents for pulmonary fibrosis, including toxins, particles, autoimmune reactions, drugs, and radiation, the etiology of the majority of cases of pulmonary fibrosis is unknown. Several studies have suggested that oxidant-antioxidant imbalance in the airways plays a critical role in the pathogenesis of PF [35]. In addition, oxidants can also modulate the production of cytokines and growth factors such as TGF-β, a key regulator of aberrant repair mechanisms that are characteristic of many fibrotic diseases. TGF-β not only induces ROS production by activation of NADPH oxidases and/or mitochondrial dysfunction but also decreases normal cellular antioxidant production through decreased expression of both catalase and mitochondrial SOD [36].

ROS are considered as carcinogenic species that facilitate *lung cancer* promotion and progression. As DNA molecule is one of the main targets of free radicals in the cell, the modifications cause loss of cellular homeostasis. Reactive oxygen species have also been proposed as activator of oncogenes such as Jun and Fos. It was found that overexpression of Jun is directly associated with lung cancer [37]. Modification of proteins and lipids also increases the risk of mutagenesis, for example, lipid peroxidative by-products can react with DNA, modify DNA polymerase, or inhibit DNA repair enzymes upon oxidation [38].

3.7 Kidney Disease

Urolithiasis is one of the most common diseases of the urinary tract. Recently, association of urolithiasis and free radicals has been reported. Different experiments performed on animals, cell cultures and human sera have also revealed the presence of enhanced oxidative stress during stone-forming conditions. But a combined study relating peroxidative stress and antioxidant capacity in stone-forming conditions in humans has not been cited yet [39, 40].

Diabetic nephropathy is one of the major causes of end-stage kidney disease. In diabetic patients, high blood glucose leads to increase in glomerular filtration as a result increased glomerular pressure. This high pressure damage to glomerular cells, which ultimately causes focal and segmental glomerulosclerosis. Increased blood glucose promotes glycosylation of circulatory and cellular proteins and initiates a series of autoxidation reactions that culminate in the formation and accumulation of advanced glycosylated end products (AGEs) in tissues. The AGEs have oxidizing potential and promote tissue damage by oxygen and nitrogen free radicals [41, 42].

3.8 Joint Disorders

Free radicals damage both chondrocytes and extracellular matrix (ECM) components of articular cartilage. ROS and RNS damage articular cartilage directly or indirectly by upregulating the mediators of the ECM degradation. All of these have important consequences on the etiology of rheumatoid arthritis (RA), pathogenesis of joint tissue injury and chronic inflammation, which may ultimately lead to periarticular deformities [43]. It also causes immunomodulation, which may lead to autoimmune diseases such as systemic lupus erythematosus (SLE) and antiphospholipid syndrome [44].

An increase in reactive oxygen species (ROS) plays an important role in the pathogenesis of *rheumatoid arthritis (RA)*. Persistent inflammation results in destruction of the cartilage and bone. This occurs through a number of mechanisms, including oxidative and proteolytic breakdown of collagen and proteoglycans. Once sequestered within the joint space, neutrophils degranulate and release a variety of potentially harmful enzymes and peptides. They may also undergo a respiratory burst and generate several reactive oxygen species, including superoxide, hydrogen peroxide, hypohalous acids, and possibly hydroxyl radical. In RA, ROS act as important intracellular signaling molecules of the immune system that amplify the synovial inflammatory-proliferative response [45]. T cells are exposed to increased oxidative stress and become refractory to growth and death stimuli, which further contribute to the perpetuation of the immune response. Interestingly, antioxidants and antioxidative enzymes have been shown to reduce this cartilage damage in animal models of RA. But unfortunately, increased ROS levels in RA result in a pro-oxidation environment, which in turn could result in decreased antioxidant activity and increased malondialdehyde (MDA) levels (an oxidative stress parameter). This ultimately results chronic inflammation [46].

It has been reported that mechanical stresses can lead to ROS-induced oxidative stress in the temporomandibular joint (TMJ), another important joint disorder. Excessive production of ROS/RNI in the TMJ thus results in tissue damage, which further propagates to *temporomandibular disorder (TMD)* [47]. ROS can be generated in the TMJ by several pathways: they include (1) direct mechanical injury, (2) hypoxia-reperfusion, and (3) arachidonic acid catabolism to the articular tissues. These ROS affect various molecular species of the TMJ and deteriorate the TMJ function: which includes (1) reduction of synovial fluid viscosity by depolymerization and/or molecular configuration of hyaluronic acid (HA); (2) reduction of lubrication of the articular surface by deterioration of the surface-active phospholipid (SAPL) layer, which acts as an extremely efficient boundary lubricant and protector of articular surfaces; (3) breakdown of collagen proteoglycans; (4) activation of cartilage-degrading enzymes such as matrix metalloproteinases. It was found that lysis of SAPL layer by phospholipase A2 (sPLA2) together with the depolymerization of HA caused by free radicals may result in a deteriorated lubrication of the articular surface, thus further proceeding to the internal derangement (ID) of the TMJ [48]. In addition, ROS, especially $HO^•$, is responsible for lipid peroxidation and disruption of cellular homeostasis during TMJ. The $HO^•$ also degrades collagen and proteoglycans (Pgs) into low molecular masses, which act as immunogens for synoviocytes or chondrocytes. These denatured Pgs may induce proinflammatory cytokines from various cells in the TMJ compartment, which deteriorate these conditions further [49].

3.9 Reproductive System Diseases

Cellular ROS and their control by antioxidants are involved in the physiology of the reproductive system. Physiological ROS levels play an important regulatory role, through various signal transduction pathways, during spermatogenesis, testis function, folliculogenesis, oocyte maturation, corpus luteum and uterine function, embryogenesis, embryonic implantation, and fetoplacental development. Imbalances between antioxidants and ROS production are considered to be responsible for the initiation or development of pathological processes affecting both male and female reproductive processes.

Oxidative stress (OS) has been identified as one of the factors that also affects fertility status. In one study scientists showed that smoking can induce OS, and this OS has a significant positive influence on sperm DNA fragmentation, axonemal damage, and with decreased sperm count [50]. Sperm from smokers has been found to be significantly more sensitive to acid-induced DNA denaturation than those from nonsmokers because the smokers' sperm have been shown to contain higher levels of DNA strand breaks.

Several other studies have also reported the role of oxidative stress and pregnancy complications. It was found that reduced antioxidant status increases risk of spontaneous abortion [51]. Moreover, ROS and RNS have been implicated in the development of premature rupture of the fetal membranes and with preeclampsia [52].

3.10 Skin Diseases

Because of its direct interaction with oxygen and other environmental factors, the skin is considered as one of the most vulnerable target organs to oxidative damage. UV irradiation interacts with the skin and produces free radicals and lipid peroxides, which ultimately damage the skin. Besides direct absorption of UVB photons by DNA and subsequent structural changes, generation of ROS following irradiation with UVA and UVB requires the absorption of photons by endogenous short-lived free radical molecules called photosensitizers. Several cellular constituents (e.g., porphyrins, flavins, quinones, and others) and biologically active drugs (e.g., tetracyclines, thiazides) can act as photosensitizers within skin cells. Because most photosensitized reactions are oxygen dependent, UV irradiation absorption results in the generation of ROS. Free radicals can also be produced by neutrophils (white blood cells having immune functions) that are increased in photodamaged skin and contribute to the overall pro-oxidant state. Thus, UV-induced generation of ROS in the skin develops OS and ultimately causes damage when their formation exceeds the antioxidant defense ability of the target cell [53]. Fortunately, the skin possesses a wide range of interlinked antioxidant mechanisms including melanin and carotenoids, which act as a UV-absorbing optical filter as well as a free radical scavenger.

4 Antioxidants

The toxic effect of reactive oxygen and nitrogen species in human is balanced by the neutralizing action of nonenzymatic antioxidants, as well as by antioxidant enzymes. Such antioxidant defenses are extremely important as they represent the direct removal of free radicals (pro-oxidants), thus providing maximal protection of biological sites. These systems not only assert with the problem of oxidative damage but also play a crucial role in wellness, health maintenance, and prevention of chronic and degenerative diseases [54]. Table 2 gives a short summary of different diseases caused by ROS in human and role of antioxidant enzymes against these circumstances.

5 Conclusion

The investigation of oxidative stress is inevitable for better understanding of aerobe organism function. There are several diseases where the reactive oxidative species and antioxidant mechanism play key role in pathogenesis. Evaluation of physiological as well as environmental factors, which affect molecular pathways via oxidative stress, can provide possible solutions for different organ malfunctions. The antioxidant supplementation, avoidance of different environmental factors, may lead to decrease in disease rate and incidence of mortality.

Table 2 Different antioxidant enzymes protect human being against reactive species-mediated diseases [55]

Disease	Specification	Main key antioxidant enzyme/s that gives protection
Allergy	Intolerance to aspirin	GPX
	Intolerance to other drugs	SOD
	Intolerance to some foods	GPX
	Reaction in skin tests	SOD
Cancer	Bowel	CAT, GPX, SOD
	Breast	GPX
	Colorectal	COX-2
	Kidney	CAT, GPX, SOD
	Leukemia	CAT, GPX, SOD
	Liver	CAT, GPX, SOD
	Skin	GPX
Cardiological and vascular injuries	Ischemia	SOD
	Atherosclerosis	SOD
Infectious disease	Arthritis	COX-2
	Helicobacter pylori	SOD
	Hepatitis	GPX
	HIV	GPX
	Influenza virus	CAT, GPX, SOD
Genetic disorder	Chronic granulomatous disease	CAT
	Down's syndrome	SOD
Metabolic malfunction	Diabetes	CAT, SOD
Neurodegenerative disease	Allergic encephalomyelitis	NOS
	Alzheimer's disease	SOD
	Amyotrophic lateral sclerosis	SOD
	Huntington's disease	SOD
	Parkinson's disease	GPX
	Prion disease	SOD
Ophthalmologic problem	Cataract	CAT, SOD

References

1. Lushchak VI (2011) Environmentally induced oxidative stress in aquatic animals. Aquat Toxicol 101:13–30
2. Elahi MM, Matata BM (2006) Free radicals in blood: evolving concepts in the mechanism of ischemic heart disease. Arch Biochem Biophys 450:78–88
3. Nathan C (2003) Specificity of a third kind: reactive oxygen and nitrogen intermediates in cell signaling. J Clin Invest 111:769–778
4. Kunwar A, Priyadarsini KI (2011) Free radicals, oxidative stress and importance of antioxidants in human health. J Med Allied Sci 1:53–60
5. Halliwell B, Gutteridge J (2007) Free radicals in biology and medicine, 4th edn. Clarendon Press, Oxford, pp 1–677
6. Chance B, Sus H, Boveris A (1979) Hydroperoxide metabolism in mammalian organs. Physiol Law 59:527–605
7. Chaitanya KV, Pathan AAK, Mazumdar SS et al (2010) Role of oxidative stress in human health: an overview. J Pharm Res 3:1330–1333
8. Arlt S, Beisiegel U, Kontush A (2002) Lipid peroxidation in neurodegeneration: new insights into Alzheimer's disease. Curr Opin Lipidol 13:289–294
9. Selley ML, Close DR, Stern SE (2002) The effect of increased concentrations of homocysteine on the concentration of (E)-4-hydroxy-2-nonenal in the plasma and cerebrospinal fluid of patients with Alzheimer's disease. Neurobiol Aging 23:383–388
10. Lotharius J, Barg S, Wiekop P et al (2002) Effect of mutant α-synuclein on dopamine homeostasis in a new human mesencephalic cell line. J Biol Chem 277:38884–38894
11. Bruijn LI, Houseweart MK, Kato S et al (1998) Aggregation and motor neuron toxicity of an ALS-linked SOD1 mutant independent from wild-type SOD1. Science 281:1851–1854

12. Valentine JS, Hart PJ (2003) Misfolded CuZnSOD and amyotrophic lateral sclerosis. Proc Natl Acad Sci 100:3617–3622
13. Klein JA, Ackerman SL (2003) Oxidative stress, cell cycle, and neurodegeneration. J Clin Invest 111:785–793
14. Ohira A, Ueda T, Ohishi K et al (2008) Oxidative stress in ocular disease. Nippon Ganka Gakkai Zasshi 112:22–29
15. Vinson JA (2006) Oxidative stress in cataract. Pathophysiology 13:151–162
16. Chalam KV, Khetpal V, Rusovici R et al (2011) A reviews: role of ultraviolet radiation in age-related macular degeneration. Eye Contact Lens 37:225–232
17. Drobek-Slowik M, Karczewicz D, Safranow K (2007) The potential role of oxidative stress in the pathogenesis of the age-related macular degeneration (AMD). Postepy Hig Med Dosw 61:28–37
18. Feldmen EL (2003) Oxidative stress and diabetic retinopathy: a new understanding of an old problem. J Clin Invest 111:431–433
19. Yue KKM, Chung WS, Leung AWN et al (2003) Redox changes precede the occurrence of oxidative stress in eyes and aorta, but not kidneys of diabetic rats. Life Sci 73:2557–2570
20. de Champlain J, Wu R, Girouard H et al (2004) Oxidative stress in hypertension. Clin Exp Hypertens 26:593–601
21. Linke A, Adams V, Schulze PC et al (2005) Antioxidative effects of exercise training in patients with chronic heart failure: increase in radical scavenger enzyme activity in skeletal muscle. Circulation 111:1763–1770
22. Devasagayam TPA, Tilak JC, Boloor KK et al (2004) Free radicals and antioxidants in human health: current status and future prospects. J Assoc Physicians India 52:794–804
23. Alexandrova M, Bochev P, Markova V et al (2004) Dynamics of free radical processes in acute ischemic stroke: influence on neurological status and outcome. J Clin Neurosci 11:501–506
24. Coppack SW (2001) Pro-inflammatory cytokines and adipose tissue. Proc Nutr Soc 60:349–356
25. Dandona P, Mohanty P, Ghanim H et al (2001) The suppressive effect of dietary restriction and weight loss in the obese on the generation of reactive oxygen species by leukocytes, lipid peroxidation, and protein carbonylation. J Clin Endocrinol Metab 86:355–362
26. Pavlova LE, Savov VM, Petkov HG et al (2007) Oxidative stress in patients with beta-thalassemia major. Prilozi 28:145–154
27. El-Sabagh ME, Ramadan KS, El-slam IMA et al (2011) Antioxidants status in acute lymphoblastic leukemic patients. Am J Med Med Sci 1:1–6
28. Irshad M (2002) Oxidative stress in liver diseases. Trop Gastroenterol 23:6–8
29. Zhu R, Wang Y, Zhang L et al (2012) Oxidative stress and liver disease. Hepatol Res 42:741–749
30. Tiedge M, Lortz S, Drinkgern J et al (1997) Relation between antioxidant enzyme gene expression and antioxidative defense status of insulin-producing cells. Diabetes 46:1733–1742
31. Drews G, Krippeit-Drews P, Dufer M (2010) Oxidative stress and beta-cell dysfunction. Pflugers Arch 460:703–718
32. Murata Y, Shimamura T, Hamuro J (2002) The polarization of T(h)1/T(h)2 balance is dependent on the intracellular thiol redox status of macrophages due to the distinctive cytokine production. Int Immunol 14:201–212
33. Rahman I, Yang SR, Biswas SK (2006) Current concepts of redox signaling in the lungs. Antioxid Redox Signal 8:681–689
34. Ercan H, Birben E, Dizdar EA et al (2006) Oxidative stress and genetic and epidemiologic determinants of oxidant injury in childhood asthma. J Allergy Clin Immunol 118:1097–1104
35. Cantin AM, North SL, Fells GA et al (1987) Oxidant-mediated epithelial cell injury in idiopathic pulmonary fibrosis. J Clin Invest 79:1665–1673
36. Koli K, Myllarniemi M, Keski-Oja J et al (2008) Transforming growth factor-ß activation in the lung: focus on fibrosis and reactive oxygen species. Antioxid Redox Signal 10:333–342
37. Szabo E, Riffe ME, Steinberg SM et al (1996) Altered cJUN expression: an early event in human lung carcinogenesis. Cancer Res 56:305–315
38. Marnett LJ (2000) Oxyradicals and DNA damage. Carcinogenesis 21:361–370
39. Menon M, Resnick MI (2002) Urinary lithiasis: etiology, diagnosis and medical management. In: Walsh PC (eds) Campbell's urology 8th edn. Saunder's, London, 4, pp 3229–3305
40. Singh PP, Barjatia MK (2002) Peroxidative stress and antioxidant status in relation to age in normal population and renal stone formers. Indian J Nephrol 12:10–15
41. Robert C, Stanton T (2011) Oxidative stress and diabetic kidney disease. Curr Diab Rep 11:330–336
42. Anderson S, Brenner BM (1988) Pathogenesis of diabetic glomerulopathy: hemodynamic considerations. Diabetes Metab Rev 4:163–177
43. Kalpakcioglu B, Senel K (2008) The interrelation of glutathione reductase, catalase, glutathione peroxidase, superoxide dismutase, and glucose-6-phosphate in the pathogenesis of rheumatoid arthritis. Clin Rheumatol 27:141–145
44. Khan F, Siddiqui AA (2006) Prevalence of anti-3-nitrotyrosine antibodies in the joint synovial fluid of patients with rheumatoid arthritis, osteoarthritis and systemic lupus erythematosus. Clin Chim Acta 370:100–107
45. Hitchon CA, El-Gabalawy HS (2004) Oxidation in rheumatoid arthritis. Arthritis Res Ther 6:265–278
46. Griffiths HR (2005) ROS as signaling molecules in T cells – evidence for abnormal redox signaling in the autoimmune disease, rheumatoid arthritis. Redox Rep 10:273–280
47. Kawai Y, Kubota E, Okabe E (2000) Reactive oxygen species participation in experimentally induced arthritis

of the temporomandibular joint in rats. J Dent Res 79:1489–1495
48. Yoshiaki K, Lee MC, Kubota E (2008) Oxidative stress and temporomandibular joint disorders. Japan Dent Sci Rev 44:145–150
49. Xie D, Homandberg GA (1993) Fibronectin fragments bind to and penetrate cartilage tissue resulting in proteinase expression and cartilage damage. Biochim Biophys Acta 1182:189–196
50. Makker K, Agarwal A, Sharma R (2009) Oxidative stress and male infertility. Indian J Med Res 129:357–367
51. Vural P, Akgul C, Yildirim A et al (2000) Antioxidant defence in recurrent abortion. Clin Chim Acta 295:169–177
52. Bilodeau JF, Hubel CA (2003) Current concepts in the use of antioxidants for the treatment of preeclampsia. J Obstet Gynaecol Can 25:742–750
53. Park G, Kim HG, Sim Y et al (2013) Sauchinone, a lignan from Saururus chinensis, protects human skin keratinocytes against ultraviolet B-induced photoaging by regulating the oxidative defense system. Biol Pharm Bull 36:1134–1139
54. Taibur Rahman T, Islam MMT, Shekhar HU (2012) Oxidative stress and human health. Adv Biosci Biotechnol 3:997–1019
55. Matés JM, Sánchez-Jiménez F (1999) Antioxidant enzymes and their implications in pathophysiologic processes. Front Biosci 4:339–345

Gene–Environment Interaction in Oxidative Stress-Induced Pathologies

Shanthi Sri Vundru, Nupoor Prasad, Rohit Patel, Vibha Rani, and Umesh Chand Singh Yadav

Abstract

Interaction between the environment and genes and resultant phenotype determine the onset as well as severity of many complex diseases such as cardiovascular, respiratory, age-related diseases and cancers. However, the relative contributions of either environmental or genetic component in these diseases are difficult to assess. Addition of oxidative stress dimension to this equation further complicates the scenario. The role of oxidative stress in the initiation and propagation of various diseases has been extensively investigated which has enhanced our understanding of how interaction between environment and genetic components leads to manifestation of these pathologies. The genetic predisposition to a disease may or may not be evident depending upon exposure to a particular environment. Similarly, exposure to a particular environment may determine the genetic background of an individual. These possible determinants and their constant interactions shape the molecular machinery that regulates human health and diseases. It is therefore required to consider the genes, environment, as well as their interaction in order to understand the etiology of various complex diseases. The better understanding of gene–environment interactions will enhance our knowledge about the emergence of various complex diseases and pave way to design novel preventive and therapeutic strategies.

Keywords

Environment • Genetic factor • Epigenetics • Diseases • Oxidative stress • Gene–environment interaction

S.S. Vundru • N. Prasad • R. Patel • U.C.S. Yadav (✉)
School of Life Sciences, Central University of Gujarat
Sector 30, Gandhinagar, Gujarat 382030, India
e-mail: umeshyadav@cug.ac.in

V. Rani
Department of Biotechnology, Jaypee Institute of Information Technology, A-10, Sector 62, Noida 201307, Uttar Pradesh, India

V. Rani and U.C.S. Yadav (eds.), *Free Radicals in Human Health and Disease*,
DOI 10.1007/978-81-322-2035-0_6,

1 Introduction

Oxidative stress represents deregulation of the redox homeostasis, i.e., loss of balance between oxidants and antioxidants in cells due to many factors including either an excess of oxidative insult or depletion of antioxidants [1]. Increased oxidative stress is associated with a number of pathologies including lung inflammation, chronic obstructive pulmonary disease (COPD), cancer, cardiovascular diseases, diabetes and its complications, and age-related diseases [2]. This indicates that oxidative stress affects the cellular function and regulation of gene expression, and the environment in which the cell or the organism exists plays an important role in contributing to these effects. The manner in which the environment regulates gene integrity and expression is determined by how oxidative stress modulates the cell function. This modulation either thwarts the impact of oxidative stress by enhancing antioxidant capacity of the cells or cells undergo oxidative stress, which may result in cellular toxicity leading to cell death, cell proliferation, and altered cell physiology. An important response to oxidative stress is upregulation of protective antioxidant genes. Upon exposure to oxidants or oxidative stress-inducing agents, mammalian cells upregulate stress-responsive genes, many of which encode antioxidant defense enzymes such as catalase (CAT), superoxide dismutase (SOD), and glutathione peroxidase-1 (GPx) [3]. Simultaneously, oxidative machinery is also activated that causes inflammation and further promotes oxidative stress. Thus, along with oxidative stress-induced inflammatory cascade, activation of antioxidant system ensures a proper damage control and repair mechanism. Out of these two mechanisms, one that prevails depending upon the outcome of oxidative stress-modulated gene–environment interactions determines the fate of the cells. In this chapter we have discussed with the help of few specific pathogenesis how environment regulates the expression of genes modulated by oxidative stress.

2 Oxidative Stress Induction by Environment

The environment surrounding the cells determines the redox-status and in many cases is the inducer of oxidative stress. Upon being exposed to different environmental cues, cells respond by producing certain mediators and intermediates that propagate the environmental signal to the cell nucleus, where gene regulation and gene transcription occur. These intermediates are mostly oxygen free radical and/or their derivatives which are together referred to as reactive oxygen species (ROS). With the help of ROS generation, the environment interacts with genes in the cells and induces varied phenotypic responses which appear in the form of inflammation and pathogenesis. Different types of environmental factors such as ultraviolet (UV) radiation, pesticides, and air pollution/smoke act in different ways on the cellular machinery to bring out different kinds of responses. Radiation constitutes major exogenous source of ROS. Nonionizing irradiation such as UV-C (<290 nm), UVB (290–320 nm), and UVA (320–400 nm) can indirectly produce a variety of ROS [4]. Prolonged exposure to UV radiation may lead to immunosuppression, tumor development, and photoaging. ROS generated by exposure to UV radiation attack the proteins, lipids, and DNA. To combat day-to-day exposure, antioxidant machinery present in cells such as ascorbic acid, glutathione (GSH), catalase (CAT), and superoxide dismutase (SOD) quells the effects of exposures and maintains the pro-oxidant/antioxidant balance and cellular homeostasis. However, prolonged exposure results in excess production of ROS that causes disequilibrium leading to oxidative stress. It has been reported that UVA produces more oxidative stress in the skin than UVB by inducing ROS in the cellular environment [5]. Photoaging of the skin includes one of the various deleterious effects of UV radiation [6].

Solar UV radiation is considered as one of the most important environmental factors responsible for skin aging. Aging, a cumulative process, involves enhanced oxidative damage to cellular components including proteins [7]. ROS induces the formation of protein carbonyls either by oxidative cleavage of proteins or by direct oxidation of amino acid residues such as lysine, arginine, proline, and threonine [8]. The oxidized proteins are usually degraded by proteasome. However, during the aging process, compromised function of proteasome leads to accumulation of oxidized proteins in the cell. In addition, lipofuscin, a highly cross-linked and modified protein aggregate formation in the cytosol, also inhibits the proteasome [9]. Further, ROS activates pro-inflammatory transcription factors such as nuclear factor kappa-light-chain-enhancer of activated B cells (NF-kB) and activator protein-1 (AP-1) which transcribe many inflammatory genes including matrix metalloproteinases (MMPs) that play a major role in the pathogenesis of photoaging. MMPs are zinc-dependent endopeptidases that show proteolytic activity and degrade matrix proteins such as collagen and elastin. Each MMP degrades different dermal matrix proteins, e.g., MMP-1 cleaves collagen type I, II, and III, whereas MMP-9 degrades collagen type IV, V, and gelatin, eventually leading to collagen degradation and wrinkle formation which is the major manifestation of photoaging (Fig. 1) [10]. UV radiations also induce mutation of the mitochondrial DNA (mtDNA) that promotes photoaging [11]. Induction of excessive ROS in mitochondria generates mtDNA mutations, which lead to a defective respiratory chain and produce more ROS forming a vicious cycle that induces further mutations suggesting important role of mitochondria in photoaging.

Further, the relationship between environment, e.g., UV radiation and gene expression leading to pathogenesis of photoaging has been suggested in a study by Takeuchi and Runger [12] who showed that UVA exposed skin cells increased the expression of progerin, a protein associated with aging. The authors demonstrated that induction of progerin was mediated by ROS-induced alternate splicing of Lamin A gene [12].

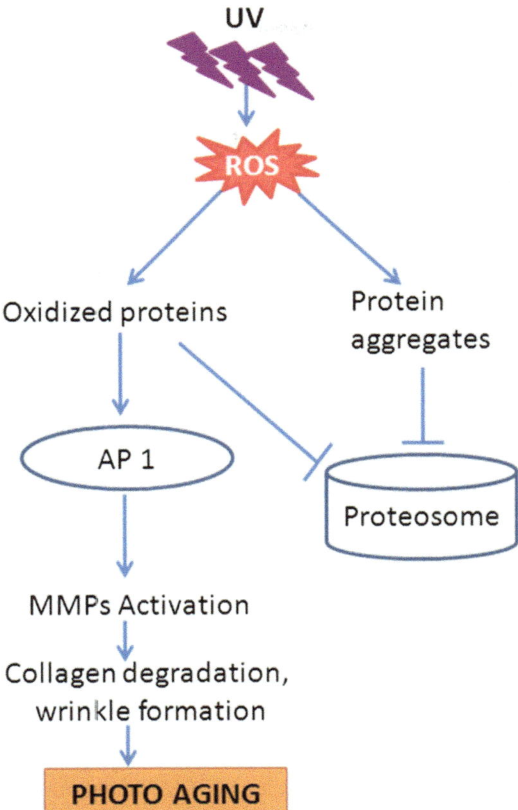

Fig. 1 UV-induced ROS mediates photoaging in skin cells

Polluted air is another major exogenous environmental factor that causes oxidative stress in cells, especially lung cells leading to various lung disorders, e.g., COPD, emphysema, and cancer. Cigarette smoke and air pollution both affect the lungs and modulate gene expression in exposed individuals [13]. Cigarette smoke, known to contain thousands of chemicals, is a major source of free radicals including both ROS and reactive nitrogen species (RNS) such as hydrogen peroxide, superoxide, and nitric oxide [14]. Cigarette smoke, an aerosol, contains particulate and gaseous phases. The particulate phase, a minority fraction, constitutes 4–9 % (w/w) of the total smoke, while the gaseous phase makes majority fraction and comprises 91–96 % (w/w) [15]. Gaseous phase contains a variety of chemicals such as formaldehyde, acrolein, and hydrogen cyanide, whereas chemicals in particulate phase include polycyclic aromatic hydrocarbons and tobacco-specific nitrosamines. These constituents are potential oxidants and are responsible for ROS production by inflammatory and epithelial cells in the lung as part of an inflammatory-immune response towards the irritant, cigarette smoke. Enhanced ROS production in cells activates a molecular cascade that leads to synthesis of cytokines and chemokines by various immune cells. One of the key mechanisms behind cigarette smoke – induced activation of inflammatory cells – is induction of NF-kB pathway. In resting cells NF-kB subunits, p65 and p50, remain sequestered in the cytoplasm in an inactive form through association with intracellular inhibitor, IkB. In response to environmental stimuli such as cigarette smoke, the activation of redox-sensitive molecular cascade phosphorylates IkB. This phosphorylation leads to poly-ubiquitination of IkB and subsequent degradation by proteasome. Degradation of IkB allows the translocation of NF-kB to the nucleus, where it induces transcription of various genes involved in inflammation and immune regulation such as IL-8, tumor necrosis factor alpha (TNF-α), and Cyclooxygenase-2 (COX-2) which further activates NF-kB and inflammation in autocrine and paracrine manner [16].

Toxic aldehydes present in cigarette smoke are also known to activate NADPH oxidase-2 (Nox2) in macrophages, neutrophils, and epithelial cells and generate $O_2^{\bullet -}$ (superoxide radical). They form highly reactive peroxynitrite ($ONOO^-$) radicals by reacting with NO and trigger inflammation, DNA damage, protein denaturation, and lipid peroxidation [17]. In addition, neutrophils are a rich cellular source of ROS, and neutrophil-derived myeloperoxidase metabolizes H_2O_2 in the presence of Cl^- to generate the strong oxidant hypochlorous acid [17], which modifies and denature proteins.

To protect cells from harmful effects of free radicals, antioxidant system including GPx and CAT present in the cells converts these toxic H_2O_2 into water and molecular oxygen. The net effect of ROS activity is greater in smokers who have been shown to possess higher levels of exhaled ROS than nonsmokers which may explain development of pathological condition including lung inflammation and COPD in smokers [18].

3 Oxidative Stress and Gene Expression

Oxidative stress involves imbalance between oxidant-producing systems and antioxidant mechanisms which result in the generation of excessive ROS which then modulates gene expression. An elevated ROS level modulates the transcription of genes that regulates the cell antioxidant defense, detoxification, immune response, cell proliferation, and cell differentiation. Upon exposure to oxidants, cells induce the expression of various defensive genes as protection measure from oxidative damage. Studies suggest that in prokaryotic cells, transcription factors OxyR and SoxRS sense cell's redox condition and induce the expression of ~80 defensive genes during oxidative stress [19]. Similarly, in eukaryotic cells, several antioxidant genes are expressed during onset of oxidative stress which helps in the neutralization of free radicals. In eukaryotes, nuclear factor erythroid 2-related factor (Nrf2) pathway gets activated in response to oxidative stress and transcribes many antioxidative genes, and resultant protein products help in removal of ROS [20]. Under normal redox condition, Nrf2 is associated with its inhibitor, keap1 protein, in the cytoplasm which helps in ubiquitin-mediated degradation of Nrf2. However, during oxidative stress, Nrf2 gets phosphorylated and dissociates from Keap 1 and translocates to nucleus, where it binds with small adaptor proteins such as musculoaponeurotic fibrosarcoma (Maf) and finally targets antioxidant response element (ARE) of genes including that of NADPH quinine oxidoreductase 1 (NQO1), glutathione-S-transferase Ya subunit,

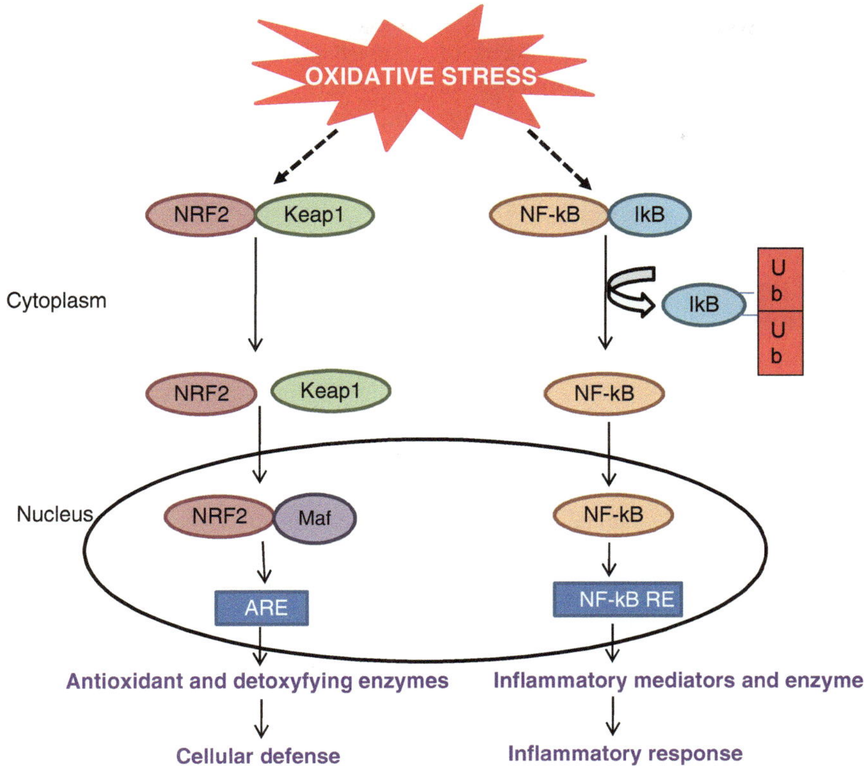

Fig. 2 Molecular regulation of Nrf2 and NF-kB transcription factors under oxidative stress

hemeoxygenase1 (HO-1), and ϒ-glutamyl cysteine synthase [21] (Fig. 2). These gene products play an important role in eliminating ROS and establishing cellular homeostasis.

NF-kB, an important redox-sensitive transcription factor, also gets activated during oxidative stress, translocates to nucleus, and transcribes various prooxidant and inflammatory genes including cytokines such as TNF-α, IL-1, and IL-6; chemokines such as monocyte chemoattractant protein-1(MCP-1), IL-8, and macrophage inflammatory protein-1(MIP-1α); adhesion molecules including intracellular adhesion molecule-1 (ICAM-1) and vascular cell adhesion molecule (VCAM-1); and inflammatory enzymes such as COX-2 and iNOS [22]. NF-kB also induces transcription of c-myc and Ras oncogenes resulting in increased cell proliferation that may cause tumorigenesis and cancer [23]. NF-kB may also induce the expression of many antioxidant genes in response to oxidative stress such as manganese superoxide dismutase (MnSOD), NQO1, HO-1, and glutathione peroxidase-1 [24].

AP-1, another redox-sensitive transcriptional factor, also gets activated under prooxidant conditions. Oxidative stress induces the levels of c-fos and c-Jun, which activate AP-1 transcription factor that leads to expression of several pro-inflammatory as well as antioxidant genes. Thus, oxidative stress in cells modulates the gene expression of several prooxidant, pro-inflammatory, as well as antioxidant genes depending upon the intensity and duration of oxidative stress [25].

Several investigators have worked extensively on various oxidative stress-induced pathways that lead to various diseases and have identified many molecular targets. However, drugs against these targets have not been successful in clinical trials due to severe side effects. The focus of investigations has now shifted on enhancing the antioxidative capacity of cells by using various phytochemicals with some initial success [26].

4 Oxidative Stress and Epigenetic Changes in Genes

Epigenetics is a major mechanism that accommodates changes in gene expression in response to gene–environment interactions without any changes in DNA sequences. The oxidative stress induced by various environmental stimuli is associated with the disruption of epigenetic homeostasis. Oxidative stress causes damage to DNA either by addition of hydroxyl radical to double bond of bases or by eliminating H atom from methyl group of thymine and C–H bonds of 2-deoxyribose resulting in formation of DNA lesions such as 8-hydroxy guanine, O^6-methyl guanine, 8-hydroxyl-2-deoxyguanosine, and thymine glycol [27]. This results in formation of apurinic/apyrimidinic sites, base deletions, single-strand breaks, double-strand breaks, mutations, and chromosomal rearrangements [28]. DNA lesion further induces epigenetic modification in cell by altering the pattern of DNA methylation and histone modification.

DNA methylation regulates many cellular processes such as maintaining chromatin structure and remodeling, chromosome stability, genomic imprinting, and gene transcription [29]. Under normal conditions, DNA is methylated symmetrically on both strands. Immediately following DNA replication, the newly synthesized double-stranded DNA contains hemimethylated sites, catalyzed by a family of DNA methyltransferases (Dnmts) that transfer a methyl group from S-adenyl methionine (SAM) to the fifth carbon of a cytosine residue to form five methyl cytosines [30]. Three members of the Dnmt family (Dnmt1, Dnmt3a, and Dnmt3b) directly catalyze the addition of methyl groups onto DNA. Among them Dnmt3a and Dnmt3b, the de novo Dnmts, transfer methyl groups onto naked DNA [31], whereas the Dnmt 1 maintains DNA methylation pattern during replication. Dnmt1 binds with the replication foci and precisely replicates the original DNA methylation pattern by adding methyl groups onto the newly formed daughter strand [32].

The majority of DNA methylation occurs on cytosine base that precedes a guanine nucleotide such as CpG sites. CpG islands are roughly 1,000 base pair long stretches of DNA that have a higher CpG density than the rest of the genome. In eukaryotes, about 50–60 % gene promoters reside within CpG island. In normal condition most of the CpG island remains unmethylated which ensures active gene transcription [33]. Methylation of CpG islands can impair transcription factor binding, recruit repressive methyl-binding proteins, and may stably silence gene expression [34].

Oxidative DNA damage can interfere with the ability of methyl transferases to interact with DNA. During oxidative stress the formation of DNA lesion such as 8-oxodG inhibits methyl CpG-binding protein (MBP) which recruits Dnmts enzyme and prevents methylation of adjacent C residues leading to global hypomethylation [35]. Oxidative stress also causes depletion of SAM molecule which is important for DNA methylation. During oxidative stress there is increase in production of GSH due to enhanced demand. Homocysteine is a common precursor for both GSH and SAM synthesis. Hence, increased utilization of homocysteine for GSH synthesis during oxidative stress decreases the production of SAM leading to global hypomethylation [36]. Global hypomethylation is associated with expression of oncogenes and genomic instability leading to diseases such as cancer. Oxidative stress may also cause DNA hypermethylation of CpG island in promoter region which in turn may cause transcriptional repression of target genes. In hepatocellular carcinoma cell line oxidative stress induction resulted in hypermethylation of E-cadherin promoter by increasing Snail expression which in turn recruits HDAC and Dnmt-1 [37]. Hypermethylation is also associated with the gene silencing of several tumor suppressor genes such as CDKN2A, Rb, and BRAC1 which increases risk of cancer [38].

Oxidative stress is also known to induce histone acetylation and methylation of promoter region leading to change in gene expression. The lysine residue in amino termini of histone3 (H3) and histone4 (H4) can undergo several modifications including acetylation and methylation. Histone acetylation induces transcriptional activation whereas decreased acetylation causes transcriptional repression [39]. Oxidative stress affects the activity and function of histone deacetylase enzyme (HDAC) which plays an important role in histone acetylation. Several studies showed that oxidative stress induces HDAC enzyme resulting in overexpression of genes in cancer [37]. Oxidative stress also affects histone methylation either by inducing the activity of DNA methylase enzyme or by decreasing production of SAM. Methylation of histone is associated with both transcription activation and repression of gene [40]. Hence, oxidative stress leads to changes in epigenetics by histone modification and DNA methylation which in turn modulates the expression of genes leading to chronic diseases such as cancer.

5 Gene–Environment Interaction in Oxidative Stress-Induced Pathologies

Oxidative stress induced by different environmental stimuli modulates the transcription of various genes which results in different chronic diseases in human beings including cardiovascular diseases, COPD, cancer, and neurodegenerative diseases. Some of these diseases are discussed below.

5.1 Pesticide-Induced Pathologies

Pesticides are commonly used chemicals intended to kill the pests, vectors of diseases, in agricultural field and enhance the agricultural production. However, excessive and indiscriminate use has led to dissemination in the environment – soil, water, and air – which adversely affects the human health. Several studies have reported that long-term exposure to even lower amount of pesticides leads to chronic diseases such as cancer [41], cardiovascular [42], and neurodegenerative diseases [43]. The toxic effect of pesticides such as organophosphates, organochlorines, pyrethroids, thiazines, and paraquat is mainly due to their ability to induce oxidative stress [44].

Pesticides induce oxidative stress in exposed cells by the following mechanisms:

1. Pesticides cause production of ROS either as by-products during pesticide metabolism or due to mitochondrial dysfunction. When pesticide enters inside the body, they get metabolized by several detoxifying enzymes such as NADPH-cytochrome P450 reductase, xanthine oxidase, and NADH-ubiquinone oxidoreductase. The by-product of this metabolism results in production of secondary free radicals, e.g., paraquat (PQ) metabolism leads to generation of paraquat mono-cation which further gets reoxidized to generate paraquat di-cation and oxygen radicals as shown in the equation resulting in redox imbalance [45]:

$$PQ^{2+} + e^{-}CYP450 \longrightarrow PQ^{+}$$

$$PQ^{+} + O_2 \longrightarrow PQ^{2+} + O_2^{\bullet -}$$

   Some of the pesticide such as rotenone causes mitochondrial dysfunction by blocking ubiquinone binding site of complex-I of electron transport system and prevents transport of electrons from complex-I to ubiquinone, and free electrons react with oxygen generating free radicals [46].
2. Pesticides also cause change in antioxidant homeostasis. Long-term exposure to pesticides such as organophosphates leads to decrease in antioxidant enzymes such as SOD and CAT which are first line of defense against oxidative stress [47]. As a result exposed cells fail to neutralize generated ROS, leading to enhanced stress.
3. Glutathiones (GSH) are the most important intrinsic defense system that directly scavenges free radicals. Decrease in GSH level has been observed in liver, brain, kidney, and spleen of rat exposed to organophosphate pesticides [48].

Pesticides cause decrease in level of GSH either by direct oxidation of GSH to glutathione disulfide (GSSG), by causing decreased level of glutathione reductase (GR) enzyme, or by oxidation of NADPH. GR needs NADPH as a substrate for reduction of oxidized GSSG to its reduced and active form, GSH:

$$GSSG + NADPH + H^{+} \xrightarrow{GR} 2GSH + NADP^{+}$$

Paraquat herbicide is reported to cause oxidation of NADPH and hence prevent regeneration of GSH in the cells resulting in decrease GSH/GSSG ratio and redox imbalance. Although depletion of NADPH activates pentose phosphate pathway by feedback mechanism that leads to restoration of NADPH, this results in continuous redox cycling between paraquat and oxygen, resulting in formation of $O_2^{\bullet -}$ [48]. Further, since GSH plays important role in regeneration of other antioxidants such as ascorbic acid and alpha-tocopherol, decreased GSH level may prevent their regeneration as well. Pesticides thus decrease antioxidant defense of the body leading to oxidative stress.

4. Pesticides such as organophosphates, synthetic pyrethroid, and carbamates may also increase the lipid peroxidation activity in erythrocytes cell [49]. Initial generation of ROS by detoxifying enzyme results in lipid peroxidation which further generates lipid-free radicals resulting in increased oxidative stress. Organophosphates are reported to induce peroxidative damage of biological membrane resulting in accumulation of lipid peroxidation products [48].
5. Pesticide-induced oxidative stress is also determined by extent of exposure and genetic polymorphism in pesticide metabolizing enzymes such as paraoxanase-1 (PON-1), pseudo or butyryl cholinesterase (BCHE), and GST, which play an important role in metabolizing organophosphates. Different polymorphic forms of PON-1 enzyme such as PON1-192RR, PON1–108TT, and PON1–909CC occur which are associated with lower PON-1 activity, which lead to pesticide toxicity and induction of oxidative stress. Short-term exposure to organophosphates in PON1-192RR genotype leads to increase in GPx and CAT as an adaptive mechanism against oxidative stress. Long-term exposure of organophosphate in PON1-192RR genotype results in lower SOD activity making them susceptible to oxidative stress. GSTs are another class of polymorphic enzyme that play protective role by deactivating oxygen free radical upon exposure to pesticide. Short-term exposure in GSTM1 null genotype was found to be associated with decrease in SOD level making them susceptible to oxidative stress. Hence, polymorphic gene encoding PON1 and GSTs are important determinants of organophosphate pesticide-induced oxidative stress [44]. These evidences indicate that the extent of pesticide-induced oxidative stress depends on the type of polymorphism in metabolizing enzyme genes, which in turn impacts the expression as well as activity of antioxidant enzymes.

Pesticides cause generation of oxidative stress in the cells which activates redox-sensitive transcription factors such as NF-kB which in turn transcribes numerous genes of pro-inflammatory enzymes, chemokines, and cytokines [45]. These mediators increase inflammation and tissue damage which results in diseases such as cancer, neurodegeneration, and cardiovascular diseases (Fig. 3). Increased peroxidation of membrane lipids, membrane receptor, and membrane-bound enzyme can also alter function, structure, and fluidity of membrane leading to pathogenesis. Brain tissues are more prone to pesticide-induced oxidative stress since they are rich in polyunsaturated fatty acids which easily undergo peroxidation leading to neural degeneration diseases such as Parkinson [50].

Several investigations are underway to establish the link between pesticides and diseases and their causative mechanism [51]. Such studies may lead to novel approaches for prevention and treatment of pesticide- induced diseases. Nevertheless, owing to high toxic effect of pesticides on human health, a number of pesticides have been banned and use of less toxic pesticides is encouraged.

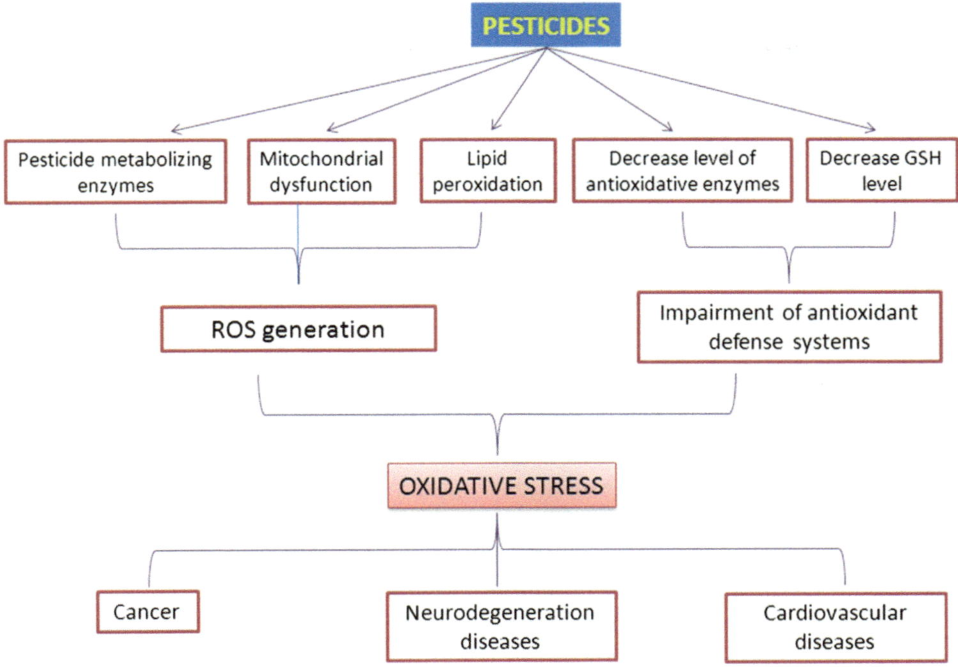

Fig. 3 Summary of pesticide-induced oxidative stress leading to chronic diseases

5.2 Cardiovascular Disease

Cardiovascular diseases (CVD) are the leading cause of mortality and morbidity worldwide [52]. CVD include many diseases related to heart and blood vessels, e.g., coronary heart disease, ischemic heart disease, atherosclerosis, hypertension, cardiac hypertrophy, cardiomyopathies, and congestive heart failure [53]. A disturbance in redox homeostasis in the body may result in failure at genetic transcription level due to the exposure to various environmental factors that could lead to development of CVD [54]. Although it is well known that environmental factors such as smoke, poor diet, diabetes, and aging increase oxidative stress, the relationship between oxidative stress, genes, and CVD remains complex. The possible mechanisms of gene–environment interaction-induced oxidative stress in CVD are important to understand for better therapeutic strategies.

Cigarette smoke, a common environmental factor associated with increased oxidative stress, is linked with oxidation of low-density lipoprotein (LDL) resulting in formation of oxidized low-density lipoprotein (oxLDL), which is implicated in plaque formation in coronary artery [54–56]. Smoking also causes vascular endothelial damage which leads to inflammation followed by an increase in the expression of cell adhesion molecules. The important cellular sources of oxidative stress in cardiovascular system upon exposure to environmental oxidants include enzymes Nox, xanthine oxidoreductase (XOR) and nitric oxide synthetase (iNOS), and mitochondrial enzyme such as mitochondrial uncoupling protein (UCP) [57, 58].

Tobacco smoke containing metabolites are converted into ROS by enzyme such as myeloperoxidase [59]. ROS besides causing oxidation of LDL forming oxLDL also activate more ROS-producing machinery such as Nox which further lead to generation of more oxLDL [60]. In endothelial cells, oxLDL bind to lectin-like oxidized low-density lipoprotein receptor 1 (LOX-1) encoded by OLR 1 gene [61, 62]. An adaptor protein, TRAF3IP2, interacts with active LOX-1 and binds to inhibitor of nuclear factor kappa-B kinase subunit beta (IKKβ) upon activation [63, 64]. Binding of TRAF3IP2 and IKK causes a conformational change in IKK which facilitates the binding of JNK (c-jun N-terminal kinase) [63, 65]. Activated IKK-JNK detaches from TRAF3IP2 and activates NF-kB and AP-1 transcription factors which transcribe many inflammatory and cell adhesion marker genes including VCAM, ICAM [64]. Increased expression of ICAM, VCAM adhesion molecules on endothelial cells (EC) attracts monocytes and platelets towards the arterial walls which adhere to EC and migrate in subendothelial space and differentiate into macrophage [66]. The oxLDL is taken up by these monocytes/macrophages with the help of scavenger receptors expressed on the cell surface during differentiation leading to accumulation of oxLDL in macrophage which results in foam cell formation. The latter are known to promote the progression of atherosclerosis (Fig. 4) [67]. Although the biochemical mechanism is clear, the molecular events emanating from gene–environment interaction remain unclear. More studies are required for a proper understanding of the molecular events that cause pathogenic effect in cardiovascular diseases that is critical. Identification of key modulator of this pathway could pave way for novel and potential therapeutic targets to prevent atherosclerosis.

5.3 Chronic Obstructive Pulmonary Disease (COPD)

COPD is a major incurable global health burden and projected to become the fifth largest cause of death worldwide by 2020 [68]. COPD leads to airway obstruction, lung tissue destruction, and emphysema. The most important risk factors for COPD are cigarette smoke, exposure to dust, fumes, and air pollution particles [69]. Apart from these factors genetic predisposition also plays a major role in the development of COPD [70].

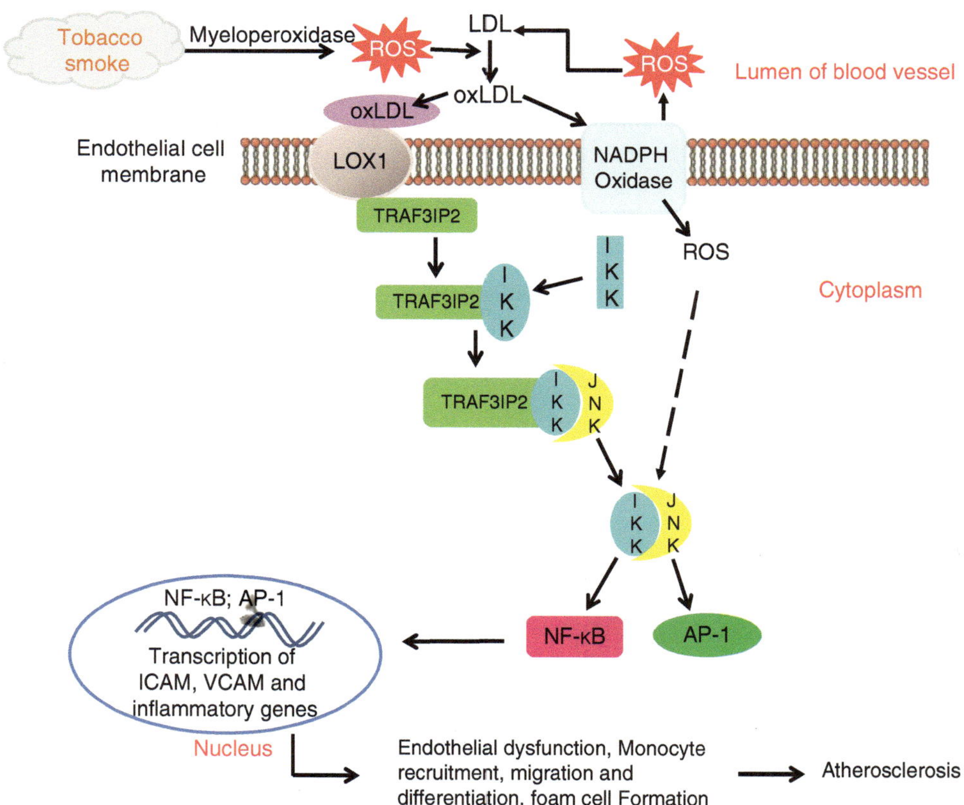

Fig. 4 Tobacco smoke-induced gene expression of adhesion molecules (ICAM and VCAM) in endothelial cell causes monocyte attachment, and their migration in subendothelial space leads to differentiation in macrophage and foam cell formation during atherosclerosis

Oxidative stress plays an important role in COPD development reflecting the increased oxidant burden in smokers. Cigarette smoke is a rich source of oxidants that lead to the recruitment of macrophages and neutrophils. Once activated by cigarette smoke, neutrophils release $O_2^{\bullet-}$ radical that can either react with NO to form highly reactive peroxynitrite molecule ($ONOO^-$) or be immediately converted to H_2O_2 under the influence of SOD. This in turn can result in the nonenzymatic production of more damaging hydroxide radical (·OH) from H_2O_2 in the presence of Fe^{2+} through the Fenton reaction [71].

The subsequent oxidation of Fe^{2+} to Fe^{3+} in turn generates ·OH from $O_2^{\bullet-}$ and Fe^{2+} regenerated through the Haber–Weiss reaction [72]. This redox cycling of Fe^{2+} and Fe^{3+} can therefore rapidly result in the formation of more damaging hydroxyl radical (·OH) from the initial supply of $O_2^{\bullet-}$. This oxidative process is particularly relevant to COPD as smokers have higher levels of iron in their lungs which increases ROS burden [73]. Macrophages also employ other enzymes including heme peroxidases and myeloperoxidase to produce ROS. These enzymes catalyze the formation of potent and damaging oxidants such as hypochlorous acid (HOCl) and hypobromous acid (HOBr) from H_2O_2, in the presence of chloride (Cl^-) and bromide (Br^-) ions, which leads to destruction of lung tissue [74].

In COPD, ROS-induced tissue damage leads to protein nitration where NO and $O_2^{\bullet-}$ contribute to peroxynitrite anions ($ONOO^-$), a highly reactive oxidant species. $ONOO^-$ attack sulfhydryl groups of proteins, a process called nitrosylation, and form nitrosothiols. $ONOO^-$ has also been shown to induce hyperresponsiveness in lung tissue, inhibit pulmonary surfactant, and induce membrane lipid peroxidation that causes damage in pulmonary epithelial cells (Fig. 5) [75]. $ONOO^-$ can also transfer its nitro group to the hydroxyl group of tyrosine and produce a stable product nitro tyrosine. Tyrosine nitration of proteins such as HDAC can affect enzyme activity

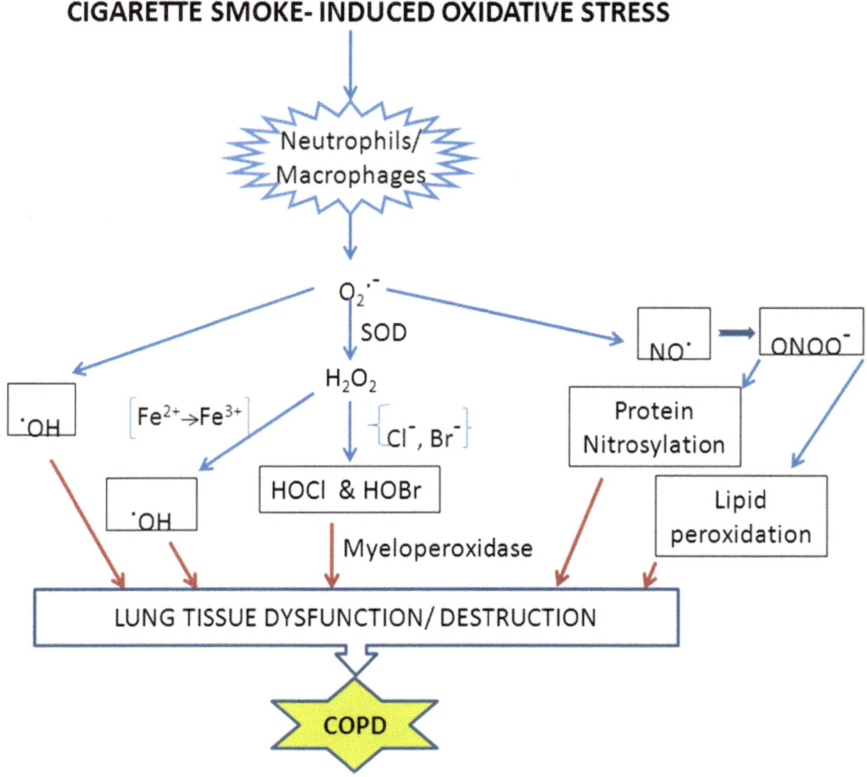

Fig. 5 Cigarette smoke-induced oxidative stress resulting in lung inflammation and destruction leading to emphysema and COPD

leading to the altered gene expression and gene regulation [76].

ROS may also oxidize membrane phospholipids by lipid peroxidation, which leads to impairment of membrane function [74]. Membrane lipid peroxidation can form highly reactive aldehydes such as acrolein and 4-hydroxy-2-nonenal [77]. These reactive aldehydes have high affinity towards cysteine, histidine, and lysine residues of cellular proteins and form adducts with them that alter their function, e.g., histone deacetylase HDAC-2 has been shown to form adduct with reactive aldehydes which alter its activity [78].

Oxidative stress-induced epigenetic changes also regulated DNA repair and cell proliferation. Transcriptional co-activators possess intrinsic histone acetyl transferase (HAT) activity leading to histone acetylation which plays a major role in inflammatory gene expression. Several genes responsible for inflammatory response in the lung are thought to be regulated by histone acetylation and deacetylation mechanisms [79]. DNA is tightly wrapped around a tetrameric set of core histone proteins (H2A, H2B, H3, and H4) to form nucleosome, a constituent unit of chromatin. Activated transcription factors recruit co-activators containing histone acetylase (HAT) activity which results in histone acetylation causing the DNA to uncoil around the histone core. As a consequence RNA polymerases can gain access to DNA to begin transcription. In contrast, gene transcription is shut down by histone deacetylases (HDACs) that remove acetyl groups from the histones, thereby facilitating condensation of DNA around the histone core.

ROS promote histone acetylation while at the same time inactivating certain histone deacetylases: HDAC-2, HDAC-5, and HDAC-8 [76]. Inactivation of HDAC-2 by ROS is achieved through increased nitration leading to altered transcription of pro-inflammatory and antioxidant gene expression. The impact of ROS on HDAC-2 is particularly important as it has also been shown to be required for corticosteroid-mediated inhibition of the inflammatory response [80].

Since lungs act as barrier as they lie at the interface between environment and the body, they are endowed with ample antioxidant defense for protection from oxidant-induced tissue damage. Among antioxidants, GSH a tripeptide (l-γ-glutamyl-l-cysteinyl-glycine), has an important protective role in the lung. GSH contains a thiol (-SH) group which functions as an antioxidant by acting as a sacrificial target for ROS and products of lipid peroxidation such as reactive carbonyls. In this reaction GSH is oxidized to its dimeric form (GSSG) or forms adducts with reactive carbonyls and other reactive xenobiotics (GS-X). It has been reported that GSH levels are increased in the epithelial lining fluid of both asthmatics and chronic cigarette smokers which suggests that lung activates GSH synthesis in the presence of oxidative stress as a protective measure [81]. Oxidative stress also causes upregulation of glutamate cysteine ligase (GCL), an important enzyme involved in the synthesis of GSH [82]. The expression of GCL mRNA is elevated in smokers' lungs, even more pronounced in smokers with COPD [83]. Similarly, bronchial epithelial cells of rats exposed to cigarette smoke have shown increased expression of antioxidant genes such as MnSOD, metallothionein, and GPx [84]. These evidences suggest the importance of an adaptive antioxidant gene response against the injurious effects of cigarette smoke and how these enzymes neutralize and remove the excess of ROS in the lung environment and prevent lung tissue from further damage. A very complex and diverse nature of gene regulation in different individuals may be an explanation why only a fraction (~10 %) of smokers develop COPD even though a majority (~90 %) of COPD patients are smokers [85].

The antioxidant genes are regulated by transcription factors, for example, Nrf2, which is a redox-sensitive transcription factor containing sulfhydryl groups. It binds to the antioxidant response element (ARE) in DNA regulating a variety of antioxidant genes. Approximately 50 antioxidant and cytoprotective genes in the lungs are transcriptionally controlled by Nrf2, which work to overcome the effects of cigarette smoke [86]. It has been shown that disruption of Nrf2 gene in mice leads to an early and intense emphysema in response to cigarette smoke [87]. These evidences indicate that Nrf2 is an important transcription factor that regulates the antioxidant genes, and any disruption in it may result in loss of antioxidant defense leading to various lung pathogenesis including COPD.

The preceding discussion thus indicates that environmental factors such as cigarette smoke induce both oxidative and antioxidant response. Thus, targeting the molecular cascade of either pathway may be effective in regulating the pathological outcome. Another approach could be to identify epigenetic regulation of environmental oxidant-induced pathogenesis which may result in a novel therapeutic and prevention paradigm against oxidative stress-induced disease including COPD.

6 Summary and Future Perspective

Oxidative stress occurs when free radical production exceeds the body's ability to neutralize them. Free radicals are generated in the body by many endogenous and exogenous factors. Environmental factors such as UV radiation, air pollution, and pesticides are involved in ROS production leading to oxidative stress. The survival depends upon the ability of cells and tissues to adapt or resist the stress and repair or remove damaged molecules or cells. As a consequence, a number of defense systems have evolved to balance ROS. Interaction between genes and environment results in induction of many factors which include transcription factors, replication proteins, proteases, inflammatory marker genes, and various antioxidants. However, when antioxidant defense mechanisms fail to counteract ROS, it results in oxidative damage of biological molecules such as DNA, proteins, and lipids. The oxidative damage results in altered cell functions leading to many pathological complications like cardiovascular diseases, photoaging, lung inflammation (COPD), and cancer (Fig. 6). Understanding the effects of ROS in basic cellular functions and molecular mechanisms such as pro-inflammatory responses, defective repair mechanism and signaling path-

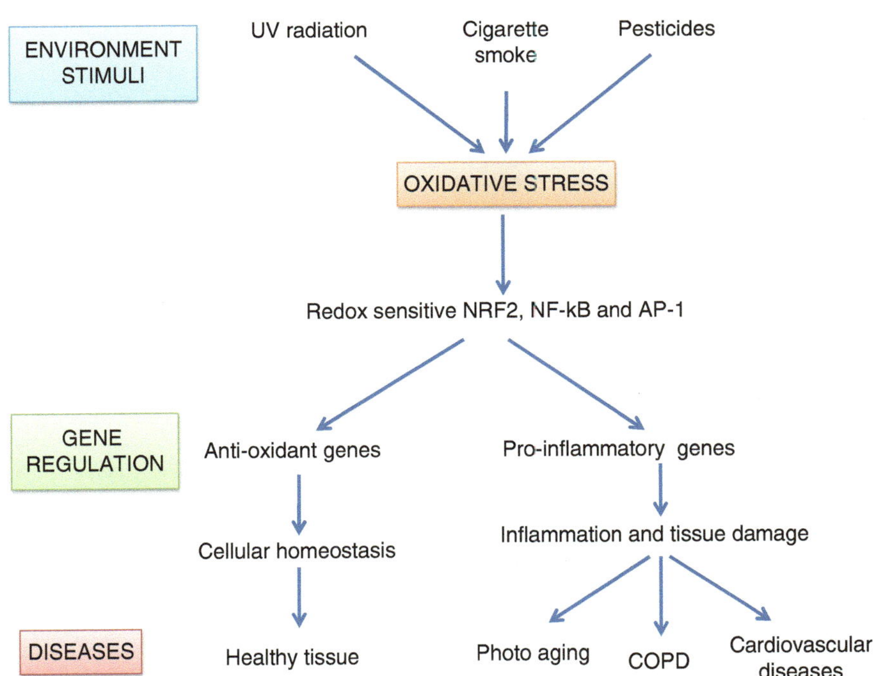

Fig. 6 Overview of oxidative stress-mediated gene–environment interaction and pathologies

ways will provide valuable information regarding basic pathologies. The identification of new genes or molecules that are altered during oxidative stress can be an important molecular target to treat the oxidative stress-induced diseases.

Acknowledgement The award of Ramanujan Fellowship and financial support from Department of Science and Technology (DST), Government of India, to Dr. Umesh C. S. Yadav is acknowledged. SSV, NP and RP acknowledge financial support from UGC.

References

1. Morel Y, Barouki R (1999) Repression of gene expression by oxidative stress. Biochem J 342:481–496
2. Pham-Huy LA, He H, Huy CP et al (2008) Free radicals, antioxidants in disease and health. Int J Biomed Sci 4:89–96
3. Noori S (2012) An overview of oxidative stress and antioxidant defensive system 1:413
4. Shadyro OI, Yurkova IL, Kisel MA (2002) Radiation-induced peroxidation and fragmentation of lipids in a model membrane. Int J Radiat Biol 78:211–217
5. de Gruijl FR (2002) Photocarcinogenesis: UVA vs. UVB radiation. Skin Pharmacol Appl Skin Physiol 15:316–320
6. Berneburg M, Plettenberg H, Krutmann J (2000) Photoaging of human skin. Photodermatol Photoimmunol Photomed 16:239–244
7. Sander CS, Chang H, Salzmann S et al (2002) Photoaging is associated with protein oxidation in human skin *in vivo*. J Invest Dermatol 118:618–625
8. Levine RL, Stadtman ER (2001) Oxidative modification of proteins during aging. Exp Gerontol 36:1495–1502
9. Hohn A, Jung T, Grimm S et al (2011) Lipofuscin inhibits the proteasome by binding to surface motifs. Free Radic Biol Med 50:585–591
10. Wenk J, Brenneisen P, Meeves C et al (2001) UV-induced oxidative stress and photoaging. Curr Probl Dermatol 29:83–94
11. Yang JH, Lee HC, Wei YH (1995) Photoageing-associated mitochondrial DNA length mutations in human skin. Arch Dermatol Res 287:641–648
12. Takeuchi H, Runger TM (2013) Longwave UV light induces the aging-associated progerin. J Invest Dermatol 133:1857–1862
13. Conrad B, Antonarakis SE (2007) Gene duplication: a drive for phenotypic diversity and cause of human disease. Annu Rev Genomics Hum Genet 8:17–35
14. Rodgman A, Perfetti TA (2013) The chemical components of tobacco and tobacco smoke. CRC Press, Boca Raton
15. Clunes LA, Bridges A, Alexis N et al (2008) *In vivo* versus *in vitro* airway surface liquid nicotine levels following cigarette smoke exposure. J Anal Toxicol 32:201–207
16. Ahn KS, Aggarwal BB (2005) Transcription factor NF-kappaB: a sensor for smoke and stress signals. Ann N Y Acad Sci 1056:218–233
17. Rahman I, Adcock IM (2006) Oxidative stress and redox regulation of lung inflammation in COPD. Eur Respir J 28:219–242
18. Maio S, Baldacci S, Martini F et al (2014) COPD management according to old and new GOLD guidelines: an observational study with Italian general practitioners. Curr Med Res Opin 30:1–33
19. Zheng M, Storz G (2000) Redox sensing by prokaryotic transcription factors. Biochem Pharmacol 59:1–6
20. Giudice A, Arra C, Turco MC (2010) Review of molecular mechanisms involved in the activation of the Nrf2-ARE signaling pathway by chemopreventive agents. Methods Mol Biol 647:37–74
21. Kensler TW, Wakabayashi N, Biswal S (2007) Cell survival responses to environmental stresses via the Keap1-Nrf2-ARE pathway. Annu Rev Pharmacol Toxicol 47:89–116
22. Siomek A (2012) NF-kappaB signaling pathway and free radical impact. Acta Biochim Pol 59:323–331
23. Ma Q (2010) Transcriptional responses to oxidative stress: pathological and toxicological implications. Pharmacol Ther 125:376–393
24. Morgan MJ, Liu ZG (2011) Crosstalk of reactive oxygen species and NF-κB signaling. Cell Res 21:103–115
25. Sen CK, Packer L (1996) Antioxidant and redox regulation of gene transcription. FASEB J 10:709–720
26. Espín JC, García-Conesa MT, Tomás-Barberán FA (2007) Nutraceuticals: facts and fiction. Phytochemistry 68:2986–3008
27. Cooke MS, Evans MD, Dizdaroglu M et al (2003) Oxidative DNA damage: mechanisms, mutation, and disease. FASEB J 17:1195–1214
28. Freitas AA, De Magalhaes JP (2011) A review and appraisal of the DNA damage theory of ageing. Mutat Res 728:12–22
29. Baccarelli A, Bollati V (2009) Epigenetics and environmental chemicals. Curr Opin Pediatr 21:243–251
30. Hergersberg M (1991) Biological aspects of cytosine methylation in eukaryotic cells. Experientia 47:1171–1185
31. Goto K, Numata M, Komura JI et al (1994) Expression of DNA methyltransferase gene in mature and immature neurons as well as proliferating cells in mice. Differentiation 56:39–44
32. Ramsahoye BH, Biniszkiewicz D, Lyko F et al (2000) Non-CpG methylation is prevalent in embryonic stem cells and may be mediated by DNA methyltransferase 3a. Proc Natl Acad Sci U S A 97:5237–5242
33. Ziech D, Franco R, Pappa A et al (2011) Reactive Oxygen Species (ROS)—induced genetic and epigenetic alterations in human carcinogenesis. Mutat Res 711:167–173
34. Mohn F, Weber M, Rebhan M et al (2008) Lineage-specific polycomb targets and de novo

DNA methylation define restriction and potential of neuronal progenitors. Mol Cell 30:755–766
35. Nishida N, Kudo M (2013) Oxidative stress and epigenetic instability in human hepatocarcinogenesis. Dig Dis 3:447–453
36. Hitchler MJ, Domann FE (2007) An epigenetic perspective on the free radical theory of development. Free Radic Biol Med 43:1023–1036
37. Lim SO, Gu JM, Kim MS et al (2008) Epigenetic changes induced by reactive oxygen species in hepatocellular carcinoma: methylation of the E-cadherin promoter. Gastroenterology 135:2128–2140
38. Mah WC, Lee CG (2014) DNA methylation: potential biomarker in hepatocellular carcinoma. Biomarker Res 2:1–13
39. Rahman I, Marwick J, Kirkham P (2004) Redox modulation of chromatin remodeling: impact on histone acetylation and deacetylation, NF-kappaB and pro-inflammatory gene expression. Biochem Pharmacol 68:1255–1267
40. Pinney SE, Simmons RA (2010) Epigenetic mechanisms in the development of type 2 diabetes. Trends Endocrinol Metab 21:223–229
41. Dich J, Zahm SH, Hanberg A et al (1997) Pesticides and cancer. Cancer Causes Control 8:420–443
42. Mills KT, Blair A, Freeman LE et al (2009) Pesticides and myocardial infarction incidence and mortality among male pesticide applicators in the Agricultural Health Study. Am J Epidemiol 170:892–900
43. Baldi I, Lebailly P, Mohammed-Brahim B et al (2003) Neurodegenerative diseases and exposure to pesticides in the elderly. Am J Epidemiol 157:409–414
44. Hernandez AF, Lacasana M, Fernando G et al (2013) Evaluation of pesticide-induced oxidative stress from a gene-environment interaction perspective. Toxicology 307:95–102
45. Gawarammana IB, Buckley NA (2011) Medical management of paraquat ingestion. Br J Clin Pharmacol 72:745–757
46. Li N, Ragheb K, Lawler G et al (2003) Mitochondrial complex I inhibitor rotenone induces apoptosis through enhancing mitochondrial reactive oxygen species production. J Biol Chem 278:8516–8525
47. Lopez O, Hernandez AF, Rodrigo L et al (2007) Changes in antioxidant enzymes in humans with long-term exposure to pesticides. Toxicol Lett 171:146–153
48. Ojha A, Yaduvanshi SK, Srivastava N (2011) Effect of combined exposure of commonly used organophosphate pesticides on lipid peroxidation and antioxidant enzymes in rat tissues. Pestic Biochem Physiol 99:148–156
49. Ogut S, Gultekin F, Kisioglu AN et al (2011) Oxidative stress in the blood of farm workers following intensive pesticide exposure. Toxicol Health 27:820–825
50. Lukaszewicz-Hussain A (2010) Role of oxidative stress in organophosphate insecticide toxicity – short review. Pestic Biochem Physiol 98:145–150
51. Mostafalou S, Abdollahi M (2013) Pesticides and human chronic diseases: evidences, mechanisms, and perspectives. Toxicol Appl Pharmacol 268:157–177
52. Sun C, Burgner DP, Ponsonby AL et al (2013) Effects of early-life environment and epigenetics on cardiovascular disease risk in children: highlighting the role of twin studies. Pediatr Res 73:523–530
53. Kukreja RC, Hess ML (1992) The oxygen free radical system: from equations through membrane-protein interactions to cardiovascular injury and protection. Cardiovasc Res 26:641–655
54. Libby P (2002) Inflammation in atherosclerosis. Nature 420:868–874
55. Ross R (1999) Atherosclerosis–an inflammatory disease. N Engl J Med 340:115–126
56. Charo IF, Taub R (2011) Anti-inflammatory therapeutics for the treatment of atherosclerosis. Nat Rev Drug Discov 10:365–376
57. Berry CE, Hare JM (2004) Xanthine oxidoreductase and cardiovascular disease: molecular mechanisms and pathophysiological implications. J Physiol 555:589–606
58. Hare JM, Stamler JS (2005) NO/redox disequilibrium in the failing heart and cardiovascular system. J Clin Invest 115:509–517
59. Rudolph TK, Rudolph V, Baldus S (2008) Contribution of myeloperoxidase to smoking-dependent vascular inflammation. Proc Am Thorac Soc 5:820–823
60. Heinloth A, Heermeier K, Raff U et al (2000) Stimulation of NADPH oxidase by oxidized low-density lipoprotein induces proliferation of human vascular endothelial cells. J Am Soc Nephrol 11:1819–1825
61. Mehta JL, Li DY (1998) Identification and autoregulation of receptor for OX-LDL in cultured human coronary artery endothelial cells. Biochem Biophys Res Commun 248:511–514
62. Sawamura T, Kume N, Aoyama T et al (1997) An endothelial receptor for oxidized low-density lipoprotein. Nature 386:73–77
63. Leonardi A, Chariot A, Claudio E et al (2000) CIKS, a connection to Ikappa B kinase and stress-activated protein kinase. Proc Natl Acad Sci U S A 97:10494–10499
64. Valente AJ, Irimpen AM, Siebenlistd U et al (2014) OxLDL induces endothelial dysfunction and death via TRAF3IP2: inhibition by HDL3 and AMPK activators. Free Radic Biol Med 70C:117–128
65. Valente AJ, Clark RA, Siddesha JM et al (2012) CIKS (Act1 or TRAF3IP2) mediates Angiotensin-II-induced Interleukin-18 expression, and Nox2-dependent cardiomyocyte hypertrophy. J Mol Cell Cardiol 53:113–124
66. Moore KJ, Tabas I (2011) The cellular biology of macrophages in atherosclerosis. Cell 145:341
67. Kzhyshkowska J, Neyen C, Gordon S (2012) Role of macrophage scavenger receptors in atherosclerosis. Immunobiology 217:492–502

68. Lopez AD, Murray CC (1998) The global burden of disease, 1990–2020. Nat Med 4(11):1241–1243
69. Dennis RJ, Maldonado D, Norman S et al (1996) Woodsmoke exposure and risk for obstructive airways disease among women. Chest J 109:115–119
70. Brebner JA, Stockley RA (2013) Recent advances in alpha-1-antitrypsin deficiency-related lung disease. Expert Rev Respir Med 7:213–229
71. Mohamadin AM (2001) Possible role of hydroxyl radicals in the oxidation of dichloroacetonitrile by Fenton-like reaction. J Inorg Biochem 84:97–105
72. Halliwell B, Gutteridge JM (1990) Role of free radicals and catalytic metal ions in human disease: an overview. Methods Enzymol 186:1–85
73. Ghio AJ, Pritchard RJ, Dittrich KL et al (1997) Non-heme (Fe3+) in the lung increases with age in both humans and rats. J Lab Clin Med 129:53–61
74. Gutteridge JM (1995) Lipid peroxidation and antioxidants as biomarkers of tissue damage. Clin Chem 41:1819–1828
75. Ichinose M, Sugiura H, Yamagata S et al (2000) Increase in reactive nitrogen species production in chronic obstructive pulmonary disease airways. Am J Respir Crit Care Med 162:701–706
76. Ito K, Hanazawa T, Tomita K et al (2004) Oxidative stress reduces histone deacetylase 2 activity and enhances IL-8 gene expression: role of tyrosine nitration. Biochem Biophys Res Commun 315:240–245
77. Parola M, Bellomo G, Robino G et al (1999) 4-Hydroxynonenal as a biological signal: molecular basis and pathophysiological implications. Antioxid Redox Signal 1:255–284
78. Yao H, Hwang JW, Moscat J et al (2010) Protein kinase C zeta mediates cigarette smoke/aldehyde- and lipopolysaccharide-induced lung inflammation and histone modifications. J Biol Chem 285: 5405–5416
79. Adcock IM, Ito K, Barnes PJ et al (2005) Histone deacetylation: an important mechanism in inflammatory lung diseases. COPD 2:445–455
80. Marwick JA, Kirkham PA, Stevenson CS et al (2004) Cigarette smoke alters chromatin remodeling and induces proinflammatory genes in rat lungs. Am J Respir Cell Mol Biol 31:633–642
81. Smith LJ, Houston M, Anderson J et al (1993) Increased levels of glutathione in bronchoalveolar lavage fluid from patients with asthma. Am Rev Respir Dis 147:1461–1464
82. Stover SK, Gushansky GA, Salmen JJ et al (2000) Regulation of γ-Glutamate-cysteine ligase expression by oxidative stress in the mouse preimplantation embryo. Toxicol Appl Pharmacol 168:153–159
83. Rahman I, van Schadewijk AA, Hiemstra PS et al (2000) Localization of gamma-glutamylcysteine synthetase messenger rna expression in lungs of smokers and patients with chronic obstructive pulmonary disease. Free Radic Biol Med 28:920–925
84. Gilks CB, Price K, Wright JL et al (1998) Antioxidant gene expression in rat lung after exposure to cigarette smoke. Am J Pathol 152:269–278
85. Tashkin DP, Murray RP (2009) Smoking cessation in chronic obstructive pulmonary disease. Respir Med 103:963–974
86. Rangasamy T, Misra V, Biswal S (2009) Cigarette smoke-induced emphysema in A/J mice is associated with pulmonary oxidative stress, apoptosis of lung cells and global alterations in gene expression. Am J Physiol Lung Cell Mol Physiol 296:L888–L900
87. Rangasamy T, Cho CY, Thimmulappa RK et al (2004) Genetic ablation of Nrf2 enhances susceptibility to cigarette smoke-induced emphysema in mice. J Clin Invest 114:1248–1259

Oxidative Stress-Induced Molecular and Genetic Mechanisms in Human Health and Diseases

Neeraj Dholia, Prerna Ramteke, Johnna F. Varghese, Vibha Rani, and Umesh Chand Singh Yadav

Abstract

Oxidative stress is a common denominator in many inflammatory diseases. A number of genetic and molecular factors are known to regulate oxidative stress by modulating cellular redox imbalance. These events lead to molecular and biochemical changes within the cells resulting in a myriad of biological phenomena including cell growth, initiation and progression of inflammation, programmed cell death, and cellular senescence. Reactive oxygen species (ROS) are well known to act as important secondary messengers which regulate gene expression through signal transduction pathways especially the MAP kinase and NF-κB pathway. Additionally, other molecular cascades are also regulated by oxidants. These molecular and genetic alterations lead to several changes such as genetic mutations, mitochondrial stress, and changes in NADH/NAD$^+$ ratio that further augment and aggravate oxidative stress leading to health deterioration and disease development. Antioxidants, on the other hand, regulate molecular pathways and gene expression within cells as part of the cellular defense system by destroying ROS through enzymatic and nonenzymatic mechanisms. In the present chapter, we discuss the genetic and molecular mechanisms that regulate oxidative stress and play a crucial role in human health and diseases.

Keywords

Oxidative stress • Reactive oxygen species • MAP kinase • Gene regulation • Inflammation

N. Dholia • P. Ramteke • J.F. Varghese
U.C.S. Yadav (✉)
School of Life Sciences, Central University of Gujarat, Sector 30, Gandhinagar, Gujarat 382030, India
e-mail: umeshyadav@cug.ac.in

V. Rani
Department of Biotechnology, Jaypee Institute of Information Technology, A-10, Sector 62, Noida 201307, Uttar Pradesh, India

V. Rani and U.C.S. Yadav (eds.), *Free Radicals in Human Health and Disease*,
DOI 10.1007/978-81-322-2035-0_7, © Springer India 2015

1 Introduction

Oxidative stress resulting from exposure to various stress conditions constantly impact human health. Oxidative stress can be caused by infections, exposure to chemical agents, xenobiotics, or radiations. Under stress conditions cells tend to produce reactive oxygen species (ROS) [1, 2]. ROS can be generated either from the extracellular environment or from within the cell's own biochemical events. Exogenous sources include γ and UV radiations, xenobiotics, pollutants, toxins and food; while endogenous sources include immune cells, ROS-producing enzymes, mitochondria-mediated metabolism, and metal toxicity. Increased levels of ROS promote deleterious actions in cells, with potential to cause dysfunction in the biochemical system and alteration in normal physiological processes. These deregulations lead to a wide variety of degenerative processes and cause diseases such as neoplastic transformations and cancer, diabetic complications, chronic inflammatory diseases such as rheumatoid arthritis, psoriatic arthritis and inflammatory bowel diseases, atherosclerosis, acute inflammatory responses, and neurodegenerative diseases such as Alzheimer's disease and Parkinson's disease [3, 4]. ROS act as signal intermediates and messengers and play an important role in the propagation of molecular signal transduction and altered gene regulation. Contrary to this, antioxidative mechanisms in cells tend to maintain redox equilibrium by activating alternate genetic and molecular machinery. Cells promote defensive mechanisms in the presence of oxidative challenge to curb the deleterious effects produced by oxidants. In both the responses the cell's genetic and molecular machinery play an important role in regulating the final outcome. The molecular signaling pathways induced upon oxidant exposure ultimately lead to transmission of signals to the nucleus where it regulates the expression of target genes and promotes appropriate physiological response. There are several pathways that are sensitive to the presence and exposure of oxidants, the most prominent and well studied among them is the MAP kinase pathway that eventually activates redox-sensitive transcription factor such as NF-κB which in turn regulates the expression of various proinflammatory genes [5]. In this chapter, the genetic and molecular mechanisms with regard to oxidative stress response specifically the MAP kinase/NF-κB pathway as well as pathways involving antioxidant response genes have been discussed. In addition, enzymatic and nonenzymatic mechanisms of the antioxidants have also been discussed.

1.1 Mitogen-Activated Protein Kinases (MAPKs)

The MAP kinase pathway, evolutionarily conserved and unique to eukaryotes, is one of the most studied signal transduction pathways [6]. The MAP kinase signaling pathway has three major families of kinases: extracellular signal-regulated kinase (ERK1/2), p38 kinase, and c-Jun N-terminal kinases/stress-activated protein kinases (JNK/SAPK). MAP kinases comprise a group of serine/threonine kinases that are activated in response to various signaling stimulus, e.g., exposure to oxidants, pollutants, radiations, as well as inflammatory cytokines. It is a vital signaling cascade which controls the gene expression related to cell growth, proliferation, differentiation, apoptosis, and immune response. Oxidative stress-induced activation of MAP kinase has been studied in various types of cells including T lymphocytes, hepatocytes, cardiomyocytes, fibroblasts, epithelial cells, endothelial cells, smooth muscle cells, and pleural cells [7]. ROS activates the ERK pathway through growth factor receptors or by direct activation of the Ras, a GTP-binding molecule involved in signal transduction, leading to the sequential activation of Raf kinase, MEK1/2(MAP kinase kinase), and finally ERK1/2. In the case of JNK/SAPK pathway, ROS activates tumor necrosis factor (TNF) receptors or alternatively activates either MAP kinase kinase kinase (MEKK1) or ASK1 proteins, which then activate MKK4/7 (MAP kinase kinase) and finally JNK/SAPK. JNK/SAPK activates various downstream transcription factors such as Jun, activating transcription factor 2 (ATF-2), nuclear factor erythroid 2-related factor

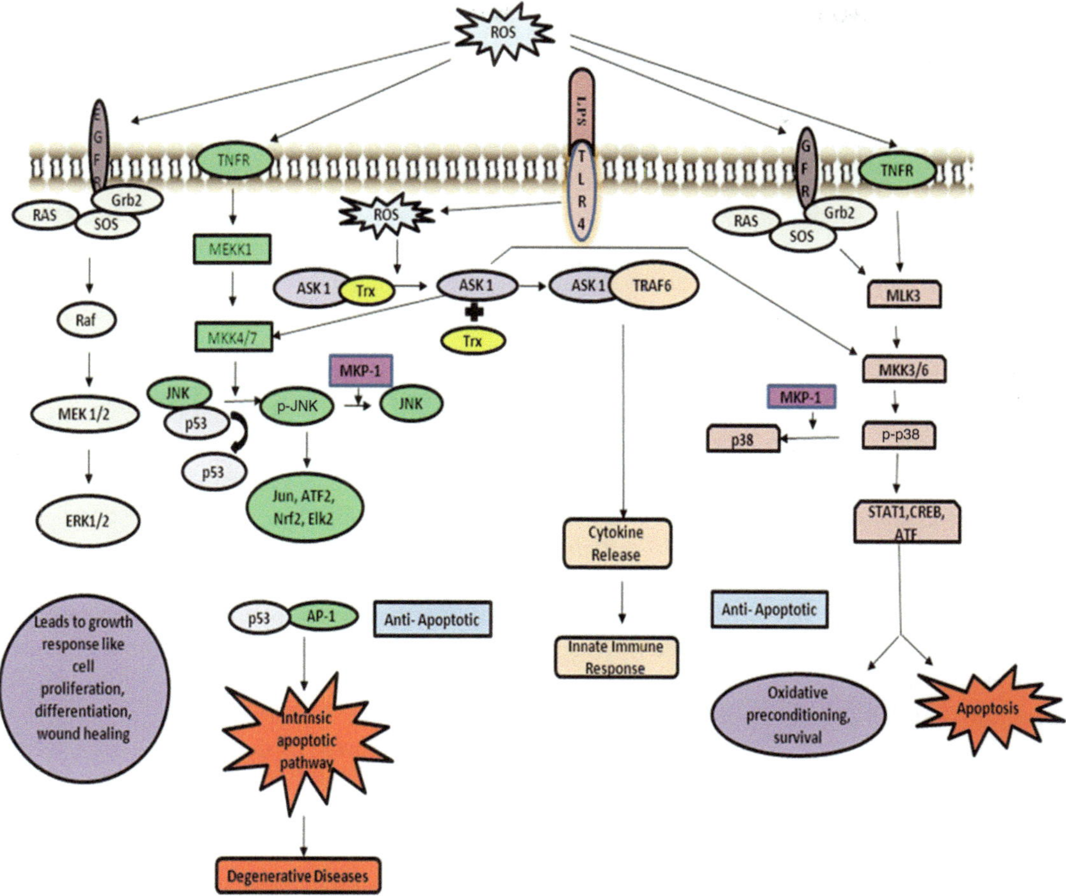

Fig. 1 The schematic diagram showing the activation of different MAPK pathways activated in response to exogenous and endogenous ROS produced during oxidative stress

(Nrf2), and Elk2 – a transcription factor belonging to the E-26 transformation-specific (Ets) family. The p38 pathway is activated either by the TNF receptor or EGF receptor activating downstream molecules such as cell division control protein 42 (cdc42), a GTP-binding protein, and Rac1 (Ras-related C3 botulinum toxin substrate 1), a GTP-binding protein of the Rho family, respectively, followed by activation of MKKKs and MKK3/6 which activate p38. The active p38 phosphorylates its corresponding target molecules such as transcription factors ATF and cAMP response element-binding protein (CREB) [7] (Fig. 1). Upon exposure to γ rays, X-ray, or UV irradiation, heat/osmotic shock and oxidative/nitrosative stress cells produce ROS and reactive nitrogen species (RNS) which act as key messengers and activate p38 and JNK/SAPK, which preferentially promote apoptosis [8]. Experimental evidences suggest the upregulation of MAP kinase pathways when exposed to oxidative agents such as H_2O_2 [9]. Endogenously produced through the process of respiratory burst, H_2O_2 induces the ERK1/2 pathway whereas exogenous source of H_2O_2 activates the p38 MAP kinase rather than ERK [10]. Hence, a balance among the three MAP kinase pathways is important to manifest a particular response by the cells to a specific type of stimulus. The ERK1/2 pathway mainly regulates cell proliferation and survival, whereas a decreased ERK and an increased JNK/SAPK lead to cell apoptosis [9]. The p38 could manifest either cell survival or cell death depending upon the nature of the stimulus.

1.2 Role of ROS-Induced MAP Kinases in Growth, Survival, and Homeostasis

ROS are mostly known for negative impacts on cells however, many studies have implicated the role of ROS in development and growth. Growth hormones, which play an important role in growth and development, are known to act through ROS. For example, PDGF induces ROS production in vascular smooth muscle cells and in lens epithelial cells [11], while TGFβ1 stimulates ROS production in lung fibroblast cells [12] and EGF in corneal epithelial cells [13] finally leading to promotion of growth in these cells. One of the important physiological functions in response to ROS observed in cells is to produce mitogenic stimuli and induce cell proliferation [13]. H_2O_2 is a type of ROS which possesses the ability to stimulate tyrosine phosphorylation of epidermal growth factor receptor (EGFR) and Src homology 2 domain-containing (SHC) adapter protein in a time-dependent manner. This is followed by the induction of a complex formation between EGFR and SHC-Grb2-SOS, thereby leading to Ras and ERK activation (Fig. 1). Hence, in vascular smooth muscle cells H_2O_2 mediates proliferation through EGFR activation via the MAP kinase pathway [14]. This suggests that besides their role in oxidant-induced cytotoxicity, MAPKs also mediate cell survival and homeostasis in the presence of growth factor-induced ROS leading to growth and development.

MAPKs play a vital role in immune response. ROS generated by phagocytic cells as part of an immune response against a pathogenic attack is an important aspect of survival and homeostasis strategy. MAPKs role in precipitating appropriate immune response is preceded by Toll-like receptor (TLR)-mediated pathogenic recognition. TLRs play a crucial role in the recognition of different pathogen-associated molecular patterns (PAMPs) such as lipids, proteins, lipoproteins, etc., of microorganisms and initiate an innate immune response followed by cytokine production. There are various types of TLRs (TLR1 to 11) depending upon specificity toward binding different ligands. Among these TLR4 is highly specific for gram-negative bacterial cell wall component, lipopolysaccharide (LPS). When bound to LPS, TLR4 recruits different adapter molecules such as TNF receptor-associated factor (TRAFF2/6), myeloid differentiation primary response gene 88 (MyD88), and Toll–interleukin 1 receptor (TIR) domain-containing adapter protein (TIRAP) and initiates a cascade of signals which eventually activate JNK and p38 MAP kinase pathway [15]. One of the mechanisms that underlie this innate immune response is ROS-mediated ASK1–MAP kinase activation. Apoptosis signal-regulating kinase 1 (ASK1) is a MAP kinase kinase kinase (MKKK), ubiquitously expressed in mammalian cells. LPS-induced TLR4 mediates ROS production through NADPH oxidase enzyme, associated with the membrane-bound TLR [16]. ROS then triggers the dissociation of thioredoxin (Trx) from ASK1 which is followed by the association of TRAF6 to ASK1. Binding of TRAF6 to ASK1 is followed by the activation of p38 MAP kinase leading to the transcription of cytokines and chemokines, which cause inflammation [17] (Fig. 1).

The role of p38 MAP kinase during oxidative stress has always been controversial. H_2O_2 challenged retinal pigment epithelial cells showed apoptosis through p38 MAP kinase pathway whereas the same cells when exposed to an oxidant *tert*-butyl hydroperoxide (t-BOOH), upon activation of p38 MAP kinase, portrayed a protective role [18]. The activation of p38 MAP pathway has been shown to play an important role in oxidative preconditioning, a process that protects cells upon subsequent exposure to oxidative agents for a long duration and at a higher dose [19]. Thus, ROS at low levels play a crucial role in maintaining a balanced redox state within the cells and also activate signal transduction pathways responsible for growth, differentiation, survival, and homeostasis. Further, insufficient levels of ROS can hinder the cells normal functioning such as growth and immune response, while overwhelming levels tend to propel the cell toward cell senescence and death.

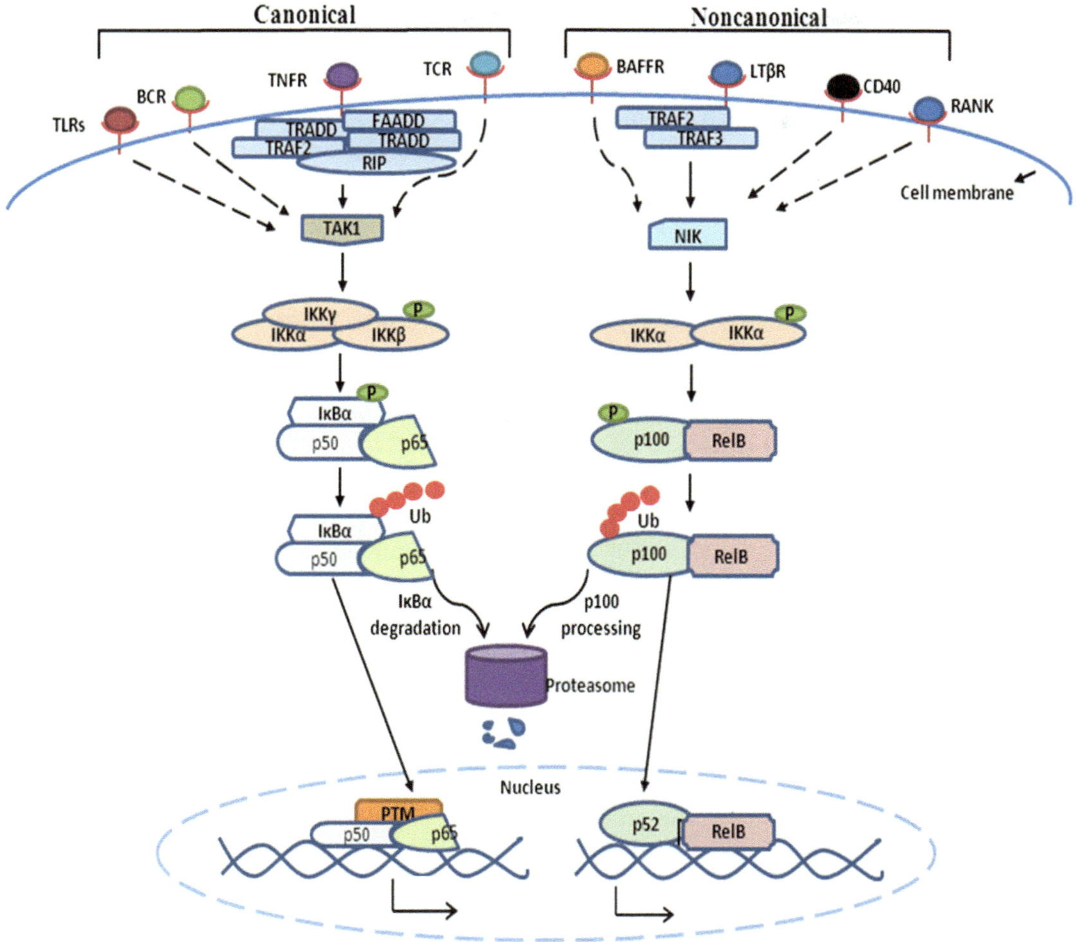

Fig. 2 Activation of canonical and noncanonical NF-κB pathways through different cell receptors

1.3 Role of Oxidative Stress-Induced MAP Kinases in Apoptosis

Apoptosis or programmed cell death is a crucial phenomenon required for development as well as a tool to promote cell death in the presence of incessant oxidative environment. As discussed above, the MAP kinase pathway is one of the main pathways through which oxidants regulate apoptosis through transduction of different apoptosis-inducing stimuli. ROS-induced protein oxidation in cells leads to induction of apoptosis through ASK1-mediated MAP kinase pathway. Upon activation ASK1 triggers the apoptotic response by phosphorylating JNK/SAPK and p38 MAP kinases, which subsequently phosphorylate and activate c-Jun transcription factor and ATF-2. The active c-Jun protein forms either a homodimer or a heterodimer with ATF-2 and forms active transcription factor AP-1 [8], which transcribes proinflammatory and apoptotic proteins. JNK/SAPK also regulates tumor-suppressor protein p53. Unphosphorylated JNK forms a complex with p53 leading to its ubiquitination and proteasomal degradation. Upon phosphorylation JNK phosphorylates p53 and stabilizes it by dissociating from the complex [20]. The active p53 then forms a complex with AP-1 and activates the intrinsic apoptotic pathway. AP-1-mediated pathway has been shown to induce apoptosis in mesangial cells [21]. On the contrary, the MAP kinase/

AP-1 pathway has also been shown to induce the expression of MAP kinase phosphatase-1 (MKP-1) in mesangial cells in a dose-dependent manner when exposed to H_2O_2. MKP-1, an oxidative stress-inducible enzyme, inactivates MAP kinases by dephosphorylating the specific tyrosine and threonine residues and protects against H_2O_2-induced apoptosis. All the three MAP kinases, ERK1/2, JNK/SAPK, and p38, need to be activated, and they cooperate to induce the expression of MKP-1. Although there are evidences of MKP-1 being regulated by pathways other than the MAP kinase, such as PI-3 kinase/Akt pathway in different cell types in response to H_2O_2 exposure, MAP kinase stands out to be a key regulator of the MKP-1, a vital antiapoptotic gene [22].

1.4 Role of Oxidative Stress-Induced MAP Kinases in Diseases

The cellular level of oxidants is an important aspect that determines the cells survival, proliferation, differentiation, or apoptosis. An imbalance in the redox status of cells is an important reason behind various types of human diseases. Oxidative stress has been implicated in various pathological conditions such as cardiovascular diseases, diabetes, neurological disorders, cancer, and age-related diseases. Heart diseases are a leading cause of death worldwide. The pathophysiology of cardiac hypertrophy is mediated by ASK1–MAP kinase pathway. Superoxide ions produced by angiotensin II (AngII) through its G protein-coupled receptor activates ASK1–MAPK pathway that leads to prominent heart diseases including cardiac hypertrophy, high blood pressure, and heart valve stenosis [23]. Alzheimer's disease, a neurodegenerative disease found among older people, is characterized by the presence of amyloid-β peptide, plaques, and neurofibrillary tangles. Amyloid-β deposition has been shown to activate ASK1 and subsequently JNK, leading to neuronal cell death [24]. Similarly, exposure to H_2O_2 and diethyl maleate, oxidative stress-inducing agents, activate p38, and ERK1/2 MAP kinase pathway leads to increase in damage of sensory neuron cells [9]. Further, diabetes, a metabolic disease affecting more than 300 million people worldwide who develop secondary diabetic complications such as retinopathy, nephropathy, and neuropathy, is caused by oxidative stress. Inflammation-induced ROS activates ASK1-mediated pathway leading to a proinflammatory condition followed by JNK activation that phosphorylates insulin receptor substrate 1(IRS1), a key mediator of insulin signaling. These steps lead to insulin resistance and an increase in blood sugar levels [25]. Hyperglycemic conditions established during diabetes lead to superoxide generation in endothelial cells in the arterial walls, promoting low-density lipoprotein (LDL) peroxidation [26]. The oxidized LDL alters the redox equilibrium within the cell followed by the activation of specific transcription factors AP-1 and NF-κB through the MAP kinase pathway, thereby causing an increase in the release of various growth factors like cytokines, which promote vascular smooth cell proliferation resulting in progressive plaque formation in the arterial walls [27]. This shows the interlink between diabetes and atherosclerosis and how oxidative stress and MAP kinase act as a common denominator [28].

The skin, being constantly exposed to oxidative agents such as UV radiations, pollutants, and pathogens, produces ROS exogenously and/or endogenously. A prolonged exposure to ROS could lead to a variety of skin disorders such as skin cancer, eczema, and psoriasis. Oxidants and ROS activate the redox signaling pathway including the MAP kinase, NF-κB, and Janus kinase/signal transducer and activator of transcription (JAK-STAT) signaling pathway which contribute toward the progression of skin diseases. For example, ERK1/2 levels have been shown to increase in lesional skin in comparison to the normal skin. Similarly, JNK and p38 MAPK pathways have been linked to over activation of TNF-α and keratinocyte hyper-proliferation by activating c-Jun in psoriasis [7].

MAP kinase pathway is an extremely complex signal transduction pathway involving numerous targets, which leads to different types of responses under oxidative stress. Under oxidative stress conditions all three MAP kinases, ERK1/2, JNK/

SAPK, and p38, are activated and cause oxidative injury and pathogenesis. Mostly ERK1/2 and JNK are associated with antiapoptotic response and lead to cell proliferation and differentiation, whereas p38 MAP kinase is linked with proapoptotic action. Thus, the MAP kinase pathway is an important oxidative stress-stimulated cascade that can be regulated and controlled by the presence of specific antioxidant agents and serves as important molecular target to prevent or treat oxidative stress-induced pathogenesis [9].

1.5 Role of Oxidative Stress-Induced MAP Kinases in Wound Healing and Repair

Signal transduction pathways are also implicated in wound healing and repair subsequent to oxidative injury and damage. Wound healing involves various important phases such as inflammation, cell proliferation, cell migration, and tissue remodeling. Different immune cells including macrophages and neutrophils get recruited in the wound site where they produce ROS which mediates the release of cytokines to protect the organism against the invading pathogen. Several studies demonstrate that inflammation caused by wound results in the release of H_2O_2, which activates p38 and MEK pathway leading to cytoplasmic translocation and active release of high-mobility group protein B1 (HMGB1) in monocytes/macrophages [29]. HMGB1 is a nuclear protein released passively in damaged cells. This protein triggers inflammation, attracts inflammatory cells, recruits stem cells, and promotes their proliferation ultimately leading to tissue repair [30]. Reepithelialization or cell migration of proliferated wound-related cells to the damaged site is of utmost importance. The extracellular matrix of the cells required for reepithelialization is degraded by matrix metalloproteinase (MMPs). A redox-dependent MMP production is observed during wound healing [31]. MMPs expression is activated by JNK through the Ets or AP-1 transcription factor or through the activation of NOX-4 (NADPH oxidase family), an enzyme that produces large amounts of ROS and activates MAPK and NF-κB in the downstream [32]. An apt example for wound healing is corneal wound healing. Cornea, the outermost transparent covering of the eye, is under constant exposure of UV-light, chemicals, bacteria, and fungus which injure or damage it. In a study, intracellular ROS produced in human corneal epithelial cells, upon exposure to epidermal growth factor, stimulated cell proliferation, adhesion, migration, and corneal surface wound healing [13]. In another study, the initial oxidative injury led to the release of keratinocyte growth factor and hepatocyte growth factor which activated p44/42 MAPK and p38 MAP kinase pathways leading to initiation of cell migration and proliferation that resulted in corneal wound healing [33].

Besides the different roles of the MAP kinase pathway during oxidative stress, as discussed in preceding sections the MAP kinase pathway also plays an important role in the antioxidative balance during oxidative stress. Homeostasis and redox balance within the cells need to be maintained by establishing an intricate equilibrium between oxidant production and destruction. The cells possess the ability to scavenge or quench the excess ROS produced within the cell by activating various antioxidant enzymes. The JNK and p38 pathways are known to activate an important transcription factor nuclear factor erythroid 2-related factor (Nrf2). Nrf2 is kept in an inactive state by interaction with the actin-anchored protein Kelch ECH-associating protein 1 (Keap 1) [34]. However, in oxidative stress Nrf2 is released and translocated into the nucleus where they bind to antioxidant response elements for the induction of cellular defense genes which include glutathione S-transferase, glutathione reductase, glutathione peroxidase, heme-oxygenase and NADPH quinone oxidoreductase-1. These gene products help in maintaining the redox balance by ROS disruption/quenching.

The identification of oxidative stress-induced molecular cascade has led to the investigation of various modulators, which may become important drug molecules against oxidative stress-induced diseases. Several phytochemicals possess the ability to ameliorate the oxidative damage caused by ROS. They act as exogenous antioxidants and mediate their action through modulating oxidative stress-induced signal transduction pathways. For example, phytochemicals such as oleanolic acid has been shown to activate ERK and JNK pathway followed by activation of transcription factor Nrf2 [35]. Nrf2 regulates the levels of thioredoxin peroxidase, an H_2O_2-scavenging enzyme, and heme-oxygenase 1 (HO-1), a stress-inducible antioxidant enzyme localized in non-neural tissues, plays a key role in defense against inflammation and oxidative injury [36]. Quercetin, an antioxidant, showed Nrf2-induced expression of HO-1 in esophageal epithelial cells through the activation of MAP kinase and PI-3K pathways [37].

2 NF-κB Pathway

NF-κB, a redox-sensitive transcription factor, coordinates the different regulators of immunity, inflammation, cell survival, proliferation, and development [38]. Dysregulation of NF-κB activity is associated with inflammatory disorders, metabolic and autoimmune diseases, as well as cancer [39]. The NF-κB or Rel family of transcription factors is a downstream target of oxidative stress-induced MAPK pathway that is activated due to bacterial and viral infection, inflammatory cytokines, and upon ligand binding of antigen receptors, which are the sources of ROS. This suggests that NF-κB plays a crucial role in orchestrating immune and inflammatory response through regulation of different oxidant and antioxidant systems. The active NF-κB is a dimer made from monomer units p50, p52, p65 (Rel A), Rel B, and c-Rel proteins [5]. The members of the NF-κB family have a conserved region which is known as NF-κB/Rel/dorsal region or Rel homology domain. This conserved region spans the first 300 N-terminal amino acids and comprises three different regions such as a nuclear localization sequence, a DNA-binding domain, and a dimerization region.

In eukaryotes two alternates NF-κB-activating pathways occur: canonical and noncanonical pathways. Most of the NF-κB-activating physiological stimuli for instance signals from cytokine receptors, such as tumor necrosis factor receptor (TNFR) and interleukin-1 receptor (IL-1R), and antigen receptors such as B cell receptor (BCR), T cell receptor (TCR), and Toll-like receptor 4 (TLR4) induce the canonical pathway. The canonical pathway is characterized by the activation of IκB kinase (IKKβ) and IKKγ/NF-κB essential modulator (NEMO) cascade that leads to phosphorylation of IκBα, which then degrades and facilitates nuclear translocation of p65-containing heterodimer which transcribes target genes. The noncanonical pathway relies on IKKα-mediated phosphorylation of p100 associated with RelB, which leads to processing of p100 and the formation of p52–RelB complex. Specific members of the TNF cytokine family such as CD40 ligand, B cell-activating factor belonging to the TNF family (BAFF), receptor activator for nuclear factor κB (RANK) [40], and lymphotoxin-β2 (LT-β) induce noncanonical NF-κB signaling, although they may also regulate canonical pathway [41]. Different members of TNFR-associated factors (TRAF), TNF receptor-associated death domain (TRADD), and Fas-associated protein with death domain (FADD) adaptor proteins recruited at the intracellular portion of different receptors may activate both canonical and noncanonical signal transduction pathways [39] (Fig. 2).

In unstimulated healthy cells IκB, an NF-κB inhibitory protein which includes IκBα, IκBβ, IκBγ, and probably B cell lymphoma-3 (Bcl-3) is found bounded to a nuclear localization sequence of NF-κB and retains it in an inactive form within the cytosol [5]. ROS, generated by different oxidative mechanisms transduces hyperphosphorylation of IκB, which leads to its degradation by 26s proteasome complex via ubiquitination. After degradation of IκB, NF-κB proteins dimerize and translocate into the nucleus and transcribe various target genes.

2.1 Role of Oxidative Stress in NF-κB Pathway Activation

An antioxidant system activated through NF-κB can be expected to regulate the redox state when the cell is under oxidative stress. There are two genes that could be induced by NF-κB, inducible Mn-dependent mitochondrial form of superoxide dismutase (MnSOD). MnSOD is expressed after stimulation of cells with TNF, LPS, and PMA. The second one is thioredoxin, which is induced by viral transactivator Tax through NF-κB [42].

ROS have paradoxical effects on regulation of NF-κB. A moderate increase in ROS leads to the activation of NF-κB, while severe increase inactivates NF-κB that could result in cell death. Moderately increased ROS-mediated activation of NF-κB increases cell survival by enhancing (a) expression of antiapoptotic Bcl-2 family members, such as Bcl-xL and A1/Bfl-1; (b) caspase inhibitors, such as IAPs; (c) $FLIP_L$, the active homologue of caspase-8; (d) GADD45, which inhibits JNK-mediated cell death; (e) TRAF1, TNF receptor-associated factor; and (f) antioxidants such as MnSOD and ferritin heavy chain (FHC) [43, 38]. The DNA-binding activity of NF-κB was found to increase with age, which reflects increased oxidative stress in aged tissues and cells [44].

The direct evidence that ROS regulates NF-κB has come from the exposure of different cells to H_2O_2 such as Wurzburg subclone of T cells, human breast mammary epithelial cell (MCF-7), skeletal muscle myotubes L6, and 70Z/3 pre-B cells and HeLa cells [45, 46]; however, some other studies could not detect NF-κB activation by H_2O_2 in HeLa cells [45]. It indicates that NF-κB activation may be cell type specific [5]. Various indirect evidences also suggest a role of ROS as a common and critical intermediate for NF-κB activation. This is based on the fact that antioxidants such as butylated hydroxyanisole (BHA) and pyrrolidine dithiocarbamate (PDTC) as well as overexpression of antioxidant enzymes including SOD, thioredoxin, and GSH peroxidase inhibit/prevent NF-κB activation, although the extent of inhibition also depends on type of cell and stimulus [5].

Even though IκB degradation is one of the phenomena observed during oxidative stress, oxygen radicals have the capability to covalently modify several amino acids in proteins such as, cysteine, methionine, and histidine residues. Thus, direct oxidative damage can also cause dissociation of the NF-κB–IκB complex [42]. Further, few compounds are generated upon lipid peroxidation, such as malondialdehyde or 4-hydroxynonenol, that could covalently alter and dissociate the NF-κB–IκB complex [42]. However, much work is needed to confirm molecules and steps that are exactly responsible for oxidative stress-induced damage. Some studies suggest activation of IKK as the key regulatory step in NF-κB activation by oxidative stress and proinflammatory stimuli (TNF, IL-1, LPS, and ds-RNA). Thus, directly or indirectly ROS play a general signaling role in NF-κB activation [45].

2.2 Role of Oxidative Stress in NF-κB-Mediated Inflammatory Gene Induction

Inflammatory cytokines and chemokines promote NF-κB activation. These inflammatory mediators together with antimicrobial molecules and ROS cooperate to kill pathogens, eliminate infection, and remove dead cells. The end of inflammation is vital for maintaining good health while overwhelming inflammation is associated with various inflammatory diseases like rheumatoid arthritis (RA), type 2 diabetes, and atherosclerosis [47].

In the development of RA, the activation of NF-κB-dependent genes plays a key role. The receptor activator of NF-κB [40], CD40, B cell-activating factor, lymphotoxin β receptor, and TLRs are found to be upregulated in RA [40, 48]. Evidences show involvement of free radicals and NF-κB in the destruction of β cells and disease progression of insulin-dependent diabetes mellitus or type I diabetes [49]. Pancreas-specific ROS production plays a critical role in the autoimmune or inflammatory response by activating NF-κB [51]. Proinflammatory cytokines such as IL-1 have been observed to activate NF-κB in rat insulinoma cell lines [50]. The NF-κB pathway also seems to be involved in the insulin resistance of type II diabetes [51]. These evidences indicate that NF-κB is crucial in the development of varied inflammatory pathologies and thus could be an important target in designing therapeutic approaches. Indeed, targeted disruption of IKK-β or salicylate inhibition of NF-κB has been shown to improve insulin sensitivity and reduction of obesity in mice [52].

3 Role of Oxidative Stress in Gene Regulation

Oxidative stress, an important modulator of cellular redox status, regulates the expression of proinflammatory/prooxidant genes as well as anti-inflammatory/antioxidative genes. The fine balance and eventual tilt between these two seemingly opposite mechanisms determine the fate of cells. As discussed in preceding sections, oxidative stress stimulates a molecular cascade that activates transcription factors, which transcribe various target genes including cytokines, chemokines, growth factors, and other inflammatory marker genes. It is therefore evident that gene expression is finely regulated in and by oxidative stress. The variation in oxidative status of cells alters the cellular environment, which affects transcription factors, and ultimately gene expression is affected.

Binding of transcription factors to a specific region of negatively charged DNA sequences requires accumulation of positively charged amino acids, resulting in stabilization of deprotonated thiol groups. In the presence of ROS, thiol residues such as cysteine get oxidized affecting DNA-binding efficiency. The signaling proteins behave differently to oxidation, depending upon the level of cysteine moiety in the cells during oxidative stress and intensity of oxidative stress. The signaling cascade of NF-kB and activator protein (AP-1), leading to gene regulation of glutathione reductase, serves as an example of oxidation of conserved cysteine groups, which is a reversible phenomenon [53].

Some of the gene products also turn on the transcription of various detoxifying enzymes and proteins such as glutathione S-transferase (GST) and NADPH quinone reductase that act as an antioxidant system during oxidative stress [54]. The cis-acting antioxidant response elements [55], responsible for inducible as well as constitutive gene expression of these two enzymes, mediate their transcriptional activation and are induced by sulfhydryl groups containing compounds such as diethyl maleate, isothiocyanates, and dithiothiones [56]. Similarly, oxidative stress also affects other transcription factors such as nuclear factor 1 (NF1), Sp1, and p53 NF1, activator of DNA replication, has a DNA-binding domain and a transactivation domain. The presence of oxidative stress affects cysteine residues of DNA-binding domain which hampers its DNA-binding activity through mutation to serine residues [54]. Regulation of redox imbalance by glutaredoxine or thioltransferase prevents mutation of its cysteine residues [54]. Sp1 is a ubiquitous transcription factor, rich in GC content, and has zinc finger motifs, which are crucial for DNA-binding activity. A decrease in glutathione level and increase in oxidants such as H_2O_2, represses the activity of Sp1-driven genes. Oxidative stress also represses the function of p53 by modulating its conformation as shown by decreased reporter gene expression [54, 57]. PDTC, which acts as prooxidant, increases the intracellular level of redox-active copper that downregulates p53. This would prevent the activation of p53 by stimuli such as UV or temperature shift, which are known to activate PDTC during oxidative stress. Oxidation of cysteine residue in p53 also inhibits or alters expression of apoptotic and DNA repair genes that leads to carcinogenesis [57].

The cytochrome P450 (CYP) genes encode for ubiquitous enzymes responsible for metabolism of xenobiotics. The NF1 binding site, located on CYP promoter, is the target for oxidation during redox imbalance. ROS is known to downregulate the isoforms of CYP, e.g., CYP1A1 by inflammatory cytokines, due to depletion of glutathione [54]. A study by Morel and Barouki [58] reported that increased TNF-α mRNA during oxidative stress downregulates this isoform of CYP through redox-sensitive mechanism involving NF1. This could also affect the NF1 and DNA interaction which in turn alters the redox-sensitive gene transcription [58].

Oxidative stress-induced alteration in gene expression is a well known cause of age-related disorders, e.g., Alzheimer's disease (AD) [59]. Various superoxide anion-producing enzymes such as NADPH oxidase (NOX) and xanthine oxidase (XO) along with mitochondrial dysfunction are major players in the oxidative stress-induced neurodegenerative disease. *NOX*, present in microglial cells, neurons, and astrocytes get activated by cytosolic subunit of G protein Ras. After activation, NOX translocates to the membrane and forms active NAPDH oxidase complex [16, 54]. The enzyme transfers protons across the membrane through anion channel to compensate the charges across the membrane and results in the formation of superoxide radicals. In astrocytes, amyloid β, an oxidant and inducer of NADPH oxidase, also activates the expression of *NOX* genes including *NOX1*, *NOX2*, and *NOX3* through CD36 receptor. The increased expression of NOX isoforms leads to an enhanced oxidative stress which depolarizes the mitochondrial membrane and induces opening of mitochondrial pore permeability and activates apoptosis that results in neuronal degeneration. Apart from neuronal death, aggregation of amyloid β protein leads to hyperphosphorylation of tau (τ) protein which is a primary or early stage biomarker in AD. A neuroprotective hormone melatonin acts as an antioxidant for NADPH enzyme complex and prevents the formation of amyloid β protein aggregation [60].

Oxidative stress-induced mitochondrial dysfunction, loss of function of DJ-1 gene, and A53T α-synuclein protein aggregation have been implicated in the pathogenesis of Parkinson's disease [61, 62]. DJ-1, a neuroprotective molecule, has an antioxidative mechanism in neurons. It increases the synthesis of glutathione, cellular redox buffer system, by regulating a rate-limiting enzyme glutathione–cysteine ligase (GCL). However, elevated level of ROS may cause allelic mutations in DJ-1 gene from cysteine to serine residues, leading to altered gene expression and progression of pathogenesis [62]. Additionally, under stress a nonfunctional GCL promotes A53T α-synuclein protein aggregation in dopamine neuronal cells which are responsible for cell death [60]. Glutathione improves survival of primary dopamine neurons and prevents dopamine-induced neuronal cell death [61, 62]. Although implicated in neuroinflammation, ischemia, age-related neurodegeneration, and Parkinson's disease, a detailed mechanism involving DJ-1/PARK7 gene function for ROS-mediated pathogenesis remains unclear [61, 63]. A better understanding of this pathway would be helpful in devising novel neuroprotective and anti-neurodegenerative therapy.

4 Conclusion

The emerging concepts regarding ROS and its effects on cellular physiology have gained enormous attention. This chapter focused on the molecular and genetic mechanisms and the different responses manifested by the cell when exposed to ROS. Deleterious effects of oxidative stress or the resultant antioxidant response to the same depend on the nature of the agents causing oxidative stress, their dosage and duration of exposure, and the signal transduction pathway activated resulting in specific transcription factor activation. Hence, it becomes difficult to determine the exact mechanism underlying ROS-induced stress at the molecular level. Nonetheless, several pathways have been identified, which get activated in response to ROS, especially the MAP kinase and NF-κB pathway. Many studies have implicated different upstream mediators that activate NF-κB, but how exactly cytosolic NF-κB complex senses oxidative stress is not well understood. Our knowledge is far from complete when it comes to identifying the precise upstream and downstream regulatory molecules in the oxidative stress-induced pathway and its regulation through gene expression. Future studies are required to clearly understand the molecular and genetic mechanism(s) that regulates oxidative stress-induced pathogenesis, which will lay the foundation for better preventive and therapeutic approaches for human health and diseases.

Acknowledgement Dr. Umesh C. S. Yadav would like to acknowledge department of science and technology (DST), Government of India for the award of Ramanujan Fellowship. Financial support from UGC to ND, PR and JFV is also acknowledged.

References

1. Blokhina O, Virolainen E, Fagerstedt KV (2003) Antioxidants, oxidative damage and oxygen deprivation stress: a review. Ann Bot 91:179–194
2. Petrosillo G, Ruggiero FM, Pistolese M et al (2004) Ca^{2+}-induced reactive oxygen species production promotes cytochrome c release from rat liver mitochondria via mitochondrial permeability transition (MPT)-dependent and MPT-independent mechanisms: role of cardiolipin. J Biol Chem 279:53103–53108
3. Imlay JA (2013) The molecular mechanisms and physiological consequences of oxidative stress: lessons from a model bacterium. Nat Rev Microbiol 11:443–454
4. Davies KJ (1995) Oxidative stress: the paradox of aerobic life. Biochem Soc Symp 61:1–31
5. Allen RG, Tresini M (2000) Oxidative stress and gene regulation. Free Radic Biol Med 28:463–499
6. Chang L, Karin M (2001) Mammalian MAP kinase signalling cascades. Nature 410:37–40
7. Zhou Q, Mrowietz U, Rostami-Yazdi M et al (2009) Oxidative stress in the pathogenesis of psoriasis. Free Radic Biol Med 47:891–905
8. Sumbayev VV, Yasinska IM (2005) Regulation of MAP kinase-dependent apoptotic pathway: implication of reactive oxygen and nitrogen species. Arch Biochem Biophys 436:406–412
9. Valko M, Leibfritz D, Moncol J et al (2007) Free radicals and antioxidants in normal physiological functions and human disease. Int J Biochem Cell Biol 39:44–84
10. Iles KE, Forman HJ (2002) Macrophage signaling and respiratory burst. Immunol Res 26:95–105
11. Chen KC, Zhou Y, Zhang W et al (2007) Control of PDGF-induced reactive oxygen species (ROS) generation and signal transduction in human lens epithelial cells. Mol Vis 13:374–387
12. Junn E, Lee KN, Ju HR et al (2000) Requirement of hydrogen peroxide generation in TGF-beta 1 signal transduction in human lung fibroblast cells: involvement of hydrogen peroxide and Ca2+ in TGF-beta 1-induced IL-6 expression. J Immunol 165: 2190–2197
13. Huo Y, Qiu WY, Pan Q et al (2009) Reactive oxygen species (ROS) are essential mediators in epidermal growth factor (EGF)-stimulated corneal epithelial cell proliferation, adhesion, migration, and wound healing. Exp Eye Res 89:876–886
14. Rao GN (1996) Hydrogen peroxide induces complex formation of SHC-Grb2-SOS with receptor tyrosine kinase and activates Ras and extracellular signal-regulated protein kinases group of mitogen-activated protein kinases. Oncogene 1:713–719
15. Symons A, Beinke S, Ley SC (2006) MAP kinase kinase kinases and innate immunity. Trends Immunol 27:40–48
16. Jiang F, Zhang Y, Dusting GJ (2011) NADPH oxidase-mediated redox signaling: roles in cellular stress response, stress tolerance, and tissue repair. Pharmacol Rev 63:218–242
17. Matsuzawa A, Ichijo H (2008) Redox control of cell fate by MAP kinase: physiological roles of ASK1-MAP kinase pathway in stress signaling. Biochim Biophys Acta 1780:1325–1336
18. Pocrnich CE, Liu H, Feng M et al (2009) p38 mitogen-activated protein kinase protects human retinal pigment epithelial cells exposed to oxidative stress. Can J Ophthalmol 44:431–436
19. Han H, Wang H, Long H et al (2001) Oxidative preconditioning and apoptosis in L-cells. Roles of protein kinase B and mitogen-activated protein kinases. J Biol Chem 276:26357–26364
20. Fuchs SY, Adler V, Pincus MR et al (1998) MEKK1/JNK signaling stabilizes and activates p53. Proc Natl Acad Sci U S A 95:10541–10546
21. Saitoh M, Nishitoh H, Fujii M et al (1998) Mammalian thioredoxin is a direct inhibitor of apoptosis signal-regulating kinase (ASK) 1. EMBO J 17:2596–2606
22. Xu Q, Konta T, Nakayama K et al (2004) Cellular defense against H2O2-induced apoptosis via MAP kinase-MKP-1 pathway. Free Radic Biol Med 36:985–993
23. Izumiya Y, Kim S, Izumi Y et al (2003) Apoptosis signal-regulating kinase 1 plays a pivotal role in angiotensin II-induced cardiac hypertrophy and remodeling. Circ Res 93:874–883
24. Kadowaki H, Nishitoh H, Urano F et al (2005) Amyloid beta induces neuronal cell death through ROS-mediated ASK1 activation. Cell Death Differ 12:19–24
25. Imoto K, Kukidome D, Nishikawa T et al (2006) Impact of mitochondrial reactive oxygen species and apoptosis signal-regulating kinase 1 on insulin signaling. Diabetes 55:1197–1204
26. Graier WF, Simecek S, Kukovetz WR et al (1996) High D-glucose-induced changes in endothelial Ca2+/EDRF signaling are due to generation of superoxide anions. Diabetes 45:1386–1395
27. Maziere C, Auclair M, Djavaheri-Mergny M et al (1996) Oxidized low density lipoprotein induces activation of the transcription factor NF kappa B in fibroblasts, endothelial and smooth muscle cells. Biochem Mol Biol Int 39:1201–1207
28. Kunsch C, Medford RM (1999) Oxidative stress as a regulator of gene expression in the vasculature. Circ Res 85:753–766
29. Tang D, Shi Y, Kang R et al (2007) Hydrogen peroxide stimulates macrophages and monocytes to actively release HMGB1. J Leukoc Biol 81:741–747
30. Bianchi ME, Manfredi AA (2007) High-mobility group box 1 (HMGB1) protein at the crossroads between innate and adaptive immunity. Immunol Rev 220:35–46
31. Yoon SO, Park SJ, Yoon SY et al (2002) Sustained production of H(2)O(2) activates pro-matrix metal-

loproteinase-2 through receptor tyrosine kinases/phosphatidylinositol 3-kinase/NF-kappa B pathway. J Biol Chem 277:30271–30282
32. Sen CK, Roy S (2008) Redox signals in wound healing. Biochim Biophys Acta 1780:1348–1361
33. Imayasu M, Shimada S (2003) Phosphorylation of MAP kinase in corneal epithelial cells during wound healing. Curr Eye Res 27:133–141
34. Kensler TW, Wakabayashi N, Biswal S (2007) Cell survival responses to environmental stresses via the Keap1-Nrf2-ARE pathway. Annu Rev Pharmacol Toxicol 47:89–116
35. Wang X, Ye XL, Liu R et al (2010) Antioxidant activities of oleanolic acid in vitro: possible role of Nrf2 and MAP kinases. Chem Biol Interact 184:328–337
36. Takahashi T, Shimizu H, Morita K (2009) Heme oxygenase-1 is an essential cytoprotective component in oxidative tissue injury induced by hemorrhagic shock. J Clin Biochem Nutr 44:28–40
37. Kim JS, Song HJ, Ko SK et al (2010) Quercetin-3-O-beta-d-glucuronopyranoside (QGC)-induced HO-1 expression through ERK and PI3K activation in cultured feline esophageal epithelial cells. Fitoterapia 81:85–92
38. Trachootham D, Lu W, Ogasawara MA et al (2008) Redox regulation of cell survival. Antioxid Redox Signal 10:1343–1374
39. Oeckinghaus A, Hayden MS, Ghosh S (2011) Crosstalk in NF-kappaB signaling pathways. Nat Immunol 12:695–708
40. Frank M, Duvezin-Caubet S, Koob S et al (2012) Mitophagy is triggered by mild oxidative stress in a mitochondrial fission dependent manner. Biochim Biophys Acta 1823:2297–2310
41. Sun SC (2011) Non-canonical NF-kappaB signaling pathway. Cell Res 21:71–85
42. Schreck R, Albermann K, Baeuerle PA (1992) Nuclear factor kappa B: an oxidative stress-responsive transcription factor of eukaryotic cells (a review). Free Radic Res Commun 17:221–237
43. Finkel T, Holbrook NJ (2000) Oxidants, oxidative stress and the biology of ageing. Nature 408:239–247
44. Poynter ME, Daynes RA (1998) Peroxisome proliferator-activated receptor alpha activation modulates cellular redox status, represses nuclear factor-kappaB signaling, and reduces inflammatory cytokine production in aging. J Biol Chem 273:32833–32841
45. Li N, Karin M (1999) Is NF-kappaB the sensor of oxidative stress? FASEB J 13:1137–1143
46. Wang X, Martindale JL, Liu Y et al (1998) The cellular response to oxidative stress: influences of mitogen-activated protein kinase signalling pathways on cell survival. Biochem J 333:291–300
47. Baker RG, Hayden MS, Ghosh S (2011) NF-kappaB, inflammation, and metabolic disease. Cell Metab 1:11–22
48. Van Loo G, Beyaert R (2011) Negative regulation of NF-kappaB and its involvement in rheumatoid arthritis. Arthritis Res Ther 13:221
49. Kumar A, Takada Y, Boriek AM et al (2004) Nuclear factor-kappaB: its role in health and disease. J Mol Med (Berl) 82:434–448
50. Kwon G, Corbett JA, Rodi CP et al (1995) Interleukin-1 beta-induced nitric oxide synthase expression by rat pancreatic beta-cells: evidence for the involvement of nuclear factor kappa B in the signaling mechanism. Endocrinology 136:4790–4795
51. Yuan M, Konstantopoulos N, Lee J et al (2001) Reversal of obesity- and diet-induced insulin resistance with salicylates or targeted disruption of Ikkbeta. Science 293(5535):1673–1677
52. Kumar A, Takada Y, Boriek AM et al (2004) Nuclear factor-kB: its role in health and disease. J Mol Med 82:434–448
53. Droge W (2002) Free radicals in the physiological control of cell function. Physiol Rev 82(1):47–95
54. Morel Y, Barouki R (1999) Repression of gene expression by oxidative stress. Biochem J 342:481–496
55. Irani K, Xia Y, Zweier JL et al (1997) Mitogenic signaling mediated by oxidants in Ras-transformed fibroblasts. Science 275:1649–1652
56. Nguyen T, Nioi P, Pickett CB (2009) The Nrf2-antioxidant response element signaling pathway and its activation by oxidative stress. J Biol Chem 284:13291–13295
57. Verhaegh GW, Richard MJ, Hainaut P (1997) Regulation of p53 by metal ions and by antioxidants: dithiocarbamate down-regulates p53 DNA-binding activity by increasing the intracellular level of copper. Mol Cell Biol 17:5699–5706
58. Morel Y, Barouki R (1998) Down-regulation of cytochrome P450 1A1 gene promoter by oxidative stress. Critical contribution of nuclear factor 1. J Biol Chem 273:26969–26976
59. Gandhi S, Abramov AY (2012) Mechanism of oxidative stress in neurodegeneration. Oxid Med Cell Longev 2012:428010
60. Pappolla MA, Sos M, Omar RA et al (1997) Melatonin prevents death of neuroblastoma cells exposed to the Alzheimer amyloid peptide. J Neurosci 17:1683–1690
61. Kahle PJ, Waak J, Gasser T (2009) DJ-1 and prevention of oxidative stress in Parkinson's disease and other age-related disorders. Free Radic Biol Med 47:1354–1361
62. Zhou W, Freed CR (2005) DJ-1 up-regulates glutathione synthesis during oxidative stress and inhibits A53T alpha-synuclein toxicity. J Biol Chem 280:43150–43158
63. Billia F, Hauck L, Grothe D et al (2013) Parkinson-susceptibility gene DJ-1/PARK7 protects the murine heart from oxidative damage in vivo. Proc Natl Acad Sci U S A 110:6085–6090

Hydrogen Peroxide Sensing and Signaling

Vibha Rani, Shivangi Mishra*, Tanuja Yadav*, Umesh Chand Singh Yadav, and Shrey Kohli

Abstract

H_2O_2 has been found to act as a signaling molecule and secondary messenger in many signal transduction pathways like in insulin signaling, cell proliferation, and apoptosis. There are biological sensors which sense the presence of H_2O_2 and trigger downstream signaling events which in turn activate complex disease pathways. A balance in the H_2O_2 levels is achieved by its compartmentalization in different cellular compartments and level is maintained by the antioxidant enzymes like catalase, glutathione peroxidase, and thioredoxin peroxidase. In this chapter we describe the way H_2O_2 is sensed in the biological system and further explain the downstream signaling events. We also explain the role of H_2O_2 signaling during specific biological events and disease conditions.

Keywords

Hydrogen peroxide (H_2O_2) • SUMOylation • Cardiac diseases • Antioxidants • Signaling

1 Introduction

Hydrogen peroxide (H_2O_2) produced by the reduction of O^{2-}, formed in the electron transport chain, has many beneficial roles in both plants and animals. However, due to its cytotoxic nature, its production is tightly regulated in the cell, since a slight increase in its level may lead to toxicity. Maintaining its gradient across the membrane helps in maintenance of cellular redox status and thus the vital cell functions. Its steady-state concentration varies

*Both the authors have contributed equally for this chapter.

V. Rani (✉) • S. Mishra • T. Yadav
Department of Biotechnology, Jaypee Institute of Information Technology, A-10, Sector-62, Noida 201307, Uttar Pradesh, India
e-mail: vibha.rani@jiit.ac.in

U.C.S. Yadav
School of Life Sciences, Central University of Gujarat, Sector 30, Gandhinagar, Gujarat 382030, India

S. Kohli
Universitatsklinkum, Medical Faculty at Otto Von Guericke University, Magdeburg, Germany

V. Rani and U.C.S. Yadav (eds.), *Free Radicals in Human Health and Disease*,
DOI 10.1007/978-81-322-2035-0_8, © Springer India 2015

with the type of cell and the subcellular compartment. Mitochondria, the major source of H_2O_2 in the cell, maintain its cytosolic steady state with the help of antioxidant enzymes. H_2O_2 acts as a secondary messenger in the cell thereby regulating the signal transduction, for example, in immune cell activation and vascular remodeling, cell proliferation, and apoptosis [1]. To act as a signaling molecule, its steady-state concentration should be maintained at a certain level with regulated synthesis and metabolism. H_2O_2 is produced by a wide range of cell and tissue types, including the immune and nonimmune cells, and by NADPH oxidases leading to growth and differentiation, for example, in insulin signaling and angiogenesis [2]. H_2O_2 is the most stable reactive oxygen species with the ability to penetrate cell membrane as an uncharged molecule [3]. Its property to be used as a signaling molecule owes to its structure and nature; being small in size, diffusible, specific in its reactivity and responses, rapid generation, and controlled degradation. These properties make it useful as a secondary messenger in both local and distant cell signaling [4].

H_2O_2 is known to increase the cell's tolerance level to any stress in the second encounter [5]. Stress encounter increases its level leading to a stress response, thus participating in signal transduction pathways and activating gene expression, enzyme activation, or apoptosis, i.e., programmed cell death.

In plants also H_2O_2 generation is known to play an important role. It has been found that the compound is responsible for defense responses and is known to play an important role in bacterial and fungal responses in plants. The genes *AtrbohD* and *AtrbohF*, responsible for the H_2O_2 generation in *Arabidopsis*, when knocked out, made the plant more susceptible to infections. It also provides protection in plants against pathogens by signaling stomatal opening or closing and abscission [6]. In this chapter, we have discussed the mechanisms of H_2O_2 sensing and how it acts as a secondary messenger in the cell leading to signal transduction in different signaling pathways.

2 H_2O_2 Concentration Effects on Cell

An exogenous H_2O_2 has less impact on the signal transduction process as compared to those generated endogenously as signals from H_2O_2 get transmitted to only shorter distances. Cells respond to the differing levels of H_2O_2 differently which is specific to the concentration. A low concentration of H_2O_2 triggers the activity of antioxidants that lowers the ROS level and hence leads to cell survival, while a higher concentration activates prooxidants, thus increasing the ROS level causing the cell to undergo apoptosis. As H_2O_2 is a by-product of a variety of enzymatic reactions, its quantitative determination is necessary for biological purposes. For the purpose of quantification, sensors can be used. One such sensor is HyPer, discussed later in the chapter. For a normal physiological functioning of the cell, concentration of H_2O_2 should be in the range of 1–15 μM. Exposure to ≥50 μM H_2O_2 is cytotoxic to a wide range of cells in culture, and when it exceeds 100 μM, it might result into serious diseases [7]. Level of H_2O_2 is elevated beyond 100 μM in response to stimulation by several growth factors including insulin, platelet-derived growth factor (PDGF), transforming growth factor β (TGF β), and fibroblast growth factor (FGF). For instance, in brain acute exposure to H_2O_2 blocks neurotransmitters like dopamine and norepinephrine affecting neurotransmission negatively. H_2O_2 concentration in cell is involved in the regulation of protein modification such as addition of the SUMO (small ubiquitin-like modifiers) group to the protein. When present in lower concentration, it acts as a secondary messenger activating different sensory proteins and playing a role in signal transduction in the body.

3 H_2O_2 Sensing

3.1 Efficient Sensor Characterization

Sensing is the change in the activity of a gene or a protein including transcription factors on exposure to H_2O_2. The activity can increase or decrease with an increase or decrease in the concentration of H_2O_2. A good sensor should interact selectively with the molecule and transmit the signal downstream and should not interfere with the structure or property of the molecule. The sensor should be able to transmit the signal at a very low concentration, and it should be sensitive towards rapid slight changes in the concentration of the chemical. It should be dynamic enough such that it gets activated before the signal gets terminated [8].

3.2 H_2O_2 Sensors in Humans

There are at least two sensors: one located in pulmonary neuroepithelial bodies (NEB) and the other in carotid bodies in blood vessels. H_2O_2 production causes constriction of the lungs and blood vessels. Pulmonary NEBs resembling chemoreceptors are distributed in airway mucosa of the lungs [9]. Under normal oxygen condition, when NEB cells are exposed to hydrogen peroxide, there occurs an increase in outward K + current [10]. An increase in blood oxygen level causes activation of NADPH oxidase further increasing the production of H_2O_2 inhibiting the secretion of serotonin, resulting in the opening of K + channel. This results into constriction of the airways of lungs, decreasing oxygen supply to the body. In carotid body, glutathione peroxidase is responsible for scavenging H_2O_2. The cytochrome b in these bodies acts as a sensor in chemoreception controlling the degradation of H_2O_2 by glutathione to control the K + conductivity of carotid body cells. The opening of the channel leads to release of catecholamines and hence dilation of the blood vessels [11, 12]. These sensors thus activated by H_2O_2 play an important role in antioxidant defense of the organism (Fig. 1).

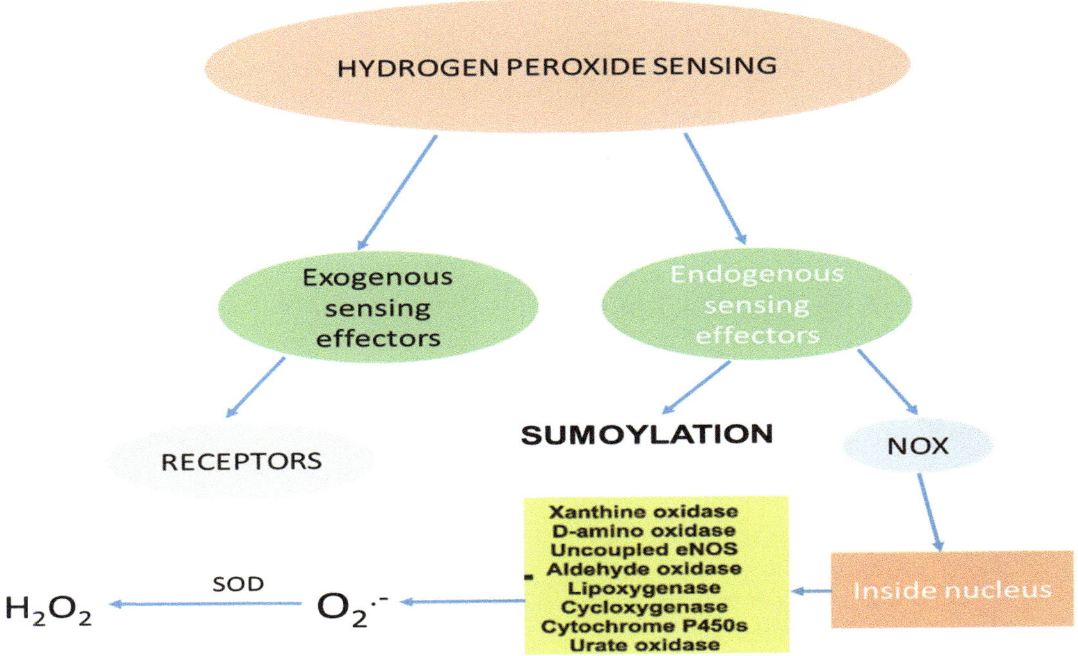

Fig. 1 Sensing of H_2O_2

3.3 HyPer

Other than the previously discussed natural sensors, a highly specific genetically encoded probe HyPer has been designed to detect the presence of H_2O_2 inside the cell. It determines the submicromolar quantities of H_2O_2 without being sensitive to other oxidants. It is based on circularly permuted Yellow Fluorescent Protein (cpYFP) which is inserted into the regulatory domain of the H_2O_2-sensing protein oxyR. OxyR is a transcription factor in eukaryotes, which senses H_2O_2 by the formation of intramolecular disulfide bonds. It is a prototype of proteins whose function is activated by H_2O_2 via cysteine modification [13]. HyPer is used to detect fast changes in H_2O_2 level in the different cellular compartments, without generating any artifactual ROS. When no H_2O_2 is present, HyPer gives two excitation maxima at 500 nm and 420 nm. In the presence of H_2O_2, there occurs a proportional decrease in the excitation peak at 420 nm with an increase in the excitation peak at 500 nm [14].

4 Hydrogen Peroxide Signaling

4.1 Biological Factors of H_2O_2 Production

The enzymatic reaction involved in the production of H_2O_2 varies with the cell organelle. In chloroplasts it is produced via Mehler reaction, in mitochondria through electron transport chain, and in peroxisomes via photorespiration [15]. H_2O_2 acts as a mediator in various pathways. But for it to act as a signaling molecule there must be a regulated synthesis, a specific response, and a specific cellular target. Also H_2O_2 must be able to evade from antioxidants so as to act as a signaling molecule (Fig. 2). Major biological sources of H_2O_2 involves the catalytic or spontaneous breakdown of superoxide anions that are produced by the partial reduction of oxygen during aerobic respiration or due to exposure of cells to other stresses [16].

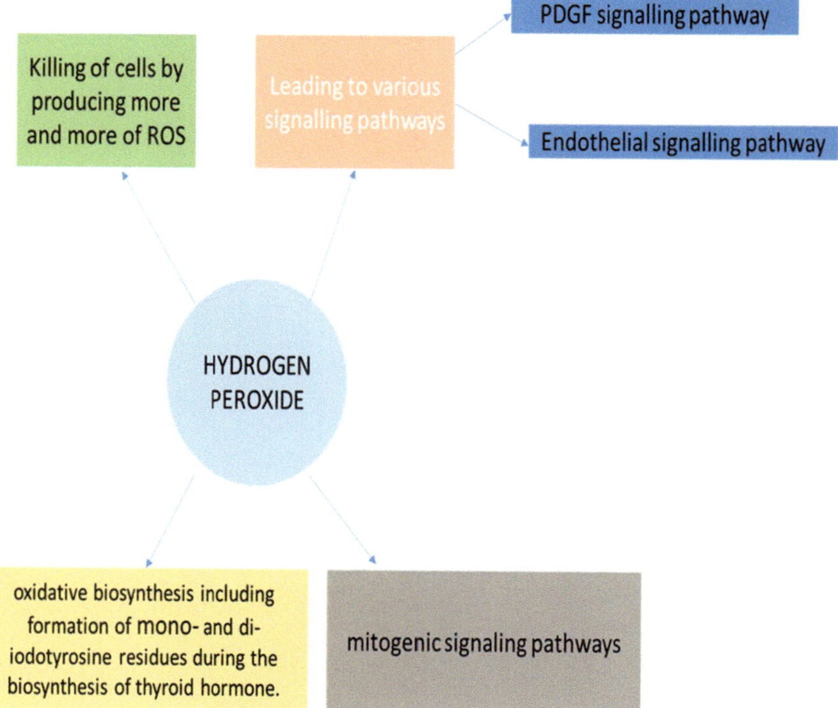

Fig. 2 Hydrogen peroxide function in the biological system

4.2 Regulated Formation of H_2O_2

In response to various stimuli, NOX complexes are involved in the generation of H_2O_2. These complexes present in peroxisomes are involved in generating O_2 when there is an attack by microorganisms. There are multiple isoforms of the catalytic units of these complexes that provide a highly regulated signaling pathway by utilizing H_2O_2. gp91phox (NOX2), which is a predominantly characterized NOX complex [17], consists of a catalytic subunit and a GTP-binding protein Rac. The catalytic subunit is bound to multiple regulatory subunits p22phox and p40phox. C-terminal is the cytoplasmic domain containing an NADPH binding site and a bound FAD. The N-terminal portion consists of six membrane-spanning helices that are bound with two iron-containing heme prosthetic groups. The main function of this complex is O_2 generation, but some isoforms particularly DUOX1 and DUOX2 may also release H_2O_2 as the primary product rather than O_2 [18].

In humans there are seven isoforms of gp91phox. NOX1, NOX2, NOX3, and NOX4 are similar to gp91phox as described above. But the NOX5 isoform contains an additional calmodulin-like domain present on the N-terminal domain. This calmodulin-like domain contains four calcium binding sites, allowing the isoform to be regulated by intracellular calcium levels. gp91phox is activated by the help of three distinct components: phosphorylation of p47phox, Rac guanine nucleotide exchange, and metabolism of phospholipids. Initially phosphorylation of the regulator p47phox leads to the translocation of cytoplasmic p47phox, along with the regulatory components p67phox and p40phox, to plasma membrane to interact with flavocytochrome b558 (composed of gp91phox and p22phox) [19]. Activation is further enhanced by guanine nucleotide exchange on the GTP-binding protein RAC which promotes dissociation of Rac from RhoGDI and its association with the plasma membrane. Activated Rac binds to p67phox, enabling the formation of active NOX complex. These enzyme complexes then serve to generate H_2O_2 in a regulated fashion for signal transduction.

H_2O_2 signaling in mammals can be understood by focusing on the signaling events actually stimulated by the low physiologic concentrations of H_2O_2 that stimulate cell growth and proliferation. The low concentrations of H_2O_2 that stimulate proliferation of mammalian cells have been demonstrated to encourage the induction of specific genes and upregulation of specific proteins.

4.3 H_2O_2 Production Facilitated by Receptors

The best studied receptor is the PDGF (platelet-derived growth factor) receptor. When PDGF binds to its receptor, it results in autophosphorylation of the tyrosine residues. These phosphotyrosine residues then initiate the signaling cascade. There are seven autophosphorylation sites in the PDGF-β receptor which are specific sites for Src (Tyr579 and Tyr581), PtdIns 3-kinase (PI3K, Tyr740, and Tyr751), GAP (Tyr771), SH2 domain-containing protein tyrosine phosphatase-2 (SHP-2, Tyr1009), and PLC-γl (Tyr1021). Experimentally it has been shown that only the PI3K binding site is sufficient for PDGF-induced hydrogen peroxide production [20].

4.4 H_2O_2-Induced Biological Responses

H_2O_2 production in the body activates a complex network of genes and proteins with different H_2O_2 concentrations activating different pathways. The primary response towards H_2O_2 production is the activation of antioxidants along with the transcription of repair proteins.

Cysteine residues of most proteins present in the cytosol, are protonated due to low pH of the cytosol so that they are unable to be sensed by H_2O_2. Hence it is necessary that sensor proteins have low pKa to get utilized by H_2O_2 as a signaling molecule [21]. H_2O_2 may also directly oxidize some transcription factors (TFs) in either the cytoplasm or nucleus. These TFs further interact with similar H_2O_2 response elements and modulate gene expression. One best example of such factors is the oxyR transcriptional activator [22]. This activator gets directly oxidized in response to hydrogen peroxide. Although both oxidized and reduced oxyR can bind to the DNA, only the oxidized form of oxyR can activate transcription of antioxidant genes. When H_2O_2 interacts with cysteine residues, it generates many modified species that are the cysteine sulfenic acid and sulfonic acids. Sulfenic acids are reversible and most common cysteine alterations while the sulfonic acids are irreversible alterations [23]. Apart from this, H_2O_2 can also be altered or interconverted to other oxidative species such as hypochlorous acids and superoxide. In normal human plasma it has been found that H_2O_2, generated by phagocytes and activated neutrophils ranges from 1 to 8 mM. But phagocytes are found to generate higher levels of H_2O_2 than other cells. Increased levels of H_2O_2 have been found in various diseases including lung diseases and cardiac diseases. The level of H_2O_2 is lowered down by NO which competes with superoxide dismutase for the superoxide anion to form peroxynitrite. Intracellularly, superoxide is generated by one electron reduction of oxygen by NADPH oxidase complexes that are present in the plasma and nuclear membranes.

In case of eukaryotes, protein tyrosine phosphatases (PTPs) play a major role in controlling signaling events initiated in response to many driving factors, including growth factors and cytokines. PTPs become susceptible to inactivation by H_2O_2 in the presence of deprotonated cysteine residues. In case of some PTPs, for example, Cdc25C, LMW-PTP, and PTEN PIP3 phosphatase, disulfide bond formation with a neighboring cysteine residue protects the catalytic cysteine residue from further getting oxidized. H_2O_2 also reacts with Fe^{2}+ and its cofactors found in some proteins that are potentially susceptible to oxidation. For example, PerR transcription repressor found in *Bacillus subtilis* contains two metal binding sites: a structural Zn^{2}+ binding site and a regulatory Fe^{2}+ binding site [24]. PerR containing Fe^{2}+ at the regulatory site, binds to DNA, and represses the expression of target genes encoding antioxidant enzymes, such as catalase [25]. In presence of H_2O_2, derepression of PerR-regulated gene expression occurs due to oxidation of two Fe^{2}+ coordinating histidine residues. The involvement of Fe in this mechanism means that the response of PerR to hydrogen peroxide is modulated by the levels of both hydrogen peroxide and Fe (Fig. 3).

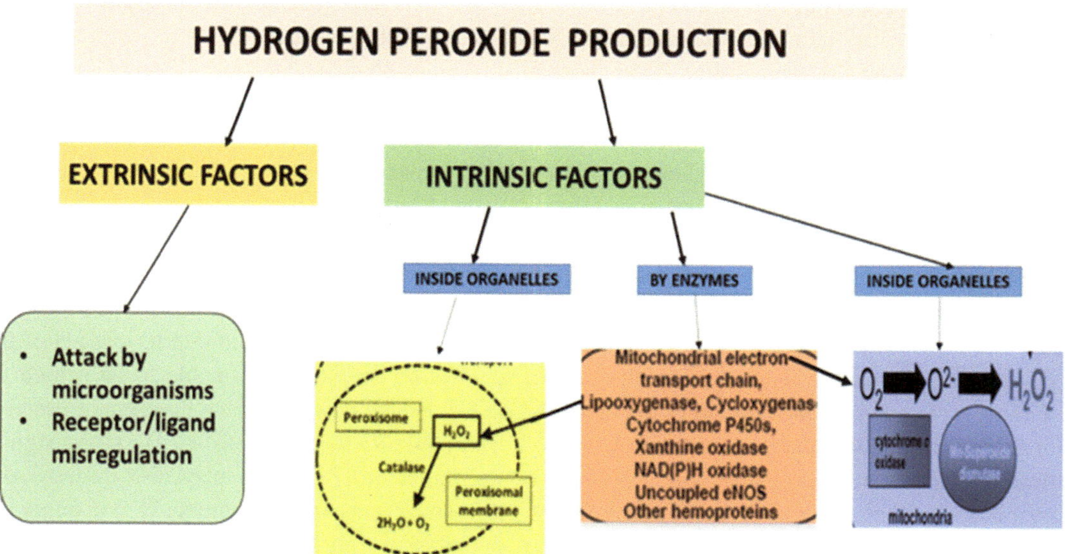

Fig. 3 Production of H_2O_2 in the cell and outside the cell

YAP1 or YAP2 are the genes which get transcribed in response to H_2O_2 produced during oxidative stress. These genes make sure that the oxidation-preventing enzymes in the body gets activated [26]. Yap1 transcription factor normally resides in the cytoplasm due to the nuclear export by the nuclear export receptor Crm1/Xpo1. Upon H_2O_2 exposure, disulfide bonds are formed between two cysteine residues (C303 and C598) in the C-terminal domain. This changes the conformation of the Yap1 and prevents its recognition by the Crm1 receptor [27]. Resulting in the accumulation of Yap1 in the nucleus.

Other transcriptional factors regulated by H_2O_2 are SP1, NOTCH, AP1, CREB, TP53, and HIF1. As these are the factors playing a role in the cell cycle regulation, apoptosis, cell survival and development tight regulation of these factors is required. In the cell, there occurs a gradient of H_2O_2 across different compartments, making the DNA less susceptible to damages by the toxicity and suggesting a role of H_2O_2 as a secondary messenger. It brings about the changes in the structure and function of proteins through posttranscriptional modifications or other processing [28].

H_2O_2 exposure of the cell leads to ADP-ribose formation in the nucleus and mitochondria by the enzyme poly ADP-ribose polymerase (PARP), which activates and opens transient receptor potential melastatin (TRPM) 2 channels [29, 30]. In mammals, six TRP receptors are found: TRPC, TRPV, TRPM, TRPML, TRMV, and TRPA. These channels are present ubiquitously, indicating a diverse range of functions in the body. Although the major physiological role of these proteins is not clear, they are known to play a role in sensory transduction for stimuli like touch, osmolarity, temperature, and pheromones including other stimuli inside and outside the cell [31]. TRPM 2 is a member of cation channels, which is mostly expressed in brain, lungs, heart, and in the cells of the immune system [32, 33]. It has an important role in the release of insulin from pancreatic cells. Loss of this receptor is also known to negatively influence the activity of macrophages. The binding of ADP-ribose to the Nudix box sequence motif (NUDT9-H) in the C-terminal domain of TRPM2 changes its conformation resulting in its opening. This results in the calcium entry into the cell, eliciting the calcium-dependent signaling processes [34]. The exact mechanism of the activation of the TRPM2 is not well understood.

Apart from this, H_2O_2 is also involved in oxidative biosynthesis. One such example involves the formation of tyr-try protein cross-link formed during fertilization envelope of sea urchin eggs [35]. These cross-links are formed oxidatively by the heme-dependent oxyperoxidase by utilizing H_2O_2 that is generated by NADPH oxidase Udx1. In mammals H_2O_2-mediated oxidative biosynthesis includes the formation of mono- and diiodotyrosine residues, catalyzed by heme-dependent thyroid peroxidases, by utilizing H_2O_2 formed by NADPH oxidases DUOX1 and DUOX2 during the biosynthesis of thyroid hormone.

4.5 Antioxidant Enzyme Involvement in Response Regulation of H_2O_2

Generation of H_2O_2 is not only dependent on the above factors but its production is also limited to the presence of enzymatic and nonenzymatic cellular antioxidants. Antioxidant enzymes like catalase, glutathione peroxidase, and thioredoxin peroxidase are mainly involved in its regulation. Catalase is highly expressed in peroxisomes where the level of ROS remains high. It utilizes a heme prosthetic group for the catalytic reduction of H_2O_2, while thioredoxin utilizes the cysteine residues and glutathione peroxidase utilizes the selenocysteine residues [36]. These enzymes are found to be expressed in various cellular locations so as to provide better protection against H_2O_2.

Several antioxidant enzymes also undergo posttranslational modifications to regulate the peroxide detoxifying activity of the enzyme. Several eukaryotic antioxidant enzymes, such as 2-Cys peroxiredoxins (PRX), thioredoxin, glutathione peroxidase (GPX), and catalase are regulated by oxidation, phosphorylation (P), ubiquitination (Ub), and/or protein-protein interactions (e.g., thioredoxin with TXNIP). Regulation of tyrosine phosphorylation of catalase provides a mechanism for initiating distinct responses to low and high levels of peroxide [37]. On the other hand, thioredoxin peroxidase activity of a mammalian typical 2-Cys Prx is inhibited by phosphorylation by the Cdc2 cyclin-dependent kinase (Cdk) during mitosis [38]. When the level of H_2O_2 is low, thioredoxin peroxidase activity is found to be inactivated in eukaryotic 2-Cys Prx by its over-oxidation at the peroxidatic cysteine to sulfinic acid [39]. This increased susceptibility to over-oxidation was linked to the presence of specific amino acids, inserted near the C terminus, that reduces the rate of formation of the disulfide between the peroxidatic and resolving cysteines of typical 2-Cys Prx, further increasing oxidation of the sulfenic intermediate by hydrogen peroxide [40].

Fe is often present at the active site of proteins, rendering these proteins exposed as targets for hydrogen peroxide-induced damage [41]. The presence of abundant enzymatic and nonenzymatic antioxidants helps prevent this damage by removing hydrogen peroxide before it reaches susceptible proteins. As well as increased sequestration of Fe, another common bacterial response inhibit the uptake of extracellular Fe and increase uptake of Mn (and to a lesser extent Zn), possibly by replacing Fe in the active sites of many Fe proteins and thus protecting them from metal-catalyzed oxidation (MCO) by hydrogen peroxide.

5 Specific H_2O_2 Responses

5.1 SUMOylation

SUMO (small ubiquitin-like modifiers) are the proteins that are attached to other proteins in their posttranslational modification. SUMOylation is an ATP-dependent process similar to ubiquitination, directed by a cascade of enzymes but here the protein is not directed toward its degradation, rather SUMO binding can modify a protein's activity, localization, and its ability to bind to other proteins or DNA. These are expressed under stress conditions. There are four SUMO isoforms in humans: SUMO-1, SUMO-2, SUMO-, and SUMO4. SUMO1, SUMO2, and SUMO 3 are majorly expressed in all types of tissues whereas SUMO 4 is expressed in the liver, kidney, and lymph nodes. The process of SUMOylation involves three enzymes: SUMO-activating enzyme E1 (Uba2-Aos1), SUMO-conjugating enzyme E2 (Ubc9), and the third enzyme E3 which facilitates conjugation [42]. Hydrogen peroxide is known to play an important role in the SUMOylation process. It has been found that higher level (concentration >10 mM) of H_2O_2 is linked to an increased protein SUMOylation while a reduced level (1 mM concentration) is responsible for de-SUMOylation [43]. The inhibition of SUMOylation occurs by the formation of disulfide bonds with the catalytic cysteine residue of the SUMO E1 subunit Uba2 and the E2-conjugating enzyme Ubc9 while its activation occurs by the formation of SUMO E1 and SUMO E2 thioester intermediates [44, 45]. This regulation determines the biological response with different H_2O_2 concentration in the cell.

5.2 Angiotensin II and Thrombin-Mediated H_2O_2 Signal Transduction

Angiotensin II (AngII) activation of p38 MAP kinase in vascular smooth muscle cells is the most studied oxidation-dependent signaling pathway. In the absence of AngII, levels of H_2O_2 (100–200 μ*M*) increase due to which phosphorylation of p38 MAP kinase is stimulated. AngII, a major factor stimulating the growth of smooth muscle cell in vascular systems, acts through the G-protein-coupled AT1 receptor [46]. It triggers the release of O_2/H_2O_2 from vascular smooth muscle cells. This generation of the ROS is inhibited by binding of antisense RNA to the p22phox component of the NOX complexes [47]. A similar pathway is that of thrombin stimulation of p38 MAP kinase phosphorylation in vascular smooth muscle cells. Thrombin is a component of the coagulation cascade that helps in the stimulation of smooth muscle cell proliferation [48]. It also stimulates the release of H_2O_2 and O_2 from vascular smooth muscle cells. Thrombin-induced proliferation is found to be inhibited by extracellular catalase [49].

AngII also stimulates the activation of Ras and the formation of Ras–SSG which is an oxidized form of Ras, in which Cys118 is in a mixed disulfide form with glutathione. This generation of Ras–SSG in response to AngII is inhibited by the overexpression of catalase. While in the absence of AngII, when H_2O_2 is present in higher concentration (250 μM), it results in the activation of Ras. Activation of Ras leads to the eventual phosphorylation of p38 MAP kinase as discussed above.

The TDPs (thiolate-dependent phosphatases) including tyrosine phosphatases, dual-specificity phosphatases, and lipid phosphatases represent oxidation-dependent step in multiple signal transduction pathways. In humans there are approximately 103 distinct TDPs present which are involved in various signaling pathways [50]. The thiolate moiety at the active sites of TDPs reacts with H_2O_2 to generate the sulfenic acid. The sulfenic acid then typically reacts with a second nucleophile, the nature of which is dependent on the specific environment of the individual enzyme's active site. The sulfenic acid reacts with a backbone of amide nitrogen to generate the sulfenyl amide in PTP-1B and with a second Cys-thiol to generate a disulfide bond in the Cdc25B dual-specificity phosphatase [51].

5.3 Regulation of Endothelial Inflammatory Responses by Hydrogen Peroxide

When endothelial cells are exposed to H_2O_2, various inflammatory proteins such as VCAM (vascular cell adhesion molecule-1), ICAM (intercellular adhesion molecule-1), and MCP (monocyte chemoattractant protein-1) get activated. H_2O_2 also requires extracellular stimuli such as TNFα for the activation of ICAM-1 and VCAM-1. All these molecules including ICAM-1, VCAM-1, MCP-1, PAF (platelet-activating factor), and P-selectin then further mediate neutrophil adhesion to the endothelium [52]. PAF and P-selectin also facilitate platelet activation and endothelium–platelet interaction (Fig. 4).

$O2^-$and H_2O_2 are produced in vascular cells by multiple enzymatic systems including vascular NAD(P)H oxidases, mitochondrion, xanthine oxidase, and uncoupled eNOS as explained above. Some of the $O2^-$ are degraded by reacting with NO, but some are preserved by dismutation into H_2O_2 that results in the prolonged signaling. Selectively overproducing or removing H_2O_2 significantly alters atherogenesis [53].

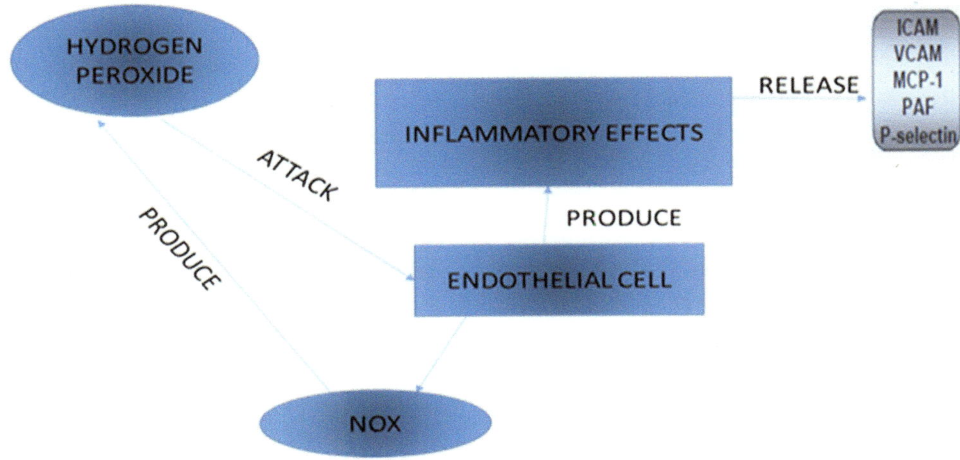

Fig. 4 Inflammatory responses by H_2O_2 on endothelial cells

5.4 Regulation of Endothelial Cell Growth and Proliferation by H_2O_2

NAD(P)H oxidase-derived H_2O_2 is also generated in vascular cells including endothelial cells, vascular smooth muscle, and fibroblasts. NAD (P)H oxidase-derived H_2O_2 is capable of amplifying ROS production in vascular cells. Propagation of these cells also occurs via other mechanisms including intracellular iron uptake, activation of sources of mitochondria, NAD (P)H oxidases, xanthine oxidase, and uncoupled eNOS. These mechanisms amplify and sustain H_2O_2 production in large amounts, contributing to continued pathological signaling. H_2O_2 generated by xanthine oxidase helps in enhanced endothelial cell growth [54]. It leads to various downstream processes including the Flk1/KDR that further activates the ERK1/2 [55]. The effector protein (p90RSK) of this ERK1/2 is activated by H_2O_2 in the endothelial cells. When there are mitogenic stimuli such as vascular endothelial growth factor (VEGF), vascular NAD (P)H oxidases get activated to form superoxide and hydrogen peroxide. This mediates activation of p38 MAPK, ERK1/2, and transcriptional factors that are involved in growth signaling. H_2O_2 also upregulates the endothelial expression of VEGF. Higher level of H_2O_2 (exceeding 200 Amol/L) leads to endothelial apoptosis. This is also regulated by the transferrin receptor (TfR)-dependent intracellular iron uptake and activations of mitochondrion, FAS, and JNK/c-Jun pathway. Other signaling pathways like activation of eIF4E and eukaryotic initiation factor of protein synthesis are also enhanced that is effective in promoting endothelial cell tube formation [56, 57].

5.5 Enzyme Metabolism of H_2O_2 in Cardiac Diseases

Heart has significant amounts of catalase, which helps in the decomposition of H_2O_2. Mammalian cells contain myeloperoxidase enzyme and multiple peroxiredoxin isoforms, which help in controlling basal cellular levels of H_2O_2. These enzymes have Km for H_2O_2 in the tens of micromolar range [58]. Peroxiredoxins may act as checkers of oxidant signaling, as higher H_2O_2 induces hyperoxidation of the peroxiredoxin catalytic cysteine to sulfinic acid, causing inhibition. Hyperoxidation has been demonstrated in response to peroxides generated endogenously by an effector molecule such as glucose oxidase or TNFα, supporting peroxiredoxin hyperoxidation as a physiological event [59]. This leads to the possibility that cellular H_2O_2 concentration can rise beyond the Km of these proteins for H_2O_2, exerting biological effects (oxidant signaling). Although exogenous H_2O_2 concentration of 0.1–1 mM is required for peroxiredoxin sulfinic acid formation, the intracellular concentration at this time remains unknown.

6 Conclusion

Increased levels of hydrogen peroxide in cells can lead to oxidative stress ultimately leading to cellular damage. However, in higher eukaryotes it has been found that hydrogen peroxide also works as a signaling molecule to regulate many different cellular processes. Hence, hydrogen peroxide can have both positive (signaling) and negative (damaging) effects on the cell. However for it to act as a signaling molecule, there must be a regulated synthesis, specific response, and specific cellular target. Also, it must be able to evade from antioxidants so as to act as a signaling molecule. Levels of H_2O_2 should be maintained, and different levels regulate different functions. A lower concentration of H_2O_2 triggers the activity of antioxidants that lower the ROS level leading to cell survival, while a higher concentration activates prooxidants, thus increasing the ROS level causing the cell to undergo apoptosis. H_2O_2 interact with cysteine residues within proteins to start the signaling cascade. A deprotonated cysteine residue present in protein makes it susceptible to undergo oxidation by H_2O_2. The cysteine residues of most protein present in the cytosol are protonated due to the low pH of the cytosol making them unable to be sensed by H_2O_2. So it is necessary that sensor proteins have low pKa to get utilized by H_2O_2 as a signaling molecule. H_2O_2 is also involved in direct oxidation of some transcription factors in either the cytoplasm or nucleus. These TFs further interact with H_2O_2 response elements and modulate gene expression. One best example of such factors is the oxyR transcriptional activator. Apart from these transcriptional factors, NOX complexes are also involved in the generation of H_2O_2 in response to various stimuli. So H_2O_2 as a signaling molecule is becoming important from the point of view of research, and further exploring it will lead to the eradication of various diseases.

References

1. Alberto B, Enrique C (2000) Mitochondrial production of hydrogen peroxide regulation by nitric oxide and the role of ubisemiquinone. IUBMB Life 50:245–250
2. Elizabeth AV, Alison MD, Brian AM (2007) Hydrogen peroxide sensing and signalling. Mol Cell 26:1–14
3. Olga BB, Tamara VC, Kurt VF (2001) Anoxic stress leads to hydrogen peroxide formation in plant cells. J Exp Bot 52:1179–1190
4. Ewald S, Philip E (2008) Hydrogen peroxide as an endogenous mediator and exogenous tool in cardiovascular research: issues and considerations. Curr Opin Pharmacol 8:153–159
5. Neill SJ, Desikan R, Clarke A et al (2002) Hydrogen peroxide and nitric oxide as signalling molecules in plants. J Exp Bot 53:1237–1247
6. Zhang P (1997) Thioredoxin peroxidase is a novel inhibitor of apoptosis with a mechanism distinct from that of Bcl-2. J Biol Chem 272:30615–30618
7. Graziella S, Antonella A, Carmine C et al (2013) Electrical characterization and hydrogen peroxide sensing properties of gold/nafion: polypyrrole/MWCNTs electrochemical devices. Sensors 13:3878–3888
8. Marinho HS, Carla R, Fernando A et al (2014) Hydrogen peroxide sensing, signaling and regulation of transcription factors. Redox Biol 2:535–562
9. Youngson C, Colin N, Herman Y et al (1993) Oxygen sensing in airway chemoreceptors. Nature 365:153–155
10. Wang D, Youngson C, Wong V et al (1996) NADPH-oxidase and a hydrogen peroxide-sensitive K^+ channel may function as an oxygen sensor complex in airway chemoreceptors and small cell lung carcinoma cell lines. Proc Natl Acad Sci U S A 93:13182–13187
11. Skulachev VP (2001) H2O2 sensors of lungs and blood vessels and their role in the antioxidant defense of the body. Biochemistry (Mosc) 66:1153–1156
12. Acker H, Bolling B, Delpiano MA et al (1992) The meaning of H2O2 generation in carotid body cells for PO2 chemoreception. J Auton Nerv Syst 41:41–51
13. Hee JC (2001) Structural basis of the redox switch in the OxyR transcription factor. Cell 105:103–113
14. Belousov VV, Fradkov AF, Lukyanov KA et al (2006) Genetically encoded fluorescent indicator for intracellular hydrogen peroxide. Nat Methods 3:281–286
15. Babior BM (1999) NADPH oxidase: an update. Blood 93:1464–1476
16. Geiszt M, Leto TL (2004) The Nox family of NAD (P) H oxidases: host defense and beyond. J Biol Chem 279:51715–51718
17. Brandes RP, Kreuzer J (2005) Vascular NADPH oxidases: molecular mechanisms of activation. Cardiovasc Res 65:16–27
18. Ameziane EHR, Morand S, Boucher JL et al (2005) Dual oxidase-2 has an intrinsic Ca2+-dependent H2O2-generating activity. J Biol Chem 280:30046–30054
19. Mackay DJ, Hall A (1998) Rho GTPases. J Biol Chem 273:20685–20688
20. Bae YS, Sung JY, Kim OS et al (2000) Platelet-derived growth factor-induced H_2O_2 production requires the activation of phosphatidylinositol 3-kinase. J Biol Chem 275:10527–10531
21. Lambeth JD (2002) Nox/Duox family of nicotinamide adenine dinucleotide (phosphate) oxidases. Curr Opin Hematol 9:11–17
22. Storz G, Tartaglia LA, Ames BN (1990) Transcriptional regulator of oxidative stress-inducible genes: direct activation by oxidation. Science 248:189–194
23. Kim SO, Merchant K, Nudelman R et al (2002) OxyR: a molecular code for redox-related signaling. Cell 109:383–396
24. Lee JW, Helmann JD (2006) The PerR transcription factor senses H_2O_2 by metal-catalysed histidine oxidation. Nature 440:363–367
25. Herbig AF, Helmann JD (2001) Roles of metal ions and hydrogen peroxide in modulating the interaction of the *Bacillus subtilis* PerR peroxide regulon repressor with operator DNA. Mol Microbiol 41:849–859
26. Derek JJ (1998) Oxidative stress responses of the Yeast *Saccharomyces cerevisiae*. Yeast 14:1511–1527
27. Delaunay A, Isnard AD, Toledano MB (2000) H_2O_2 sensing through oxidation of the Yap1 transcription factor. EMBO J 19:5157–5166
28. Branco MR, Marinho HS, Cyrne L et al (2004) Decrease of H_2O_2 plasma membrane permeability during adaptation to H_2O_2 in *Saccharomyces cerevisiae*. J Biol Chem 279:6501–6506
29. Claudie MH, Gias UA, Stephen MV et al (2008) Role of TRPM2 channel in mediating H_2O_2-induced Ca^{2+} entry and endothelial hyperpermeability. Circ Res 102:347–355
30. Kühn FJ, Heiner I, Luckhoff A (2005) TRPM2: a calcium influx pathway regulated by oxidative stress and the novel second messenger ADP-ribose. Pflugers Arch 451:212–219

31. Togashi K, Yuji H, Tomoko T et al (2006) TRPM2 activation by cyclic ADP-ribose at body temperature is involved in insulin secretion. EMBO J 25: 1804–1815
32. Gurling H (1998) Chromosome 21 workshop. Psychiatr Genet 8:109–114
33. Sano Y, Inamura K, Miyake A et al (2001) Immunocyte Ca^{2+} influx system mediated by LTRPC2. Science 293:1327–1330
34. Makiko K, Takaaki S, Kenji S et al (2012) Redox signal-mediated sensitization of transient receptor potential melastatin 2 (TRPM2) to temperature affects macrophage functions. PNAS 25:1–6
35. Elizabeth V, Alison D (2011) Hydrogen peroxide as a signalling molecule. Antioxid Redox Signal 2(10):e213
36. Sablina AA, Budanov AV, Ilyinskaya GV (2005) The antioxidant function of the p53 tumor suppressor. Nat Med 11:1306–1313
37. Cao C, Leng Y, Liu X et al (2003) Catalase is regulated by ubiquitination and proteasomal degradation. Role of the c-Abl and Arg tyrosine kinases. Biochemistry 42:10348–10353
38. Chang TS, Jeong W, Choi SY et al (2002) Regulation of peroxiredoxin I activity by Cdc2-mediated phosphorylation. J Biol Chem 277:25370–25376
39. Rabilloud T, Heller M, Gasnier F et al (2002) Proteomics analysis of cellular response to oxidative stress. Evidence for in vivo overoxidation of peroxiredoxins at their active site. J Biol Chem 277:19396–19401
40. Wood ZA, Poole LB, Karplus PA (2003) Peroxiredoxin evolution and the regulation of hydrogen peroxide signalling. Science 300:650–653
41. Faulkner MJ, Helmann JD (2011) Peroxide stress elicits adaptive changes in bacterial metal ion homeostasis. Antioxid Redox Signal 15:175–189
42. Marco F (2011) Cross-talk between JNK and SUMO signaling pathways: deSUMOylation is protective against H_2O_2- induced cell injury. PLoS ONE 6:1–9
43. Vibha S, Marina P, Eliana G et al (2010) SUMO proteins are involved in the stress response during spermatogenesis and are localized to DNA double-strand breaks in germ cells. Reproduction 139:999–1010
44. Bossis G, Melchior F (2006) Regulation of SUMOylation by reversible oxidation of SUMO conjugating enzymes. Mol Cell 21:349–357
45. Xu Z, Lam LS, Lam LH et al (2008) Molecular basis of the redox regulation of SUMO proteases: a protective mechanism of intermolecular disulfide linkage against irreversible sulfhydryl oxidation. FASEB J 22:1–11
46. Touyz RM, Schiffrin EL (2000) Signal transduction mechanisms mediating the physiological and pathophysiological actions of angiotensin II in vascular smooth muscle cells. Pharmacol Rev 52:639–672
47. Viedt C, Soto U, Krieger BHI (2000) Differential activation of mitogen activated protein kinases in smooth muscle cells by angiotensin II: involvement of p22phox and reactive oxygen species. Arterioscler Thromb Vasc Biol 20:940–948
48. McNamara CA, Sarembock IJ, Gimple LW et al (1993) Thrombin stimulates proliferation of cultured rat aortic smooth muscle cells by a proteolytically activated receptor. J Clin Invest 91:94–98
49. Patterson C, Ruef J, Madamanchi NR et al (1999) Stimulation of a vascular smooth muscle cell NAD (P) H oxidase by thrombin. Evidence that p47phox may participate in forming this oxidase in vitro and in vivo. J Biol Chem 274:19814–19822
50. Alonso A, Sasin J, Bottni S et al (2004) Protein tyrosine phosphatases in the human genome. Cell 117:699–711
51. Salmeen A, Andersen JN, Myers MP et al (2003) Redox regulation of protein tyrosine phosphatase 1B involves a sulphenyl-amide intermediate. Nature 423:769–773
52. True AL, Rahman A, Malik AB (2000) Activation of NF-kappaB induced by H(2)O(2) and TNF-alpha and its effects on ICAM-1 expression in endothelial cells. Am Physiol Lung Cell Mol Physiol 279:L302–L311
53. Cai H (2005) Hydrogen peroxide regulation of endothelial function: origins, mechanisms, and consequences. Cardiovasc Res 68:26–36
54. Ruiz-Gines JA, Lopez-Ongil S, Gonzalez-Rubio M (2000) Reactive oxygen species induce proliferation of bovine aortic endothelial cells. J Cardiovasc Pharmacol 35:109–113
55. Colavitti R, Pani G, Bedogni B et al (1998) Reactive oxygen species as downstream mediators of angiogenic signaling by vascular endothelial growth factor receptor- 2/KDR. J Biol Chem 277:3101–3108
56. Lin SJ, Shyue SK, Liu PL et al (1999) Adenovirus-mediated overexpression of catalase attenuates oxLDL induced apoptosis in human aortic endothelial cells via AP-1 and C-Jun N-terminal kinase/ extracellular signal-regulated kinase mitogen- activated protein kinase pathways. J Mol Cell Cardiol 36:129–139
57. Chen K, Thomas SR, Albano A (2002) Mitochondrial function is required for hydrogen peroxide-induced growth factor receptor transactivation and downstream signaling. J Biol Chem 279:35079–35086
58. Brennan JP, Eaton P (2006) Oxidized proteins in cardiac ischemia and reperfusion. In: Dalle-Donne I, Scaloni A, Allan Butterfield D (eds) Redox proteomics: from protein modifications to cellular dysfunction and diseases. Wiley, Hoboken, pp 605–649

Part III

Oxidative Stress and Diseases

Oxidative Stress-Induced Lipid Peroxidation: Role in Inflammation

Umesh Chand Singh Yadav

Abstract

Oxidative stress-induced lipid peroxidation is known to produce mediators that are implicated in the pathophysiology of a wide range of inflammatory diseases. In many inflammatory diseases, various lipid-derived aldehydes (LDAs) including 4-hydroxy-trans-2-nonenal (HNE), acrolein, and malondialdehyde (MDA) have been identified. The reactive oxygen species (ROS), generated by various oxidants, attack the membrane lipids resulting into lipid peroxidation which forms a number of potentially toxic lipid aldehydes. The lipid aldehydes activate upstream signaling kinases and subsequently alter the redox signaling pathway resulting in cytotoxicity such as excessive cell proliferation or cell death. Further, LDAs also cause posttranslational modification of various other cellular proteins and genetic material, deoxyribonucleic acid (DNA) and ribonucleic acid (RNA), leading to cytotoxicity and genotoxicity. HNE, one of the highly reactive lipid aldehyde, directly as well as indirectly in the form of glutathione conjugate, has been implicated in the activation of protein kinase cascades leading to the activation of redox-sensitive transcription factors such as NF-κB and AP-1. The activated transcription factors translocate to the nucleus and transcribe several inflammatory marker genes including cytokines, chemokines, and various cellular proteins involved in cell survival, cell differentiation, and cell death eventually resulting in the pathogenesis of various diseases. Thus, a clear understanding of oxidative stress-generated LDA-induced alteration in cellular physiology would provide opportunities to prevent or ameliorate a number of inflammatory diseases.

U.C.S. Yadav (✉)
School of Life Sciences, Central University of Gujarat, Sector 30, Gandhinagar, Gujarat 382030, India
e-mail: umeshyadav@cug.ac.in

V. Rani and U.C.S. Yadav (eds.), *Free Radicals in Human Health and Disease*,
DOI 10.1007/978-81-322-2035-0_9, © Springer India 2015

Keywords
Oxidative stress • Lipid peroxidation • Oxidants • Redox signaling • Inflammation

1 Introduction

A number of oxidative stress-induced inflammatory pathologies including obesity, diabetes, cardiovascular complications, neural degeneration and aging, hepatotoxicity, nephrotoxicity, carcinogenesis, smoke-induced lung inflammation, retinopathy, and others have been linked with oxidative stress and subsequent generation of lipid-derived aldehydes (LDAs) [1–10]. Oxidative stress is known to generate different types of free radicals which then attack the membrane lipids, particularly the polyunsaturated fatty acids (PUFA) such as arachidonic acid. Because of the presence of unsaturation, PUFAs are highly susceptible to the free radical attack and are the primary source of LDAs. Attack by free radicals leads to the peroxidation of membrane lipids and generation of highly reactive and electrophilic α, β-unsaturated aldehydes. LDAs are highly toxic because of their reactivity with nucleophilic cellular biomolecules and ability to easily form adducts with protein and DNA without prior metabolic activation [11]. Although LDAs are relatively stable than their parental free radicals such as oxygen and hydroxyl free radicals, they are highly reactive molecules due to electrophilic nature [12]. Among various LDAs, HNE and MDA are produced abundantly upon lipid peroxidation, whereas acrolein is the most reactive of all LDAs [13]. Because of tendency to attract electrons and presence of unsaturation, LDAs have the tendency to readily react with nucleophiles including thiols and amines containing cellular macromolecules such as proteins and nucleic acids. LDAs form adducts (Michael adduct and Schiff base, respectively) with these important biomolecules which is known to cause cellular damage and accumulation of chemically altered macromolecules [14]. Such conjugates and adducts of LDAs with proteins and nucleic acids have been identified in many diseases; for example, HNE-protein adducts were detected in mitotic, necrotic, and apoptotic cells in brain tissue [15]. A number of studies have also revealed association of LDAs with the progression of inflammatory diseases such as chronic obstructive pulmonary disease (COPD) [9], atherosclerosis [16], Alzheimer's disease [4, 5], and stroke [17] based on the presence of LDAs in the affected tissues. LDAs such as MDA, acrolein, HNE, and 4-hydroxyhexenal (HHE) have been identified in the diseased tissue [5, 15, 18, 19]. The concentration of HNE, the most abundant LDA, has been reported to be up to 10 nanomols per gram tissue [12]. Since LDAs are known to disrupt the cellular homeostasis by modifying and/or damaging cellular proteins and nucleic acids besides altering cellular signaling events leading to cytotoxicity and pathogenesis, in this chapter, we have discussed the mechanism of LDA production and their roles in regulating inflammatory signaling pathways that activate transcription factors and other modulators which in turn cause a number of inflammatory pathogenesis.

2 Mechanism of Reactive Oxygen Species (ROS) Production and Lipid Peroxidation

In eukaryotic system, especially aerobes, the importance of oxygen in cell growth and maintenance is well understood. Although oxygen is indispensible for life in most organisms, some of its metabolic derivatives can be toxic. A variety of highly reactive chemical species, called free radicals or reactive oxygen species (ROS), are formed as the by-product of the oxygen utilization. It is interesting here to note that on the one hand, ROS are considered important bimolecular entity regulating normal cellular functions such as growth factor-induced growth, development, and morphogenesis; on the other hand, their toxic by-products are detrimental to the cells. This noticeable paradox regarding the roles of ROS could be, at least in part, related to differences in the ROS concentrations. ROS are recognized as natural players in the normal metabolism of oxygen and thus have significant roles in cell signaling and homeostasis [20]. However, in stressful condition such as exposure to endogenous or exogenous oxidants, carcinogens, and radiation, the levels of ROS can increase dramatically, compromising the cellular structure and function that lead to oxidative stress, inflammation, and disease. Different types of ROS include superoxide anions (O_2^-), hydroxyl radicals ($^{\cdot}OH$), nitric oxide radicals ($NO^{\cdot}$), and their by-products such as lipid peroxides and hydrogen peroxide, H_2O_2.

There are several ways in which ROS can be formed. In the context of cells, the source of ROS could be exogenous such as environmental pollutants, tobacco smoke, drugs and ionizing radiation, or endogenous such as via mitochondria or nicotinamide adenine dinucleotide phosphate (NADPH) oxidase (NOX). Ionizing radiation causes water in the cells to lose an electron and convert into hydroxyl radical ($^{\cdot}OH$), which then in sequential steps converted into hydrogen peroxide (H_2O_2), superoxides ($O^{\cdot}{}_2^-$), and finally into molecular oxygen (O_2). The hydroxyl radical being highly reactive can react with any molecule and extract an electron converting them into free radical and thus continue a chain reaction. Endogenously, excessive ROS can be produced by dysfunctional mitochondria [21]. Electron leaks from the mitochondrial membrane may reduce oxygen, a terminal electron accepter, into superoxide anion ($O^{\cdot}{}_2^-$). Further, the NOX family of NADPH oxidases shares the capability to transport electrons across the plasma membrane and generate superoxide and other reactive oxygen species in the cells, e.g., allergic pollens have been shown to produce ROS in the lungs by this mechanism [22], and it is a well-known mechanism (oxidative burst) of professional phagocytes to kill invading pathogens. Indeed, the phagocytic cells are endowed with a well-characterized $O^{\cdot}{}_2^-$ generating NOX family of proteins in plasma membrane capable of producing large amounts of ROS that they use in host defense [23]. Emerging evidence also suggests the role of smooth endoplasmic reticulum (without ribosomes), which contains enzymes that detoxify lipid-soluble drugs and other harmful metabolic products, in superoxide production [24]. The cytochrome *p-450* and b_5 families of enzymes can oxidize unsaturated fatty acids and xenobiotics and reduce molecular O_2 to produce superoxide ($O^{\cdot}{}_2^-$) [25, 26]. Peroxisomes, endowed with a number of H_2O_2-generating enzymes including glycolate oxidase, D-amino acid oxidase, urate oxidase, L-α-hydroxy acid oxidase, and fatty acyl-CoA oxidase, could be yet another source of cellular peroxides [27].

Generated in excess, the superoxides are capable of causing oxidative damage to biomolecules such as peroxidation of membrane lipids, oxidation of amino acid side chains (especially cysteine), formation of protein-protein cross-links, oxidation of polypeptide backbones resulting in protein fragmentation, and DNA damage [28, 29]. The oxidation of membrane lipid is highly damaging because it on the one hand disrupts the membrane structure and function compromising the cellular homeostasis; on the other hand, it leads to the formation of toxic lipid peroxidation products leading to loss of cell function and severe cytotoxicity causing either uncontrolled cell growth (neoplasia) or cell death (apoptosis) [30, 31].

The lipid peroxidation includes a series of sequential chemical reactions that include initiation, propagation, and termination (Fig. 1) [32]. In the initiation, free radicals such as hydroxyl radical ($^{\cdot}OH$), alkoxyl radical ($RO^{\cdot}$), peroxyl radical ($ROO^{\cdot}$), and $HO^{\cdot}{}_2$ abstract hydrogen atom from the lipid molecule and produces a lipid free radical ($L^{\cdot}$), which in turn reacts with molecular oxygen and forms a lipid peroxyl radical ($LOO^{\cdot}$). In the propagation, which is a chain reaction, lipid peroxyl ($LOO^{\cdot}$) free radical species removes hydrogen atom from another lipid molecule and produces a lipid hydroperoxide (LOOH) and generate a second lipid free radical ($L^{\cdot}$). The LOOH can further be reduced by metals, such as Fe^{++}, forming a lipid alkoxyl radical ($LO^{\cdot}$). Both alkoxyl and peroxyl radicals attack the adjacent lipid molecules and cause excessive

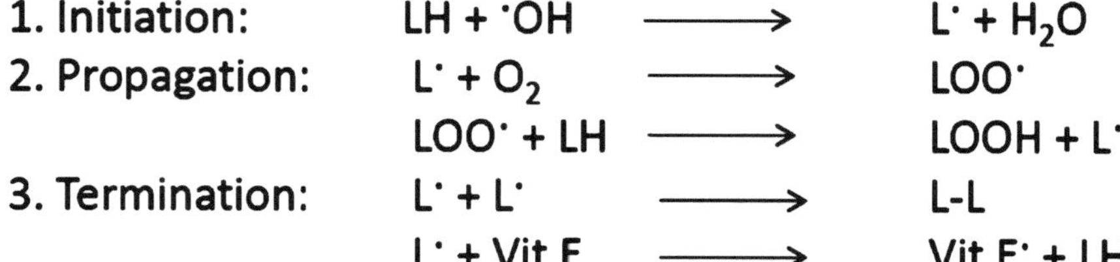

Fig. 1 Three reaction steps in free radical-initiated lipid (LH) peroxidation reaction that lead to the formation of highly reactive lipid molecules (LOO·) that subsequently form lipid aldehydes

lipid peroxidation by abstracting additional hydrogen atoms [33]. This results in the disruption of membrane lipids which disturbs the cell membrane causing alterations in membrane fluidity and permeability, alterations of ion transport, and suppression of normal metabolic processes [34]. In addition, LOOH may break down into highly reactive lipid-derived aldehydes (LDAs) such as MDA, crotonaldehyde, HNE, HHE, and acrolein in the presence of reduced metals or ascorbate [12, 35]. In the termination, formation of a hydroperoxide is achieved by the reaction of a peroxyl radical with α-tocopherol, a lipophilic chain-breaking molecule found in the cell membrane [33]. Termination can also result when a lipid radical (L·) reacts with lipid peroxide (LOO·) or when two peroxide molecules combine together and result in a nonreactive species LOOL or hydroxylated derivative (LOH), respectively, which are chemically more stable. Some of the lipid peroxides could also react with the membrane proteins and terminate the reaction [36, 37]. The propagation step of lipid peroxidation is self-sustaining and continues until either substrate is consumed or the antioxidants or free radical quenchers terminate the reaction [37]. Thus, lipid peroxidation, being a self-sustaining reaction, amplifies the effects of the oxygen free radical and cause extensive tissue damage.

3 Redox Signaling and Inflammation

Redox signaling involves signaling mechanisms that utilize ROS to initiate processes which allow cells to survive exposure to oxidative stress within certain limits, and also to ensure cell death when stress and damage become too great [38]. Redox signaling can also be defined as intracellular signaling pathways that are activated via changes in intracellular metabolic oxidation/reduction (redox) reactions involving ROS. Different ROS including superoxide and hydrogen peroxide have been implicated in activating signaling pathway that result in inflammation and cause inflammatory disease including cancer, neurodegeneration, and cardiovascular disorders. Although several signal circuitries regulate ROS-inducible oxidative stress responses, the corresponding upstream ROS receptors are not known as yet. In stressed cells, ROS is generated as part of a defense reaction, initially intended to clear infectious and environmental threats, including microbial agents. However, excessively formed ROS such as ·OH and H_2O_2 species can facilitate further oxidation and participate in intracellular signaling events that lead to inflammatory response and subsequently to pathogenesis [39, 40]. The ROS-induced upstream signaling pathways which include stress response kinases and MAP kinases activate redox-sensitive transcription factors, of which nuclear factor-kappa B (NF-κB) is a prominent member [41]. A clear understanding of the various facets of redox regulation of NF-κB and its target genes has advanced our understanding of inflammatory processes [42]. Various pathogens, oxidants, cytokines, chemokines, and growth factors either via specific receptors or general oxidative stress induce the molecular signals that eventually lead to activation of a redox-sensitive transcription factor NF-κB [43]. NF-κB comprise a family of transcription factors that regulate expression of numerous genes that play important roles in immune and stress responses, inflammation and apoptosis [44, 45]. The NF-κB family consists of five well-characterized proteins p50, p52, RelA or p65, c-Rel, and RelB, which homo- or heterodimerize through Rel homology domain. These NF-κB-dimerized subunits, ubiquitously expressed in mammalian cells and highly conserved across the species, form biologically active molecule of NF-κB, which translocates to the nucleus and transcribes various genes including inflammatory cytokine and chemokines genes [46]. The expression of a large number of genes involved in apoptosis, cell growth, survival, differentiation, and immune response is regulated by NF-κB, which are associated with an array of inflammatory diseases [45].

4 Role of Oxidative Stress-Derived Lipid Aldehydes in Mediating NF-κB Redox Signaling

Cumulative evidences suggest that lipid aldehydes generated endogenously during ROS-induced process of lipid peroxidation are involved in several pathophysiological effects associated with oxidative stress in cells and tissues [47]. Although previously it was believed that lipid peroxidation products only elicit cellular damage, recent evidence suggests that more varied effects dependent upon factors including their species, concentration, and protein target involved [48]. A growing number of studies now implicate LDAs in the cellular signaling [48–51]. The LDAs affect cellular processes in various ways including covalent and non-covalent interactions with proteins [52, 53]. For example, LDAs interact with nucleophilic protein constituents, including amino acid residues such as cysteine, lysine, and histidine, and form protein adducts through Schiff base formation [54]. Another mechanism of protein adduct formation by LDAs is Michael addition, where thiolate groups of cysteine residues react with electrophilic carbons present in α, β-unsaturated carbonyls [52, 53].

LDAs activate cell-signaling pathways mostly by non-covalent mechanisms that involves binding to a protein receptor or by covalently modifying protein kinases or by activating pathways that cause calcium influx and further intracellular ROS formation. For example, HNE, besides being a cytotoxic lipid aldehyde, triggers phosphorylation of epidermal growth factor receptor (EGFR) at lower doses and activates downstream signaling intermediates such as ERK1/2 and protein kinase B (PKB) or Akt which are known to be involved in cell proliferation [10]. Similarly, HNE could evoke signaling mechanisms for cellular defense and can self-regulate its toxicity and simultaneously may affect multiple signaling pathways through its interactions with membrane receptors and transcription factors/repressors [55]. Studies suggest that a constitutive level of HNE is needed for normal cell functions and any decrease in this level may promote proliferative machinery, while an increase may promote apoptotic signaling suggesting that HNE has concentration-dependent effects on key signaling components (e.g., protein kinase C, adenylate cyclase) which may be opposite to their final effects on cells [56]. Further, HNE has been shown to modulate ligand-independent signaling by membrane receptors such as epidermal growth factor receptor (EGFR) or Fas (CD95) and as a sensor of external stimuli for eliciting stress response [56]. Biasi et al. showed that a significant cross-talk between HNE-activated signaling pathways such as SAPK/JNK/AP-1 pathway and TGFβ1 activated Smad 2/3/4 pathway cooperated to increase apoptotic process in human colon cells [57]. Additionally, glutathione conjugates of LDAs (GS-LDAs) have been shown to act as secondary messengers and activate signaling molecules that eventually leads to activation of redox-sensitive transcription factors NF-κB and AP-1 and elicit various pathophysiological responses [58, 59]. A summary of regulation of potential signaling targets by HNE that modulate redox-sensitive transcription factors and in turn various cell functions is included in Fig. 2 [60].

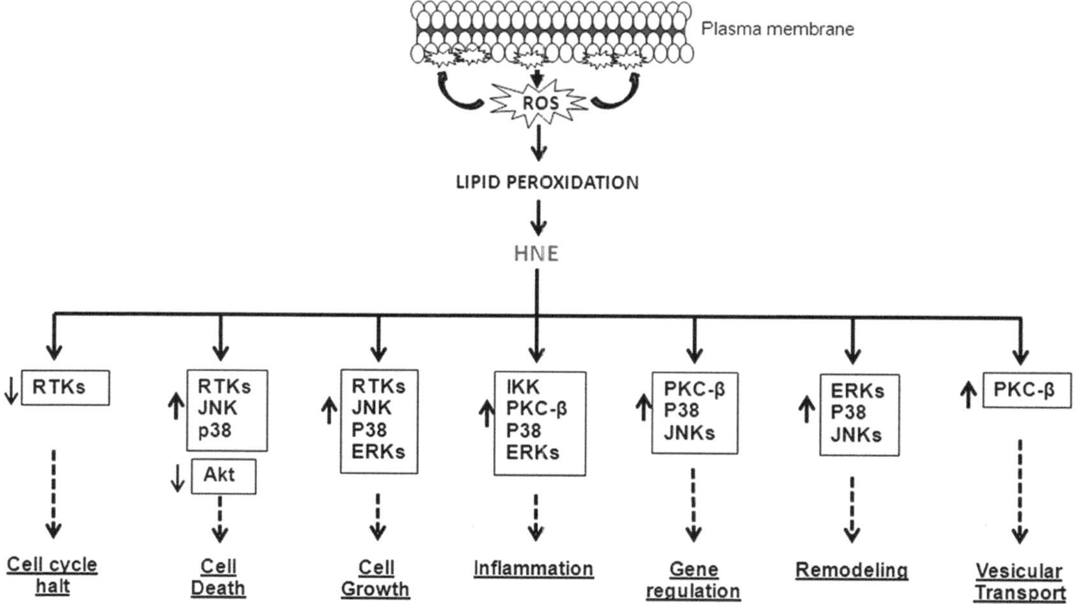

Fig. 2 HNE regulates different potential signaling targets of various cell functions

5 Activation of Signaling in Oxidative Stress-Induced Inflammation

Increased oxidative stress is a hallmark of inflammatory diseases, and ROS has been suggested as an essential mediator of intracellular signaling leading to inflammatory conditions [61–63]. It is well understood that ROS mediate the activation of redox-sensitive transcription factors NF-κB and AP-1, which in turn transcribe an array of inflammatory cytokines, chemokines, and other inflammatory marker genes [64]. Excessive and unrestrained production of inflammatory mediators causes cytotoxicity in an autocrine and paracrine manner leading to tissue damage. Among various ROS-regulated and redox-sensitive transcription factors, NF-κB has been extensively studied and implicated in many oxidative stress-induced inflammatory diseases. It is well known that phosphorylation of upstream protein kinases including protein kinase C (PKC), mitogen-activated protein kinase (MAPK), and IκB kinase (IKK) is activated by oxidative stress and in turn phosphorylates IκB and the subsequent ubiquitination of IκB activates NF-κB leading to transcription of inflammatory mediators (Fig. 3) [45]. However, how exactly ROS activates these kinases is not clearly understood.

This missing link in the ROS-induced activation of signaling cascade and activation of NF-κB has been provided by many evidences which implicate ROS-induced lipid peroxidation products such as lipid aldehydes in the activation of signaling cascade [58, 59]. Indeed, ROS-induced lipid peroxidation products have been suggested to be major contributor in the pathophysiology of many inflammatory disorders. HNE, a peroxidation product of arachidonic acid in the cell membrane, has been shown to be generated by bacterial infection, xenobiotics, environmental pollutants, and autoimmune disorders [65]. Acrolein, another highly reactive species generated both exo- and endogenously, is also implicated in many diseases including COPD, atherosclerosis, and Alzheimer's diseases [4, 5, 9]. Depending upon their respective concentrations, both HNE and acrolein elicit varied physiological outcomes and are known to activate upstream kinases including MAPK and PKC [66–68]. These evidences suggest that cellular enzymes that either regulate or metabolize LDAs could mediate oxidative stress signals. Accordingly,

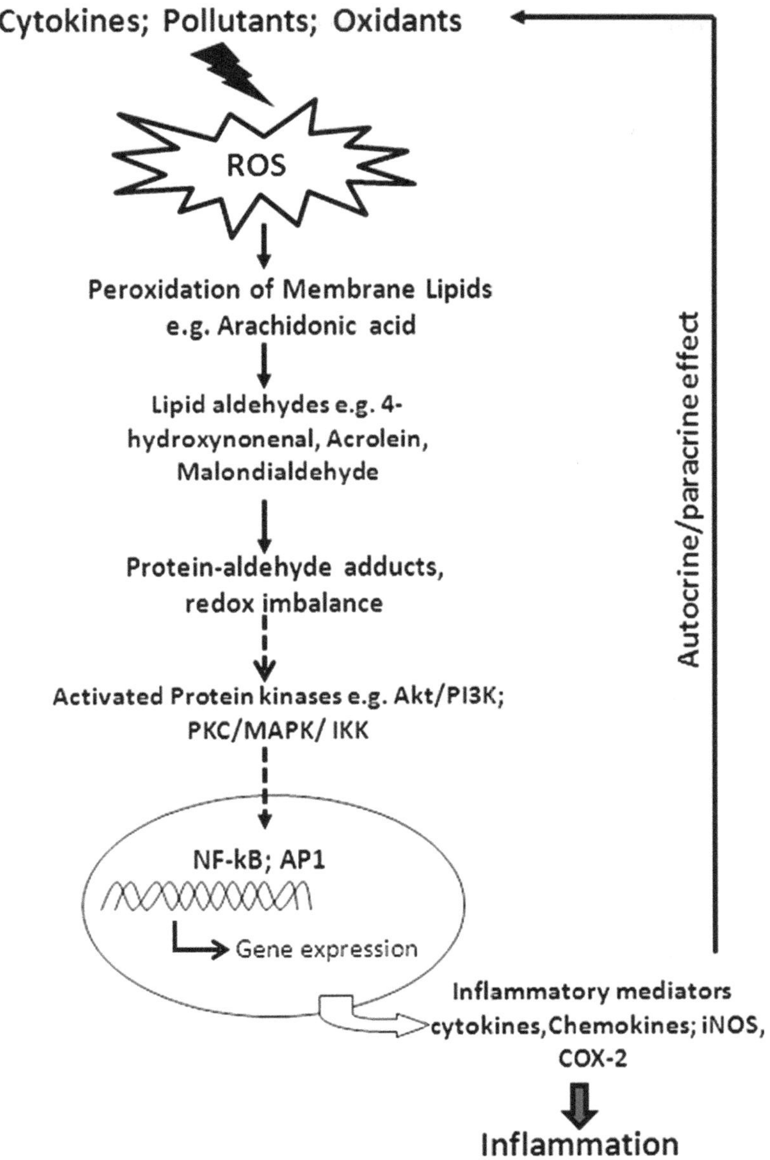

Fig. 3 Activation of signaling cascade by lipid aldehyde leading to NF-κB activation and inflammation

HNE-metabolizing enzymes have been shown to modify and activate MAPKs and ROS-sensitive transcription factors [59]. The overexpression of antioxidants such as catalase or superoxide dismutase (SOD) or treatment with antioxidants such as N-acetylcysteine (NAC) is known to diminish growth factor and cytokine-stimulated cell growth, which suggests that increased ROS production by growth factors and cytokines is an essential step in their effects on cell growth [69, 70]. Studies have indeed indicated that ROS act as secondary messengers that activate NF-κB and influence the cellular functions of growth factors, cytokines, and other stimulants [71, 72]. These evidences clearly suggest that oxidative stress-induced signaling is mediated by ROS-induced generation of other secondary messengers such as lipid-derived aldehydes that activate redox-sensitive transcription factors.

6 Cellular Regulators of Lipid Aldehyde Signals

In the oxidative stress-induced inflammatory diseases, cellular antioxidant status undergoes severe alteration leaving the cells overexposed to excessive free radicals. The oxidative free radicals then cause lipid peroxidation which continues to be unabated leading to generation of more free radicals. This leads to an unending chain of lipid peroxidation reactions generating toxic LDAs which cause inflammation resulting in the establishment and progression of the diseases [73]. However, in healthy cells a robust antioxidant system regulates the toxicity of ROS-derived toxic by-products [74]. The cellular antioxidant system includes superoxide dismutase (SOD), which catalyzes the breakdown of the superoxide anion into oxygen and hydrogen peroxide; catalase, which converts hydrogen peroxide to water and oxygen; and peroxiredoxin, which reduces hydrogen peroxide, organic hydroperoxides, as well as peroxynitrite [75]. Additionally, the glutathione system which includes glutathione, glutathione reductase, glutathione peroxidases, and glutathione *S*-transferases maintains redox balance and protects cells from oxidative challenges [76, 77]. Moreover, exogenous antioxidants and free radical scavengers can also protect the cells from oxidative insults. The supplementation of antioxidants in the experimental models of oxidative stress-generated inflammatory pathologies has been shown to successfully block the inflammation [78]. Several antioxidants are inhibitors of oxidative stress signaling intermediates and thus are effective in halting the inflammation in experimental models. For example, vitamins C and E, N-acetylcysteine (NAC), lipoic acid, glutathione, carotenoids and flavonoids from plants, and melatonins have been demonstrated to effectively ameliorate inflammatory diseases including cardiovascular, neurodegenerative, autoimmune and allergic diseases, infection, and diabetic complications in experimental animals. A number of antioxidants have also gone under clinical trial [79–85] but could not be successfully translated for the clinical use as they act as prooxidants and result in serious side effects at the clinically effective doses among other relevant factors [86, 87]. This has restricted the use of antioxidants as therapeutic drugs and allows them to be used only as prophylactic or preventive drugs. The foregoing discussion thus suggests that a drug which can be antioxidant and also anti-inflammatory that scavenges ROS and inhibits ROS-derived LDAs-induced activation of signaling intermediates would be suitable for clinical use.

7 Conclusion and Future Direction

LDAs have been identified as mediators in many inflammatory complications including cardiovascular disorders, bacterial sepsis, cancer, and asthma and present enormous pathological challenges worldwide. Investigators have demonstrated immense importance of oxidative stress-induced generation of LDAs in pathophysiology of many diseases, which suggests that urgent development of new therapeutic strategies is needed for clinical intervention in LDAs-mediated inflammatory diseases. In order to achieve this goal, a clear understanding of lipid aldehyde-induced signaling events in the inflammatory diseases is crucial. Based on the precise elucidations of molecular signals, better therapeutic interventions could be developed to contain the inflammatory responses in patients. Extensive research in the past few decades has identified that oxidative stress-induced lipid aldehydes play a major role in the mediation of oxidative stress-induced inflammatory signals via PLC/PKC/MAPK/IKK pathway eventually activating transcription factors NF-κB and AP-1. Further, these studies have delineated a novel mechanism regulating inflammation and have laid foundation for future studies that could result in identifying biomarkers and novel molecular target(s) that can be used in diagnosis and clinical intervention.

Acknowledgment Dr. Umesh C. S. Yadav acknowledges the award of Ramanujan Fellowship and financial support from Department of Science and Technology (DST), Government of India.

References

1. Singh SP, Niemczyk M, Saini D et al (2008) Role of the electrophilic lipid peroxidation product 4-hydroxynonenal in the development and maintenance of obesity in mice. Biochemistry 47(12):3900–3911
2. Pillon NJ, Croze ML, Vella RE et al (2012) The lipid peroxidation by-product 4-hydroxy-2-nonenal (4-HNE) induces insulin resistance in skeletal muscle through both carbonyl and oxidative stress. Endocrinology 153(5):2099–2111
3. Mattson MP (2009) Roles of the lipid peroxidation product 4-hydroxynonenal in obesity, the metabolic syndrome, and associated vascular and neurodegenerative disorders. Exp Gerontol 44(10):625–633
4. Butterfield DA, Bader Lange ML, Sultana R (2010) Involvements of the lipid peroxidation product, HNE, in the pathogenesis and progression of Alzheimer's disease. Biochim Biophys Acta 1801(8):924–929
5. Bradley MA, Markesbery WR, Lovell MA (2010) Increased levels of 4-hydroxynonenal and acrolein in the brain in preclinical Alzheimer disease. Free Radic Biol Med 48(12):1570–1576
6. Kang SC, Kim HW, Kim KB et al (2011) Hepatotoxicity and nephrotoxicity produced by 4-hydroxy-2-nonenal (4-HNE) following 4-week oral administration to Sprague-Dawley rats. J Toxicol Environ Health A 74(12):779–789
7. Karihtala P, Kauppila S, Puistola U et al (2011) Divergent behaviour of oxidative stress markers 8-hydroxydeoxyguanosine (8-OHdG) and 4-hydroxy-2-nonenal (HNE) in breast carcinogenesis. Histopathology 58(6):854–862
8. Uno K, Kato K, Kusaka G et al (2011) The balance between 4-hydroxynonenal and intrinsic glutathione/glutathione S-transferase A4 system may be critical for the epidermal growth factor receptor phosphorylation of human esophageal squamous cell carcinomas. Mol Carcinog 50(10):781–790
9. Moretto N, Volpi G, Pastore F (2012) Acrolein effects in pulmonary cells: relevance to chronic obstructive pulmonary disease. Ann NY Acad Sci 1259:39–46
10. Vatsyayan R, Chaudhary P, Sharma A et al (2011) Role of 4-hydroxynonenal in epidermal growth factor receptor-mediated signaling in retinal pigment epithelial cells. Exp Eye Res 92(2):147–154
11. Lee SE, Park YS (2013) Role of lipid peroxidation-derived α, β-unsaturated aldehydes in vascular dysfunction. Oxid Med Cell Longev 2013:1–7
12. Esterbauer H, Schaur RJ, Zollner H (1991) Chemistry and biochemistry of 4-hydroxynonenal, malonaldehyde and related aldehydes. Free Radic Biol Med 11:81–128
13. Loidl-Stahlhofen A, Hannemann K, Spiteller G (1994) Generation of alpha-hydroxyaldehydic compounds in the course of lipid peroxidation. Biochim Biophys Acta 1213:140–148
14. LoPachin RM, Gavin T, Petersen DR et al (2009) Molecular mechanisms of 4-hydroxy-2-nonenal and acrolein toxicity: nucleophilic targets and adduct formation. Chem Res Toxicol 22(9):1499–1508
15. Juric-Sekhar G, Zarkovic K, Waeg G et al (2009) Distribution of 4-hydroxynonenal-protein conjugates as a marker of lipid peroxidation and parameter of malignancy in astrocytic and ependymal tumors of the brain. Tumori 95(6):762–768
16. Vindis C, Escargueil-Blanc I, Uchida K et al (2007) Lipid oxidation products and oxidized low-density lipoproteins impair platelet-derived growth factor receptor activity in smooth muscle cells: implication in atherosclerosis. Redox Rep 12(1):96–100
17. Lee WC, Wong HY, Chai YY et al (2012) Lipid peroxidation dysregulation in ischemic stroke: plasma 4-HNE as a potential biomarker? Biochem Biophys Res Commun 425(4):842–847
18. Bradley MA, Xiong-Fister S, Markesbery WR et al (2012) Elevated 4-hydroxyhexenal in Alzheimer's disease (AD) progression. Neurobiol Aging 33(6):1034–1044
19. Begenik H, Soyoral YU, Erkoc R et al (2013) Serum malondialdehyde levels, myeloperoxidase and catalase activities in patients with nephrotic syndrome. Redox Rep 18(3):107–112
20. Devasagayam TP, Tilak JC, Boloor KK et al (2004) Free radicals and antioxidants in human health: current status and future prospects. J Assoc Physicians India 52:794–804
21. Turrens JF (2003) Mitochondrial formation of reactive oxygen species. J Physiol 552(Pt 2):335–344
22. Bacsi A, Choudhury BK, Dharajiya N et al (2006) Subpollen particles: carriers of allergenic proteins and oxidases. J Allergy Clin Immunol 118(4):844–850
23. Minakami R, Sumimotoa H (2006) Phagocytosis-coupled activation of the superoxide-producing phagocyte oxidase, a member of the NADPH oxidase (nox) family. Int J Hematol 84(3):193–198
24. Santos CX, Tanaka LY, Wosniak J, Laurindo FR (2009) Mechanisms and implications of reactive oxygen species generation during the unfolded protein response: roles of endoplasmic reticulum oxidoreductases, mitochondrial electron transport, and NADPH oxidase. Antioxid Redox Signal 11(10):2409–24027
25. Capdevila J, Parkhill L, Chacos N et al (1981) The oxidative metabolism of arachidonic acid by purified cytochromes P-450. Biochem Biophys Res Commun 101:1357–1363
26. Puntarulo S, Cederbaum AI (1996) Role of cytochrome P-450 in the stimulation of microsomal production of reactive oxygen species by ferritin. Biochim Biophys Acta 1289(2):238–246
27. Tolbert NE, Essner E (1981) Microbodies: peroxisomes and glyoxysomes. J Cell Biol 91:271–283

28. Vuillaume M (1987) Reduced oxygen species, mutation, induction and cancer initiation. Mutat Res 186(1):43–72
29. Berlett BS, Stadtman ER (1997) Protein oxidation in aging, disease, and oxidative stress. J Biol Chem 272(33):20313–20316
30. Circu ML, Aw TY (2010) Reactive oxygen species, cellular redox systems, and apoptosis. Free Radic Biol Med 48(6):749–762
31. Bartsch H, Nair J (2004) Oxidative stress and lipid peroxidation-derived DNA-lesions in inflammation driven carcinogenesis. Cancer Detect Prev 28(6):385–391
32. Catalá A (2006) An overview of lipid peroxidation with emphasis in outer segments of photoreceptors and the chemiluminescence assay. Int J Biochem Cell Biol 38(9):1482–1495
33. Buettner GR (1993) The pecking order of free radicals and antioxidants: lipid peroxidation, alpha-tocopherol, and ascorbate. Arch Biochem Biophys 300(2):535–543
34. Nigam S, Schewe T (2000) Phospholipase A(2)s and lipid peroxidation. Biochim Biophys Acta 1488(1–2):167–181
35. Parola M, Bellomo G, Robino G et al (1999) 4-Hydroxynonenal as a biological signal: molecular basis and pathophysiological implications. Antioxid Redox Signal 1(3):255–284
36. Rubbo H, Parthasarathy S, Barnes S et al (1995) Nitric oxide inhibition of lipoxygenase-dependent liposome and low-density lipoprotein oxidation: termination of radical chain propagation reactions and formation of nitrogen-containing oxidized lipid derivatives. Arch Biochem Biophys 324(1):15–25
37. Jongberg S, Carlsen CU, Skibsted LH (2009) Peptides as antioxidants and carbonyl quenchers in biological model systems. Free Radic Res 43(10):932–942
38. Soberman RJ (2003) The expanding network of redox signaling: new observations, complexities, and perspectives. J Clin Invest 111(5):571–574
39. Thannickal VJ, Fanburg BL (2000) Reactive oxygen species in cell signaling. Am J Physiol Lung Cell Mol Physiol 279(6):L1005–L1028
40. Wall SB, Oh JY, Diers AR et al (2012) Oxidative modification of proteins: an emerging mechanism of cell signaling. Front Physiol 3:369
41. Pantano C, Reynaert NL, van der Vliet A et al (2006) Redox-sensitive kinases of the nuclear factor-kappaB signaling pathway. Antioxid Redox Signal 8(9–10):1791–1806
42. Srivastava SK, Ramana KV (2009) Focus on molecules: nuclear factor-kappa B. Exp Eye Res 88(1):2–3
43. Grossmann M, Nakamura Y, Grumont R et al (1999) New insights into the roles of ReL/NF-kappa B transcription factors in immune function, hemopoiesis and human disease. Int J Biochem Cell Biol 31(10):1209–1219
44. Baeuerle PA, Baltimore D (1996) NF-kappa B: ten years after. Cell 87(1):13–20
45. Barnes PJ, Karin M (1997) Nuclear factor-kappaB: a pivotal transcription factor in chronic inflammatory diseases. N Engl J Med 336(15):1066–1071
46. Ghosh S, May MJ, Kopp EB (1998) NF-kappa B and Rel proteins: evolutionarily conserved mediators of immune responses. Annu Rev Immunol 16:225–260
47. Poli G, Biasi F, Chiarpotto E et al (1989) Lipid peroxidation in human diseases: evidence of red cell oxidative stress after circulatory shock. Free Radic Biol Med 6(2):167–170
48. Higdon A, Diers AR, Oh JY et al (2012) Cell signalling by reactive lipid species: new concepts and molecular mechanisms. Biochem J 442(3):453–464
49. Levonen AL, Landar A et al (2004) Cellular mechanisms of redox cell signalling: role of cysteine modification in controlling antioxidant defences in response to electrophilic lipid oxidation products. Biochem J 378(Pt 2):373–382
50. Ceaser EK, Moellering DR et al (2004) Mechanisms of signal transduction mediated by oxidized lipids: the role of the electrophile-responsive proteome. Biochem Soc Trans 32(Pt 1):151–155
51. Zmijewski JW, Landar A, Watanabe N et al (2005) Cell signalling by oxidized lipids and the role of reactive oxygen species in the endothelium. Biochem Soc Trans 33(Pt 6):1385–1389
52. Dickinson DA, Darley-Usmar VM, Landar A (2006) The covalent advantage: a new paradigm for cell signaling by thiol reactive lipid oxidation products. In: Dalle-Donne I, Scalone A, Butterfield DA (eds) Redox proteomics: from protein modifications to cellular dysfunction and diseases. Wiley, Indianapolis, pp 345–367
53. Doorn JA, Petersen DR (2002) Covalent modification of amino acid nucleophiles by the lipid peroxidation products 4-hydroxy-2-nonenal and 4-oxo-2-nonenal. Chem Res Toxicol 15(11):1445–1450
54. Isom AL, Barnes S, Wilson L et al (2004) Modification of cytochrome c by 4-hydroxy- 2-nonenal: evidence for histidine, lysine, and arginine-aldehyde adducts. J Am Soc Mass Spectrom 15(8):1136–1147
55. Chaudhary P, Sharma R, Sharma A et al (2010) Mechanisms of 4-hydroxy-2-nonenal induced pro- and anti-apoptotic signaling. Biochemistry 49(29):6263–6275
56. Dwivedi S, Sharma A, Patrick B et al (2007) Role of 4-hydroxynonenal and its metabolites in signaling. Redox Rep 12(1):4–10
57. Biasi F, Vizio B, Mascia C et al (2006) c-Jun N-terminal kinase upregulation as a key event in the proapoptotic interaction between transforming growth factor-beta1 and 4-hydroxynonenal in colon mucosa. Free Radic Biol Med 41(3):443–454
58. Ramana KV, Bhatnagar A, Srivastava S et al (2006) Mitogenic responses of vascular smooth muscle cells to lipid peroxidation-derived aldehyde 4-hydroxy-trans-2-nonenal (HNE): role of aldose reductase-catalyzed reduction of the HNE-glutathione conjugates in regulating cell growth. J Biol Chem 281(26):17652–17660

59. Ramana KV, Fadl AA, Tammali R et al (2006) Aldose reductase mediates the lipopolysaccharide-induced release of inflammatory mediators in RAW264.7 murine macrophages. J Biol Chem 281(44): 33019–33029
60. Leonarduzzi G, Robbesyn F, Poli G (2004) Signaling kinases modulated by 4-hydroxynonenal. Free Radic Biol Med 37(11):1694–1702
61. Segovia J, Sabbah A et al (2012) TLR2/MyD88/NF-κB pathway, reactive oxygen species, potassium efflux activates NLRP3/ASC inflammasome during respiratory syncytial virus infection. PLoS ONE 7(1), e29695
62. Auerbach A, Hernandez ML (2012) The effect of environmental oxidative stress on airway inflammation. Curr Opin Allergy Clin Immunol 12(2):133–139
63. Profumo E, Buttari B, Riganò R (2011) Oxidative stress in cardiovascular inflammation: its involvement in autoimmune responses. Int J Inflamm 2011:295705
64. Ray PD, Huang BW, Tsuji Y (2012) Reactive oxygen species (ROS) homeostasis and redox regulation in cellular signaling. Cell Signal 24(5):981–990
65. Srivastava SK, Yadav UC et al (2011) Aldose reductase inhibition suppresses oxidative stress-induced inflammatory disorders. Chem Biol Interact 191(1–3):330–338
66. Zhang H, Forman HJ (2008) Acrolein induces heme oxygenase-1 through PKC-delta and PI3K in human bronchial epithelial cells. Am J Respir Cell Mol Biol 38(4):483–490
67. Tanel A, Averill-Bates DA (2007) P38 and ERK mitogen-activated protein kinases mediate acrolein-induced apoptosis in Chinese hamster ovary cells. Cell Signal 19(5):968–977
68. Zhang H, Forman HJ (2009) Signaling pathways involved in phase II gene induction by alpha, beta-unsaturated aldehydes. Toxicol Ind Health 25(4–5):269–278
69. Fu YQ, Fang F, Lu ZY et al (2010) N-acetylcysteine protects alveolar epithelial cells from hydrogen peroxide-induced apoptosis through scavenging reactive oxygen species and suppressing c-Jun N-terminal kinase. Exp Lung Res 36(6):352–361
70. Umar S, Zargan J, Umar K et al (2012) Modulation of the oxidative stress and inflammatory cytokine response by thymoquinone in the collagen induced arthritis in Wistar rats. Chem Biol Interact 197(1):40–46
71. Müller JM, Rupec RA, Baeuerle PA (1997) Study of gene regulation by NF-kappa B and AP-1 in response to reactive oxygen intermediates. Methods 11(3):301–312
72. Meyer M, Pahl HL, Baeuerle PA (1994) Regulation of the transcription factors NF-kappa B and AP-1 by redox changes. Chem Biol Interact 91(2–3):91–100
73. Bartsch H, Nair J (2006) Chronic inflammation and oxidative stress in the genesis and perpetuation of cancer: role of lipid peroxidation, DNA damage, and repair. Langenbecks Arch Surg 391(5):499–510
74. Matés JM, Sánchez-Jiménez F (1999) Antioxidant enzymes and their implications in pathophysiologic processes. Front Biosci 4:D339–D345
75. Cerutti P, Ghosh R, Oya Y et al (1994) The role of the cellular antioxidant defense in oxidant carcinogenesis. Environ Health Perspect 102:123–129
76. Limón-Pacheco JH, Gonsebatt ME (2010) The glutathione system and its regulation by neurohormone melatonin in the central nervous system. Cent Nerv Syst Agents Med Chem 10(4):287–297
77. Ballatori N, Krance SM, Notenboom S et al (2009) Glutathione dysregulation and the etiology and progression of human diseases. Biol Chem 390(3):191–214
78. Ozkanlar S, Akcay F (2012) Antioxidant vitamins in atherosclerosis – animal experiments and clinical studies. Adv Clin Exp Med 21(1):115–123
79. Hopkins MH, Fedirko V, Jones DP et al (2010) Antioxidant micronutrients and biomarkers of oxidative stress and inflammation in colorectal adenoma patients: results from a randomized, controlled clinical trial. Cancer Epidemiol Biomarkers Prev 19(3):850–858
80. Mazloom Z, Hejazi N, Dabbaghmanesh MH et al (2011) Effect of vitamin C supplementation on postprandial oxidative stress and lipid profile in type 2 diabetic patients. Pak J Biol Sci 14(19):900–904
81. Goodman M, Bostick RM, Kucuk O et al (2011) Clinical trials of antioxidants as cancer prevention agents: past, present, and future. Free Radic Biol Med 51(5):1068–1084
82. Ienco EC, LoGerfo A, Carlesi C et al (2011) Oxidative stress treatment for clinical trials in neurodegenerative diseases. J Alzheimers Dis 24:111–126
83. Greenberg ER, Baron JA et al (1994) A clinical trial of antioxidant vitamins to prevent colorectal adenoma. Polyp Prevention Study Group. N Engl J Med 331(3):141–147
84. Mathew MC, Ervin AM, Tao J et al (2012) Antioxidant vitamin supplementation for preventing and slowing the progression of age-related cataract. Cochrane Database Syst Rev 6, CD004567
85. Yadav UC, Kalariya NM, Ramana KV (2011) Emerging role of antioxidants in the protection of uveitis complications. Curr Med Chem 18(6): 931–942
86. Steinhubl SR (2008) Why have antioxidants failed in clinical trials? Am J Cardiol 101(10A):14D–19D
87. Cochemé HM, Murphy MP (2010) Can antioxidants be effective therapeutics? Curr Opin Investig Drugs 11(4):426–431

Oxidative Stress and Its Biomarkers in Cardiovascular Diseases: An Overview

Vibha Rani and Aditi Jain

Abstract

Cardiovascular diseases (CVD) are today's largest single contributor toward global mortality and will continue to dominate mortality trends in the coming future. Growing evidences indicate the role of reactive oxygen species (ROS) and the dysregulation of oxidant–antioxidant pathways under pathophysiologic conditions in the development of CVDs. ROS has several potential sources within the cardiovascular system which lead to oxidative stress at the cellular level by interrupting with different signaling pathways within the cell. These may include the oxidation and nitration of cellular proteins, lipids, nucleic acids and the formation of aggregates of oxidized molecules leading to the loss of cellular function and the inability of cells to withstand physiological stresses. Evidences confirm that sources of ROS, physiological and pathophysiological conditions, and cellular oxidant targets of ROS determine the characteristic nature of a disease process and resultant outcomes. Diverse free radical-mediated responses observed in different cardiovascular conditions are discussed in detail in this chapter. Also, in recent times, important milestones have been reached in studying the role of ROS in CVDs with the finding of explicit biomarkers of oxidative stress involved in CVDs. These biomarkers are also discussed in detail.

Keywords

Cardiovascular disease (CVD) • Coronary heart disease (CHD) • Oxidative stress • Biomarkers • Antioxidants

V. Rani (✉) • A. Jain
Department of Biotechnology, Jaypee Institute of Information Technology, A-10, Sector-62, Noida 201307, Uttar Pradesh, India
e-mail: vibha.rani@jiit.ac.in

V. Rani and U.C.S. Yadav (eds.), *Free Radicals in Human Health and Disease*,
DOI 10.1007/978-81-322-2035-0_10, © Springer India 2015

1 Introduction

Cardiovascular diseases (CVD) are caused by disorders of the heart and blood vessels and include coronary heart disease (heart attacks), cerebrovascular disease (stroke), raised blood pressure (hypertension), peripheral artery disease, rheumatic heart disease, congenital heart disease, and heart failure. The World Health Organization has declared CVD a global epidemic as it is the number one cause of death in the world and it creates a plethora of complications. The major risk factors for CVDs include high blood pressure, high cholesterol, overweight and obesity, physical inactivity, unhealthy diet, smoking, and harmful use of alcohol. At the molecular level, important factors include reactive oxygen species (ROS) and oxidative stress which play a very crucial role in the pathophysiology of different CVDs and other metabolic syndromes. Numerous studies in the past four decades have confirmed the cardiac abnormalities in response to oxidative stress and studied various pathways in understanding the role of oxidative stress in cardiac dysfunctions under different CVDs. Oxidative stress is also a common mediator in the pathogenicity of conventional cardiovascular risk factors and associated with a marked increase in vascular ROS production (Fig. 1) [1].

As we have already studied in the previous chapters, ROS is involved as a signaling molecule in various important cellular processes, has a very short half-life, highly reactive in nature, and, if not neutralized, can hamper cellular functions rigorously. Also, inbuilt antioxidant enzymatic machinery and other antioxidants are present in cells to counterbalance elevated levels of ROS (Fig. 2). Decrease in this antioxidant reserve in response to excessively elevated levels of ROS results in oxidative stress which further leads to the onset of different pathophysiological conditions in different tissues and organs.

Several studies suggest the role of ROS in different growth processes associated with vascular injury and remodeling like integrin-linked kinase 1/βPIX/Rac-1 pathway, nuclear factor-κB pathway, etc. [2, 3]. Studies also indicate that many harmful cellular phenotypes identified in a hypertrophied and deteriorating myocardium are attributed to oxidative stress [4]. Thus, we can say that any acute or chronic cardiac stress condition, resulting in a relative deficit in the myocardial antioxidant reserve, is associated with an increase in myocardial oxidative stress.

ROS leads to the increased entry of Ca^{2++} into myocytes and results in the activation of diverse chemical reactions inside the cells, upregulates the growth factor gene expression in vascular endothelial cells, and modifies cellular proteins which lead to cellular dysfunction [5]. Cardiomyocytes exposed to H_2O_2 or glucose oxidase have shown a reduction in mRNA levels for α-actin, troponin I, myosin light chain 2, and M isoform of creatinine kinase [6]. ROS also induce specific posttranslational modifications that alter the function of important cellular proteins and signaling pathways in the heart [7]. Oxygen radicals are capable of reacting with unsaturated lipids of cellular membranes and initiate the chain reactions of lipid peroxidation. Early in vivo studies have examined the role of free radical-mediated increase in lipid peroxide activity which results in catecholamine-induced heart disease as a result of autoxidation of catecholamines [8].

Changes in myocardial antioxidants as well as oxidative stress have been observed in vivo suggesting the role of free radicals in the pathogenesis of heart failure subsequent to myocardial infarction (MI). These changes have also been correlated with the cardiac function at different stages of failure [9]. The direct involvement of oxidative stress in apoptosis has been demonstrated in a variety of cell types and also been documented in the myocardium of MI and heart failure patients [10]. The assessment of oxidative stress in various CVDs is done by studying various biomarkers to assess the response to oxidative challenges.

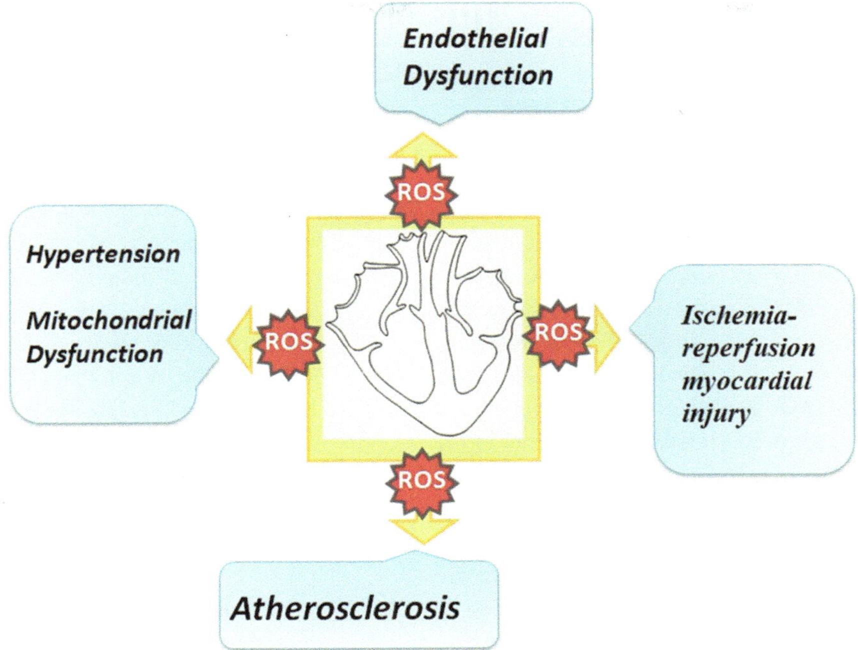

Fig. 1 Overproduction of ROS in cardiac system plays an integral role in the development of CVD and leads to different pathophysiologic conditions

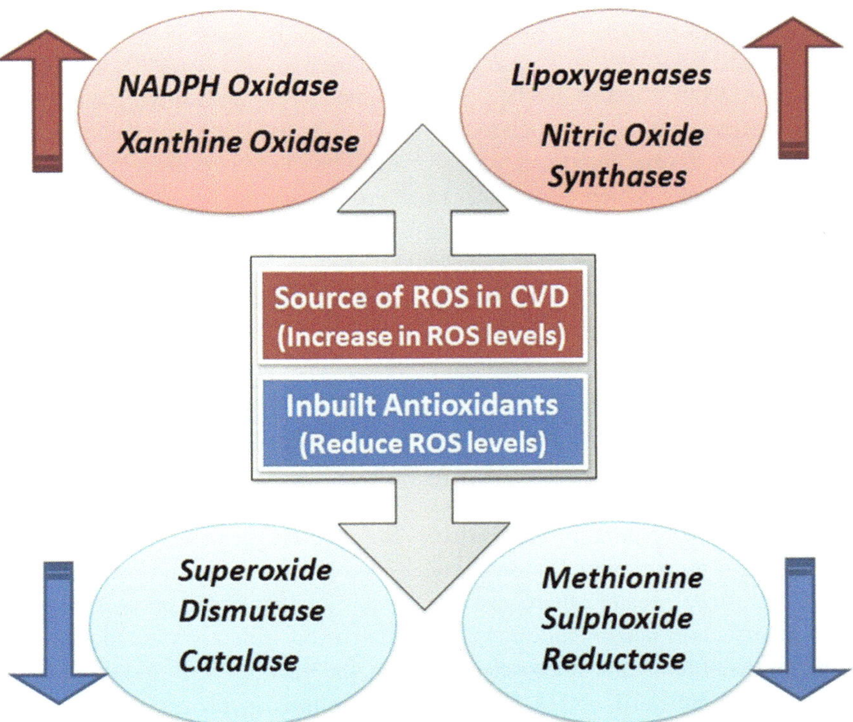

Fig. 2 Source of free radicals in cardiovascular system which lead to oxidative stress and the progression of CVD. An inbuilt antioxidant enzymatic machinery to counterbalance the elevated ROS in the cell

2 Major Sources of ROS and Their Effects in Cardiovascular Diseases

ROS differs in their source, effect, and chemical reactivity based on their radical nature. ROS is generated by dysfunction of the mitochondrial electron transport system, the xanthine oxidase reaction, neutrophil activation, arachidonic acid metabolism, and the autooxidation of catecholamines. ROS is also generated by endothelial dysfunction (ED) as well as from the exposure to radiation or air pollution, whereas decline in the antioxidant reserve is due to the exhaustion and/or changes in gene expression [11]. Clearly, many cellular enzymes as well as nonenzymatic processes within the cardiovascular system are potential sources of free radicals which lead to oxidative stress and the progression of CVD.

2.1 ROS Derived from the Mitochondrial Respiratory Chain

The mitochondrial electron transport chain is a rich source of free radicals, and mitochondria-generated ROS plays a very important role in many signaling pathways which contribute to CVD. These ROS also get modified, further amplifying its deleterious effect. Manganese superoxide dismutase (Mn-SOD) enzyme is mainly responsible for the conversion of $O2^{\bullet-}$ to H_2O_2, and it has been observed that individuals null for the Mn-SOD allele exhibit perinatal lethality due to cardiac dysfunction, and cardiac-specific Mn-SOD deletion produces progressive congestive heart failure with specific molecular defects in mitochondrial respiration [12]. Also, targeted deletion of a mitochondrial uncoupling protein 2 gives rise to a greater atherosclerotic lesion as well as increased macrophage accumulation and apoptosis. Premature atherosclerosis has been reported in complex III- and IV-deficient individuals, and premature atherosclerosis and other vascular complications are associated with mitochondrial dysfunction [13].

2.2 NADPH Oxidase-Derived ROS

NADPH oxidase is a multiple subunit, membrane-associated complex and a source of ROS in CVD which modulates both vascular physiology and pathophysiology. Its increased activity contributes strongly to the pathogenesis in experimental models of vascular disease, including cholesterol-induced atherosclerosis. Several studies indicate the roles of NADPH oxidase isoforms in cardiac fibrosis, cardiac remodeling, post-MI, and angiotensin II-dependent cardiac hypertrophy [14–16]. Within the vasculature, it may also contribute to atherosclerosis, aortic aneurysm formation, and the response to arterial injury [17].

2.3 Xanthine Oxidase-Derived ROS

Xanthine oxidase or xanthine oxidoreductase catalyzes the conversion of hypoxanthine to xanthine and xanthine to uric acid and generates $O^{2\bullet-}$ and H_2O_2. Its function is upregulated by NADPH oxidase, and hence the factors regulating it also influence the activity of xanthine oxidase [18]. Xanthine oxidase has been implicated as a source of ROS after reperfusion of ischemic tissue in the cardiac system.

2.4 Lipoxygenase-Derived ROS

They are nonheme, iron-containing enzymes which catalyze the incorporation of molecular oxygen into polyunsaturated fatty acids thereby modifying the fatty acids, and ROS is generated as a by-product during the lipoxygenase activity. Some lipoxygenase isoforms promote atherosclerosis by generating ROS and oxidatively modifying lipids and lipoproteins such as LDL. In the period from early to advanced atherosclerosis, the quantity of 5-lipoxygenase (5-LO) positive cells is increased in human atherosclerotic plaque specimens, and elevated 5-LO activity has been linked to plaque instability [19]. In human abdominal aortic aneurysm tissue, 5-LO is expressed in macrophage-rich adventitial areas and associated with intraluminal thrombus.

2.5 Nitric Oxide Synthases

They catalyze the conversion of L-arginine to L-citrulline with the production of NO, and its isoforms are constitutively present in most cardiovascular tissues. Under specific circumstances, nitric oxide synthases may become uncoupled and reduce molecular oxygen rather than transferring electrons to L-arginine, thereby generating oxygen free radicals and hence act as an important source of ROS. In the pressure overload induced by aortic constriction, eNOS, an isoform of nitric oxide synthases, leads to more severe left ventricular (LV) hypertrophy, LV dysfunction, and myocardial fibrosis [20].

3 Free Radical-Mediated Responses in Cardiovascular Complications

3.1 Initiation of Endothelial Dysfunction

Endothelial dysfunction (ED) is a strong predictor of future cardiovascular events in patients with cardiovascular risk factors. Increased ROS has been implicated as an important mechanism that contributes to ED by modulating the cellular pathways. Oxidative stress is a major cause of ED in experimental and clinical atherosclerosis, as observed in many studies [21, 22]. Polymorphism has been observed in a variety of genes whose products have been implicated in ED, for example, in the genes of methylenetetrahydrofolate reductase, angiotensin-converting enzyme, glutathione S-transferase, cytochrome P450, and nitric oxide synthase gene [23]. Various pro-atherogenic stimuli increase oxidative stress in endothelial cells resulting in endothelial dysfunction. This may include the activation of MMPs by ROS, proinflammatory signaling in vascular smooth muscle cells, and enhanced expression of oxidative stress marker NF-κB [24, 25]. Few studies demonstrate that oxidative stress is one of the most potent inductors of endothelial dysfunction and is involved in all stages of atherosclerotic plaque evolution [26, 27].

3.2 Atherosclerosis

Atherosclerosis generates from ED and inflammation and characterized by elevated plasma cholesterol levels, smooth muscle cell proliferation, and alteration in arterial walls. Transport of oxidized low-density lipoprotein (LDL) across the endothelium into the artery wall may be an important initiating event for atherosclerosis and likely to occur at the sites of endothelial damage which are caused by oxidized LDL. The involvement of ROS in atherosclerotic heart disease is very well accepted which results in the rupture of a lipid-rich atherosclerotic plaque and may lead to MI or sudden death [28].

3.3 Oxidation of Low-Density Lipoprotein

LDL, also known as "the bad cholesterol," gets easily oxidized by the free radicals produced and results in the formation of oxidized LDL which is a critical factor of atherogenesis and has many proinflammatory properties [29]. Peroxidation of long-chain polyunsaturated fatty acids within LDL in the presence of hydroxyl radicals gives rise to conjugated dienes and lipid hydroperoxy radicals. These radicals further attack the next fatty acid in a self-propagating manner, and the fragmentation of whole chains takes place. As a result of this, the accumulation of highly reactive products like malondialdehyde and lysophosphatides in LDL takes place and modifies the LDL. These modified oxidized LDL is readily ingested and scavenged by subendothelial macrophages and forms fatty streaks in the arterial endothelium, the earliest histopathological evidence for the development of atherosclerotic plaque and fibrosis [30].

3.4 Oxidative Stress and Mitochondrial Dysfunction

As we know, intracellular ROS is derived from the mitochondria primarily from complex I and complex III of the electron transport chain; under pathophysiological conditions, the electron transport chain may become uncoupled and result in the increase of ROS inside the cells. First and foremost, mitochondrial DNA gets affected by this ROS as it is in close proximity of the newly produced ROS and lacks histone proteins [31]. This DNA damage can lead to several functional changes in the pathways associated with the mitochondria. Increased mitochondrial DNA damage has been observed in the vascular tissues of CVD patients. 8-oxoG, a product of oxidative DNA damage, is highly immunogenic and has been found present in the plaques of the carotid artery [32].

3.5 Oxidative Stress in Hypertension

Increased production of ROS in hypertension has been demonstrated in many studies [33]. Hypertension is associated with elevated superoxide anion and H_2O_2 production as well as decreased antioxidant capacity. Also, increased levels of lipid peroxides and decreased concentrations of antioxidant vitamin E has been observed in the plasma of hypertensive patients. ROS has dual adverse effect in hypertensive patients. *Firstly*, free radicals inactivate NO and convert it into peroxynitrile, thereby causing arteriolar vasoconstriction and elevation of peripheral hemodynamic resistance; *secondly*, ROS initiate the mechanism for the oxidative damage of numerous macromolecules including LDL. Increased peroxynitrile causes cytokine-induced myocardial contractile failure by inactivating sarcoplasmic Ca^{2+} ATPase and dysregulating Ca^{2+} homeostasis. This condition may further lead to heart failure. Endothelial cells, the source of NO, are considered as the primary source of ROS in hypertension [34]. Also, activation of the renin–angiotensin system is a major mediator of NAD(P)H oxidase activation and ROS production in human hypertension [35].

3.6 Heart Failure

Increased oxidative stress has been observed in heart failure and plays a critical role in the pathophysiology of cardiac dysfunction. This may contribute many changes that characterize the disease progression. The increased level of malondialdehyde, the end product of lipid peroxidation, is present in the plasma of patients with heart failure. The mitochondria is considered as a major source of ROS production in the failing heart, which may further decrease the NO bioavailability and impair diastolic functions [36]. Xanthine oxidase and NADPH oxidase activity is increased in the failing heart and results in ROS generation in the myocardium and plasma, respectively [37].

3.7 Ischemia–Reperfusion Myocardial Injury

Ischemia occurs when myocardial oxygen demand exceeds the oxygen supply, and reperfusion of an ischemic myocardium restores oxygen which in turn may lead to reperfusion injury. This reperfusion injury induced by the cycles of ischemia and reperfusion may be, in part, due to the generation of ROS. The involvement of ROS in the ischemia–reperfusion damage has been studied by measuring the end products of free radical-mediated damages like malondialdehyde and the cellular antioxidant capacity. In patients undergoing cardiopulmonary bypass, a prominent increase in ROS levels has been observed, following aortic declamping suggesting an impairment of antioxidant mechanisms in the ischemic tissues. Reperfusion leads to the burst of ROS generation which contributes to oxygen tension in the myocardium [38]. ROS can also be formed as a result of increased intracellular Ca^{2+} levels as it leads to the conversion of xanthine dehydrogenase to xanthine oxidase and leads to tissue damage by ROS-mediated events. As the cardiac muscles are exposed to ischemia–reperfusion, mitochondrial dysfunction may also lead to increased ROS levels. Superoxides either inactivate NO and decrease its bioavailability or react with NO and form peroxynitrile which is an important determinant of postischemic myocardial function [39]. ROS-mediated complications worsen the efficiency of interventions used in the treatments of coronary heart diseases.

4 Biomarkers of Oxidative Stress and Their Role in Cardiovascular Diseases

Oxidative species are highly active and short-lived but leave a detectable trace of modified oxidative products at the site of atherosclerotic lesions [40]. These modified oxidative products are measured as biomarkers of oxidative stress and help establish that ROS and oxidative events are part of the pathophysiology of CVD. Various biomarkers have been studied and summarized in Fig. 3.

4.1 Biomarkers for Lipid Peroxidation

Isoprostanes and malondialdehyde are two well-studied biomarkers for lipid peroxidation. Other lipid oxidation products that have been explored as biomarkers include lipid hydroperoxides, fluorescent products of lipid peroxidation, and oxysterols. Isoprostanes are generated from the peroxidation of arachidonic acid of cellular membranes. Isoprostane levels are increased in human atherosclerotic lesions and may participate in the actual pathogenesis of atherosclerosis through its harmful effects on vasoconstriction, platelet aggregation, and proliferation of vascular smooth muscle cells [41]. Malondialdehyde, on the other hand, is generated by the peroxidation of polyunsaturated fatty acids, and it itself is potentially atherogenic. It promotes atherosclerosis by forming lysine–lysine cross-links with the oxidized LDL and impairs its interaction with the macrophages [42]. Various studies suggest the potential role of lipid oxidation in predicting

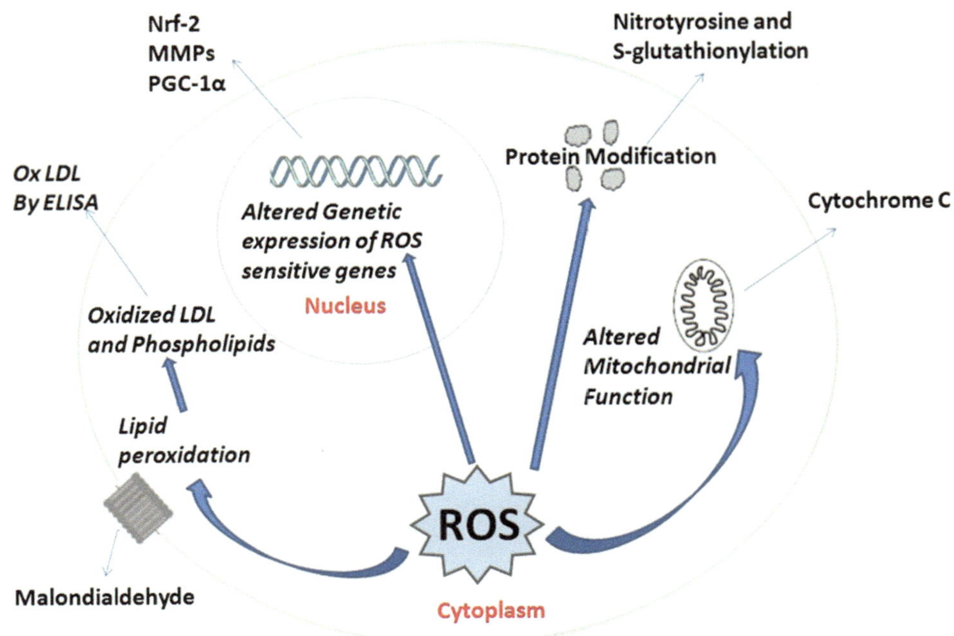

Fig. 3 Biomarkers of ROS and Oxidative stress generated in various cardiovascular abnormalities

the progression of CVD. Malondialdehyde is quantified with the colorimetric assay known as thiobarbituric acid reactive substances (TBARS) assay and has been applied as an indicator of oxidative stress in a number of CVD models.

4.2 Biomarkers for Oxidative Protein Modifications

Nitrotyrosine and S-glutathionylation are the two commonly used biomarkers for studying oxidative protein modifications. Free nitrotyrosine corresponds to nitrated proteins and can be measured by tandem mass spectrometry coupled with HPLC as the current gold standard technique [43]. Its presence has been observed in the vascular and myocardial tissue in both healthy individuals and those with CVD. Nitration of proteins and lipoproteins plays a direct pathophysiological role, for example, nitrated LDL is taken up by macrophages leading to plaque formation [44]. Despite the importance of nitration, there are several challenges in applying nitrotyrosine as a CVD biomarker because circulating nitrated proteins and lipoproteins may not accurately reflect the degree of nitration of key proteins in the case of atherosclerosis. Hence another marker used is S-glutathionylation, which is the formation of a disulfide bridge between a reactive cysteine residue and the abundant cellular tripeptide glutathione. Quantification of S-glutathionylation of target proteins with important functional consequences is a promising biomarker for CVD processes. The impact of glutathionylation of various membrane proteins has been reported in the myocardium and vascular tissue with altered function resulting in alterations in intracellular Na^+ and Ca^{2+} handling and other key signaling pathways particularly relevant to cardiovascular function [45].

4.3 Enzymatic Biomarkers for Oxidative Stress

Myeloperoxidase is a key enzyme in ROS generation by catalyzing the conversion of hydrogen peroxide to various free radical species including $^{\cdot}OH$, $ONOO^-$, hypochlorous acid, and NO_2. Myeloperoxidase-derived ROS can then modify lipids, lipoproteins, and proteins which promote ED by reducing the bioavailability of NO and generating atherogenic oxidized LDL [46]. The activity of this enzyme can be measured by peroxidase activity assays and quantified spectrophotometrically. Myeloperoxidase is enriched within atheromatous plaques and hence implicated in the pathogenesis of atherosclerosis. Elevated circulating myeloperoxidase levels have been found to be associated with the presence of coronary artery diseases, and its measurement can help us in predicting the risk of developing coronary artery disease in healthy individuals and myocardial infarction. These applications make myeloperoxidase levels one of the most promising biomarkers of oxidative stress for clinical cardiologists [47]. The activity of other antioxidant enzymes like catalase, glutathione peroxidase 1, and SOD is also measured for assessing the antioxidant status. Patients with suspected coronary artery disease have shown to have elevated levels of catalase and glutathione peroxidase and hence inversely associated with the incidence of cardiovascular events [48]. The commercial availability of antioxidant enzyme assay kits allows us to use these potential biomarkers to be evaluated in a large-scale high-throughput screening.

4.4 Oxidized LDL and Phospholipids

The oxidation of LDL and phospholipids plays a central role in the pathogenesis of atherosclerosis [49]. Oxidized LDL is generally measured by using specific monoclonal antibodies that directly recognize unique oxidation-specific epitopes known as OxLDL ELISA. Levels of oxidized LDL are higher in patients with CVD. Also, the increasing levels of LDL correlate with increasing severity of disease and appear to be a tool for the prediction of future CAD in apparently healthy men [50].

4.5 Change in the Genetic Expression of ROS-Sensitive Genes

ROS influence the expression of various genes involved in regulating critical cellular and systemic oxidative stress in different pathophysiological conditions. Major ROS-sensitive genes include genes for nuclear factor like Nrf-2 which results in increased expression of cellular antioxidant enzymes, peroxisome proliferator-activated receptor gamma coactivator 1-alpha (PGC-1α) gene, and matrix metalloproteinase genes [51]. Profiling the expression of these ROS-sensitive genes by high-throughput methods like microarray technology may be a helpful tool, particularly relevant to assessing the cardiovascular redox status. Low level of expression of these genes may reflect a low level of oxidative stress in the relevant system or individual variation in response and may result in a higher level of oxidative cellular damage. This approach remains to be further explored for clinical applicability of ROS-responsive genes as biomarkers in various diseases.

5 Antioxidant Therapies Used in Cardiovascular Oxidative Stress

Attempts have been made to decrease the oxidative stress produced in different cardiovascular abnormalities by the application of different antioxidants and free radical scavengers in vitro. Studies have implicated the role of antioxidants in decreasing the ROS levels as well as bringing potentially atheroprotective changes in biochemical and functional disease-specific markers [52]. Antioxidant therapies for heart failure in animal models also suggest a benefit from reduced generation of free radicals, improves cardiac function after reperfusion, attenuates remodeling and cardiomyopathy [53]. Observational studies of vitamin E supplementation suggest that vitamin E is associated with a lower risk of coronary heart diseases [54]. Theoretically, antioxidant therapies are expected to be most effective in secondary prevention of CVD but there is no consistent evidence of benefit for prevention of Coronary heart disease from the meta analysis of the trials. There have been multiple justifications for the fact including that the ROS effects are complex and make the outcome predictions difficult and that the types of antioxidants, their dosage and duration of action have been inadequate. It is also known that genetic factors influence the response to antioxidants raising the possibility that only some patients get benefits from antioxidant treatments [55].

6 Conclusions

There are substantial data linking oxidative stress and its role in the physiology and pathophysiology of CVD. There are several sources of reactive oxygen species that are known to be active in the cardiovascular system, and there is an evident linking of each of the sources with CVD pathology. With the discovery of specific ROS biomarkers in CVD, the assessment of oxidative stress has become a very helpful method in assessing an individual's response to oxidative challenges. Initial attempts to improve manifestations of CVD with simple antioxidant strategies have not proven helpful, probably because ROS have important and diverse physiological roles. Clearly, atherosclerosis is a complex multifactorial and multicellular disease, and oxidative stress responds to antioxidant therapy differently in different cell types and at different stages of disease progression. Thus, correct trial design is required for the assessment of the potential of different antioxidants in reducing atherosclerosis and other CVDs.

Acknowledgement We acknowledge Jaypee Institute of Information Technology, deemed to be the university for providing the infrastructural support. This work was supported by a research grant awarded to Dr. Vibha Rani by the Department of Biotechnology (DBT), Government of India (BT/PR3978/17/766/2011).

References

1. Ho E, Galougahi KK, Chia-Chi L et al (2013) Biological markers of oxidative stress: applications to cardiovascular research and practice. Redox Biol 1:483–491
2. Vecchione C, Carnevale D, Di Pardo A et al (2009) Pressure-induced vascular oxidative stress is mediated through activation of integrin linked kinase 1/βPIX/Rac-1 pathway. Hypertension 54:1028–1034
3. Djordjevic T, Hess J, Herkert O et al (2004) Rac regulates thrombin induced tissue factor expression in pulmonary artery smooth muscle cells involving the nuclear factor-κB pathway. Antioxid Redox Signal 6:713–720
4. Sawyer DB, Siwik DA, Xiao L et al (2002) Role of oxidative stress in myocardial hypertrophy and failure. J Mol Cell Cardiol 34:379–388
5. Dhalla NS, Golfman L, Takeda S et al (1999) Evidence for the role of oxidative stress in acute ischemic heart disease: a brief review. Can J Cardiol 15:587–593
6. Torti SV, Akimoto H, Lin K et al (1998) Selective inhibition of muscle gene expression by oxidative stress in cardiac cells. J Mol Cell Cardiol 30:1173–1180
7. Rasmussen HH, Hamilton EJ, Liu CC et al (2010) Reversible oxidative modification: implications for cardiovascular physiology and pathophysiology. Trends Cardiovasc Med 20:85–90
8. Singal PK, Beamish RE, Dhalla NS (1983) Potential oxidative pathways of catecholamines in the formation of lipid peroxides and genesis of heart disease. Adv Exp Med Biol 161:391–401
9. Hill MF, Singal PK (1996) Antioxidant and oxidative stress changes during heart failure subsequent to myocardial infarction in rats. Am J Pathol 148:291–300
10. Singala PK, Khapera N, Palacea V et al (1998) The role of oxidative stress in the genesis of heart disease. Cardiovasc Res 40:426–432
11. Dhalla NS, Temsah RM, Netticadan T (2000) Role of oxidative stress in cardiovascular diseases. J Hypertens 18:655–673
12. Nojiri H, Shimizu T, Funakoshi M et al (2006) Oxidative stress causes heart failure with impaired mitochondrial respiration. J Biol Chem 281:33789–33801
13. Finsterer J (2007) Is atherosclerosis a mitochondrial disorder? Vasa 36:229–240
14. Johar S, Cave AC, Narayanapanicker A et al (2006) Aldosterone mediates angiotensin II-induced interstitial cardiac fibrosis via a Nox2-containing NADPH oxidase. FASEB J 20:1546–1548
15. Looi YH, Grieve DJ, Siva A et al (2008) Involvement of Nox2 NADPH oxidase in adverse cardiac remodeling after myocardial infarction. Hypertension 51:319–325
16. Satoh M, Ogita H, Takeshita K et al (2006) Requirement of Rac1 in the development of cardiac hypertrophy. Proc Natl Acad Sci U S A 103:7432–7437
17. Judkins CP, Diep H, Broughton BR et al (2009) Direct evidence of a role for Nox2 in superoxide production, reduced nitric oxide bioavailability, and early atherosclerotic plaque formation in ApoE−/−mice. Am J Physiol Heart Circ Physiol 298:H24–H32
18. Landmesser U, Spiekermann S, Preuss C et al (2007) Angiotensin II induces endothelial xanthine oxidase activation: role for endothelial dysfunction in patients with coronary disease. Arterioscler Thromb Vasc Biol 27:943–948
19. Cipollone F, Mezzetti A, Fazia ML et al (2005) Association between 5-lipoxygenase expression and plaque instability in humans. Arterioscler Thromb Vasc Biol 25:1665–1670
20. Ichinose F, Bloch KD, Wu JC et al (2004) Pressure overload-induced LV hypertrophy and dysfunction in mice are exacerbated by congenital NOS3 deficiency. Am J Physiol Heart Circ Physiol 286:H1070–H1075
21. Cai H, Harrison DG (2000) Endothelial dysfunction in cardiovascular disease: the role of oxidant stress. Circ Res 87:840–844
22. Landmessar U, Dikalov S, Price SR et al (2003) Oxidation of tetrahydrobiopterin leads to uncoupling of endothelia cell nitric oxide synthase in hypertension. J Clin Invest 111:1201–1209
23. Durier S, Fassot C, Laurent S et al (2003) Physiological genomics of human arteries: quantitative relationship between gene expression and arterial stiffness. Circulation 108:1845–1851
24. Galis ZS, Sukhova GK, Lark MW et al (1994) Increased expression of matrix metalloproteinases and matrix degrading activity in vulnerable regions of human atherosclerotic plaques. J Clin Invest 94:2493–2503
25. Foncea R, Carvajal C, Almarza C, Leighton F (2000) Endothelial cell oxidative stress and signal transduction. Biol Res 33:89–96
26. Harrison D, Griendling KK, Landmesser U et al (2003) Role of oxidative stress in atherosclerosis. Am J Cardiol 91:7A–11A
27. Glass CK, Witztum JL (2001) Atherosclerosis: the road ahead. Cell 104:503–516
28. Kunsch C, Medford RM (1999) Oxidative stress as a regulator of gene expression in the vasculature. Circ Res 85:753–766
29. Chisolm GM, Steinberg D (2000) The oxidized modification hypothesis of atherogenesis: an overview. Free Radic Biol Med 28:1815–1826
30. Fruchart JC, Duriez P (1994) Free radicals and atherosclerosis. In: Rice-Evans CA, Burdon RH (eds) Free radical damage and its control. Elsevier Science, Amsterdam, pp 257–281
31. Wei YH (1998) Oxidative stress and mitochondrial DNA mutations in human aging. Proc Soc Exp Biol Med 217:53–63
32. Martinet W, Knaapen MW, De Meyer GR et al (2002) Elevated levels of oxidative DNA damage and DNA repair enzymes in human atherosclerotic plaques. Circulation 106:927–932

33. Zalba G, San JG, Moreno MU et al (2001) Oxidative stress in arterial hypertension: role of NAD(P)H oxidase. Hypertension 38:1395–1399
34. Brigelius-Flohe R, Banning A, Kny M et al (2004) Redox events in interleukin-1 signalling. Arch Biochem Biophys 423:66–73
35. Schiffrin EL, Touyz RM (2003) Multiple actions of angiotensin II in hypertension: benefits of AT1 receptor blockade. J Am Coll Cardiol 42:911–913
36. Lakshmi SV, Padmaja G, Kuppusamy P et al (2009) Oxidative stress in cardiovascular disease. Indian J Biochem Biophys 46:421–440
37. Saavedra WF, Paolocci N, St John ME et al (2002) Imbalance between xanthine oxidase and nitric oxide synthase signalling pathways underlies mechanoenergetic uncoupling in the failing heart. Circ Res 90:297–304
38. Angelos MG, Kutala VK, Torres CA et al (2006) Hypoxic reperfusion of the ischemic heart and oxygen radical generation. Am J Physiol Heart Circ Physiol 290:H341–H347
39. Wang P, Zweier JL (1996) Measurement of nitric oxide and peroxynitrite generation in the postischemic heart. Evidence for peroxynitrite-mediated reperfusion injury. J Biol Chem 271:29223–29230
40. Schnabel R, Blankenberg S (2007) Oxidative stress in cardiovascular disease: successful translation from bench to bedside? Circulation 116:1338–1340
41. Gniwotta C, Morrow JD, Roberts LJ et al (1997) Prostaglandin F2-like compounds, F2-isoprostanes, are present in increased amounts in human atherosclerotic lesions. Arterioscler Thromb Vasc Biol 17:3236–3241
42. Uchida K (2000) Role of reactive aldehyde in cardiovascular diseases. Free Radic Biol Med 28:1685–1696
43. Duncan MW (2003) A review of approaches to the analysis of 3-nitrotyrosine. Amino Acids 25:351–361
44. Bartesaghi S, Ferrer-Sueta G, Peluffo G et al (2007) Protein tyrosine nitration in hydrophilic and hydrophobic environments. Amino Acids 32:501–515
45. Figtree GA, Karimi GK, Liu CC et al (2012) Oxidative regulation of the Na(+)–K(+) pump in the cardiovascular system. Free Radic Biol Med 53:2263–2268
46. Abu-Soud HM, Hazen SL (2005) Nitric oxide modulates the catalytic activity of myeloperoxidase. J Biol Chem 275:5425–5430
47. Schindhelm RK, Zwan LPV, Teerlink T et al (2009) Myeloperoxidase: a useful biomarker for cardiovascular disease risk stratification? Clin Chem 55:1462–1470
48. Blankenberg S, Rupprecht HJ, Bickel C et al (2003) Glutathione peroxidase 1 activity and cardiovascular events in patients with coronary artery disease. N Engl J Med 349:1605–1613
49. Tsimikas S (2006) Oxidized low-density lipoprotein biomarkers in atherosclerosis. Curr Atheroscler Rep 8:55–61
50. Meisinger C, Baumert J, Khuseyinova N et al (2005) Plasma oxidized low-density lipoprotein, a strong predictor for acute coronary heart disease events in apparently healthy, middle-aged men from the general population. Circulation 112:651–657
51. Singh S, Vrishni S, Singh BK et al (2010) Nrf2-ARE stress response mechanism: a control point in oxidative stress-mediated dysfunctions and chronic inflammatory diseases. Free Radic Res 44:1267–1288
52. Fearon IM, Faux SP (2009) Oxidative stress and cardiovascular disease: novel tools give (free) radical insight. J Mol Cell Cardiol 47:372–381
53. Redout EM, Van der Toorn A, Zuidwijk MJ et al (2010) Antioxidant treatment attenuates pulmonary arterial hypertension-induced heart failure. Am J Physiol Heart Circ Physiol 298:H1038–H1047
54. Sesso HD, Buring JE, Christen WG et al (2008) Vitamins E and C in the prevention of cardiovascular disease in men: the Physicians' Health Study II randomized controlled trial. JAMA 300:2123–2133
55. Sugamura K, Keaney JF (2011) Reactive oxygen species in cardiovascular disease. Free Radic Biol Med 51:978–992

Free Radicals and Oxidative Stress in Neurodegenerative Disorders

Darshika Nigam

Abstract

Oxidative stress (OS) leading to free radical attack on neural cells contributes calamitous role to neurodegeneration. Free radicals are produced either endogenously (by metabolism and antioxidant system of the body) or exogenously (by environmental sources) by imbalance between antioxidant and prooxidant in the body. In-built antioxidant system of body plays critical role in prevention of any loss due to free radicals. However, imbalanced defense mechanism of antioxidants, overproduction, or incorporation of free radicals from environment to living system leads to serious penalty leading to neurodegeneration. This chapter discusses the role of free radicals in pathophysiology of some common neurodegenerative disorders like Alzheimer's disease, Parkinson's disease, Huntington's disease, and amyotrophic lateral sclerosis.

Keywords

Free radicals • Oxidative stress • Neurodegenerative disorders • Alzheimer's disease • Parkinson's disease • Huntington's disease • Amyotrophic lateral sclerosis

1 Introduction

Neurodegenerative diseases are heterogeneous group of central nervous system and peripheral nervous system disorders with different etiologies. Neurodegenerative diseases affect many physiological functions of the human body, such as balance, movement, breathing, talking, etc. Cellular antioxidants are known to change the redox state, and they can be targeted for destruction, regulate oxidative processes involved in signal transduction, and effect gene expression and pathways of cell proliferation and death. The increased incidence of neurodegenerative diseases may be attributed to a prooxidative environment caused by UV and γ irradiation, smoking, alcoholism, air pollution, toxins, drugs, viruses as well as inappropriate nutrition [1, 2]. Neurodegeneration is the progressive loss of

D. Nigam (✉)
Department of Biochemistry, School of Life Sciences, Dr. B. R. Ambedkar University, Agra, India
e-mail: drdarshikanigam@rediffmail.com

V. Rani and U.C.S. Yadav (eds.), *Free Radicals in Human Health and Disease*,
DOI 10.1007/978-81-322-2035-0_11, © Springer India 2015

Table 1 Some neurodegenerative disorders caused by aggregation of abnormal proteins

Neurodegenerative disorders	Aggregated protein	Lesion and accumulated sites	Reference
Alzheimer's disease	Amyloid-β peptide	Amyloid plaques extracellularly	[19]
	Tau	NFT in intracytoplasmic neurons	
	α-Synuclein	Lewy bodies variants in intracytoplasmic neurons	
Parkinson's disease	α-Synuclein	Lewy bodies in intracytoplasmic neurons	[7]
Huntington's disease	Huntingtin (polyglutamine repeats)	Neuronal inclusions in nuclei of neurons	[20]
Amyotrophic lateral sclerosis	SOD1	Hyaline inclusions in intracytoplasmic neurons	[21]
Dementia with Lewy bodies	α-Synuclein	Lewy bodies in intracytoplasmic neurons	[22]
Multiple system atrophy	α-Synuclein	Glial cytoplasmic inclusions in cytoplasm of oligodendrocytes	[23]
Supranuclear palsy	Tau	Tau inclusions in intracytoplasmic neurons, astrocytes, and oligodendrogliocytes	[10]
Spinocerebellar ataxia	Ataxin (polyglutamine repeats)	Neuronal inclusions in nuclei of neurons	[6]
Prion diseases	Protease-resistant prion proteins	Prion plaques extracellularly	[24]
Pick's disease	Tau	Pick bodies in intracytoplasmic neurons	[11]

structure or function of neurons, including death of neurons. Many diseases including Parkinson's, Alzheimer's, Huntington's, amyotrophic lateral sclerosis, Friedreich's ataxia, Lewy body disease, and spinal muscular atrophy occur as a result of neurodegenerative processes. Neurodegeneration may be found in many different levels of neuronal circuitry ranging from molecular to systemic. These diseases have many similarities on a subcellular level [3, 4]. These similarities offer hope for therapeutic advances that could ameliorate many diseases simultaneously.

1.1 Pathogenesis of Neurodegenerative Disorders

1.1.1 Genetics

Pathogenesis of several neurodegenerative diseases is genetic mutations. Most of these mutations are located in completely unrelated genes. In many of different neurodegenerative diseases, the mutated gene has a common feature, for example, a repeat of the CAG tri nucleotide (encodes for the amino acid glutamine) and results in a polyglutamine (polyQ) tract. Diseases showing this feature are known as polyglutamine diseases. Extra glutamine residues exert toxic effects in many ways, including irregular protein folding and degradation pathways, altered subcellular localization, and abnormal interactions with other cellular proteins. Nine inherited neurodegenerative diseases, for example, Huntington's disease, spinocerebellar ataxia, and others, are caused by the expansion of the CAG nucleotide triplet [5, 6].

1.1.2 Protein Misfolding

In several neurodegenerative diseases, aggregation of misfolded proteins (proteopathies or proteinopathies) occurs, some of which are as follows (see Table 1).

1.1.2.1. Alpha-Synuclein

α-Synuclein is the primary structural protein of insoluble Lewy bodies formed in pathological conditions such as in Parkinson's disease, dementia with Lewy bodies, and multiple system atrophy. α-Synuclein fragment, known as the non-abeta component (NAC), is also found in amyloid plaques in Alzheimer's disease [7, 8].

1.1.2.2. Tau

Tau is a microtubule-associated protein. Hyperphosphorylated tau protein is the main component of neurofibrillary tangles present in intracytoplasmic neurons in Alzheimer's disease. It is also found in progressive supranuclear palsy and Pick's disease [9–11].

1.1.2.3. Beta-Amyloid

It is an abnormal protein-polysaccharide complex deposited extracellularly in various tissues or organs. It is the major component of senile plaques in Alzheimer's disease [12].

1.1.3 Mitochondrial Dysfunction

The intrinsic mitochondrial apoptotic pathway is the most common form of cell death in neurodegeneration. This pathway regulates the activation of caspase-9 by release of cytochrome c from the mitochondrial intermembrane space. Reactive oxygen species (ROS) are usually generated during electron transport chain (ETC) process whose concentration is mediated by mitochondrial antioxidants such as manganese superoxide dismutase (SOD2) and glutathione peroxidase. Excess production of ROS (or oxidative stress) is a key feature of all neurodegenerative disorders. In addition to the ROS generation, mitochondria play significant role in calcium homeostasis, programmed cell death, mitochondrial fission and fusion, maintaining lipid contents of mitochondrial membranes, and alteration in mitochondrial permeability [13, 14].

1.1.4 Membrane Damage

Misfolded proteins cause damage to the membranes of cell organelles and may also contribute to neurodegenerative diseases. For example, α-synuclein may damage membranes by inducing membrane curvature, and extensive vesiculation and tubulation were observed when these proteins were incubated with artificial phospholipid vesicles [15].

1.1.5 Axonal Transport

Axonal swelling has been observed in many neurodegenerative diseases due to accumulation of organelles. Axonal transport may be disrupted by a variety of mechanisms including damage to motor proteins such as kinesin and dynein, microtubules, cargoes, and mitochondria. When axonal transport is severely disturbed, a degenerative pathway which is known as Wallerian-like degeneration is triggered [16].

1.1.6 Protein Degradation Pathways

There are two main pathways by which abnormal proteins or organelles are removed:

1.1.6.1. Ubiquitin–Proteasome

This is the primary route to degrade cellular proteins. Protein ubiquitin along with enzymes is important for the degradation of many proteins that cause proteinopathies including polyQ expansions and α-synucleins. If these irregular proteins are not correctly cleaved by proteasome enzymes, then they may possibly result in a more toxic species which may in turn cause neurodegeneration [17, 18].

1.1.6.2. Autophagy–Lysosome Pathways

It is a form of programmed cell death (PCD); this becomes the favorable route when a protein is a poor proteasome substrate. This may be divided into two forms of autophagy: macroautophagy and chaperone-mediated autophagy (CMA). Macroautophagy is involved with nutrient recycling of macromolecules under conditions of starvation. If it is absent, this leads to the formation of ubiquitinated inclusions. CMA pathway receptors present on lysosomal membrane binds to mutant proteins and block their own degradation as well as the degradation of other substrates [25, 26].

1.1.7 Programmed Cell Death

Programmed cell death (PCD) is death of a cell in any form, mediated by an intracellular program. These PCD pathways may also be artificially stimulated due to injury or disease.

1.1.7.1. Type I PCD

Type I PCD or apoptosis is one of the main types of programmed cell death (PCD). Apoptosis is triggered by intrinsic or extrinsic stress signals. In extrinsic apoptotic pathway, a series of biochemical events leads to cell death through activation of cell surface death receptors (e.g., Fas, TNF-R1). This results in the activation of caspase-8 or caspase-10. In intrinsic pathway, death signals trigger mitochondria to release cytochrome c or endoplasmic reticulum malfunctions both of which lead to the activation of caspase-9. The nucleus and Golgi apparatus are other organelles that have damage sensors (intrinsic apoptotic pathway) [26–28].

1.1.7.2. Type II PCD

Type II PCD or autophagy is basically a form of intracellular phagocytosis in which a cell actively consumes damaged organelles or misfolded proteins by enclosing them into an autophagosome, which fuses with a lysosome to degrade the autophagosomal contents. Defect in autophagy shows unusual protein aggregates which are common in many neurodegenerative diseases [26].

1.1.7.3. Type III PCD

Type III or cytoplasmic cell death, a non-apoptotic process, is the least understood PCD mechanism. Type III PCD either might be caused by hyperactivation of trophic factor receptors (trophotoxicity) or by other cytotoxins that induce PCD at low concentrations but act to cause necrosis or aponecrosis (the combination of apoptosis and necrosis) when in higher concentrations [29].

2 Role of Free Radicals and Oxidative Stress in Neurodegenerative Disorders

Many neurodegenerative diseases are late onset. Neurons gradually lose function as the disease progresses with age. Mitochondrial DNA mutations and oxidative stress both contribute to aging. The mitochondrial electron transport chain continually generates ROS by which mitochondrial DNA is exposed leading to mutations. This may accumulate exponentially with age (Fig. 1). The simultaneous increase in oxidation of mitochondrial lipids and proteins adds to the oxidative stress effects and initiates the vicious cycle of molecular degeneration [13, 14, 30]. All aerobic organisms including humans consume appreciable amount of O_2 for various metabolic processes. It readily reacts with other radicals because O_2 is a diradical. Free radicals and partially reduced

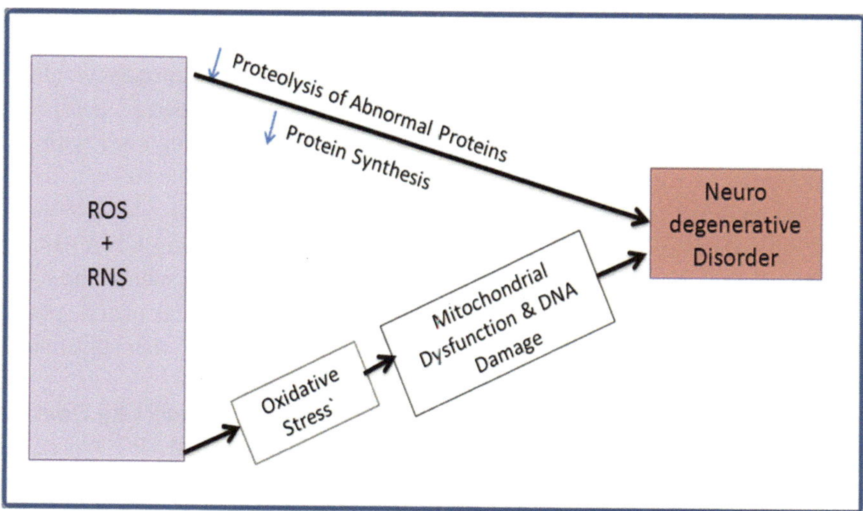

Fig. 1 Interplay of reactive species with proteolysis and mitochondrial function in neurodegenerative disorders

species are often generated from O_2 itself during normal metabolism in the body. The most common cellular free radicals are hydroxyl radical ($OH^{\bullet}$), superoxide anion radical (SAR) ($O_2^{\bullet-}$), and nitric oxide (NO•). Other molecules, such as hydrogen peroxide (H_2O_2) and peroxynitrite anion ($ONOO^-$), are not free radicals but may lead to the generation of free radicals through various chemical reactions. Due to the presence of an unpaired electron, free radicals are highly unstable and tend to react with cellular elements [31, 32].

The brain is susceptible to oxidative stress due to its high-energy demand and the specialized redox activities of neurons. Free radicals and other purported reactive species are constantly produced in the brain in vivo. Some arise accidentally and may be the leakage of electrons from the mitochondrial electron transport chain to generate SAR. Other radicals are generated to carry out normal physiological activities, such as the role of nitric oxide in neurotransmission and the production of SAR by activated microglia. Other sources of the SAR include the short electron chain in the endoplasmic reticulum, cytochrome P450, and the enzyme nicotinamide adenine dinucleotide phosphate (NADPH) oxidase, which generates substantial quantities especially during early pregnancy, and other oxidoreductases. The brain contains multiple antioxidant defenses, out of which the mitochondrial manganese-containing superoxide dismutase and reduced glutathione are important [31–33].

Oxidative stress occurs when the balance between ROS generation and antioxidant activity is disturbed or the rate of formation of free radicals exceeds the antioxidant capacity of the system and complicates the underlying disease [32]. It is reported in literature that the best-characterized cellular damage caused by •OH is its ability to stimulate the free radical chain reaction known as lipid peroxidation. Unsaturated lipids are predominantly susceptible to oxidative modification. Lipid peroxidation is the result of attack by free radicals on double bond of unsaturated fatty acid and arachidonic acid. It leads to generation of highly reactive lipid peroxyl radicals that initiate a chain reaction of further attacks on other unsaturated fatty acid [34]. Transition metal ions also affect lipid peroxidation by decomposing peroxides. The transition metal ions, when added to lipid systems containing peroxides, decompose these peroxides into alkoxyl (LO•) and peroxyl (LOO•) radicals which in turn abstract hydrogen and propagate the chain reaction of lipid peroxidation. Lipid hydroperoxides formed in membrane of nerve cell alter its fluidity causing influx of various ions as Ca^{2+} in neurons and disrupts normal functioning of cell. In addition to polyunsaturated fatty acids of membrane lipids, the free radicals attack proteins. Neurofilaments and proteins constituting the cytoskeleton of the nerve cell are particularly rich in lysine residues. In the oxidative stress, the amino groups of these residues react with aldehydes derived from lipid oxidation and lose their electric charge. This leads to distortions of its tertiary and quaternary structure, and a number of enzymes in the neurons lose their activity [35].

Increased levels of oxidative damage to DNA, lipids, and proteins have been detected in autopsy tissues from patients with Parkinson's disease, Alzheimer's disease, Huntington's disease, amyotrophic lateral sclerosis, and many others neurodegenerative disorders [36–39]. Some of these changes may occur early during progression of these diseases. The accumulation and precipitation of proteins that occur in these diseases may be aggravated by oxidative damage and may in turn cause more oxidative damage by interfering with the function of the proteasome. Proteasomal inhibition increases levels of oxidative damage not only to proteins but also to other biomolecule [26].

2.1 Alzheimer's Disease and Oxidative Stress

Alzheimer's disease (AD) is a neurodegenerative disease in which progressive impairment of cognitive function occurs and is accompanied by behavioral dysfunctions. AD is the commonest cause of dementia (a progressive decline in intellectual functions that substantially interferes with individual's activities). The usual onset age is over 65 years, but AD can also appear in the early 40s [40].

AD often results in the clinical manifestations including impaired memory, impaired thought processes that involve visual and spatial awareness, behavior changes, confusion, restlessness, impaired judgment, unable to recognize friends and family members, and impaired communication. Motor function is often unaffected in AD. Alzheimer's disease is characterized by loss of neurons and synapses in the certain areas of the brain (cerebral cortex and hippocampus). This loss results in degeneration of temporal lobe and parietal lobe and parts of the frontal cortex and cingulate gyrus regions [40].

Two principal characteristic features of AD are extracellular senile plaques (cluster of nerve endings surrounding the core of extracellular beta-amyloid peptide) [12, 41] and intracellular neurofibrillary tangles (filamentous protein aggregates mainly composed of hyperphosphorylated microtubule-associated protein, tau) [9]. Many genes involved with the disease have been identified including amyloid precursor protein (*APP*) gene located on chromosome 21 and presenilin genes 1 and 2 (*PS1* and *PS2*) located on chromosomes 14 and 1, respectively. *PS1* and *PS2* genes are responsible for early onset of the disease [42]. The apolipoprotein E (*apoE*) gene located on chromosome 19, the α2-macroglobulin gene located on chromosome 12, and many unidentified genes may determine susceptibility in late-onset forms. Mutations on the *APP*, *PS1*, or *PS2* gene all feature increased production of beta-amyloid peptide. The majority of cases are sporadic, involving both genetic and environmental factors [43–45].

Senile plaques are made up of small peptides of 39–43 amino acid residues called beta-amyloid (Aβ). A transmembrane neuronal protein, APP, is divided enzymatically into smaller fragments through proteolysis by an unknown process. One of these fragments gives rise to Aβ fibrils, which form clumps and deposit outside neurons in dense formations known as senile plaques. Various complementary factors, including cytokines, transforming growth factor β1, and interleukin 1, seem to be involved in triggering the process of amyloidogenesis. APP is also important for neuronal growth and post-injury repair [46, 47].

2.1.1 Role of Free Radicals and Oxidative Stress in AD

The oxidation of mitochondrial DNA (lesser extent of nuclear DNA) has been observed in the parietal cortex of AD patients. Protein oxidation has also been observed in AD patients in the cerebral cortex and hippocampal regions of the brain. These oxidations are related to the apoE genotype. Individuals with E4 allele are likely to be more susceptible to peroxidation than are those without this allele [44]. Moreover, within the brains of AD patients, particularly in the neurofibrillary tangles, malondialdehyde, peroxynitrite anions, carbonyls, advanced glycosylation end products (AGEs), superoxide dismutase-1, and heme oxygenase-1 (a cellular enzyme that are upregulated in the brain and in other tissues in response to an oxidative stress or other harmful stimuli) levels are high [48–50]. ROS causes alterations to membrane phospholipids which may be specific to the pathogenesis of AD. Lipid peroxidation is a major cause of depletion of membrane phospholipids in AD. One of the products of lipid peroxidation 4-hydroxynonenal (4-HNE) found in high concentrations in hippocampal cells of AD patients is a highly reactive aldehyde and is thought to cause neuronal death by altering the ATPases involved in ionic transfers and calcium homeostasis [51]. The increased intracellular calcium ion concentration could cause a cascade of intracellular events. It results in increased ROS and cell death. The glutamate-dependent flux of calcium is associated with the production of free radicals by the mitochondria [52]. Nitric acid and peroxynitrite play a crucial role in the excitotoxicity related to *N*-methyl-D-aspartate glutamate receptor activation [53]. It has been reported that cerebrospinal fluid F2-isoprostane concentrations are elevated in AD patients. It is produced by free radical-catalyzed peroxidation of arachidonic acid (independent of the cyclooxygenase enzyme) [54].

β-Amyloid is also crucial for the generation of oxidative stress. It interacts with vascular endothelial cells and produces excess of SAR that can scavenge the endothelium-derived relaxing factor and produce oxidizing agents which causes lipid peroxidation. Mass spectrometric and electron paramagnetic resonance spin trapping indicate that in aqueous solution, β-amyloid fragments and generates free radical peptides. Some of these free radical peptides have a potent lipoperoxidizing effect on the synaptosomal membranes in the neocortex [55]. β-Amyloid also activates protein tyrosine kinase-dependent signaling and superoxide production in the microglia [56].

2.1.2 Role of Metals in Free Radicals Generation in AD

Iron, copper, aluminum, and zinc play a major catalytic role in the production of free radicals in AD [57, 58]. The concentration of iron is elevated in the brain of AD patient. Iron is involved in the formation of the free hydroxyl radical by Fenton's and Haber–Weiss's reactions, which has deleterious effects. Iron, transferrin, and ferritin have been found in senile plaques and neurofibrillary tangles. The iron-binding protein p97 is elevated in blood and cerebrospinal fluid in AD patients. p97 protein concentrations could be used as a marker of the disease for following the course of the disease [59]. Aluminum has been suggested as a causal factor in AD, in part because of reports showing the toxicity of aluminum. The aluminum content is not elevated in the brain regions of AD patients that are selectively vulnerable to the neuropathologic changes associated with the disease [60]. Copper may act as a catalyst in the production of ROS. The APP molecule contains a copper-binding site. The binding of Cu(II) leads to the modification of APP via the oxidation of cysteines 144 and 158, which leads to the formation of cysteine and Cu(I) [61]. This metal is essential for many enzyme activities, including cytochrome *c* oxidase and Cu/Zn superoxide dismutase [62]. Contrary to this, lower concentrations of copper in five zones of the brains of AD patients, particularly in the hippocampus, are observed [57]. Possible role of zinc has also been observed in AD. Zinc induces a rapid amyloid formation. APP binds with Zn(II) which modulates the functional properties of APP by inhibiting the cleavage of APP by α-secretase and increases binding to heparin [63, 64]. Zinc, the most abundant trace metal in the brain, has numerous functions, both in health and in disease. Zinc is released into the synaptic cleft of glutamatergic neurons along glutamate from where it interacts and modulates NMDA and AMPA receptors (both are glutamate receptors) and causes increased influx of calcium as discussed earlier in this chapter. In addition, zinc has multifactorial functions in Alzheimer's disease (AD). Zinc is critical in the enzymatic nonamyloidogenic processing of the APP and in the enzymatic degradation of the β-amyloid peptide. Zinc binds to Aβ promoting its aggregation into neurotoxic species, and zinc dyshomeostasis in the brain results in synaptic and memory deficits [65].

2.2 Parkinson's Disease and Oxidative Stress

Parkinson's disease (PD) is the second most common degenerative disorder (old age) of the central nervous system after AD and manifests as bradykinesia, rigidity, tremor at rest, and postural instability. Parkinson's disease is characterized by the loss of dopaminergic neurons that arise in the substantia nigra, a region of the midbrain (mesencephalon), and project to the putamen and caudate regions (the striatum) of the brain (the areas concerned with the control of motor movements) [66]. Ninety to ninety-five percent of PD cases are sporadic form, and the rest is familial cases [35].

Two principal hallmarks of PD are intracellular inclusion bodies called Lewy bodies [7] and accumulation of neuromelanin pigment [67]. An abnormal accumulation of the protein α-synuclein is bound to ubiquitin in the damaged cells. The alpha-synuclein–ubiquitin complex cannot be directed to the proteosome. This is due to a defect in the machinery that transports proteins between two major cellular organelles, the endoplasmic reticulum and the Golgi apparatus. These proteins accumulate and form proteinaceous cytoplasmic inclusions [68]. Lewy bodies are abnormal intracytoplasmic neuronal inclusions that consist of neurofilaments (intermediate cytoskeletal filaments) and other amorphous material (vesicular and granular materials). They are also seen in pigmented brain stem nuclei in various disorders and in normal aging brain. Lewy bodies occur in two characteristic forms. The first is classical or subcortical (brain stem) type consisting of a single or multiple round or oval eosinophilic structures with a central core surrounded by a less dense peripheral zone and an outermost pale halo that is sharply demarcated from the neuronal cytoplasm. The cortical type is not sharply delineated from the cytoplasm, and the outline of the central part is rather indistinct [7, 38].

The second characteristic feature of the neurons within the substantia nigra is the age-dependent accumulation of neuromelanin, an insoluble polymer related to melanin of skin. Neuromelanin is a dark brown pigment that accumulates metal ions, particularly iron. In PD, these neuromelanin-containing cells are most likely to be lost [69]. Dopamine is a precursor of neuromelanin. Neuromelanin is present in pigment-bearing dopaminergic neurons of four deep brain nuclei. These are the substantia nigra pars compacta region, the locus coeruleus, the dorsal motor nucleus of the vagus nerve, and the median raphe nucleus of the pons. In addition to the degeneration of the nigrostriatal dopaminergic pathway, a variety of other neuronal systems are involved, causing complex patterns of functional deficits in PD [70]. This degeneration also affects the dopaminergic neurons containing areas of the brain including mesolimbic and mesocortical system, the noradrenergic locus coeruleus (oral parts) and motor vagal nucleus, the serotonergic raphe nuclei, the cholinergic nucleus basalis of Meynert, pedunculopontine nucleus, Westphal–Edinger nucleus, and many peptidergic brain stem nuclei containing somatostatin, cholecystokinin, substance P, neuropeptide Y, and met-enkephalin [70, 71].

2.2.1 Role of Free Radicals and Oxidative Stress in Parkinson's Disease

In Parkinson's disease, oxidative stress induced by free radicals damages neuronal membrane lipids, proteins, and other components of brain tissue. Oxidative stress is implicated as a major factor for nigral neuronal cell death. The biological targets of oxidants include membrane proteins, unsaturated lipids, and DNA. Oxidative stress promotes aggregation and accumulation of α-synuclein in sporadic PD. There are several potential sources of the increased free radical production in Parkinson's disease including increased dopamine metabolism, formation of neuromelanin, mitochondrial dysfunction, increased free iron levels, and low ferritin concentration. These sources are specific only for Parkinson's disease, and not found in other neurodegenerative diseases associated with degeneration of dopaminergic neurons [36, 72].

2.2.2 Increased Oxidation of Dopamine and Formation of Neuromelanin

Catecholamines and particularly dopamine (DA) are an important source of free radicals in the brain. As long as dopamine is stored in synaptic vesicles, it is stable. DA is an essential neurotransmitter, and as it is a catechol, it is also a good metal chelator [70]. However, when it is in excess in cytosol, it is oxidized either enzymatically by monoamine oxidase (MAO) (see reaction 1) or autoxidized to generate hydrogen peroxide (see reaction 2). Autoxidation of dopamine or L-dopa via quinone formation generates free radicals such as superoxide radical and hydrogen peroxide. DA reduces the oxidation state of the transition metals such as Cu^{2+} and Fe^{3+} and subsequently stimulates production of H_2O_2 [73]. By Fenton reaction, hydrogen peroxide and reduced metal ions produce hydroxyl radicals (see reaction 3) [3].

Reaction 1: DA + O_2 + H_2O →3,4 dihydroxyphenyl acetaldehyde + NH_3+ H_2O_2

Reaction 2: DA + O2 → semiquinone radical+ $O_2^{\bullet-}$ + H^+DA + $O_2^{\bullet-}$ + $2H^+$→ semiquinone radical + H_2O_2

Reaction 3: H_2O_2 + Fe^{2+}/ Cu^+→ •OH + OH^- + Fe^{3-}/Cu^{2+}

Moreover, dopamine and L-dopa quinone are easily oxidized to aminochromes and finally polymerize to form melanin. In PD patients, metabolism of dopamine is greatly enhanced and in unaffected dopaminergic neurons tyrosine hydroxylase activity is increased. This may be the compensatory mechanism to fill dopamine deficiency [74].

The cause for degeneration of dopamine neurons is not well understood. However, it can be assumed that the interactions between external toxins arise from environmental, dietary, and lifestyle factors, and internal toxins arising from normal metabolism, genetic factor, and epigenetic (mitochondria, membranes, and proteins) components of neurons occur continuously [75, 76].

In PD, the dopaminergic neurons containing the largest quantities of neuromelanin are damaged first. Although, the exact function of neuromelanin in the brain is unknown, but the pigment is made from oxyradical metabolites of neurotransmitters including dopamine and norepinephrine. Neuromelanin may also be seen as a kind of free radicals, which is able to catalyze the dismutation of superoxide radical to hydrogen peroxide. In addition, neuromelanin may bind with free radicals (such as superoxide anion radicals and hydroxyl radicals) and some metal ions (such as iron ions). At low iron concentrations, neuromelanin is known to have antioxidant properties and protects against dopamine-induced free radical generation, but at high metal burden, melanins are prooxidant [73]. However, neuromelanin is an iron storage molecule in substantia nigra and has both high- and low-affinity Fe^{3+}-binding sites [76].

ROS may also a cause of DNA damage and altered expression of susceptible genes including α-synuclein, leucine-rich repeat kinase 2 (LRRK-2), and glucocerebrosidase. This shows that genetic predisposition is another important factor of PD [77].

2.2.3 Increased Iron Concentration and Low Concentration of Ferritin

Iron is found in high concentrations in several parts of the basal ganglia including the substantia nigra, the globus pallidus, and the putamen. Iron contributes to generate free radicals and may therefore increase the vulnerability of dopaminergic neurons to toxic oxygen radicals, especially in the substantia nigra of parkinsonian patients where iron content is increased. Iron is important for developing the activity of tyrosine hydroxylase and monoamine oxidase enzymes [74]. The tyrosine hydroxylase catalyzes the conversion of L-tyrosine to L-dihydroxyphenylalanine, and the monoamine oxidase catalyzes the oxidative degradation of dopamine. The translocation of iron across the blood–brain barrier is mediated by specific transferrin receptors located on the brain microvasculature. In the cells most of the iron binds to ferritin, others formed chelate compounds with phosphate groups of membrane components and some included in iron micronutrient enzymes. Ferritin serves to store iron (Fe^{3+}) in a nontoxic form, to deposit it in a safe form, and to transport it to areas where it is required. By linking the iron from some biochemical reactions, the ferritin limited its ability to stimulate oxidative processes [78, 79]. Iron mediates oxidative damage to cellular components through the one-electron transfer called the Fenton reaction, which leads to production of the unstable hydroxyl radical (OH•) that oxidizes lipid, protein, nucleic acid, and carbohydrate, whichever is proximate [30, 80]. Furthermore, dopaminergic neurons may express transferrin receptors on their cell surface to facilitate the uptake of iron bound to transferrin. If the intracellular iron pool is regulated by receptor-mediated transferrin uptake, then an upregulation of transferrin receptor number may play a role in the pathogenesis of nigral cell damage in Parkinson's disease. Early in the disease process, surviving dopaminergic neurons may increase the number of transferrin receptors in order to meet the increased metabolic demand associated with compensatory changes in dopamine synthesis and turnover.

The uptake of ferrotransferrin by dopaminergic neurons may result in a progressive elevation in the cellular iron load that exceeds the regulatory capacity for increased ferritin expression in the aging brain. Increased iron stimulates the formation of free radicals, and changes in the ratio of Fe^{2+}/Fe^{3+} confirmed the presence of oxidative stress [81, 82].

Antioxidant protection of the brain is provided by superoxide dismutase (SOD), catalase (CAT), and glutathione peroxidase (GSH-Px). Glutathione peroxidase is one of the most potent enzymes that protects from ROS-mediated toxicity by scavenging H_2O_2 generated during cellular metabolism [83, 84]. GSH-Px is detected exclusively in glial cells of the midbrain. Under normal conditions, there is a balance between the rate of formation and decomposition of H_2O_2, which prevents the occurrence of oxidative stress. It is assumed that the amount of reduced glutathione (GSH) is the limiting factor for the removal of H_2O_2 and other obtained by membrane oxidation phospholipid peroxides. The recovery of reduced glutathione is performed by reduction of oxidized glutathione (GSSG) by glutathione reductase (GSSG-R) [81]. The level of GSH in substantia nigra is decreased and of GSSG level is high [84, 85]. The depletion of reduced glutathione in the substantia nigra in Parkinson's disease could be the result of neuronal loss. A decrease in the availability of reduced glutathione would impair the capacity of neurons to detoxify hydrogen peroxide and increase the risk of free radical formation and lipid peroxidation. Indeed, the nigra contains increased levels of malondialdehyde and hydroperoxides. An increase in the activity of mitochondrial SOD in the substantia nigra in Parkinson's disease may indicate a compensatory mechanism to nullify the augmented oxidative stress which is not observed in other tissues [81, 84].

2.3 Huntington's Disease and Oxidative Stress

Huntington's disease (HD) or Huntington chorea is an autosomal-dominant inherited progressive neurodegenerative disorder, characterized by brief involuntary movements (chorea) and progressive deterioration of higher neural functions including cognition along with the development of psychological symptoms [86]. Although the disease has the potential to present itself at any time from childhood to old age, it usually begins between 35 and 50 years of age and progress relentlessly. Symptoms include weakening of mental abilities leading to a change in personality such as depression and suicidal tendencies. Loss of psychomotor functions due to lack of muscle coordination and abnormal jerky involuntary movements, violent behavior, and development of dementia are rare symptoms of HD [87, 88].

HD is caused by an abnormal expansion of normal CAG trinucleotide polyglutamine repeats (polyQ repeats) on the N-terminus of the IT 15 (*HTT*) gene, as it codes for the protein huntingtin (Htt) and is located near the tip of chromosome 4 (4p16.3), close to the telomere [20]. Normally, Htt is abundantly expressed in the brain and is tested with moderate expression observed in other organs such as the liver, heart, and lungs. Although the complete function of Htt is yet to be discovered, it has been observed to be involved in embryogenesis and development (cytoskeletal anchoring and transport of mitochondria along with vesicle trafficking to mediate endocytosis). Htt is also important during the postembryonic development of some brain regions and in the survival of neurons. The active participation of Htt in brain development and maintenance therefore illustrates the importance of Htt in the CNS [89, 90]. In HD patient, the abnormally amplified CAG repeats in the *HTT* gene lead to transcription of the mutant huntingtin. In a normal person, the number of CAG repeats is approximately 10–30, whereas in patients with HD have 38–120 repeats. The intensity of the disease progression and earlier onset is directly related to the number of these CAG repeats. The CAG repeat may result in aggregation of proteins within the striatal neurons which further gets broken up into long and short pieces. When short toxic pieces misfold and clump together and accumulate in the nerve cells, they disrupt cell function in a number of ways which appear to be the main cause of neurodegeneration in the disease [20, 91]. A wide range of possible mechanisms for neurotoxicity have been suggested, including inhibition of proteasome activity, caspase activation, dysregulation of transcriptional pathways, and increased production of ROS [92–95].

The areas affected in HD are mainly in the corpus striatum (caudate nucleus and putamen) and cortex (frontal and temporal cortices). However, many other nuclei including the globus pallidus, thalamus, hypothalamus, subthalamic nucleus, substantia nigra, and cerebellum also are affected [88, 94, 96, 97]. The neuropathological hallmark of HD is the degeneration of the caudate nuclei of the basal ganglia situated in the lateral ventricle brain, while it is less prominent in the putamen. As the disease progresses, there is an increase in neuronal loss from caudate, and they are partly replaced by both astrocytes and microglia (called gliosis) in the gray matter of the caudate. In early stages, no significant gliosis is observed [88]. In the neostriatum, mutant Htt is found in the cell bodies and synaptic processes of surviving neurons and glial cells [92]. The weaker signals from striatum's subthalamic nuclei thus cause reduced initiation and modulation of movement, resulting in the characteristic movements of the disorder [91, 93].

Biphasic changes in DA function may occur in HD, with early increase followed by late decreases. Presynaptic activation of the nigrostriatal DA pathway induces chorea, while loss of DA inputs induces akinesia [98]. Overactivity in the nigrostriatal region might arise from a deficiency in GABA. Normally, GABA inhibits DA release by activating GABA receptors on nigrostriatal somata and terminals [99].

2.3.1 Role of Free Radicals and Oxidative Stress in Huntington's Disease

The neurodegeneration initially caused by mutant Htt may be further exacerbated by free radical production followed by oxidative stress. Like in AD and PD, disruption of iron regulation also plays a key role in the etiology of HD. The most frequent is the toxic role of iron linked with the catalytic production of OH• from H_2O_2 by Fenton reaction and causing damage to various biomolecules [3, 31, 32, 100].

Expression of mutant huntingtin results in early impairment of mitochondrial function and axonal transport. Glutamate binds a variety of metabotropic and ionotropic receptors. Among these receptors, NMDA receptor plays a key role in mediating neuronal toxicity. Overactivity of the NMDA-type glutamate receptors (NMDA) causes excitotoxicity, especially neuronal dysfunction and death particularly in the striatum region of brain which is an early effect of mutant Htt. Excitotoxicity stands for neuronal death due to prolonged exposure to the neurotransmitter glutamate. This leads to overactivation of glutamate receptors with depolarization of neuronal membrane, Ca^{2+} influx, and mitochondrial energy depletion. Mutant Htt destabilizes the mitochondrial outer membrane, which increases the permeability of mitochondria to calcium ions and other apoptotic stimuli [101, 102]. Studies on transgenic mouse model R6/2 suggest complex IV deficiency in mitochondria and elevated nitric oxide and superoxide anion radical generation followed by neuronal death which contributes to pathogenesis of HD [103]. It has been observed that nuclear localization and proteolysis of mutant Htt occur in early stage, whereas the transcriptional dysregulation and ubiquitinated aggregates of mutant Htt appear as later stage of HD [102].

2.4 Amyotrophic Lateral Sclerosis

Amyotrophic lateral sclerosis (ALS) or Lou Gehrig's disease is a devastating neurodegenerative disorder characterized by death of motor neurons leading to muscle wasting, dyspnea (difficulty in breathing), dysphagia (difficulty in swallowing), and dysarthria (difficulty in speaking), and death occurs usually within 2–3 years of symptom onset [104]. The causes of ALS are not completely understood as the neurodegenerative processes involved in disease progression are diverse and complex. There is substantial evidence implicating oxidative stress as a central mechanism by which motor neuron death occurs.

It includes elevated markers of oxidative damage in ALS patient's spinal cord and cerebrospinal fluid and mutations in the antioxidant enzyme Cu/Zn superoxide dismutase (SOD1) causing approximately 20 % of familial ALS cases. However, the precise mechanism by which mutant SOD1 leads to motor neuron degeneration has not been defined with certainty in literature [105].

2.4.1 Role of Free Radicals and Oxidative Stress in Amyotrophic Lateral Sclerosis

SOD1 is a powerful antioxidant enzyme that protects the body from damage caused by SAR generated in the mitochondria. Mutations of the SOD gene may reduce its superoxide dismutase activity, thereby elevating free radical levels that cause damage to both mitochondrial and nuclear DNA as well as proteins within cells. In addition, the mutant SOD protein may function as a peroxidase to oxidize cellular components, and it may also react with peroxynitrite to form nitrate proteins [38]. However, other reports documented that the levels of hydrogen peroxide and the hydroxyl radical are significantly higher, and the level of the SAR is significantly lower in ALS implying that the mutant enzyme detoxifies $O_2^{\bullet-}$ into H_2O_2 to further trigger formation of $OH^{\bullet}$ by Fenton's reaction. Furthermore, free radicals trigger oxidative damage to proteins, membrane lipids, and DNA, followed by destroying neurons. The selective degeneration of motor neurons in ALS may be caused by the high level of SOD1 present in large number of glutamatergic synapses projecting to these motor neurons [39].

More than 110 different mutations in SOD1 have been linked with ALS. The primary cellular sites where SOD1 mutations act are located in astrocytes which exert toxic effects on the motor neurons. Autopsy and laboratory studies in ALS indicate that oxidative stress plays a major role in motor neuron degeneration and astrocyte dysfunction. As the disease progresses, nutritional deficiency, wasting syndrome (cachexia), and psychological stress respiratory failure may further increase oxidative stress [106].

3 Conclusion

The maintenance of redox balance within cells is a primary component of homeostasis underlying neuronal survival. Any process that leads to a disruption of the redox balance can extremely interfere with several biochemical processes and result in neuronal deficits and dysfunction. The brain is more vulnerable to oxidative stress due to its high oxygen consumption, high content of polyunsaturated fatty acids, and low levels of antioxidant enzymes than other organs of the body. Despite that neuronal cells are endowed with a range of protective mechanisms, when additional oxidative load overwhelms, a failure of protective mechanisms may allow endogenous oxidative processes to damage cells and result in the pathophysiology of neurodegenerative disorders. There is growing evidence supporting increased oxidative stress in neurodegenerative disorders in addition to environmental, genetic, and immunological factors. However, the exact molecular mechanisms are yet to be determined. The evidence to date for oxidative stress in PD, AD, ALS, HD, and other neurodegenerative diseases is more convincing. Increase in energy metabolism by aerobic pathways increases the intracellular concentration of free oxygen radicals, which in turn enhance the rate of the autocatalytic process of lipid peroxidation, inducing damage to brain structures, especially when physiological defenses become insufficient or depleted. In AD, amyloid plaques are deposited by binding of transition metal ions such as Fe^{3+}, Cu^{2+}, and Zn^{2+} with β-amyloid peptide. This produces H_2O_2 catalytically in the presence of transition metals and finally gives toxic $OH^{\bullet}$ radicals. PD is characterized by deposition of inclusion bodies (Lewy bodies) of α-synuclein in substantia nigra that is ubiquitously expressed in brain. Dopamine is a neurotransmitter as well as very potent metal chelator and electron donor that set in vivo conditions for redox metal chemistry to generate toxic free radicals. It has high tendency to coordinate with Cu^{2+} and Fe^{3+} and reducing metals to initiate Fenton's chemistry to generate H_2O_2. HD is characterized by involuntary movements and progressive deterioration of cognition along with the development of psychological symptoms. It is caused by an abnormal expansion of normal CAG trinucleotide polyglutamine repeats in *Htt* gene and formation of mutant huntingtin. The neurodegeneration is initially caused by mutant Htt that may be further exacerbated by free radical production followed by oxidative stress and excitotoxicity caused by glutamate. In ALS, lower motor neurons from spinal cord and cerebral cortex are lost due to deposition of a misfolded protein in neuronal tissue in relation with toxic gain of function by mutated SOD1 that leads to conversion of SOD itself in prooxidant protein that participates in ROS generation.

References

1. Thompson LM (2008) Neurodegeneration: a question of balance. Nature 452(7188):707–708
2. Barnham K, Masters C, Bush A (2004) Neurodegenerative diseases and oxidative stress. Nature 3:205–214
3. Halliwell B (2006) Oxidative stress and neurodegeneration: where are we now? J Neurochem 97:1634–1658
4. Ailton M, Larissa M, Rute MFL et al (2010) Oxidative stress in neurodegenerative diseases: mechanisms and therapeutic perspectives. Free Radic Biol Med 48(5):629–641
5. Orr HT (2009) Unstable nucleotide repeat minireview series: a molecular biography of unstable repeat disorders. J Biol Chem 284(12):7405
6. Zoghbi HY, Orr HT (2009) Pathogenic mechanisms of a polyglutamine-mediated neurodegenerative disease, spinocerebellar ataxia type 1. J Biol Chem 284(12):7425–7429
7. Sigleton AB, Farrer M, Johnson J et al (2003) Alpha-synuclein locus triplication causes Parkinson's disease. Science 302:841
8. Hashimoto M, Rockenstein E, Crews L et al (2003) Role of protein aggregation in mitochondrial dysfunction and neurodegeneration in Alzheimer's and Parkinson's diseases. Neuromolecular Med 4(1–2):21–36
9. Yan SD, Chen X, Schmidt AM et al (1994) Glycated tau protein in Alzheimer disease: a mechanism for induction of oxidant stress. Proc Natl Acad Sci U S A 91:7787–7791
10. Houlden H, Baker M, Morris HR et al (2001) Corticobasal degeneration and progressive supranuclear palsy share a common tau haplotype. Neurology 56:1702–1706
11. Goedert M, Jakes R (2005) Mutations causing neurodegenerative tauopathies. Biochim Biophys Acta 1739:240–250
12. Tiraboschi P, Hansen LA, Thal LJ et al (2004) The importance of neuritic plaques and tangles to the development and evolution of AD. Neurology 62(11):1984–1989
13. Boveris A, Navarro A (2008) Brain mitochondrial dysfunction in aging. Life 60(5):308–314
14. Lin MT, Beal MF (2006) Mitochondrial dysfunction and oxidative stress in neurodegenerative diseases. Nature 443(7113):787–795
15. Varkey J, Isas JM, Mizuno N et al (2010) Membrane curvature induction and tubulation are common features of synucleins and apolipoproteins. J Biol Chem 285(42):32486–32493
16. De Vos KJ, Grierson AJ, Ackerley S et al (2008) Role of axonal transport in neurodegenerative diseases. Ann Rev Neurosci 31:151–173
17. Nijholt DAT, De Kimpe L, Elfrink HL et al (2010) Removing protein aggregates: the role of proteolysis in neurodegeneration. Curr Med Chem 18:2459–2476
18. Rubinsztein DC (2006) The roles of intracellular protein-degradation pathways in neurodegeneration. Nature 443(7113):780–786
19. Duda JE, Giasson BI, Mabon ME et al (2002) Concurrence of alpha-synuclein and tau brain pathology in the Contursi kindred. Acta Neuropathol 104:7–11
20. Penney JB Jr, Vonsattel JP, MacDonald ME et al (1997) CAG repeat number governs the development rate of pathology in Huntington's disease. Annals Neurol 41(5):689–692
21. Bruijn L, Houseweart M, Kato S et al (1998) Aggregation and motor neuron toxicity of an ALS-linked SOD1 mutant independent from wild-type SOD1. Science 281(5384):1851–1854
22. Spillantini MG, Crowther RA, Jakes R et al (1998) Alpha-Synuclein in filamentous inclusions of Lewy bodies from Parkinson's disease and dementia with Lewy bodies. Proc Natl Acad Sci U S A 95:6469–6473
23. Tu PH, Galvin JE, Baba M et al (1998) Glial cytoplasmic inclusions in white matter oligodendrocytes of multiple system atrophy brains contain insoluble alpha-synuclein. Ann Neurol 44:415–422
24. Laurén J, Gimbel DA, Nygaard HB et al (2009) Cellular prion protein mediates impairment of synaptic plasticity by amyloid-beta oligomers. Nature 457(7233):1128–1132
25. Ravikumar B, Duden R, Rubinsztein DC (2002) Aggregate-prone proteins with polyglutamine and polyalanine expansions is degraded by autophagy. Hum Mol Genet 11(9):1107–1117
26. Ghavamia S, Shojaeid S, Yeganehb B et al (2014) Autophagy and apoptosis dysfunction in neurodegenerative disorders. Prog Neurobiol 112:24–49
27. Hashemi M, Ghavami S, Eshraghi M et al (2007) Cytotoxic effects of intra and extracellular zinc chelation on human breast cancer cells. Eur J Pharmacol 557:9–19
28. Fuchs Y, Steller H (2011) Programmed cell death in animal development and disease. Cell 147:742–758

29. Lee EW, Seo J, Jeong M et al (2012) The roles of FADD in extrinsic apoptosis and necroptosis. BMB Rep 45:496–508
30. Navarro A, Boveris A, Bández MJ et al (2009) Human brain cortex: mitochondrial oxidative damage and adaptive response in Parkinson's disease and in dementia with Lewy bodies. Free Radic Biol Med 46:1574–1580
31. Kohen R, Nyska A (2002) Oxidation of biological system: oxidative stress phenomena, antioxidants, redox reaction, and methods for their quantification. Toxicol Pathol 33(6):620–650
32. Sies H (1991) Role of reactive oxygen species in biological processes. Wien Klin Wochenschr 69:965–968
33. Halliwell B, Gutteridge C (2006) Free radicals in biology and medicine. 4 edn, Oxford University Press, UK
34. Sies H (1985) Oxidative stress. Academic, New York
35. Farooqui T, Farooqui A (2011) Lipid-mediated oxidative stress and inflammation in the pathogenesis of Parkinson's disease. Park Dis 2011:247467
36. Nikolova G (2012) Oxidative stress and Parkinson disease. Trakia J Sci 10(1):92–101
37. Smith MA, Perry G, Richey PL et al (1996) Oxidative damage in Alzheimer's. Nature 382:120–121
38. Liu D (1996) The roles of free radicals in amyotrophic lateral sclerosis. J Mol Neurosci 7(3):159–167
39. Barber SC, Shaw PJ (2009) Oxidative stress in ALS: key role in motor neuron injury and therapeutic target. Curr Neuropharmacol 7(1):65–74
40. Wenk GL (2003) Neuropathologic changes in Alzheimer's disease. J Clin Psychiatry 64(9):7–10
41. Hensley K, Carney J, Mattson M et al (1994) A model for b-amyloid aggregation and neurotoxicity based on free radical generation by the peptide: relevance to Alzheimer's disease. Proc Natl Acad Sci U S A 91:3270–3274
42. Younkin SM, Tanzi RE, Christen Y (eds) (1998) Presenilins and Alzheimer's disease. Springer-Verlag, Heidelberg
43. Christen Y (2000) Oxidative stress and Alzheimer's disease. Am J Clin Nutr 71:621S–629S
44. Ramassamy C, Krzywokowski P, Bastianetto S et al (1998) Apolipoprotein E, oxidative stress and EGb 761 in Alzheimer's disease brain. In: Packer L, Christen Y (eds) Ginkgo biloba extract (EGb 761) study: lesson from cell biology. Elsevier, Paris, pp 69–83
45. Leininger-Muller B, Jolivalt C, Bertrand P et al (1998) Oxidation of human apolipoprotein E: isoforms susceptibility and protection with Ginkgo biloba EGb 761 extract. In: Packer L, Christen Y (eds) Ginkgo biloba extract (EGb 761) study: lessons from cell biology. Elsevier, Paris, pp 57–68
46. Wyss-Coray T, Masliah E, Mallory M et al (1997) Amyloidogenic role of cytokine TGF-b1 in transgenic mice and in Alzheimer's disease. Nature 389:603–606
47. Patterson P, Kordon C, Christen Y (eds) (1999) Neuro-immune interactions and neurological and psychiatric disorders. Springer-Verlag, Heidelberg
48. Smith MA, Richey Harris PL et al (1997) Widespread peroxynitrite-mediated damage in Alzheimer's disease. J Neurosci 17:2653–2657
49. Vitek MP, Bhattacharya K, Glendening JM et al (1994) Advanced glycation end products contribute to amyloidosis in Alzheimer disease. Proc Natl Acad Sci U S A 91:4766–4770
50. Smith MA, Kutty RK, Richey PL et al (1994) Heme oxygenase-1 is associated with the neurofibrillary pathology of Alzheimer's disease. Am J Pathol 145:42–47
51. Markesbery WR (1997) Oxidative stress hypothesis in Alzheimer's disease. Free Radic Biol Med 23:134–147
52. Dugan LL, Sensi SL, Canzoniero LMT et al (1995) Mitochondrial production of reactive oxygen species in cortical neurons following exposure to N-methyl-D-aspartate. J Neurosci 15:6377–6388
53. Dowson VL, Kizushi VM, Huang PL et al (1996) Resistance to neurotoxicity in cortical cultures from neuronal nitric oxide synthase deficient mice. J Neurosci 16:2463–2478
54. Montine TJ, Markesbery WR, Morrow JD et al (1998) Cerebrospinal fluid F2-isoprostane levels are increased in Alzheimer's disease. Ann Neurol 44:410–413
55. Pogocki D (2003) Alzheimer's beta-amyloid peptide as a source of neurotoxic free radicals: the role of structural effects. Acta Neurobiol Exp 63:131–145
56. Meda L, Cassatella MA, Szendrei GI et al (1995) Activation of microglial cells by b-amyloid protein and interferon gamma. Nature 374:647–650
57. Deibel MA, Ehmann WD, Markesbery WR (1997) Copper, iron, and zinc imbalances in severely degenerated brain regions in Alzheimer's disease: possible relation to oxidative stress. J Neurol Sci 143:137–142
58. Crapper DR, Quittkat S, Krishnan SS et al (1980) Intranuclear aluminum content in Alzheimer's disease, dialysis encephalopathy, and experimental aluminum encephalopathy. Acta Neuropathol (Berl) 50:19–24
59. Kennard ML, Feldman H, Yamada T et al (1996) Serum levels of the iron binding protein p97 is elevated in Alzheimer's disease. Nat Med 2:1230–1235
60. Bjertness E, Candy JM, Torvik A et al (1996) Content of brain aluminum is not elevated in Alzheimer disease. Alzheimer Dis Assoc Disord 10:171–174
61. Multhaup G, Masters CL, Veyreuther K (1998) Oxidative stress in Alzheimer's disease. Alzheimer Rep 1:147–154
62. Linder MC, Hazegh-Azam M (1996) Copper biochemistry and molecular biology. Am J Clin Nutr 63:797S–811S

63. Bush AI, Pettingell WH, Multhaup G et al (1994) Rapid induction of Alzheimer Aβ amyloid formation by zinc. Science 265:1464–1467
64. Choi DW, Koh JY (1998) Zinc and brain injury. Annu Rev Neurosci 21:347–375
65. Szewczyk B (2013) Zinc homeostasis and neurodegenerative disorders. Free Radic Biol Med 21(65C):509–527
66. Jellinger K (1986) An overview of morphological changes in Parkinson's disease. Adv Neurol 45:1–16
67. Zecca L (2003) The neuromelanin of human substantia nigra: structure, synthesis and molecular behavior. J Neural 65(1):145–155
68. Galvin JE, Uryu K, Lee VM et al (1999) Axon pathology in Parkinson's disease and Lewy body dementia hippocampus contains alpha-, beta-, and gamma-synuclein. Proc Natl Acad Sci U S A 96:13450–13455
69. Marsden D (1983) Neuromelanin and Parkinson's disease. J Neural 19(1):121–141
70. Cadet L, Brannock C (1998) Free radicals and the pathobiology of brain dopamine systems. Neurochem Int 32:117–131
71. Dauer W, Przedborski S (2003) Parkinson's disease: mechanisms and models. Neurology 39(6):889–909
72. Jenner P, Olanow W (1996) Oxidative stress and the pathogenesis of Parkinson's disease. Neurology 47(3):161–170
73. Faucheux A, Martin M, Beaumont C et al (2003) Neuromelanin associated redox-active iron is increased in the substantia nigra of patients with Parkinson's disease. J Neurochem 86:1142–1148
74. Asanuma M, Miyazaki I, Ogawa N (2003) Dopamine- or L-DOPA-induced neurotoxicity: the role of dopamine quinone formation and tyrosinase in a model of Parkinson's disease. Neurotox Res 5:165–176
75. Habibi E, Masoudi-Nejad A, Abdolmaleky HM et al (2011) Emerging roles of epigenetic mechanisms in Parkinson's disease. Funct Integr Genomics 11(4):523–537
76. Mash C, Singer J, Pablo J et al (1993) Iron storage and transport markers in Parkinson's disease and MPTP treated mice. Iron Cent Nerv Syst Disord :103–116
77. Wang C, Cai Y, Gu Z et al (2014) Clinical profiles of Parkinson's disease associated with common leucine-rich repeat kinase 2 and glucocerebrosidase genetic variants in Chinese individuals. Neurobiol Aging 35(3):725
78. Dexter T, Wells R, Lees J et al (1989) Increased nigra iron content and alterations in other metal ions occurring in brain in Parkinson's disease. J Neurochem 52:1830–1836
79. Riederer P, Sofic E, Rauseh D et al (1989) Transition metals, ferritin, glutathione, and ascorbic acid in Parkinsonian brain. J Neurochem 52:515–520
80. Cohen G (1999) Oxidative stress and Parkinson's disease. In: Gilbert DL, Colton CA (eds) Reactive oxygen species. Biological systems: an interdisciplinary approach, pp 593–608
81. Repetto MG, Boveris A (2011) Transition metals: bioinorganic and redox reactions in biological systems. In: Nova Science Publishers Inc (ed) Transition metals: uses and characteristics. New York
82. Dalle-Donne I, Scaloni Giustarini A, Cavarra D et al (2005) Proteins as biomarkers of oxidative stress in diseases: the contribution of redox proteomics. Mass Spectrom Rev 24:55–99
83. Saggu H, Cooksey J, Dexter D (1989) A selective increase in particulate superoxide dismutase activity in Parkinsonian-substantia nigra. J Neurochem 53:692–697
84. Sofic E, Lange W, Jellinger K, Riederer P (1992) Reduced and oxidized glutathione in the substantia nigra of patients with Parkinson's disease. Neurosci Lett 142:128–130
85. Perry L, Godin V, Hansen S (1982) Parkinson's disease: a disorder due to nigral glutathione deficiency? Neurosci Lett 33:305–310
86. Cattaneo E, Zuccato C, Tartari M (2005) Normal huntingtin function: an alternative approach to Huntington's disease. Nat Rev Neurosci 6:919–930
87. Walker FO (2007) Huntington's disease. Lancet 369(9557):218–228
88. Milnerwood J, Raymond LA (2010) Early synaptic pathophysiology in neurodegeneration: insights from Huntington's disease. Trends Neurosci 33(11):513–523
89. Reiner A, Dragatsis I, Zeitlin S, Goldowitz D (2003) Wild-type huntingtin plays a role in brain development and neuronal survival. Mol Neurobiol 28(3):259–275
90. Reiner A, Del Mar N, Meade CA et al (2001) Neurons lacking huntingtin differentially colonize brain and survive in chimeric mice. J Neurosci 21(19):7608–7619
91. Frederick CN, Masayuki S, Matthew FP et al (2001) Interference by huntingtin and atrophin-1 with CBP-mediated transcription leading to cellular toxicity. Science 291(5512):2423–2428
92. Rubinsztein DC, Carmichael J (2003) Huntington's disease: molecular basis of neurodegeneration. Expert Rev Mol Med 5(20):1–21
93. Pavese N, Gerhard A, Tai YF et al (2006) Microglial activation correlates with severity in Huntington disease: a clinical and PET study. Neurology 66(11):1638–1643
94. DiFiglia M, Sapp E, Chase KO et al (1997) Aggregation of huntingtin in neuronal intranuclear inclusions and dystrophic neurites in brain. Science 277(5334):1990–1993
95. Borlongan CV, Kanning K, Poulos SG et al (1996) Free radical damage and oxidative stress in Huntington's disease. J Fla Med Assoc 83(5):335–341
96. Kremer HP, Roos RA, Dingjan G et al (1990) Atrophy of the hypothalamic lateral tuberal nucleus in Huntington's disease. J Neuropathol Exp Neurol 49:371–382

97. Heinsen H, Rub U, Gangnus D, Jungkunz G et al (1996) Nerve cell loss in the thalamic centromedian-parafascicular complex in patients with Huntington's disease. Acta Neuropathol 91:161–168
98. Spokes EGS (1981) The neurochemistry of Huntington's chorea. TINS 4:115–118
99. Paladini CA, Tepper JM (1999) GABAA and GABAB antagonists differentially affect the firing pattern of substantia nigra dopaminergic neurons in vivo. Synapse 32:165–176
100. Jomova K, Valko M (2011) Importance of iron chelation in free radical-induced oxidative stress and human disease. Curr Pharm Des 17(31): 3460–3473
101. Di Figlia M (1990) Excitotoxic injury of the neostriatum: a model for Huntington's disease. Trends Neurosci 13:286–289
102. Cowan CM, Raymond LA (2006) Selective neuronal degeneration in Huntington's disease. Curr Top Dev Biol 75:25–71
103. Tabrizi SJ, Workman J, Hart PE et al (2000) Mitochondrial dysfunction and free radical damage in the Huntington R6/2 transgenic mouse. Ann Neurol 47(1):80–86
104. Boillée S, Vande Velde C, Cleveland D (2006) ALS: a disease of motor neurons and their nonneuronal neighbors. Neuron 52(1):39–59
105. Carri MT, Cozzolino M (2011) SOD1 and mitochondria in ALS: a dangerous liaison. J Bioenerg Biomembr 43:593–599
106. D'Amico E, Factor-Litvak P, Santella RM et al (2010) Clinical perspective on oxidative stress in sporadic amyotrophic lateral sclerosis. Indian J Psychiatry 52(1):21–27

Thyroid Gland in Free Radical-Induced Oxidative Stress

Mohan K. Kale

Abstract

This oxidative stress has been implicated in a variety of pathological conditions such as diabetes mellitus, inflammation, cancer, ageing, ischemia, atherosclerosis, liver damage, etc. The present study revealed that all the stressors invariably increase oxidative stress in all the tissues as evident from the markers of oxidative stress, i.e., LPO, SOD, CAT, and GSH, in various tissues like blood, muscle, and liver. It was also found that chemical stress produces maximum oxidative stress as compared to physiological and psychological stress. Similarly, changes in the markers of oxidative stress in blood parallel with changes in muscle and liver. In all the stressed conditions, there was an increase in T_3 and T_4 and decrease in TSH. There was a concurrent increase in LPO and decrease in the SOD and CAT activity and reduction in the reduced glutathione content in blood. The data on oxidative stress and blood levels of thyroid hormones T3 and T4 condition exhibited a linear correlation. The changes in thyroid hormone levels correlate with the parameters of oxidative stress. Hence, it can be contemplated that thyroid hormones may play a pivotal role in the induction of oxidative stress in stress-exposed subjects.

Keywords

Stress • Free radicals • Thyroid gland • Oxidative stress • Thyroid function • Antioxidants

1 Thyroid Gland in Free Radical-Induced Oxidative Stress

Generalized oxidative stress is a common consequence of most stressful conditions, and prolonged oxidative stress is known to cause

M.K. Kale (✉)
Department of Pharmacology, Konkan Gyanpeeth Rahul Dharkar College of Pharmacy and Research Institute, Dist. Raigarh, 410201 Karjat, India
e-mail: kalemkpharm@gmail.com

V. Rani and U.C.S. Yadav (eds.), *Free Radicals in Human Health and Disease*,
DOI 10.1007/978-81-322-2035-0_12, © Springer India 2015

degenerative changes in the body leading to a pathological state. Despite the fact that many research articles have been written about stress, stress-related diseases, oxidative stress, etc., very little work has been done indicating the role of thyroid in induction of oxidative stress.

1.1 Stress

Stress is a force that disrupts equilibrium and affects homeostasis. Therefore, it is also defined as a state of threatened homeostasis. Stressor may be viewed as a stimulus for a classical stress response particularly when a person cannot handle stress. Prolonged and repeated exposure to stress had been shown to produce a variety of pathological conditions ranging from diabetes to cardiovascular diseases. During stress, an adaptive compensatory specific response of the organism is activated to sustain homeostasis. The adaptive response reflects the activation of specific neuronal circuits. It is genetically programmed and constantly modified by environmental factors. The neuropeptide – CRH – is a central regulator of the hypothalamic-pituitary-adrenal (HPA) stress response and is implicated in various stress-related conditions [1, 2].

The commonly identified stress conditions can be classified as follows:

1. Psychological stress due to anxiety, fear, frustration, restrain, isolation, etc.
2. Physiological stress or metabolic stress due to abnormal metabolism
3. Environmental stress due to pollution, radiation, electromagnetic field, etc.
4. Physical stress due to cold, heat, intense radiation, noise, vibration etc.
5. Chemical stress due to poisons, chemicals, drugs, medicines, etc.

The literature has demonstrated that most of the above stressors and subsequent stress generate free radicals [3–6] and produce oxidative stress, the underlying cause of degenerative type of pathological conditions.

1.2 Oxidative Stress

Oxidative stress arises when highly reactive free radicals produce oxidative damages to the macromolecular structures of the cell. When generation of free radicals and other reactive oxygen species overwhelms the endogenous antioxidant defense of the body, then such condition is called as oxidative stress. Oxidative stress is a general term used to describe the steady-state level of oxidative damage in a cell, tissue, or organ, caused by the reactive oxygen species (ROS). This damage can affect a specific molecule or the entire organism. ROS, such as free radicals and peroxides, represent a class of molecules that are derived from the metabolism of oxygen and exist inherently in all aerobic organisms. Oxidative stress is imposed on cells as a result of one of three factors such as an increase in oxidant generation, a decrease in antioxidant protection, and a failure to repair oxidative damage. Cell damage is induced by ROS. ROS are either free radicals, reactive anions containing oxygen atoms, or molecules containing oxygen atoms that can either produce free radicals or are chemically activated by them. The main source of ROS in vivo is aerobic respiration, although ROS are also produced by peroxisomal oxidation of fatty acids, microsomal cytochrome P450 metabolism of xenobiotic compounds, stimulation of phagocytosis by pathogens or lipopolysaccharides, arginine metabolism, and tissue-specific enzymes. In normal conditions, ROS are cleared from the cell by the action of superoxide dismutase, catalase, or glutathione peroxidase [7–9].

1.3 Free Radicals

Radicals are molecules that are over-reactive because they are unbalanced electronically. Radicals are mainly generated in the mitochondria with a purpose to provide energy to activate the intracellular enzymes that they can act on their substrate. However, mitochondria, if excessively stimulated, generate the radicals in excess.

The additional unpaired electron on radical often makes it highly unstable and reactive, and hence it tries to gain stability by capturing an additional electron from the molecule of the surrounding structures, so that the unpaired electron can become paired. The process of capturing an electron involves reacting with "donor" molecule, which loses an electron and is said to have been "oxidized." This oxidized donor molecule then has a capacity to oxidize other molecules, which sets up a chain reaction that potentially leads to damaging surrounding tissues. The unutilized radical is scavenged by intracellularly available antioxidants. However, in the event of overwhelming generation of radicals, the antioxidants fall short and the unscavenged radicals remain free. Such radical is called as "free radical."

A free radical can be defined as a chemical species possessing an unpaired electron. It can also be considered as a fragment of a molecule. As such, free radical can be formed in three ways:

1. By homolytic cleavage of a covalent bond of a normal molecule, with each fragment retaining one of the paired electrons
2. By the loss of a single electron from a normal molecule
3. By the addition of single electron from a normal molecule

The latter, electron transfer, is a far more common process in biological system than is homolytic fission, which generally requires high-energy input from either high temperature, UV light, or ionizing radiation. Heterolytic fission, in which the electrons of the covalent bond are retained by only one of the fragments of the parent molecule, does not result in free radical but in ions, which are changed. Free radical can be positively charged, negatively charged, or electrically neutral. The unpaired electron and the radical nature of a species are conventionally indicated by writing it with a heavy superscript dot.

Since free radical generation is the outcome of imbalanced metabolism, and thyroid gland secretions are known for activating the basal metabolic rate, it was contemplated that the same may be playing a central role in the induction of generalized oxidative stress.

2 Thyroid Gland

The thyroid gland consists of two lobes that lie on each side of the trachea, just below the Adam's apple. It is the largest and most sensitive endocrine glands in the body (Fig. 1). It produces the thyroid hormones thyroxine (T_4) and triiodothyronine (T_3). Both hormones are classified as biogenic amines and are derived from the amino acid tyrosine. The thyroid gland is the only gland, which stores its secretory product in large quantities, normally about a 10-day supply. The synthesis of the thyroid hormones requires the amino acid tyrosine and the trace mineral iodine. T_3 and T_4 are synthesized by attaching iodines to the amino acid tyrosine, stored for some period of time and then secreted into the blood. Thyroid-stimulating hormone (TSH) stimulates most of the steps. The thyroid secretes about ten times as much T_4 as T_3; however, T_3 is roughly two to three times more potent. Thyroxine is converted into the more active triiodothyronine with the selenium-dependent enzyme 5′-deiodinase. T3 and T4 are lipid soluble and combine with special transport proteins upon release into the blood serum, called thyroxine-binding globulins (TBG). Less than one percent of thyroid hormones travel unattached in their free state.

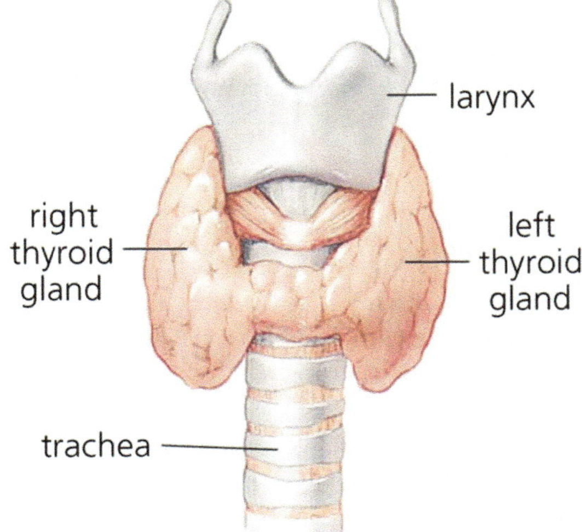

Fig. 1 Thyroid gland

2.1 Oxidative Stress and Thyroid Gland

Oxidative stress has been thought to contribute to the general decline in cellular functions that are associated with many human diseases. Increased cellular level of ROS due to oxidative stress can result in an increased steady-state level of oxidative DNA damage. The oxidative stress has been implicated in a variety of pathological conditions such as diabetes mellitus, inflammation, cancer, ageing, ischemia, atherosclerosis, liver damage, etc. [10, 11].

The thyroid gland is the body's primary regulator of metabolism. Because it controls the body's metabolic rate and the rate at which energy is produced, imbalances of thyroid hormones can have a profound effect on an individual's energy levels. Thyroid hormones accelerate cellular reactions and increase oxidative metabolism. By stimulating enzymes that control active transport pumps, demand for cellular oxygen increases, and as ATP production goes up, heat is produced. This creates a thermoregulatory effect, which increases body temperature. Basal metabolic rate (BMR) is directly influenced by thyroid hormone chemistry. Thyroid hormones can target, influence, and alter the metabolism of virtually every cell in the body. Thyroid hormones are among the most important humoral factors involved in setting the basal metabolic rate on a long-term basis in target tissues such as liver, heart, kidney, and brain [12]. Oxygen free radicals can develop during several steps of normal metabolic events. Although free radicals have the potential to damage the organism, their generation is inevitable for some metabolic processes. The main endogenous sources of free radicals are the mitochondrial electron transport chain, the microsomal membrane electron transport chain, reactions of oxidant enzymes, and auto-oxidation reactions [12, 13]. Both hydrogen peroxide and superoxide anion produce highly reactive hydroxyl radical (OH) by the Haber-Weiss reaction. The hydroxyl radical can initiate lipid peroxidation, which is a free radical chain reaction leading to damage of membrane structure and function [14, 15].

Thyroid hormones are key regulators of growth, development, and metabolism. The development of a hyperthyroid state in vertebrates leads to an enhancement in their basal metabolic rate (BMR) due to an increase in the rate of O_2 consumption in most tissues, excluding the spleen, testis, and adult brain [16]. Current available data indicate that thyroid calorigenesis is achieved by both:

1. A short-term signaling mechanism mediated by 3,5-diiodothyronine (3,5-T_2) and 3,3,5-triiodothyronine (T_3) with the allosteric activation of cytochrome c oxidase ([17, 18]
2. A long-term pathway inducing nuclear and mitochondrial gene transcription through T_3 signaling [19, 20], thus stimulating basal thermogenesis

The T_3-dependent long-term signaling mechanism induces the synthesis of the enzymes involved in energy metabolism and the components of the respiratory chain apparatus, leading to a higher capacity of oxidative phosphorylation [21, 22]. The consequent increase in ATP production is partially balanced by the concomitant diminution in the efficiency of ATP synthesis due to intrinsic uncoupling, afforded by induction of uncoupling proteins (UCPs) by T_3 [23], with the consequent enhancement in mitochondrial O_2 uptake. Although these short- and long-term pathways are mainly responsible for the increased cellular respiration induced by hyperthyroid state, other processes may also play a role, namely:

1. Energy expenditure due to higher active cation transport
2. Loss of energy from futile cycles due to increases in catabolic and anabolic pathways of intermediary metabolism
3. Higher activity of membrane-bound enzymes associated with electron transfer and metabolite carriers due to changes in the lipid composition of mitochondrial membranes [21]
4. O_2 equivalents related to oxidative stress [22, 24], a redox imbalance that leads to various pathophysiological events in the liver [25, 26]

The relation between thyroid calorigenesis and oxidative stress has been studied extensively [22] in line with the significant correlation established for BMR and the lipid peroxidative potential of tissues from several mammalian species [27]. Experimental animals made hyperthyroid by T_3 administration and exhibit a thermogenic response that coincides with increases in the rate of O_2 consumption by the liver [28]. Acceleration of hepatic respiration during thyroid calorigenesis leads to a marked elevation in the rate of superoxide production by liver submitochondrial particles in the presence of NADH (142 %) or succinate (152 %), with higher rates of hydrogen peroxide (H_2O_2) generation, either under basal conditions or in the succinate-supported process, both in the absence and presence of antimycin-A [29, 30]. Enhancement in liver mitochondrial H_2O_2 production also occurs in the transition from hypothyroid to hyperthyroid state as a function of the content of auto-oxidizable electron carriers [31], an effect that is mimicked by cold-induced hyperthyroidism [32]. Development of a hyperthyroid state in rats results in the proliferation of the smooth endoplasmic reticulum, with higher activities of NADPH-cytochrome P_{450} reductase [33] and NADPH oxidase [28]. The latter enzymatic activity represents the oxidase activity of cytochrome P_{450} responsible for O_2 and H_2O_2 production [34], which has been recently related to the induction of the highly prooxidant cytochrome P_{4502E1} isoform by T_3 [35]. These changes, and the increase in the activity of the NADPH-generating system glucose-6-phosphate dehydrogenase [36], are likely to determine high rates of cytochrome P_{450} reduction in hyperthyroid state, thus explaining the increase in:

1. NADPH-dependent antioxidant-sensitive rates of O_2 uptake [37].
2. NADPH-supported O_2 generation [28]. and
3. The biotransformation of a variety of xenobiotics [22]. In addition to the T_3-induced liver mitochondrial and microsomal capacity of reactive oxygen species (ROS) generation, cytosolic enzymatic mechanisms are also increased, namely, the well-known ROS generator xanthine oxidase [38] and the reactive nitrogen species (RNS) producing system nitric oxide synthase (NOS) [39].

These changes are presumed to occur primarily at the parenchymal cell level, with Kupffer cells playing a secondary role, as evidenced by the enhancement in the respiratory burst activity after T_3 administration [40]. The latter process is mainly due to the activity of NADPH oxidase, with a smaller contribution by NOS [41] being the T_3-induced increase in liver NOS activity partially inhibited by the Kupffer cell inactivator gadolinium chloride (GdC13) [39]. T_3-induced liver free radical generation occurs in concomitance with enhanced respiratory burst activity and chemiluminescent response in polymorphonuclear leukocytes, both in experimental and human hyperthyroidism [42]. Thyroid hormone-induced liver free radical activity is associated with a diminution in antioxidant defenses, namely:

1. Reduction in the activity of superoxide dismutase (SOD) and catalase [43], probably due to enzyme inactivation by the ROS/RNS produced
2. Depletion of reduced glutathione (GSH) [38, 43, 44], α-tocopherol, β-carotene, and lycopene [36]

In conclusion, a higher prooxidant activity is developed in the liver as result of the functional interdependence established between thyroid calorigenesis, hepatic respiration, and ROS/RNS generation, which accounts for 16–25 % of the net increase in the total rate of O_2 consumption by T_3 [30], including O_2 equivalents used in the oxidation of hepatic biomolecules. This T_3-induced liver free radical activity is paralleled by a decrease in antioxidant defenses, leading to oxidative stress [24] in liver, as well as in extrahepatic tissues exhibiting a calorigenic response [22].

At the cellular level, oxidative stress leads to a wide spectrum of responses, depending on the cell type, the level of ROS achieved, and the duration of the exposure [45]. Under conditions of thyrotoxicosis, T_3-induced liver oxidative stress triggers different molecular changes associated with either cell dysfunction or adaptive responses to injury. As a consequence of the enhanced oxidative stress imposed on the liver by thyroid calorigenesis, damages to polyunsaturated fatty acids, proteins, and DNA have been evidenced by the increases in biochemical indicators of:

1. Lipid peroxidation [thiobarbituric acid reactants (TBARs)] [28, 31, 46, 47], hydroperoxide formation, and chemiluminescence [28, 37, 48]
2. Protein oxidation [content of protein hydrazone derivatives] [40]
3. DNA oxidation [8-oxo-deoxyguanosine levels] [49], in experimental animals

In man, hyperthyroidism is characterized by significant changes in circulating parameters related to oxidative stress, including:

1. Increased levels of TBARs [50–57] and conjugated dienes [54, 56]
2. Elevated levels of H_2O_2 and lipid hydroperoxides [57]
3. Reduced levels of thiols [54, 58], ascorbic acid, α-tocopherol [52], and coenzyme Q [52]

The above changes are either reduced or normalized by thyrostatic therapy or antioxidant supplementation, correlate with the elevation in urinary TBARs levels [50] and chemiluminescent response [59], and occur concomitantly with a higher susceptibility of erythrocytes to an oxidant challenge [50]. In the liver, T_3-induced oxidative stress exacerbates hepatic injury caused by other harmful stimuli, probably due to potentiation of the prooxidant state and increased Kupffer cell functioning due to macrophage hyperplasia and hypertrophy [22, 40].

Variations of the levels of thyroid hormones can be one of the main physiological modulators of in vivo cellular oxidative stress due to their known effects on mitochondrial respiration [11]. In particular, it has been suggested that the increase in reactive oxygen species induced by thyroid hormones can lead to an oxidative stress condition in the liver and in the heart and some skeletal muscles with a consequent lipid peroxidative response [12, 60].

Both oxygen consumption and free radical production take place in mitochondrial membranes. The sensitivity to oxidative damage of these membranes is strongly dependent on their unsaturated fatty acid content (PUFAs) which are among the more susceptible cellular macromolecules to oxidative stress. At the same time, the lipid environment can directly affect membrane function, including mitochondrial electron transport and possibly oxygen free radical production. Moreover, changes in lipid composition of cellular membranes appear to be well-established features of altered thyroid states in rat tissues [61].

In aerobic cells, active oxygen species, e.g., superoxide and hydrogen peroxide, are generated as by-products of oxidative metabolism in mitochondria. These species are toxic to biomembranes and eventually lead to peroxidation of lipids unless they are removed by free radical scavenging enzymes, such as superoxide dismutase (SOD), glutathione peroxidase (GPx), and catalase. Antioxidant enzymes act to scavenge free radicals by converting them to less harmful molecules [62].

Thyroid hormones stimulate protein synthesis and increase the rate at which triglycerides are broken down (lipolysis). This is why they are sometimes taken by athletes in sports where physical appearance is judged, especially during the final stages of pre-contest dieting. At appropriate levels, these hormones help to preserve muscle and reduce body fat, but when used incorrectly or excessively, they are highly destructive to muscle (catabolic). Low levels of T_3 stimulate the hypothalamus to secrete thyrotropin-releasing hormone (TRH) and the anterior pituitary to secrete TSH. TRH also stimulates the anterior pituitary to secrete TSH. Then TSH stimulates virtually all aspects of thyroid gland activity. The thyroid gland releases T_3 and T_4 until the metabolic rate returns to normal.

During growth, thyroid hormones provide an anabolic influence on protein metabolism. This is due to their influence on insulin secretion. T4 and insulin also connect in the liver, where they mutually affect IGF activity. IGF (insulin growth factors) are powerful muscle-building control agents. In the absence of adequate levels of thyroid hormones, human growth hormone (hGH) also loses its growth-promoting action and is not secreted normally.

Thyroid hormones are associated with the oxidative and antioxidative status of the organism. Thyroid hormones regulate oxidative metabolism and thus play an important role in free radical production [63]. In hypothyroidism, a decrease in free radical production is expected because of the metabolic suppression brought about by the decrement in thyroid hormone levels [63, 64]. However, there are some studies reporting oxidative stress in hypothyroidism [65–67]; paraoxonase (PON1) is an antioxidant enzyme that hydrolyzes lipid peroxides in oxidized lipoproteins [68]. PON1 exerts paraoxonase and arylesterase activities, since the enzyme hydrolyzes both organophosphates (such as paraoxonase) and aromatic esters (such as phenyl acetate). Paraoxonase activity was suggested to be inversely associated with oxidative stress in serum and macrophages [69]. Reduced paraoxonase activities have been reported in several groups of patients with diabetes, hypercholesterolemia, and cardiovascular disease who are under increased oxidative stress [70, 71]. Vitamin E is a fat-soluble vitamin that exists in eight different forms. It is also a powerful biological antioxidant. Vitamin E has also been shown to play a role in immune function, in DNA repair, and in other metabolic processes. Vitamin E is the major fat-soluble free radical chain-breaking antioxidant found in plasma, red blood cell, and tissues [72]. Vitamin E is a widely studied antioxidant molecule and several authors have proposed that vitamin-E supplementation could be useful in disturbances that are associated with oxidative stress [73].

Thyroid hormones play a crucial role in the regulation of mitochondrial oxidative metabolism; the synthesis and degradation of proteins and vitamins, such as vitamin E, vitamin A, and β-carotene; the sensitivity of tissues to catecholamines; and the regulation of antioxidant enzyme levels [61]. Hyperthyroidism and hypothyroidism represent opposite clinical conditions characterized, respectively, by enhanced oxidative metabolism and by reduced oxidative metabolism. The hypermetabolic state that characterizes hyperthyroidism accelerates free radical production in the mitochondria and induces changes in the antioxidant defense system [63, 74]. In contrast, the metabolic suppression brought about by hypothyroidism is associated with a decrease in free radical production, and it has also been suggested that hypothyroidism protects tissues against acceleration of lipid peroxidation [63, 64]. Increasing experimental and epidemiological evidence shows that high oxidative stress status favors oxidative modifications of LDL and plays an important role in the development of atherosclerosis [75, 76]. Nevertheless, hypothyroidism but not hyperthyroidism represents an important risk factor for atherosclerosis and coronary heart disease.

High concentrations of thyroid hormones may change the metabolism of oxygen in the cells and stimulate the production of free radicals [77]. In the course of hyperthyroidism, oxidative stress and the peroxidation of lipids can be generated [47]. Acceleration of the mitochondrial respiration [78] and basal metabolic rate and the energy metabolism of tissues in several mammalian species represent some of the major functions of thyroid hormones [79]. Accumulating evidence has suggested that the hypermetabolic state in hyperthyroidism is associated with increase in free radical production and lipid peroxide levels [28, 74], whereas the hypometabolic state induced by hypothyroidism is associated with a decrease in free radical production [80] and in lipid peroxidation products [64]. The changes in the levels of the scavengers α-tocopherol [81], glutathione [37, 82] and coenzyme Q [81], and activities of antioxidant enzymes [74] in various tissues were found to be imbalanced and often opposite. It is worth mentioning that some of the antithyroid drugs have antioxidant effects [83, 84]. It was shown that both methimazole and propylthiouracil abolished or reduced the oxygen radical production by complement-attacked thyroid cells and decreased cytokine production [77]. The role of thyroid hormones in metabolic pathways is well known. However, their involvement in lipid peroxidation and antioxidant enzyme activities is not known.

Reports suggest that high concentration of thyroid hormones can affect the metabolism of oxygen in aerobic condition and stimulate free radical generation in mitochondria [85]. Oxidative stress produces immunosuppression [86]. Reactive oxygen types play an important role in physiological mechanism; however, extremely reactive oxygen types lead to oxidative stress [87].

Reports suggest that hypothyroidism reduces oxidative stress in kidney and testis tissue, but short-term, high dose of thyroxine administration in addition to hypothyroidism increased oxidative stress in same tissue of rats [88]. Other studies showed that hypothyroidism reduced oxidative damage in cerebral, hepatic, and cardiac tissues of rats and high dose of thyroxine increased oxidative damage in tissues [89]. It has been proposed that hypothyroidism provides in vivo protection against free radical-induced damage, and this cellular defense mechanism may be acting differently from antioxidant defense system [47]. Basedow disease patients exhibited increased T3 and T4, and their treatment with propylthiouracil resulted in decreased catabolism and lowered oxidant generation [73]. These evidences suggest that oxidative stress in any diseased condition or in infected condition or in stressed condition may be mediated through thyroid gland. Hence, it is contemplated that thyroid gland plays a central role in generating generalized oxidative stress in diseased condition. There are merits and demerits of free radical generation. Thyroid hormones play a crucial role in the regulation of mitochondrial oxidative metabolism [61].

Despite the fact that extensive research has been conducted about stress, stress-related diseases, oxidative stress, etc., very little work has been done indicating the role of thyroid in induction of oxidative stress. The body generates free radical by cellular mechanisms and/or endocrine mechanism. Many scientific studies suggest that corticotropin-releasing factor (CRF) supports neuronal system; increasing the neuronal effects may generate more free radicals. Most scientists view stress as the situation when the hypothalamo-pituitary-adrenocortical (HPA) axis, represented mainly by elevated adrenocorticotropic hormone (ACTH) levels, is activated [90]. Others suggest that activating other systems with or without an elevation in ACTH may reflect stress-induced disturbed homeostasis [2]. Apart from other factors, the role of neuroendocrine response in coping with stress is well recognized. During stress response the physiological processes are playing a vital role to redirect energy utilization among various organs. The thyroid gland is the body's primary regulator of metabolism. Thyroid-stimulating hormone (TSH) affects metabolism and may be affected by the thyroxin secretions. Therefore, it is postulated that thyroxin may play an important role in induction of oxidative stress [91]. Variations of the levels of thyroid hormones can be one of the main physiological modulators of in vivo cellular oxidative stress due to their known effects on mitochondrial respiration [92]. The thyroid hormones may play a crucial role in inducing the generation of generalized oxidative stress.

It is known that stress causes the release of corticotropin-releasing factor, which activates the sympathetic system as well as adrenals glands. This characteristic component of stress response influences several organs, which consequently affects the equilibrium and eventually disturbs the homeostasis. Therefore, it is also defined as a state of threatened homeostasis. A stress response particularly appears when a person cannot handle stress. Psychological stress, physiological or metabolic stress, environmental stress, physical stress, and chemical stress conditions implicate degenerative changes in many organs leading to variety of pathological conditions. It is believed that the degeneration of the tissues is subsequent to the cytotoxic effects of the free radicals, the underlying cause for the oxidative stress. Oxidative stress is a condition when the cell-bound macromolecular structures such as DNA, proteins, and lipids undergo oxidative damages due to chemical influences of the free radicals. Radicals are electronically unbalanced moieties and, hence, extremely reactive. Radicals are mainly generated in the mitochondria with a purpose to provide energy to activate the intracellular enzymes. However, in the event of excessive stimulation of mitochondria, the radicals are generated in excess. Due to an unpaired electron, the radical tries to gain an additional electron or donate same to any molecule in its surrounding. The molecule which loses an electron gets oxidized. This oxidized donor molecule sometimes sets up a chain reaction which potentially causes tissue damages. The unutilized radical is normally inactivated by intracellular available antioxidants.

However, in the event of overwhelming generation of radicals, the antioxidants fall short and the unscavenged radicals remain free. Such radicals are called as "free radicals."

The free radicals are oxygen or nitrogen derived, leading to formation of reactive oxygen species (ROS) and reactive nitrogen species (RNS), respectively. Chief danger of free radicals comes from the oxidative damages to the cellular component such as DNA, lipids, and protein. Probably, therefore, increased lipid peroxidation has been seen in a wide range of tissue injuries and diseases. Lipids appear more susceptible to radical attacks as their oxidation leads to formation of lipid peroxides. When free radicals excessively oxidize the intracellular components, the ensuing state of cell is called as "oxidative stress." This oxidative stress has been implicated in a variety of pathological conditions. Plethora of evidences has demonstrated the generation of oxidative stress on exposure to the stressor, and in addition, an antioxidant therapy has been shown to prevent the same. Probably, therefore, antioxidants are clinically found effective in many stress-related pathologies, which suggest that they are subsequent to the oxidative stress. Why prolonged induction of stress leads to oxidative stress is not yet properly delineated.

As mentioned earlier, exposure to any stressor initiates release of CRF from hypothalamus which increases the metabolic and functional status of many organs via sympathetic system and adrenal steroids. The glucocorticoids are known to mobilize the metabolic substrates so that cells can perform more work, as demanded by neuronal stimulus. The activation is normally proportional to the intensity of stressor. However, why should it produce oxidative stress is unclear because mobilization of metabolic substrates does not lead to free radicals generation. As mentioned earlier, the primary site for radical generation is the mitochondria, and this cell structure and its processes are more influenced by thyroxin.

Incidentally, it is important to note that hypothalamus also causes the release of thyrotrophic releasing hormones (TRH) during the stress response so that the mitochondria can provide the radicals to activate the enzymes required to derive the energy from the metabolic substrates, mobilized indirectly by CRF via glucocorticoids. In view of these evidences, it is likely that thyroxin, released in higher amount in response to stress, would activate the mitochondria to generate more radicals which could lead to oxidative stress.

The literature suggests the basic purpose of the endocrine component of stress response is to derive energy from the metabolic substrates for the increased organ functions. Plethora of evidences suggests that free radicals are often generated in stress-exposed subjects. However, it is not clear whether radical generation is the cause of enhanced metabolic activities or they are subsequent due to increased metabolic rate. It is known that:

(a) Stress causes the release of CRF and TRH from hypothalamus.
(b) Thyrotropin-releasing hormone (TRH) causes the release of thyroxin.
(c) Thyroxin influences the mitochondria to generate the radicals.
(d) Overdose of thyroxin generates the oxidative stress.
(e) Oxidative stress is often evident in stress-related pathologies.

These evidences suggest that the thyroid hormones released in response to stress may be playing a pivotal role in generation of oxidative stress. Hence, it was proposed to find experimental evidences to support the probable involvement of thyroid hormones in the oxidative stress generated by exposing rats to various types of stressors.

Hence, the investigations were carried out to define the role of thyroid hormones in the generation of oxidative stress. In vivo studies were conducted, in which the generalized oxidative stress was generated by a stress exposure, which initiated a supraspinal stress response involving the role of thyroid hormones. The literature has documented several models, which can induce stress response. However, there is no equivocal agreement on their abilities to generate generalized oxidative stress. Therefore, experiments were conducted on a number of models to test their ability to generate oxidative stress in various tissues such as liver, muscle, and blood.

The study revealed that all the stressors invariably increase oxidative stress in all the tissues as evident from the markers of oxidative stress, i.e., LPO, SOD, CAT, and GSH, in various tissues like blood, muscle, and liver. It was also found that chemical stress produces maximum oxidative stress as compared to physiological and psychological stress. Similarly, changes in the markers of oxidative stress in blood parallel with changes in muscle and liver. Hence, it appears that mere estimation of markers in blood can represent generalized oxidative stress which prevails all over the body. In all the stressed conditions, there was an increase in T3 and T4 and decrease in TSH. There was a concurrent increase in LPO and decrease in the SOD and CAT activity and reduction in the reduced glutathione content in blood. The data on oxidative stress and blood levels of thyroid hormones T3 and T4 condition exhibited a linear correlation. The changes in thyroid hormone levels correlate with the parameters of oxidative stress. In unstressed condition, there was a decrease in T3 and T4 levels and increase in TSH in hypothyroid state and the levels of LPO decreased, whereas the SOD and CAT activity and the reduced glutathione content increased significantly as compared to normal. In all the stressed conditions, there were an increase in T3 and T4 levels and a decrease in TSH in hyperthyroid state and levels of LPO increased, whereas SOD and CAT activity and the reduced glutathione content decreased significantly as compared to normal. It was further observed that the oxidative stress was higher in those rats which received exogenous thyroxin. The study revealed an intense oxidative stress in hyperthyroid state on exposure to any one of the employed stressor where in the chemical stress was found to be most effective. In hypothyroid state there was no indication of any oxidative stress on exposure to either of the stressors. Daily Vitamin-E treatment significantly prevented oxidative stress as compared to vehicle-treated rats in control (unstressed) as well as in experimental (stressed) group of rats having different functional states of thyroid. Vitamin-E treatment did not affect the levels of T3, T4, and TSH in any of the groups (Tables 1 and 2).

Thus, the investigations have provided an experimental evidence to show that the exposure to stressor implicates thyroid hormones in the generation of oxidative stress and antithyroid treatment in such situation prevents the same whereas the antioxidant treatment attenuates the oxidative stress more effectively in hypo- or normal functional state of thyroid, but its activity is lowered in hyperthyroid state. Hence, assessing the thyroid function in stressed individuals and then using antithyroid agents appear to be a better therapeutic approach to prevent the generation of stress-induced oxidative stress and prevent the oxidative degeneration of the body. The present study demonstrated that the subjection of the rats to either of the stressor caused the rise in the levels of thyroid hormones and also produced significant oxidative stress in all tested tissues. The oxidative stress was however not seen in rats, which were treated with methimazole and had low levels thyroid hormones. On the contrary, those who received exogenous thyroxin exhibited further rise in the oxidative stress. This shows that thyroid hormones play a pivotal role in the generation of oxidative stress on exposure to stressor. It was further observed that vitamin-E treatment eliminated the stress-induced oxidative stress while the raised levels of T_3 and T_4 remained unaffected.

3 Conclusion

The present study demonstrated that the subjection of the rats to either of the stressors caused the rise in the levels of thyroid hormones and also produced significant oxidative stress in all tested tissues. The oxidative stress was however not seen in rats, which were treated with methimazole and had low levels of thyroid hormones. On the contrary, those who received exogenous thyroxin exhibited further rise in the oxidative stress. This shows that thyroid hormones play a pivotal role in the generation of oxidative stress on exposure to stressor. It was further observed that vitamin-E treatment eliminated the stress-induced oxidative stress while the raised levels of T_3 and T_4 remained unaffected. The present investigations

Table 1 Influence of stress on the markers of oxidative stress in blood of rats having hypo- and hyperfunctional state of thyroid gland

	Control (unstressed)							Experimental (stressed)						
	LPO	SOD	GSH	CAT	$T_{3,}$	T_4	TSH	LPO	SOD	GSH	CAT	$T_{3,}$	T_4	TSH
Normal	2.27±0.21	30.07±0.49	10.80±0.24	268±1.78	38.46±0.92	2.76±0.12	1.82±0.04	2.80±0.26	27.93±0.98	10.60±0.35	246±8.94	43.5±0.76	2.95±0.15	1.70±0.33
Hypothyroid	2.07±0.04	32.8±0.34	11.30±0.17	298±2.89	30.33±0.57	1.28±0.15	12.55±0.43	1.86±0.89	28.4±0.56	11.05±1.73	268±7.15	31.26±0.51	1.29±0.10	9.80±0.03
Hyperthyroid	2.93±0.42	29.20±0.45	9.84±0.35	254±2.68	56.93±0.45	3.32±0.12	0.86±0.44	4.48±0.72	23.02±0.10	8.96±0.10	225.7±2.13	69.83±0.73	4.98±0.17	0.98±0.05

Values are mean±SEM ($n=6$). *$P<0.001$, #$P<0.05$, and **$P<0.01$, when compared to respective control (unstressed)
T3 (ng/dL)
T4 (µg/dL)
TSH (µIU/mL)
LPO (nmMDA/mg protein)
SOD (U/mg protein)
GSH µM/mg protein
CAT (nM of H_2O_2 decomposed/min/mg protein) of H_2O_2 decomposed/min/mg protein)

Table 2 Influence of stress on the markers of oxidative stress in blood of rats having hypo- and hyper- functional state of thyroid gland in normal (vehicle treated) and in vitamin-E-treated rats

State of thyroid	Treatments	Control (unstressed) LPO	SOD	GSH	CAT	$T_{3,}$	T_4	TSH	Experimental (stressed) LPO	SOD	GSH	CAT	T_{3Q}	T_4	TSH
Normal	Vehicle	2.23±0.56	33.07±0.49	11.23±0.31	240.3±4.03	35.46±0.92	2.26±0.09	1.80±0.05	3.00±0.35	32.47±0.76	9.57±0.33	238±2.68	39.88±0.87	2.76±0.11	1.65±0.05
	Vit-E	2.14±0.22	35.03±0.76	11.90±0.08	255±3.57	35.01±1.08	2.34±0.01	1.83±0.01	2.26±0.16	34.10±0.35	9.98±0.26	267±2.69	35.01±1.9	2.69±0.01	2.43±0.01
Hypothyroid	Vehicle	1.89±0.54	33.66±0.13	11.43±0.07	261.3±6.34	32.33±0.59	1.28±0.15	12.55±0.43	2.69±0.03	32.30±0.17	10.42±0.35	248.6±0.17	32.33±0.59	1.29±0.17	12.55±0.43
	Vit-E	1.74±0.21	35.17±0.81	12.30±0.08	295±2.62	31.68±0.14	1.24±0.26	13.2±0.38	1.88±0.26	35.70±0.53	12.70±0.17	289±1.78	32.18±0.14	1.21±0.26	13.2±0.38
Hyperthyroid	Vehicle	3.13±0.32	25.37±0.36	10.61±0.09	210.7±5.94	56.93±0.14	4.29±0.12	1.02±0.43	3.64±0.24	24.60±0.35	9.10±0.26	191.3±0.26	58.93±0.32	4.54±0.32	0.99±0.56
	Vit-E	2.95±0.36	25.97±1.18	10.70±0.17	212.7±4.03	54.97±0.42	4.26±0.38	1.03±0.24	3.45±0.44	25.10±0.44	10.30±0.17	214±2.68	55.75±0.75	4.13±0.43	1.01±0.24

Values are mean ± SEM ($n=6$). $*P<0.001$, $\#P<0.05$, and $**P<0.01$, when compared to respective control (unstressed)
T3 (ng/dL), T4 (µg/dL), TSH (µIU/mL), LPO (nmMDA/mg protein)
SOD (U/mg protein), GSH µM/ mg protein, CAT (nM of H_2O_2, decomposed/min/mg protein of H_2O_2 decomposed/min/mg protein)

provided enough evidence to show that the supraspinal response to stress basically elevates the thyroid hormone levels, which onward causes the oxidative stress. Such conclusion further suggests the use of antithyroid agent to prevent the stress-induced oxidative stress and subsequent tissue degeneration. However, it is cautioned that the levels of thyroid hormones must be first assessed in stressed-exposed subjects before any antithyroid agent is used to prevent the oxidative stress as the hypothyroidism after the inappropriate dose of antithyroid agent is likely to impair the stress-sustenance ability.

References

1. Behan DP, Grigoriadis DE, Lovenberg T et al (1996) Neurobiology of corticotropin releasing factors (CRH) receptors and CRH-binding protein: implications for the treatment of CNS disorders. Mol Psychol 1:265–277
2. lsboer F, Bardan N (1996) Antidepressant and hypothalamic pituitary adrenocortical regulation. Endocr Rev 17:187–205
3. Reagon LP, Magarinos AM, McEwen BS (1999) Neurological changes induced by stress in streptozotocin diabetic rats. Ann N Y Acad Sci 893:126–137
4. Pekarkova I, Parara S, Holecek V et al (2001) Does exogenous melatonin influences the free radicals metabolism and pain sensation in rat? Physiol Res 50(6):595–602
5. Mileva M, Bekalova R, Tancheva l et al (2002) Effect of immobilization, cold and cold restraint stress on liver monooxygenase activity and lipid peroxidation of influenza virus – infected mice. Arch Toxicol 76(2):96–103
6. Gupta YK, Sharma M, Chaudhari G (2002) Pyrogallol-induced hepatotoxicity in rats: a model to evaluate antioxidant hepatoprotective agents. Exp Clin Pharmacol 24(8):497–500
7. Fiers W, Beyaert R, Declercq W et al (1999) More than one way to die: apoptosis, necrosis and reactive oxygen damage. Oncogene 18:7719–7730
8. Nicolas MG, Fujiki K, Murayama K et al (1996) Exp Eye Res 62:399–408
9. Hayes JD (1999) Glutathione and glutathione-dependent enzymes represent a coordinately regulated defense against oxidative stress. Free Radic Res 31:273–300
10. Cheesseman KH, Slater TF (1993) Free radicals in medicine. Br Med Bull 49:481–491
11. Halliwell B, Gutteridge JMC (1990) Role of free radicals and catalytic metal ions in human disease: an overview. Methods Enzymol 186:1–85
12. Guerrero A, Pamplona R, Postero-Otin M et al (1999) Effect of thyroid status on lipid composition and peroxidation in the mouse liver. Free Radic Biol Med 26:73–80
13. Freeman BA, Crapo JD (1982) Biology of disease, free radicals and tissue injury. Lab Invest 47:412–426
14. Mates JM, Perez-Gomez C, Castro IN (1999) Antioxidant enzymes and human diseases. Clin Biochem 32:595–603
15. Hauck JS, Bartke A (2000) Effects of growth hormone on hypothalamic CAT and Cu/Zn superoxide dismutase. Free Radic Biol Med 28:970–978
16. Baker S, Klitgaar HM (1952) Metabolism of tissues excised from thyroxine-injected rats. Am J Physiol 170:81–86
17. Arnold S, Goglia F, Kadenbach B (1998) 3,5-Diiodothyronine binds to subunit Va of cytochrome-c oxidase and abolishes the allosteric inhibition of respiration by ATP. Eur J Biochem 252:325–330
18. Moreno M, Lombardi A, Beneduce L et al (2002) Are the effects of T3 on resting metabolic rate in euthyroid rats entirely caused by T3 itself? Endocrinology 143:504–510
19. Goglia F, Moreno M, Lanni A (1999) Actions of thyroid hormones at the cellular level: the mitochondrial target. FEBS Lett 452:115–120
20. Ram PA, Waxman DJ (1992) Thyroid hormone stimulation of NADPH P450 reductase expression in liver and extrahepatic tissues. J Biol Chem 267:3294–3301
21. Soboll S (1993) Thyroid hormone action on mitochondrial energy transfer. Biochim Biophys Acta 1144:1–6
22. Videla LA (2000) Energy metabolism, thyroid calorigenesis, and oxidative stress: functional and cytotoxic consequences. Redox Rep 5:265–275
23. Lanni A, Moreno M, Lombardi A et al (2003) Thyroid hormone and uncoupling proteins. FEBS Lett 543:5–10
24. Sies H (1986) Biochemistry of oxidative stress. Angew Chem Int Ed Engl 25:1058–1071
25. Videla LA, Ferna'ndez V, Carrio'n Y (1995) Biochemical mechanisms in hepatotoxicity: oxidative stress induced by xenobiotics and hormonal changes. J Braz Assoc Adv Sci 47:385–394
26. Jaeschke H, Gores GJ, Cederbaum AI et al (2002) Mechanisms of hepatotoxicity. Toxicol Sci 65:166–176
27. Cutler RG (1985) Peroxide-producing potential of tissues: correlation with longevity of mammalian species. Proc Natl Acad Sci U S A 87:1620–1624
28. Ferna'ndez V, Barrientos X, Kipreos K et al (1985) Superoxide radical generation, NADPH oxidase activity, and cytochrome P-450 content of rat liver microsomal fractions in an experimental hyperthyroid state: relation to lipid peroxidation. Endocrinology 117:496–501
29. Ferna'ndez V, Videla LA (1993) Influence of hyperthyroidism on superoxide radical and hydrogen

peroxide production by rat liver submitochondrial particles. Free Radic Res Commun 18:329–335

30. Ferna'ndez V, Videla LA (1993) 3,3,5-Triiodothyronine-induced hepatic respiration: effects of desferrioxamine and allopurinol in the isolated perfused rat liver. Toxicol Lett 69:205–210
31. Venditti P, De Rosa R, Di Meo S (2003) Effect of thyroid state on H_2O_2 production by rat liver mitochondria. Mol Cell Endocrinol 205:185–192
32. Venditti P, De Rosa R, Di Meo S (2004) Effect of cold-induced hyperthyroidism on H2O2 production and susceptibility to stress conditions of rat liver mitochondria. Free Radic Biol Med 36:348–358
33. Tata JR, Ernster L, Lindberg O (1962) Control of basal metabolic rate by thyroid hormones and cellular function. Nature 193:1058–1060
34. Goeptar AR, Scheerens H, Vermeulen NPE (1995) Oxygen and xenobiotic reductase activities of cytochrome P450. Crit Rev Toxicol 25:25–65
35. Ferna'ndez V, Massa L, Quin˜ones L et al (2003) Effects of g-hexachlorocyclohexane and 1-3,3V,5-triiodothyronine on rat liver cytochrome P4502E1-dependent activity and content in relation to microsomal superoxide radical generation. Biol Res 36:359–365
36. Simon-Giavarotti KA, Giavarotti L, Gomes LF et al (2002) Enhancement of lindane-induced liver oxidative stress and hepatotoxicity by thyroid hormone is reduced by gadolinium chloride. Free Radic Res 36:1033–1039
37. Ferna'ndez V, Llesuy S, Solari L et al (1988) Chemiluminescent and respiratory responses related to thyroid hormone-induced liver oxidative stress. Free Radic Res Commun 5:77–84
38. Ferna'ndez V, Simizu K, Barros SBM et al (1991) Effects of hyperthyroidism on rat liver glutathione metabolism: related enzymes, activities, efflux, and turnover. Endocrinology 129:85–91
39. Huh K, Kwon TH, Kim JS et al (1998) Role of the hepatic xanthine oxidase in thyroid dysfunction: effect of thyroid hormones in oxidative stress in rat liver. Arch Pharm Res 21:236–240
40. Fernandez V, Cornejo P, Tapia G et al (1997) Influence of hyperthyroidism on the activity of liver nitric oxide synthase in the rat. Nitric Oxide Biol Chem 1:463–468
41. Tapia G, Pepper I, Smok G et al (1997) Kupffer cell function in thyroid hormone-induced liver oxidative stress in the rat. Free Radic Res 26:267–279
42. Wang JF, Komarov P, de Groot H (1993) Luminol chemiluminescence in rat macrophages and granulocytes: the role of NO, O_{2}_/H_2O_2, and HOCl. Arch Biochem Biophys 304:189–196
43. Videla LA, Correa L, Rivera M et al (1993) Zymosan-induced luminol-amplified chemiluminescence of whole blood phagocytes in experimental and human hyperthyroidism. Free Radic Biol Med 14:669–675
44. Fernandez V, Llesuy S, Solari L et al (1988) Chemiluminescence and respiratory responses related to thyroid hormone-induced liver oxidative stress. Free Radic Res Commun 5:77–84
45. Droge W (2002) Free radicals in the physiological control of cell function. Physiol Rev 82:47–95
46. Fernandez V, Videla LA (1996) Effect of hyperthyroidism on the biliary release of thiobarbituric acid reactants in the rat. Toxicol Lett 84:149–153
47. Venditti P, Balestrieri M, Meo S et al (1997) Effect of thyroid state on lipid peroxidation, antioxidant defences and susceptibility to oxidative stress in rat tissues. J Endocrinol 155:151–157
48. Marzoev AI, Kozlov AV, Andryushchenko AP et al (1982) Activation of lipid peroxidation in liver mitochondria of hyperthyroid rabbits. Bull Exp Biol Med 93:269–272
49. Andican G, Gelisgen R, Civelek S et al (2004) Oxidative damage to nuclear DNA in hyperthyroid rat liver: inability of vitamin C to prevent the damage. J Toxicol Environ Health 67:413–420
50. Videla LA, Sir T, Wolff C (1988) Increased lipid peroxidation in hyperthyroid patients: suppression by propylthiouracil treatment. Free Radic Res Commun 5:1–10
51. Seven R, Gelisgen R, Seven A et al (2001) Influence of propylthiouracil treatment on oxidation stress and nitric oxide in Basedow diseases. Toxic Environ Health A 62:495–503
52. Bianchi G, Solaroli E, Zaccheroni V et al (1999) Oxidative stress and anti-oxidant metabolites in patients with hyperthyroidism: effect of treatment. Horm Metab Res 31:620–624
53. Sewerynek J, Wiktorska J, Nowak D et al (2000) Methimazole protection against oxidative stress induced by hyperthyroidism in Graves disease. Endocr Regul 34:83–89
54. Komosinska-Vessev K, Olczyk K, Kucharz EJ et al (2000) Free radical activity and antioxidant defense mechanisms in patients with hyperthyroidism due to Graves' disease during therapy. Clin Chim Acta 300:107–117
55. Guerra LN, Moiguer S, Karner M et al (2001) Antioxidants in the treatment of Graves disease. IUBMB Life 51:105–109
56. Yavuz DG, Yuksel M, Deyneli O et al (2004) Association of serum paraoxonase activity with insulin sensitivity and oxidative stress in hyperthyroid and TSH-suppressed nodular goiter patients. Clin Endocrinol 61:515–521
57. Bednarek J, Wysocki H, Sowinski J (2004) Oxidation products and antioxidant markers in plasma of patients with Graves, disease and toxic multinodular goiter: effect of methimazole treatment. Free Radic Res 38:659–664
58. Wilson R, Chopra M, Bradley H et al (1989) Free radicals and Graves, disease: the effect of therapy. Clin Endocrinol 30:429–433
59. Lissi EA, Salim-Hanna M, Sir T et al (1992) Is spontaneous urinary visible chemiluminescence a reflection of in vivo oxidative stress? Free Radic Biol Med 12:317–322
60. McCord JM, Fridovich I (1969) Superoxide dismutase, an enzymic function of erythrocuprein. J Biol Chem 244:6049–6055

61. Asayama K, Kato K (1990) Oxidative muscular injury and its relevance to hyperthyroidism. Free Radic Biol Med 8:293–303
62. Fernandez V, Videla LA (1989) Thyroid hormone, active oxygen, and lipid peroxidation. In: Miquel J, Quintanilha AT, Weber H (eds) Handbook of free radicals and antioxidants in biomedicine. CRC Press Inc., Boca Raton, pp 105–115
63. Pereira B, Rosa LF, Safi DA et al (1994) Control of superoxide dismutase, CAT and glutathione peroxidase activities in rat lymphoid organs by thyroid hormones. J Endocrinol 140:73–77
64. Paller MS (1986) Hypothyroidism protects against free radical damage in ischemic acute renal failure. Kidney Int 29:1162–1166
65. Dumitriu L, Bartoc R, Ursu H et al (1988) Significance of high levels of serum malonyl dialdehyde and ceruloplasmin in hyper- and hypothyroidism. Endocrinology 26:35–38
66. Yilmaz S, Ozan S, Benzer F et al (2003) Oxidative damage and antioxidant enzyme activities in experimental hypothyroidism. Cell Biochem Funct 21(4):325–330
67. Costantini F, Pierdomenico SD, De Cesare D (1998) Effect of thyroid function on LDL oxidation. Arterioscler Thromb Vasc Biol 18:732–737
68. Aviram M, Rosenblat M, Bisgaier CL et al (1998) Paraoxonase inhibits high-density lipoprotein oxidation and preserves its functions: a possible peroxidative role for paraoxonase. J Clin Invest 101:1581–1590
69. Rozenberg O, Rosenblat M, Coleman R et al (2003) Paraoxonase (PON1) deficiency is associated with increased macrophage oxidative stress: studies in PON1-knockout mice. Free Radic Biol Med 34:774–784
70. Ayub A, Mackness MI, Arrol S et al (1999) Serum paraoxonase after myocardial infarction. Arterioscler Thromb Vasc Biol 19:330–335
71. Mackness MI, Harty D, Bhatnagar D et al (1991) Serum paraoxonase activity in familial hypercholesterolaemia and insulin dependent diabetes mellitus. Atherosclerosis 86:193–199
72. Burton GW, Traber MG (1990) Vitamin E: antioxidant activity, bio-kinetics, and bioavailability. Annu Rev Nutr 10:357–382
73. Chow CK (1991) Vitamin E and oxidative stress. Free Radic Biol Med 1:215–232
74. Seven A, Seymen O, Hatemi S et al (1996) Antioxidant status in experimental hyperthyroidism: effect of vitamin E supplementation. Clin Chim Acta 256:65–74
75. Asayama K, Dobashi K, Hayashibe H et al (1987) LPO and free radical scavengers in thyroid dysfunction in the rat: a possible mechanism of injury to heart and skeletal muscle in hyperthyroidism. Endocrinology 121:2112–2118
76. Steinberg D, Parthasarathy S, Carew TE et al (1989) Beyond cholesterol: modifications of low density lipoprotein that increase its atherogenicity. N Engl J Med 320:915–924
77. Steinbrecher UP, Zhang H, Lougheed M (1990) Role of oxidatively modified LDL in atherosclerosis. Free Radic Biol Med 9:155–168
78. Weetman AP, Tandon N, Morgan BP (1992) Antithyroid drugs and release of inflammatory mediators by complement attacked thyroid cells. Lancet 340:633–636
79. Nishiki K, Ericinska M, Wilson DF et al (1978) Evaluation of oxidative phosphorylation in hearts from euthyroid, hypothyroid and hyperthyroid rats. Am J Physiol 235:C212–C219
80. Schwartz HL, Oppenheimer JH (1978) Physiologic and biochemical actions of thyroid hormone. Pharmacol Ther 3:349–376
81. Swaroop A, Ramasarma T (1985) Heat exposure and hypothyroid conditions decrease hydrogen peroxide generation in liver mitochondria. Biochem J 226:403–408
82. Mano T, Sinohora R, Sawai Y et al (1995) Changes in lipid peroxidation and free radicals scavengers in the brain of hyper and hypothyroid aged rats. J Endocrinol 147:361–365
83. Morini P, Casalino E, Sblano C et al (1991) The response of rat liver lipid peroxidation, antioxidant enzyme activities and glutathione concentration to the thyroid hormone. Int J Biochem 23:1025–1030
84. Heufelder AE, Wenzel BE, Bahn RS (1992) Methimazole and propylthiouracil inhibit the oxygen free radical-induced expression of a 72 kilodalton heat shock protein in Graves' retroocular fibroblasts. J Clin Endocrinol Metab 40:720–723
85. Hicks M, Wong LS, Day RO (1992) Antioxidant activity of propylthiouracil. Biochem Pharmacol 4:439–444
86. Sewerynek E, Wiktorska J, Lewinski A (1999) Effect of melatonin on the oxidative stress induced by thyrotoxicosis in rats. Neuroendocrinology 20:157–161
87. Karel P, Miklos P (2001) Stressor specifically of central neuroendocrine responses: implication of stress related disorders. Endocr Rev 224:502–548
88. Karbowni KM, Lewinski A (2003) The role of oxidative stress in physiological and pathological processes in the thyroid gland, possible involvement in pineal-thyroid interaction. Neuroendocrinol Lett 24:293–303
89. Mogulkoc R, Baltaci AK, Aydin L et al (2005) Short term administration leads LPO in renal and testicular tissues of rats with hypothyroidism. Acta Biol Hung 56:225–232
90. Mogulkoc R, Baltaci AK, Aydin L et al (2005) The effect of thyroxine administration on LPO in different tissues of rats with hypothyroidism. Acta Physiol Hung 92:39–46
91. Owen MJ, Nemeroff CP (1991) Physiology and pharmacology of corticotropin releasing factor. Pharmacol Rev 43:425–473
92. Umathe SN, Kale MK, Bhusari KP (2006) Oxidative stress and the thyroid, positive health Portsmouth, London UK 119:24–28

Oxidative Stress Events and Neuronal Dysfunction in Alzheimer's Disease: Focus on APE1/Ref-1-Mediated Survival Strategies

Navrattan Kaur, Bibekananda Sarkar, Sunil Mittal, Monisha Dhiman, Gulio Taglialatela, Regino J. Perez-Polo, and Anil K. Mantha

Abstract

Alzheimer's disease (AD) is an important public health problem which affects millions of people worldwide. The major pathological hallmarks associated with AD are the accumulation of amyloid beta (Aβ) in senile plaques and neurofibrillary tangles (NFT) made up of hyperphosphorylated tau proteins. New findings suggest that oligomeric Aβ is a more toxic species than fibrillar Aβ relevant to AD pathology. Although the molecular mechanism(s) underlying the disease is not identified completely, various factors have been implicated in the development of AD. Accumulating evidences point towards the role of oxidative stress and mitochondrial dysfunction in the pathogenesis of AD and recognise them as an early event in AD development. Ageing is considered the greatest risk factor for AD and is linked to oxidative stress which causes accumulation of somatic mutations in mitochondrial DNA (mtDNA) over time and leads to genome

N. Kaur • B. Sarkar
Centre for Biosciences, School of Basic and Applied Sciences, Central University of Punjab, Bathinda 151 001, Punjab, India

S. Mittal
Centre for Environmental Science and Technology, School of Environmental and Earth Sciences, Central University of Punjab, Bathinda, Punjab, India

M. Dhiman
Centre for Genetic Diseases and Molecular Medicine, School of Emerging Life Science Technologies, Central University of Punjab, Bathinda 151 001, Punjab, India

G. Taglialatela
Department of Neurology, University of Texas Medical Branch, Galveston, TX 77555, USA

R.J. Perez-Polo
Department of Biochemistry and Molecular Biology, University of Texas Medical Branch, Galveston, TX 77555, USA

A.K. Mantha (✉)
Centre for Biosciences, School of Basic and Applied Sciences, Central University of Punjab, Bathinda 151 001, Punjab, India

Department of Biochemistry and Molecular Biology, University of Texas Medical Branch, Galveston, TX 77555, USA
e-mail: anilmantha@gmail.com; Anil.kumar@cup.ac.in

V. Rani and U.C.S. Yadav (eds.), *Free Radicals in Human Health and Disease*, DOI 10.1007/978-81-322-2035-0_13, © Springer India 2015

instability and mitochondrial dysfunction. Recent studies on AD patients and transgenic mouse models suggest that amyloid precursor protein (APP) and Aβ localise to mitochondria, interact with mitochondrial proteins, disrupt electron transport chain (ETC), increases reactive oxygen species (ROS) level, impair axonal mitochondrial trafficking, thus leading to synaptic damage and cognitive decline associated with AD. It is not known whether accumulation of Aβ is the cause or outcome of declining mitochondrial function in AD. In order to counteract oxidative stress and maintain genome integrity, various DNA repair pathways exist, base excision repair (BER) pathway being the predominant pathway for repairing oxidised base lesions in neuronal cells. APE1 is the central enzyme of the BER pathway, having both repair and redox activities and shown to enhance neuronal survival after oxidative stress. Newer studies are revealing the role of APE1 in maintenance of mitochondrial genome repair and function. In this scenario, antioxidant-based therapy, which could reduce oxidative stress and modulate the activities of APE1, can serve as effective treatment providing neuroprotection in AD. This book chapter summarises some recent developments in understanding the pathogenesis of AD linking Aβ-induced oxidative stress, mitochondrial dysfunction, role of APE1 and phytochemicals toward AD therapeutics.

Keywords
Alzheimer's disease • Oxidative stress • Amyloid beta • Mitochondria • APE1/Ref-1 • Phytochemicals

1 Introduction

In 1907, Alois Alzheimer, a German psychiatrist and neuropathologist, described the hallmark lesions associated with 'presenile dementia', which later came to be known as Alzheimer's disease AD [1], a progressive and always fatal disorder characterised clinically by memory loss and behavioural abnormalities. The hallmarks described were extracellular plaques composed of amyloid beta (Aβ) and intracellular neurofibrillary tangles (NFTs) made up of a protein called tau (τ). Since then, it has been more than 100 years that the neuropathological hallmarks of the disease have been described, the underlying molecular mechanism(s) of the pathogenesis of AD are yet to be elucidated. Since 1992, the 'Aβ cascade hypothesis' has been the main model describing the pathogenesis of AD. According to it, accumulation of Aβ, owing to increased processing of amyloid precursor protein (APP), induces biochemical, histological and clinical changes associated with the disease [2]. Various modifications of the model have taken place over the time and currently, oligomers, the soluble form of Aβ, rather than insoluble fibrillar Aβ, are considered to be the more toxic species and relevant to AD pathogenesis [3]. Many studies have proposed that oxidative stress plays an important role in the pathogenesis of AD and consider it to be one of the early changes associated with AD [4]. Also, mitochondrial dysfunction is seen as an early event in the pathogenesis of AD [5]. Thus, these studies have led to the formulation of a new hypothesis, viz. 'mitochondrial cascade hypothesis', indicating the role of the mitochondria and its dysfunction initiating late-onset AD pathologies particularly in relation to sporadic AD [6].

Ageing is considered to be the greatest risk factor for AD. Linking ageing with neurodegenerative diseases, the free radical theory of ageing suggests that oxidative imbalance, i.e. elevated levels of reactive oxygen species (ROS), has a role in the pathogenesis of many neurodegenerative diseases like AD [7]. The brain is particularly vulnerable to oxidative stress due to a low level of antioxidant system and high consumption of oxygen [7]. There are a number of exogenous and endogenous sources of ROS, which increase the oxidative stress and lead to genome instability. Mitochondria are considered to be the major internal source of ROS. One study has showed that overexpression of antioxidant catalase, targeted to the mitochondria reduces oxidative damage, thus highlighting the role of the mitochondria as a source of these radicals [8]. In line with this evidence, many studies have shown the mitochondria as a source as well as a target of ROS. The link between Aβ and free radical generation ultimately leading to increased ROS in neuronal cells has been well established [9–12]. There is an emerging body of evidence revealing that Aβ enters the mitochondria and induces generation of free radicals causing oxidative damage, thus providing a link between Aβ and mitochondrial dysfunction in the pathogenesis of AD [13].

Postmitotic cells like neurons are terminally differentiated cells which cannot be replaced. Neurons are particularly sensitive to oxidative damage. Accumulation of DNA base modifications especially 8-oxo-7,8-dihydro-2′-deoxyguanosine (8-OHdG) are the major factors leading to genome instability [14, 15]. Base excision repair (BER) is the predominant pathway in the nucleus as well as in the mitochondria that removes these oxidised base lesions [16, 17]. Amongst the BER enzymes, apurinic/apyrimidinic endonuclease (APE1) is a multifunctional enzyme involved in DNA repair and redox regulation of various transcription factors (TFs), thus is also known as redox effector factor 1, Ref-1 [17–19]. Elevated nuclear expression of APE1/Ref-1 in cerebral cortical regions of AD patients highlights the role of APE1 in neurons in response to oxidative stress [20]. But mitochondrial APE1's function is still not clear. Recently, it was shown that nuclear APE1 regulates nuclear-encoded mitochondrial-related genes by modulating the DNA-binding activity of nuclear respiratory factor-1 (NRF1) in a redox-dependent manner [21]. This indicates that APE1 regulates the expression of some nuclear-encoded mitochondrial constitutive genes and consequently modulates mitochondrial functions in response to oxidative stress.

There are a number of phytochemicals which have shown their neuroprotective abilities in rescuing the Aβ-induced oxidative stress in vitro and in vivo. Some of them have also shown modulation of APE1/Ref-1's repair and redox activities, making them effective molecules for prevention and treatment of cancers and degenerative diseases [22–24]. Thus, phytochemicals may serve as effective treatments providing neuroprotection in AD. This book chapter summarises the available literature in the field and suggests a link between oxidative stress, mitochondrial dysfunction, APE1 and phytochemical-based interventions towards AD therapeutics.

2 Neurodegenerative Diseases

Neurodegenerative diseases represent a range of diseases affecting the central nervous system (CNS) characterised by selective neuronal vulnerability and degeneration in specific regions of the brain. This causes disabling and debilitating conditions involving either impairment of memory (dementia) or movement-related disabilities (ataxia), ultimately leading to death. The degeneration is caused due to abnormal accumulation and aggregation of proteins in specific parts of the brain intracellular or extracellular as insoluble or soluble forms. These diseases are also known as protein conformational diseases as the aggregated proteins bear a β-sheet conformation that aid in protein aggregation and fibril formation [25, 26] which are the pathological hallmarks associated with these diseases. The exact reason as to why these proteins begin to accumulate is not known but research has revealed that this may happen due to a number of reasons, majorly due to disruption of ubiquitin-proteosome machinery, autophagy failure and oxidative stress. Some of the common neurodegenerative diseases include AD, Parkinson's disease (PD), Huntington's disease (HD) and amyotropic lateral sclerosis (ALS). Although the mechanism of the pathogenesis of these diseases is not understood to date, some genetic and environmental factors have been implicated which confer susceptibility to these diseases.

2.1 Alzheimer's Disease

AD is the most common neurodegenerative disease and leading cause of dementia in people above 65 years of age. It is a progressive brain disorder in which a person's memory, thinking ability and behaviour is affected with the deterioration worsening with time. Around 27 million people are estimated to be affected by this disease [27]. AD is characterised by the presence of two major neuropathological hallmarks which are extracellular Aβ plaques and intracellular NFTs composed of tau causing neuronal dysfunction [28]. Ageing is the greatest risk factor for AD with majority of patients above 65 years of age [29]. But cases of early-onset of AD are also present in those 35–60 years of age [30].

Genetic mutations are the known cause of early-onset familial AD with a prevalence of less than 1 % [29]. Thus, three forms of early-onset AD which are inherited as autosomal dominant traits involving three genes, viz. APP on chromosome 21, presenilin 1 (PS1) on chromosome 14 and presenilin 2 (PS2) on chromosome 1 [30], are recognised. Mutations in any of these three genes can make an individual susceptible to AD. Apolipoprotein E, type ϵ4 (APO ϵ4), is a risk factor for late-onset familial and sporadic AD, which accounts for majority (>99 %) of the AD cases [31]. Several genome-wide association studies (GWAS) have identified and implicated many genes in the aetiology of AD [32–34].

Accumulating evidences have shown that oxidative stress and mitochondrial dysfunction are important events occurring during the development of AD. In relation to this, decline in mitochondrial function with advancing age owing to accumulation of somatic mutations in mtDNA is reported [35]. Aβ is also known to induce free radicals [10, 11, 36] and cause decline in mitochondrial function. Oligomeric Aβ is considered to be more toxic than the insoluble fibrillar Aβ [3], which localises to the mitochondria and interacts with a number of mitochondrial proteins, causing synaptic damage and leading to memory impairment and cognitive decline, associated with AD [37, 38]. But the underlying mechanism(s) of the pathogenesis of AD is not known as yet. At present, there is no cure for AD but symptomatic treatments are available to relieve the symptoms and slowdown the impairment of memory associated with AD.

2.2 Factors Governing/Responsible for AD Pathology

AD, a disabling and fatal disease, has become an important public health problem and poses an enormous economic and social burden for affected individuals, their caregivers and society. Though the neuropathological hallmarks of AD are known, the etiological factors involved in the pathogenesis of the disease are unknown. A number of studies have indicated some risk factors for onset of disease, advancing age being the most important risk factor. Other potential risk factors for AD are described in the following sections.

2.2.1 Genetic Factors

First, an individual having a family history of AD, is susceptible to develop it, especially if he is a first-degree relative of an affected person [29]. In comparison to males, females are more prone to develop AD. Persons with Down's syndrome (trisomy 21) are at a greater risk of developing AD [39]. Additionally, an individual's chance for developing AD increases if he inherits an *APOE* ϵ4 allele, one of the three common alleles (ϵ2, ϵ3, ϵ4) of *APOE* gene, from his parents and has an increased risk if two *APOE* ϵ4 alleles are inherited [29]. Mutations in *APP*, *PS1* and *PS2* genes are recognised to make a person susceptible to early-onset of familial AD [30]. Apart from the *APOE* gene, a number of new genes are implicated in the pathogenesis of late-onset AD

(LOAD). Bridging integrator 1 (BIN1), also known as amphiphysin 2, is now recognised as an important genetic risk factor after *APOE4* for LOAD [40]. BIN1 transcripts levels were observed to be elevated in AD brains, showing BIN1 as a genetic susceptibility locus in AD [32]. Also, decreased expression of drosophila BIN1 ortholog *Amph* suppressed tau-mediated neurotoxicity, highlighting the role of BIN1 in mediating AD risk and its role in modulating AD pathogenesis at the level of the tau pathway [32]. Thus, BIN1 can be thought of as a target for treatment of AD. In addition to *APOE* and *BIN1*, other susceptibility gene loci identified include phosphatidylinositol-binding clathrin assembly protein (*PICALM*), ATP-binding cassette transporter (*ABCA7*), CD2-associated protein (*CD2AP*), clusterin (*CLU*), complement receptor 1 (*CR1*), CD33 antigen (*CD33*), ephrin receptor Eph-A1 (*EPHA1*) and a cluster of membrane-spanning 4A (*MS4A*) genes [33, 34, 40, 41].

2.2.2 Vascular Factors

Drinking alcohol poses a threat of developing AD at a later life. Heavy drinkers who are *APOE4* ϵ4 allele carriers are at a greater risk of developing AD at a later life than moderate alcohol drinkers [42]. Many studies have found association between AD and cigarette smoking. For example, *APOE4* noncarriers are at a greater risk of developing AD [43]. People with high blood pressure in middle age are more susceptible to develop AD at a later stage in their life [44]. Obesity is also a risk factor for AD [45]. Higher body mass index (BMI) or obesity in middle age can increase the risk of developing AD at a later life [46]. Higher serum cholesterol levels are also associated with development of AD at a later stage [47, 48]. Use of statins, i.e. cholesterol lowering drugs, poses a lower risk of AD [49]. In line with this, it was seen that people having a diet rich in polyunsaturated fatty acids and fish have a lower risk of AD [50] as compared to those consuming saturated fatty acids in their diet [51, 52]. An association between diabetes and AD has been observed. Diabetic people are more prone to develop AD, with an increased risk if diabetes occurs in middle age [53]. All these factors are modifiable risk factors that can be modulated by adopting healthy eating habits and an active life thus, lowering the risk of dementia and AD.

2.2.3 Head Trauma and Head Injury

Increased risk of AD development is associated with head injury and head trauma [29]. People with moderate head injuries are at an increased risk of developing dementia than those without any head injuries. Those with severe head injuries are at the greatest risk of developing dementia at a later life. Susceptibility of developing AD increases if a person carries an APOE ϵ4 allele and has suffered any head injury [54]. Thus, boxers and football players are at greater risk of AD at a later life [29].

2.2.4 Mild Cognitive Impairment

Mild cognitive impairment (MCI) is said to be a transitional stage between ageing and development of AD [55]. It is a condition in which a person has memory impairment though it doesn't affect a person's daily activity. It is shown that those who have MCI are at a greater risk of developing AD at a later life [55].

2.2.5 Autophagy Failure

Accumulation of Aβ plaques and tau protein in neurons shows that the neuronal housekeeping and protein quality control systems are impaired in AD. One of these is autophagy, which is a lysosomal degradative process involving removal of toxic proteins and preventing protein aggregation. Recent studies have indicated beclin-1 to be involved in autophagy regulation and modulation of APP metabolism [56]. Beclin-1 expression was also seen to be reduced in the AD brain [57]. Rapamycin is emerging as a potential neuroprotective agent for AD and functions via enhancement of autophagy [58]. A recent study has shown that Aβ(1-42)-induced beclin-1 expression was upregulated by rapamycin and that the beclin-1-dependent autophagy can prevent neuronal cell death before occurrence of AD in PC12 cells [59]. Inhibition of beclin-1-dependent autophagy was shown to speed up neuronal cell death. Thus, beclin-1 dysfunction is associated with

AD. Therefore, autophagy failure is a potential factor leading to accumulation of toxic Aβ plaques and tau in AD.

2.2.6 Brain Inflammation

Chronic brain inflammation is associated with AD. Activated microglia and reactive astrocytes are seen in close proximity to neuritic Aβ plaques in the AD brain [60]. Studies have shown that complement system is activated in the AD brains [61, 62]. The microglia, astrocytes along with complement system components, cytokines and chemokines are involved in inflammatory responses against Aβ. Microglias are cells of CNS involved in the protection of the brain as first-line defence against any invading pathogen and pathological conditions. On the other hand, astrocytes provide trophic support to neurons and are involved in Aβ clearance [63]. In order to clear Aβ, the activated microglia and astrocytes secrete ROS, nitric oxide (NO) and proteolytic components which further increase APP production and proteolytic processing of APP, thus causing neuronal dysfunction [63]. β-site APP-cleaving enzyme 1 (BACE1), a membrane-bound aspartic proteinase, is primarily expressed by the neurons and is associated with the generation of Aβ peptides from APP owing to its β-secretase activity [64]. BACE1 expression has been observed in reactive astrocytes in the AD brains while resting astrocytes do not display BACE1 at detectable levels, thus showing that activation of astrocytes may lead to the development of AD [65]. Another study showed that the TF NF-κB acts as a repressor in neuronal and nonactivated astrocytes, while NF-κB acts as activator of BACE1 transcription in activated astrocytes and Aβ-exposed neuronal cells. Also, the presence of increased level of activated astrocytes with ageing is well demonstrated, which may lead to increased processing of BACE1 causing increased Aβ resulting into chronic inflammation and subsequently astrocyte activation ending up forming a feedback loop [66]. Thus, inflammation plays a major role in AD pathology.

2.2.7 Hormones

Levels of reproductive hormones change with advancing age and are considered as risk factors for AD. Studies have shown that an elevated level of luteinising hormone (LH) increases the risk for developing AD [67]. Subsequently, using a transgenic mouse model of AD, it was shown that leuprolide {a gonadotropin-releasing hormone (GnRH) agonist} lowered serum LH levels, improved working memory and decreased Aβ deposition [68]. A study has also suggested that spatial memory impairment observed in postmenopausal women or female rats after ovariectomy is attributed to high LH levels [69]. Thus, reduction in LH levels may serve as potential therapeutics for AD. Also, in males there is a significant reduction in testosterone and elevation in LH levels with ageing. This is a potential risk factor for development of AD in males. A study showed that gonadectomised (GDX) mice had increased levels of Aβ, and the levels of Aβ were significantly lowered on treatment with testosterone [70]. This suggests potential of androgen therapy for treating AD in hypogonadal men.

2.2.8 Pathogens

Intracellular bacterial pathogen *Chlamydophila (Chlamydia) pneumoniae* is a risk factor for AD. Studies have shown presence of *Chlamydia* in the brains of LOAD patients [71]. Also, the presence of herpes simplex virus type 1 (HSV1) in the brain of *APOE4*-є4 carriers is a risk factor for development of AD [72]. In a recent finding, association between periodontal disease and AD was established by the presence of elevated serum antibodies against periodontal infection, caused by bacteria, in individuals who later developed AD as compared to serum antibody levels in control individuals [73]. The mechanisms leading to the pathogenesis of AD due to the presence of these pathogens in the brain need to be understood to a greater extent.

2.2.9 Metal Exposure

Exposure to metals has been linked to AD from a very long time. A recent study has shown that elevated magnesium (Mg^{2+}) in the brain has a synapto-protective effect and improves cognition deficits by reducing Aβ plaques and stabilising BACE1 expression in a transgenic mouse model of AD [74]. An earlier study had shown that reduced Mg^{2+} levels are present in AD patients

[75]. This shows that restoring brain Mg^{2+} levels has a potential to treat AD. Elevated levels of aluminium (Al) are seen in the serum of AD patients [76]. In line with this study, recently it was shown that Al may mediate liver toxicity in AD patients and lead to free copper (Cu^{2+}) in serum, as seen in AD patients [77]. This study has put forward a likely mechanism showing that Al toxicity to liver leads to abnormal synthesis of ceruloplasmin and ATPase7B causing increase in Cu^{2+} levels in serum, and these free Cu^{2+} may cause accumulation of Aβ and neuronal dysfunction associated with AD [77]. Thus, reducing Al levels and reviving normal liver function can be thought of as a treatment for AD. Another significant study has showed that drinking silicon (Si)-rich mineral water reduced Al levels, leading to the removal of Al via urine and thus improved cognition in AD patients [78]. Lead (Pb) exposure has also been linked to the pathogenesis of AD. Pb is shown to increase APP expression and change methylation patterns of *APP* gene, making it hypomethylated in PC12 cells [79]. Another study showed that Pb exposure facilitates Aβ fibril formation and increases Aβ deposition in a transgenic mouse brain [80]. Another evidence supporting the role of Pb in the pathogenesis of AD showed increased fibrillisation of tau protein on exposure to Pb via interaction with His330 and His362, the His mutants of wild-type tau [81]. Inorganic Cu^{2+}, present in drinking water and Cu^{2+} supplements, has also been linked to AD and cognitive impairment in the elderly [82]. A recent study has demonstrated that exposure to iron (Fe^{2+}) leads to upregulation of a disintegrin and metalloproteinase 10 (ADAM10) and increased transcription levels of BACE1, and these were reported to be associated with increased expression of APP-α-CTF and APP-β-CTF, respectively, in PC12 cells, suggesting that Fe^{2-} induces enhanced expression of ADAM10/BACE1 leading to altered APP carboxyl-terminal processing [83]. Increased intracellular calcium (Ca^{2+}) levels brought about by oligomeric Aβ are associated with impaired synaptic plasticity. A study pointed towards the synaptic loss of phosphorylated (active) Ca^{2+}/calmodulin-dependent protein kinase II-α[p(Thr286)CaMKII], a critical enzyme mediating synaptic events, in the MCI and AD hippocampal regions. The loss of p(Thr286) CaMKII was also observed in mice hippocampal regions on treatment with oligomeric Aβ. This was shown to be prevented by inhibiting the phosphatase calcineurin (CaN). Thus, dysregulated Ca^{2+} signaling is associated with AD and MCI [84]. Altered Zinc (Zn^{2+}) homeostasis is implicated in AD. Higher concentrations of releasable Zn^{2+} are present in synaptic vesicles as compared to extracellular fractions in AD hippocampus [85]. In line with this, Bjorklund et al. also showed that in the case of non-demented with AD neuropathology (NDAN) individuals, Aβ oligomers are absent in hippocampal postsynapses along with lower total Zn^{2+} levels, thus leading to intact cognitive function in NDAN individuals [86]. Excess exposure to Zn^{2+} leads to enhanced APP processing, Aβ accumulation and memory impairment as seen in transgenic mice and SH-SY5Y human neuroblastoma cells overexpressing the Swedish mutant form of human *APP* (*APPsw*). Thus, Zn^{2+} overload has a toxic role in AD pathogenesis [87].

2.2.10 Air Pollution

There is a possible association between air pollution and development of AD. Studies have shown that air pollution causes accumulation of Aβ(1-42), increase in cyclooxigenase-2 (COX-2) expression and brain inflammation that cause neuronal dysfunction leading to pathological hallmarks that are associated with AD [88]. Damage to olfactory bulb, olfactory mucosa and frontal regions of the brain, as observed in AD brains, is associated with exposure to air pollution [89]. Ozone (O_3) is a powerful gaseous air pollutant and oxidising agent. It is shown that memory impairment, motor deficiency and lipid peroxidation are caused by exposure to different doses of O_3 in rats [90]. Dysregulation of inflammatory processes, progressive neurodegeneration, chronic loss of brain repair in the hippocampus and brain plasticity changes occurred in rats on exposure to low doses of O_3 [91]. Thus, air pollution is an important environmental hazard which may play a role in development of AD. Additional studies are needed at the population level to clearly understand the role of air pollution in pathophysiology of AD.

2.3 Amyloid Proteins (Oligomers and Fibrils)

The Aβ protein is central to the amyloid cascade hypothesis, the prevailing hypothesis for more than 20 years, which explains the pathogenesis of AD. Though it does not pinpoint towards a specific Aβ species in the aetiology of AD, several studies have implicated soluble oligomers of Aβ, rather than monomers or insoluble amyloid fibrils, in causing neuronal dysfunction in AD [2, 92, 93].

Abnormal processing of APP in the neurons leads to generation and accumulation of Aβ. APP is a type I, single-pass transmembrane protein with large extracellular domains and belongs to a family of proteins which include amyloid precursor-like proteins (APLP1 and APLP2) in mammals and the amyloid precursor protein-like (APPL) in drosophila [94]. APP contains a 40- or 42-amino acid sequence, i.e. Aβ(1-40) or Aβ(1-42), and 3 sites for cleavage by various proteinases, which are designated as α, β and γ secretases [95]. APP is expressed in all tissues ubiquitously and present on the plasma membrane, majority of which localises to endoplasmic reticulum (ER), Golgi apparatus (GA) and mitochondria [96]. Two pathways exist for processing of APP, viz. amyloidogenic pathway and non-amyloidogenic pathway as illustrated in Fig. 1. In the amyloidogenic pathway, the activity of β-secretase (also known as BACE1) at the beginning of Aβ domain of APP results into a soluble N-terminal fragment (sAPPβ) and an amyloidogenic C-terminal fragment of 99 amino acids (CTF99). Further γ-secretase cleaves this C-terminal fragment to mainly Aβ(1-40) and Aβ(1-42), which may accumulate in the mitochondria and other cellular compartments affecting the cellular functions and leading to mitochondrial dysfunction and hyperphosphorylation of tau. Non-amyloidogenic pathway is the major APP processing pathway in which the activity of α-secretase generates a large soluble N-terminal fragment (sAPPβ) and a non-amyloidogenic C-terminal fragment of 83 residues (CTF83) owing to the cleavage within the Aβ domain. This C-terminal fragment is then cleaved by γ-secretase resulting into non-amyloidogenic peptide (P3) and APP intracellular domain (AICD) which are non-toxic and degraded rapidly and occurs in majority of individuals including non-demented and healthy individuals [97]. In case of early-onset AD, mutations in *APP, PS1* and *PS2* are known to activate β- and γ-secretases leading to generation of Aβ, but in case of sporadic AD, it is proposed that oxidative stress activates β-secretase which increases the production of Aβ [97]. Although the mechanism of APP trafficking is not known, APP is said to be axonally transported, endocytosed and sorted to different compartments of the cell, thus leading to Aβ generation and accumulation in different cellular compartments thus, impairing normal cellular function. A recent study pointed towards the role of huntingtin-associated protein 1 in regulating APP trafficking to the non-amyloidogenic pathway resulting into reduced Aβ levels [98]. As excessive BACE1 expression and APP processing can lead to uncontrolled production of Aβ, different regulatory mechanisms are present [99, 100]. Apart from the transcriptional control, a complex network of neurotransmitter systems and translational regulation is present. Amongst the different systems, viz. glutamatergic, adrenergic, serotonergic, cholinergic and dopaminergic systems, cholinergic system is known to be affected in the early stages of AD. A study showed that downregulation of M2 acetylcholine receptor in the brains of AD patients affects a number of AD-relevant genes including BACE1 [99].

In the amyloidogenic pathway, different lengths of Aβ are produced ranging from 37 to 43 amino acids. Aβ(1-40) is the most dominant species produced from the processing of APP. In comparison to Aβ(1-40), only a small amount of Aβ(1-42) is produced in the human brain in the ratio of approximately 99 to 1. But Aβ(1-42) is the main component of the amyloid deposits associated with AD and has a tendency to form aggregates spontaneously and, thus, may form oligomers. Therefore, Aβ(1-42) is considered more neurotoxic than Aβ(1-40) [101]. While both Aβ(1-40) and Aβ(1-42) are capable of forming

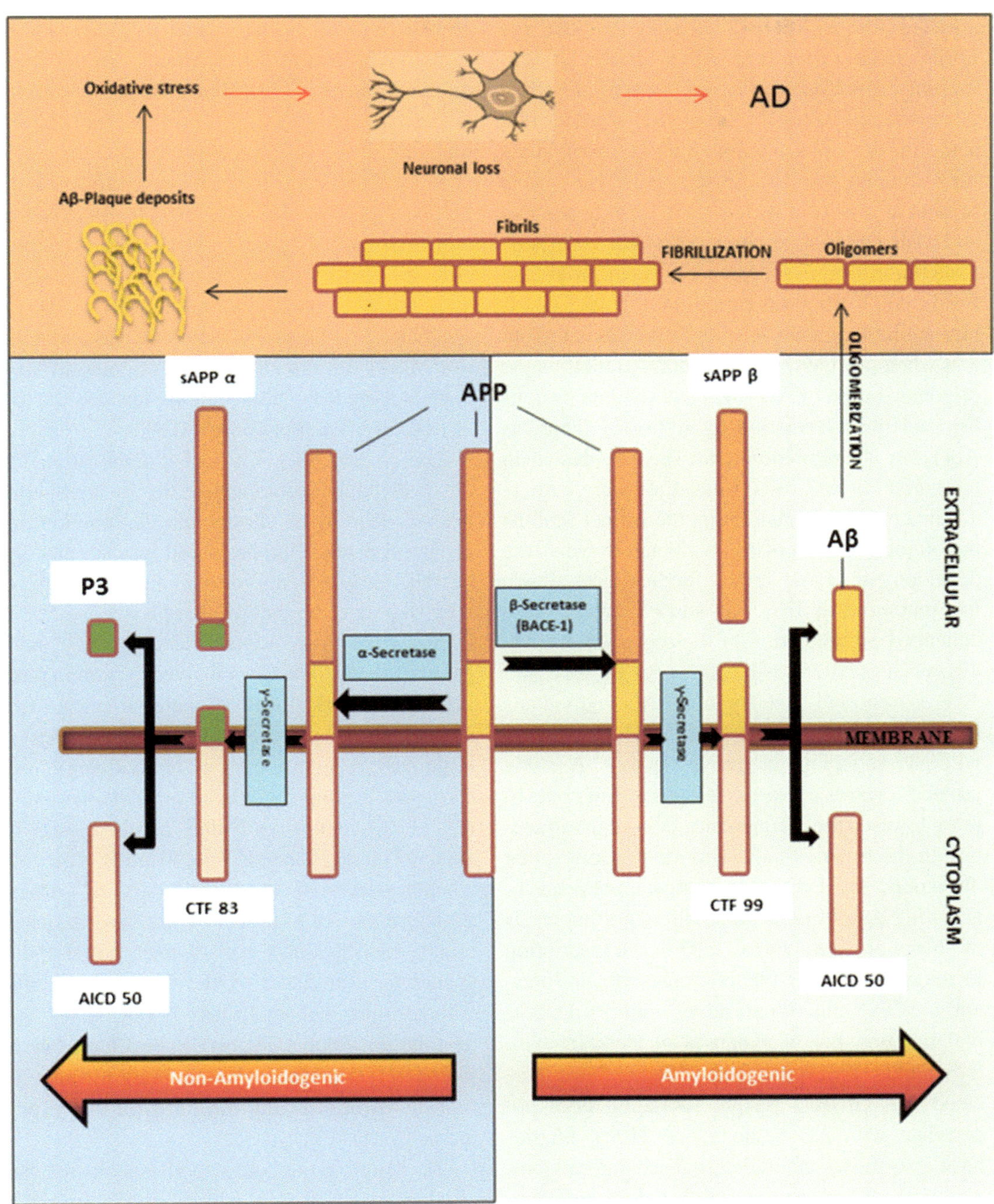

Fig. 1 APP processing leads to the generation of cleavage products via the amyloidogenic and non-amyloidogenic pathways. Both oligomerisation and aggregation of Aβ lead to the formation of senile plaques, a hallmark of AD phenotype. APP processing occurs via two pathways: amyloidogenic and non-amyloidogenic pathway. In the amyloidogenic pathway, β-secretase cleaves APP at the beginning of Aβ domain generating a soluble N-terminus fragment (*sAPPβ*) and amyloidogenic C-terminal fragment of 99 residues (*CTF 99*). Next, γ-secretase acts on CTF 99 and generates Aβ and APP intracellular domain (*AICD 50*). While in the α-secretase-based non-amyloidogenic pathway, α-secretase cleaves within the Aβ domain of APP and generates a soluble N-terminal fragment (*sAPPα*) and a non-amyloidogenic C-terminal fragment of 83 amino acid residues (*CTF 83*). This C-terminal fragment, i.e. CTF 83, is further cleaved by γ-secretase and generates a non-amyloidogenic peptide (*P3*) and APP intracellular domain (*AICD 50*). The amyloidogenic pathway, which generates the toxic Aβ, occurs in 10 % of individuals who have chances of developing AD in the later life. The generated Aβ(1-40) or Aβ(1-42) may further undergo different conformations including oligomerisation resulting into generation of oligomers which are considered to be the most toxic species of Aβ implicated in AD. Further fibrillisation may take place leading to fibril formation and deposition in the neurons, a characteristic feature of AD

amyloid fibrils, Aβ(1-42) is said to make the fibrils much faster than Aβ(1-40). Using ion-mobility spectrometry, it was shown that when Aβ(1-40) and Aβ(1-42) are present together in a solution, the Aβ(1-40) and Aβ(1-42) monomers form dimers, trimers and tetramers. But, when present alone, Aβ(1-42) tends to form pentamers and hexamers (paranuclei) which on self-association form dodecamers, protofibrils and fibrils. Aβ(1-40) alone produces oligomer distribution up to tetramer level only. This pointed out that when present together, Aβ(1-40) inhibits the oligomerisation of Aβ(1-42) and inhibits protofibril and fibril formation by Aβ(1-42). Thus, as Aβ(1-40) is the predominant species present in the human brain in ~10 times the level of Aβ(1-42) in a healthy human brain, the former inhibits the oligomerisation of Aβ(1-42) and prevents the development of AD [102]. Studies have shown that mutations in *APP*, *PS1* and *PS2* may lead to enhanced accumulation of the more toxic Aβ(1-42) species [101, 103–105].

It is reported that some NDAN individuals have significant amounts of amyloid plaques without displaying cognitive decline. A study pointed towards a resistance mechanism contributing towards the maintenance of cognitive function in these individuals. This study showed that oligomeric Aβ is absent in hippocampal postsynapses in NDAN brains. In addition, normal levels of phosphorylated (active) CREB, a transcription factor important for synaptic plasticity, are present in NDAN individuals advocating for the normal functionality of synapses in these in these individuals [86]. A number of studies have indicated that amyloid plaque formation does not correlate with AD pathogenesis [106]. Studies have also shown that deleterious changes associated with AD occur very early before the accumulation of amyloid plaques [107]. Accumulating evidences reveal that oligomers, the soluble form of Aβ, are the more toxic species and associated with cognitive decline in AD, rather than the insoluble fibrillar deposits [108]. Further, different approaches have been employed which have supported the above findings and implicated soluble Aβ in AD. Hippocampal long-term potentiation (LTP) is one such approach, which is correlated with learning and memory, and is shown to be inhibited by synthetic and naturally secreted human Aβ oligomers. Behavioural studies performed on living wild-type rats showed that rats developed learning and memory deficiency after human oligomers were infused in the hippocampus, which were readily observed in morris water maze models [101]. Recently, a possible role of cellular prion protein (PrPc) acting as a neuronal receptor for oligomeric Aβ was speculated. It was shown that there was no inhibition of LTP after treatment with oligomeric Aβ in mice lacking PrPc, thus pointing towards the role of PrPc in Aβ neurotoxicity [109].

Oligomeric Aβ(1-42) is also demonstrated to be involved in inducing toxicity in cholinergic neurons leading to cholinergic dysfunction and progressive basal forebrain cell loss, assumed to be an early event in the pathogenesis of AD. Heinitz et al. showed that on treating SN56.B5.G4 cells with oligomeric Aβ(1-42), many genes of the ER and GA involved in protein modification and degradation were affected. This indicated a possible role of ER-mediated stress in oligomeric Aβ(1-42) toxicity in cholinergic neurons and leading to cholinergic dysfunction in AD [110]. In line with this, another study by Joerchel et al. showed that SN56.B5.G4 cells when treated with oligomeric Aβ(1-42) affects the expression of a number of proteins, viz. calreticulin, MAPK kinase 6c, γ-actin, Rho-GDP dissociation inhibitor (Rho-GDI), ubiquitin carboxyl-terminal hydrolase-1 (UCHL-1) and α6-tubulin, which are known to be affected in the brains of AD patients, thus pointing towards the role of Aβ in affecting the integrity of the proteome in AD [111].

The amyloid cascade hypothesis has reached another level of complexity with recent studies revealing the appearance of multiple types of Aβ oligomers and their role in the pathogenesis of AD [2, 112, 113]. Various studies have recognised four endogenously produced Aβ oligomeric assemblies, viz. dimers, trimers, Aβ*56 and APFs (annular protofibrils), which may alter neuronal function in human and transgenic mice and have different consequences on neuronal survival [114]. In case of AD, soluble Aβ monomers may form higher-order assemblies ranging from low-molecular weight oligomers (dimers and trimers) to Aβ*56, then to APFs and fibrils, which are the primary components of amyloid plaques, characteristic of AD.

Taken together, recent studies provide the evidence that Aβ oligomers are the more toxic species than insoluble fibrillar deposits relevant to AD pathology (Fig. 1). Also, soluble monomers can form higher-order assemblies finally leading to fibril formation which is the characteristic component of amyloid plaques. Thus, development of antibodies against specific oligomeric species can be an effective approach for treating AD.

2.4 Role of Mitochondria in AD Pathology

The 'Mitochondrial cascade hypothesis' was formulated in 2004, which provides an explanation of the aetiology of sporadic AD [115]. According to the 'mitochondrial cascade hypothesis', ageing and sporadic AD are two convergent events, and the etiological factors for autosomal dominant and sporadic AD are not the same. Also, mitochondrial dysfunction has been viewed as the common element between autosomal dominant and sporadic AD forms. In addition to sporadic AD, this hypothesis also predicts the aetiology of autosomal dominant AD forms. In case of autosomal dominant forms, it points out that excessive Aβ causes mitochondrial dysfunction and this Aβ-induced dysfunction further initiates histopathologies associated with AD. For sporadic AD cases, it is believed that age-related mitochondrial changes cause mitochondrial dysfunction and activate downstream cellular changes as observed in sporadic AD which include processing of APP to Aβ, tau phosphorylation, synaptic loss and finally neurodegeneration [6]. Thus, mitochondrial cascade believes Aβ accumulation as a downstream event in the cascade.

Mounting evidences are present which point to an association between ageing and mitochondrial dysfunction. An earlier study pointed out that somatic mutations in mtDNA can lead to premature onset of ageing phenotypes [116]. But it is not known whether the decline in mitochondrial function is a cause or outcome of ageing.

Of particular interest, a large number of evidences from studies involving experimental models and human samples suggest an association of APP and Aβ with the mitochondria. In particular, a study involving a mouse model showed that Aβ, particularly Aβ (1-42), accumulates in the mitochondria in the presence of a mutant *APP* gene and there is a decline in the activity of respiratory complexes III and IV. Also, Aβ was shown to accumulate in the mitochondria very early before the extensive extracellular deposition takes place [117]. These point towards the involvement of the mitochondria in the pathogenesis of AD (Fig. 2).

In addition, it is reported that age-related accumulation of somatic mutations in mtDNA causes an increase in ROS production in the mitochondria [118, 119]. It is known that oxidative stress induces an increased expression of BACE1 [120, 121]. Thus, it is speculated that with advanced age, an increase in ROS levels occur which increases Aβ levels owing to increased processing of APP by BACE1. Also, this generated Aβ induces an increased production of free radicals (ROS and RNS), causing decline in mitochondrial function. Thus, a feedback loop exists in which age-related ROS levels cause increased production of Aβ, and this Aβ further leads to increased production of ROS, resulting in memory impairment and cognitive decline associated with AD [36].

Several evidences point out that Aβ interacts with mitochondrial proteins and increase ROS levels, finally leading to synaptic damage. Mounting evidences show that Aβ, in its soluble oligomeric form, disrupts the axonal mitochondrial transport and disturbs the mitochondrial fission-fusion balance [122, 123]. Mitochondrial trafficking/transport is a phenomenon in which synaptic mitochondria, synthesised in the cell body, are transported down the axon/dendrite to areas of high energy demand, thus serving cellular energy demands [97]. During the transport, the

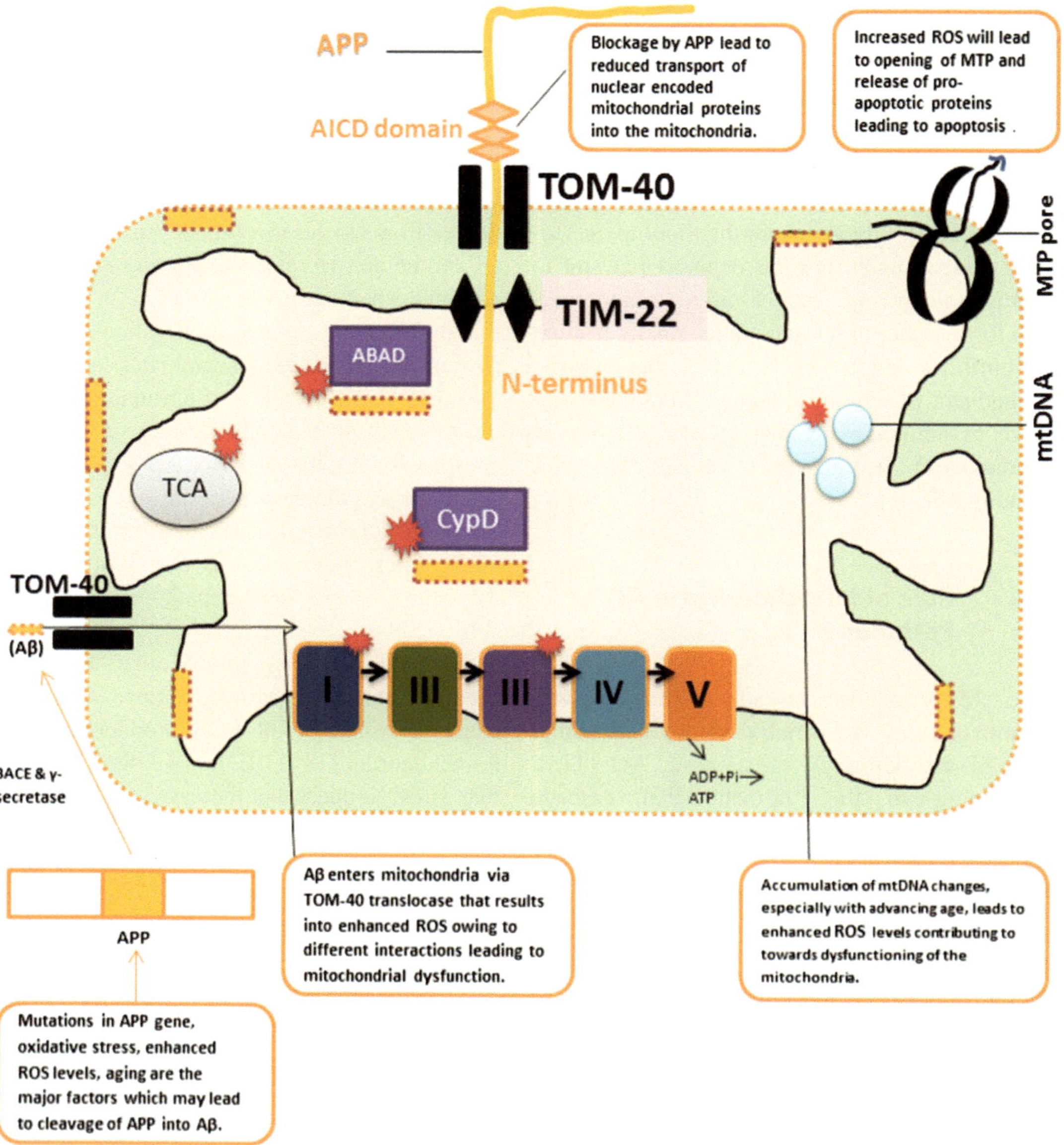

Fig. 2 Aβ induces mitochondrial dysfunction that results in neurodegeneration and AD. A mitochondrion is surrounded by two lipid membranes: outer membrane and inner membrane. The outer membrane is porous, while the inner membrane restricts ionic flow and harbours the electron transport chain (*ETC*). ETC is composed of complexes I–V and is responsible for the generation of mitochondrial ATP via oxidative phosphorylation. Electron leaks in complexes I and III are responsible for the generation of free radicals in the mitochondria. Components of tricarboxylic acid cycle (*TCA*), viz. α-ketoglutarate dehydrogenase and beta oxidation, are present in the mitochondrial matrix and generates superoxide radicals. These generated radicals thus cause lipid peroxidation and protein and DNA oxidation. In case of early-onset AD, genetic mutations in *APP*, *PS 1* and *PS 2* activate beta- and gamma-secretase and lead to increased processing of APP to Aβ. This Aβ is found to accumulate in the outer, inner mitochondrial membrane and the matrix. When associated with the outer membrane, it causes blockage of the entry of nuclear-encoded mitochondrial proteins. On being localised to inner membrane, it directly induces free radical production, decreases cytochrome oxidase activity, interferes with complex activities and impairs ATP production. Once inside, Aβ interacts with mitochondrial matrix proteins (ABAD and cyclophilin D) which increases the oxidative stress and worsens the mitochondrial damage. In case of late-onset AD, ageing is one of the major factors contributing to increased free radical production owing to mutations in mtDNA. APP is also said to be localised to mitochondrial outer membrane and forms complexes with the translocases of outer (*TOM*) and inner membrane (*TIM*). But the importation of C-terminus of APP is blocked due to the AICD domain of APP and thus blocks the mitochondrial pores. This prevents the entry of nuclear-encoded mitochondrial proteins causing impairment of mitochondrial enzyme activities, increased oxidative stress, neuronal damage and cognitive decline

mitochondria can encounter each other, leading to fusion (aided by fusion proteins) and subsequently resulting in exchange of mitochondrial content. This is necessary to maintain genome stability as it allows exchange of highly pathogenic mtDNA and helps to maintain mitochondrial function [124]. Mitochondrial fission plays an important role in apoptosis. Unbalanced fission and fusion results in disruption of mitochondrial dynamics, causing mitochondrial fragmentation and contributing to mitochondrial dysfunction [125], which has been speculated to be the underlying mechanism(s) causing synaptic degeneration in AD. Recently, it was shown that both monomers and oligomers of Aβ interact with mitochondrial fission protein, dynamin-related protein 1 (Drp1) and may be involved in abnormal mitochondrial dynamics causing mitochondrial fragmentation leading to neuronal dysfunction in AD. Also, expression levels of mitochondrial fission genes *Drp1* and *Fis1* (Fission 1) were increased and those of mitochondrial fusion genes *Mfn1* (mitofusin 1), *Mfn2* (mitofusin 2), *Opa1* (optic atrophy 1) and *Tom40* were decreased in AD patients [37]. This study also provided evidence that mitochondrial fragmentation may be initiated by interaction of Aβ with Drp1 causing abnormal mitochondrial dynamics, which increases as AD progresses. Thus, targeting these abnormal interactions can serve to minimise the neuronal damage caused by AD.

It has been shown that interaction of Aβ with Aβ-binding alcohol dehydrogenase (ABAD) in the mitochondria causes enhanced ROS production and apoptosis in AD patients and transgenic mice [126]. Some recent studies have implicated mitochondrial permeability transition pore (mPTP) formation in Aβ-mediated mitochondrial dysfunction [127]. The mPTP consists of cyclophilin D (CypD) in the mitochondrial matrix, voltage-dependent anion channel (VDAC) in outer mitochondrial membrane and adenine nucleotide translocase (ANT) in inner mitochondrial membrane. CypD has a role in opening of mPTP by binding with ANT and VDAC after its release from the matrix [38]. In relation to VDAC1, a recent study has shown that interaction of Aβ and phosphorylated tau with VDAC1 blocks mitochondrial pore leading to mitochondrial dysfunction in AD patients and transgenic APP mice [12]. Another mitochondrial protein CypD is shown to enhance mitochondrial and neuronal stress by interacting with Aβ. This interaction promotes ROS which leads to recruitment of CypD to the inner mitochondrial membrane and results in opening of mPTP causing cell death [128]. In addition, this study showed that CypD deficiency improved learning and memory in a mouse model of AD. A recent study has provided a new insight into the role of CypD in the disruption of axonal mitochondrial transport [38]. It was shown that depletion of CypD protects the Aβ-induced axonal mitochondrial transport damage and improves mitochondrial motility and dynamics. An increase in axonal mitochondrial density and bidirectional transport of the axonal mitochondria was observed on CypD depletion. CypD in the presence of Aβ promoted the opening of mPTP and, thus, disrupted Ca^{2+} homeostasis and increased accumulation of ROS, further activating P38 MAPK signal transduction pathway, causing synaptic injury. It was also demonstrated that CypD-mediated mPTP blockade improved synaptic function against Aβ toxicity using CypD-deficient mice. This study suggests a relation between CypD-mediated disruption of axonal mitochondrial trafficking and Aβ-induced mitochondrial dysfunction leading to synaptic injury, thus speculates a mechanism of mitochondrial dysfunction involved in the pathogenesis of AD (Fig. 2).

Another study has shown that age-related accumulation of Aβ occurs to a larger extent in the synaptic mitochondria as compared to the non-synaptic mitochondria and this accumulation in the synaptic mitochondria occurs very early as compared to its accumulation in the non-synaptic mitochondria. This led to altered mitochondrial transport in murine primary axons [129]. Thus, it can be ascertained that the synaptic mitochondria are more probable targets of Aβ-induced oxidative stress. Sirtuin 3 (SIRT3), a deacetylase has an essential role in maintaining mitochondrial function. Studies have highlighted its neuroprotective role and its role in energy homeostasis by maintaining basal ATP levels for survival of the cells [130, 131]. A recent study has shown upregulation of SIRT3 in relation to Aβ-accumulation in AD patients and transgenic AD mouse model. Also, it was shown that mitochondrial ROS levels regulated SIRT3 expression

[132]. This speculates that upregulation of SIRT3, seen in AD patients and mouse model, may be due to Aβ-induced mitochondrial oxidative stress. This suggests that in response to Aβ-induced oxidative stress in the mitochondria, upregulation of SIRT3 occurs to promote neuronal survival.

Recently, the RanBP9-cofilin pathway has been implicated in AD when it was shown that RanBP9, a scaffolding protein, generates Aβ and promotes Aβ-induced neurotoxity along with activating cofilin, having a key role in regulating actin dynamics and mitochondria-mediated apoptosis in a mouse model of AD pathology [133]. Also, suppression of Aβ and RanBP9-induced apoptosis by siRNA knockdown of cofilin confirmed the role of RanBP9-cofilin pathway in AD and can be the probable therapeutic target for lowering Aβ-induced neurotoxicity.

A study by Kopeikina et al. showed that soluble tau species are more toxic than tau aggregates and cause mitochondrial distribution deficiencies in a mouse model of tautopathy and in the human AD brains, possibly due to axonal transport deficiencies resulting in mitochondrial and neuronal dysfunction [134]. This suggests that soluble tau like the oligomeric Aβ are more toxic than aggregated form and involved in neuronal dysfunction associated with AD.

Poly (ADP-ribose) polymerase-1 (PARP-1), a predominantly nuclear enzyme responsible for genome stability and transcriptional regulation, is thought to be involved in the pathogenesis of AD. Activation of PARP-1 by oxidative stress (induced by Aβ) is believed to be an early event in the pathogenesis of AD. Enhancement of PARP-1 activity and accumulation of PAR is observed in the brains of AD patients [135]. A recent study has established the role of PARP-1 in microglial activation by its interaction with NF-kB [136]. Recently, the mitochondrial localisation of PARP-1 and its interaction with mitochondrial protein, mitofilin, has been established [137]. It was shown that in the absence of PARP-1, there is an accumulation of mtDNA damage, suggesting that mitochondrial PARP-1 has a role in mtDNA damage repair/signaling. Overexpression of PARP-1, Bax and p53 and altered mitochondrial function in the presence of oxidative stress induced by Aβ has also been shown recently [138]. Additional studies are needed to ascertain the role of PARP-1 and mitochondrial dysfunction in AD pathogenesis.

Taken together, these studies suggest that mitochondrial dysfunction is associated with ageing and AD. Along with Aβ, age-related accumulation of somatic mutations in mtDNA increases mitochondrial ROS levels leading to a decline in mitochondrial function. Recent studies have pointed out the role of oligomeric Aβ in the pathogenesis of AD. Oligomeric Aβ and APP are said to localise to mitochondria, mainly synaptic mitochondria, interact with mitochondrial proteins and disrupt axonal mitochondrial trafficking, causing synaptic injury and cognitive impairment. In relation to this, the use of mitochondria-targeted antioxidants can be seen as important approaches to treat AD.

3 Apurinic/Apyrimidinic Endonuclease (APE1): An Emerging Neuroprotective Enzyme

There are many external and internal agents, which bring human genome under stress and finally bring modification in the genomic stability. These threats are mainly produced internally from mitochondrial electron transport chain (ETC) or externally by different biological, chemical and physical agents like ultraviolet (UV) rays, ionising radiation (IR), chemotherapeutic agents, pollutants and heavy metals [139]. ROS attacks DNA readily and generates a variety of DNA base lesions [140]. DNA damage is a continuous process and ~10^4 DNA lesions are estimated to be produced in a mammalian genome each day as a result of spontaneous decay, errors in replication and cellular metabolism. To maintain genomic integrity, a cell has an internal regulatory mechanism which maintains DNA damage and repair in a balanced condition. There are two pathways by which cells maintain genome integrity: (i) antioxidants which quench the ROS/RNS, nonenzymatic antioxidants (e.g. α-tocopherol, β-carotene, lycopene and ascorbic acid) and enzymatic antioxidants (e.g. superoxide dismutase (SOD), glutathione (GSH), peroxidases and catalase) and (ii) DNA repair by different processes. The DNA repair system comprises of base excision repair (BER), transcription-coupled repair (TCR), global genome repair (GGR), mismatch repair (MMR), homologous recombination (HR) and nonhomologous end joining [NHEJ] [141].

In mammals and higher organisms, different organs consist of various cell types; some of them are dividing while others are nondividing. In adults, cell types such as myocytes, adipocytes, skin cells and neurons are nondividing cells, i.e. terminally differentiated [142, 143]. BER is the major pathway for oxidative DNA base damage caused by ROS/RNS as well as for abasic (AP) sites and single-strand breaks (SSBs). Apurinic/apyrimidinic endonuclease (APE1) is a primary BER enzyme and responsible for repair and removal of AP sites and strand breaks [16, 17, 144].

Human *APE gene* (~3 kb in size) is localised on chromosome 14q11.2-12 and consists of four introns and five exons [145]. The human APE *cDNA* is about 1.4 kb in length and encompasses a coding region of 954 nucleotides and encodes a protein comprising of 318 amino acids. APE1 is abundant (~10^5 copies per cell) in eukaryotic cells and has a relatively long half-life [~8 h] [146]. APE1 is a dual function protein. Its C-terminus displays repair activity and its N-terminal contains a bipartite nuclear localisation signal, NLS [18, 147, 148] and displays redox activity responsible for transcriptional regulation through redox-based mechanisms [18, 149, 150].

3.1 Role of APE1 in Oxidative DNA Damage Repair

The ROS-induced damage to the DNA is implicated in a number of human diseases including neurodegenerative diseases like AD, PD, HD and cancers [16, 17, 151]. It is thus very important to repair the ROS-induced DNA damage in order to maintain the genomic integrity. BER, an evolutionarily conserved process, is responsible for repairing most endogenous lesions like oxidised bases, AP sites and SSBs in both nuclear DNA and mitochondrial DNA. The basic BER pathway involves enzymes, viz. DNA glycosylase, APE1, DNA polymerase and DNA ligase. APE1 is involved in the repair of oxidised base lesions generated in the DNA as a result of oxidative damage. Attempts to generate APE1-null mice were not successful and lead to an early embryonic death [152, 153]. Further attempts to generate cell lines from APE1-null embryos failed, showing the essentiality of APE1 in maintaining cell viability. A study pointing towards the role of APE1 in neuronal cell survival showed that overexpression of APE1 in hippocampal and sensory cells exposed to H_2O_2 lead to an increase in cell viability [154]. Upregulation of APE1 in cerebral cortical region of AD patients was also seen [155]. An immunohistochemical study pointing towards the role of APE1/Ref-1 in regulating cellular response towards oxidative stress showed that increased nuclear expression of APE1/Ref-1 is present in cerebral cortical regions of AD patients [20]. Another study showed the colocalisation of APE1/Ref-1 with Aβ in the senile plaques in AD hippocampus [156]. This study also showed that varying concentrations of Aβ(1-42) regulates APE1/Ref-1 expression, thus pointing towards the neuroprotective role of APE1/Ref-1 in response to oxidative stress. A number of evidences point towards the role of cyclin-dependent kinase 5 (Cdk5) in mediating neuronal loss. In line with this, it was shown that Cdk5 complexes with p35 and phosphorylates APE1 at Thr232, causing reduction in APE1's endonuclease activity and leading to accumulation of DNA damage and neuronal loss [157]. It can be interpreted that APE1 has a major role in overcoming the oxidative stress and maintaining neuronal cell viability and integrity.

3.1.1 Nuclear BER Pathway

A number of DNA repair pathways operate in the nucleus. Amongst them, BER pathway is the most versatile repair pathway operating in the nucleus in response to oxidative damage for repairing alkylated and oxidised DNA lesions, AP sites and SSBs. Two models of BER are present: short-patch BER (SN-BER) and long-patch BER (LP-BER). SN-BER involves removal of a DNA lesion and incorporation of a single nucleotide, while a patch size of 2–8 nucleotides is associated with LP-BER [17, 144, 158]. The choice of the pathway depends on factors like type of lesions, AP sites and nature of 5′ terminus. The first step of the BER pathway is recognition and excision of a damaged base by DNA glycosylase. Two types of DNA glycosylases are present: monofunctional and bifunctional. Monofunctional DNA glycosylases (M-DG) include thymine DNA glycosylase (TDG), uracil-DNA glycosylase 1 (UDG1) and MutY homolog (MUTYH) and excise the substrate base, e.g. alkylated bases and uracil, generating an AP site which is later processed by APE1. Bifunctional DNA glycosylases (B-DG) which include 8-oxoguanine DNA glycosylase (OGG1) and *Nei*-like-1 (NEIL1), *Nei*-like-2 (NEIL2) and endonuclease III-like 1 (NTH1) have an additional lyase activity specific for oxidised bases and incise the DNA backbone 3′ to the AP site via β or β,γ elimination [144, 159–161]. The second step of BER involves processing of the generated AP site by APE1 that generates a nick containing a 3′OH residue and dRP at 5′ end, by cleaving the phosphodiester bond 5′ to the AP site. During the third step of the BER pathway, repair of the AP site is catalysed by polymerase (pol) β. If an unaltered group in deoxyribose is present, then pol β owing to its dRP lyase activity can carry SN-BER. The LP-BER occurs when AP sites are further oxidised by ROS and pol β cannot remove the 5′ blocking groups. 5′ flap-endonuclease-1 (FEN1), part of DNA replication machinery, is shown to displace and cleave this 5′ blocking group along with 4–6 nucleotides as a single-stranded DNA flap. PCNA also has a role to play by acting as a sliding clamp in LP-BER. The last step of BER involves nick sealing by DNA ligase which in case of LP-BER involves DNA pol ϵ/δ and DNA ligase I, and in case of SN-BER involves DNA ligase IIIα/XRCC1 complex [144, 160]. PARP-1 is known to modulate the capacity of BER and efficiently recognise and repair SSBs, thus acts as a DNA damage sensor and signal transducer [17, 162]. It was shown that Aβ is involved in the activation of PARP-1 through NO cascade in adult hippocampus [163]. Another protein XRCC1 also acts a SSB sensor protein and acts as a scaffold for BER proteins for SSB repair [144]. XRCC1 also interacts and stimulates APE1 [164]. Thus, it can be said that BER pathway is indeed a versatile pathway involved in maintaining genome integrity and involving a number of enzymes and interactions.

3.1.2 Mitochondrial BER Pathway

It is well known that the mitochondria have a role in the ageing process. Mitochondrial DNA (mtDNA) is more prone to oxidative DNA damage due to its close proximity to the site of ATP production in the inner mitochondrial membrane [165]. Thus, DNA damage repair in the mitochondria appears to be very important for maintaining proper functioning of the cell, especially during ageing. Various studies have shown that DNA repair actively takes place in the mitochondria, which was earlier thought to be present only in the nucleus. Recent studies have identified new DNA repair enzymes that participate in the DNA repair pathways operating in the mitochondria. Amongst the different repair pathways, BER is considered to be the major DNA repair pathway taking place in the mitochondria (Fig. 3). BER in the mitochondria helps to cope up with the oxidised DNA lesions generated due to the presence of free radicals and thus maintains mtDNA stability. The basic mechanism by which mitochondrial BER acts remains the same as the nuclear BER, but some specific BER enzymes are present in the mitochondria and these are coded by nuclear genes [166]. For a very long time, it was considered that only SN-BER occurs in the mitochondria which include removal of a DNA lesion and incorporation of a single nucleotide. But now it is believed that owing to the rate at which oxidised base lesions are generated in mtDNA, LP-BER may also take place in the mitochondria [17, 160, 167, 168]. The first step of the mtBER pathway involving recognition of a damaged base is

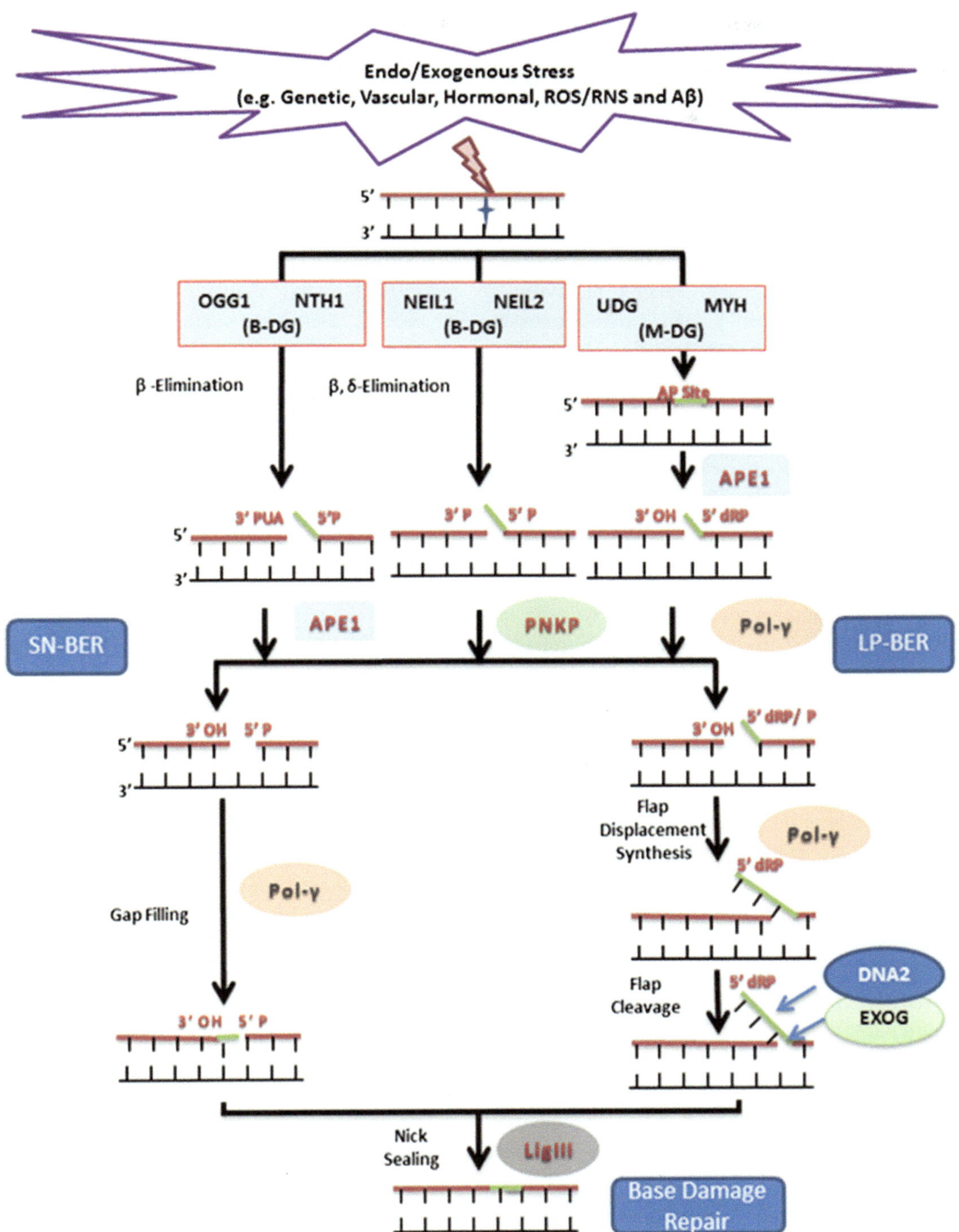

Fig. 3 Mitochondrial BER has more or less similar repair machinery as nuclear BER. DNA damage repair in the mitochondria is believed to be important in maintaining genomic integrity, especially during ageing. The basic mechanism of mitochondrial BER (*mtBER*) remains the same as nuclear-BER pathway, but some specific enzymes are present in the mtBER pathway. Both SN-BER and LP-BER are believed to take place in the mitochondria. Two M-DG, i.e. UDG1 and MYH, and four B-DG, viz. OGG1, NTH1, NEIL1 and NEIL2, are present in the mitochondria. APE1 in the mitochondria is found to be an N-terminal truncated product of APE1. In the mitochondria, DNA polγ is the only polymerase both in LP-BER and SN-BER pathways. Ligation of single-strand nick is performed by DNA ligase III. In addition, EXOG is an essential component of BER/SSBR pathway unique to the mitochondria and forms complex with APE1, DNA polγ and DNA ligase III and is involved in repairing endogenous SSBs in the mtDNA

performed by DNA glycosylase. Two M-DG, i.e. UDG1 and MYH, and four B-DG, viz. OGG1, NTH1, NEIL1 and NEIL2 are present in the mitochondria [169, 170]. Next step of the mtBER pathway involves processing of the AP site by APE1, which is the main AP endonuclease of the mammalian cell, and this mtAPE1 is believed to be an N-terminal truncated product of APE1. It is also shown that deletion of the 33 N-terminal residues increases the specific activity of mtAPE1 by threefold [171]. The next step involves insertion of correct nucleotides by DNA polγ which is followed by ligation of single-strand nick by DNA ligase III (Fig. 3). This mitochondrial ligase III is derived from *LIG3* gene and is known to be independent of XRCC1 while the nuclear variant of DNA ligase III interacts with XRCC1 [165, 172]. A recent study has shown that EXOG is an essential component of BER/SSBR pathway unique to the mitochondria and forms complex with APE1, DNA polγ and DNA ligase III and is involved in repairing endogenous SSBs in the mtDNA. Also, it was shown that depletion of EXOG increases ROS levels and induces apoptosis in normal cells [173]. FEN1 is involved in repairing oxidative DNA damage via LP-BER in the mitochondria [174]. Cellular death and embryonic lethality in presence of gamma radiation-induced DNA damage was observed in *FEN1* gene knockout mice [175]. The hDNA2, possessing nuclease, helicase and ATPase activities, is also involved in DNA replication and repair in the mitochondria. The hDNA2 forms a complex with polγ and stimulates the polymerase activity. It is also involved in RNA primer removal during mtDNA replication. The hDNA2, owing to its nuclease property, can also process flap LP-BER intermediates. This points towards the synergistic roles of FEN1 and hDNA2 to process the 5′flap intermediates during DNA replication and repair in the mitochondria [176]. Thus, all through these years we have gained knowledge about some of the repair pathways occurring in the mitochondria and identified different repair enzymes present in the mitochondria but much more needs to be understood towards establishing pathophysiologies which overtakes these repair processes.

3.2 Role of APE1 in Redox and Transcriptional Regulation

Owing to APE1's N-terminal domain which contains the NLS and redox regulatory domain, APE1/Ref-1 is considered to be an important mammalian redox regulator of transcription. It is identified that Cys65 is the redox active site in APE1/Ref-1 and that Cys93 interacts with Cys65 via disulphide bond formation and thus these two Cys residues contribute to alter the redox state of a number of TFs [177]. A number of studies have shown that APE1/Ref-1 modifies the DNA-binding ability of several TFs, such as activator protein-1 (AP-1), Fos and Jun, NF-κB, p53, Myb, early growth response-1 (Egr-1), polyoma virus enhancer-binding protein-2 (PEBP-2), activating transcription factor/cAMP response element-binding protein (ATF-CREB), hypoxia-inducible factor (HIF-1α) and HIF-like factor [18, 149, 178]. The reduction of Cys in the DNA-binding domains of the TFs by APE1/Ref-1 enhances the DNA-binding ability of TFs. It was earlier shown that reduction of the conserved Cys residue in the DNA-binding domain of c-Jun by APE1/Ref-1 enhances the DNA-binding activity of AP-1 in vitro [179]. Also, the ability to reactivate Fos-Jun DNA-binding declines on oxidation of APE1, which can be restored on treatment with thioredoxin, TRX [177]. Thus, it can be said that alterations in the redox state could lead to alterations in gene expression of key cellular signaling and other regulatory proteins. While APE1 is considered as a redox activator of several TFs like AP-1, p53, HIF-1α, it also acts as a trans-acting factor which causes Ca^{2+}-dependent repression of parathyroid hormone (PTH) and renin genes [180, 181]. An increase in extracellular Ca^{2+} causes binding of APE1 to negative Ca^{2+} response element (nCaRE: nCaRE-A and nCaRE-B) in the respective gene promoters causing repression of PTH and renin gene expression. Acetylation of APE1 is also shown to modulate APE1's transcriptional regulatory function. In addition, APE1 interacts stably with other trans-acting factors like HIF 1-α, STAT 3 and CBP/P300 [18, 153, 182]. Thus, it points towards the redox-independent functions of APE1 in regulating transcription.

3.3 Other Functions

Apart from being the major DNA repair enzyme of the BER pathway and redox activator of several TFs, newer studies have shown that APE1 serves some other important functions as discussed below.

3.3.1 APE1 as Endoribonuclease

For a very long time, RNA decay in eukaryotes was considered to be an exoribonucleolytic process, while in prokaryotes it was considered to be an endoribonucleolytic process. But numerous evidences in the recent past have demonstrated that endoribonucleases also have a significant role in eukaryotic RNA metabolism and contribute to RNA turnover in eukaryotes [183]. Recently, APE1 was identified as an endoribonuclease that cleaves within the UA and CA dinucleotides of *c-myc* (a proto-oncogene) mRNA and regulates the *c-myc* mRNA levels [7, 147]. It was shown that APE1 knockdown in HeLa cells led to increased *c-myc* mRNA levels and its half-life [147]. In line with this, a study identified the active site of APE1 and found that common active site is shared for endoribonuclease and nuclease activities of APE1 but the mechanisms of cleavage of RNA and DNA are not identical [184]. Thus, the role of APE1 in controlling the levels and turnover of other mRNAs in the neuronal as well as other cell types need to be understood.

3.3.2 Maintenance of Cellular Homeostasis

Maintenance of proper cellular redox balance is an essential prerequisite for proper functioning of biological systems. An increase in the ROS/RNS levels beyond the normal physiologic limits could disturb the redox homeostasis leading to cell death and disease development. To cope up with the increased oxidative stress build-up in the cell, different enzymatic/nonenzymatic antioxidant systems are present. APE1/Ref-1 is known to act as an important redox regulator of the cell which helps in maintaining proper levels of ROS/RNS for cell survival and proliferation. An earlier study showed that APE1/Ref-1 helps in regulating oxidative stress by inhibiting ROS production and NF-κB activation by modulating the activation of rac1 GTPase and inhibits apoptosis [185, 186]. Overexpression of APE1/Ref-1 was shown to increase SH-SY5Y cell viability following exposure to H_2O_2 [150]. A recent study of Mantha et al. has identified several key neuronal proteins those are involved in various cellular functions are interacting with APE1 in response to Aβ(25-35)-induced stress in PC12 and SH-SY5Y cells [187]. Thus, it can be interpreted that APE1/Ref-1 has a role in providing protection against oxidative stress and helps to maintain cellular redox balance.

4 Importance of Phytochemicals in Modulating Functions of APE1/Ref-1 Towards AD Therapeutics

Human beings have used plant extracts for centuries for treating various types of ailments. For the last few years, scientists are trying to figure out the active ingredients present in the plant extracts responsible for the specific action and decipher the molecular mechanism(s) by which these phytochemicals exert their action (Fig. 4). Phytochemicals like resveratrol, isoflavones, curcumin, decursin, EGCG, L-carnitine and *Ganoderma lucidum* extract have shown potent properties against neuronal disorders. In addition, some of them have also shown their effect on modulation of APE1/Ref-1 repair as well as redox activity in relation to cancer. Limited studies warrant additional studies to see how these phytochemicals may affect APE1/Ref-1 functions in neurodegenerative diseases like AD. The following sections describe these phytochemicals as having potential therapeutic effect against AD.

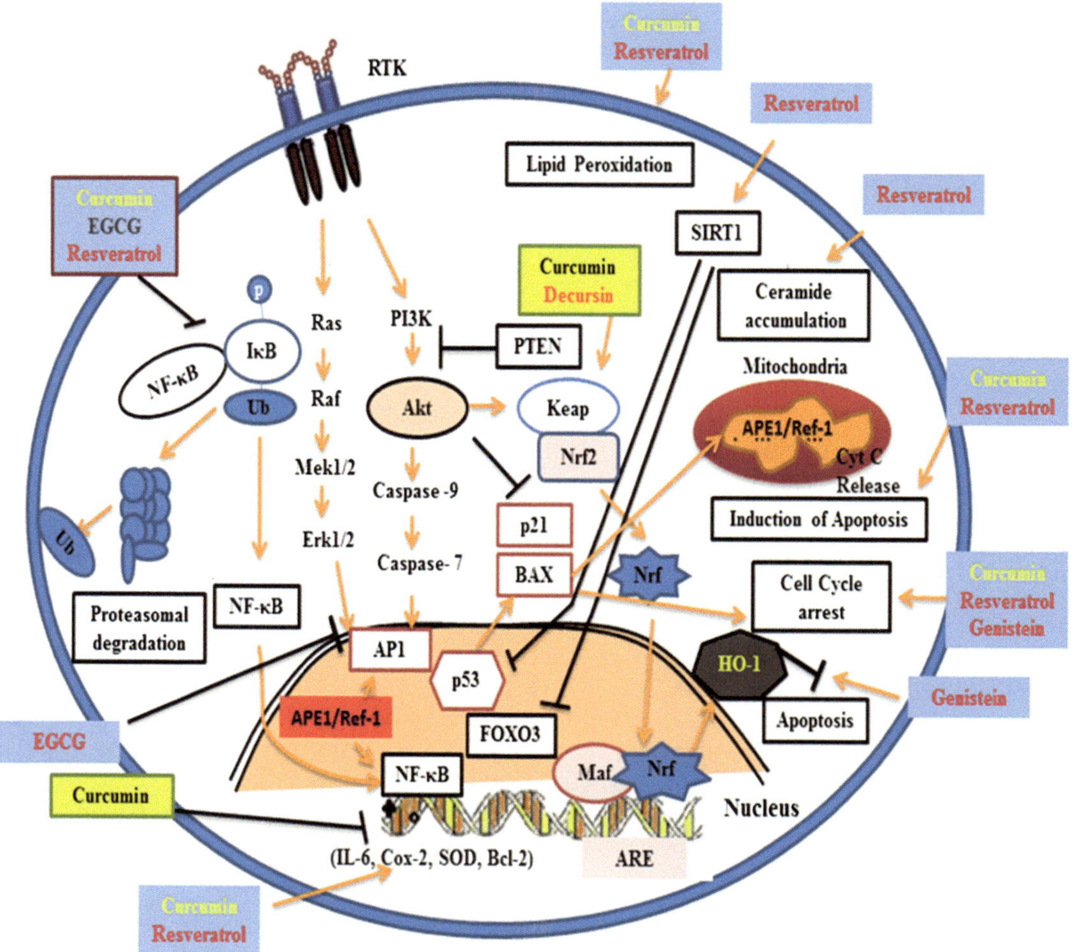

Fig. 4 Phytochemicals regulate different cellular interactions and modulates APE1/Ref-1 activity towards AD therapeutics. The phytochemical resveratrol triggers apoptosis through ceramide accumulation. Resveratrol activates SIRT1 and inhibits the transcription regulator p53 and FOXO3. Curcumin and decursin activate Nrf and rescue cell from oxidative stress. EGCG is an active polyphenol in green tea and inhibits multiple signal transduction pathways, including AP-1 and NF-κB, whereas curcumin and resveratrol activate different cellular factors like IL-6, Cox-2, SOD and Bcl-2. Curcumin, genistein and resveratrol also induce apoptosis and cell cycle arrest via functionally activated p53

4.1 Resveratrol

Resveratrol (3,5,4′-trihydroxy-*trans*-stilbene) is a phytoalexin and a polyphenolic compound found in the seeds and wine made from grape cultivars which provides natural protection to the plant against environmental stresses such as UV radiation and fungal infections. Resveratrol is the possible explanation for the French Paradox. According to French Paradox, France has a low rate of coronary heart disease (CHD) in spite of high intake of saturated fats, thus presents a situation which is paradoxical when compared with other countries having comparable diet rich in saturated fats and subsequently high CHD [188]. Wine intake is highest in France and studies have pointed out that drinking red wine confers cardioprotection and this is attributable to resveratrol present in it along with other polyphenols [189, 190], thus explaining the paradox. Therefore, resveratrol has assumed a great importance over time. The protective abilities of resveratrol have been attributable to its antioxidant properties [191]. Several lines of studies have shown that drinking red wine also confers neuroprotection and reduces the incidence of neurological diseases like AD [192, 193]. It has been shown that resveratrol exerts protective effects against Aβ-induced neurotoxicity in rat hippocampal cells with the involvement of PKC [194]. Resveratrol showed anti-apoptotic effect and interference in cell cycle progression in SH-SY5Y neuroblastoma cells [195]. The anti-Aβ potential of resveratrol in clearing Aβ via a mechanism involving activation of proteasome was also identified [196]. In addition, resveratrol oligomers from *V. amurensis* were shown to rescue Aβ-mediated oxidative stress in PC12 cells by inhibiting ROS production [197]. Resveratrol was found to stimulate NO production and reduce the oxidative stress after a focal cerebral ischaemia (FCI) injury in rats [198]. Downregulation of iNOS and enhancement of HO-1 expression by resveratrol rescues the Aβ-induced neurotoxicity [199]. Upregulation of iNOS is associated with Aβ levels [200], indicating a connection between iNOS and Aβ in the progression of AD. Resveratrol is also shown to be a SIRT-1 activator protecting the neuroblastoma cells from oxidative damage caused by Aβ [201].

SIRT-1 has an important role in maintaining genome integrity through regulation of BER pathway. An increase in association of APE1 with XRCC1 under genotoxic stress is reported, while the knockdown of SIRT-1 decreases this association. Resveratrol has been shown to promote binding of APE1 to XRCC1 by a mechanism involving activation of SIRT-1 [202]. Resveratrol is also shown to regulate the redox activity of APE1/Ref-1 and is identified as a potent APE1/Ref-1 inhibitor [24]. In this study, an increase in expression of Ref-1 was seen in human melanoma cells which may be partly due to mitochondrial dysfunction owing to high ROS levels and presence of oxidised melanin in these cells. Overexpressing Ref-1 led to increase in basal NF-κB transcription activities. In addition, in response to APE1/Ref-1 antibody, reduced AP-1 and NF-κB DNA-binding activities were observed. Thus, resveratrol seems to act as an APE1/Ref-1 inhibitor upregulating AP-1 and NF-κB DNA-binding activities, highlighting its anti-melanoma potential [24]. In a recent finding, it was shown that resveratrol mitigates the $AlCl_3$-induced direct neuroinflammation in rats [203]. Also, an increase in APE1 level and decrease in β-secretase and Aβ levels were observed. In addition, a decrease in expression of TNF-α, IL-6 and iNOS in the rat brain was seen on treatment with resveratrol, thus revealing the anti-inflammatory effects of resveratrol [203]. Taken together, these findings suggest resveratrol as a potent phytochemical for treating oxidative stress-induced mitochondrial dysfunction and inflammation in neurodegenerative diseases like AD and which might alter the APE1/Ref-1 function to mediate neuronal cell viability to counter AD.

4.2 Curcumin

Curcumin is the main active flavonoid derived from the rhizome of *Curcuma longa* (Zingiberaceae). Curcumin has potent anti-inflammatory property due to its antioxidant activity resulting in the scavenging of the ROS generated inside the body under stress conditions [204]. Curcumin owing to its antioxidant and anti-inflammatory action suppresses the oxidative damage and decreases amyloid deposition [205].

Curcumin acts as a strong metal chelator and has the ability to repress the inhibition of DG, NEIL1 caused by divalent metals like Cu and Fe in SH-SY5Y neuroblastoma cells [206, 207]. Curcumin acts as a potential therapeutic agent owing to its two effects – reduction of oxidative stress and acting as a metal chelator [160, 207]. Curcumin is shown to increase the heme oxygenase1 (HO-1) expression in cultured hippocampal neurons in response to glucose oxidase (GO)-mediated oxidative damage [208]. Curcumin has also shown to reduce the formation of Aβ and decrease plaque burden in transgenic AD mice [209]. Moreover, curcumin has a strong ability to cross blood-brain barrier (BBB) and shown to reduce aggregation of Aβ(1-40) and cause disaggregation of Aβ(1-40). In addition, curcumin prevented Aβ(1-42) oligomer formation and toxicity, making it an effective molecule for prevention and treatment of AD [210]. Thus, this curry spice has a great potential in alleviating oxidative stress and improving cognitive decline in AD.

4.3 Decursin

Decursin (D) and decursinol angelate (DA) are the major coumarins present in the roots of *Angelica gigas* Nakai (Umbelliferae). The roots of *Angelica gigas* Nakai have been used in traditional Korean medicine for treating anaemia and as a sedative and an anodyne agent [211]. Many reports highlight the antitumour [211], antibacterial [211], anti-nematodal [212] and antioxidant [213] properties of *Angelica gigas* Nakai, mainly due to the presence of D and DA. A study demonstrated the anti-amnesic property of D which rescued the impairment induced by scopolamine through the inhibition of acetylcholinesterase (AChE) in the hippocampus of treated mice [211]. Decursin was shown to cross the BBB [212], thus showing a potential to intervene the CNS to treat disorders like AD. The neuroprotective role of D and DA in rescuing the glutamate-induced oxidative stress in primary cortical cells was highlighted in a study [214]. Another study showed the neuroprotective ability of D and DA and its role in nuclear factor erythroid 2-related factor (Nrf2) activation and elevation of antioxidant levels in rescuing Aβ-mediated oxidative stress in PC12 cells [213]. Both D and DA were shown to inhibit Aβ fibrillation. This study indicated that D and DA can be utilised as an important antioxidant to help reduce the oxidative stress induced by Aβ in AD. In a recent finding, it was shown that in response to Aβ(23-35)-induced oxidative stress, treatment with D leads to decreased ROS levels and activation of mitogen-activated protein kinases (MAPK) signal pathways, leading to Nrf2 activation and upregulation of HO-1 expression, thus protecting the PC12 cells from Aβ-mediated neurotoxicity [215]. Taken together, these findings suggest that D and DA can protect neurons from Aβ-mediated oxidative stress. Further studies are needed to show the potential of D and DA in modulating APE1/Ref-1's functions to limit neurodegeneration and increase cell survival in AD.

4.4 Soy Isoflavones

Soy isoflavones are the major flavonoids found in soybean, which have been a traditional food in Asia for a very long time. Apart from isoflavones, soy is also rich in phytic acid, trypsin inhibitors and saponins [216]. But soy isoflavones have dragged attention in the recent past due to its numerous health benefits particularly its neuroprotective effects. Soy isoflavones are also referred to as phytoestrogens due to their beneficial effects on estrogenic problems [217]. Soy isoflavones include genistein, daidzein and glycitein [218]. An earlier study showed that genistein could attenuate the oxidative stress induced by Aβ(25-35) and reduce the ROS levels and inhibit cell apoptosis possibly through Nrf/HO-1signal pathway in PC12 cells [219]. Another study showed that genistein improves the short-term spatial memory in rats by mitigating Aβ(1-40)-induced impairment via an estrogenic pathway [219]. Soy isoflavones suppressed the production of inflammatory cytokines and downregulated NF-κB activity, which was induced by Aβ(1-42) and improved the learning and memory impairment in rats [220]. Another study showed that isoflavones, specifically genistein and glycitein, have an anti-fibrillisation, anti-oligomerisation and fibril-destabilising potential on Aβ(1-40) and Aβ(1-42) in vitro and that glycitein, in particular, binds directly to Aβ monomers, oligomers and fibrils and exhibit highest affinity for Aβ(25-35) [221]. Thus, isoflavones can be employed towards effective therapy to directly target amyloid assemblies for the treatment of AD. In addition, genistein showed neuroprotection and increased cell viability and protein kinase C (PKC) activity in PC12 cells which were treated with Aβ(25-35) and this involved PKC signaling pathway, which is known to regulate neuronal survival in AD [222]. In addition, downregulation of *PS1*, involved in Aβ generation, by treatment with genistein was shown [223]. A recent finding showed that soy isoflavones reduced the oxidative stress in the mitochondria induced by Aβ(1-42) in the rat brain and increased the mitochondrial membrane potential (MMP) and antioxidant function [217]. As a result, isoflavones help to maintain redox balance in the brain. Together, these findings show that isoflavones with potential can improve mitochondrial function and maintain redox balance for neuronal survival.

4.5 Epigallocatechin-3-Gallate (EGCG)

Epigallocatechin-3-gallate (EGCG) isolated from the leaves of green tea (*Camellia sinensis*) and a type of catechin [224]. It has a number of beneficial health effects owing to its neuroprotective, anticarcinogenic and anti-inflammatory property [225, 226]. The consumption of green tea and incidence of dementia, AD and PD are inversely correlated [227, 228]. Numerous animal model studies have suggested that EGCG exerts neuroprotective effects against age-related cognitive decline and neurodegenerative diseases. An earlier study using in vitro and in vivo models has shown that EGCG elevates the levels of soluble APP-α (an N-terminal cleavage product) and promotes the cleavage of α-C-terminal fragment of APP. This shows that EGCG promotes α-secretase activity leading to decreased Aβ levels and plaque formation [229, 230]. EGCG is shown to bind to the β-sheet-rich aggregates and bring about a conformational change remodelling mature α-synuclein and Aβ fibrils into smaller amorphous nontoxic protein aggregates and reduce cellular toxicity [231]. Another in vivo study involving passive avoidance and water maze tests showed that EGCG reduces the Aβ(1-42)-induced memory dysfunction dose-dependently and suppresses the activities of β- and γ-secretase. In addition, an inhibition in the activation of extracellular signal-regulated protein kinase (ERK) and NF-κB by EGCG was observed in the Aβ(1-42)-injected mouse brains [232]. EGCG has also emerged as a mitochondrial restorative compound which was demonstrated to restore MMP, ROS levels and ATP levels in a double-transgenic mouse model of AD. Thus, EGCG was shown to lessen the Aβ-induced mitochondrial dysfunction, which is implicated during the onset and progression of AD [230]. These studies point out that EGCG, owing to its anti-amyloidogenic and mitochondrial restorative property, has a tremendous potential in AD therapy.

4.6 L-Carnitine

L-carnitine is a derivative of the amino acid, lysine. Its name is derived from the fact that it was first isolated from meat (*carnus*). Acetyl L-carnitine (ALCAR), an L-carnitine ester of acetic acid, crosses the BBB and modifies acetylcholine production in the brain [233]. ALCAR is involved in regulation of mitochondrial energetics and oxidative stress associated with ageing [234]. An earlier study pointed towards induction of HO-1 and upregulation of Nrf-2, a redox-sensitive TFs, on treating astrocytes with ALCAR [235]. ALCAR was also shown to increase the synthesis of nerve growth factor receptors (NGFR) in PC12 cells and thus elicits neurite outgrowth by stimulating NGF uptake in these cells [236]. ALCAR is a physiological activator of the mitochondrial fatty acid metabolism and has been reported to improve cognitive deficits in aged animals and to slow deterioration in AD patients [237]. A study showed that ALCAR promotes α-secretase activity and physiological APP metabolism by facilitating the delivery of ADAM10 to the postsynaptic compartment regulating α-secretase activity towards APP, leading to release of a non-amyloidogenic product [238]. A study by Abdul et al. showed that ALCAR displayed neuroprotective effect towards Aβ(1-42)-induced oxidative stress in cortical neurons by upregulating the levels of glutathione (GSH) and heat shock proteins [HSPs] [239]. Thus, ALCAR displays neuroprotection and modulates mitochondrial function and oxidative stress, thus has a potential and can be employed in AD therapy upon further studies.

4.7 Ganoderma Lucidum

G. lucidum is a medicinal fungus used clinically in many Asian countries for health and longevity. A study showed the neuroprotective effect of *G. lucidum* in which the extract induced the neuronal differentiation of PC12 cells and prevented NGF-dependent PC12 neurons from apoptosis. This effect was thought to be mediated by the activation of ERK and CREB signaling pathways that maintained the survival of the NGF-dependent neurons [240]. An earlier study by Pillai et al. had demonstrated that an aqueous extract of *G. lucidum* protected against the radiation-induced nuclear DNA damage [241]. *G. lucidum* polysaccharides (GLP) was shown to reduce the expression of Caspase-3 and FasL leading to improved cognition and learning ability in Aβ(25-35)-injected mice [242]. Another study provided evidence that *G. lucidum* increased the non-amyloidogenic protein secretion, i.e. sAPPα secretion, in SH-SY5Y cells involving phosphatidylinositol 3 kinase (PI3K) and ERK signaling pathways [243]. A study pointing towards the antioxidant properties of *G. lucidum* has shown that the activities of heart TCA enzymes and mitochondrial complex (I-IV) improved on treating aged mice with an ethanolic extract of *G. lucidum* [244]. In line with this, another study showed that *G. lucidum* elevated the activities of mitochondrial dehydrogenases, i.e. succinate dehydrogenase (SDH), malate dehydrogenase (MDH), α-ketoglutarate dehydrogenase (α-KGDH) and pyruvate dehydrogenase (PDH), as well as complex I and II activities in the mitochondria of aged Wistar rat brains. Also, the level of lipid peroxidation was decreased in the *G. lucidum*-treated rats [245]. A recent in vivo study involving Sprague-Dawley rats showed that *G. lucidum* spore (GLS) improved mitochondrial functioning, alleviated oxidative stress and protected the hippocampal neurons from apoptosis, improving cognition in these rats [86]. Another recent study showed that *G. lucidum* promoted neurite outgrowth in differentiating N2a cells [246]. Thus, *G. lucidum* seems to have a great therapeutic importance in reviving brain and cognitive health in AD patients.

Although some of the phytochemicals described here are not studied directly with relation to APE1/Ref-1's functions, their beneficial effects as discussed further suggest testing them to understand their role in modulating repair, redox and other newly discovered roles of APE1/Ref-1 towards neuronal cell survival. It is a prerequisite for the neuronal cell to counter the oxidative stress responses elicited by different agents and mechanisms as discussed in this review and, further, APE1-/Ref-1-mediated intervention along with phytochemicals, thus, emerges a new field of study to tackle the AD.

5 Conclusions and Future Perspectives

AD is a disabling and debilitating disease affecting millions worldwide and is projected to affect many more. The pathological hallmarks of the disease are known from a very long time, but the molecular mechanism(s) underlying the AD is not known to date. Researchers have tried to understand the various factors responsible for the progression of this fatal disease. Some risk factors have been associated with the disease. These are mutations in the *APP*, *PS1* and *PS2* genes, which are responsible for the accumulation of Aβ, the main culprit, in the neurons leading to development of early-onset AD. A significant number of studies pointing out that Aβ oligomers are the more toxic species rather than the insoluble fibrillar deposits. Other risk factors for AD include the presence of *APOE* ϵ4 allele. Ageing is the greatest risk factor for AD. In the recent past, studies have implicated oxidative stress and mitochondrial dysfunction in the pathogenesis of AD. It is also observed that mitochondrial dysfunction occurs very early in AD pathogenesis. Studies have pointed out that accumulation of somatic mtDNA mutations over time causes genome instability and mitochondrial dysfunction. It is well established that Aβ causes oxidative stress. Recent studies have shown that Aβ and APP localise to the mitochondria, interact with mitochondrial proteins, increase ROS/RNS levels and cause mitochondrial dysfunction. Recent studies have provided evidence that Aβ accumulates more in the synaptic terminals of neurons and interferes with the axonal mitochondrial transport, leading to synaptic damage and cognitive decline associated with AD. During the mitochondrial trafficking, fusion process occurs which is essential for exchange of pathogenic mtDNA. Unbalanced fusion/fission has been implicated in various neurodegenerative diseases like AD. BER is the predominant pathway responsible for repairing oxidised base lesions in the nucleus as well as in the mitochondria, with APE1 being the central repair enzyme of this pathway. Importance of APE1/Ref-1 for the cell can be discerned by the fact that attempts to generate APE1-null mice failed as it led to their early embryonic death. APE1/Ref-1 has been shown to play a major role in overcoming the oxidative stress and maintaining neuronal cell viability and integrity owing to various roles played by it, viz. as a redox and transcriptional regulator, as an endoribonuclease and as a regulator of cellular homeostasis. Phytochemicals like soy isoflavones, resveratrol and curcumin have been shown to modulate APE1/Ref-1 activity both in vitro and in vivo. In addition to these, decursin, L-carnitine, *Ganoderma lucidum* and EGCG have shown to lower the oxidative stress induced by Aβ in various studies. Thus, these phytochemicals have a potential to reduce the oxidative stress and modulate functions of APE1/Ref-1 and can be used as an effective approach to treat AD by protecting APE1/Ref-1's functions.

Acknowledgement A.K.M is supported by Alzheimer's Association, USA, NIRG-11-203527 grant. N.K. acknowledge financial support in the form of JRF from ICMR, New Delhi. We thank Dr. Alpna Saini, Centre for Comparative Literature, School of Languages, Central University of Punjab, Bathinda, India, for proofreading the manuscript for typographic and linguistic errors. Because of the limited focus of the article, many relevant and appropriate references could not be included, for which the authors apologise. The CUPB institutional number provided for this book chapter is P-136.

References

1. Stelzmann RA, Norman Schnitzlein H, Reed Murtagh F (1995) An English translation of Alzheimer's 1907 paper, "Über eine eigenartige Erkankung der Hirnrinde". Clin Anat 8(6):429–431
2. Hardy J, Selkoe DJ (2002) The amyloid hypothesis of Alzheimer's disease: progress and problems on the road to therapeutics. Science 297(5580):353–356
3. Lambert MP, Barlow AK, Chromy BA et al (1998) Diffusible, nonfibrillar ligands derived from Abeta1-42 are potent central nervous system neurotoxins. Proc Natl Acad Sci 95(11):6448–6453
4. Zhu X, Raina AK, Lee HG et al (2004) Oxidative stress signalling in Alzheimer's disease. Brain Res 1000(1):32–39
5. Leuner K, Muller WE, Reichert AS (2012) From mitochondrial dysfunction to amyloid beta formation: novel insights into the pathogenesis of Alzheimer's disease. Mol Neurobiol 46(1):186–193
6. Swerdlow RH, Burns JM, Khan SM (2010) The Alzheimer's disease mitochondrial cascade hypothesis. J Alzheimers Dis 20(2):S265–S279
7. Clark TA, Lee HP, Rolston RK et al (2010) Oxidative stress and its implications for future treatments and management of Alzheimer disease. Int J Biomed Sci 6(3):225–227
8. Schriner SE, Linford NJ, Martin GM et al (2005) Extension of murine life span by overexpression of catalase targeted to mitochondria. Science 308(5730):1909–1911
9. Behl C, Davis JB, Lesley R et al (1994) Hydrogen peroxide mediates amyloid beta protein toxicity. Cell 77(6):817–827
10. Hensley K, Carney J, Mattson M et al (1994) A model for beta-amyloid aggregation and neurotoxicity based on free radical generation by the peptide: relevance to Alzheimer disease. Proc Natl Acad Sci 91(8):3270–3274
11. Kadowaki H, Nishitoh H, Urano F et al (2005) Amyloid β induces neuronal cell death through ROS-mediated ASK1 activation. Cell Death Differ 12(1):19–24
12. Manczak M, Reddy PH (2012) Abnormal interaction of VDAC1 with amyloid beta and phosphorylated tau causes mitochondrial dysfunction in Alzheimer's disease. Hum Mol Genet 21(23):5131–5146
13. Reddy PH (2006) Amyloid precursor protein-mediated free radicals and oxidative damage: implications for the development and progression of Alzheimer's disease. J Neurochem 96(1):1–13
14. de la Monte SM, Luong T, Neely TR et al (2000) Mitochondrial DNA damage as a mechanism of cell loss in Alzheimer's disease. Lab Invest 80(8):1323–1335
15. Yang JL, Weissman L, Bohr VA et al (2008) Mitochondrial DNA damage and repair in neurodegenerative disorders. DNA Repair 7(7):1110–1120
16. Hegde ML, Mantha AK, Hazra TK et al (2012) Oxidative genome damage and its repair: implications in aging and neurodegenerative diseases. Mech Ageing Dev 133(4):157–168
17. Mantha AK, Sarkar B, Tell G (2013) A short review on the implications of base excision repair pathway for neurons: relevance to neurodegenerative diseases. Mitochondrion. doi:10.1016/j.mito.2013.10.007

18. Bhakat KK, Mantha AK, Mitra S (2009) Transcriptional regulatory functions of mammalian AP-endonuclease (APE1/Ref-1), an essential multifunctional protein. Antioxid Redox Signal 11(3):621–638
19. Xanthoudakis S, Curran T (1992) Identification and characterization of Ref-1, a nuclear protein that facilitates AP-1 DNA-binding activity. EMBO J 11(2):653
20. Marcon G, Tell G, Perrone L et al (2009) APE1/Ref-1 in Alzheimer's disease: an immunohistochemical study. Neurosci Lett 466(3):124–127
21. Li M, Vascotto C, Xu S et al (2012) Human AP endonuclease/redox factor APE1/ref-1 modulates mitochondrial function after oxidative stress by regulating the transcriptional activity of NRF1. Free Radic Biol Med 53(2):237–248
22. Singh-Gupta V, Zhang H, Yunker CK et al (2010) Daidzein effect on hormone refractory prostate cancer in vitro and in vivo compared to genistein and soy extract: potentiation of radiotherapy. Pharm Res 27(6):1115–1127
23. Silva JP, Gomes AC, Coutinho OP (2008) Oxidative DNA damage protection and repair by polyphenolic compounds in PC12 cells. Eur J Pharmacol 601(1–3):50–60
24. Yang IK, Heffron SE et al (2005) Alterations in the expression of the apurinic/apyrimidinic endonuclease-1/redox factor-1 (APE/Ref-1) in human melanoma and identification of the therapeutic potential of resveratrol as an APE/Ref-1 inhibitor. Mol Cancer Ther 4(12):1923–1935
25. Takalo M, Salminen A, Soininen H et al (2013) Protein aggregation and degradation mechanisms in neurodegenerative diseases. Am J Neurodegener Dis 2(1):1–14
26. Soto C, Kindy MS, Baumann M et al (1996) Inhibition of Alzheimer's amyloidosis by peptides that prevent β-sheet conformation. Biochem Biophys Res Commun 226(3):672–680
27. Wimo A, Jonsson L, Winblad B (2006) An estimate of the worldwide prevalence and direct costs of dementia in 2003. Dement Geriatr Cogn Disord 21(3):175–181
28. Irvine GB, El-Agnaf OM, Shankar GM et al (2008) Protein aggregation in the brain: the molecular basis for Alzheimer's and Parkinson's diseases. Mol Med 14(7–8):451
29. Thies W, Bleiler L (2011) 2011 Alzheimer's disease facts and figures. Alzheimers Dement 7(2):208–244
30. Roses M, Allen D (1996) Apolipoprotein E alleles as risk factors in Alzheimer's disease. Annu Rev Med 47(1):387–400
31. Saunders A, Strittmatter W, Schmechel D et al (1993) Association of apolipoprotein E allele ϵ4 with late-onset familial and sporadic Alzheimer's disease. Neurology 43(8):1467–1467
32. Chapuis J, Hansmannel F, Gistelinck M et al (2013) Increased expression of BIN1 mediates Alzheimer genetic risk by modulating tau pathology. Mol Psychiatry. doi:10.1038/mp.2013.1
33. Hollingworth P, Harold D, Sims R et al (2011) Common variants at ABCA7, MS4A6A/MS4A4E, EPHA1, CD33 and CD2AP are associated with Alzheimer's disease. Nat Genet 43(5):429–435
34. Lambert JC, Heath S, Even G et al (2009) Genome-wide association study identifies variants at CLU and CR1 associated with Alzheimer's disease. Nat Genet 41(10):1094–1099
35. Wei YH, Lee HC (2002) Oxidative stress, mitochondrial DNA mutation, and impairment of antioxidant enzymes in aging. Exp Biol Med 227(9):671–682
36. Reddy PH, Tripathi R, Troung Q et al (2012) Abnormal mitochondrial dynamics and synaptic degeneration as early events in Alzheimer's disease: implications to mitochondria-targeted antioxidant therapeutics. Biochim Biophys Acta 1822(5):639–649
37. Manczak M, Calkins MJ, Reddy PH (2011) Impaired mitochondrial dynamics and abnormal interaction of amyloid beta with mitochondrial protein Drp1 in neurons from patients with Alzheimer's disease: implications for neuronal damage. Hum Mol Genet 20(13):2495–2509
38. Guo L, Du H, Yan S et al (2013) Cyclophilin D deficiency rescues axonal mitochondrial transport in Alzheimer's neurons. PLoS One 8(1):e54914
39. Wilcock DM, Griffin WS (2013) Down's syndrome, neuroinflammation, and Alzheimer neuropathogenesis. J Neuroinflammation 10:84
40. Tan MS, Yu JT, Tan L (2013) Bridging integrator 1 (BIN1): form, function, and Alzheimer's disease. Trends Mol Med 19(10):594–603
41. Schellenberg GD, Montine TJ (2012) The genetics and neuropathology of Alzheimer's disease. Acta Neuropathol 124(3):305–323
42. Anttila T, Helkala EL, Viitanen M et al (2004) Alcohol drinking in middle age and subsequent risk of mild cognitive impairment and dementia in old age: a prospective population based study. BMJ 329(7465):539
43. Ott A, Slooter AJ, Hofman A et al (1998) Smoking and risk of dementia and Alzheimer's disease in a population-based cohort study: the Rotterdam Study. Lancet 351(9119):1840–1843
44. Launer LJ, Ross GW, Petrovitch H et al (2000) Midlife blood pressure and dementia: the Honolulu-Asia aging study. Neurobiol Aging 21(1):49–55
45. Whitmer RA, Gustafson DR, Barrett-Connor E et al (2008) Central obesity and increased risk of dementia more than three decades later. Neurology 71(14):1057–1064
46. Kivipelto M, Ngandu T, Fratiglioni L et al (2005) Obesity and vascular risk factors at midlife and the risk of dementia and Alzheimer disease. Arch Neurol 62(10):1556–1560
47. Solomon A, Kareholt I, Ngandu T et al (2007) Serum cholesterol changes after midlife and late-life cogni-

tion: twenty-one-year follow-up study. Neurology 68(10):751–756

48. Solomon A, Kivipelto M, Wolozin B et al (2009) Midlife serum cholesterol and increased risk of Alzheimer's and vascular dementia three decades later. Dement Geriatr Cogn Disord 28(1):75–80
49. Haag MD, Hofman A, Koudstaal PJ et al (2009) Statins are associated with a reduced risk of Alzheimer disease regardless of lipophilicity. The Rotterdam Study. J Neurol Neurosurg Psychiatry 80(1):13–17
50. Huang TL, Zandi PP, Tucker KL et al (2005) Benefits of fatty fish on dementia risk are stronger for those without APOE epsilon4. Neurology 65(9):1409–1414
51. Laitinen MH, Ngandu T, Rovio S et al (2006) Fat intake at midlife and risk of dementia and Alzheimer's disease: a population-based study. Dement Geriatr Cogn Disord 22(1):99–107
52. Povova J, Ambroz P, Bar M et al (2012) Epidemiological of and risk factors for Alzheimer's disease: a review. Biomed Pap 156(2):108–114
53. Xu W, Qiu C, Gatz M et al (2009) Mid- and late-life diabetes in relation to the risk of dementia: a population-based twin study. Diabetes 58(1):71–77
54. Crawford FC, Vanderploeg RD, Freeman MJ et al (2002) APOE genotype influences acquisition and recall following traumatic brain injury. Neurology 58(7):1115–1118
55. Petersen RC, Smith GE, Waring SC et al (1999) Mild cognitive impairment: clinical characterization and outcome. Arch Neurol 56(3):303–308
56. Salminen A, Kaarniranta K, Kauppinen A et al (2013) Impaired autophagy and APP processing in Alzheimer's disease: the potential role of Beclin 1 interactome. Prog Neurobiol 106–107:33–54
57. Pickford F, Masliah E, Britschgi M et al (2008) The autophagy-related protein beclin 1 shows reduced expression in early Alzheimer disease and regulates amyloid beta accumulation in mice. J Clin Investig 118(6):2190–2199
58. Cai Z, Yan LJ (2013) Rapamycin, Autophagy, and Alzheimer's disease. J Biochem Pharmacol Res 1(2):84–90
59. Xue Z, Zhang S, Huang L et al (2013) Upexpression of Beclin-1-dependent autophagy protects against beta-amyloid-induced cell injury in PC12 cells. J Mol Neurosci 51(1):180–186
60. Meda L, Baron P, Scarlato G (2001) Glial activation in Alzheimer's disease: the role of Abeta and its associated proteins. Neurobiol Aging 22(6):885–893
61. Strohmeyer R, Shen Y, Rogers J (2000) Detection of complement alternative pathway mRNA and proteins in the Alzheimer's disease brain. Mol Brain Res 81(1):7–18
62. Webster S, Lue LF, Brachova L et al (1997) Molecular and cellular characterization of the membrane attack complex, C5b-9, in Alzheimer's disease. Neurobiol Aging 18(4):415–421
63. Rubio-Perez JM, Morillas-Ruiz JM (2012) A review: inflammatory process in Alzheimer's disease, role of cytokines. Sci World J 2012:756357
64. Sinha S, Anderson JP, Barbour R et al (1999) Purification and cloning of amyloid precursor protein beta-secretase from human brain. Nature 402(6761):537–540
65. Rossner S, Lange-Dohna C, Zeitschel U et al (2005) Alzheimer's disease beta-secretase BACE1 is not a neuron-specific enzyme. J Neurochem 92(2): 226–234
66. Bourne KZ, Ferrari DC, Lange-Dohna C et al (2007) Differential regulation of BACE1 promoter activity by nuclear factor-κB in neurons and glia upon exposure to β-amyloid peptides. J Neurosci Res 85(6):1194–1204
67. Barron AM, Verdile G, Martins RN (2006) The role of gonadotropins in Alzheimer's disease: potential neurodegenerative mechanisms. Endocrine 29(2):257–269
68. Casadesus G, Milliken EL, Webber KM et al (2007) Increases in luteinizing hormone are associated with declines in cognitive performance. Mol Cell Endocrinol 269(1–2):107–111
69. Ziegler SG, Thornton JE (2010) Low luteinizing hormone enhances spatial memory and has protective effects on memory loss in rats. Horm Behav 58(5):705–713
70. Rosario ER, Carroll JC, Pike CJ (2012) Evaluation of the effects of testosterone and luteinizing hormone on regulation of beta-amyloid in male 3xTg-AD mice. Brain Res 1466:137–145
71. Dreses-Werringloer U, Bhuiyan M, Zhao Y et al (2009) Initial characterization of Chlamydophila (Chlamydia) pneumoniae cultured from the late-onset alzheimer brain. Int J Med Microbiol 299(3):187–201
72. Itzhaki RF, Wozniak MA, Appelt DM et al (2004) Infiltration of the brain by pathogens causes Alzheimer's disease. Neurobiol Aging 25(5):619–627
73. Stein SMJ, Smith C et al (2012) Serum antibodies to periodontal pathogens are a risk factor for Alzheimer's disease. Alzheimers Dement 8(3):196–203
74. Li W, Yu J, Liu Y et al (2013) Elevation of brain magnesium prevents and reverses cognitive deficits and synaptic loss in Alzheimer's disease mouse model. J Neurosci 33(19):8423–8441
75. Andrasi E, Pali N, Molnar Z et al (2005) Brain aluminum, magnesium and phosphorus contents of control and Alzheimer-diseased patients. J Alzheimers Dis 7(4):273–284
76. Zapatero MD, Garcia de Jalon A, Pascual F et al (1995) Serum aluminum levels in Alzheimer's disease and other senile dementias. Biol Trace Elem Res 47(1–3):235–240
77. Brenner S (2013) Aluminum may mediate Alzheimer's disease through liver toxicity, with aberrant hepatic synthesis of ceruloplasmin and

ATPase7B, the resultant excess free copper causing brain oxidation, beta-amyloid aggregation and Alzheimer disease. Med Hypotheses 80(3):326–327
78. Davenward S, Bentham P, Wright J et al (2013) Silicon-rich mineral water as a non-invasive test of the 'aluminum hypothesis' in Alzheimer's disease. J Alzheimers Dis 33(2):423–430
79. Li YY, Chen T, Wan Y et al (2012) Lead exposure in pheochromocytoma cells induces persistent changes in amyloid precursor protein gene methylation patterns. Environ Toxicol 27(8):495–502
80. Gu H, Robison G, Hong L et al (2012) Increased beta-amyloid deposition in Tg-SWDI transgenic mouse brain following in vivo lead exposure. Toxicol Lett 213(2):211–219
81. Zhu HL, Meng SR, Fan JB et al (2011) Fibrillization of human tau is accelerated by exposure to lead via interaction with His-330 and His-362. PLoS One 6(9):e25020
82. Brewer GJ (2009) The risks of copper toxicity contributing to cognitive decline in the aging population and to Alzheimer's disease. J Am Coll Nutr 28(3):238–242
83. Kim CH, Yoo YM (2013) Altered APP carboxyl-terminal processing under ferrous iron treatment in PC12 cells. Korean J Physiol Pharmacol 17(3):189–195
84. Reese LC, Laezza F, Woltjer R et al (2011) Dysregulated phosphorylation of Ca2+/calmodulin-dependent protein kinase II-α in the hippocampus of subjects with mild cognitive impairment and Alzheimer's disease. J Neurochem 119(4):791–804
85. Bjorklund NL, Sadagoparamanujam VM, Taglialatela G (2012) Selective, quantitative measurement of releasable synaptic zinc in human autopsy hippocampal brain tissue from Alzheimer's disease patients. J Neurosci Methods 203(1):146–151
86. Bjorklund NL, Reese LC, Sadagoparamanujam V et al (2012) Absence of amyloid β oligomers at the postsynapse and regulated synaptic Zn2+ in cognitively intact aged individuals with Alzheimer's disease neuropathology. Mol Neurodegener 7(1):1–13
87. Wang WT, Zheng W, Zhao BL et al (2010) Zinc overload enhances APP cleavage and Abeta deposition in the Alzheimer mouse brain. PLoS One 5(12):e15349
88. Calderon-Garciduenas L, Reed W, Maronpot RR et al (2004) Brain inflammation and Alzheimer's-like pathology in individuals exposed to severe air pollution. Toxicol Pathol 32(6):650–658
89. Moulton PV, Yang W (2012) Air pollution, oxidative stress, and Alzheimer's disease. J Environ Public Health 2012:472751
90. Dorado-Martinez C, Paredes-Carbajal C, Mascher D et al (2001) Effects of different ozone doses on memory, motor activity and lipid peroxidation levels, in rats. Int J Neurosci 108(3–4):149–161
91. Rivas-Arancibia S, Guevara-Guzman R, Lopez-Vidal Y et al (2010) Oxidative stress caused by ozone exposure induces loss of brain repair in the hippocampus of adult rats. Toxicol Sci 113(1):187–197
92. Lesne S, Koh MT, Kotilinek L et al (2006) A specific amyloid-beta protein assembly in the brain impairs memory. Nature 440(7082):352–357
93. Pham E, Crews L, Ubhi K et al (2010) Progressive accumulation of amyloid-beta oligomers in Alzheimer's disease and in amyloid precursor protein transgenic mice is accompanied by selective alterations in synaptic scaffold proteins. FEBS J 277(14):3051–3067
94. O'Brien RJ, Wong PC (2011) Amyloid precursor protein processing and Alzheimer's disease. Annu Rev Neurosci 34:185
95. Nunan J, Small DH (2000) Regulation of APP cleavage by α-, β-and γ-secretases. FEBS Lett 483(1):6–10
96. Pagani L, Eckert A (2011) Amyloid-Beta interaction with mitochondria. Int J Alzheimer Dis 2011:925050
97. Reddy PH, Beal MF (2008) Amyloid beta, mitochondrial dysfunction and synaptic damage: implications for cognitive decline in aging and Alzheimer's disease. Trends Mol Med 14(2):45–53
98. Yang GZ, Yang M, Lim Y et al (2012) Huntingtin associated protein 1 regulates trafficking of the amyloid precursor protein and modulates amyloid beta levels in neurons. J Neurochem 122(5):1010–1022
99. Zuchner T, Schliebs R, Perez-Polo JR (2005) Down-regulation of muscarinic acetylcholine receptor M2 adversely affects the expression of Alzheimer's disease-relevant genes and proteins. J Neurochem 95(1):20–32
100. Tonelli DDP, Mihailovich M, Di Cesare A et al (2004) Translational regulation of BACE-1 expression in neuronal and non-neuronal cells. Nucleic Acids Res 32(5):1808–1817
101. Lublin AL, Gandy S (2010) Amyloid-β oligomers: possible roles as key neurotoxins in Alzheimer's disease. Mt Sinai J Med 77(1):43–49
102. Murray MM, Bernstein SL, Nyugen V et al (2009) Amyloid β protein: Aβ40 inhibits Aβ42 oligomerization. J Am Chem Soc 131(18):6316–6317
103. Vassar R, Bennett BD, Babu-Khan S et al (1999) Beta-secretase cleavage of Alzheimer's amyloid precursor protein by the transmembrane aspartic protease BACE. Science 286(5440):735–741
104. Hutton M (2004) Presenilin mutations associated with fronto-temporal dementia. Ann Neurol 55(5):604–606
105. Gandy S (2005) The role of cerebral amyloid beta accumulation in common forms of Alzheimer disease. J Clin Inv 115(5):1121–1129
106. Glabe CG (2008) Structural classification of toxic amyloid oligomers. J Biol Chem 283(44):29639–29643
107. Billings LM, Oddo S, Green KN et al (2005) Intraneuronal Abeta causes the onset of early Alzheimer's disease-related cognitive deficits in transgenic mice. Neuron 45(5):675–688

108. Cleary JP, Walsh DM, Hofmeister JJ et al (2005) Natural oligomers of the amyloid-beta protein specifically disrupt cognitive function. Nat Neurosci 8(1):79–84
109. Lauren J, Gimbel DA, Nygaard HB et al (2009) Cellular prion protein mediates impairment of synaptic plasticity by amyloid-beta oligomers. Nature 457(7233):1128–1132
110. Heinitz K, Beck M, Schliebs R et al (2006) Toxicity mediated by soluble oligomers of β-amyloid (1–42) on cholinergic SN56. B5. G4 cells. J Neurochem 98(6):1930–1945
111. Joerchel S, Raap M, Bigl M et al (2008) Oligomeric β-amyloid (1-42) induces the expression of Alzheimer disease-relevant proteins in cholinergic SN56. B5. G4 cells as revealed by proteomic analysis. Int J Dev Neurosci 26(3):301–308
112. Kokubo H, Kayed R, Glabe CG et al (2009) Amyloid beta annular protofibrils in cell processes and synapses accumulate with aging and Alzheimer-associated genetic modification. Int J Alzheimer Dis 2009 pii: 689285
113. Shafrir Y, Durell SR, Anishkin A et al (2010) Beta-barrel models of soluble amyloid beta oligomers and annular protofibrils. Proteins Struct Funct Bioinforma 78(16):3458–3472
114. Larson ME, Lesné SE (2012) Soluble Aβ oligomer production and toxicity. J Neurochem 120(s1):125–139
115. Swerdlow RH, Khan SM (2004) A "mitochondrial cascade hypothesis" for sporadic Alzheimer's disease. Med Hypotheses 63(1):8–20
116. Trifunovic A, Wredenberg A, Falkenberg M et al (2004) Premature ageing in mice expressing defective mitochondrial DNA polymerase. Nature 429(6990):417–423
117. Caspersen C, Wang N, Yao J et al (2005) Mitochondrial Abeta: a potential focal point for neuronal metabolic dysfunction in Alzheimer's disease. FASEB J 19(14):2040–2041
118. Larsson NG (2010) Somatic mitochondrial DNA mutations in mammalian aging. Annu Rev Biochem 79:683–706
119. Greaves LC, Turnbull DM (2009) Mitochondrial DNA mutations and ageing. Biochim Biophys Acta 1790(10):1015–1020
120. Tamagno E, Bardini P, Obbili A et al (2002) Oxidative stress increases expression and activity of BACE in NT2 neurons. Neurobiol Dis 10(3):279–288
121. Tan JL, Li QX, Ciccotosto GD et al (2013) Mild oxidative stress induces redistribution of BACE1 in non-apoptotic conditions and promotes the amyloidogenic processing of Alzheimer's disease amyloid precursor protein. PLoS One 8(4):e61246
122. Pigino G, Morfini G, Atagi Y et al (2009) Disruption of fast axonal transport is a pathogenic mechanism for intraneuronal amyloid beta. Proc Natl Acad Sci 106(14):5907–5912
123. Wang X, Perry G, Smith MA et al (2010) Amyloid-beta-derived diffusible ligands cause impaired axonal transport of mitochondria in neurons. Neurodegener Dis 7(1–3):56–59
124. Heller A, Brockhoff G, Goepferich A (2012) Targeting drugs to mitochondria. Eur J Pharm Biopharm 82(1):1–18
125. Wang X, Su B, Siedlak SL et al (2008) Amyloid-β overproduction causes abnormal mitochondrial dynamics via differential modulation of mitochondrial fission/fusion proteins. Proc Natl Acad Sci 105(49):19318–19323
126. Lustbader JW, Cirilli M, Lin C et al (2004) ABAD directly links Abeta to mitochondrial toxicity in Alzheimer's disease. Science 304(5669):448–452
127. Du H, Yan SS (2010) Mitochondrial permeability transition pore in Alzheimer's disease: cyclophilin D and amyloid beta. Biochim Biophys Acta 1802(1):198–204
128. Du H, Guo L, Fang F et al (2008) Cyclophilin D deficiency attenuates mitochondrial and neuronal perturbation and ameliorates learning and memory in Alzheimer's disease. Nat Med 14(10):1097–1105
129. Du H, Guo L, Yan S et al (2010) Early deficits in synaptic mitochondria in an Alzheimer's disease mouse model. Proc Natl Acad Sci 107(43):18670–18675
130. Ahn BH, Kim HS, Song S et al (2008) A role for the mitochondrial deacetylase Sirt3 in regulating energy homeostasis. Proc Natl Acad Sci 105(38):14447–14452
131. Kim SH, Lu HF, Alano CC (2011) Neuronal Sirt3 protects against excitotoxic injury in mouse cortical neuron culture. PLoS One 6(3):e14731
132. Weir HJ, Murray TK, Kehoe PG et al (2012) CNS SIRT3 expression is altered by reactive oxygen species and in Alzheimer's disease. PLoS One 7(11):e48225
133. Woo JA, Jung AR, Lakshmana MK et al (2012) Pivotal role of the RanBP9-cofilin pathway in Abeta-induced apoptosis and neurodegeneration. Cell Death Differ 19(9):1413–1423
134. Kopeikina KJ, Carlson GA, Pitstick R et al (2011) Tau accumulation causes mitochondrial distribution deficits in neurons in a mouse model of tauopathy and in human Alzheimer's disease brain. Am J Pathol 179(4):2071–2082
135. Strosznajder JB, Czapski GA, Adamczyk A et al (2012) Poly(ADP-ribose) polymerase-1 in amyloid beta toxicity and Alzheimer's disease. Mol Neurobiol 46(1):78–84
136. Kauppinen TM, Suh SW, Higashi Y et al (2011) Poly(ADP-ribose)polymerase-1 modulates microglial responses to amyloid beta. J Neuroinflammation 8:152
137. Rossi MN, Carbone M, Mostocotto C et al (2009) Mitochondrial localization of PARP-1 requires interaction with mitofilin and is involved in the maintenance of mitochondrial DNA integrity. J Biol Chem 284(46):31616–31624

138. Turunc Bayrakdar E, Uyanikgil Y et al (2014) Nicotinamide treatment reduces the levels of oxidative stress, apoptosis, and PARP-1 activity in Abeta(1-42)-induced rat model of Alzheimer's disease. Free Radic Res 48(2):146–158
139. Altieri F, Grillo C, Maceroni M et al (2008) DNA damage and repair: from molecular mechanisms to health implications. Antioxid Redox Signal 10(5):891–937
140. Bergamini CM, Gambetti S, Dondi A et al (2004) Oxygen, reactive oxygen species and tissue damage. Curr Pharm Des 10(14):1611–1626
141. Kruman II (2004) Why do neurons enter the cell cycle? Cell Cycle 3(6):767–771
142. Iyama T, Wilson DM 3rd (2013) DNA repair mechanisms in dividing and non-dividing cells. DNA Repair (Amst) 12(8):620–636
143. Crescenzi M, Soddu S, Tatò F (1995) Mitotic cycle reactivation in terminally differentiated cells by adenovirus infection. J Cell Physiol 162(1):26–35
144. Hegde M, Hazra T, Mitra S (2008) Early steps in the DNA base excision/single-strand interruption repair pathway in mammalian cells. Cell Res 18(1):27–47
145. Fritz G (2000) Human APE/Ref-1 protein. Int J Biochem Cell Biol 32(9):925–929
146. Tell G, Zecca A, Pellizzari L, Spessotto P et al (2000) An 'environment to nucleus' signaling system operates in B lymphocytes: redox status modulates BSAP/Pax-5 activation through Ref-1 nuclear translocation. Nucleic Acids Res 28(5):1099–1105
147. Barnes T, Kim WC, Mantha AK et al (2009) Identification of Apurinic/apyrimidinic endonuclease 1 (APE1) as the endoribonuclease that cleaves c-myc mRNA. Nucleic Acids Res 37(12):3946–3958
148. Zaky A, Busso C, Izumi T et al (2008) Regulation of the human AP-endonuclease (APE1/Ref-1) expression by the tumor suppressor p53 in response to DNA damage. Nucleic Acids Res 36(5):1555–1566
149. Tell G, Quadrifoglio F, Tiribelli C et al (2009) The many functions of APE1/Ref-1: not only a DNA repair enzyme. Antioxid Redox Signal 11(3):601–619
150. Jiang Y, Guo C, Fishel ML et al (2009) Role of APE1 in differentiated neuroblastoma SH-SY5Y cells in response to oxidative stress: use of APE1 small molecule inhibitors to delineate APE1 functions. DNA Repair 8(11):1273–1282
151. Thakur S, Sarkar B, Cholia RP et al (2014) APE1/Ref-1 as an emerging therapeutic target for various human diseases: phytochemical modulation of its functions. Exp Mol Med 46:e106
152. Xanthoudakis S, Smeyne RJ, Wallace JD et al (1996) The redox/DNA repair protein, Ref-1, is essential for early embryonic development in mice. Proc Natl Acad Sci 93(17):8919–8923
153. Izumi T, Brown DB, Naidu C et al (2005) Two essential but distinct functions of the mammalian abasic endonuclease. Proc Natl Acad Sci U S A 102(16):5739–5743
154. Vasko MR, Guo C, Kelley MR (2005) The multifunctional DNA repair/redox enzyme Ape1/Ref-1 promotes survival of neurons after oxidative stress. DNA Repair 4(3):367–379
155. Davydov V, Hansen LA, Shackelford DA (2003) Is DNA repair compromised in Alzheimer's disease? Neurobiol Aging 24(7):953–968
156. Tan Z, Shi L, Schreiber SS (2009) Differential expression of redox factor-1 associated with beta-amyloid-mediated neurotoxicity. Open Neurosci J 3:26–34
157. Huang E, Qu D, Zhang Y et al (2010) The role of Cdk5-mediated apurinic/apyrimidinic endonuclease 1 phosphorylation in neuronal death. Nat Cell Biol 12(6):563–571
158. Frosina G, Fortini P, Rossi O et al (1996) Two pathways for base excision repair in mammalian cells. J Biol Chem 271(16):9573–9578
159. Jacobs AL, Schär P (2012) DNA glycosylases: in DNA repair and beyond. Chromosoma 121(1):1–20
160. Hegde ML, Hegde PM, Rao K, Mitra S (2011) Oxidative genome damage and its repair in neurodegenerative diseases: function of transition metals as a double-edged sword. J Alzheimers Dis 24:183–198
161. Krokan H, Standal R, Slupphaug G (1997) DNA glycosylases in the base excision repair of DNA. Biochem J 325:1–16
162. Ariumi Y, Turelli P, Masutani M et al (2005) DNA damage sensors ATM, ATR, DNA-PKcs, and PARP-1 are dispensable for human immunodeficiency virus type 1 integration. J Virol 79(5):2973–2978
163. Strosznajder JB, Jesko H, Strosznajder RP (2000) Effect of amyloid beta peptide on poly(ADP-ribose) polymerase activity in adult and aged rat hippocampus. Acta Biochima Pol 47(3):847–854
164. Vidal AE, Boiteux S, Hickson ID et al (2001) XRCC1 coordinates the initial and late stages of DNA abasic site repair through protein-protein interactions. EMBO J 20(22):6530–6539
165. Gredilla R, Bohr VA, Stevnsner T (2010) Mitochondrial DNA repair and association with aging–an update. Exp Gerontol 45(7):478–488
166. Bohr VA (2002) Repair of oxidative DNA damage in nuclear and mitochondrial DNA, and some changes with aging in mammalian cells. Free Radic Biol Med 32(9):804–812
167. Akbari M, Visnes T, Krokan HE et al (2008) Mitochondrial base excision repair of uracil and AP sites takes place by single-nucleotide insertion and long-patch DNA synthesis. DNA Repair (Amst) 7(4):605–616
168. Szczesny B, Tann AW, Longley MJ et al (2008) Long patch base excision repair in mammalian mitochondrial genomes. J Biol Chem 283(39):26349–26356
169. Alexeyev M, Shokolenko I, Wilson G et al (2013) The maintenance of mitochondrial DNA integrity—critical analysis and update. Cold Spring Harb Perspect Biol 5(5):a012641
170. Banerjee D, Mandal SM, Das A et al (2011) Preferential repair of oxidized base damage in the

transcribed genes of mammalian cells. J Biol Chem 286(8):6006–6016
171. Chattopadhyay R, Wiederhold L, Szczesny B et al (2006) Identification and characterization of mitochondrial abasic (AP)-endonuclease in mammalian cells. Nucleic Acids Res 34(7):2067–2076
172. Lakshmipathy U, Campbell C (2000) Mitochondrial DNA ligase III function is independent of Xrcc1. Nucleic Acids Res 28(20):3880–3886
173. Tann AW, Boldogh I, Meiss G et al (2011) Apoptosis induced by persistent single-strand breaks in mitochondrial genome: critical role of exog (5′exo/endonuclease) in their repair. J Biol Chem 286(37):31975–31983
174. Liu P, Qian L, Sung J-S et al (2008) Removal of oxidative DNA damage via FEN1-dependent long-patch base excision repair in human cell mitochondria. Mol Cell Biol 28(16):4975–4987
175. Larsen E, Gran C, Sæther BE et al (2003) Proliferation failure and gamma radiation sensitivity of Fen1 null mutant mice at the blastocyst stage. Mol Cell Biol 23(15):5346–5353
176. Zheng L, Zhou M, Guo Z et al (2008) Human DNA2 is a mitochondrial nuclease/helicase for efficient processing of DNA replication and repair intermediates. Mol Cell 32(3):325–336
177. Walker L, Robson C, Black E et al (1993) Identification of residues in the human DNA repair enzyme HAP1 (Ref-1) that are essential for redox regulation of Jun DNA binding. Mol Cell Biol 13(9):5370–5376
178. Liu H, Colavitti R, Rovira II et al (2005) Redox-dependent transcriptional regulation. Circ Res 97(10):967–974
179. Xanthoudakis S, Miao G, Wang F et al (1992) Redox activation of Fos-Jun DNA binding activity is mediated by a DNA repair enzyme. EMBO J 11(9):3323–3335
180. Okazaki T, Ando K, Igarashi T et al (1992) Conserved mechanism of negative gene regulation by extracellular calcium. Parathyroid hormone gene versus atrial natriuretic polypeptide gene. J Clin Investig 89(4):1268–1273
181. Fuchs S, Philippe J, Corvol P et al (2003) Implication of Ref-1 in the repression of renin gene transcription by intracellular calcium. J Hypertens 21(2):327–335
182. Bhakat KK, Izumi T, Yang SH et al (2003) Role of acetylated human AP-endonuclease (APE1/Ref-1) in regulation of the parathyroid hormone gene. EMBO J 22(23):6299–6309
183. Tomecki R, Dziembowski A (2010) Novel endoribonucleases as central players in various pathways of eukaryotic RNA metabolism. RNA 16(9):1692–1724
184. Kim WC, Berquist BR, Chohan M et al (2011) Characterization of the endoribonuclease active site of human apurinic/apyrimidinic endonuclease 1. J Mol Biol 411(5):960–971
185. Ozaki M, Suzuki S, Irani K (2002) Redox factor-1/APE suppresses oxidative stress by inhibiting the rac1 GTPase. FASEB J 16(8):889–890
186. Angkeow P, Deshpande S, Qi B, Liu Y et al (2002) Redox factor-1: an extra-nuclear role in the regulation of endothelial oxidative stress and apoptosis. Cell Death Differ 9(7):717–725
187. Mantha AK, Dhiman M, Taglialatela G et al (2012) Proteomic study of amyloid beta (25-35) peptide exposure to neuronal cells: impact on APE1/Ref-1's protein-protein interaction. J Neurosci Res 90(6):1230–1239
188. Das S, Das DK (2007) Resveratrol: a therapeutic promise for cardiovascular diseases. Recent Pat Cardiovasc Drug Discov 2(2):133–138
189. Rotondo S, Rajtar G, Manarini S et al (1998) Effect of trans-resveratrol, a natural polyphenolic compound, on human polymorphonuclear leukocyte function. Br J Pharmacol 123(8):1691–1699
190. Tsai SH, Lin-Shiau SY, Lin JK (1999) Suppression of nitric oxide synthase and the down-regulation of the activation of NFkappaB in macrophages by resveratrol. Br J Pharmacol 126(3):673–680
191. Belguendouz L, Fremont L, Linard A (1997) Resveratrol inhibits metal ion-dependent and independent peroxidation of porcine low-density lipoproteins. Biochem Pharmacol 53(9):1347–1355
192. Orgogozo JM, Dartigues JF, Lafont S et al (1997) Wine consumption and dementia in the elderly: a prospective community study in the Bordeaux area. Rev Neurol (Paris) 153(3):185–192
193. Bastianetto S, Zheng WH, Quirion R (2000) Neuroprotective abilities of resveratrol and other red wine constituents against nitric oxide-related toxicity in cultured hippocampal neurons. Br J Pharmacol 131(4):711–720
194. Han YS, Zheng WH, Bastianetto S et al (2004) Neuroprotective effects of resveratrol against beta-amyloid-induced neurotoxicity in rat hippocampal neurons: involvement of protein kinase C. Br J Pharmacol 141(6):997–1005
195. Rigolio R, Miloso M, Nicolini G et al (2005) Resveratrol interference with the cell cycle protects human neuroblastoma SH-SY5Y cell from paclitaxel-induced apoptosis. Neurochem Int 46(3):205–211
196. Marambaud P, Zhao H, Davies P (2005) Resveratrol promotes clearance of Alzheimer's disease amyloid-beta peptides. J Biol Chem 280(45):37377–37382
197. Jang MH, Piao XL, Kim HY et al (2007) Resveratrol oligomers from Vitis amurensis attenuate beta-amyloid-induced oxidative stress in PC12 cells. Biol Pharm Bull 30(6):1130–1134
198. Tsai SK, Hung LM, Fu YT et al (2007) Resveratrol neuroprotective effects during focal cerebral ischemia injury via nitric oxide mechanism in rats. J Vasc Surg 46(2):346–353
199. Huang TC, Lu KT, Wo YY et al (2011) Resveratrol protects rats from Abeta-induced neurotoxicity by the reduction of iNOS expression and lipid peroxidation. PLoS One 6(12):e29102
200. Wang Q, Rowan MJ, Anwyl R (2004) Beta-amyloid-mediated inhibition of NMDA receptor-dependent

long-term potentiation induction involves activation of microglia and stimulation of inducible nitric oxide synthase and superoxide. J Neurosci 24(27):6049–6056
201. Albani D, Polito L, Batelli S et al (2009) The SIRT1 activator resveratrol protects SK-N-BE cells from oxidative stress and against toxicity caused by α-synuclein or amyloid-β (1-42) peptide. J Neurochem 110(5):1445–1456
202. Yamamori T, DeRicco J, Naqvi A et al (2010) SIRT1 deacetylates APE1 and regulates cellular base excision repair. Nucleic Acids Res 38(3):832–845
203. Zaky A, Mohammad B, Moftah M et al (2013) Apurinic/apyrimidinic endonuclease 1 is a key modulator of aluminum-induced neuroinflammation. BMC Neurosci 14(1):1–12
204. Kunchandy E, Rao M (1990) Oxygen radical scavenging activity of curcumin. Int J Pharm 58(3):237–240
205. Zhang C, Browne A, Child D et al (2010) Curcumin decreases amyloid-β peptide levels by attenuating the maturation of amyloid-β precursor protein. J Biol Chem 285(37):28472–28480
206. Hegde M, Hegde P, Holthauzen L et al (2010) Specific inhibition of NEIL-initiated repair of oxidized base damage in human genome by copper and iron potential etiological linkage to neurodegenerative diseases. J Biol Chem 285(37):28812–28825
207. Baum L, Ng A (2004) Curcumin interaction with copper and iron suggests one possible mechanism of action in Alzheimer's disease animal models. J Alzheimers Dis 6(4):367–377
208. Scapagnini G, Colombrita C, Amadio M et al (2006) Curcumin activates defensive genes and protects neurons against oxidative stress. Antioxid Redox Signal 8(3–4):395–403
209. Lim G, Chu T, Yang F et al (2001) The curry spice curcumin reduces oxidative damage and amyloid pathology in an Alzheimer transgenic mouse. J Neurosci 21(21):8370–8377
210. Yang F, Lim G, Begum A et al (2005) Curcumin inhibits formation of amyloid β oligomers and fibrils, binds plaques, and reduces amyloid in vivo. J Biol Chem 280(7):5892–5901
211. Kang SY, Lee KY, Park MJ et al (2003) Decursin from Angelica gigas mitigates amnesia induced by scopolamine in mice. Neurobiol Learn Mem 79(1):11–18
212. Shiomi K, Hatano H, Morimoto H et al (2007) Decursin and decursinol angelate selectively inhibit NADH-fumarate reductase of Ascaris suum. Planta Med 73(14):1478–1481
213. Li L, Li W, Jung S-W et al (2011) Protective effects of decursin and decursinol angelate against amyloid β-protein-induced oxidative stress in the PC12 cell line: the role of Nrf2 and antioxidant enzymes. Biosci Biotechnol Biochem 75(3):434–442
214. Kang SY, Kim YC (2007) Decursinol and decursin protect primary cultured rat cortical cells from glutamate-induced neurotoxicity. J Pharm Pharmacol 59(6):863–870
215. Li L, Du JK, Zou LY et al (2013) Decursin isolated from Angelica gigas Nakai Rescues PC12 cells from amyloid beta-protein-induced neurotoxicity through Nrf2-mediated upregulation of heme oxygenase-1: potential roles of MAPK. Evid Based Complement Alternat Med 2013:467245
216. Anderson RL, Wolf WJ (1995) Compositional changes in trypsin inhibitors, phytic acid, saponins and isoflavones related to soybean processing. J Nutr 125(3 Suppl):581S–588S
217. Ding J, Yu HL, Ma WW et al (2013) Soy isoflavone attenuates brain mitochondrial oxidative stress induced by beta-amyloid peptides 1-42 injection in lateral cerebral ventricle. J Neurosci Res 91(4):562–567
218. Lee Y-B, Lee HJ, Sohn HS (2005) Soy isoflavones and cognitive function. J Nutr Biochem 16(11):641–649
219. Bagheri M, Joghataei MT, Mohseni S et al (2011) Genistein ameliorates learning and memory deficits in amyloid beta(1-40) rat model of Alzheimer's disease. Neurobiol Learn Mem 95(3):270–276
220. Ding BJ, Ma WW, He LL et al (2011) Soybean isoflavone alleviates beta-amyloid 1-42 induced inflammatory response to improve learning and memory ability by down regulation of Toll-like receptor 4 expression and nuclear factor-kappaB activity in rats. Int J Dev Neurosci 29(5):537–542
221. Hirohata M, Ono K, Takasaki J et al (2012) Anti-amyloidogenic effects of soybean isoflavones in vitro: fluorescence spectroscopy demonstrating direct binding to Abeta monomers, oligomers and fibrils. Biochim Biophys Acta 1822(8):1316–1324
222. Luo S, Lan T, Liao W et al (2012) Genistein inhibits Abeta(25-35)-induced neurotoxicity in PC12 cells via PKC signaling pathway. Neurochem Res 37(12):2787–2794
223. Okumura N, Yoshida H, Nishimura Y et al (2012) Genistein downregulates presenilin 1 and ubiquilin 1 expression. Mol Med Rep 5(2):559–561
224. Katiyar SK, Agarwal R, Wang ZY et al (1992) (−)-Epigallocatechin-3-gallate in Camellia sinensis leaves from Himalayan region of Sikkim: inhibitory effects against biochemical events and tumor initiation in Sencar mouse skin. Nutr Cancer 18(1):73–83
225. Weinreb O, Amit T, Mandel S et al (2009) Neuroprotective molecular mechanisms of (−)-epigallocatechin-3-gallate: a reflective outcome of its antioxidant, iron chelating and neuritogenic properties. Genes Nutr 4(4):283–296
226. Zaveri NT (2006) Green tea and its polyphenolic catechins: medicinal uses in cancer and noncancer applications. Life Sci 78(18):2073–2080
227. Hu G, Bidel S, Jousilahti P et al (2007) Coffee and tea consumption and the risk of Parkinson's disease. Mov Disord 22(15):2242–2248

228. Mandel S, Amit T, Kalfon L et al (2008) Cell signaling pathways and iron chelation in the neurorestorative activity of green tea polyphenols: special reference to epigallocatechin gallate (EGCG). J Alzheimers Dis 15(2):211–222
229. Rezai-Zadeh K, Shytle D, Sun N et al (2005) Green tea Epigallocatechin-3-Gallate (EGCG) modulates amyloid precursor protein cleavage and reduces cerebral amyloidosis in Alzheimer transgenic mice. J Neurosci 25(38):8807–8814
230. Dragicevic N, Smith A, Lin X et al (2011) Green tea epigallocatechin-3-gallate (EGCG) and other flavonoids reduce Alzheimer's amyloid-induced mitochondrial dysfunction. J Alzheimers Dis 26(3):507–521
231. Bieschke J, Russ J, Friedrich RP et al (2010) EGCG remodels mature α-synuclein and amyloid-β fibrils and reduces cellular toxicity. Proc Natl Acad Sci 107(17):7710–7715
232. Lee JW, Lee YK, Ban JO et al (2009) Green tea (−)-epigallocatechin-3-gallate inhibits β-amyloid-induced cognitive dysfunction through modification of secretase activity via inhibition of ERK and NF-κB pathways in mice. J Nutr 139(10):1987–1993
233. Carta A, Calvani M, Bravi D et al (1993) Acetyl-l-Carnitine and Alzheimer's disease: pharmacological considerations beyond the cholinergic spherea. Ann N Y Acad Sci 695(1):324–326
234. Hagen TM, Ingersoll RT, Wehr CM et al (1998) Acetyl-L-carnitine fed to old rats partially restores mitochondrial function and ambulatory activity. Proc Natl Acad Sci 95(16):9562–9566
235. Calabrese V, Ravagna A, Colombrita C et al (2005) Acetylcarnitine induces heme oxygenase in rat astrocytes and protects against oxidative stress: involvement of the transcription factor Nrf2. J Neurosci Res 79(4):509–521
236. Taglialatela G, Angelucci L, Ramacci M et al (1991) Acetyl-L-carnitine enhances the response of PC12 cells to nerve growth factor. Dev Brain Res 59(2):221–230
237. Pettegrew J, Levine J, McClure R (2000) Acetyl-L-carnitine physical-chemical, metabolic, and therapeutic properties: relevance for its mode of action in Alzheimer's disease and geriatric depression. Mol Psychiatry 5(6):616–632
238. Epis R, Marcello E, Gardoni F et al (2008) Modulatory effect of acetyl-L-carnitine on amyloid precursor protein metabolism in hippocampal neurons. Eur J Pharmacol 597(1):51–56
239. Abdul HM, Calabrese V, Calvani M et al (2006) Acetyl-L-carnitine-induced up-regulation of heat shock proteins protects cortical neurons against amyloid-beta peptide 1–42-mediated oxidative stress and neurotoxicity: Implications for Alzheimer's disease. J Neurosci Res 84(2):398–408
240. Cheung WMW, Hui WS, Chu PWK et al (2000) Ganoderma extract activates MAP kinases and induces the neuronal differentiation of rat pheochromocytoma PC12 cells. FEBS Lett 486(3):291–296
241. Pillai TG, Salvi VP, Maurya DK et al (2006) Prevention of radiation-induced damages by aqueous extract of Ganoderma lucidum occurring in southern parts of India. Curr Sci 91(3):341–347
242. Yueping Z, Hua Y, Li L et al (2008) Effects of Ganoderma Lucidum polysaccharides on caspase-3 and FasL expressions in the hippocampus of Alzheimer disease model rats [J]. Chin J Histochem Cytochem 5:016
243. Pinweha S, Wanikiat P, Sanvarinda Y et al (2008) The signaling cascades of Ganoderma lucidum extracts in stimulating non-amyloidogenic protein secretion in human neuroblastoma SH-SY5Y cell lines. Neurosci Lett 448(1):62–66
244. Sudheesh N, Ajith T, Janardhanan K (2009) Ganoderma lucidum (Fr.) P. Karst enhances activities of heart mitochondrial enzymes and respiratory chain complexes in the aged rat. Biogerontology 10(5):627–636
245. Ajith T, Sudheesh N, Roshny D et al (2009) Effect of Ganoderma lucidum on the activities of mitochondrial dehydrogenases and complex I and II of electron transport chain in the brain of aged rats. Exp Gerontol 44(3):219–223
246. Phan CW, David P, Naidu M et al (2013) Neurite outgrowth stimulatory effects of culinary-medicinal mushrooms and their toxicity assessment using differentiating Neuro-2a and embryonic fibroblast BALB/3T3. BMC Complement Altern Med 13(1):261

ROS in Carcinogenesis and Anticancerous Drug-Induced Toxicity

Vibha Rani and Khushboo Gupta

Abstract

Equilibrium between the cell proliferation and cell death helps in maintaining cell number within a tissue and the imbalance between the two leads to cancer. Both endogenous and exogenous factors influence DNA damage, cell growth, and cell death and contribute to carcinogenesis. There are experimental evidences to support the role of reactive oxygen species in the cancer process. Increase in reactive oxygen species in the cell, through either biological modification or chemical exposure to carcinogen, contributes to the process of carcinogenesis. As already stated, reactive oxygen species can arise through a variety of factors and pathways. Oxidative stress due to these can directly lead to production of single- or double-stranded DNA breaks, purine, pyrimidine, or deoxyribose modifications, and DNA cross-links. Persistent DNA damage can result in either arrest or induction of transcription, induction of signal transduction pathways, replication errors, and genomic instability, all of which are seen in carcinogenesis. Oxidative stress also plays a dual role in inducing toxicity due to administration of drugs given to treat cancer aggregation which further lead to toxicity in various organs and tissues such as cardiotoxicity, neurotoxicity.

Keywords

Reactive oxygen species • Carcinogenesis • Cardiotoxicity • Oxidative stress • Mutation • Doxorubicin

V. Rani (✉) • K. Gupta
Department of Biotechnology, Jaypee Institute of Information Technology, A-10, Sector-62, Noida 201307, Uttar Pradesh, India
e-mail: vibha.rani@jiit.ac.in

1 Introduction

Cancer is one of the leading causes of death in most developed countries. Excess generation of reactive oxygen species can be associated with the process of carcinogenesis. Oxidative stress leads to DNA damage and modifications, leading

V. Rani and U.C.S. Yadav (eds.), *Free Radicals in Human Health and Disease*,
DOI 10.1007/978-81-322-2035-0_14, © Springer India 2015

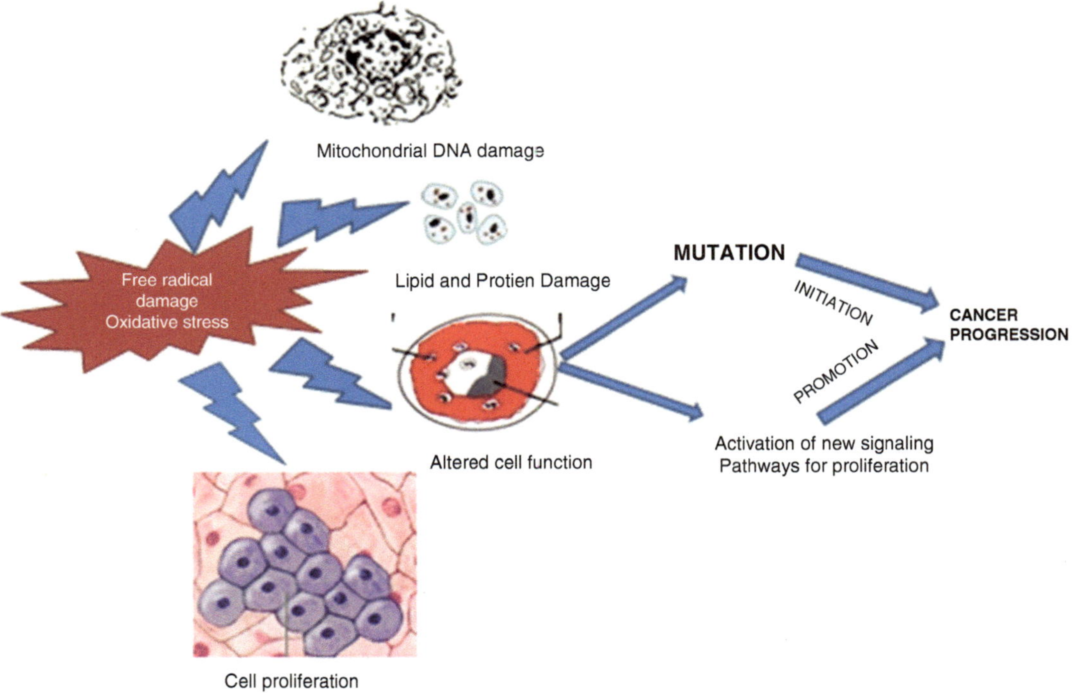

Fig. 1 The overall mechanism of oxidative stress-induced toxicity leading to cancer. Free radicals lead to deteriorating effects like mitochondrial DNA damage, damage to lipids and proteins, changes in cell function, and ultimately cell proliferation. All these effects lead to changes in signaling pathways. This mutation and altered cellular function leads to initiation followed by progression of cancer

to changes in the genomic information, which is the most critical aspect of cancer progression. Several studies suggested that ROS can act as secondary messengers controlling various signaling cascades. Reactive oxygen species can be involved in initiation and promotion of carcinogenesis via activation of proto-oncogenes and inactivation of tumor-suppressing genes. The unregulated and prolonged production of free radicals can lead to mutations as well as modification of gene expression by stimulation of signal transduction pathways and activation of key transcription factors such as Nrf2 and NF-*k*B. The induction of cancer by chemicals is a multistep process which involves multiple molecular and cellular events that transform a normal cell to a malignant one. Firstly, the initiation process occurs when a normal cell sustains DNA mutation, and after a round of DNA synthesis, fixation of mutation occurs which forms an initiated cell. This mutation can be induced by genotoxic agents such as UV radiation and other chemical carcinogens present in the environment. Progression of these tumor cells occurs during normal cell proliferation and DNA synthesis wherein mutations may be acquired through misrepair of damaged DNA resulting in spontaneous initiated mutated cells. Drug-induced oxidative stress is implicated as a mechanism of toxicity in numerous tissues and organ systems, including the liver, kidney, ear, and cardiovascular and nervous systems. This chapter highlights the various mechanisms like oxidative DNA damage, mitochondrial DNA damage, modulation of DNA methylation, production of lipid peroxides, and changes in cell growth and gene expression which occur during the process of carcinogenesis induced by reactive oxygen species (Fig. 1). The role of oxidative stress in Drug-induced toxicity has also been discussed in the chapter.

2 Oxidative DNA Damage and Carcinogenesis

Oxidative DNA damage is a major cause of mutation in living organisms, with more than hundred oxidative DNA adducts (purine, pyrimidine, and the deoxyribose backbone) identified [1]. The estimated frequency of oxidative DNA damage in human cells is 10^4 lesions/cell/day [1]. Reactive nitrogen species, such as peroxynitrites and nitrogen oxides, have been implicated in cancer formation. Many forms of reactive oxygen species are capable of forming oxidized bases such as hydroxyl radical which in particular has been shown to produce a number of oxidized DNA lesions [2]. For the hydroxyl radical to react and oxidize DNA, it must be generated adjacent to the nucleic acid material. H_2O_2, a precursor to hydroxyl radical, being less reactive and readily diffusible is more likely to be involved in the formation of oxidized bases [3, 4]. Peroxynitrite is a strong cellular oxidant, which is formed from the coupling of nitric oxide and superoxide and is easily diffusible between cells and taken up by active transport mechanisms into cells [5]. Nitric oxide and superoxide are produced in activated macrophages, and it is likely that peroxynitrite may be formed in proximity to these cells so thus they are equally important to the induction of mutation. The DNA-damaging capability of peroxynitrite therefore helps to explain the association between inflammation and mutation [2].

Oxidized DNA bases appear to be mutagenic and capable of inducing mutations that are commonly observed in neoplasia. Neoplasia is a multistep process involving DNA damage and cell proliferation. Chemical carcinogens control various stages of this process and function through modification of cellular and molecular events. Genotoxic agents are the chemicals that directly damage genomic DNA, which in turn can result in mutation and/or clastogenic changes. These chemicals when frequently activated in the target cell produce a dose-dependent increase in neoplasm formation. A second category of carcinogenic compounds (nongenotoxic) appears to function through non-DNA-reactive or indirect DNA-reactive mechanisms. Although much less is known about the exact mode of action of nongenotoxic carcinogens, they modulate cell growth and cell death. Changes in gene expression and cell growth parameters are paramount in the action of nongenotoxic carcinogens. These agents frequently function during the promotion stage of the cancer process [6, 7] (Fig. 2).

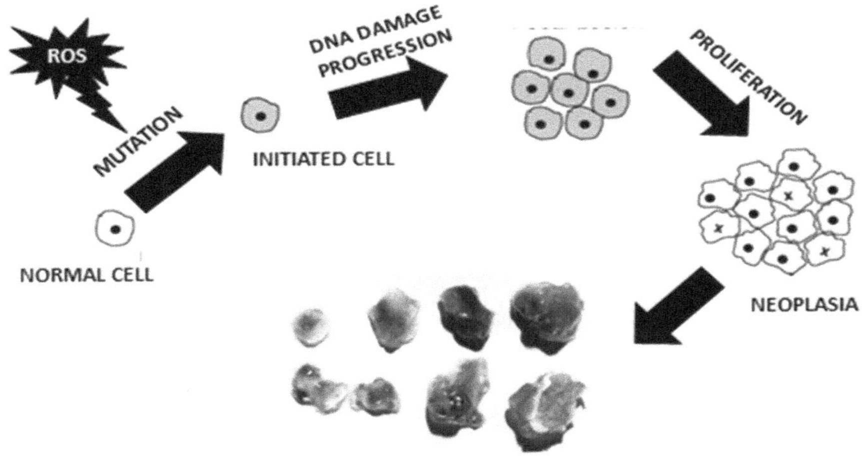

Fig. 2 Process of tumor formation. ROS induces mutation in normal cell which leads to formation of an initiated cell. DNA damage occurs in this cell and further progression leads to proliferation of these cells and formation of neoplasia. Further proliferation results in the formation of tumor at various stages

3 Mitochondrial DNA Damage and Carcinogenesis

Although the role of oxidative nuclear DNA damage has been implicated in the formation of neoplasia, there are some evidences that show the involvement of mitochondrial oxidative DNA damage in the carcinogenesis process [8]. The sustained oxidative burden in mitochondria has been linked to the induction of mutation. It has been found that tumor cells appear to be more glycolytic than the normal cells as the mitochondria are under the burden of sustained oxidative stress and increased mutation frequency [9]. Mitochondrial DNA mutations and alterations in mitochondrial genomic function appear to be causally related to the development of neoplasia. Mitochondrial DNA mutations have been identified in a number of cancers [10]. Altered expression and/or mutations in mitochondrial genes encoding for complexes I, III, IV, and V, and in the hypervariable regions of mitochondrial DNA, have been identified in human tumors. Mutagenesis occurs more rapidly in rodents as compared to humans although mitochondrial genes have been found in human tumors [11]. Compared to the nuclear genome, the mitochondrial genome appears to be more susceptible to oxidative base damage [12, 13]. Furthermore; the mutation rate in mitochondrial DNA has been reported to be at least two orders of magnitude higher than that of the nuclear DNA [11]. Increased susceptibility of the mitochondrial genome to oxidative damage can be because of the following three factors:

(a) Mitochondrial DNA and the electron transport system, a major source of reactive oxygen species, are in close proximity to each other. Under physiological conditions, mitochondria convert 4–5 % of oxygen consumed into superoxide anion and subsequently hydrogen peroxide [14].
(b) Mitochondrial DNA is not protected by histones.
(c) Due to lack of nucleotide excision repair, DNA repair capacity is limited in the mitochondria [15, 16].

Collectively, these findings may help in explaining the increased frequency of mitochondrial mutations seen in tumor cells.

Although the extent to which mitochondrial DNA alterations participate in the cancer process and the portion of tumor cells that possess mutated mitochondrial DNA has not been fully recognized, significant information exists that supports the involvement of the mitochondria in carcinogenesis. Cellular ATP production and an overall cellular energy balance can be affected by mutations in genes encoding oxidative phosphorylation. Decreases in ATP can disturb the cell cycle by delaying progression through the cell cycle [17]. Fragments of mitochondrial DNA have been found to be inserted into nuclear DNA, and it has been suggested as a mechanism for activation of oncogenes [18]. Hydrogen peroxide and other reactive oxygen species have been associated with the activation of nuclear genes that can lead to mitochondrial biogenesis, transcription, and replication of the mitochondrial genome. The stimulation of mitochondrial biogenesis may be a cellular response to compensate dysfunctional oxidative phosphorylation associated with mutated mitochondrial DNA. Low levels of hydrogen peroxide appear to cause stimulation of mitogenesis in a variety of mammalian cell types [19]. As observed with oxidative genomic DNA modification, oxidative damage and the induction of mutation in mitochondrial DNA may participate at multiple stages of the process of carcinogenesis, involving mitochondria-derived reactive oxygen species, induction of mutations in mitochondrial genes, and perhaps the insertion of mitochondrial genes into nuclear DNA [20].

4 Cell Growth Regulation and Carcinogenesis

Production of reactive oxygen species and oxidative stress play a role both in the stimulation of cell proliferation and cell removal by apoptosis [22, 23]. The mechanisms for the involvement of oxidative stress in the induction of cell proliferation and apoptotic processes are not known. The effects of reactive oxygen species and oxidative stress are cell specific and dependent upon the form as well as the intercellular concentration of reactive oxygen species. Thus, the involvement of reactive oxygen species in cell growth regulation is a complex process which depends on a number of cellular and biochemical parameters. Reactive oxygen species function to induce cell proliferation during the tumor promotion stage of carcinogenesis [24]. Both H_2O_2 and superoxide anion have been found to induce mitogenesis and cell proliferation in several mammalian cell types [25]. Oxidative stress also has a role in inducing apoptosis. High concentrations of reactive oxygen species trigger an apoptotic signaling pathway which results in cell loss [26]. A number of endogenous substances (prostaglandins and lipid hydroperoxides), redox cycling compounds (quinones, adriamycin), and growth factors (transforming growth factor-β and tumor necrosis factor-α) induce apoptosis via the generation of reactive oxygen species [27, 28]. Antioxidants such as *N*-acetyl cysteine (NAC), glutathione, and dithiothreitol help in scavenging the reactive oxygen species thereby inhibiting the apoptotic process [28]. Xenobiotics may differentially interact within the cell to elicit biological responses. Lipophilic and esterified compounds can freely cross the plasma membrane and yield effects within a cell; other compounds gain entry to the cell only through specific channels or energy-dependent pumps, whereas membrane-impermeable chemicals may stimulate cellular responses by acting on receptors that initiate signaling cascades in the cell. As a result, each chemical may provide a unique stimulus that sets in motion specific signaling pathways. Although no single mechanism explains the increased cell proliferation and/or inhibition of apoptosis observed following conditions that favor increased cellular oxidants, mounting evidence is emerging that links reactive oxygen species with altered expression of growth regulatory genes [21].

5 Cellular Transformation and Carcinogenesis

Cellular transformation in cancer biology is a process whereby normal cells acquire properties of malignant cells and start behaving in a different manner. Gain-of-function mutations in oncogenes and the loss-of function mutations in tumor suppressor genes are the main causes of malignant transformation of cells [29]. Oxidative stress has a very important role in both genotoxic and nongenotoxic mechanisms. Accordingly, rather than structural molecular changes in a specific gene or gene cluster, there are "hallmarks of cancer" that enable a cell to change its transcription machinery, become malignant, and metastasize. The reason for these early changes is closely related to aberrant DNA methylation, which is considered to be amongst the earliest changes to occur in carcinogenesis. Aberrant DNA methylation is extensively present in cancer cells as global hypomethylation and focal, specific hypermethylation in promoters of tumor suppressor genes, leading to their silencing [30]. The mutations lead to agitations of a number of signaling molecules, including p53, Raf, retinoblastoma (Rb), protein phosphatase 2A, telomerase, RalGEFs, phosphatidylinositol 3-kinase (PI3K), Ras, Rac, cellular v-myc myelocytomatosis viral oncogene homolog (c-Myc), STAT3, NF-jB, and HIF-1a (Fig. 3). Chemicals, viruses, radiation, hypoxia, and nutrient deprivation can also induce mutations in these genes, thereby giving rise to cancer cells [31]. Evidences accumulated over the past several years have indicated an association between reactive oxygen species and malignant transformation [32–34]. The mechanism behind raised ROS levels leading to oncogene activation is still not clearly understood, but DNA damage is known to play a role. For instance, the oncogenic transformation of ovarian epithelial cells with *H*-Ra^{v12} or tyrosine kinase *Bcr-Abl* in hematopoietic cells was associated with an increase in reactive oxygen species [35]. In another study, transformation of fibroblasts with constitutively active isoforms of Rac and Ras was

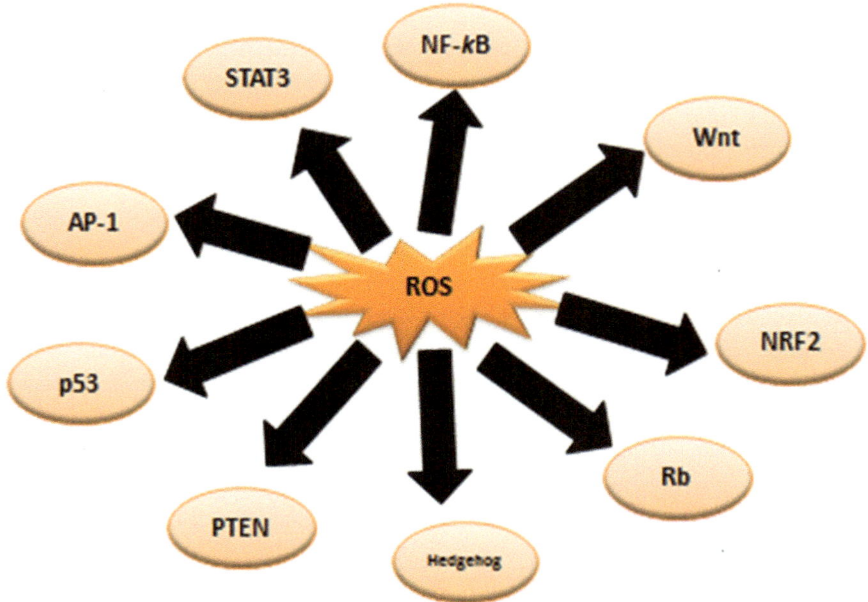

Fig. 3 Altered cellular signaling leads to changes in the function of various transcription factors which further leads to changes in cell functionality

associated with production of superoxide; further study revealed that transformation could be suppressed by treatment with antioxidants [36]. Mox1 is a phagocytic NOX, the overexpression of which has been shown to increase superoxide generation in mouse fibroblasts [37]. The cells expressing Mox1 exhibited a transformed appearance and produced tumors in athymic mice [37]. In a study, cells genetically transformed to express the cancer phenotype were able to generate ROS in response to the small-molecule piperlongumine; normal cells, on the other hand, could rarely be induced to generate ROS [38]. The inflammatory cytokine tumor necrosis factor-α (TNF-α) has been shown to play a role in the transformation of mouse fibroblasts into malignant cells; this effect was partially suppressed by antioxidants [39]. Apurinic/apyrimidinic endonuclease/redox effector factor-1 (APE/Ref-1) is a multifunctional protein involved in both DNA repair and redox regulation. Ref-1 was shown to induce malignant transformation in JB6 mouse epithelial cells through the mediation of reactive oxygen species [40]. Matrix metalloproteinase (MMP)-3, a stromal enzyme that is upregulated in many breast tumors, has been shown to induce ROS, DNA damage, genomic instability, and the transformation of mouse mammary epithelial cells into malignant cells [41]. Therefore, it can be said that reactive oxygen species appear to play a role in the transformation of normal cells into cancer cells, wherein transformed cells seem to have greater ROS levels than normal cells. However, how ROS transform normal cells is not exactly known.

6 Lipid Damage and Carcinogenesis

Damage to lipid bilayer due to oxidative stress also plays an important role in chemical carcinogenesis. Damage to cellular biomembrane generates a variety of products including reactive electrophiles such as epoxides and aldehydes [42]. Malondialdehyde (MDA), a by-product of lipid degradation, is both highly electrophilic and nucleophilic. This characteristic allows not only the reaction with cellular nucleophiles but also the formation of MDA oligomers [43]. MDA was shown to induce thyroid tumors in chronically treated rats [44]. MDA reacts with several nucleic acid bases to form dG, dA, and dC adducts [45]. The identification of MDA-DNA adducts in humans may be significant as MDA-DNA adducts have been detected in the genome of healthy humans in quantities comparable to those generated by exogenous chemicals in rodent carcinogenesis studies. The observed MDA-DNA adducts appear to be promutagenic as they induce mutations in oncogenes and tumor suppressor genes seen in human tumors. MDA-DNA adduct levels also appear to correlate with altered cell cycle control and gene expression in cultured cells [46].

7 Gene Modulation and Carcinogenesis

Besides understanding the effects of ROS on DNA damage and induction of mutations, the role of ROS on epigenetic effects should also be considered [47]. An upregulation of genes in response to stress is most commonly seen in mammalian cells upon exposure to oxidants or oxidative stress-inducing agents. While high levels of ROS production may lead to the induction of apoptosis or necrosis, increasing evidence demonstrates that low or transient ROS exposure increases cell proliferation likely through altered expression of growth factors and proto-oncogenes [48, 49]. Hence there are many chemical and biological agents that increase cellular levels of ROS through varied mechanisms. ROS-induced alteration of gene expression can occur through a modulation of a host of signaling pathways including cAMP-mediated cascades, calcium-calmodulin pathways, and intracellular signal transducers such as nitric oxide [50–52]. Studies have also shown that ROS-induced release of calcium stored inside the cells results in the activation of kinases, such as protein kinase C (PKC), which regulates a variety of cell functions including proliferation, cell cycle, differentiation, cytoskeletal organization, cell migration, and apoptosis [53]. It is interesting to know that the activation of PKC seems to be differentially regulated by cellular oxidants: oxidation at the NH2-terminal regulatory domain activates PKC, whereas oxidation at the COOH terminal inactivates PKC [54]. Similarly, H_2O_2 activates protein kinases such as extracellular signal-regulated kinase (ERK) 1/2 [55], phosphoinositide 3-kinase/serine-threonine kinase (P13K/Akt) [56], protein kinase B (PKB) [57], and protein tyrosine phosphatases (PTPs) [53]. Because these pathways regulate cellular migration, proliferation, survival, death responses, their irregular activation have been suggested to be a possible mechanism of ROS-induced carcinogenesis. The upregulation of free radicals has also led to activation of transcription factors such as nuclear factor erythroid 2-related factor 2 (NF-E2/rf2 or Nrf2) [58], mitogen-activated protein (MAP) kinase/AP-1 [59], and NF-kB pathways [60] as well as hypoxia-inducible transcription factor HIF-1a [61]. The level of ROS can trigger the upregulation of these transcription factors which may help to determine that either cell death or cell proliferation may result from exposure to oxidative stress.

AP-1 is a collection of dimeric basic region-leucine zipper (bZIP) proteins that belong to the Jun (c-Jun, JunB, JunD), Fos (FosB, Fra-1, Fra-2), Maf, and ATF subfamilies, all of which can bind TPA or cAMP response elements [62]. c-Jun, an important transcriptional regulator, often forms stable heterodimers with Jun proteins, which assist the binding of Jun to DNA [63]. AP-1 activity is triggered in response to H_2O_2 as well as several cytokines and other physical and chemical stresses. In addition, in vitro transcriptional activity of AP-1, regulated by the redox state of a specific cysteine located at the interface between the two c-Jun subunits, highlight the importance of redox status on gene transcription [64]. The free radicals invoke a signal cascade that begins with the stimulation of MAP kinases [65].

MAP kinases, a family of serine/threonine kinases, have been seen to regulate processes important in carcinogenesis including proliferation, differentiation, and apoptosis. Three major subfamilies of MAP kinases, extracellular signal-regulated kinases (ERK), c-Jun N-terminal kinases (JNK), and the p38 kinases, have been identified [66]. MAP kinases are involved in modulation of gene expression through phosphorylation of a wide array of transcription factors. The ERK pathway among the three subfamilies has most commonly been associated with the regulation of cell proliferation. Activation of the ERK, JNK, and p38 subfamilies has been observed in response to changes in the cellular redox balance. The balance between ERK and JNK activation is a key determinant for cell survival as both a decrease in ERK and a rise in JNK are required for the initiation of apoptosis [67]. The induction of AP-1 by H_2O_2, cytokines, and other stressors is mediated mainly by JNK and p38 MAP kinase cascades [65]. Once activated, JNK proteins translocate to the nucleus and phosphorylate c-Jun and ATF2, enhancing transcriptional activities [68]. H_2O_2 can activate MAP kinases and thereby AP-1 in several manners, one of which involves an MAP kinase kinase, apoptosis signal regulating kinase (ASK1) [69]. It has been seen that ASK1 activity is inhibited by thioredoxin and oxidation of thioredoxin by H_2O_2 interrupts ASK1 inhibition, resulting in ASK1 activation [69]. The other mechanism involves oxidant-mediated inhibition of MAP kinase phosphatases, which leads to increased MAP kinase activation. In either mechanism, activation of MAP kinases directly leads to increased AP-1 activity. The significance of AP-1 activation, and the role of cellular oxidants in the activation in the carcinogenesis, has been supported by a number of observations.

Increased cell proliferation is a common consequence of AP-1 activation. In particular, several evidences have demonstrated that c-fos and c-jun are positive regulators of cell proliferation [70]. One of the genes regulated by AP-1 is cyclin D1. AP-1 binding sites have been identified in the cyclin D1 promoter and AP-1 activates this promoter, resulting in activation of cyclin-dependent kinase, which helps in the entry into the cell division cycle [71]. c-Jun also stimulates the progression into the cell cycle by both induction of cyclin D1 and suppression of p21waf, a protein that inhibits cell cycle progression [72]. JunB, considered a negative regulator of c-jun-induced cell proliferation, represses c-jun-induced cyclin D1 activation by the transcription of p16 ^{INK4a}, a protein that inhibits the G1 to S phase transition [73]. Expression of c-jun and c-fos can be induced by a variety of compounds including nongenotoxic and tumor-promoting compounds (carbon tetrachloride, phenobarbital, TPA, TCDD, cadmium, alcohol, ionizing radiation, asbestos), many of which generate reactive oxygen species [74]. In addition to affecting cell proliferation, AP-1 proteins also function as positive and negative regulators of apoptosis. Whether AP-1 induces or inhibits apoptosis depends upon the balance between the pro- and antiapoptotic target genes, and this may vary from one cell type to another, the stimulus used to activate AP-1, the developmental stage, and the duration of the stimulus [67]. Finally, important to neoplastic development, through increased production of growth factors as well as modulation of cell cycle regulators, AP-1 proteins participate in oncogenic transformation through interaction with activated oncogenes such as Ha-ras [75].

NF·*k*B is an inducible and ubiquitously expressed transcription factor for genes involved in cell survival, differentiation, inflammation, and growth [76]. Active NF·*k*B complexes are dimers of proteins from the Rel family of proteins consisting of p50 (NF·*k*B1), p52 (NF·*k*B2), c-Rel, v-Rel, Rel A (p65), and Rel B [77]. NF·*k*B are present in inactive state in the cytoplasm because of binding to inhibitory I·*k*B proteins. Response to an extracellular stimuli leads to activation of NF·*k*B by dissociation of I·*k*B, which exposes the nuclear localization sequence thereby allowing the entry of NF·*k*B into the nucleus and binds *k*B-regulatory elements [78]. The presence of free radicals has also been shown to influence NF·*k*B regulation. Besides experimental evidence for redox regulation of transcription factors that has been demonstrated in vitro, in vivo evidence has also been suggested in vivo. It was seen in a study that oxidative stimuli promoted the oxidation of p53 to a mixed disulfide with GSH and that this modification inhibited transcriptional activity and nuclear localization of NF·*k*B [79]. Activation of NF·*k*B has been linked to the process of carcinogenesis through its role in inflammation, differentiation, and cell growth. NF-*k*B regulates several genes which play important part in cell transformation, proliferation, and angiogenesis [80]. Carcinogens and tumor promoters including UV radiation, phorbol esters, NNK, asbestos, alcohol, and benzo(a)pyrene are among the external stimuli that activate NF·*k*B [81]. Through complex pathways that are still being clarified, NF·*k*B activation is involved in cell survival. The expression of several genes like bcl-2, bcl-x_L, TRAF1, TRAF2, SOD, and A20, regulated by NF·*k*B, is related to cell survival to some extent through inhibition of apoptotic pathways. Expression of NF·*k*B has been shown to promote cell proliferation, whereas inhibition of NF·*k*B activation blocks cell proliferation [82]. Additionally, tumor cells from blood neoplasms and colon, breast, pancreas, and squamous cell carcinoma cell lines have all been reported to constitutively express activated NF·*k*B [83]. The mechanism for activation of NF·*k*B by reactive oxygen species is not clear, but they have been implicated as second messengers involved in activation of NF·*k*B via tumor necrosis factor (TNF) and interleukin-1 [83]. Suppression of TNF and interleukin-1 was shown to downregulate the expression of active NF·*k*B and inhibit proliferation of lymphoma and myelogenous leukemia cells [84].

The fact that protein kinases are also involved in cell response mediated by the TNF superfamily should be noted. In fact, the binding of TNF to its receptor is associated with protein-protein disulphide bond formation [85]. Oxidative changes may augment the TNF receptor-mediated signal, H_2O_2, and can function either to activate protein kinases (e.g., stress-activated protein kinase (SAPK), extracellular signal-regulated kinase (ERK), and p38) or inhibit transcription factors, such as AP-1 and NF·*k*B [86]. Therefore, cell death or cell survival will in part depend on the strength and duration of oxidant exposure and on the cell type involved. The importance of reactive oxygen species on NF·*k*B activation is further supported by studies demonstrating that activation of NF·*k*B by nearly all stimuli can be obstructed by antioxidants, including L-cysteine, NAC, thiols, green tea polyphenols, and vitamin E [87]. NF·*k*B activation appears to be selectively mediated by peroxides as activation was observed only following exposure to H_2O_2 or butyl peroxide, and not superoxide or hydroxyl radicals [88]. Likewise, NF·*k*B activity was increased in cells that had higher levels of superoxide dismutase and decreased in cells overexpressing catalase [89]. Collectively these findings support the linkage of NF·*k*B activation with reactive oxygen species and the carcinogenesis process. Activation of transcription factors is clearly triggered by signal transduction pathways and activated by H_2O_2 and other cellular oxidants. Transcription factors can mediate effects of both physiological and pathological exposure to reactive oxygen species through their ability to stimulate cell proliferation and regulation of apoptosis.

8 DNA Methylation and Carcinogenesis

Modulation of DNA methylation in the cell often influences gene expression [90]. Altered methylation process does not involve a modification or miscoding of DNA base-coding sequence, but rather leads to abnormal gene expression, in part, by affecting the ability of methylated DNA-binding proteins to interact with *cis* elements [91]. Under normal conditions, DNA is methylated symmetrically on both strands. Immediately following DNA replication, the newly synthesized double-stranded DNA contains hemimethylated sites that signal for DNA maintenance methylases to transfer methyl groups from S-adenosylmethionine to cytosine residues on the new DNA strand [92]. If a cell is signaled to undergo DNA synthesis prior to maintenance methylation, then double-stranded DNA with hypomethylated regions will be disseminated in subsequent cell division cycles, giving rise to possibly heritable genetic changes. 5 mC in DNA is known to affect gene expression and alteration of cellular processes such as development and differentiation and appears to be an important mechanism in carcinogenesis [93–95]. During the carcinogenesis process, both hypomethylation and hypermethylation can occur [93, 94]. Hypermethylation of genes may inhibit transcription of tumor suppressor genes [96] and is associated with decreased gene expression or gene silencing. Important to the cancer process, tumor suppressor genes are known to be hypermethylated and subsequently inactivated [93–95].

The degree of methylation within a gene inversely correlates with the expression of that gene. Progressive increases in methylation of CpG islands have been observed in bladder cancer, and specific tumor suppressor genes have been reported to be methylated in tumors, e.g., the retinoblastoma gene, p16 ^{ink4a}, and p14 ^{ARF} [92–95]. Inactivation of p16 ^{ink4a} by hypermethylation [97] of the promoter region appears to be an early event in lung cancer [98]. Furthermore, in nickel carcinogenesis, hypermethylation of p16^{ink4a} is apparently induced by reactive oxygen species and MAP kinase pathways [99]. Regional hypermethylation may impart molecular changes associated with genetic instability and may participate in the progression of neoplasia. Conversely, hypomethylation is considered an early and frequent event in the carcinogenesis process [100]. A hypomethylated gene is considered to possess an increased potential for expression as compared to a hypermethylated gene [101]. In addition, hypomethylation has been associated with increased mutation rates. Most metastatic neoplasms in humans have significantly lower 5 MeC than normal tissue [102]. Oncogenes gets hypomethylated as a result their expression is amplified [98]. Dietary constituents containing choline and methionine provide the methyl groups used in methylation reactions. Exposure of rats to a choline-/methionine-deficient diet results in hepatocellular proliferation and neoplasia [103, 104]. The induction of cell proliferation by a methyl-deficient diet appears to function through decreased hepatic levels of S-adenosyl methionine and, thus, promotes hypomethylation and subsequent expression of oncogenes. Prolonged administration of a diet deficient in choline or methyl donor groups resulted in hypomethylation of c-myc, c-fos, and c-H-ras proto-oncogenes and was associated with the induction of hepatocarcinogenesis in rodents [103, 104]. Also consistent with the role of methylation of DNA in the promotion stage of carcinogenesis process, the induction of hepatocarcinogenesis by methyl-deficient diets was shown to be reversible by the administration of S-adenosyl methionine [105, 106]. Among the agents and situations that can alter methylation status, reactive oxygen species can modify DNA methylation patterns. In particular, oxidative DNA damage [106] can result in decreased DNA methylation. Several chemical carcinogens modify DNA methylation, methyltransferase activity, and chromosomal structure. Of particular importance, the formation of oxidative DNA lesions has been linked to changes in DNA methylation profiles and the carcinogenesis process. Oxidative DNA damage can interfere with the ability of methyltransferases to interact with DNA, thus resulting in a generalized hypomethylation of cytosine residues at CpG sites. Hence, oxidative DNA damage may be an important contributor to the carcinogenesis process brought about by the loss of DNA methylation, following the expression of normally inactive genes [107]. The abnormal methylation pattern observed in cells transformed by chemical oxidants may contribute to an overall anomalous gene expression and promote the tumor process. The roles of oxidative DNA damage, mutation, and altered gene expression induced by cellular oxidants and/or altered methylation status have been discussed in relation to the carcinogenesis process.

9 Tumor Proliferation and Carcinogenesis

Uncontrolled proliferation is one of the chief characteristics of tumor cells [108, 109]. A specific set of cell cycle regulators such as cyclins and cyclin-dependent kinases (CDKs) control the progression of cell cycle events. CDK activity is controlled by the opposing effects of cyclins and CDK inhibitors. CDK inhibitors such as p21 and p27 negatively regulate CDK activity, whereas cyclins are required for CDK activity and cell cycle progression. Another protein, c-Myc, is required for the G-to-S-phase transition [109]. The expression of c-Myc, in turn, is regulated by cdc25, a phosphatase that activates CDKs. Intracellular ROS produced by extracellular stimuli as well as exogenous administration of ROS have been shown to enhance cell proliferation. In a study, administration of H_2O_2 was shown to enhance the proliferation of hepatoma cells by increasing protein kinase B and extracellular signal-regulated kinase (ERK) activities [110]. In another study, transformed bladder urothelial cells were found to be hyper-proliferative and produced elevated ROS levels in the presence of monomethylarsonous acid; the upregulation in cyclooxygenase-2 (COX-2) expression observed in these cells was found to be ROS dependent [111]. ROS produced by low concentrations of arsenite has been shown to enhance the proliferation of breast cancer cells by recruiting cells into the S phase of the cell cycle, enhancing the expression of c-Myc and heme oxygenase-1 and increasing NF-*k*B activity [112]. ROS produced by endogenous sources can also enhance cancer cell proliferation. For example, ROS produced by Romo1, a mitochondria-localized protein [113, 114], was shown to be indispensable to the proliferation of lung cancer cells [115]. Such an induction in cell proliferation was found to be ERK dependent [115]. Endogenous production of superoxide has also been shown to enhance tumor proliferation in hepatoma cells that was mediated through AKT8 virus oncogene cellular homolog (AKT) phosphorylation [116]. Similarly, an increase in endogenous ROS due to reduction in the antioxidant defense system has been correlated with an increase in the proliferation of breast [117] and ovarian [118] cancer cells. In breast cancer cells, ROS-mediated tumor proliferation was found to be dependent on the activation of PI3K pathway and reduction of PTEN activity [117].

The role of ROS in promoting tumor proliferation is further supported by observations that certain agents that have the potential to inhibit ROS generation can also inhibit tumor cell proliferation. In another study, exogenous catalase inhibited the proliferation of numerous cancer types [119]. Consistent with these observations, stable expression of human catalase in MCF-7 cells inhibited proliferation and reverted malignant features [119]. Curcumin has been shown to inhibit the proliferation of lymphoma cells by increasing endogenous antioxidant enzyme activity and by inhibiting NF-*k*B activity [120]. Inhibition of ROS generation by N-acetyl-L-cysteine (NAC), one of the most widely used ROS scavengers, has been correlated with decreased proliferation of cancer cells. For example, treatment of glioma cells with NAC inhibited cell proliferation by arresting cells in the G phase; this inhibition was correlated with a decrease in the activities of AKT, ERK1/2, and NF-*k*B [121].

10 Anticancerous Drug-Induced Oxidative Stress

Drug-induced oxidative stress has been associated with mechanism of toxicity in numerous tissues and organ systems, including the liver, kidney, ear, and cardiovascular and nervous systems. The level to which mechanisms of drug-induced oxidative stress have been characterized varies among different types of drugs. Metabolism of a drug may generate a reactive intermediate that can reduce molecular oxygen directly to generate ROS thus triggering its production and leading to oxidative stress. Chlorpromazine is an interesting example as photoactivation in the skin is considered likely to cause cutaneous phototoxicity (sunburn-like reaction and hyperpigmentation), which is a well-known adverse event associated with this compound [122]. Photodechlorination converts chlorpromazine to an excited state with subsequent energy transfer to molecular oxygen and generation of both excited singlet oxygen and superoxide species. These species may then react with DNA and macromolecules as described above and trigger adaptive or toxic responses in the skin. For other drugs, there is evidence of increase in cellular ROS in response to the exposure to anticancerous drugs, and this indicates association of ROS and oxidative stress with toxicity even if the mechanisms by which ROS are generated are not fully characterized. In this section, we discuss further the evidence for involvement of oxidative stress in drug-induced toxicities, using the examples of doxorubicin.

10.1 Doxorubicin

Doxorubicin (DOX) is an anthracycline drug used in numerous chemotherapy regimens to treat hematological and solid tumors. Though effective as an anticancer drug, dose-dependent cardiotoxicity is a well-described side effect of DOX therapy and is a major limitation to its use. Considerable research has been conducted to explore the mechanism(s) of DOX-induced cardiomyopathy [123]. The drug has been demonstrated to cause a number of potentially pathological effects. However, it is still uncertain what role these varied events play in the pathogenic process. Despite the fact that the precise mechanism(s) of toxicity is unknown, it is widely accepted that the drug stimulates the generation of oxygen free radicals and that oxidative stress is a central feature of DOX-induced cardiac toxicity [124, 125]. It is not known, however, whether oxidative injury is the cause or a consequence of the toxicity.

The mechanisms underlying effects on cardiac tissue have been investigated intensively. Free radical formation, lipid peroxidation, mitochondrial dysfunction, altered calcium handling, DNA damage, p53 accumulation, and activation of pro-apoptotic signaling cascades/inhibition of survival signaling have all been implicated. Although these mechanisms are not fully elucidated and are multifactorial, there is substantial evidence to support a key role for DOX-induced oxidative stress in clinically relevant cardiotoxicity.

10.2 Formation of ROS by DOX

DOX may generate ROS by more than one mechanism [126, 127]. Reduction of DOX by one electron via mitochondrial reductases may generate anthracycline semiquinone free radicals [128]. Under aerobic conditions, these are unstable and readily reduce molecular oxygen to the ROS superoxide anion and H_2O_2 [129]. Reactions between iron and DOX may also generate ROS. Redox reactions subsequent to interaction of DOX with iron (III) may generate an iron II-DOX free radical, capable of reducing molecular oxygen.

The oxidative stress resulting from increased free radical generation in cardiomyocytes may lead to multiple adverse effects including energetic imbalance, perturbations in mitochondrial function, activation of stress-related signaling pathways (such as p38 and JNK), p53 accumulation, and, ultimately, cell death. To illustrate, studies in mice have shown that DOX induces elevations in ROS, DNA damage, activation of ataxia telangiectasia-mutated (ATM) kinase signaling, accumulation of p53, and cardiomyocyte death. Similar DOX-induced events, when measured in cultured mouse cardiomyocytes, are not seen in the presence of a free radical scavenger. It is interesting to note that inhibition of ATM kinase (which signals DNA damage) reduces DOX-induced accumulation of p53, suggesting a link between DNA damage and apoptosis in response to DOX. Furthermore, in transgenic mice deficient in p53 or overexpressing Bcl-2 in cardiac tissue, DOX cardiac damage, including contractile dysfunction and myocyte apoptosis, was attenuated [130]. Mitochondrial swelling, depolarization, perturbations of energetics, and dysregulation of mitochondrial calcium signaling have all been reported following exposure to DOX in vitro *or* in vivo [131–133]. The consequent disruption of calcium signaling pathways and calcium-dependent ATP synthesis that could result from perturbations in mitochondrial structure and function may be key contributors to toxicity in cardiomyocytes, via induction of apoptosis [134] (Fig. 4).

11 Conclusion

It can be elucidated that cellular reactive oxygen radicals are linked with the pathogenesis of several chronic diseases including cancer. When the internal antioxidant fails to combat the excessive production of exogenous free radicals, condition of oxidative stress occurs which in turn leads to damage of cellular nucleic acids, lipid, or protein. Unrepaired damage to DNA may result in mutation which leads to cancer progression. Although significance of the role of oxidative nuclear DNA damage in the formation of neoplasia has been established, mitochondrial DNA damage, mutation, and alteration of the mitochondrial genomic function also appear to contribute to the process of carcinogenesis. ROS are important components of cell signaling pathways and have been shown to regulate cellular transformation, survival, proliferation, invasion, angiogenesis, and metastasis.

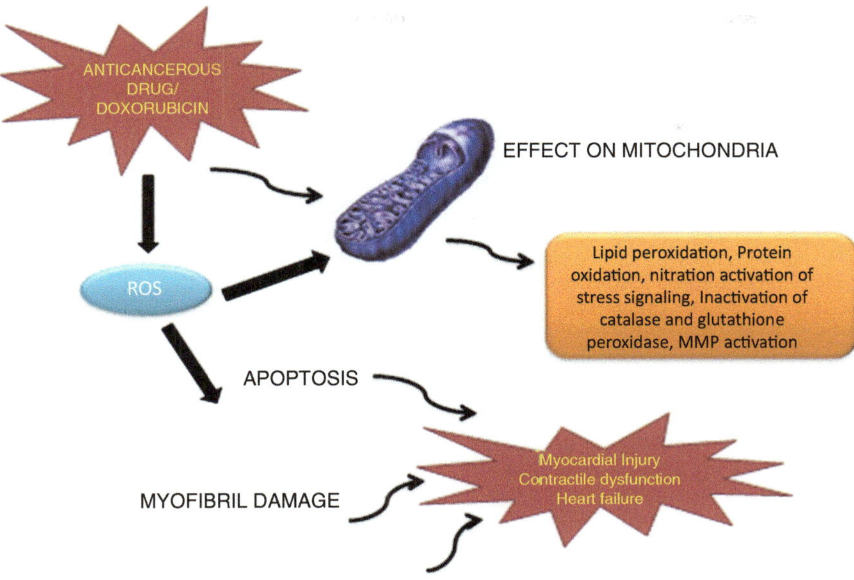

Fig. 4 Doxorubicin-induced toxicity. Doxorubicin is responsible for various cardiac injuries due to damaging effects on mitochondria which leads to apoptosis, myofibril damage and disruption of calcium signaling due to lipid peroxidation, protein oxidation, nitration activation of stress signaling, inactivation of catalase and glutathione peroxidase, MMP activation, etc.

Besides role in cancer progression, toxicity due to reactive oxygen species is also considered as an implication induced by anticancerous drugs. Further antioxidative strategies should be developed to scavenge these free radicals, stabilize the redox state to avoid the mechanisms leading to cancer, and also to minimize the toxic effects of anticancerous drugs. In this regard, exogenous, natural antioxidants can play a positive role by scavenging ROS and stabilizing redox status being less toxic and safe.

References

1. Lu AL, Li X, Gu Y et al (2001) Repair of oxidative DNA damage: mechanisms and functions. Cell Biochem Biophys 35:141–170
2. Marnett LJ (2000) Oxyradicals and DNA damage. Carcinogenesis 21:361–370
3. Guyton KZ, Kensler TW (1993) Oxidative mechanisms in carcinogenesis. Br Med Bull 49:523–544
4. Barber DA, Harris SR (1994) Oxygen free radicals and antioxidants: a review. Am Pharm NS34:26–35
5. Radi R (1998) Peroxynitrite reactions and diffusion in biology. Chem Res Toxicol 11:720–721
6. Pitot HC, Goldsworthy T, Moran S (1981) The natural history of carcinogenesis: implications of experimental carcinogenesis in the genesis of human cancer. J Supramol Struct Cell Biochem 17:133–146
7. Kolaja KL, Klaunig JE (1996) Selective dieldrin promotion of hepatic focal lesions in mice. Carcinogenesis 17:1243–1250
8. Tamura G, Nishizuka S, Maesawa C et al (1999) Mutations in mitochondrial control region DNA in gastric tumors of Japanese patient. Eur J Cancer 35:316–319
9. Horton TM, Petros JA, Heddi A et al (1996) Novel mitochondrial DNA deletion found in renal cell carcinoma. Genes Chromosomes Cancer 15:95–101
10. Cavalli LR, Liang BD (1998) Mutagenesis, tumorigenicity and apoptosis: are the mitochondria involved? Mutat Res 398:19–26
11. Wang E, Wong A, Cortopassi G (1997) The rate of mitochondrial mutagenesis is faster in mice than in humans. Mutat Res 377:157–166
12. Yakes FM, Van Houten B (1997) Mitochondrial DNA damage is more extensive and persists longer than nuclear DNA damage in human cells following oxidative stress. Proc Natl Acad Sci U S A 94:514–519
13. Zastawany TH, Dabrowska M, Jaskolski T et al (1998) Comparison of oxidative base damage in mitochondrial and nuclear DNA. Free Radic Biol Med 24:722–725
14. Albig AR, Neil JR, Schiemann WP (2006) Fibulins 3 and 5 antagonize tumor angiogenesis *in vivo*. Cancer Res 66:2621–2629

15. Sawyer DE, Van Houten B (1999) Repair of DNA damage in mitochondria. Mutat Res 434:161–176
16. Bohr VA, Dianov GL (1999) Oxidative DNA damage processing in nuclear and mitochondrial DNA. Biochimie 81:155–160
17. Van den Bogert C, Muus P, Haanen C et al (1988) Mitochondrial biogenesis and mitochondrial activity during the progression of the cell cycle of human leukemic cells. Exp Cell Res 178:143–153
18. Shay JW, Werbin H (1992) New evidence for the insertion of mitochondrial DNA into the human genome: significance for cancer and aging. Mutat Res 275:227–235
19. Davies KJ (1999) The broad spectrum of responses to oxidants in proliferating cells: a new paradigm for oxidative stress. IUBMB Life 48:41–47
20. Gupta S, Hevia D, Patchva S et al (2012) Upsides and downsides of reactive oxygen species for cancer: the roles of reactive oxygen species in tumorigenesis, prevention, and therapy. Antioxid Redox Signal 16:1295–1322
21. Klaunig EJ, Kamendulis ML (2004) The role of oxidative stress in carcinogenesis. Annu Rev Pharmacol Toxicol 44:239–267
22. Burdon RH (1995) Superoxide and hydrogen peroxide in relation to mammalian cell proliferation. Free Radic Biol Med 18:775–794
23. Slater AF, Stefan C, Nobel I et al (1995) Signaling mechanisms and oxidative stress in apoptosis. Toxicol Lett 83:149–153
24. Cerutti PA (1985) Prooxidant states and tumor promotion. Science 227:375–381
25. D'Souza RJ, Phillips EM, Jones PW et al (1993) Interactions of hydrogen peroxide with interleukin-6 and platelet-derived growth factor in determining mesangial cell growth: effect of repeated oxidant stress. Clin Sci 86:747–751
26. Dypbukt JM, Ankarcrona M, Burkitt M et al (1994) Different prooxidant levels stimulate growth, trigger apoptosis, or produce necrosis of insulin-secreting RINm5F cells the role of intracellular polyamines. J Biol Chem 269:30553–30560
27. Aoshima H, Satoh T, Sakai J et al (1997) Generation of free radicals during lipid hydroperoxide triggered apoptosis in PC12h cells. Biochim Biophys Acta 1345:35–42
28. Sandstrom PA, Mannie MD, Buttke TM (1994) Inhibition of activation-induced death in T cell hybridomas by thiol antioxidants: oxidative stress as a mediator of apoptosis. J Leukoc Biol 55:221–226
29. Wang JC (2010) Good cells gone bad: the cellular origins of cancer. Trends Mol Med 16:145–151
30. Jaganjac M, Čačev T, Čipak A et al (2012) Even stressed cells are individuals: second messengers of free radicals in pathophysiology of cancer. Croat Med J 53:304–309
31. Ralph SJ, Rodriguez-Enriquez S, Neuzil J et al (2010) The causes of cancer revisited: "mitochondrial malignancy" and ROS-induced oncogenic transformation—why mitochondria are targets for cancer therapy. Mol Aspects Med 31:145–170
32. Jackson AL, Loeb LA (2001) The contribution of endogenous sources of DNA damage to the multiple mutations in cancer. Mutat Res 477:7–21
33. Valko M, Rhodes CJ, Moncol J et al (2006) Free radicals, metals and antioxidants in oxidative stress induced cancer. Chem Biol Interact 160:1–40
34. Wang J, Yi J (2008) Cancer cell killing via ROS: to increase or decrease that is the question. Cancer Biol Ther 7:1875–1884
35. Trachootham D, Zhou Y, Zhang H et al (2006) Selective killing of oncogenically transformed cells through a ROS-mediated mechanism by beta-phenylethyl isothiocyanate. Cancer Cell 10:241–252
36. Irani K, Xia Y, Zweier JL et al (1997) Mitogenic signaling mediated by oxidants in Ras-transformed fibroblasts. Science 275:1649–1665
37. Suh YA, Arnold RS, Lassegue B et al (1999) Cell transformation by the superoxide-generating oxidase Mox1. Nature 401:79–82
38. Raj L, Ide T, Gurkar AU et al (2011) Selective killing of cancer cells by a small molecule targeting the stress response to ROS. Nature 475:231–234
39. Wondrak GT (2009) Redox-directed cancer therapeutics: molecular mechanisms and opportunities. Antioxid Redox Signal 11:3013–3069
40. Yang S, Misner BJ, Chiu RJ et al (2007) Redox effector factor-1, combined with reactive oxygen species, plays an important role in the transformation of JB6 cells. Carcinogenesis 28:2382–2390
41. Radisky DC, Levy DD, Littlepage LE et al (2005) Rac1b and reactive oxygen species mediate MMP-3-induced EMT and genomic instability. Nature 436:123–127
42. Janero DR (1990) Malondialdehyde and thiobarbituric acid reactivity as diagnostic indices of lipid peroxidation and peroxidative tissue injury. Free Radic Biol Med 9:515–540
43. Golding BT, Patel N, Watson WP (1989) Dimer and trimer of malondialdehyde. J Chem Soc Perkin 1:668–669
44. NTP toxicology and carcinogenesis studies of malonaldehyde, sodium salt (3-Hydroxy-2-propenal, sodium salt) (CAS No. 24382-04-5) in F344/N Rats and B6C3F1 Mice (Gavage Studies). NTP Tech Rep 331:513
45. Stone K, Ksebati M, Marnett LJ (1990) Investigation of the adducts formed by reaction of malondialdehyde with adenosine. Chem Res Toxicol 3:33–38
46. Ji C, Rouzer CA, Marnett LJ (1998) Induction of cell cycle arrest by the endogenous product of lipid peroxidation, malondialdehyde. Carcinogenesis 19: 1275–1283
47. Evans MD, Dizdaroglu M, Cooke MS (2004) Oxidative DNA damage and disease: induction, repair and significance. Mutat Res 567:1–61
48. Frenkel K (1992) Carcinogen-mediated oxidant formation and oxidative DNA damage. Pharmacol Ther 53:127–166

49. Fiorani M, Cantoni O, Tasinato A et al (1995) Hydrogen peroxide-and fetal bovine serum-induced DNA synthesis in vascular smooth muscle cells: positive and negative regulation by protein kinase C isoforms. Biochim Biophys Acta 1269:98–104
50. Shi H, Shi X, Liu KJ (2004) Oxidative mechanism of arsenic toxicity and carcinogenesis. Mol Cell Biochem 255:67–78
51. Liu G, Zhou W, Wang LI et al (2004) MPO and SOD2 polymorphisms, gender, and the risk of non-small cell lung carcinoma. Cancer Lett 214:69–79
52. Bertin G, Averbeck D (2006) Cadmium: cellular effects, modifications of biomolecules, modulation of DNA repair and genotoxic consequences (a review). Biochimie 88:1549–1559
53. Wu WS (2006) The signaling mechanism of ROS in tumor progression. Cancer Metastasis Rev 25: 695–705
54. Gopalakrishna R, Anderson WB (1989) Ca2+- and phospholipid-dependent activation of protein kinase C by selective oxidative modification of the regulatory domain. Proc Natl Acad Sci U S A 86:6758–6762
55. Zhou S, Kachhap S, Sun W et al (2007) Frequency and phenotypic implications of mitochondrial DNA mutations in human squamous cell cancers of the head and neck. Proc Natl Acad Sci U S A 104: 7540–7545
56. Liu LZ, Hu XW, Xia C et al (2006) Reactive oxygen species regulate epidermal growth factor-induced vascular endothelial growth factor and hypoxia-inducible factor-1alpha expression through activation of AKT and P70S6K1 in human ovarian cancer cells. Free Radic Biol Med 41:1521–1533
57. Mehdi MZ, Azar ZM, Srivastava AK (2007) Role of receptor and nonreceptor protein tyrosine kinases in H2O2-induced PKB and ERK1/2 signaling. Cell Biochem Biophys 47:1–10
58. Kensler TW, Wakabayashi N, Biswal S (2007) Cell survival responses to environmental stresses via the Keap1-Nrf2-ARE pathway. Annu Rev Pharmacol Toxicol 47:89–116
59. Benhar M, Engelberg D, Levitzki A (2002) ROS, stress-activated kinases and stress signaling in cancer. EMBO Rep 3:420–425
60. Pantano C, Reynaert NL, van der Vliet A et al (2006) Redox-sensitive kinases of the nuclear factor-kappaB signaling pathway. Antioxid Redox Signal 8:1791–1806
61. Rankin EB, Giaccia AJ (2008) The role of hypoxia-inducible factors in tumorigenesis. Cell Death Differ 15:678–685
62. Chinenov Y, Kerppola TK (2001) Close encounters of many kinds: Fos-Jun interactions that mediate transcription regulatory specificity. Oncogene 20: 2438–2452
63. Kouzarides T, Ziff E (1988) The role of leucine zipper in the fos-jun interaction. Nature 336:646–651
64. Klatt P, Molina EP, DeLacoba MG et al (1999) Redox regulation of c-Jun binding by reversible glutathiolation. FASEB J 13:1481–1490
65. Chang L, Karin M (2001) Mammalian MAP kinase signalling cascades. Nature 410:37–40
66. Martindale JL, Holbrook NJ (2002) Cellular response to oxidative stress: signaling for suicide and survival. J Cell Physiol 192:1–15
67. Xia Z, Dickens M, Raingeaud J (1995) Opposing effects of ERK and JNK-p38 MAP kinases on apoptosis. Science 270:1326–1331
68. Karin M (1995) The regulation of AP-1 activity by mitogen-activated protein kinases. J Biol Chem 270:16483–16486
69. Tobiume K, Matsuzawa A, Takahashi T et al (2001) ASK1 is required for sustained activations of JNK/p38 MAP kinases and apoptosis. EMBO Rep 2:222–228
70. Shaulian E, Karin M (2001) AP-1 in cell proliferation and survival. Oncogene 20:23390–23400
71. Brown JR, Nigh E, Lee RJ et al (1998) Fos family members induce cell cycle entry by activating Cyclin D1. Mol Cell Biol 18:55609–55619
72. Bakiri L, Lallemand D, Bossy-Wetzel E et al (2000) Cell cycle-dependent variations in c-jun and Jun B phosphorylation: a role in the control of cyclin D expression. EMBO J 19:2056–2068
73. Passague E, Wagner EF (2000) JunB suppresses cell proliferation by transcriptional activation of p16 (INK4a) expression. EMBO J 19:2969–2979
74. Amstad PA, Krupitza G, Cerutti PA (1992) Mechanism of c-fos induction by active oxygen. Cancer Res 52: 3952–3960
75. Schutte J, Minna JD, Birer MI (1989) Deregulated expression of human c-jun transforms primary rat embryo cells in cooperation with an activated c-Ha-ras gene and transforms rat-la cells as a single gene. Proc Natl Acad Sci U S A 86:2257–2261
76. Chen F, Castranova V, Shi X (2001) New insights into the role of nuclear factor- β in cell growth regulation. Am J Pathol 159:387–397
77. Beauerle PA, Lenardo M, Pierce JW et al (1988) Phorbol-ester-induced activation of the NF--B transcription factor involved dissociation of an apparently cytoplasmic Nf-*k*B/Inhibitor complex. Cold Spring Harb Symp Quant Biol 53:789–798
78. Molina EP, Klatt P, Vasquez J et al (2001) Glutathionylation of the p50 subunit of NF-*k*B: a mechanism for redox-induced inhibition of DNA binding. Biochemistry 40:14134–14142
79. Wu HH, Thomas JA, Momand J et al (2000) p53 protein oxidation in cultured cells in response to pyrrolidine dithiocarbamate: a novel method for relating the amount of p53 oxidation *in vivo* to the regulation of p53-repressive genes. Biochem J 35:87–93
80. Baldwin AS (1996) The NF-*k*B and I-*k*B proteins: new discoveries and insights. Annu Rev Immunol 14:649–683
81. Li N, Karin M (1998) Ionizing radiation and short wavelength UV activate NF-*k*B through two distinct mechanisms. Proc Natl Acad Sci U S A 95: 13012–13017
82. Rath PC, Aggarwal BB (2001) Antiproliferative effects of IFN-α correlate with the down regulation

of nuclear factor-*k*B in human Burkitt lymphoma Daudi cells. J Interferon Cytokine Res 21:523–528
83. Schulze-Oshoff K, Ferrari D, Los M et al (1998) Apoptosis signaling by death receptors. Eur J Biochem 254:439–459
84. Giri DK, Aggarwal BB (1998) Constitutive activation of NF-*k*B causes resistance to apoptosis in human cutaneous T cell lymphoma HuT-78 cells Autocrine role of tumor necrosis factor and reactive oxygen intermediates. J Biol Chem 273:14008–14014
85. Sullivan DM, Wehr NB et al (2000) Identification of oxidant-sensitive proteins: TNF-α induces protein glutathiolation. Biochemistry 39:11121–11128
86. Nebreda AR, Porrai A (2000) P38 MAP kinases: beyond the stress response. Trends Biochem Sci 25:257–260
87. Nomura M, Ma W, Chen N (2000) Inhibition of 12-O-tetradecanoylphorbol-13-acetate-induced NF·*k*B activation by tea polyphenols, (¡) epigallocatechin gallate, and theaflavins. Carcinogenesis 21:1885–1890
88. Schreck R, Rieber P, Baeuerle PA (1991) Reactive oxygen intermediates as apparently widely used messengers in the activation of the NF-*k*B transcription factor and HIV-1. EMBO J 10:2247–2258
89. Schmidt KN, Armstad P, Cerutti P et al (1995) The roles of hydrogen peroxide and superoxide as messengers in the activation of transcription factor NF-*k*B. Biol Chem 2:13–22
90. Holliday R (1990) Mechanisms for the control of gene activity during development. Biol Rev 65:431–471
91. Samiec PS, Goodman JI (1999) Evaluation of methylated DNA binding protein-1 in mouse liver. Toxicol Sci 49:255–262
92. Hergersberg M (1991) Biological aspects of cytosine methylation in eukaryotic cells. Experientia 47:1171–1185
93. Counts JL, Goodman JI (1995) Alterations in DNA methylation may play a variety of roles in carcinogenesis. Cell 83:13–15
94. Baylin SB (1997) Tying it all together: epigenetics, genetics, cell cycle, and cancer. Science 277: 1948–1949
95. Jones PA, Laird PW (1999) Cancer epigenetics comes of age. Nat Genet 21:163–167
96. Greger V, Debus N, Lohmann D et al (1994) Frequency and paternal origin of hypermethylated RB1 alleles in retinoblastoma. Hum Genet 94:491–496
97. Esteller M, Cordon-Cardo C, Corn PG et al (2001) p14 silencing by promoter hypermethylation mediates abnormal intracellular localization of MDM2. Cancer Res 61:2816–2821
98. Belinsky SA, Nikula KJ, Palmisano WA et al (1998) Aberrant methylation of p16 ^{INK4a} is an early event in lung cancer and a potential biomarker for early diagnosis. Proc Natl Acad Sci U S A 95:11891–11896
99. Govindarajan B, Klafter R, Miller MS et al (2002) Reactive oxygen-induced carcinogenesis causes hypermethylation of p16 ^{INK4a} and activation of MAP kinase. Mol Med 8:1–8
100. Goelz SE, Vogelstein B, Hamilton SR et al (1985) Hypomethylation of DNA from benign and malignant human colon neoplasms. Science 228:187–190
101. Vorce RL, Goodman JI (1989) Altered methylation of ras oncogenes in benzidine-induced B6C3F1 mouse liver tumors. Toxicol Appl Pharmacol 100:398–410
102. Gama-Sosa MA, Slagel VA, Trewyn RW et al (1983) ARF The 5-methylcytosine content of DNA from human tumors. Nucleic Acids Res 11:6883–6894
103. Abanobi SE, Lombardi B, Shinozuka H (1982) Stimulation of DNA synthesis and cell proliferation in the liver of rats fed a choline-devoid diet and their suppression by Phenobarbital. Cancer Res 42:412–415
104. Wainfan E, Poirier LA (1992) Methyl groups in carcinogenesis: effects on DNA methylation and gene expression. Cancer Res 52:S2071–S77173
105. Pascale RM, Marras V, Simile MM et al (1992) Chemoprevention of rat liver carcinogenesis by S-adenosyl L-methionine: a long-term study. Cancer Res 52:4979–4986
106. Simile MM, Saviozzi M, De Miglio MR et al (1996) Persistent chemopreventive effect of Sadenosyl-L-methionine on the development of liver putative preneoplastic lesions induced by thiobenzamide in diethylnitrosamine-initiated rats. Carcinogenesis 17:1533–1537
107. Weitzman SA, Turk PW, Milkowski DH et al (1994) Free radical adducts induce alterations in DNA cytosine methylation. Proc Natl Acad Sci U S A 91:1261–1264
108. Hanahan D, Weinberg RA (2011) Hallmarks of cancer: the next generation. Cell 144:646–674
109. Hanahan D, Weinberg RA (2000) The hallmarks of cancer. Cell 100:57–70
110. Heikkila R, Schwab G, Wickstrom E et al (1987) A c-myc antisense oligodeoxynucleotide inhibits entry into S phase but not progress from G0 to G1. Nature 328:445–449
111. Liu SL, Lin X, Shi DY et al (2002) Reactive oxygen species stimulated human hepatoma cell proliferation via cross-talk between PI3-K/PKB and JNK signaling pathways. Arch Biochem Biophys 406:173–182
112. Eblin KE, Jensen TJ, Wnek SM et al (2009) Reactive oxygen species regulate properties of transformation in UROtsa cells exposed to monomethyl arsonous acid by modulating MAPK signaling. Toxicology 255:107–114
113. Ruiz-Ramos R, Lopez-Carrillo L, Rios-Perez AD et al (2009) Sodium arsenite induces ROS generation, DNA oxidative damage, HO-1 and c-Myc proteins, NF-kappa B activation and cell proliferation in human breast cancer MCF-7 cells. Mutat Res 674:109–115
114. Chung YM, Kim JS, Yoo YD (2006) A novel protein, Romo1, induces ROS production in the mitochondria. Biochem Biophys Res Commun 347:649–655
115. Hwang IT, Chung YM, Kim JJ et al (2007) Drug resistance to 5-FU linked to reactive oxygen species modulator 1. Biochem Biophys Res Commun 359: 304–310

116. Na AR, Chung YM, Lee SB et al (2008) A critical role for Romo1-derived ROS in cell proliferation. Biochem Biophys Res Commun 369:672–678
117. Dong-Yun S, Yu-Ru D, Shan-Lin L et al (2003) Redox stress regulates cell proliferation and apoptosis of human hepatoma through Akt protein phosphorylation. FEBS Lett 542:60–64
118. De Luca A, Sanna F, Sallese M et al (2010) Methionine sulfoxide reductase A down-regulation in human breast cancer cells results in a more aggressive phenotype. Proc Natl Acad Sci U S A 107:18628–18633
119. Hu Y, Rosen DG, Zhou Y et al (2005) Mitochondrial manganese-superoxide dismutase expression in ovarian cancer: role in cell proliferation and response to oxidative stress. J Biol Chem 280:39485–39492
120. Policastro L, Molinari B, Larcher F et al (2004) Imbalance of antioxidant enzymes in tumor cells and inhibition of proliferation and malignant features by scavenging hydrogen peroxide. Mol Carcinog 39:103–113
121. Das L, Vinayak M (2012) Anticarcinogenic action of curcumin by activation of antioxidant defence system and inhibition of NF-kappa B signaling in lymphoma bearing mice. Biosci Rep 32:161–170
122. Moore DE (2002) Drug-induced cutaneous photosensitivity: incidence, mechanism, prevention and management. Drug Saf 25:345–372
123. Wallace KB (2003) Doxorubicin-induced cardiac mitochondrionopathy. Pharmacol Toxicol 93: 105–115
124. Benchekroun MN, Catroux P, Montaudon D et al (1990) Development of mechanisms of protection against oxidative stress in doxorubicin-resistant rat tumoral cells in culture. Free Radic Res Commun 11:137–144
125. Olson RD, Boerth RC, Gerber JG et al (1981) Mechanism of adriamycin cardiotoxicity: evidence for oxidative stress. Life Sci 29:1393–1401
126. Šimůnek T, Štěrba M, Popelová O et al (2009) Anthracycline-induced cardiotoxicity: overview of studies examining the roles of oxidative stress and free cellular iron. Pharmacol Rep 61:154–171
127. Menna P, Recalcati S, Cairo G et al (2007) An introduction to the metabolic determinants of anthracycline cardiotoxicity. Cardiovasc Toxicol 7:80–85
128. Davies AJK, Doroshow HJ (1986) Redox cycling of anthracyclines by cardiac mitochondria I Anthracycline radical formation by NADH dehydrogenase. J Biol Chem 261:3060–3067
129. Doroshow HJ, Davies AJK (1986) Redox cycling of anthracyclines by cardiac mitochondria II Formation of superoxide anion, hydrogen peroxide, and hydroxyl radical. J Biol Chem 261:3068–3074
130. Yoshida M, Shiojima I, Ikeda H et al (2009) Chronic doxorubicin cardiotoxicity is mediated by oxidative DNA damage-ATM-p53-apoptosis pathway and attenuated by pitavastatin through the inhibition of Rac1 activity. J Mol Cell Cardiol 47:698–705
131. Pereira CG, Silva MA, Diogo VC et al (2011) Drug-induced cardiac mitochondrial toxicity and protection: from doxorubicin to carvedilol. Curr Pharm Des 17:2113–2129
132. Nithipongvanitch R, Ittarat W, Cole PM et al (2007) Mitochondrial and nuclear p53 localization in cardiomyocytes: redox modulation by doxorubicin (Adriamycin)? Antioxid Redox Signal 9:1001–1008
133. Wallace KB (2007) Adriamycin-induced interference with cardiac mitochondrial calcium homeostasis. Cardiovasc Toxicol 7:101–107
134. Zhang WY, Shi J, Li JY et al (2009) Cardiomyocyte death in doxorubicin-induced cardiotoxicity. Arch Immunol Ther Exp 57:435–445

Oxidative Stress in Low Birth Weight Newborns

Reena Negi, Deepti Pande, Kanchan Karki, Ashok Kumar, Ranjana S. Khanna, and Hari D. Khanna

Abstract

Free radical damage has been recognized to be a common pathogenic mechanism of many neonatal diseases. An excessive and/or sustained increase in free radical production associated with diminished efficacy of the antioxidant defense systems results in oxidative stress, which occurs in many pathologic processes and contributes significantly to disease and can be the key link between size at birth and increased morbidity later in life. Low birth weight is closely associated with fetal and neonatal mortality and morbidity, inhibited growth and cognitive development, and chronic diseases later in life. Low birth weight is considered the primary factor associated with a poor perinatal outcome of maternal preeclampsia/eclampsia and premature birth. High oxidative stress and low level of enzymatic antioxidants and antioxidant nutrients such as vitamins A, E, and C, zinc, copper, and selenium and antioxidant status might be increasing the risk of pathogenesis of major complications in low birth weight newborns.

Keywords

Free radical • Low birth weight newborn • Oxidative stress • Antioxidants

1 Introduction

Reactive oxygen species (ROS) and nitrogen species are considered to play a major role in the pathogenesis of a wide range of neonatal diseases. Delivery constitutes a significant oxidative stress, and the gestation of the newborn and circumstances of delivery will affect the overall burden. The medical significance of oxidative stress has become increasingly recognized to the

R. Negi • D. Pande • K. Karki • H.D. Khanna (✉)
Department of Biophysics, Banaras Hindu University, Varanasi 221005, Uttar Pradesh, India
e-mail: hdkhanna@yahoo.co.in

A. Kumar
Department of Pediatrics, Institute of Medical Sciences, Banaras Hindu University, Varanasi 221005, India

R.S. Khanna
Department of Chemistry, Faculty of Science, Banaras Hindu University, Varanasi 221005, India

DOI 10.1007/978-81-322-2035-0_15, © Springer India 2015

point that it is now considered to be a component of virtually every disease process. There is equilibrium between a free radical (FR)/reactive oxygen species (ROS) formation and endogenous antioxidant defense mechanisms, but if this balance is disturbed, it can produce oxidative stress. This state of oxidative stress can result in injury to all the important cellular components like proteins, DNA, and membrane lipids which can cause cell death. Free radicals have a very short half-life and are therefore difficult to measure. Direct means of measuring free radicals include electron spin resonance and spin trapping methods [1]. Most commonly ROS have been tracked by measuring stable metabolites (e.g., nitrate/nitrite) and/or concentrations of their oxidation target products, including lipid peroxidation/DNA oxidation end products and oxidized proteins [2–4]. Emphasis is now being placed on biomarkers of oxidative stress, which are objectively measured and evaluated as indicators of normal biological processes, pathogenic processes, or pharmacologic responses to therapeutic intervention.

An antioxidant constitutes "any substance that delays, prevents or removes oxidative damage to a target molecule." Numerous antioxidants exist to counteract the oxidative stress and maintain cell and tissue homeostasis. Enzymatic antioxidants – superoxide dismutase (SOD), catalase, and glutathione peroxidase (GPx) – are important parts of the defense system. Antioxidant vitamins (A, E, C), with the ability to stabilize highly reactive free radicals, act as the first line of defense against free radical attack. Cells differ profoundly in their resistance to oxidative stress, which may be due to differences in their antioxidant capacity or in the balance between oxidants and antioxidants in these cells.

Many antioxidant defense systems depend on micronutrients [5]. Another way is to identify trace element deficiencies via compromised activities of dependent enzymes. SOD contains as cofactors Zn and Cu and GPx depends on selenium (Se). Selenium regulation of GPx activity is mediated by selenium (Se) regulation of GPx mRNA [6].

2 Clinical Implications of Oxidative Stress in Low Birth Weight Newborn

The low birth weight (LBW) newborns are predisposed to a number of neonatal problems. LBW has been associated with a higher risk of chronic conditions such as hypertension, diabetes mellitus, hypercholesterolemia, and other cardiovascular diseases later in life [7]. In normal-term pregnancies, the newborn weighs ≥2,500 g. Ninety-five percent of fetal weight gain occurs during the last 20 weeks of gestation, mainly during the third trimester [8]. Thus, it follows that low birth weight is a potential outcome of maternal risk factors such as oxidative stress during the fetal period.

An association between oxidative stress and low birth weight has been indicated [9]. Risk of oxidative stress in the fetus depends upon the mother antioxidant state [10] which highly protects the fetus by increasing the intrauterine growth rate and birth weight. The newborn infant is extremely vulnerable to oxidative stress and to the toxic effect of free radicals [11, 12]. At birth, the newborn is exposed to a relatively high hyperoxic environment caused by increased oxygen bioavailability [13, 14], which results in greatly enhanced free radicals generation, especially when supplemental oxygen is used for resuscitation and mechanical ventilation [15, 16].

2.1 LBW/Prematurity

Low birth weight in a newborn infant results due to intrauterine growth restriction (IUGR) or prematurity. LBW is the most significant factor contributing to neonatal mortality and morbidity. The basic underlying feature of the preterm LBW infant is immaturity of their organ systems. A delicate oxidant–antioxidant balance exists in the fetus and newborn. This balance can tip toward oxidant injury in the setting of preterm birth. This is due to the immature antioxidant enzyme system that is still underdeveloped in the early third trimester and the premature infant's limited capacity to clear oxygen free radicals during the transition to extrauterine life. Therefore oxidative stress-related diseases mostly affect neonates with preterm low birth weight.

2.1.1 Oxidative Stress

The maternal or fetal oxidative stress plays an important role in the pathophysiology of low birth weight [17]. Low birth weight newborns and especially preterm infants are probably more prone to oxidative stress than are children and young adults. There are some special reasons for this. These infants very often (1) are exposed to high oxygen concentrations, (2) have infections or inflammation, (3) have reduced antioxidant defense, and (4) have free iron which enhances the Fenton reaction leading to production of highly toxic hydroxyl radicals [18, 19]. In the perinatal period, there are many mechanisms leading to free radicals overproduction, including ischemia–reperfusion, arachidonic acid cascade, free iron, nitric oxide cascade, phagocyte activation, hypoxia, and hyperoxia [20, 21].

Oxidative stress is an imbalance between free radicals (FRs) production and antioxidant systems which plays a key role in pathogenesis of so-called free radical diseases such as retinopathy of prematurity (ROP), bronchopulmonary dysplasia (BPD), necrotizing enterocolitis (NEC), and intraventricular hemorrhage (IVH).

Post hypoxic reoxygenating injury caused by ROS may be a key factor [22]. Ischemia–reperfusion injury is now recognized as a probable contributing factor to much of the morbidity of premature infants. Very low birth weight (VLBW) infants have an increased susceptibility to brain damage as 5–15 % develop cerebral palsy and an additional 25–50 % have less severe neurological deficits [23] due to higher partial pressure of oxygen (pO_2) dissolved in blood. Furthermore, due to the developmental immaturity of their free radical defenses, the brain appears to be susceptible to oxidative stress [24].

Lung injury and eye damage due to hyperoxia and free radical-mediated damage are well established. Respiratory distress due to ROS occurs because of a surfactant deficiency at the gas–liquid interface. About 14 % of all infants <2,500 g at birth and 60 % of infants born at 29 weeks or less develop respiratory distress syndrome. Chronic lung disease (CLD) or bronchopulmonary dysplasia (BPD) is a form of chronic lung disease that usually occurs in preterm infants receiving respiratory support with mechanical ventilation or prolonged oxygen supplementation [25].

Necrotizing enterocolitis (NEC) is an inflammation of the small intestine and bowel surface, with infiltration of epithelial cells by bacteria. While its etiology is unclear, it may be precipitated by ischemia–reperfusion injury [13].

Intraventricular–periventricular hemorrhage (IVH-PVH) is a brain injury, usually occurring within 24–48 h after birth (95 % occur by 3 days) or may occur later. It is believed to be an ischemic injury followed by reperfusion injury. IVH-PVH in its severest form carries a high risk of poor neurodevelopmental outcome (>50 %). Retinopathy of prematurity (ROP) is oxygen-induced damage to blood vessels in the retina that are undergoing neovascularization. It is characterized by abnormal vascularization of the retina, causing a range of vision impairment, and remains a major cause of morbidity for premature neonates [26]. It has been proposed that the above diseases are expressions of an inability to cope with an overexposure to ROS [13].

2.1.2 Oxidative Damage Markers in Preterm LBW Newborns

Estimation of lipid peroxidation in cord blood has been proposed as a reliable marker of reactive oxygen species (ROS) activity in the fetus and a measure of perinatal outcome [27]. Malondialdehyde (MDA) appears to be a useful measure to continue to explore the role of free radical-mediated disease in the LBW infant. Urinary malondialdehyde concentration was also found highest in the extremely low birth weight newborns in previous study [28]. The oxygen-derived free radical damage resulting in lipid peroxidation is widely believed to play a role in the etiology of many of the "diseases of prematurity."

Oxidative injury is associated with the development of the long-term complications of the preterm infant. Previous study proves that there is a strong evidence of oxidative stress in the small for gestational age (SGA) babies as evidenced by increased lipid peroxidation and reduced free oxygen radical scavenger system. The protective mechanisms against oxidative damage are more efficient in appropriate for gestational age (AGA) babies of healthy mothers [29]. Other investigators have noted the presence of lipid peroxidation among SGA neonates [30, 31]. Both mothers and babies are exposed to oxidative stress during and after delivery, which is more pronounced and persistent in the perinatal period of the SGA group, while lipid peroxidation in placenta could play a role in SGA pathophysiology [32].

Oxidative protein damage measured as plasma carbonyl levels was increased in preterm low birth weight newborns [33] and could be interpreted as indicating increased oxidative stress associated with prematurity. Higher levels of circulating protein carbonyl in small for gestational age (SGA) at birth than average weight babies was reported [34]. Oxidative stress was induced both in small for gestational age (SGA) newborns and their mothers which is manifested as increased protein oxidant damage [31].

Urinary 8-OHdG, taken as a marker of oxidative DNA damage, was also reported to be increased in infants with very low birth weight [17]. It indicates that low birth weight (LBW) infants can be subjected to oxidative stress. The preterm newborns had higher level of 8-OHdG compared to full-term newborns [35]. There are limited studies of oxidative DNA damage in low birth weight newborns. Previous studies have described the relation between prematurity and urinary 8-OHdG concentrations in low birth weight infants [36]. Significant increase in urinary 8-OHdG excretion in women giving preterm birth (<37 completed weeks) or giving birth to a child with a body weight below 2,500 g or growth restriction (<10th percentile) was also reported [37]. The levels of urinary 8-OHdG in pregnant women with fetal growth retardation have been reported higher in women with normal pregnancies [38]. The concentration of 8-OHdG in cord blood can be considered as momentary steady-state levels of oxidative DNA damage, which is sustained by many factors like nature and concentration of ROS, the overall efficiency of the antioxidant defense systems, and the efficiency of the DNA repair system directed against 8-OHdG [39].

Even though it does not indicate a cause–effect relationship, oxidative stress in intrauterine life strongly represents a risk factor for the development of neonatal free radical diseases in preterm low birth weight newborns [40, 41]. Preterm infants are more exposed to oxidative damage than term infants; this is due to different reasons: organ's structural and functional immaturity, overloading of aerobic tissue metabolism with rapidly growing energy demand, conditions leading to excessive free radicals production, and lack of antioxidant system.

2.1.3 Antioxidant Defense System in Preterm LBW Newborns

Antioxidant capacity of blood has shown a more variable result in relation to the development of oxygen radical diseases of the preterm infant. The body on account of susceptibility to oxidative insult is naturally provided with an efficient antioxidant system. The intracellular oxidative defense seems to be lowered in preterm babies. Thus, the more premature the infant, the lower the defense [42]. It was speculated that preterm babies have prompter involvement of antioxidant defenses than term babies [43]. On the other hand, a highly reductive intracellular environment implies a decline in the output of free radicals to extracellular space. This effect contributes to prevent the endothelial dysfunction associated with oxidative stress [44] and its impaired effects on maternal–fetal blood flow in pregnancies complicated by IUGR and low birth weight [45]. Appropriate for gestational age (AGA) newborns possess antioxidant defense capable of resisting the physiological oxidative stress at birth [10].

The body has developed a complex defense strategy to minimize the damaging effects of various oxidants. Central to this defense are the antioxidant enzymes of the blood, which include SOD, GPx, and catalase. A reduction in SOD, the primary enzyme that inactivates the superoxide radical, and in the glutathione peroxidase and catalase activity, which is involved in the detoxification of H_2O_2, would lead to increased numbers of free radicals, and this could thereafter be responsible for the increased levels of MDA, protein carbonyl, 8-OHdG, and reactive nitrogen species in preterm low birth weight newborns.

Nonenzymatic antioxidant components consist of various molecules such as vitamin A, E, and C that react with activated oxygen species and thereby prevent the propagation of free radical chain reactions. Low levels of these vitamins (A, E, C) could reflect low intake, which results in decreased antioxidant defense system in preterm low birth weight newborns.

The presence of an association between oxidative stress markers and enzymatic/nonenzymatic antioxidants in preterm low birth weight newborns suggests that increased oxidative stress may be the result of changes in the level of antioxidants due to the cause or the effect of oxidative damage occurring at the molecular level [40, 41].

2.2 Low Birth Weight Neonates Born to Preeclamptic/Eclamptic Mothers

Preeclampsia/eclampsia is a disorder that occurs during only pregnancy and affects both the mother and the fetus. Preeclampsia (PE) is a great challenge to obstetricians because its cause is unknown, its pathophysiology is complex and incompletely understood, its diagnosis may be difficult to determine, there are no effective treatments, and antenatal care involves a difficult balance between the risks for women to continue pregnancy and those for the baby's early birth.

Affecting at least 5 % of all pregnancies, it is a rapidly progressive condition characterized by high blood pressure, swelling of lower extremities, and protein in the urine. According to the World Health Organization, preeclampsia is a major cause of both maternal and fetal/neonatal morbidity and mortality [46].

Worldwide, approximately 3 % of all pregnant women develop preeclampsia, of whom 1.9 % will develop eclampsia, although its greatest impact is in the developing world, where >90 % of the most serious preeclampsia-related maternal and fetal morbidity and mortality occurs [47]. Preeclampsia is responsible for ≤15 % of preterm births and consequently increases infant mortality and morbidity [48].

Preeclampsia is a two-stage disorder that begins with poor placentation and reduced uteroplacental blood supply, resulting in placental hypoxia. This first stage, with silent placental events, is followed by the release of several mediators: growth factors and their soluble receptors, inflammatory cytokines, placental debris, and products of placental oxidative stress. Such mediators cause endothelial cell dysfunction and the systemic inflammatory syndrome, leading to the clinical manifestation of PE (second stage) [49]. Thus, there is increasing evidence that oxidative stress plays an important role in the pathogenesis of preeclampsia, perhaps acting as the link in the two-stage model of preeclampsia [50].

Epidemiologic studies have reported alarmingly high rates of preterm births, predominantly due to increasing indications for preterm delivery, and PE is one of the most common of these indications [51, 52]. Preterm newborns of women who have PE are of great concern because strong evidence shows that they are exposed to increased oxidative stress, which has been implicated in the pathogenesis of serious diseases in neonates [53]. Thus, an unfavorable outcome could be expected in infants of women who have PE [54]. Several studies indicate that PE is associated with a higher incidence of newborns with low birth weight [55, 56].

2.2.1 Oxidative Stress

The oxidative stress theory suggests that these stresses in the fetus lead to increased lipid peroxidation but, more importantly, to modifications in gene expression that lead to adverse perinatal programming. The susceptibility of an individual fetus will depend on the time during gestation when metabolic programming occurs and on the genetic susceptibility of the fetus. Thus, the effect of preeclampsia could extend beyond the mother to involve the fetus [57].

2.2.2 Oxidative Damage Markers

Although a limited number of studies have been carried out on fetal oxidative status and umbilical cord blood, the results have consistently shown an upregulation of oxidative stress markers and altered antioxidant status in fetal circulation [58]. The fetus is affected by oxidative stress in PE, and this raises a concern about fetal and neonatal outcomes [59, 60]. The level of lipid peroxidation product (MDA) in blood is elevated in preeclamptic pregnancy, and it has been suggested that it plays a role in the etiology of the disease [61, 62]. This provides further evidence that appropriate or excessive lipid peroxidation may play an important role in the pathophysiology of preeclamptic/eclamptic pregnancies. Significant elevation of MDA levels in cord blood of pair-matched preeclamptic mother has been reported [63, 64]. This may result in a greater potential for endothelial damage ultimately leading to enhanced diastolic pressure [65] which further aggravates the condition of preeclamptic/eclamptic patients.

The direct damage of proteins during oxidative stress can give rise to the formation of protein carbonyls, which may serve as biomarkers for general oxidative stress, in addition to data provided by lipid peroxidation. This is an indication of oxidative stress in preeclampsia/eclampsia. Significant elevation of protein carbonyl level in maternal circulation [66, 67] and cord blood [60] of preeclamptic mothers has been reported. The previous study showed higher plasma protein carbonyl levels that are markers of oxidative protein damage in preeclamptic pregnant women than those in healthy pregnant women [68].

Oxygen free radical-induced DNA damage may be particularly deleterious because it can produce mutations [69]. There are only few reports on the oxidative DNA damage in preeclamptic/eclamptic pregnancies. According to one such studies, the concentration of 8-OHdG was significantly higher in the placental DNA from preeclampsia-complicated pregnancies [70]. In preeclamptic women, serum levels of 8-OHdG, a well-established marker of oxidative DNA damage, are increased [71, 72]. The generation of 8-OHdG (8-hydroxy-2-deoxyguanosine) is one of the results of DNA damage induced by oxygen free radicals [73].

2.2.3 Antioxidant Status

Significantly decreased enzymatic activities of superoxide dismutase (SOD), glutathione peroxidase (GPx), and catalase (CAT) have been reported [74, 75] in cord blood of the full-term newborns of preeclamptic mothers as compared to their pair-matched maternal blood. In normal pregnancy, by 10–12 weeks of gestation, the onset of maternal blood flow in the placenta results in a local increase in oxygen tension and parallel elevation in the expression and activity of the antioxidant enzymes [76].

The increased concentration of superoxide in the placental tissue of preeclamptic women [77] was found associated with decreased SOD activity and mRNA expression, for CuZn-SOD in trophoblast cells isolated from preeclamptic placentas [78]. Also, it was recently found that these women show a decrease in plasma levels of SOD [65]. Therefore, a decreased degradation of superoxide anion should be expected.

Many studies have reported significant reduction in maternal plasma [79, 80] and placental GPx activities in preeclamptic women [81, 82]. Various and sometimes conflicting results were reported for total GPx activities in the pathology of preeclampsia. For example, total GPx activity in whole maternal blood was found higher in the previous study [83], and the total erythrocyte GPx activity was shown to be higher [84, 85] in preeclampsia.

Deficiency in antioxidant vitamins would be associated with the development of preeclampsia [86]. Vitamin E, vitamin C, and beta-carotene (provitamin A) are known to be powerful antioxidants [87, 88], while some reports documented an increase in vitamin levels in preeclampsia and eclampsia [89, 90].

2.3 Newborn with Respiratory Distress Syndrome (RDS)

Infant respiratory distress syndrome (IRDS), also called neonatal respiratory distress syndrome or respiratory distress syndrome of newborn, previously called hyaline membrane disease (HMD), is a syndrome in premature infants caused by developmental insufficiency of surfactant production and structural immaturity in the lungs. IRDS affects about 1 % of newborn infants and is the leading cause of death in preterm infants [91]. The incidence decreases with advancing gestational age, from about 50 % in babies born at 26–28 weeks to about 25 % at 30–31 weeks.

2.3.1 Oxidative Stress

Neonatal respiratory distress syndrome is accompanied by inflammatory processes with free radical generation and oxidative stress [92, 93]. The imbalance between the oxidative forces and the antioxidant defense systems was suggested to predispose the lungs to the development of RDS [94, 95].

When phagocytes such as neutrophils are stimulated by microorganisms or other means, they become activated and increase their oxidative metabolism; as a result, toxic oxygen and nitrogen derivatives, i.e., ROS/RNS, are formed. If these toxic products are not inactivated, their high chemical reactivity leads to damage to a variety of cellular macromolecules including proteins, carbohydrates, lipids, and nucleic acid. This results in cell injury and may induce respiratory cell death [96]. Under these conditions, a surfactant deficiency may be aggravated by inactivation of the small amount of endogenous surfactant that is produced [97]. Furthermore, if exogenous surfactant is given, this may also be destroyed [98, 99].

Many studies have shown increased oxidative stress markers and/or reduced antioxidant defense in preterm infants with RDS [100, 101]. In vitro studies have shown surfactant protein SP-A and SP-D to have potent, direct antioxidant properties. The surfactant proteins protect unsaturated phospholipids and growing cells from oxidative injury at physiological concentrations [102]. Dani et al. [103] demonstrated superoxide dismutase (SOD) and catalase activities in four natural surfactants, as well as scavenger activity against hydrogen peroxide.

Hyperoxic exposure itself, although essential for survival of RDS infants, probably induces excessive production of ROS/RNS in the respiratory system. There exist, however, several potential causes of intra- and extracellular oxidant stress in the preterm newborns with RDS. The high inspiratory concentrations of oxygen required to achieve adequate arterial oxygenation, prooxidant drugs, and infections or extrapulmonary inflammation can all promote ROS accumulation and the utilization and depletion of antioxidative factors [104]. ROS/RNS also have been implicated in the molecular damage seen in the bronchoalveolar lavage (BAL) fluid of patients with RDS [105, 106].

A deficit in the precise balance between exposure to oxidants and endogenous antioxidant results obviously leads to elevated oxidative damage. The molecular damage caused by oxidative stress appears to be involved in the pathogenesis of a growing number of diseases, including RDS of the newborn [107].

2.3.2 Oxidative Damage Markers in Neonates with RDS

Oxidative damage is important in the pathogenesis of respiratory distress syndrome (RDS). It can be assumed that these newborns might be at risk of an oxidative stress. Singh found elevated lipid peroxidation product in respiratory disorders [108]. Some studies reported that prematurely born babies with RDS showed high concentrations of protein carbonyls [109, 110]. The amount of oxidatively modified protein may provide a quantitative assessment of oxygen toxicity and of pulmonary antioxidant defenses [111]. Lung proteins are attacked by oxygen reactive species. When RDS is present, pulmonary edema occurs because of increased permeability of cell membranes. The fluid in the edema is rich in proteins, which represent the ideal target for oxygen reactive species. In order to initiate the oxidative attack, oxygen reactive species inactivates alpha-1 protease thus causing an imbalance in the lung protease – antiprotease system [112]. Reactive oxygen species also interact with pulmonary surfactant as well as with other protein and lipid structures thus delaying the normal functioning of the lung. Therefore, surfactant administration before the initiation of mechanical ventilation diminishes the severity of lung lesions by providing consistent ventilation [113]. This protein oxidation process activated by oxygen reactive species has been proven to contribute to pathogenesis in newborns with RDS [114].

Oxidative DNA damage could be the crucial mechanism in the pathogenesis of respiratory disorder [115].

Oxidative damage is important in the pathogenesis of respiratory distress syndrome (RDS). It can be assumed that these newborns might be at risk of an oxidative stress. One previous study [116] also observed oxidative stress in neonates with respiratory distress syndrome. Reactive oxygen is generated by several inflammatory and structural cells of the airways. These oxidant species have important effects on a variety of lung cells as regulator of signal transduction, activators of key transcription factors, and modulators of gene expression and apoptosis. Thus, increased oxidative stress accompanied by reduced antioxidant defenses may play a role in the pathogenesis of a number of inflammatory pulmonary diseases including RDS in the newborn [117].

2.3.3 Antioxidant Defense System in Neonates with RDS

Antioxidant defenses of the immature lung will be prepared neither for the hyperoxic environment nor the inflammation found in association with respiratory distress.

The increase in SOD has clinical significance in the prevention of respiratory distress syndrome in preterm newborns [118]. Preterm birth is very likely to lead to severe respiratory distress syndrome (RDS). One study concluded that the majority of preterm infants with severe RDS do not have protective superoxide dismutase activity in tracheal fluid [119]. Depletion of glutathione, a key antioxidant, accelerates lung injury. Glutathione concentrations are reduced significantly in premature infants with respiratory distress syndrome, leaving them at greater risk of bronchopulmonary dysplasia [91, 120].

Catalase was expressed only during the later stages of lung development; catalase was the only antioxidant enzyme that increased at the level of mRNA and specific activity throughout the period of human lung development [121]. Catalase may play an important role against oxidant stress of lung and that a lack of catalase may also predispose the preterm lung of newborn with RDS to oxidant-related injury.

Many preterm infants are deficient in vitamin A at birth, and failure to correct this deficiency may contribute to the development of respiratory distress disorder. Inadequate vitamin A status could intuitively be expected to predispose to respiratory problems. Most preterm infants are born with low vitamin A stores and low plasma concentrations [122]. Vitamin E acts as a free radical scavenger, reducing peroxidation of membrane polyunsaturated fatty acids. In present study, low concentration of vitamin E was observed in the newborn with RDS compared to healthy newborns. The decrease in plasma vitamin E level was observed in infants with RDS [123]. The previous study confirmed that low plasma selenium and vitamin E levels in premature infants were significantly associated with an increased respiratory distress disorder [124]. Premature infants with RDS might need more supplemental vitamin E than premature infants without RDS [125]. Vitamin C is the most important aqueous phase chain-breaking antioxidant in plasma [126] and makes a major contribution to antioxidant protection in the lung, particularly the lower respiratory tract. This is likely to be mainly due to the direct radical scavenging properties of vitamin C but also to its capacity to recycle vitamin E [127]. Therefore, low levels of vitamin A, E, and C in premature babies with RDS may compromise antioxidant mechanisms and exacerbate oxidant damage in newborns.

3 Antioxidant Therapy in Newborns

Oxidative stress occurs at birth in all newborns as a consequence of the hyperoxic challenge after the transition from the hypoxic intrauterine environment to extrauterine life. During the perinatal period, oxidative stress can be magnified by others predisposing conditions such as premature delivery, preeclamptic/eclamptic pregnancy, and respiratory distress syndrome.

Nutrients can affect oxidative stress by increasing or decreasing free radicals or antioxidants or by providing substrate for the formation of free radicals. Poor maternal nutrition has been implicated as one of the key "adverse environmental influences in utero" which could lead to compromised placental and fetal growth and adverse long-term consequences [128].

The antioxidant defenses rely heavily on vitamins and minerals in the diet as well as essential amino acids required to synthesize glutathione and antioxidant proteins such as albumin. There have been few direct studies on the relationship between micronutrient deficiencies and oxidative stress. Administering intravenous lipid emulsions containing monounsaturated and saturated fatty acids also could minimize the risk of peroxidation [129]. The neonatal oxidant load could be further reduced by the use of low-dose multivitamins mixed with lipid emulsions, which would allow a gradual introduction of ascorbic acid and prevent excessive plasma concentrations of vitamin C [130].

Another strategy with which to protect babies from oxidative stress is to promote their antioxidant defenses, especially as regards the glutathione system. Because of physiological glutathione deficiency, loss of glutathione caused by a reduced capacity of cellular conservation at a low gestational age increases demand and consumption during hyperoxia and oxidative stimulation; the cell would need to synthesize more glutathione to achieve a positive balance. Moreover, hyperoxia is associated with lower cysteine concentrations in infants with acute respiratory failure [131]. The availability of cysteine appears to be the rate-limiting step in glutathione synthesis. It could therefore be necessary to provide the baby with adequate amino acid substrates for cellular glutathione synthesis through parenteral or enteral administration of adequate doses of amino acids immediately after birth [130].

There are diverse clinical and therapeutic possibilities of limiting oxidative damage, including the administration of corticoids (superoxide dismutase, catalase, etc.) to the mother in order to hasten the maturation of the antioxidant defense systems of the fetus, the provision of antioxidant enzymes to the neonate (these have been used to prevent bronchopulmonary dysplasia), and even nutritional intervention. Breast milk plays an important role in protecting the newborn from oxidative stress; it contains many antioxidant molecules that probably are vital for antioxidant defense at early stages of life [132]. It is conceivable that these antioxidants in breast milk help to eliminate free radicals in infants.

References

1. Baum SL, Anderson IGM, Baker RR et al (2003) Electron spin resonance and spin trap investigation of free radicals in cigarette smoke: development of a quantification procedure. Anal Chim Acta 481:1–13
2. Dalle-Donne I, Scaloni A, Giustarini D et al (2005) Proteins as biomarkers of oxidative stress in diseases: the contribution of redox proteomics. Mass Spectrom Rev 24:55–99
3. Winterbourn CC, Buss H (1999) Protein carbonyl measurement by enzyme-linked immunosorbent assay. Methods Enzymol 300:106–111
4. Levine RL, Wehr N, Williams JA et al (2000) Determination of carbonyl groups in oxidized proteins. Methods Mol Biol 99:15–24
5. Evans P, Halliwell B (2001) Micronutrients: oxidant/antioxidant status. Br J Nutr 85:S67–S74
6. Sunde RA, Weiss SL, Thompson KM et al (1992) Dietary selenium regulation of glutathione peroxidase mRNA-implications for selenium requirement. FASEB J 6:1365
7. Luo ZC, Fraser WD, Julien P et al (2006) Tracing the origins of "fetal origins" of adult diseases: programming by oxidative stress? Med Hypotheses 66:38–44
8. Largo RH, Wailli R, Duc G et al (1980) Evaluation of perinatal growth. Helv Pedaitr Acta 35:419–436
9. Pitkanen OM, Hallman M, Andersson SM (1990) Correlation of free oxygen radical-induced lipid peroxidation with outcome in very low birth weight infants. J Pediatr 116:760–764
10. Buonocore G, Perrone S (2006) Biomarkers of oxidative stress in the fetus and newborn. Hematology 2:103–107
11. Saugstad OD (1996) Mechanisms of tissue injury by oxygen radicals: implication for neonatal disease. Acta Paediatr 85:1–4
12. Halliwell B (1994) Free radicals, antioxidants, and human disease: curiosity, cause, or consequence? Lancet 344:721–724
13. Saugstad OD (1998) Oxygen radical disease in neonatology. Semin Neonatol 3:229–238
14. Buonocore G, Perrone S, Longini M et al (2000) Total hydroperoxide and advanced oxidation protein products in preterm hypoxic babies. Pediatr Res 47:221–224
15. Kondo M, Itoh S, Isobe K (2000) Chemiluminescence because of the production of reactive oxygen species

in the lungs of newborn piglets during resuscitation periods after asphyxiation load. Pediatr Res 45:524–527

16. Weinberger B, Laskin DL, Heck DE (2002) Oxygen toxicity in premature infants. Toxicol Appl Pharmacol 181:60–67
17. Matsubasa T, Uchino T, Karashima S et al (2002) Oxidative stress in very low birth weight infants as measured by urinary 8-OHdG. Free Radic Res 36:189–193
18. Saugstad OD (2005) Oxidative stress in the newborn. A 30-year perspective. Biol Neonate 88:228–236
19. Saugstad OD (2003) Oxygen toxicity at birth: the pieces are put together. Pediatr Res 54:789
20. Saugstad OD (2001) Update on oxygen radical disease in neonatology. Curr Opin Obstet Gynecol 13:147–153
21. Buonocore G, Perrone S, Longini M (2002) Oxidative stress in preterm neonates at birth and on seventh day of life. Pediatr Res 52:46–49
22. Sullivan JL, Newton RB (1988) Serum antioxidant activity in neonates. Arch Dis Child 63:748–750
23. Volpe JJ (1990) Brain injury in the premature infant: is it preventable? Pediatr Res 27(Suppl 6):S28–S33
24. Noseworthy MD, Bray TM (2000) Zinc deficiency exacerbates loss in blood-brain barrier integrity induced by hyperoxia measured by dynamic MRI. Proc Soc Exp Biol Med 223:175–182
25. Ambalavanan N, Carlo WA (2004) Bronchopulmonary dysplasia: new insights. Clin Perinatol 31:613–628
26. Shweiki D, Itin A, Soffer D et al (1992) Vascular endothelial growth factor induced by hypoxia may mediate hypoxia-induced angiogenesis. Nature 359:843–845
27. Rogers MS, Wang W, Mongelli M et al (1997) Lipid peroxidation in cord blood at birth: a marker of fetal hypoxia during labor. Gynecol Obstet Ivest 44:229–233
28. Schlenzig JS, Bervoets K, Loewenich V et al (1993) Urinary malondialdehyde concentration in preterm neonates: is there a relationship to disease entities of neonatal intensive care? Acta Paediatr 82:202–205
29. Mocatta TJ, Winterbourn CC, Inder TE et al (2004) The effect of gestational age and labour on markers of lipid and protein oxidation in cord plasma. Free Radic Res 38:185–191
30. Gupta P, Narang M, Banerjee BD et al (2004) Oxidative stress in term small for gestational age neonates born to undernourished mothers: a case control study. BMC Pediat 20020:4–14
31. Kamath U, Rao G, Kamath SU et al (2006) Maternal and fetal indicators of oxidative stress during intrauterine growth retardation (IUGR). Indian J Clin Biochem 21(1):111–115
32. Gveric-Ahmetasevic S, Sunjic SB et al (2009) Oxidative stress in small-for-gestational age (SGA) term newborns and their mothers. Free Radic Res 43:376–384
33. Saker M, Mokhtari NS, Merzouk SA et al (2008) Oxidant and antioxidant status in mothers and their newborns according to birth weight. Eur J Obstet Gynecol Reprod Biol 141:95–99
34. Sridhar MG, Setia S, John M et al (2007) Oxidative stress varies with the mode of delivery in intrauterine growth retardation: association with apgar score. Clin Biochem 40:688–691
35. Nassi N, Ponziani V, Becatti M et al (2009) Antioxidant enzymes and related elements in term and preterm newborns. Pediatr Int 51:183–187
36. Drury JA, Jeffers G, Cooke RW (1998) Urinary 8-hydroxydeoxyguanosine in infants and children. Free Radic Res 28:423–424
37. Scholl TO, Stein TP (2001) Oxidative damage to DNA and pregnancy outcome. J Matern Fetal Med 10:182–185
38. Ames BN, Shigenaga MK, Hagen TM (1993) Oxidants, antioxidants and the degenerative diseases of aging. Proc Natl Acad Sci U S A 90:7915–7922
39. Demple B, Harrison L (1994) Repair of oxidative damage to DNA: enzymology and biology. Ann Rev Biochem 63:915–948
40. Negi R, Pande D, Kumar A et al (2012) Evaluation of biomarkers of oxidative stress and antioxidant capacity in the cord blood of preterm low birth neonates. J Matern Fetal Neonatal Med 25:1338–1341
41. Negi R, Pande D, Kumar A et al (2012) *In vivo* oxidative DNA damage and lipid peroxidation as a biomarker of oxidative stress in preterm low birth weight infants. J Trop Pediatr 58:326–328
42. Inanc F, Kilinc M, Kiran G et al (2005) Relationship between oxidative stress in cord blood and route of delivery. Fetal Diagn Ther 20:450–453
43. Frosali S, Di Simplicio P, Perrone S et al (2004) Recycling and antioxidant enzyme activities in erythrocytes of term and preterm newborns at birth. Biol Neonate 85:188–194
44. Shane RT, Paul KW, Grant RD (2008) Redox control of endothelial function and dysfunction: molecular mechanisms and therapeutic opportunities. Antioxid Redox Signal 10:1713–1765
45. Gicquel C, Le Bouc Y (2006) Hormonal regulation of fetal growth. Horm Res 65:28–33
46. Roberts JM, Redman CWG (1993) Pre-eclampsia: more than pregnancy induced hypertension. Lancet 341:1447–1451
47. Villar J, Say L, Gulmezoglu AM et al (2003) Eclampsia and preeclampsia: a health problem for 2000 years. In: Critchley H, MacLean A, Poston L, Walker J (eds) Pre-eclampsia. RCOG Press, London, pp 189–208
48. Meis PJ, Goldenberg RL, Mercer BM et al (1998) The preterm prediction study: risk factors for indicated preterm births. Maternal-Fetal Medicine Units Network of the National Institute of Child Health and Human Development. Am J Obstet Gynecol 178:562–567
49. Ness RB, Roberts JM (1996) Heterogeneous causes constituting the single syndrome of preeclampsia: a

hypothesis and its implications. Am J Obstet Gynecol 175:1365–1370
50. Roberts JM, Hubel CA (1999) Is oxidative stress the link in the two-stage model of pre-eclampsia? Lancet 354:788–789
51. Sibai B, Dekker G, Kupferminc M (2005) Pre-eclampsia. Lancet 365:785–799
52. Duley L (2009) The global impact of pre-eclampsia and eclampsia. Semin Perinatol 33:130–137
53. Trindade CEP, Rugolo LMSS (2007) Free radicals and neonatal diseases. NeoReviews 8:522–532
54. Mitani M, Matsuda Y, Makino Y et al (2009) Clinical features of fetal growth restriction complicated later by preeclampsia. J Obstet Gynaecol Res 35:882–887
55. Groom K, North R, Poppe K et al (2007) The association between customised small for gestational age infants and preeclampsia or gestational hypertension varies with gestation at delivery. BJOG 114:478–484
56. Wu C, Nohr E, Bech B et al (2009) Health of children born to mothers who had preeclampsia: a population-based cohort study. Am J Obstet Gynecol 201:269.e1–269.e10
57. Steinborn A, Sohn C, Sayehli C (2001) Preeclampsia, a pregnancy-specific disease, is associated with fetal monocyte activation. Clin Immunol 100:305–313
58. Negi R, Pande D, Karki K et al (2012) Trace elements and antioxidant enzymes associated with oxidative stress in pre-eclamptic, eclamptic mother during fetal circulation. Clin Nutr 31:946–950
59. Chamy VM, Lepe J, Catalan A et al (2006) Oxidative stress is closely related to clinical severity of pre-eclampsia. Biol Res 39:229–236
60. Howlader ZH, Parveen S, Tamanna S et al (2009) Oxidative stress and antioxidant status in neonates born to preeclamptic mother. J Trop Pediatr 55:363–367
61. Maseki M, Nishigaki I, Hagihara M (1981) Lipid peroxide levels and lipid serum content of serum lipoprotein fractions of pregnant subjects with and without preeclampsia. Clin Chim Acta 155:155–161
62. McLaughlin MK (1989) Lipid peroxidation in pregnancy: new perspectives on preeclampsia. Am J Obstet Gynecol 161:1025–1034
63. Durak I (2007) Role of oxidative stress in intrauterine growth restriction. Gynecol Obstet Invest 64:187–192
64. El-Bana SM, El-Din AE, Isamil ZA (2001) Fetal and maternal oxidative stress in normal and abnormal pregnancies. Ain Shams Med J 52:421–431
65. Aydin S, Benian A, Madazli R et al (2004) Plasma malondialdehyde, superoxide dismutase, sEselectin, fibronectin, endothelin-1 and nitric oxide levels in women with preeclampsia. Eur J Obstet Gynecol Reprod Biol 113:21–25
66. Serder Z, Gur E, Colakoethullary M et al (2003) Lipid and protein oxidation and antioxidant function in women with mild and severe pre-eclampsia. Arch Gynecol Obstet 268:19–25
67. Zusterzeel PL, Mulder TP, Peters WH et al (2000) Plasma protein carbonyls in non-pregnant, healthy pregnant and pre-eclamptic women. Free Radic Res 33:471–476
68. Zusterzeel PL, Rütten H, Roelofs HM et al (2001) Protein carbonyls in decidua and placenta of pre-eclamptic women as markers for oxidative stress. Placenta 22:213–219
69. Ames BN (1989) Endogenous oxidative DNA damage, aging, and cancer. Free Radic Commun 7:121–128
70. Wiktor H, Kankofer M, Schmerold I et al (2004) Oxidative DNA damage in placentas from normal and pre-eclamptic pregnancies. Virchows Arch 445:74–78
71. Takagi Y, Nikaido T, Toki T et al (2004) Levels of oxidative stress and redox-related molecules in the placenta in preeclampsia and fetal growth restriction. Virchow Arch 444:49–55
72. Leslie M, Xiaolan C (2004) Oxidative stress in the placenta. Histochem Cell Biol 122:369–382
73. Doetsch PW, Cunningham RP (1990) The enzymology of apurinic/apyrimidinic endonucleases. Mutat Res 236:173–201
74. Orhan H, Onderoglu L, Yucel A (2003) Circulating biomarkers of oxidative stress in complicated pregnancies. Arch Gynecol Obstet 267:189–195
75. Uotila J, Tuimala R, Pyykko K et al (1993) Pregnancy induced hypertension is associated with changes in maternal and umbilical blood antioxidants. Gynecol Obstet Invest 36:153–157
76. Jauniaux E, Watson AL, Hempstock J et al (2000) Onset of maternal arterial blood flow and placental oxidative stress. A possible factor in human early pregnancy failure. Am J Pathol 157:2111–2122
77. Sikkema JM, Van Rijn BB, Franx A et al (2001) Placental superoxide is increased in pre-eclampsia. Placenta 22:304–308
78. Wang Y, Walsh SW (2001) Increased superoxide generation is associated with decreased superoxide dismutase activity and mRNA expression in placental trophoblast cells in pre-eclampsia. Placenta 22:206–212
79. Boutet M, Roland L, Thomas N et al (2009) Specific systemic antioxidant response to preeclampsia in late pregnancy: the study of intracellular glutathione peroxidases in maternal and fetal blood. Am J Obstet Gynecol 200:530.e1–530.e7
80. Karsdorp VH, Dekker GA, Bast A (1998) Maternal and fetal plasma concentrations of endothelin, lipid hydroperoxides, glutathione peroxidase and fibronectin in relation to abnormal umbilical artery velocimetry. Eur J Obstet Gynecol Reprod Biol 80:39–44
81. Mistry HD, Kurlak LO, Williams PJ et al (2010) Differential expression and distribution of placenta glutathione peroxidases 1, 3 and 4 in normal and pre-eclamptic pregnancy. Placenta 31:401–408
82. Wang Y, Walsh SW (1996) Antioxidant activities and mRNA expression of superoxide dismutase, catalase and glutathione peroxidase in normal and preeclamptic placentas. J Soc Gynecol Investig 3:179–184

83. Sharma JB, Sharma A, Bahadur A et al (2006) Oxidative stress markers and antioxidant levels in normal pregnancy and pre-eclampsia. Int J Gynaecol Obstet 94:23–27
84. Llurba E, Gratacos E, Martin-Gallan P et al (2004) A comprehensive study of oxidative stress and antioxidant status in preeclampsia and normal pregnancy. Free Radic Biol Med 37:557–570
85. Diedrich F, Renner A, Rath W et al (2001) Lipid hydroperoxides and free radical scavenging enzyme activities in preeclampsia and HELLP (hemolysis, elevated liver enzymes, and low platelet count) syndrome: no evidence for circulating primary products of lipid peroxidation. Am J Obstet Gynecol 185:166–172
86. Hubel CA, Kagan VE, Kisin ER (1997) Increased ascorbate radical formation and ascorbate depletion in plasma from women with preeclampsia – implications for oxidative stress. Free Radic Biol Med 23:597–609
87. Czerinichow S, Hercberg S (2001) International studies concerning the role of antioxidant vitamins in cardiovascular diseases: a review. J Nutr Health Aging 5:188–195
88. Diplock AT (1991) Antioxidant nutrients and disease prevention: an overview. Am J Clin Nutr 53:1893–1935
89. Wang Y, Walsh SW, Guo J et al (1991) The imbalance between thromboxane and prostacyclin in preeclampsia is associated with an imbalance between lipid peroxides and vitamin E in maternal blood. Am J Obstet Gynecol 165:695–700
90. Schiff E, Friedman SA, Stampfer M et al (1996) Dietary consumption and plasma concentrations of vitamin E in pregnancies complicated by preeclampsia. Am J Obstet Gynecol 175:1024–1028
91. Rodriguez RJ, Martin RJ, Fanaroff AA (2002) Respiratory distress syndrome and its management. In: Fanaroff AA, Martin RJ (eds) Neonatal-perinatal medicine: diseases of the fetus and infant, 7th edn. Mosby, St. Louis, pp 1001–1011
92. Nemeth I, Boda D (1994) Blood glutathione redox ratio as a parameter of oxidative stress in premature infants with IRDS. Free Rad Biol Med 16:347–353
93. Krediet TG, Cirkel GA, Vreman HJ et al (2006) End-tidal carbon monoxide measurements in infant respiratory distress syndrome. Acta Paediatr 95:1075–1082
94. Lang JD, McArdle PJ, O'Reilly PJ et al (2002) Oxidant-antioxidant balance in acute lung injury. Chest 122:314S–320S
95. Frank L, Sosenko IR (1987) Development of lung antioxidant enzyme system in late gestation: possible implications for the prematurely born infant. J Pediatr 110:9–14
96. Esteban J, Morcillo JE, Cortjo J (1999) Oxidative stress and pulmonary inflammation: pharmacological intervention with antioxidants. Pharmacol Res 40:393–404
97. Boda D, Nemeth I, Pinter S (1998) Surface tension, glutathione content and redox ratio of the tracheal aspirate fluid of premature infants with IRDS. Biol Neonate 74:281–288
98. Ikegami M, Kallapur S, Michna J et al (2000) Lung injury and surfactant metabolism after hyperventilation of premature lambs. Pediatr Res 47:398–404
99. Huertas JR, Palomino N, Ochoa JJ (1998) Lipid peroxidation and antioxidant erythrocyte membranes of full-term and preterm newborns. Biofactors 8:133–137
100. Miller NJ, Rice-Evans C, Davies MJ et al (1993) A novel method for measuring antioxidant capacity and its application to monitoring the antioxidant status in premature neonates. Clin Sci (Lond) 84:407–412
101. Ogihara T, Okamoto R, Kim HS (1996) New evidence for the involvement of oxygen radicals in triggering neonatal chronic lung disease. Pediatr Res 39:117–119
102. Bridges JP, Davis HW, Damodarasamy M (2000) Pulmonary surfactant proteins A and D are potent endogenous inhibitors of lipid peroxidation and oxidative cellular injury. J Biol Chem 275:38848–38855
103. Dani C, Buonocore G, Longini M (2009) Superoxide dismutase and catalase activity in naturally derived commercial surfactants. Pediatr Pulmonol 44:1125–1131
104. Kothecha S (2000) Lung growth: implication for the newborn infant. Arch Dis Child Fetal Neonatal Ed 82:F69–F74
105. Banks BA, Ischiropoulos H, McClelland M (1998) Plasma 3- nitrotyrosine is elevated in premature infants who develop bronchopulmonary dysplasia. Pediatrics 101:870–874
106. Dellinger RP (1999) Inhaled nitric oxide in acute lung injury and acute respiratory distress syndrome. Inability to translate physiologic benefit to clinical outcome benefit in adult clinical trials. Intens Care Med 25:881–883
107. Lamb NJ, Gutteridge JMC, Baker C (1999) Oxidative damage to proteins of bronchoalveolar lavage fluid in patients with acute respiratory distress syndrome: evidence for neutrophil-mediated hydroxylation, nitration and chlorination. Intens Care Med 25:1738–1744
108. Singh SK, Tandon A, Kumari S et al (1998) Changes in antioxidant enzymes and lipid peroxidation in hyaline membrane disease. Indian J Pediatr 65:609–614
109. Buss IH, Darlow BH, Winterbourn CC (2000) Elevated protein carbonyls and lipid peroxidation products correlating with myeloperoxidase in tracheal aspirates from premature infants. Pediatr Res 47:640–645
110. Schock BC, Sweet DG, Halliday HL et al (2001) Oxidative stress in lavage fluid of preterm infants at risk of chronic lung disease. AJP Lung Physiol 281:1386–1391
111. Gladstone IM, Levine RL (1994) Oxidation of proteins in neonatal lungs. Pediatrics 93:764–768

112. Winterbourn CC, Chan T, Buss IH et al (2000) Protein carbonyls and lipid peroxidation products as oxidation markers in preterm infant plasma: association with chronic lung disease and retinopathy and effects of selenium supplementation. Pediatr Res 48:84–90
113. Carty JL, Bevan R, Waller H (2000) The effects of Vitamin C supplementation on protein in healthy volunteers. Biochem Res Com 273:729–735
114. Levine RL, Williams JA, Stadtman ER et al (1994) Carbonyl assays for determination of oxidatively modified proteins. Methods Enzymol 233:346–357
115. Joung KE, Kim HS, Lee J et al (2011) Correlation of urinary inflammatory and oxidative stress markers in very low birth weight infants with brochopulmonary dysplasia. Free Radic Res 45:1024–1032
116. Sharda B (2006) Free radical – emerging challenge in environmental health research in childhood and neonatal disorders. Int J Environ Res Public Health 3:286–291
117. Gitto E, Reiter RJ, Karbownik M et al (2001) Respiratory distress syndrome in the newborn: role of oxidative stress. Intensive Care Med 27:1116–1123
118. Warren R, Luzminda C (1986) Endogenous antioxidant defenses in neonates. J Free Radic Biol Med 2:295–298
119. Schroder A, Herting E, Speer CP (1999) Superoxide dismutase and catalase activity in tracheobronchial secretions after surfactant treatment of newborn infants with respiratory distress syndrome. Z Geburtshilfe Neonatol 203:201–206
120. Lavoie JC, Chessex P (1997) Gender and maturation affect glutathione status in human neonatal tissues. Free Rad Biol Med 23:648–657
121. Asikainen TM, Raivio KO, Saksela M et al (1998) Expression and developmental profile of antioxidant enzymes in human lung and liver. Am J Respir Cell Mol Biol 19:942–949
122. Zachman RD (1989) Retinol (vitamin A) and the neonate: special problems of the human premature infant. Am J Clin Nutr 50:413–424
123. Simon C, Kiosz D, Hofman I (1980) Serum concentration of vitamin E in healthy infants fed commercial milk. Eur J Pediatr 133:273–276
124. Coleman M, Thompson TR (1979) A possible role of vitamin E in the prevention or amelioration of bronchopulmonary dysplasia. Am J Pediatr Hematol Oncol 1:175–178
125. Huijbers WA, Schrijver J, Speek AJ et al (1986) Persistent low plasma vitamin E levels in premature infants surviving respiratory distress syndrome. Eur J Pediatr 145:170–171
126. Frei B, England L, Ames BN (1989) Ascorbate is an outstanding antioxidant in human blood plasma. Proc Natl Acad Sci U S A 86:6377–6381
127. Niki E (1987) Interaction of ascorbate and α-tocopherol. Ann NY Acad Sci 498:186–199
128. Mathews F, Yudkin P, Neil A (1999) Influence of maternal nutrition on outcome of pregnancy: prospective cohort study. BMJ 319:339–343
129. Wiedemann M, Kontush A, Finckh B (2003) Neonatal blood plasma is less susceptible to oxidation than adult plasma owing to its higher content of bilirubin and lower content of oxidizable fatty acids. Pediatr Res 53:843–849
130. Yeung MY (2006) Influence of early postnatal nutritional management on oxidative stress and antioxidant defence in extreme prematurity. Acta Pediatr 95:153–163
131. White CW, Stabler SP, Allen RH (1994) Plasma cysteine concentrations in infants with respiratory distress. J Pediatr 125:769–777
132. Abbe LM, Friel JK (2000) Superoxide dismutase and glutathione peroxidase content of human milk from mothers of premature and full term infants during the first 3 months of lactation. J Pediatr Gastroenterol Nutr 31:270–274

Oxidative Stress and Diabetes

Rachana, Shruti Thakur, and Sujata Basu

Abstract

Increase in oxidative stress (OS) has been found to be linked with various disease conditions, including diabetes and post-diabetic complications. If there is any imbalance between the reactive oxygen species (ROS) and antioxidant species inside cells, ROS damage cellular functions directly or indirectly. Besides oxidizing the major biomolecules inside the cells, they also alter the cell signaling mechanism, cell permeability, basic genetic mechanism, etc. In case of diabetes, different types of stresses (emotional, physical, chemical, or infectious) can lead to damage to the pancreatic cells and may result in decreased production/secretion (by β-cells) or utilization (by adipocytes, skeletal muscles, hepatocytes, etc.) of insulin and so, can result into hyperglycemic conditions. For the cells, which are not insulin dependent for their glucose uptake and metabolism (retinal cells, nephrons, nerve cells, etc.), their intracellular glucose concentration rises, and as a result, an increase in oxidative stress occurs by various mechanisms. This further triggers the onset of post-diabetic complications. This chapter describes the causes and mechanisms for the onset of diabetes and post-diabetic complications.

Keywords

Type 1 diabetes • Type 2 diabetes • Hyperglycemia • Oxidative stress • Post-diabetic complications • Insulin-dependent diabetes or Juvenile-onset diabetes • Adult-onset diabetes • Neuropathy • Nephropathy • Retinopathy • Vascular dysfunction

1 Introduction

There are many reports suggesting the direct link between oxidative stress and onset of diabetes and associated complications [1–3]. In case of

Rachana (✉) • S. Thakur • S. Basu
Department of Biotechnology, Jaypee Institute of Information Technology, A-10, Sector-62, Noida 201307, Uttar Pradesh, India
e-mail: rachana.dr@iitbombay.org

V. Rani and U.C.S. Yadav (eds.), *Free Radicals in Human Health and Disease*,
DOI 10.1007/978-81-322-2035-0_16, © Springer India 2015

type 1 diabetes, ROS are shown to be involved in the destruction of pancreatic β-cells and further development of insulin-dependent diabetes mellitus (IDDM) [2]. Oxidative stress arises when a cell is unable to quench the reactive oxygen species (ROS) through the antioxidant mechanism, and when there is antioxidant deficiency, either genetically or induced by environmental factors, including infections [3]. The general mechanism of action in this pathogenesis by ROS is via oxidizing the major biomolecules in the cells and so affecting their basic structure and/or functions. Proteins get modified by oxidized or nitrosylated products of free-radical attack. These altered products lose their function and are also targeted for proteosome degradation, which further decreases their cellular functions. Damaged proteins have an effect on cell signaling, membrane permeability, transport of various molecules, and also on the energy metabolism inside the cells [4], and these ultimately lead to malfunctioning and cell death. Nucleic acids (DNA and RNA), both from the cytoplasm and nucleus, also get oxidized [1, 2], and the modified/mutated products also cannot play their normal function [5]. Similarly, lipid products also get modified and they being the main structural components of the membranes of different cellular compartments and the cell itself affect their integrity [3]. Accumulations of such injuries ultimately lead the cells to die through necrotic or apoptotic mechanisms and thus cause onset of various inflammatory reactions [6]. This chapter describes the role of reactive oxygen species (ROS) and linked pathways which might be involved in both the onset and post-diabetic complications. The entire pathway involved is shown in Fig. 1.

1.1 Effect of OS on Onset of Diabetes

For the two types of diabetes (1 and 2), the origins for the disease are different, but ultimately they result in hyperglycemia. In the first case, it is the lack of insulin in blood because of β-cell destruction or due to lack of insulin secretion caused by OS and inflammation [7]. On the other hand, in case of type 2 diabetes, it is the lack of insulin sensitivity in glucose-utilizing cells (adipose tissue, skeletal muscle, and liver cells, etc.) due to OS [8]. Following are the descriptions about the onset of both types of diabetes involving events led by oxidative stress.

1.1.1 Type 1 Diabetes

Type 1 diabetes is commonly known as "insulin-dependent" or "Juvenile-onset" diabetes as it is genetic and the onset of this disease can occur during childhood itself [7]. It is actually an autoimmune disorder which is involved in the recognition of β-cells by the T cells which are against them. This attack leads to the increase in ROS and production of many types of cytokines and hence inflammation. These further lead to the loss of β-cells present in the islets of Langerhans [9]. It has been proposed that both genetic and environmental factors might account for an increased susceptibility of β-cells to be attacked by the immune system and to dysfunction in the face of an increasing inflammatory response [10], but their exact role is still little understood. It is seen that the children, as diabetics, suffer from a rapid β-cell destruction, whereas it might be slower in case of adults [7].

It can be seen from the above discussion that type 1 diabetes is not associated with any external cause directly, but even in this case also, the onset of the disease can occur early in life if there is an increase in any type of stress/OS in the external environment by any reason.

1.1.2 Type 2 Diabetes

Type 2 diabetes is also known as the "Adult-onset diabetes." This type of diabetes is implicated with resistance to insulin with very little deficiency. It affects almost 90–95 % of patients with diabetes [8]. The patients are not required to undergo insulin treatment as β-cell destruction does not take place in the initial stages. Type 2 diabetes is often reported to be associated with the modern lifestyle which actually results in having an increase in oxidative stress [11]. The cause for this increase in OS can vary from physical, mental, psychological, chemical, to infectious load [12].

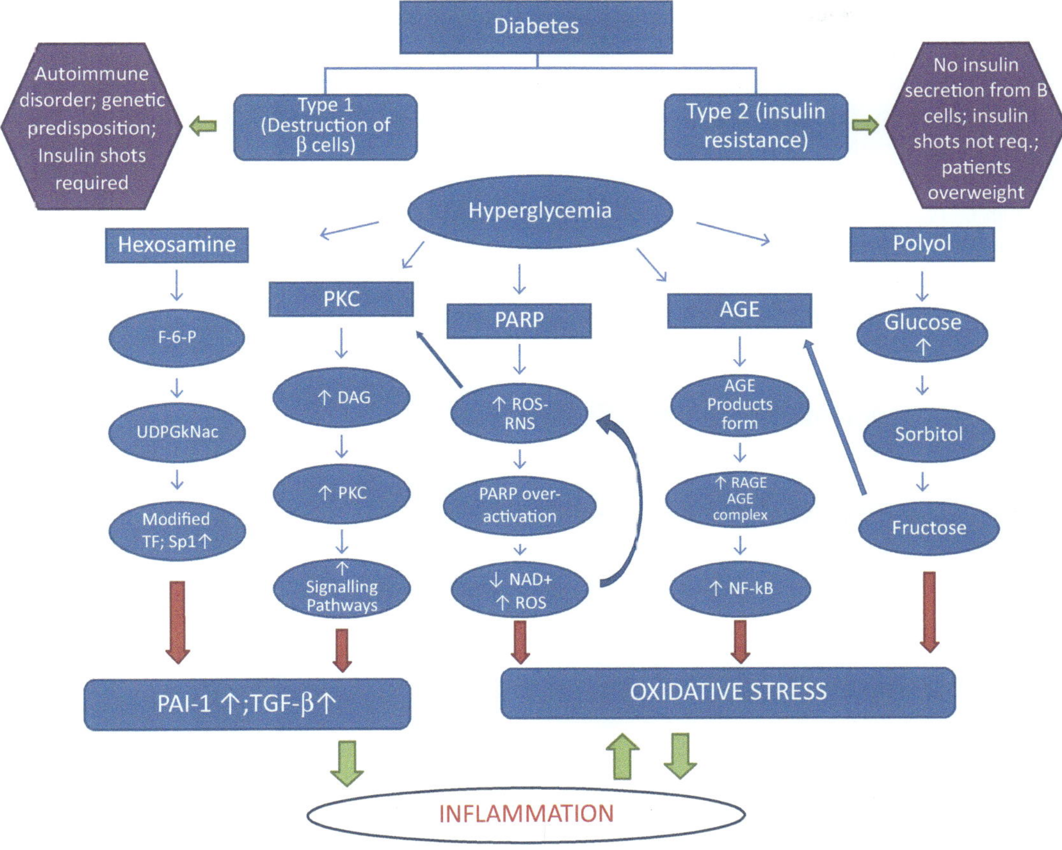

Fig. 1 OS and diabetes. Diabetes are of two types: insulin dependent and insulin independent. Type 1 or insulin dependent is due to destruction of β-cells and is prevalent in patients who already have risk of diabetes through family history and type 2 or insulin independent which is found in patients who are overweight or have sedentary. Hyperglycemia sets forth the activation of various pathways in the target cells. AGEs and polyol pathway changes the redox capacity of the cell by generation of oxidative stress. PARP, hexosamine, and PKC pathway leads to inflammation which ultimately leads to oxidative stress. Thus, there is a direct link between diabetic condition and oxidative stress. This stress then leads to diabetic complications and in turn leads to more oxidative stress. (*F-6-P*) fructose 6 phosphate, *PKC* protein kinase C, *DAG* diacyl glycerol, *PARP* poly ADP ribose polymerase, *ROS* reactive oxygen species, *AGE* advanced glycation end products, *RAGE* receptor for advanced glycation product

1.2 Post-diabetic Complications

Chronic increase in blood glucose level in both type 1 and type 2 diabetes leads to various post-diabetic complications. The major symptoms of various diabetic complications are: pain, numbness, development of Charcot's joint, difficulty in focusing objects with the eyes, paralysis in distal parts of the limbs, problem in controlling muscle while urinating, etc. [13]. Diabetic complications increase when patients are not able to control their blood glucose and have sedentary lifestyle and stressful life. Most of these complications are reported to be the result of increase in oxidative stress in the body system [14]. Increased concentration of glucose leads to increase in oxidative stress, which further deteriorates the condition of diabetes by affecting pancreatic cells for insulin secretion and insulin targets like: the skeletal muscle, for glucose uptake and its further metabolism.

In a large population-based survey for glucose intolerance, an association between stressful experiences and the diagnosis of type 2 diabetes has also been demonstrated [15]. The cells in the retina, nerves, and kidney are more fragile in comparison to other body cells as they have a very weak antioxidative state, and with the increase in sugar concentration, they lose the fight against oxidative stress [16]. This further leads to related post-diabetic complications like retinopathy, neuropathy, nephropathy, etc. This chapter also describes various causes and pathways which lead to worsening of the condition of diabetes and also towards post-diabetic complications. Causes and pathways involved in this process are shown in Fig. 1.

2 Mechanism of Action of OS on Onset of Diabetes

2.1 General Mechanism of Action for β-Cell Damage and/or Function

Stressful experiences have been implicated in the onset of diabetes in individuals already predisposed to develop the disease [16]. As early as the beginning of the seventeenth century, the onset of diabetes was linked to "prolonged sorrow" by an English physician [17]. Events like family losses and workplace stress are reported to be factors triggering the onset of diabetes – both type 1 and type 2. It is suggested that negative stressful experiences in the first 2 years of life may increase the risk of developing type 1 diabetes in children. Other factors, such as high family chaos and behavioral problems, were also implicated [18]. All these events ultimately result in increase in oxidative stress via different pathways/mechanisms.

In some other cases, where ROS production or suppressed antioxidant function is due to infections, the cells are much more prone to damage than that incurred due to genetic problems and also more prone to damage than any other tissue in the body [19].

Mitochondria of β-cells have very low levels of superoxide dismutase and glutathione peroxidase which results in its being an easy target for disruption by oxidative stress in individuals with risk to the disorder [20]. ROS derived from the mitochondria and nitrogen oxygen species (NOX) are responsible for destroying β-cells and initiating diabetes [21].

Not only the decrease in the number of β-cells but also the decrease of insulin production from the normal number of β-cells may lead to type 1 diabetic condition [20]. There are experimental evidences showing the deleterious effects of ROS on β-cell function. Maechler et al. [22] have shown that oxygen stress generated by short exposure of β-cell preparations to H_2O_2 increases production of p21 (an inhibitor of cyclin-dependent kinase); decreases insulin mRNA, cytosolic ATP, and calcium flux in the cytosol and mitochondria; and causes apoptosis [22].

2.1.1 Damage to the Macromolecules

It has been discussed earlier that ROS are known to damage the macromolecules in almost all types of cells including β-pancreatic cells. Modifications of protein, DNA damage, and peroxidation of lipids are seen when oxidative stress and ROS increase in cells of patients with diabetes [1]. Hyperglycemia or increased blood sugar level is responsible for not only generating ROS and oxidative stress but also responsible for the weakened antioxidant mechanisms in the cells. This happens when ROS destroys antioxidant enzymes and other substances helpful in the antioxidant activity [23].

Protein damage. Protein molecules are found in the cell membrane and cytosol of all cells. The function of each protein is based on the type of folding it undergoes, and if it is changed the whole function of the proteins changes. Proteins could be modified to some alternate form due to increase in ROS [24] as they can oxidize the amino acid residue side chains, modifying the cross-links between molecules of protein and also resulting in the fragmentation of proteins by oxidizing their backbone. The most susceptible amino acids for change are cysteine and methionine by ROS. Besides ROS, reactive nitrogen species (RNS) is also responsible for the damage seen in protein structure and their mutated functions. Studies have shown that cysteine and methionine residues of proteins as well as aromatic residues have high probability of getting modified through RNS [25].

Moreover, apart from just modifying protein molecules, ROS also reacts with them to form highly reactive products like ketones and aldehydes [26]. It also results in partial unfolding of proteins as surface hydrophobicity increases. Oxidation of lysine results in the formation of aminoadipic semialdehyde, and the oxidation of arginine forms glutamic semialdehyde [27].

Nucleic acid damage. NO and other ROS have been shown to damage nuclear DNA and mitochondrial DNA in β-cells [28]. In general, mitochondrial DNA is more sensitive to oxidative stress than nuclear DNA [29]. The ROS and reactive nitrogen species (RNS) cause breakage of the DNA strands and can also convert guanine to 8-oxo, 2′-deoxyguanosine (8-oxodG) which is an oxidized nucleoside and results in base substitution. This new substance 8-oxodG has been considered to be a very important biomarker of oxidative DNA damage and is important in mutagenesis. When cells of diabetic patients were investigated, a high concentration of 8-oxodG was seen in the mitochondrial DNA in islet and mononuclear cells [30].

Lipid damage. Similar to proteins and nucleic acid, ROS also oxidize lipids via lipid peroxidation. This gives rise to mutagenic lipids: epoxides, hydroperoxides, alkoxyl and peroxyl radicals, and enals (4-unsaturated aldehydes). Singlet oxygen, a high-energy and mutagenic form of oxygen, can also be produced by transfer of energy during respiratory burst from neutrophils or lipid peroxidation [30].

In addition to their ability to directly inflict macromolecular damage, ROS have also been reported to have an effect to work as signaling molecules and to activate a number of cellular stress-sensitive pathways that cause cellular damage and are also ultimately responsible for the late complications of diabetes. Furthermore, these same pathways are linked to insulin resistance and decreased insulin secretion [31].

2.1.2 Poor Antioxidative Potential of β-Cells

β-cells are responsible for sensing glucose level and secreting appropriate amount of insulin accordingly. Type 1 diabetes, or insulin-dependent diabetes mellitus (IDDM), is an autoimmune disease, but there are evidences that even if the central cause of this kind of β-cell damage is genetic and has an immunological base [16], the onset of diabetes may also be accelerated if the person is facing stressful conditions and so increase ROS in his/her life as described above. Excess production of both, ROS and NO, has been implicated to be cytotoxic to the β-cells [32]. Increase in ROS for any reason further leads to damage of β-cells, which are sensitive to ROS and RNS specifically as they are low at their antioxidative state [33]. The level of free-radical quenching (antioxidant) enzymes such as catalase, glutathione peroxidase, and superoxide dismutase is low in β pancreatic cells [34]. There are evidences for these facts as overexpression of the antioxidant enzymes in islets or transgenic mice has been shown to prevent many of the deleterious effects on the cells [35].

As the OS rises due to any of the above-said reasons, inflammatory processes are initiated and β-cell function is further affected as a result of exposure to inflammatory products [36]. Immune cells like macrophages infiltrate the islets of Langerhans. They are known to generate reactive oxygen species such as hydrogen peroxide (H_2O_2), nitric oxide, etc. [33]. So, again, as a consequence of infiltration of the immune cells, the oxidative stress increases further. Both of these are known to exert deleterious actions on the β-cells and its mitochondrial oxidative metabolism [29].

Imbalanced mitochondrial oxidative metabolism also causes type 1 diabetes as secretion of insulin is affected and might also result in decreased blood insulin level [31]. Nitric oxide (NO), the free-radical precursor, has been shown to suppress mitochondrial activity leading to a defective insulin release in response to nutrient secretagogues [37].

2.1.3 Triggers to Inflammatory Pathways

As described above, increase in ROS leads to inflammatory conditions. Recently, Tersey et al. [10] have reported on how the pathogenesis of type 1 diabetes can involve an orchestrated interplay between cell types of the immune system and the β-cell. They worked on nonobese diabetic (NOD) mice and found that the prediabetic phase of the disease can be characterized by infiltration of islets by macrophages and T cells, resulting in insulitis. Cytokines seem to have a major role in prediabetic situation as prior to overt β-cell death; the local release of cytokines interleukin 1β (IL-1β), γ-interferon (IFN-γ), and tumor necrosis factor-α (TNF-α) by infiltrating cells seems to get activated and then induces the inflammatory pathways in the β-cell leading to insulin deficiency and hyperglycemia [10].

2.1.4 ER Stress Pathway

ER stress has also been shown to be associated with the predisposal of diabetic condition in case of type 1 diabetes. Tersey et al. [10] also showed that upon exposure to a mixture of proinflammatory cytokines that mimic the microenvironment of type 1 diabetes, MIN6 β-cells have been shown to display evidence for polyribosomal runoff, a finding consistent with the translational initiation blockade characteristic of ER stress [10]. It has been suggested that β-cells of prediabetic NOD mice display dysfunction and overt ER stress that may be driven by NF-kB signaling and strategies that attenuate pathways leading to ER stress may preserve β-cell function in type 1 diabetes. The activation of these pathways may accelerate β-cell death in the prediabetic phase of the disease and thus promote further antigen exposure and T-cell activation [38].

It has been also suggested that release of a certain set of cytokines may lead to increase in ER stress; however, how it leads to cause ER stress, is not clear. Some studies suggest that nitric oxide generated via iNOS, downregulates SERCA2B [39–41] which is an ATP-dependent Ca^{2+} pump that is partially responsible for transport of Ca^{2+} into the ER lumen, thereby maintaining a steep ER:cytosolic Ca^{2+} gradient [42, 22].

2.2 Mechanism of Action for Decrease in Insulin Secretion

Along with the damage caused to β-cells, ROS also affect the process of insulin secretion [43]. This process is regulated by a combination of many events, but ultimately it is the mitochondrial glucose metabolism which links the stimulus to secretion by generating ATP to raise cytosolic Ca^{2+} concentration and many other pathways [44].

The other mechanism involved in insulin secretion is through membrane depolarization and additional mitochondrial factor(s), triggering insulin exocytosis. In the pancreatic β-cells of the mouse, H_2O_2 hyperpolarizes the cell membrane coupled with an increase of cell membrane conductance [29]. Moreover, it has recently been shown that H_2O_2 increases intracellular Ca^{2+}, decreases the ATP/ADP ratio, and inhibits glucose-stimulated insulin secretion from isolated mouse islets in response to a glucose stimulus [44]. Although, this process is complex and dependent on many factors [45], the critical importance of mitochondrial glucose metabolism in linking stimulus to secretion, is well established. Therefore, the ability of oxidative stress (H_2O_2) to damage the mitochondria and markedly blunt insulin secretion is not surprising [29].

ER stress has also been shown to be associated with insulin secretion. It can activate NF-κB which promotes defects in insulin secretion. A study conducted by Maechler et al. [29] correlates the extent of ER stress with the severity of β-cell secretory deficiency in prediabetic NOD mice [29].

3 Mechanism of Action of OS on Post-diabetic Complications

Diabetes mellitus has become a great challenge when considering health issues in the twenty-first century [14] due to its link with post-diabetic complications. There are considerable amount of data indicating that the chronic elevation of plasma glucose causes many of the major microvascular and macrovascular complications of diabetes. Recently, numerous authors have suggested that excess generation of highly reactive oxygen and nitrogen species is the key component in the development of complications invoked by hyperglycemia. However, detailed molecular mechanism remains uncertain. As said earlier the production of ROS is under tight control in healthy cells, but their overproduction during metabolic dysfunction leads to cellular injury. ROS has this ability to directly oxidize and damage DNA, protein, and lipids and so diminishing their natural metabolic functions. In addition to their ability to directly inflict macromolecular damage, ROS can function as signaling molecules to activate a number of cellular stress-sensitive pathways that cause cellular damage and are ultimately responsible for the late complications of diabetes [31]. Oxidative phosphorylation, glucose auto-oxidation, increased lipoxygenase expression, changes in the regulation and expression of nitric oxide synthase (NOS) isoforms (endothelial NOS, eNOS; inducible NOS, iNOS; neuronal NOS, nNOS) and $ONOO^-$ production, are some of the mechanisms by which hyperglycemia can generate reactive oxygen nitrogen species (RONS) [5]. The following are few mechanisms of action which are stated to be responsible for post-diabetic complications.

3.1 General Mechanism

Mesangial cells of the kidney, endothelial cells of the vascular system, neurons and neuroglia of the nervous system, and pancreatic β-cells are the types of cells which are directly linked to complications raised due to hyperglycemia. These cells are especially vulnerable to hyperglycemic conditions. These are the kind of cells which are not able to downregulate the influx of high dose of glucose in cell and thus are more exposed to hyperglycemic conditions. This condition then induces the phenomenon known as oxidative stress, this being the basic cause of diabetic complications. Studies also suggest that the increase in the level of free radicals during hyperglycemic conditions is the central reason for the complications. Thus, production of mitochondrial free radical and oxidative damage to molecular pathways, are main causes for onset, progression, and morbidity. This fact is also evident from the examination of the data from animal and cell culture models of diabetes, as well as clinical trials of antioxidants, which strongly implicate hyperglycemia-induced oxidative stress in diabetic complications such as neuropathy in diabetic patients [46].

3.1.1 Modification of Proteins

The interaction between OS and proteins may lead to oxidation of protein molecules, and due to which there may be change in confirmation and so modification in their cellular function. The general mechanism of ROS towards protein modifications has been discussed in earlier section. Besides this the following are few more specialized mechanisms of action responsible.

3.1.1.1 Modifications of Enzymes

ROS while interacting with proteins/enzymes may perturb the electron transfer among them also, thus leading to the formation of abnormal products (e.g., uncoupled endothelial nitric oxide synthase, eNOS, etc.). This is a wide concept, which directs us to imagine the complexity of such regulatory redox mechanisms pointing to an infinite number of damage pathways, since diverse redox reactions in human biological systems exist [14]. It has been reported that highly reactive peroxynitrite, when attacking proteins and lipids, inhibits their function. These can also attack iron-sulfur centers of enzymes and other proteins, to release iron atoms and consequently inhibit enzyme/protein activities. Few very important proteins are exquisitely sensitive to this type of inhibition including, complexes I–III of the electron transfer chain, aconitase of the trichloroacetic acid cycle, and biotin synthase [47].

3.1.1.2 Nonenzymatic Glycation

Metal-catalyzed glucose autoxidation is suggested to be one more mechanism for glucose toxicity [48]. It is a nonenzymatic protein glycation which takes place due to high glucose concentration in blood and cells. It is reported to be dependent on ROS (superoxide and hydroxyl) formation through transition.

3.1.1.3 Modification of Transcription Factors

ROS can attack transcription factors also, and then it can change their normal genetic mechanism in any type of cell. Oxidative modification of transcription factors not only leads to decreased expression of many proteins such as apoptosis inhibitory factor, complex I, and Bcl-2 but also results in increased expression of stress proteins that may be proapoptotic, including cyclooxygenase 2, poly-ADP ribose polymerase, and Jun kinase (JNK). This ultimately leads towards apoptotic cell death [49].

3.1.2 Modification of DNA

Understanding the DNA modifications is the most significant consequence of oxidative stress in dividing cells and may lead to genomic instability and mutations. The number of DNA strand breaks has been observed to be significantly higher in cells of diabetic patients than in cells of the control (Comet Assay and SCGE assay) [1]. Along with this, the comet tail length was found to be significantly increased in diabetic patients. Serum 8-OHdG (a modified nucleic acid due to oxidation) was found to be significantly increased in diabetic patients.

The cells which do not divide after a certain stage/age like neurons, may suffer less from oxidative damage of DNA, but it is also true that the mitochondria remain at a low antioxidant state and mitochondrial DNA is particularly sensitive to oxidative damage which further would impair energy regulation and would definitely affect the functionality of the nerve cells [47].

In a similar manner, increased oxidative stress plays an important role in the progression of diabetic nephropathy. In general, as nephrons are dividing cells, oxidative stress can affect nucleic acids and generate various modified bases in DNA. 8-OHdG is one of the most abundant oxidative products of DNA and appears to play a crucial role in mutagenesis. Once it is modified, cells do excrete it out to avoid the mutations, and it is released into the blood and urine after being excised from DNA by the repair enzyme; thus, urinary 8-OHdG can be used as biomarker for diabetic nephropathy. Various studies have revealed that cells like mononuclear cells, urine, pancreatic islet, and mitochondrial DNA of diabetic patients have higher levels of 8-OHdG. Kakimoto et al. [50] showed that the levels of 8-OHdG are increased in kidney tissues of streptozotocin-induced diabetic rats [50]. A similar increase in 8-OHdG in the urine and mononuclear cells is observed with the patients with either retinopathy or nephropathy.

3.1.3 Modification of Lipids

In general, in patients with type 2 diabetes, there is a significant inverse correlation between the fasting plasma free fatty acids (FFA) concentration and ratio of reduced/oxidized glutathione (the major endogenous antioxidant). It is evidenced by the studies where infusion of FFAs (as Intralipid) to healthy subjects causes increased oxidative stress, as judged by increased malondialdehyde levels and a decline in the plasma reduced/oxidized glutathione ratio [51]. Elevated FFA levels have numerous adverse effects on mitochondrial function evidenced by various in vitro experiments, and it includes the uncoupling of oxidative phosphorylation [52], which generates ROS, including superoxide [53]. It is suggested that generation of SOD exacerbated the situation because FFAs not only are capable of inducing oxidative stress but also impair endogenous antioxidant defenses by reducing intracellular glutathione [54, 55].

It has been reported that elevated FFA level has a link with the development of microvascular complications. It has been positively correlated with both insulin resistance [56, 57] and the deterioration of β-cell function in the context of concomitant hyperglycemia; however, their exact role in microvascular complications remains to be established [58, 59].

3.1.4 Decrease in Antioxidant Content

It is a vicious cycle in which decrease in antioxidant leads to oxidative stress and increase in oxidative stress can lead to decrease in antioxidant level in the cells. Various molecular events such as increase in FFA level lead to decrease in GSH content of the cells. GSH is maintained at a concentration of 0.2–10 mM in all mammalian cells [60]. GSH can be synthesized by many cells de novo or by γ-glutamylcysteine synthetase. Neurons do not contain the γ-glutamylcysteine synthetase enzyme and so require the dipeptide to be secreted from glial cells [61, 62], and depletion of GSH in the cell renders it susceptible to oxidative injury [63].

Direct in vitro experiments using agent 3-hydroxy-4-pentenoate to deplete mitochondrial GSH were shown to increase cell death induced by pro-oxidants such as *tert*-butyl hydroperoxide [64]. Similarly, by increasing the GSH by artificial loading particularly in the mitochondria, neuronal apoptosis was seen to be prevented which was induced by ischemia [65] and excitotoxicity [66]. Besides this, overexpression of glutathione-S-transferase in neuroblastoma cells has shown to increase their resistance to oxidative stress [67].

3.2 Pathways Involved: General

As mentioned earlier, the major diabetic complications are: neuropathy, nephropathy, retinopathy, and vascular disorders. All these complications are the result of the progression of diabetes if there is no control taken over hyperglycemia by the patient [4]. There are several molecular pathways involved which are common to most of the diabetic complications. These pathways are polyol pathway, hexamine, NF-kB, JNK/SAPK and p38 MAPK pathway, and Protein kinase C (PKC) etc. which act while glucose and FFA levels are high and lead to both insulin resistance and impaired insulin secretion [13]. So, ultimately when there is an increase in glucose level, it activates several major, well-characterized biochemical pathways that play a significant role in the etiology of diabetic complications. Data now indicate that activation of these pathways is linked not only to the development of the late complications of diabetes but also to insulin resistance and β-cell dysfunction, like a vicious cycle [31].

3.2.1 Polyol Pathways

When there is an increase in glucose level in the cells, polyol pathway gets activated. Usually, this enzyme reduces cellular toxic aldehyde into inactive alcohols. However, in the presence of high concentrations of intracellular glucose, aldose reductase (AR) reduces glucose to sorbitol, which is further converted to fructose through oxidation. The cell membrane is impermeable to sorbitol, and the increase in osmolality is compensated by moving the cell's osmolytes such as myoinositol, adenosine, etc. out [68]. The absence of myoinositol in the cytoplasm reduces formation of ATP due to exhaustion of phosphatidylinositol, and so the Na+/K+ -ATPase activity is also reduced. The conversion of glucose to sorbitol depletes NADPH which besides serving as a cofactor of this pathway is also essential to regenerate glutathione (GSH), an important cellular antioxidant [69].

Polyol pathway is also found to be the main source for RONS induced by hyperglycemia in the retina [70]. The involvement of AR in post-diabetic complication has been proved experimentally when morphological abnormalities of peripheral nerves of diabetic mice with and without AR deficiency were verified. OS was found to be less in the first group, suggesting the involvement of this pathway in the pathogenesis of acute diabetic neuropathy [71].

3.2.2 Hexosamine Pathway

The excessive flux of glucose or FFAs into a variety of cell types results in the activation of the hexosamine biosynthetic pathway. Recent data have implicated a hyperglycemia-induced increase in ROS formation in the activation of the hexosamine pathway [72].

Fructose generated through polyol pathways may get converted to fructose 6-phosphate. Fructose 6-phosphate, which is also the intermediate product of glycolysis, shifts to enter into the hexosamine pathway and gets converted to glucosamine-6-phosphate by the enzyme glutamine-fructose-6-phosphate aminotransferase (GFAT) [73]. This in turn leads to the formation of uridine diphosphate-N-acetylglucosamine, which attaches to serine and threonine residues of transcription factors, and is responsible for the increased level of expression on transcription factor Sp1. Activated Sp1 results in overexpression of transforming growth factors (TGF-α and TGF-β) and plasminogen activator inhibitor-1 (PAI-1) which increases the complications in diabetes [74].

3.2.3 NF-ĸB Pathway

NF-ĸB plays a critical role in mediating immune and inflammatory responses and apoptosis. NF-ĸB regulates the expression of a large number of genes, including several of those linked to diabetes, e.g., vascular endothelial growth factor (VEGF) and receptor for advanced glycation end products (RAGE) [75]. Many of the gene products regulated by NF-ĸB (e.g., VEGF, RAGE, etc.) in turn activate NF-ĸB, leading to a vicious circle. The aberrant regulation of NF-ĸB is associated with a number of chronic diseases, including diabetes and atherosclerosis.

NF-ĸB pathway is the most extensively studied intracellular pathway as the target of hyperglycemia, ROS, and oxidative stress [76]. NF-ĸB induction may be mediated with increase in FFA in the patient's blood. Numerous in vitro studies have reported the FFA-mediated activation of NF-κB, a likely consequence of the ability of FFAs to increase ROS formation and reduce glutathione [55]. FFA-mediated reduction of glutathione level might also be linked to FFA-mediated activation of PKC-θ [77], which has the unique ability among PKC isoforms to activate NF-ĸB [78]. FFA-induced activation of NF-κB can be prevented by vitamin E, suggesting that the alteration in cellular redox status is a contributory component of the proinflammatory effects of FFAs.

The activation of NF-ĸB involves the phosphorylation-induced, proteosome-mediated degradation of the inhibitory subunit, inhibitory protein κB (IκB). IκB is phosphorylated by an upstream serine kinase, IκB kinase β (IKK-β), which is phosphorylated and activated by additional upstream serine kinases [76].

In a healthy situation, there is a balance between the two transcription factors NF-κB and Nrf2, where Nrf2 is involved in the regulation of antioxidant defense systems [79]. During pathological conditions, the control slips in the form of over-activation of NF-κB and simultaneous suppression of Nrf2. Although Nrf2 is briefly activated by oxidative stress, extracellular related kinase (ERK) activation restrains permanent Nrf2 activation [80]. Waning in Nrf2 activity and an insistent increase in NF-κB activity can lead to enhanced nitrosative and oxidative stress [81].

3.2.4 JNK/SAPK and p38 MAPK Pathway

JNK/SAPK and p38 MAPK, the members of the complex superfamily of MAP serine/threonine protein kinases, are known as stress-activated kinases and are responsive to a variety of exogenous and endogenous stress-inducing stimuli, including hyperglycemia, ROS, oxidative stress, osmotic stress, proinflammatory cytokines, heat shock, and ultraviolet irradiation [82]. It has been shown that H_2O_2 generation, JNK/SAPK activity, and subsequent apoptosis induced by hyperglycemia could be suppressed by vitamin C. JNK/SAPK are reported to induce apoptosis in hyperglycemia-induced oxidative stress conditions in human endothelial cell [83].

Although the role of p38 MAPK pathway is not known, it is also shown to get activated in response to hyperglycemia via oxidative stress in vascular smooth muscle cells, rat aortic smooth muscle cells, glomeruli of rats (streptozotocin-induced diabetes), and nerve tissue of patients (type 1 and type 2 diabetes) [84].

3.2.5 PKC Pathway

The role of PKC has been described earlier to be linked with various other pathways. PKC is a family of 11 isoforms, out of which 9 are linked with the production of oxidative stress [85]. Hyperglycemia stimulates the formation of diacyl glycerol (DAG) which activates these nine isoforms [86]. These isoforms lead to stimulation of the expression of signaling pathways involving PAI-1, NF-κB, and TGF-β [87] which have shown to be involved in complications as above. They then lead to overproduction of cytokines and induce inflammatory response. Also, it leads to inhibition of Na+/K+ ATPase. PKC also actuates stress genes, phosphorylating transcription factors; affects the balance of gene expression; and induces oxidative stress [88].

3.2.6 Inflammation

All the above pathways ultimately are linked to apoptosis and inflammation. Two main inflammatory agents are C-reactive protein and TNF-α (tumor necrosis factor-α). As mentioned above, NF-κB is one of the transcription factors which upregulates many genes involved in inducing inflammation such as TNF-α. It also upregulates other factors such as TGF-α and TGF-β and is produced during hyperglycemia [86]. There is a two-way relation between ROS and inflammation where ROS leads to inflammation which further adds up to increase in ROS.

There are certain enzymes which get upregulated, when the factors like NF-kB are activated. Cyclooxygenase-2 is an enzyme which is upregulated by NF-κB generating prostaglandin E2 and further ROS [89]. An additional inflammatory enzyme, which is also catalyzed by NF-κB, is iNOS (inducible nitric oxide synthetase) [74]. NF-κB is the central point of activation and is the center of inflammatory response in hyperglycemic conditions. Along with it, NF-κB induces the production of cytokines in endothelial cells, Schwann cells, and neurons.

4 Few Specific Mechanisms for Specific Complications

The overall mechanism of action remains the same in all the cases of post-diabetic complications, but due to a specialized environment for/in a particular cell, the end result may vary from one to another. A sustained increase in ROS and RNS and a decrease in endogenous antioxidant defenses contribute to the establishment and maintenance of OS, leading to endothelial dysfunction (ED), insulin resistance (IR), alterations of pancreatic β-cells, and damage to all other susceptible cells. There is a cyclic relationship between DM and OS, therefore triggering deleterious cellular processes. The following are descriptions of few more specific mechanisms of action in a particular complication.

4.1 Effect on β-Cell

4.1.1 β-Cell Dysfunction

Many studies have suggested that, β-cell dysfunction is the result of prolonged exposure to high glucose, elevated FFA levels, or a combination of the two [45]. There is considerable evidence that chronic hyperglycemia in patients with type 2 diabetes contributes to impaired β-cell function (insulin production and/or release) and cell death [90].

4.1.2 Defect in Insulin Secretion

The effect of high glucose level has been described earlier to generate ROS which affects the number of β-cells and so the amount of insulin in blood. In addition to this high cell glucose inhibits insulin secretion in β-cells even if their number remains the same. It is shown by many scientists that in patients with type 2 diabetes, controlling sugar level with diet, insulin, or sulfonylureas results in improved insulin release [91]. The reverse is also reported where, in healthy individuals, high glucose infused as a clamp reduces insulin release. There are many in vitro experiments which indicate that long-term culture of pancreatic cells (e.g., HIT-T15) with elevated glucose decreases insulin release, insulin mRNA, and binding of insulin mRNA transcription factors [92, 93].

Not only at cellular or enzymatic level, glucose concentration has an effect at gene expression level also via ROS. Oxidative radicals and other species generated due to high glucose level are shown to affect indirectly on preventing glucotoxic effects on insulin gene activity [94]. When the cells are treated with the antioxidants like NAC and aminoguanidine, it is shown to markedly affect the glucose toxicity. Along with this, these antioxidants have been shown to partially prevent glucose-induced decreases in insulin mRNA, DNA-binding of pancreatic duodenal homeobox-1, insulin content, and glucose-stimulated insulin secretion [94].

4.2 Neuropathy

Around 10 % of the cases of neuropathy are associated with abnormal sensations and pain. The incidence of neuropathy increases with duration of diabetes and is accelerated by poor control of glucose level. Neurons not only are lost in diabetes, but their ability to regenerate is also impaired, particularly for the small-caliber nerve fibers [4].

The damage of lipids and proteins is more important in case of neurons as they are nondividing cells, and so their DNA is less susceptible to damage, and proteins as well as lipids are majorly responsible for their structural and functional aspects. Modified proteins are unable to perform their axonal transport and signaling and can be correlated with the alterations in brain function [95]. Loss of function in neurons rapidly promotes necrotic or apoptotic mechanisms [96, 97]. Neuropathy is suggested to be very dynamic as the process of degeneration and regeneration of the nerves both takes place simultaneously. With time, the balance between the two gets shifted towards the former, and nerves lose their function. Majorly, the Schwann cells are involved in the regeneration process, and they are reported to get affected in increased oxidative stress due to hyperglycemia. Loss of regeneration is also suggested to include lack of insulin, impairing the decrease in specific types of PKC activity and loosing the growth factor system [4].

4.3 Nephropathy

As discussed earlier oxidative stress has a major role towards the pathogenesis of diabetic nephropathy. However, detailed molecular mechanism remains uncertain [1]. One of the mechanisms suggested involved in nephropathy conditions is increase in advanced glycation end products (AGEs). It was postulated that the chemical pathways leading to advanced glycation end product formation and the renin-angiotensin systems may interact through the generation of free radicals. The renin-angiotensin system is known to block both upstream and downstream pathways leading to tissue injury in case of nephropathy [98]. AGE-dependent pathways may play a role in the development of tubulointerstitial fibrosis in the diabetic kidney. This effect is mediated through RAGE and is TGF-β and CTGF dependent [99]. AGEs and their receptors co-localize in the kidney. It is also shown that AGE binding sites are present in rat renal proximal tubules [100].

AGEs have a cross talk with other pathways also, such as VCAM-1, TGF-β, and CTGF. These molecules might mediate the effects of AGEs and angiotensin II, predominantly on the glomerulus leading to mesangial expansion. Other molecules such as RANTES and MCP-1 mediate the effects of filtered protein predominantly on renal tubules. This ultimately leads to tubulointerstitial fibrosis and inflammation. The enzyme protein kinase C is likely to be the mediator of altered intrarenal albumin processing which contributes to diabetes-related albuminuria, and its effects are exerted on both the glomerulus and the renal tubules. Both extracellular and intracellular proteins are shown to bind to AGEs. Extracellular proteins such as ezrin, radixin, and moesin (ERM) bind AGEs with their amino-terminal domain [101]. It is important to mention here that these ERM proteins function as a link between the cytoplasmic tail of membrane proteins and cytoplasmic actin filaments and also regulate kinases (such as: rho kinase, focal adhesion kinase, and phosphatidylinositol 3-kinase). Another important function of ERM proteins is that they also modulate membrane ion transport proteins. In addition, preliminary evidence suggests that endogenous ligands for RAGE such as the S100/calgranulin polypeptides promote inflammatory reactions in the diabetic kidney through a separate and additional mechanism to that described for AGEs [102]. There are evidences that the processes like increased carbonyl modification of proteins might be involved in diabetic glomerular lesions.

4.4 Retinopathy

As the number of diabetic patients is increasing, patients suffering from diabetic retinopathy are also increasing in number. According to a report published by the WHO, five million people suffer from diabetic retinopathy, and retinopathy is responsible for 5 % of blind people in the world [103].

The complications are rarely seen in early stages of diabetes, but after 20–25 years of diabetes, patients suffer different degrees of retinopathy. There are reports indicating that ROS formation in retinal mitochondria is a direct consequence of hyperglycemia. More recent studies have suggested that increased FFA levels may also result in ROS formation [31].

The mitochondria, being the main site for the generation of ROS, can damage their own DNA, as it remains in the free floating their "naked" condition. Hence, it is the first victim of increased level of ROS in the mitochondria. Mitochondrial DNA is responsible for encoding 13 subunits of ETC, and so damage to the DNA can compromise the functioning of ETC [104].

Although the cells of the retina try to reverse the damage by initiating gene expressions for the formation of repair enzymes, but generally, the enzymes fall short as compared to the multitude oxidant species [105].

The increase in reactive oxygen species and reactive nitrogen species (ROS/RNS) production in the intramitochondrial environment is deleterious to cellular functioning also as molecules like hydrogen peroxide (H_2O_2) and peroxynitrite ($ONOO^-$) may cross mitochondrial membranes and damage macromolecules in other cellular regions as well (Fig. 1) [14].

Additionally, as eyes are the most vascularized tissue in the body and full of capillaries. The capillaries are made up of endothelial cells which maintain the blood-retinal barrier. These cells are destroyed by repeated attacks of high blood glucose that results in gradual loss of sight [106].

4.5 Diabetes and Macrovascular Dysfunction

Two- to threefold increase in the incidences of macrovascular complications is found to be associated with diabetes, and most patients with diabetes develop evidence of microvascular diseases also. Despite current methods of treatment, if the duration of follow-up exceeds 10 years [99], the number of incidences also increases. The role of ROS is reported to be a major cause in this complication. Nishikawa et al. [107] showed that blockade of hyperglycemia-induced ROS production reverses the pathways implicated in diabetic angiopathy in cultured endothelial cells [107].

As depicted in the earlier section, functioning of endothelial cells is affected in high glucose concentration. The endothelial cell lining the arterial vasculature actively balances between thrombosis and fibrinolysis by modulating the relationship between the cellular elements of the blood and the vascular wall. This balance is linked to NO production by eNOS [108]. Any imbalance between the two results in peripheral arterial disease (PAD), a major risk factor for lower-extremity amputation, and is also accompanied by a high possibility for symptomatic cardiovascular and cerebrovascular disease. Hyperglycemia is reported to disturb the activity of eNOS. Diabetes and smoking are the strongest risk factors for PAD. An elevated C-reactive protein (CRP) level has been considered to be a biomarker of PAD. CRP binds to endothelial cell receptors and promotes apoptosis as well as colocalizes with oxidized LDL in atherosclerotic plaques [108]. It also catalyzes the production of procoagulant tissue factor, leukocyte adhesion molecules, and chemotactic substances and inhibits endothelial cell nitric oxide (NO) synthase (eNOS), resulting in abnormalities in the regulation of vascular tone. CRP is also reported to impair fibrinolysis as it increases the local production of compounds, such as plasminogen activator inhibitor (PAI)-1.

In addition to hyperglycemia, insulin resistance also plays a role in the loss of normal NO homeostasis [109] via excess liberation of FFAs which activate protein kinase C (PKC), inhibit phosphatidylinositol (PI)-3 kinase (an eNOS agonist pathway), and also produce reactive oxygen species.

High glucose concentration also disturbs vascular smooth muscle cell (VSMC) function via stimulating pro-atherogenic activity through mechanisms similar to that in endothelial cells and in addition linking RAGE and NF-κB. These events accelerate atherosclerosis and are also associated with plaque destabilization and precipitation of clinical events.

Improper functioning of platelets in diabetic patients are also reported to be involved as there is an increased expression of glycoprotein Ib and IIb/IIIa receptors by platelets, which are important in thrombosis via their role in adhesion and aggregation.

5 Concluding Remarks

The increase in glucose concentration, insulin resistance, release of large number of FFA, and ultimately increased production of ROS and RNS can be explained to be responsible for the onset and post-diabetic complications. The treatment of diabetes is contained by strict glycemic control and balanced lifestyle, but is not shown to be sufficient to treat it. Insulin therapy can only work in case of insulin-dependent diabetes, and it can also lead to insulin resistance later in life if the blood sugar level is not controlled by diet maintenance and regular exercise. Although many drugs targeting metabolic sites in the pathways have been identified and researched upon, crucial advances are still required to look for better and newer sites aiming for reversal of diabetes. Many antioxidants have also been reported to have positive effects, but drugs continue to be the main factor in curbing the disorder. The future aspects can be seen in the advent of therapeutic targets at the mitochondrial metabolic control level, and inflammatory pathways are most likely to diminish the incidence of diabetic complications. Other than this, there are various herbs (*Salacia*, *Syzygium*, and *Momordica*, etc.) which are reported to be antidiabetic. They are reported to contain strong antioxidants and may help to fight the complications. They also are reported to have specific mechanisms which may help to secrete more amount of insulin and may also decrease insulin resistance.

References

1. Pan HZ, Chang D, Feng LG et al (2007) Oxidative damage to DNA and its relationship with diabetic complications. Biomed Environ Sci 20(2):160–163
2. Chung SSM, Ho ECM, Lam KSL et al (2003) Contribution of polyol pathway to diabetes-induced oxidative stress. J Am Soc Nephrol 14(3):s233–s236
3. Maritim AC, Sanders RA, Watkins JB (2003) Diabetes, oxidative stress, and antioxidants: a review. J Biochem Mol Toxicol 17(1):24–38
4. Vincent AM, Russell JW, Low P et al (2004) Oxidative stress in the pathogenesis of diabetic neuropathy. Endocr Rev 25(4):612–628
5. Valko M, Leibfritz D, Moncol J et al (2007) Free radicals and antioxidants in normal physiological functions and human disease. Int J Biochem Cell Biol 39(1):44–84
6. Devasagayam TPA, Tilak JC, Boloor KK et al (2004) Free radicals and antioxidants in human health: current status and future prospects. J Assoc Physicians India 52:794–804
7. American Diabetes Association (2010) Diagnosis and classification of diabetes mellitus. Diabetes Care 33(1):S62–S69
8. Patti ME, Corvera S (2010) The role of mitochondria in pathogenesis of type 2 diabetes. Endocr Rev 31(3):364–395
9. Daneman D (2006) Type 1 diabetes. Lancet 367(9513):847–858
10. Tersey SA, Nishiki Y, Templin AT et al (2012) Islet β-cell endoplasmic reticulum stress precedes the onset of type 1 diabetes in the nonobese diabetic mouse model. Diabetes 61(4):818–827
11. Ceriello A, Motz E (2004) Is oxidative stress the pathogenic mechanism underlying insulin resistance, diabetes, and cardiovascular disease? The common soil hypothesis revisited. Arterioscler Thromb Vasc Biol 24:816–823
12. American Diabetes Association (2004) Hyperglycemic crises in diabetes. Diabetes Care 27(1):s94–s102
13. Said G (2007) Diabetic neuropathy – a review. Nat Clin Pract Neurol 3:331–340
14. Bandeira SM, Fonseca LJS, Guedes GS et al (2013) Oxidative stress as an underlying contributor in the development of chronic complications in diabetes mellitus. Int J Mol Sci 14(2):3265–3284
15. Mooy JM, Grootenhuis PA, Vries H et al (1996) Intra-individual variation of glucose, specific insulin and proinsulin concentrations measured by two oral glucose tolerance tests in a general Caucasian population: the Hoorn Study. Diabetologia 39(3):298–305
16. Lloyd C, Smith J, Weinger K (2005) Stress and diabetes: a review of the links. Diabetes Spectr 18(2):121–127
17. Willis T (1959) Pharmaceutical rational is or an excitation of the operations of medicines in human bodies. In: The works of Thomas Willis Dring. Harper, Leigh, London
18. Sepa A, Frodi A, Ludvigsson J (2005) Mothers' experiences of serious life events increase the risk of diabetes-related autoimmunity in their children. Diabetes Care 28(10):2394–2399
19. Machlin LJ, Bendich A (1987) Free radical tissue damage: protective role of antioxidant nutrients. FASEB J 1(6):441–445

20. Kolluru GK, Bir SC, Kevil CG (2012) Endothelial dysfunction and diabetes: effects on angiogenesis, vascular remodeling, and wound healing. Int J Vasc Med 2012:1–30
21. Padgett LE, Broniowska KA, Hansen PA et al (2013) The role of reactive oxygen species and proinflammatory cytokines in type 1 diabetes pathogenesis. Ann N Y Acad Sci 1281(1):16–35
22. Maechler P, Jornot L, Wollheim CB (1999) Hydrogen peroxide alters mitochondrial activation and insulin secretion in pancreatic beta cells. J Biol Chem 274:27905–27913
23. Song F, Jia W, Yao Y et al (2007) Oxidative stress, antioxidant status and DNA damage in patients with impaired glucose regulation and newly diagnosed type 2 diabetes. Clin Sci (Lond) 112(12):599–606
24. Berlett BS, Stadtman ER (1997) Protein oxidation in aging, disease, and oxidative stress. J Biol Chem 272:20313–20316
25. Naudi A, Jove M, Ayala V et al (2012) Cellular dysfunction in diabetes as maladaptive response to mitochondrial oxidative stress. Exp Diab Res 2012:1–14
26. Thorpe SR, Baynes JW (2003) Maillard reaction products in tissue proteins: new products and new perspectives. Amino Acids 25(3–4):275–281
27. Dean RT, Fu S, Stocker R et al (1997) Biochemistry and pathology of radical-mediated protein oxidation. Biochem J 324(1):1–18
28. Yakes FM, Houten BV (1997) Mitochondrial DNA damage is more extensive and persists longer than nuclear DNA damage in human cells following oxidative stress. Proc Natl Acad Sci U S A 94(2):514–519
29. Maechler P, Kennedy ED, Pozzan T et al (1997) Mitochondrial activation directly triggers the exocytosis of insulin in permeabilized pancreatic beta-cells. EMBO J 16(13):3833–3841
30. Ames BN, Shigenaga MK, Hagen TM (1993) Review: oxidants, antioxidants, and the degenerative diseases of aging. Proc Natl Acad Sci 90:7915–7922
31. Evans JL, Goldfine ID, Maddux BA et al (2003) Are oxidative stress-activated signaling pathways mediators of insulin resistance and beta-cell dysfunction? Diabetes 52(1):1–8
32. Nishida T, Nakagawa S, Manab¢ R (1984) Superoxide dismutase activity in diabetic rat retina. Jpn J Ophthalmol 28:377–382
33. Uttara B, Singh AV, Zamboni P et al (2009) Oxidative stress and neurodegenerative diseases: a review of upstream and downstream antioxidant therapeutic options. Curr Neuropharmacol 7(1):65–74
34. Hagglof B, Marklund SL, Holmgren G (1983) CuZn superoxide dismutase, Mn superoxide dismutase, catalase, and glutathione peroxidase in lymphocytes and erythrocytes in insulin-dependent diabetic children. Acta Endocrinol 102:235–239
35. Cadet J, Loft S, Olinski R et al (2012) Biologically relevant oxidants and terminology, classification and nomenclature of oxidatively generated damage to nucleobases and 2-deoxyribose in nucleic acids. Free Radical Res 46(4):367–381
36. Cnop M, Welsh N, Jonas JC et al (2005) Mechanisms of pancreatic β-cell death in type 1 and type 2 diabetes: many differences, few similarities. Diabetes 54(2):S97–S107
37. Newsholme P, Haber EP, Hirabara SM et al (2007) Diabetes associated cell stress and dysfunction: role of mitochondrial and non-mitochondrial ROS production and activity. J Physiol 583(1):9–24
38. Chambers KT, Unverferth JA, Weber SM et al (2008) The role of nitric oxide and the unfolded protein response in cytokine-induced beta-cell death. Diabetes 57:124–132
39. Oyadomari S, Takeda K, Takiguchi M et al (2001) Nitric oxide-induced apoptosis in pancreatic beta cells is mediated by the endoplasmic reticulum stress pathway. Proc Natl Acad Sci U S A 98:10845–10850
40. Cardozo AK, Ortis F, Storling J et al (2005) Cytokines downregulate the sarcoendoplasmic reticulum pump Ca^{2+} ATPase 2b and deplete endoplasmic reticulum Ca2+, leading to induction of endoplasmic reticulum stress in pancreatic beta-cells. Diabetes 54:452–461
41. Kulkarni RN, Roper MG, Dahlgren G et al (2004) Islet secretory defect in insulin receptor substrate 1 null mice is linked with reduced calcium signaling and expression of sarco (endo) plasmic reticulum Ca2+-ATPase (SERCA)-2b and -3. Diabetes 53:1517–1525
42. Meldolesi J, Pozzan T (1998) The endoplasmic reticulum Ca2+ store: a view from the lumen. Trends Biochem Sci 23:10–14
43. Brownlee M (2001) Biochemistry and molecular cell biology of diabetic complications. Nature 414:813–820
44. Meglasson MD, Matschinsky FM (1986) Pancreatic islet glucose metabolism and regulation of insulin secretion. Diabetes Metab Rev 2:163–214
45. Grodsky GM (2000) Kinetics of insulin secretion: underlying metabolic events in diabetes mellitus. In: Le Roith D, Taylor SI, Olefsky JM (eds) Diabetes mellitus: a fundamental and clinical text. Lippincott Williams & Wilkins, Philadelphia
46. Henry WL (1962) Perspectives in diabetes. J Natl Med Assoc 54(4):476–478
47. Brown GC, Borutaite V (1999) Nitric oxide, cytochrome c and mitochondria. Biochem Soc Symp 66:17–25
48. Wolff SP, Dean RT (1987) Glucose autoxidation and protein modification: the potential role of 'autoxidative glycosylation' in diabetes. Biochem J 245:243–250
49. Paschen W, Mengesdorf T, Althausen S et al (2001) Peroxidative stress selectively down-regulates the neuronal stress response activated under conditions of endoplasmic reticulum dysfunction. J Neurochem 76:1916–1924

50. Kakimoto M, Inoguchi T, Sonta T et al (2002) Accumulation of 8-hydroxy-2′-deoxyguanosine and mitochondrial DNA deletion in kidney of diabetic rats. Diabetes 51:1588–1595
51. Paolisso G, Giugliano D (1996) Oxidative stress and insulin action. Is there a relationship? Diabetologia 39:357–363
52. Wojtczak L, Schonfeld P (1993) Effect of fatty acids on energy coupling processes in mitochondria. Biochim Biophys Acta 1183:41–57
53. Bakker SJ, IJzerman RG, Teerlink T (2000) Cytosolic triglycerides and oxidative stress in central obesity: the missing link between excessive atherosclerosis, endothelial dysfunction, and beta-cell failure? Atherosclerosis 148:17–21
54. Toborek M, Hennig B (1994) Fatty acid-mediated effects on the glutathione redox cycle in cultured endothelial cells. Am J Clin Nutr 59:60–65
55. Hennig B, Meerarani P, Ramadass P et al (2000) Fatty acid-mediated activation of vascular endothelial cells. Metabolism 49:1006–1013
56. McGarry JD (2002) Banting Lecture 2001: dysregulation of fatty acid metabolism in the etiology of type 2 diabetes. Diabetes 51:7–18
57. Boden G (1997) Role of fatty acids in the pathogenesis of insulin resistance and NIDDM. Diabetes 46:3–10
58. Poitout V, Robertson RP (2002) Mini review: secondary beta-cell failure in type 2 diabetes—a convergence of glucotoxicity and lipotoxicity. Endocrinology 143:339–342
59. Harmon JS, Gleason CE, Tanaka Y et al (2001) Antecedent hyperglycemia, not hyperlipidemia, is associated with increased islet triacylglycerol content and decreased insulin gene mRNA level in Zucker diabetic fatty rats. Diabetes 50:2481–2486
60. Anderson ME (1998) Glutathione: an overview of biosynthesis and modulation. Chem Biol Interact 111–112:1–14
61. Iwata-Ichikawa E, Kondo Y, Miyazaki I et al (2002) Glial cells protect neurons against oxidative stress via transcriptional up-regulation of the glutathione synthesis. J Neuro Chem 72:2334–2344
62. Keelan J, Allen NJ, Antcliffe D et al (2001) Quantitative imaging of glutathione in hippocampal neurons and glia in culture using monochlorobimane. J Neurosci Res 66:873–884
63. Lowndes HE, Beiswanger CM, Philbert MA et al (1994) Substrates for neural metabolism of xenobiotics in adult and developing brain. Neurotoxicology 15:61–73
64. Shan X, Jones DP, Hashmi M et al (1993) Selective depletion of mitochondrial glutathione concentrations by (R, S)-3-hydroxy-4-pentenoate potentiates oxidative cell death. Chem Res Toxicol 6:75–81
65. Li L, Shen YM, Yang XS et al (2002) Effects of spiramine T on antioxidant enzymatic activities and nitric oxide production in cerebral ischemia-reperfusion gerbils. Brain Res 944:205–209
66. Kobayashi MS, Han D, Packer L (2000) Antioxidants and herbal extracts protect HT-4 neuronal cells against glutamate-induced cytotoxicity. Free Radic Res 32:115–124
67. Xie C, Lovell MA, Xiong S et al (2001) Expression of glutathione-S- transferase isozyme in the SY5Y neuroblastoma cell line increases resistance to oxidative stress. Free Radic Biol Med 31:73–81
68. Evans JL, Goldfine ID, Maddux BA et al (2002) Oxidative stress and stress-activated signaling pathways: a unifying hypothesis of type 2 diabetes. Endocr Rev 23:599–622
69. Pop-Busui R (2010) DCCT and EDIC studies in type 1 diabetes: lessons for diabetic neuropathy regarding metabolic memory and natural history. Curr Diab Rep 10:276–282
70. Lee AYW, Chung SSM (1999) Contributions of polyol pathway to oxidative stress in diabetic cataract. FASEB J 13:23–30
71. Boucek P (2006) Advanced diabetic neuropathy: a point of no return? Rev Diabet Stud 3(3):143–150
72. Marshall S, Garvey WT, Traxinger RR (1991) New insights into the metabolic regulation of insulin action and insulin resistance: role of glucose and amino acids. FASEB J 5:3031–3036
73. Callaghan BC, Cheng HT, Stables CL et al (2012) Diabetic neuropathy: clinical manifestations and current treatments. Lancet Neurol 11:521–534
74. Edwards JL, Vincent AM, Cheng HT, Feldman EL (2008) Diabetic neuropathy: mechanisms to management. Pharmacol Ther 120:1–34
75. Bierhaus A, Schiekofer S, Schwaninger M et al (2001) Diabetes-associated sustained activation of the transcription factor nuclear factor-κB. Diabetes 50:2792–2808
76. Mohamed AK, Bierhaus A, Schiekofer S et al (1999) The role of oxidative stress and NF-κB activation in late diabetic complications. Biofactors 10:157–167
77. Griffin ME, Marcucci MJ, Cline GW et al (1999) Free fatty acid-induced insulin resistance is associated with activation of protein kinase C θ and alterations in the insulin signaling cascade. Diabetes 48:1270–1274
78. Coudronniere N, Villalba M, Englund N et al (2000) NF-kappa B activation induced by T cell receptor/CD28 co-stimulation is mediated by protein kinase C-theta. Proc Natl Acad Sci U S A 97:3394–3399
79. Hosseini A, Abdollahi M (2013) Diabetic neuropathy and oxidative stress: therapeutic perspectives. Oxid Med Cell Longev 168039:1–15
80. Zheng H, Whitman SA, Wu W (2011) Therapeutic potential of Nrf2 activators in streptozotocin-induced diabetic nephropathy. Diabetes 60(11): 3055–3066
81. Palsamy P, Subramanian S (2011) Resveratrol protects diabetic kidney by attenuating hyperglycemia-mediated oxidative stress and renal inflammatory cytokines via Nrf2-Keap1 signaling. Biochim Biophys Acta 7:719–731

82. Ho FM, Liu SH, Liau CS et al (2000) High glucose-induced apoptosis in human endothelial cells is mediated by sequential activations of c-Jun NH(2)-terminal kinase and caspase-3. Circulation 101:2618–2624
83. Natarajan R, Scott S, Bai W et al (1999) Angiotensin II signaling in vascular smooth muscle cells under high glucose conditions. Hypertension 33:378–384
84. Purves T, Middlemas A, Agthong S et al (2001) A role for mitogen-activated protein kinases in the etiology of diabetic neuropathy. FASEB J 15:2508–2514
85. Geraldes P, King GL (2010) Activation of protein kinase C isoforms and its impact on diabetic complications. Circ Res 106:1319–1331
86. Cameron NE, Cotter MA (2002) Effects of protein kinase C beta inhibition on neurovascular dysfunction in diabetic rats: interaction with oxidative stress and essential fatty acid dysmetabolism. Diabetes Metab Res Rev 18:315–323
87. Cotter MA, Jack AM, Cameron NE (2002) Effects of the protein kinase C beta inhibitor LY333531 on neural and vascular function in rats with streptozotocin-induced diabetes. Clin Sci (Lond) 103:311–321
88. Rajbhandari SM, Piya MK (2005) A brief review on the pathogenesis of human diabetic neuropathy: observations and postulations. Int J Diab Metab 13:135–140
89. Busui RP, Marinescu V, Huysen VC (2002) Dissection of metabolic, vascular, and nerve conduction interrelationships in experimental diabetic neuropathy by cyclooxygenase inhibition and acetyl-L-carnitine administration. Diabetes 51:2619–2628
90. Robertson RP, Harmon JS, Tanaka Y et al (2000) Glucose toxicity of the β-cell: cellular and molecular mechanisms. In: Le Roith D, Taylor SI, Olefsky JM (eds) Diabetes mellitus: a fundamental and clinical text. Lippincott Williams & Wilkins, Philadelphia
91. Boden G, Ruiz J, Kim CJ et al (1996) Effects of prolonged glucose infusion on insulin secretion, clearance, and action in normal subjects. Am J Physiol 270:E251–E258
92. Robertson RP, Zhang HJ, Pyzdrowski KL et al (1992) Preservation of insulin mRNA levels and insulin secretion in HIT cells by avoidance of chronic exposure to high glucose concentrations. J Clin Invest 90:320–325
93. Poitout V, Olson LK, Robertson RP (1996) Chronic exposure of beta TC-6 cells to supraphysiologic concentrations of glucose decreases binding of the RIPE3b1 insulin gene transcription activator. J Clin Invest 97:1041–1046
94. Tanaka Y, Gleason CE, Tran PO et al (1999) Prevention of glucose toxicity in HIT-T15 cells and Zucker diabetic fatty rats by antioxidants. Proc Natl Acad Sci U S A 96:10857–10862
95. Butterfield DA, Koppal T, Howard B et al (1998) Structural and functional changes in proteins induced by free radical-mediated oxidative stress and protective action of the antioxidants N-tert-butyl-α -phenylnitrone and vitamin E. Ann N Y Acad Sci 854:448–462
96. Deng G, Su JH, Ivins KJ et al (1999) Bcl-2 facilitates recovery from DNA damage after oxidative stress. Exp Neurol 159:309–318
97. Aksenova MV, Aksenov MY, Payne RM et al (1999) Oxidation of cytosolic proteins and expression of creatine kinase BB in frontal lobe in different neurodegenerative disorders. Dement Geriatr Cogn Disord 10:158–165
98. Kobori H, Nangaku M, Navar LG et al (2007) The intrarenal renin-angiotensin system: from physiology to the pathobiology of hypertension and kidney disease. Pharmacol Rev 59(3):251–287
99. Jerums G, Panagiotopoulos S, Forbes J et al (2003) Evolving concepts in advanced glycation, diabetic nephropathy, and diabetic vascular disease. Arch Biochem Biophys 419:55–62
100. Youssef S, Nguyen DT, Soulis T et al (1999) Aminoguanidine ameliorates changes in the IGF system in experimental diabetic nephropathy. Kidney Int 55:907–916
101. McRobert EA, Gallicchio M, Jerums G et al (2003) The Amino-terminal domains of the Ezrin, Radixin, and Moesin (ERM) proteins bind advanced glycation end products, an interaction that may play a role in the development of diabetic complications. J Biol Chem 278:25783–25789
102. Yamamoto Y, Kato I, Doi T et al (2001) Development and prevention of advanced diabetic nephropathy in RAGE-over expressing mice. J Clin Invest 108:261–268
103. Gilbert C, Ackland P, Resnikoff S, Gilbert S et al (2007) International. Vision 2020 The right to sight, global initiative for the elimination of avoidable blindness. Action plan 2006–2011 World Health Organization
104. Indo HP, Davidson M, Yen HC et al (2007) Evidence of ROS generation by mitochondria in cells with impaired electron transport chain and mitochondrial DNA damage. Mitochondrion 7:106–118
105. Madsen-Bouterse SA, Zhong Q, Mohammad G et al (2010) Oxidative damage of mitochondrial DNA in diabetes and its protection by manganese superoxide dismutase. Free Radic Res 44(3):313–321
106. Hammes HP, Feng P, Fister et al (2011) Diabetic retinopathy: targeting vasoregression. Diabetes 60(1):9–16
107. Nishikawa T, Edelstein D, Du XL et al (2000) Normalizing mitochondrial superoxide production blocks three pathways of hyperglycaemic damage. Nature 404:787–790
108. Simionescu M, Simionescu N (2006) Implications of early structural-functional changes in the endothelium for vascular disease. Arterioscler Thromb Vasc Biol 27(2):266–274
109. Steinberg HO, Baron AD (2002) Vascular function, insulin resistance and fatty acids. Diabetologia 45:623–634

Oxidative Stress and Inflammation in Cardiovascular Diseases: Two Sides of the Same Coin

Monisha Dhiman, Shweta Thakur, Shishir Upadhyay, Amandeep Kaur, and Anil K. Mantha

Abstract

Globally, the major cause of long-term disability and death is an "epidemiologic transition" from infectious diseases and malnutrition complications to non-communicable chronic diseases like cardiovascular disease (CVD), cancer and diabetes. CVD accounts for major global mortality. Imbalance due to the generation of reactive oxygen species (ROS) levels above normal baseline levels and decreased antioxidant defence reserve makes the cardiovascular system (cardiac and vascular cells) susceptible to oxidative stress and damage. Growing evidences support the notion that oxidative stress plays a crucial role in the development and progression of CVD by altering normal functions such as inactivation of nitric oxide (NO) leading to endothelial dysfunction, intracellular Ca^{2+} overload and others. Oxidative stress also mediates inflammation through various signalling cascades such as the activation of inflammatory transcription factors (TFs) namely NF-κB, AP-1 and Nrf-1. A vicious cycle of oxidative stress-mediated inflammation and inflammation-induced oxidative stress makes the CVD-related complications worse. Therefore, it is also very important to clearly understand the role of enzymatic sources of ROS,

M. Dhiman (✉)
Centre for Genetic Diseases and Molecular Medicine, School of Emerging Life Science Technologies, Central University of Punjab, Bathinda 151 001, Punjab, India
e-mail: monisha.dhiman@gmail.com; monisha.dhiman@cup.ac.in

S. Thakur • S. Upadhyay • A. Kaur
Centre for Biosciences, School of Basic and Applied Sciences, Central University of Punjab, Bathinda 151 001, Punjab, India

A.K. Mantha
Centre for Biosciences, School of Basic and Applied Sciences, Central University of Punjab, Bathinda 151 001, Punjab, India

Department of Biochemistry and Molecular Biology, University of Texas Medical Branch, Galveston, TX 77555, USA

V. Rani and U.C.S. Yadav (eds.), *Free Radicals in Human Health and Disease*, DOI 10.1007/978-81-322-2035-0_17, © Springer India 2015

mechanisms underlying pathological conditions and link between oxidative stress and inflammation during each stage of CVD. The present chapter will elucidate the role of oxidative stress and inflammation in CVD development and progression. It is important to find the remedial measures, to develop the efficient biomarkers and to design the therapeutic strategies for CVD in the near future.

Keywords
Cardiovascular diseases • Reactive oxygen and nitrogen species (ROS/RNS) • Inflammation • Oxidative stress

1 Introduction

Cardiovascular diseases (CVDs) are disorders involving the heart and the blood vessels or both. CVD includes acute coronary syndrome, angina, arrhythmia, cardiomyopathy, congenital heart disease, coronary heart disease, heart failure, inflammatory heart disease, ischemic heart disease, rheumatic heart disease and valvular disease, brain-related cardiovascular diseases (such as cerebrovascular disease), haemorrhagic stroke/ischemic stroke and peripheral circulatory system-related cardiovascular diseases [deep vein thrombosis, hypertensive heart disease, peripheral artery disease (PAD) and pulmonary embolism] [1]. More than 80 % of the deaths from CVD occur in low- and middle-income countries, and the economic burden due to CVD remains high in developing countries [2]. Many modifiable (environment, diet and exercise) and non-modifiable (genetic, gender, age, early menopause and ethnic group) causal risk factors are involved in CVD [3]. Current preventive and treatment strategies are inadequate to prevent and cure CVD. This raises an urgent need for effective therapeutic strategies which in turn require extensive fundamental understanding of key processes involved in the development of CVD.

The human heart is an obligate aerobic organ with very high metabolic energy demand and relatively lower levels of antioxidant defence as compared to other organs. The metabolic energy generation involves the use of oxygen thereby resulting in significant production of reactive oxygen species (ROS) which in turn is implicated in processes affecting cardiac function. Atherosclerosis and hypertension are the most common causes of CVD. Atherosclerosis is the thickening of the arterial wall due to Ca^{2+} and cholesterol, which reduces the elasticity of arterial wall, reduces the blood flow and thus increases blood pressure. It is due to chronic inflammatory response elicited by the accumulation and rupture of white blood cells (T cells and macrophages) in response to cholesterol-carrying oxidised low-density lipoprotein (oxLDL) molecules, which prompts endothelial cells to produce adhesive substances that snag monocyte/macrophage precursor cells from the blood [4].

The conventional view of CVD as altered lipid storage is now changing, as the role of key mechanistic pathways of oxidative stress and inflammation in the initiation, development and progression of CVD is becoming clearer. Studies advocate the involvement of oxidative stress and inflammation in endothelial dysfunction and atherosclerosis. Blood vessels consist of three main layers—the outer layer is the connective tissue and provides structure to the layers beneath, the middle layer is the smooth muscle and controls the blood flow, and the inner lining is the thin layer of endothelial cells. The underlying cause of all vascular diseases is the dysfunction of endothelial cells that occur much before the appearance of clinical symptoms. Endothelial cells prevent entry of harmful blood-borne substances into the smooth muscle of the blood vessel. The oxidative damage to the inner layer can result into the modulation of endothelial permeability,

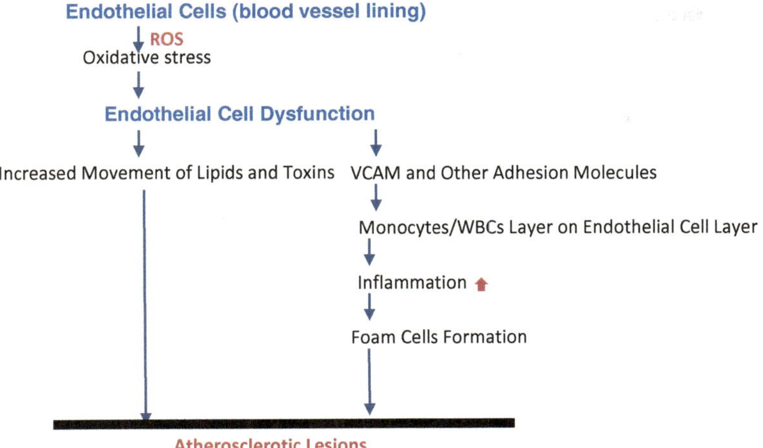

Fig. 1 Effect of oxidative stress on endothelial cell functions. Oxidative stress makes the endothelial cells dysfunctional by altering their permeability, thus allowing the entry and movement of lipids and toxins across the endothelial layer and smooth muscle cells into sub-endothelial space. It also stimulates the expression of vascular cell adhesion molecule (VCAM) and other adhesion molecules, which leads to the recruitment of monocyte/WBC layer on endothelial layer that leads to inflammation-induced foam cells and atherosclerotic lesion formation

thus allowing the movement of lipids and toxins across the endothelial layer and smooth muscle cells into the sub-endothelial space [5]. The normal endothelium resists adhesion by white blood cells. However, exposure to risk factors may result in the expression of vascular cell adhesion molecule (VCAM-1) on the surface of endothelial cells. These adhesion molecules mediate adhesion of monocytes to endothelial cells [6]. These monocytes then mature into macrophages which then form the cholesterol-laden foam cells in arterial wall forming "atheroma" plaques. Exploration of the role of inflammation and oxidative stress in CVD will not only increase our understanding of diseases but also have applications in risk stratification and the development of targeted therapeutics to control these diseases [7] (Fig. 1).

2 Pathophysiological Agents Associated with CVD

2.1 Role of ROS in CVD

ROS are highly reactive free radicals derived from molecular oxygen (O_2) that can readily oxidise other molecules. ROS have vital role in normal physiological cellular signalling pathways in various cells of the cardiovascular system (CVS) and other systems. However, pathological levels of ROS can alter structure and vital functions of cellular proteins, lipids and nucleic acids. The site and extent of ROS production have important consequences and thus determine the ultimate cell/tissue fate (Fig. 2). ROS can be formed in the heart, vascular tissue, splenocytes and blood leukocytes through the action of specific oxidases and oxygenases (xanthine oxidase, NADPH oxidase and NOX) and peroxidases (myeloperoxidase); through the Fenton reaction; and as by-products of the electron transport chain (ETC) of the mitochondria [8]. Further, cyclooxygenase, lipoxygenase and cytochrome P-450 enzymes produce ROS during arachidonic acid metabolism [9]. Nitric oxide (NO) is produced by the enzymatic activity of nitric oxide synthases (NOSs), which oxidises L-arginine, transferring electrons from NADPH [10]. The endothelium has been identified as a major source of ROS in the human blood vessels, and endothelial function is closely linked to the homoeostasis of ROS formation within the vascular wall [11]. The human system has enzymatic and nonenzymatic systems to get rid of increased ROS. Imbalance

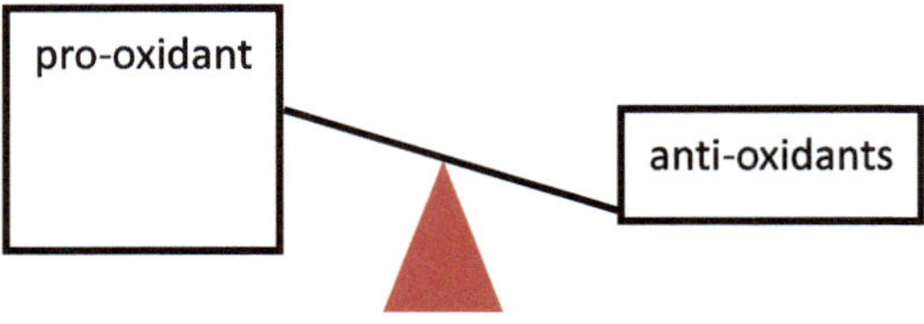

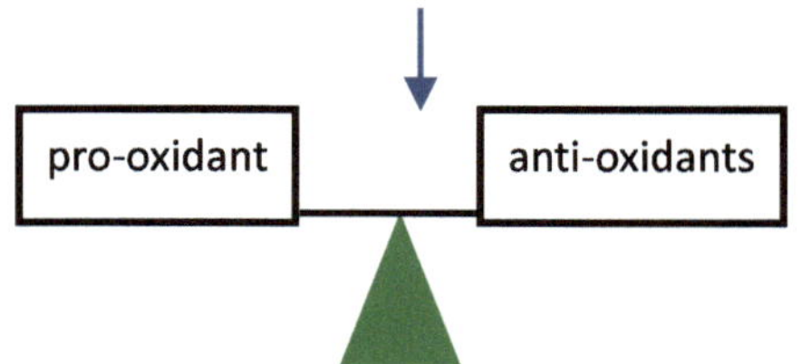

Fig. 2 (**a**) Pathophysiological ROS levels result as a consequence of imbalance between pro-oxidants and antioxidants; and (**b**) where as physiological ROS levels cause/result in balance between pro-oxidants and antioxidants

between ROS production and antioxidant defence of the cell can result in oxidative stress which in turn may cause tissue injury and diseased condition [12], which is summarised in Table 1.

2.2 Role of Mitochondrial Electron Transport Chain in CVD

The mitochondrial electron transport chain is the major endogenous source of ROS. In the mitochondria, the partial reduction of O_2 occurs as a result of leakage of electrons from the ETC, contributing one, two or three electrons to form $O_2^{\cdot}$, H_2O_2 or HO·, respectively. As much as 2–4 % of the reducing equivalents escape the respiratory chain, leading to $O_2^{\cdot-}$ formation. $O_2^{\cdot-}$ is dismutated by manganese superoxide dismutase (MnSOD) to H_2O_2 that may then be converted to highly reactive and deleterious HO· radicals. Generally, the leakage of electrons at CI flavoprotein generates $O_2^{\cdot-}$ in mitochondrial matrix while CIII ubisemiquinones (UQ–) generated at Q1 (UQ−1) and Q*o* (UQ−0) sites release $O_2^{\cdot-}$ in the matrix and intermembrane space of the mitochondria, respectively. Superoxide anions can give rise to other reactive species such as peroxynitrite ($ONOO^-$), H_2O_2, OH^- radicals and hypochlorous acid (HOCl). Increased $O_2^{\cdot-}$ levels are the result of electron loss from ETC or diminished activity of mitochondrial ROS scavengers such as MnSOD [8]. Monoamine oxidase (MAO) present in outer mitochondrial membrane contributes to oxidative stress via generation of NO or H_2O_2, thus contributing to oxidative stress [34]. Altered mitochondrial DNA as result of oxidative stress affects vital mitochondrial functions, leading to the production of more ROS [35].

The heart is highly dependent on the mitochondria for the energy required for its contractile and other metabolic activities. The mitochondria represent 30 % of the total volume of cardiomyocytes and provide >90 % of the cellular ATP energy through oxidative phosphorylation [36]. It is now documented that myopathic heart sustains mitochondrial dysfunction at gene, protein and biochemical levels. Global microarray profiling of gene expression has identified alterations in several of the mitochondrial function-related transcripts in the myocardial biopsies of humans [37] and experimental animals [38]. Further studies documented a decline in the activities of respiratory complexes and NADH-ubiquinone reductase (CI), ubiquinol-cytochrome c reductase (CIII) and ATP synthase (CV) complexes in diseased hearts [39].

A study showed that MnSOD-deficient mice developed progressive congestive heart failure with specific molecular defects in mitochondrial respiration. Studies have revealed the role of peroxiredoxin-3, a mitochondrial H_2O_2 scavenger, in the prevention of heart failure after experimental myocardial infarction (MI) in mice [40]. Numerous studies have shown that physiological stimuli such as vasoactive agents angiotensin II [41], epidermal growth factor (EGF), transforming growth factor-β (TGF-β) and tumour necrosis factor-α (TNF-α) that are involved in pathogenesis of vascular diseases can lead to the production of mitochondrial ROS. Studies have also shown

Table 1 Major sources of ROS and their respective functions in different cell types of CVS

Cell type	Major source of ROS	Functions	References
Cardiomyocytes	*Mitochondrial enzymes*		
	1. P66shc	Catalyses electron transfer from cytochrome c to oxygen	[13–16]
	2. Nicotinamide adenine dinucleotide phosphate oxidases 2 and 4 (NOX2 and 4) isoforms	Associates with subunit p22^{phox} for its activation upon stimulation by G protein-coupled receptor (GPCR) agonists and TGF-β, generates superoxide and H_2O_2	[17–21]
	3. Monoamine oxidases (MAOs)	MAO metabolises serotonin, releases H_2O_2	[22]
Endothelial cells	1. Nicotinamide adenine dinucleotide phosphate oxidase 4 (Nox4)	Constitutively active at low level, levels increase in response to pressure overload, generates H_2O_2, triggers further ROS generation	[19]
	2. Endothelial nitric oxide synthase (eNOS)	Generates NO, which affects cell functions by post-translational modification of effector proteins or by stimulating guanylate cyclase	[23–25]
	3. Heme oxygenase I (HO-1)	Upregulation of HO-1 in the endothelium protects it against inflammation via enzymatic degradation of the pro-oxidant and pro-inflammatory molecule heme and via the generation of its anti-inflammatory products bilirubin and CO	[26]
Vascular smooth muscle cells (VSMCs) and fibroblast cells	1. NADP(H) oxidase	Produce $O_2^{-\cdot}$, leads to the production of peroxynitrite, involved in the growth response of VSMCs and fibroblasts, VSMC migration and cell apoptosis	[27–30]
	2. Mitochondria	ANG II-stimulated mitochondrial reactive oxygen species production in rat cardiac fibroblasts is accompanied by a reduction in the expression of the mitochondrial antioxidant peroxiredoxin-3 (Prx-3)	[31]
Infiltrating immune cells	1. Myeloperoxidase	Neutrophils and monocytes secrete myeloperoxidases and NOXs which initiates lipid peroxidation via tyrosyl radical and nitrogen dioxide	[32, 33]
	2. NADP(H) oxidase	NOXs which initiate lipid peroxidation via tyrosyl radical and nitrogen dioxide	

that mitochondrial ROS in hypoxia induces signalling which in turn is closely associated with inflammation. Mitochondrial H_2O_2 is linked with flow-mediated dilatation in human coronary resistance arteries [42].

In our recent study, in which we utilised genetically modified mice, we found that chagasic MnSOD transgenic mice equipped with a variable capacity to scavenge mitochondrial and cellular ROS had low inflammatory infiltrate and showed significant myocardial remodelling when compared to the chronic chagasic wild-type mice [43]. Collectively, it is indicative that the mitochondria are an important source of ROS that has implications for the cardiovascular system (CVS).

2.3 Role of Oxidases and Oxygenases in CVD

Numerous oxidases and oxygenases expressed in different cell types and locations within the cell contribute to the formation of ROS. By definition, oxidases reduce O_2, whereas oxygenases (oxidoreductases) transfer O_2 to substrates. ROS produced by activated phagocytes such as macrophages and neutrophils from NADPH oxidase and/or by myeloperoxidase activity is termed as "oxidative burst" [44, 45]. This ROS production is critical to antimicrobial function, contributing either directly or indirectly to the killing of intracellular organisms. NADPH oxidase, produced by many types of phagocytes, reduces O_2 to $O_2^{\cdot-}$ [46]. Subsequently, $O_2^{\cdot-}$ and HOCl further can react to form HO· [47].

2.3.1 NADPH Oxidase (NOX)

The NADPH oxidase (nicotinamide adenine dinucleotide phosphate oxidase, NOX) is a membrane-bound enzymatic complex that remains present as a transmembrane complex and transfers electrons to oxygen across the biomembrane that converts into superoxide anion, referred as respiratory burst, that serves as the first line of host defence against microbes [48]. The prototypic NOX (gp91phox), renamed as NOX2, was first identified in phagocytes (neutrophils, macrophages). Presently, seven mammalian NOX homologues have been identified, namely, NOX1–NOX5 and dual oxidases 1 and 2 (DUOX1 and DUOX2). In the cardiovascular system, NOX1, NOX2, NOX4 and NOX5 have been identified. NOX1 is expressed mainly in vascular smooth muscle cells (VSMCs). NOX2 and NOX4 are expressed in endothelial cells, cardiomyocytes, fibroblasts and VSMC [49]. NOX5 has been reported in human endothelial cells and smooth muscle cells but is absent in rodents [50]. It has been found that some NOX family members preferred NADPH or NADH, but some can't discriminate between NADPH and NADH. NOX2 generally remains present in phagocytic cells such as neutrophils and macrophages [48] and helps vascular endothelial growth factor (VEGF) to induce angiogenesis mediating through NOX3 [51]. NOX4 protects the vasculature against inflammatory stress [52]. NOX-derived ROS plays a physiological role in the regulation of endothelial function and vascular tone and a pathophysiological role in endothelial dysfunction, inflammation, hypertrophy, apoptosis, migration, fibrosis, angiogenesis, rarefaction, important processes underlying cardiovascular and renal remodelling in hypertension and diabetes [53, 54]. It has been reviewed that monocyte/macrophage extravasation through NOX into the vessel wall is a critical step in the development of atherosclerosis. Upon activation, NOX complex of monocytes produces a burst of superoxide anion. This superoxide anion develops oxidative stress at the inflammatory sites. ROS thus generated activate an enzyme that makes the macrophages adhere to the arterial wall by polymerising actin fibres [54].

In a study by Liu and co-workers, antioxidant as a peptide inhibitor of NOX has been shown to reduce blood pressure and forestall macrophage accumulation in rats during angiotensin II infusion [55]. Similarly, work on chagasic cardiomyopathy showed that NOX-dependent ROS is a critical regulator of the splenic response (phagocytes, T cells and cytokines) which effects the heart-infiltrating phagocytes and $CD8^+$ T cells resulting in cardiac remodelling [56].

2.3.2 Xanthine Oxidase

Both xanthine oxidase and xanthine dehydrogenase, derived from xanthine oxidoreductase (XOR), produce H_2O_2 and $O_2^{\cdot-}$ while metabolising hypoxanthine and xanthine to uric acid [57]. ROS due to XOR has been implicated in endothelial dysfunction, atherosclerosis, hypertension and heart failure. Further inhibitors of xanthine oxidase such as febuxostat, allopurinol and oxypurinol showed diminished ROS, improved contractile function and myocardial efficiency [58]. Understanding of the mechanism(s) of action of XOR in CVD development can lead to the development of therapies targeting XOR. Enhanced xanthine oxidase has been shown to be associated with vascular dysfunction in animal models of hypercholesterolemia [59]. Overexpression of renin and angiotensinogen has been linked with increased xanthine oxidase activity in endothelial malfunction in transgenic rats [60].

2.3.3 Lipoxygenase

Lipoxygenase (LO) is an important enzyme involved in the conversion of arachidonic acid to leukotrienes which are pro-inflammatory lipid mediators [61]. The role of leukotrienes (LTs) as mediators in asthma is well known [62]; consequently, leukotriene inhibitors are used for treatment of asthma. Now, researchers are trying to elucidate the role of 5-lipoxygenase and leukotrienes in other chronic inflammatory diseases including atherosclerosis [63]. 5-LO is abundant in monocytes/macrophages, dendritic cells, mast cells and neutrophilic granulocytes of 5-LO^{+} cells markedly increased in advanced lesions.

Lipoxygenase or cyclooxygenase metabolised products such as eicosanoids and LTs are associated with several pathogenesis in human beings such as cancer, CVD, asthma and others [62, 64]. LTs play important role in immunity as well as inflammation [63]. Lipoxygenase induces/attracts various leukocytes including macrophages, circulatory monocytes, mast cells and foam cells. But lipoxygenase remains non-functional until 5-lipoxygenase-activating protein (FLAP) is not associated with it. Lipoxygenase products like leukotriene B4 (LTB4) is the most powerful inflammatory product, while the products of 12/15-LOs have both pro-inflammatory and anti-inflammatory responses. This inflammatory response promotes atherosclerosis, abdominal aortic aneurysm and myocardial infarction/reperfusion injury via increased leukocyte chemotaxis and vascular inflammation, enhanced permeability and subsequent tissue/matrix degeneration. Some recent studies have shown that LTB4 is a signal-relay molecule secreted by neutrophils [61] which promotes atherosclerotic initiation by inflammation through various mechanism(s) including the release of pro-inflammatory cytokines IL-6 and TNF-α.

The expression of 5-LO by activated macrophages in symptomatic plaques leads to LTB4 accumulation and enhanced synthesis and release of matrix metalloproteinases (MMPs) that can promote plaque rupture [65]. From studies on animal models, it has been observed that 12/15-LOs play a crucial role during late-phase inflammation and atherosclerosis by fixing the interactions between monocytes and endothelial tissues in vivo [66]. It has been found that 15-LOs in monocytes generate superoxide that leads to the oxidation of LDL [67]. This 15-LO protein has been observed to be localised in atherosclerotic lesions in rabbit and humans [68, 69].

2.3.4 Heme Oxygenase-1

Heme oxygenase-1 (HO-1) catalyses the oxidation of heme to generate carbon monoxide, biliverdin and iron. These reaction products of HO-1 have potent anti-inflammatory and anti-oxidative functions. Although HO-1 is expressed at low levels in most tissues under normal basal conditions, it is highly inducible in response to various pathophysiological stresses. The role of HO-1 in inflammation and several CVDs such as atherosclerosis, myocardial infarction, graft survival after heart transplantation and abdominal aortic aneurysm has been reported. HO-1 is emerging as a great potential therapeutic target for treating CVD [70]. Experimental evidence from various cell culture and animal models suggests an association of HO-1 with the complex sequence of events that cause atherosclerosis [26]. It has been demonstrated that HO-1-deficient mice develop cardiac abnormalities, thus suggesting its role in CVD [71, 72].

2.3.5 Myeloperoxidase

Myeloperoxidase (MPO) is a heme peroxidase enzyme abundantly expressed in monocytes, macrophages and activated neutrophils. It utilises co-substrate to generate other ROS/RNS [73]. MPO oxidises tyrosine to tyrosyl radical and also produces HOCl during phagocytosis from H_2O_2 and Cl^- by neutrophil respiratory bursts to kill bacteria and infectious pathogen [74]. Therefore, MPO is an enzyme that plays an important role in innate immune system. The oxidants produced by MPO are also associated with CVD in the coronary circulation or in peripheral arterial vasculature. High plasma MPO is reported to be a risk factor for early adverse cardiac events in patients with chest pain [73].

In the vascular system, MPO remains present in endothelial cells attached with cytokeratin 1 [75]. It has been elucidated that MPO-generated oxidants are one of the major causes of vascular damage during inflammation [76] and also act as biomarker of vasculitis [77], cardiac dysfunction and left vascular ejection fraction (LVEF) and a risk marker in acute coronary syndrome [78]. MPO functions as a survival signal for neutrophils and thereby contributes to prolonged inflammation [79]. The resultant inflammatory response induced by MPO leads to endothelial dysfunction, atheroma initiation and propagation, subsequent complications of plaque rupture, thrombosis and ventricular remodelling [80] which makes MPO a major player contributing to CVD.

Elevated plasma MPO levels lead to inflammatory diseases like dermatitis herpetiformis, systemic inflammatory response syndrome and anti-neutrophil cytoplasmic autoantibody (ANCA)-mediated glomerulonephritis [76]. MPO suppresses the vascular dilation and promotes smooth muscle cell proliferation, which is a major reason of endothelial dysfunction [76]. Chronic inflammatory process leads to oxidative damage of the arterial wall and its subsequent outcome is atherosclerosis [81]. The foam cells (hallmark of atherosclerosis) formed by the accumulation of cholesterol and lipids during atherosclerosis arise from oxidised LDL by MPO [40]. High levels of oxLDL are associated with increased risk of future myocardial infarction [82].

MPO, a major granule enzyme in neutrophils, accounts for 5 % of the total neutrophil proteins and is responsible for the production of oxidant HOCl [61]. The release of ROS and HOCl by neutrophils may cause damage to important biological structures, such as proteins carbohydrates, lipids and nucleic acids, and may enhance inflammatory responses. Dityrosine-containing protein cross-linking products, designated as advanced oxidation protein products (AOPPs), are formed by HOCl-induced chlorination of amines and constitute an excellent marker of MPO activation. AOPPs are found in the extracellular matrix of human atherosclerotic plaques, and increased levels of AOPP have been described as an independent risk factor for coronary artery disease [57] and in several infectious inflammatory diseases [83].

2.4 Role of Nitric Oxide in CVD

During aerobic respiration, ROS are generally produced, while in hypoxic condition, NO are produced and that in turn form RNS. Some more reactive species have been noticed such as reactive aldehydes, i.e. malondialdehyde (MDA) product of lipid peroxidation [84]. NO synthase is an enzyme that helps in the synthesis of NO from L-arginine in various types of cells and tissues. In mammals, there are three distinct types of isozymes of NOS, neuronal (nNOS or NOS-1), inducible (iNOS or NOS-2) and endothelial (eNOS or NOS-3) [84]. iNOS and nNOS are soluble and found predominantly in the cytosol, while eNOS is membrane associated. eNOS and nNOS are constitutively expressed while the expression of iNOS is activated during infection [85].

NO is a soluble gas that delivers signalling as paracrine hormones in vasorelaxation, vascular haemostasis, neurotransmission and cytotoxicity [86]. It protects blood vessel from injurious consequence of platelets and cells circulating in the blood. The NOS catalyses an NADPH- and $O_2^{\cdot-}$-dependent oxidation of arginine to generate NO and citrulline, with the formation of N-hydroxyarginine (NOHA) as an intermediate [87]. iNOSs in the macrophages are different from the others, as it is Ca^{2+} independent. Different NOSs synthesise NO in response to different stimuli such as eNOS synthesising NO in a vascular endothelial cell in response to acetylcholine, nNOS synthesising NO in a neuron in response to glutamate, and iNOS synthesising NO in macrophages following its induction by IFN-γ [88].

In the endothelium vessels, some major functions such as blood pressure, platelet aggregation, leukocyte adherence and vascular smooth muscle cell mitogenesis are regulated by eNOS. eNOS is acutely activated by agonists of diverse G protein-coupled cell surface receptors and by physical stimuli such as haemodynamic shear stress and varying oxygenation. Mutational studies have proved that unlike other NOS isoforms, eNOS shows N-myristoylation which targets its localisation to the plasma membrane [89]. Diminished NO availability contributes to systemic and pulmonary hypertension, atherosclerosis and airway dysfunction [90]. It has been seen that C-reactive protein (CRP), a prototypic marker of inflammation, decreases the eNOS level in endothelial tissues leading to CVD and atherogenesis [91].

From pharmacological studies in cultured cells, it has been observed that increased cAMP can exert opposite effects on the endotoxin- or cytokine-induced expression of NOS-2, being either stimulatory or inhibitory in macrophages; stimulatory in adipocytes, smooth muscle, skeletal muscle and brain endothelial cells; and inhibitory in pancreatic, liver and brain glial cells. The regulation of NOS-2 gene transcription appears to be the primary mechanism of action of cAMP and, whether it is stimulatory or inhibitory, hinges on the cell-specific regulation of transcription factors (TFs) including CREB, NF-κB and C/EBP. cAMP must therefore be considered a modulator rather than a suppressor of NOS-2 expression [92].

2.5 Role of Fenton Reaction in CVD

Besides oxidases and oxygenases, the "Fenton reaction" is another mechanism of ROS formation which forms the basis for CVD. The reaction results in the Fe^{+2}- or Cu^{+}-mediated conversion of H_2O_2 to HO· [41]. The relationship between iron and CVD was proposed in 1981 by Jerome Sullivan. Since then, numerous epidemiologic studies have been conducted to test this hypothesis. Increased iron levels in the body after menopause in women and adolescence in men are associated with the development of atherosclerosis and ischemia [93]. Further, reduction of iron in the body via phlebotomy may be used in the treatment of CVD. The administration of deferoxamine, a potent iron chelator, resulted in a decrease in myocyte necrosis in a random study of ischemia/reperfusion in dogs, thus indicating the role of Fenton reaction [94].

Study on the association between oxidative stress markers and iron nutrition status in humans revealed significantly higher concentrations of serum ferritin than control group. Also these subjects showed significantly lower levels of the transferrin receptor than control group. Further, significantly higher levels of oxidative stress markers including heme oxygenase activity, oxLDL and thiobarbituric acid reactive substances were reported in individuals with metabolic syndrome than in the control group. DePalma and group found a positive correlation between ferritin levels, inflammatory biomarker interleukin-6, CRP levels and mortality in patients with symptomatic PAD. This study suggested the role of iron-induced oxidative stress in the initiation and development of inflammation in PAD patients. Further, statins were found to suppress ferritin levels which were in turn associated with improved clinical outcomes [95]. On the other hand, the increased intake of vegetables, fruits, tea and coffee is associated with lower levels of oxidative stress.

3 Role of Oxidative Stress and Inflammation: Development and Progression of CVD

3.1 Oxidative Stress in CVD

Various studies have demonstrated the role of oxidative stress in the development and progression of CVD [96, 97]. It is well known that imbalance between raised ROS levels and antioxidant systems creates oxidative stress, which makes the cells prone to the damage. Antioxidant defence systems such as catalase (CAT), superoxide dismutase (SOD) and glutathione peroxidase (GPx) scavenge ROS and inhibit degradation of NO. Prostacyclin and endothelium-derived hyperpolarizing factor (EDHF) and NO are vasodilators which help in maintaining vascular homoeostasis via regulating vascular tone, reactivity and vascular smooth muscle cell proliferation [98, 99]. Oxidative stress inactivates NO inducing decreased NO bioavailability and further production of ROS species such as $ONOO^-$. Decreased NO bioavailability leads to imbalance of vessel wall, impaired endothelial-dependent vasodilation and thus endothelial dysfunction. Studies have suggested the role of ROS-mediated accelerated inactivation of NO in endothelial dysfunction in CVD [100]. Apart from reduced NO bioavailability, endothelial dysfunction also results from lipid peroxidation and inflammatory responses. ROS-mediated altered endothelial permeability prompts the entry of LDL into subendothelial space and its subsequent oxidation to form oxLDL. oxLDL molecules are seized by macrophages and transformed macrophages (i.e. foam cells) and are then added to the atherosclerotic lesion development [101]. Elevated abnormal ROS also activates inflammatory TFs such as NF-κB, AP-1 and Nrf1 involved in immune system and inflammatory processes [102].

Clinical studies have also demonstrated the role of oxidative stress in the development and progression of CVD. Oxidative stress-mediated increased lipid peroxidation has been shown to correlate with the severity of heart failure and has been found to be reduced by free-radical scavenger [103–105]. Elevated free-radical activity and low SOD/GPx levels have been reported in patients with congestive heart failure [106]. It has been observed that risk factor-stimulated enzymatic sources generate ROS leading to hypertension [107, 108]. Altogether, these studies advocate for the role of oxidative stress in CVD. Further investigations delineating the role of enzymatic sources of ROS and oxidative stress can help to develop specific therapeutic strategies to prevent the development and progression of CVD.

3.2 Inflammation in CVD

Till now, it is not well understood whether inflammation is responsible for CVD or not. But it has been found that many heart diseases and stroke are related to inflammation and it also promotes atherogenesis (formation of atheromatous lesions in arterial walls) that can directly lead to CVD. In Atlanta, on March 14 and 15, 2002, a workshop titled "CDC/AHA Workshop on Inflammatory Markers and Cardiovascular Disease: Applications to Clinical and Public Health Practice" was held, whose main purpose was to illustrate some potential markers in pro-inflammatory or inflammatory condition [109].

As discussed earlier, oxidative stress can enhance inflammation, and this inflammation promotes many diseases along with coronary diseases including the initiation and progression of atherogenesis characterised by atherosclerotic plaque, plaque rupture and thrombosis [2]. Collagen and some other factors strengthen the formation of a tough cap over plaques. Inflammatory mediators can weaken this cap by inhibiting collagen synthesis and also by enhancing the production of collagen-breaking enzymes. Inflammation is therefore responsible for not only the initiation of atherosclerosis but also the promotion of other complications. Macrophages, another potent mediator of inflammation, in the later stage of plaque rupturing, break down the clotting factors [10]. oxLDL in endothelial tissue initiates a cascade of events to cholesterol-laden macrophages and accumulates them in the arterial wall during atherosclerotic plaques [10]. CRP is a traditional inflammatory marker observed during inflammation and also independently (genetically). Its high levels have been linked with incidences of CVD and coronary heart diseases through IL-6 [110, 111]. Inflammation raises cytokine IL-6 level that has been also found to relate with MI and CVD [112]. A well-recognised signalling molecule such as NF-κB also contributes in inflammation and atherosclerosis [113].

Some common physiological behaviour such as hypertension is also associated with an increased risk of inflammation and, consequently, development of carotid heart disease [7]. Several inflammatory cytokines have been shown to contribute to cardiac dysfunction under various pathophysiological conditions associated with heart failure, including I/R injury, MI, atherosclerosis, hypertrophy and acute viral myocarditis [114].

Cytokines and chemokines implicated in the progression of heart failure include TNF-α, IL-1, IL-6, IL-8, IL-13, IL-18, IFN-γ, cardiotrophin-1, monocyte chemotactic peptide-1 (MCP-1) and macrophage inflammatory protein-1 alpha (MIP-1α), anti-inflammatory mediators transforming growth factor beta (TGF-β), IL-10 and other pro-inflammatory mediators [114]. Previously, most of the studies related to inflammation and CVD were performed in experimental animals, and there was no proof that inflammation contributes to human heart diseases until, recently, when studies by various groups started to report the role of inflammation in heart diseases. A correlation was found between inflammation, oxidative stress and persistent platelet activation in android obese women [115]. A review summarising the results from clinical studies indicated the role of vascular inflammation in CVD, and a positive correlation between upregulated inflammatory markers and cardiovascular risk has been reported [116].

All of these observations support the idea that antioxidant depletion and inefficient scavenging of ROS, resulting in sustained oxidative stress, are of pathological importance in human CVD. The reexpression of foetal genes (ANP, BNP, α-actin and β-MHC) is a hallmark of hypertrophic remodelling, and a considerable body of evidence shows the redox regulation of various signalling cascades and remodelling responses in cardiac diseases of various aetiologies. Current evidence supports the involvement of the following pathways: (i) ERK-1/ERK-2 [117] and the small GTPase Ras [118] in response to adrenergic agonist and angiotensin II stimulation [119, 120], (ii) MAPKs in pressure-overload hypertrophy and (iii) NF-κB and apoptosis signal-regulating kinase 1 (ASK-1) in response to angiotensin II infusion. ASK-1 is the upstream of p38 MAPK and JNK in the MAPK signalling cascade, and both of these have been shown to be activated by NOX/ROS [121]. The inhibition or scavenging of free radicals has been shown to modulate the ERK signalling and hypertrophic responses in neonatal and adult cardiomyocytes. Besides ROS, experimental studies have shown that the inflammatory cytokines (e.g. TNF-α, IL-1β, and MCP-1) also promote myocardial hypertrophy and contribute to the development and progression of heart failure [122]. Several recent studies have supported the concept that cytokines produced by T cells and other inflammatory cells contribute to hypertension. More recently, it has been found that the novel, pro-inflammatory cytokine IL-17 contributes to hypertension. This cytokine is produced by TH_{17} cells, a subset of $CD4^+$ cells, which are distinct from TH1 and TH2 cells [123].

3.3 Relationship Between Oxidative Stress and Inflammation

ROS enhances inflammation directly via the activation of certain inflammatory TFs such as AP-1, NF-κB and Nrf2 [102, 124], modifying the expression of gene coding for chemokines and adhesion molecules, causing accumulation of inflammatory cells [125–127]. On the other way, ROS-mediated oxidative injury augments endothelial permeability, which prompts the lipoproteins to enter sub-endothelial space, gets oxidised and intensifying inflammation [5]. These oxidised lipoproteins also interact with Toll-like receptors (TLRs) to foster vascular disease [128]. Inflating inflammatory cells further release ROS, strengthening the oxidative environment and continuing the series of events of oxidative stress-inflammation-oxidative stress.

There is substantial evidence to show that ROS modulate T-cell function and can affect T-cell polarisation and cytokine secretion [114]. Exogenously generated ROS cause apoptosis and suppression of T-cell proliferation and production of IL-2. Of note, T cells also produce ROS endogenously via a NOX2-based NADPH oxidase, promoting a TH2 phenotype.

Inflammation and oxidative stress are involved in atherosclerosis right from initiation through development to thrombotic stage. Therefore, clearance of the clogged vessels via surgery and standard medical treatments cannot be the prime treatment option. Instead, inflammatory processes and oxidative stress pathways involved in cardiac diseases need to be targeted [109]. Reckoning with the studies advocating the relationship between oxidative stress and inflammation in the matter of CVD and the general fact that overall level of cellular ROS is determined by the relative rate of generation and the rate of reduction by antioxidants, enzymatic and nonenzymatic antioxidants, scavenging myocardial ROS can decrease inflammation and thus can demote CVD.

4 Are There Ways to Prevent CVD?

Numerous clinical trials have been performed to examine the potential for preventing CVD using antioxidant therapies. Some antioxidant studies have focused on the primary prevention of CVD, meaning the prevention of CVD in patients that do not already have the disease. β-Carotene, vitamin C and vitamin E have been investigated and randomised trials of this antioxidant failed to show any effect on the risk of death from CVD. Thus, it has been difficult to demonstrate that antioxidant supplementation has significant impact in CVD.

Secondary prevention refers to inhibiting the manifestations of CVD in those patients who already have the disease. Because the risk of a second cardiovascular event (MI, stroke, angina) is high in patients that have already had a first event, established prevention measures (e.g. cholesterol lowering, smoking cessation and others) are the most effective in secondary prevention.

4.1 Enzymatic Antioxidants

Enzymatic antioxidants are expressed in response to ROS production and display function as catalyst in reactions that convert specific ROS to different and, presumably, less harmful species. The principal enzymatic antioxidants are SOD, CAT, peroxiredoxin (Prx) and GPx [129]. SOD converts $O2^{\cdot -}$ to H_2O_2. MnSOD, the mitochondrial isoform, makes up to ~70 % of the SOD activity in heart and 90 % in cardiomyocytes. The remaining fraction consists primarily of cytoplasmic CuZnSOD with <1 % extracellular SOD (ECSOD). The importance of MnSOD in regulating $O_2^{\cdot -}$ in the myocardium is demonstrated by the fact that MnSOD$^{-/-}$ mice die soon after birth with dilated cardiomyopathy [130]. GPx (isoforms GPx1–GPx5), using glutathione (GSH), reduces H_2O_2 or ROOH to H_2O or alcohols (ROH), respectively. GPx1 and GPx3 are the most abundant intracellular isoforms and GPx4 is a mitochondrial isoform. Unlike MnSOD mice deficient in GPx develop normally and show no marked pathological changes under normal physiological conditions and exhibit a pronounced susceptibility to myocardial ischemia-reperfusion injury [131]. CAT, located in peroxisomes, is highest in the liver and erythrocytes and converts H_2O_2 to H_2O and O_2. Prx reduces peroxides, including H_2O_2 and alkyl hydroperoxides (ROOH).

The importance of removing mitochondrial $O_2^{\cdot -}$ is emphasised by observations that animals null for the MnSOD allele exhibit perinatal lethality due to cardiac dysfunction, and cardiac-specific MnSOD deletion/depletion produces progressive congestive heart failure with specific molecular defects in mitochondrial respiration. It is also important to realise that MnSOD generates H_2O_2, another ROS with pathophysiological importance, as overexpression of Prx3 (a mitochondrial H_2O_2 scavenger) prevents heart failure after experimental MI in mice [132].

4.2 Nonenzymatic Antioxidants

The role of glutathione (GSH) in maintaining cellular redox state is complex. GSH cooperates with GPx in the detoxification of H_2O_2 to $2H_2O$. In addition, GSH participates in reactions with glutathione S-transferase (GST) to bind ROS such as attachment of NO to form S-nitrosoglutathione adducts. Glutathione reductase (GR) functions to regenerate antioxidant capacity, converting from glutathione disulphide (GSSG) to GSH. Vitamins and other chemical antioxidants play an important role in the control of ROS cascades. Vitamin E (α-tocopherol) is active in membranes where it functions to reduce ROS and lipid peroxy radicals. Vitamin C (ascorbate) serves predominantly as an antioxidant in plasma due to its water solubility. It functions by reducing α-tocopherol lipid peroxide radicals to normal form [129]. Uric acid, found in extracellular fluids, detoxifies HO· metal ions (Fe^{+2} or Cu^{+}) contributing to ROS-mediated peroxidation of lipids via the Fenton reaction that produces H_2O_2. Additionally, a study by Ku and group has shown that the relationship of vitamin D (calciferol) deficiency with diabetes, hypertension, inflammation and increased cardiovascular risk and also analysed the association between vitamin D supplementation and the reduction in CVD [133].

4.3 Phytochemicals

Consumption of fruits and vegetables has been associated with lower risk of CVD [134], and their cardioprotective role is not attributable to any of the macro- and micronutrients, thus indicating role of other plant components in CVD. Plant sterols, flavonoids and sulphur-containing compounds are three categories of compounds in fruits and vegetables that may have important roles which prevent the cardiac diseases in some way. Non-nutritional bioactive compounds including isoflavones, diosgenin, resveratrol, quercetin, rosmarinic acid, catechin, sulphoraphane, tocotrienols and carotenoids comes under these three classes of compounds and are proven to reduce the risk of CVD and aid in cardioprotection [135, 136]. These compounds further need to be characterised as their mechanisms of action are not yet understood. Apart from these, vitamins, phytoestrogens and trace minerals may also have roles in cardioprotection. Most of the CVD consist of multiple events. Hence, a single cardioprotectant may not be enough to combat the CVD. These phytochemicals nowadays are greatly used in various pharmacological medicines in curing of CVD as well as cardioprotective due to their properties and mechanisms involved including antioxidative, anti-hypercholesterolemic, anti-angiogenic, anti-ischemic, inhibition of platelet aggregation and anti-inflammatory activities that reduce the risk of CVD.

Low to moderate consumption of red wine is associated with decreased incidence of CVD [137, 138]. Resveratrol and quercetin, polyphenols present in red wine, are anti-proliferative, anti-mitogenic, anti-platelet and anti-inflammatory [139]. The therapeutic action of resveratrol has been reported in animal models of arterial injury [140]. Resveratrol increases vascular NO production through an oestrogen receptor modulation and also inhibits NF-κB activation [141]. Quercetin and other wine polyphenols have been shown to prevent cardiac cells from apoptosis, oxidative stress and endothelial dysfunction both in vitro and in vivo studies [142].

Organosulfides, present mainly in garlic and onion, possess antioxidant and anti-inflammatory properties. The efficacy of experimental and clinical effects of garlic preparations and constituents in CVD complications have been well studied [143]. Garlic extract prevents oxidative stress via NOX and lipid peroxidation in experimental model of metabolic syndrome [144]. The systolic blood pressure decrease is associated with intake of garlic extract and allicin in fructose-induced hyperinsulinemic, hyperlipidemic and hypertensive rats [144, 145]. The action of garlic extracts against hypertension may be exerted via prostaglandin-like effects [146] or by increasing the bioavailability of NO or scavenging oxidants [147] or by inhibiting angiotensin-converting enzyme in vitro [148]. The effect of garlic against hypertension has been demonstrated in high blood pressure human cases in randomised controlled trials and meta-analysis trial [149, 150]. Garlic exerts anti-inflammatory action via the inhibition of the expression of intercellular cell adhesion molecule 1 (ICAM1) through the down-regulation of AP-1 and c-Jun N-terminal kinase (JNK) pathway [151]. The mechanism and role of garlic and its active constituents in cardioprotection need to be verified in humans through more experimental and clinical studies, so that they can be utilised as better therapeutic agents.

Higher intake of fruits and vegetables is linked with a decrease in risk of MI in prospective cohort study of women, thus increase in consumption of fruits and vegetables may protect against CVD. Increased total flavonoid and flavone intake is associated with lower risk of mortality due to CVD in large cohort of 38,180 men and 60,289 women, while long-term supplementation with β-carotene has been proved ineffective in preventing CVD in a randomised, double-blind, placebo-controlled trial [152]. Additionally, the combination of β-carotene and vitamin A showed no positive effect on the risk of CVD [153]. Tea and coffee consumption is associated with the risk of CVD [154]. Phytochemicals present in tea and coffee may exert their cardioprotective action through the regulation of vascular tone by influencing endothelial function, enhanced reverse cholesterol transport, ameliorated glucose metabolism, restrained foam cell formation, immunomodulation, inhibition of oxidative stress and effects on platelet function or by altering gene expression [155]. Further research is required to identify the role of constituents of tea and coffee in CVD [155].

Another phytochemical that showed the capability of having therapeutic potential in the treatment of CVD is cannabidiol (CBD). CBD is an abundant constituent of *Cannabis sativa*, which have been reported to have anti-inflammatory effect in various disease models including multiple sclerosis in humans. A study investigated the effects of CBD on cardiac dysfunction, elevated oxidative stress and amplified inflammatory cell signalling pathways in a mouse model with type I diabetic cardiomyopathy. It was observed that CBD remarkably attenuates myocardial dysfunction, oxidative stress, cardiac cell death, inflammatory and other interrelated cell signalling pathways. It was also found to inhibit high-glucose-mediated elevated ROS production as well as NF-κB activation in primary human cardiomyocytes [156].

5 Conclusions and Future Perspectives

Till date CVD remains the principal cause of long-term disability and death worldwide. Numerous studies have been performed, but its efficient prevention and treatment management are still failing off. For better understanding of the pathophysiology of CVD, for identifying specific biomarkers and for determining effective therapeutic targets, it is important to resolve the complexity of biochemistry of involved important pathophysiological factors such as oxidative stress and inflammation, displaying association between, which promotes the development and progression of CVD.

Oxidative stress is the disturbance of physiological balance between oxidants (ROS/RNS generation) and antioxidant defence systems. Oxidative stress is the fundamental mechanism of cell damage, and ROS is the damaging factor leading to the onset and progression of CVD. Increased ROS production in the vascular wall promotes endothelial dysfunction, infiltration and accumulation of inflammatory cells. ROS are also involved in redox activation of certain TFs responsible for expression of genes for inflammatory responses. Elevated levels of pro-inflammatory cytokines in circulation such as CRP, TNF-α and IL-6 have been shown to be associated with CVD. Accumulating evidence shows that increased ROS-mediated oxidative damage coupled with downstream inflammatory pathways augments pathological complications associated with CVD.

Therapeutic interventions involving diet, nutrition and pharmacology may target enzymatic sources of ROS, activates antioxidant defence systems in vasculature and can prevent oxidative stress and inflammation in CVD. Evidences from studies suggest that antioxidants hold the potential to act as therapeutic intervention against CVD by reducing ROS generation, thus attenuating oxidative stress and downstream inflammatory processes. Results from various studies also indicate that phytochemicals may have therapeutic potential against CVD, by reducing oxidative stress and oxidative stress-mediated inflammation. The motive of this book chapter was to make understand the role of excessive oxidative stress and chronic inflammation in CVD and to summarise the recent scenario displaying the perspectives of antioxidant and anti-inflammatory therapeutic interventions in CVD (Fig. 3). Future studies aimed at inferring the biochemical and molecular links concerning

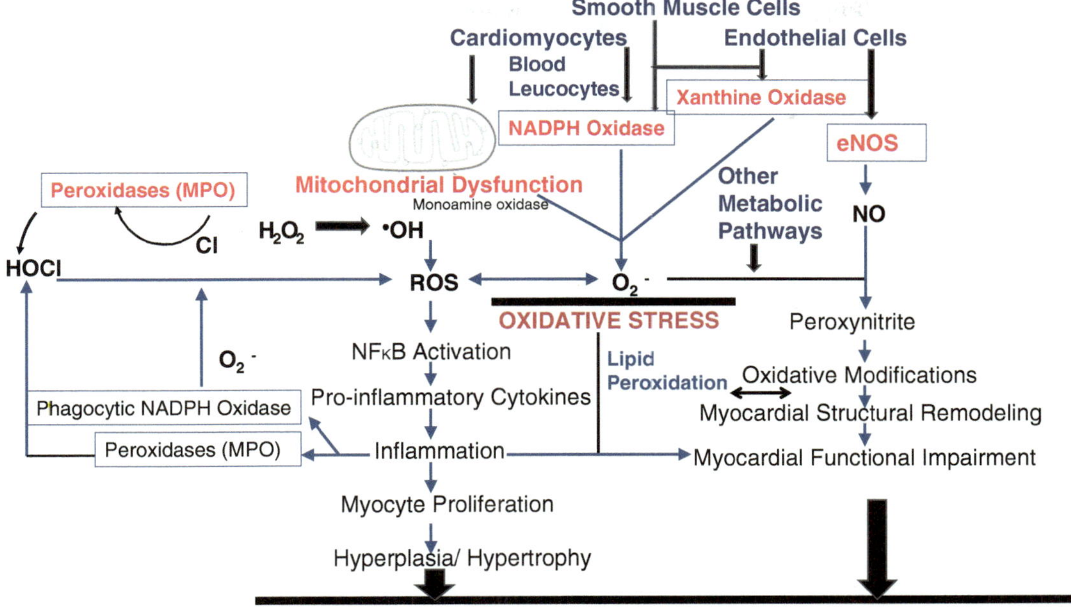

Fig. 3 Role of oxidative stress and inflammation in the development and progression of CVD. Specific enzymes of cells of the cardiovascular system, such as NADPH oxidases, xanthine oxidases and myeloperoxidases (MPO) of cardiomyocytes and endothelial cells, generate pathophysiological levels of ROS/RNS which cause oxidative stress-induced damage, promote endothelial dysfunction and accumulation of inflammatory cells and further involved in the activation of inflammatory response signalling pathways. Altogether leads to the severity of CVD

oxidative stress and inflammation and mechanisms underlying phytochemical-induced cardioprotection in CVD need to be elucidated in CVD to develop more new effective therapeutic interventions.

Acknowledgements The financial support from DST (Fast-Track) to M.D., ICMR-JRF to S.T. and CUPB institutional fellowship to S.U. and Alzheimer's Association, USA (NIRG-11-203527), grant to A.K.M is acknowledged. Because of the limited focus of the article, many appropriate references could not be included, for which the authors apologise. The CUPB publication number provided for this article is P-34/14.

References

1. Labarthe D (1998) Epidemiology and prevention of cardiovascular diseases: a global challenge. Aspen Publishers. Jones & Bartlett Learning, Gaithersburg
2. Kelly BB, Narula J, Fuster V (2012) Recognizing global burden of cardiovascular disease and related chronic diseases. Mt Sinai J Med: J Transl Personalized Med 79(6):632–640
3. Yusuf S, Reddy S, Ounpuu S et al (2001) Global burden of cardiovascular diseases part I: general considerations, the epidemiologic transition, risk factors, and impact of urbanization. Circulation 104(22):2746–2753
4. Alexander RW (1995) Hypertension and the pathogenesis of atherosclerosis oxidative stress and the mediation of arterial inflammatory response: a new perspective. Hypertension 25(2):155–161
5. Sima AV, Stancu CS, Simionescu M (2009) Vascular endothelium in atherosclerosis. Cell Tissue Res 335(1):191–203
6. Marui N, Offermann M, Swerlick R et al (1993) Vascular cell adhesion molecule-1 (VCAM-1) gene transcription and expression are regulated through an antioxidant-sensitive mechanism in human vascular endothelial cells. J Clin Invest 92(4):1866
7. Libby P, Ridker PM, Maseri A (2002) Inflammation and atherosclerosis. Circulation 105(9):1135–1143
8. Turrens JF (2003) Mitochondrial formation of reactive oxygen species. J Physiol 552(2):335–344
9. Shafaq N (2012) An overview of oxidative stress and antioxidant defensive system. Open Access Sci Rep 1(8):413, 10.4172/scientificreports
10. Andrew PJ, Mayer B (1999) Enzymatic function of nitric oxide synthases. Cardiovasc Res 43(3):521–531

11. Winterbourn CC, Buss IH, Chan TP et al (2000) Protein carbonyl measurements show evidence of early oxidative stress in critically ill patients. Crit Care Med 28(1):143–149
12. Chen K, Keaney JF Jr (2012) Evolving concepts of oxidative stress and reactive oxygen species in cardiovascular disease. Curr Atheroscler Rep 14(5):476–483
13. Carpi A, Menabo R, Kaludercic N et al (2009) The cardioprotective effects elicited by p66(Shc) ablation demonstrate the crucial role of mitochondrial ROS formation in ischemia/reperfusion injury. Biochim Biophys Acta 1787(7):774–780
14. Giorgio M, Migliaccio E, Orsini F et al (2005) Electron transfer between cytochrome c and p66^{Shc} generates reactive oxygen species that trigger mitochondrial apoptosis. Cell 122(2):221–233
15. Migliaccio E, Giorgio M, Mele S et al (1999) The p66^{Shc} adaptor protein controls oxidative stress response and life span in mammals. Nature 402(6759):309–313
16. Orsini F, Migliaccio E, Moroni M et al (2004) The life span determinant p66Shc localizes to mitochondria where it associates with mitochondrial heat shock protein 70 and regulates trans-membrane potential. J Biol Chem 279(24):25689–25695
17. Brown DI, Griendling KK (2009) Nox proteins in signal transduction. Free Radic Biol Med 47(9):1239–1253
18. Martyn K, Frederick L, von Loehneysen K et al (2006) Functional analysis of Nox4 reveals unique characteristics compared to other NADPH oxidases. Cell Signal 18(1):69–82
19. McNally JS, Davis ME, Giddens DP et al (2003) Role of xanthine oxidoreductase and NAD(P)H oxidase in endothelial superoxide production in response to oscillatory shear stress. Am J Physiol Heart Circ Physiol 285(6):H2290–H2297
20. Takac I, Schroder K, Zhang L et al (2011) The E-loop is involved in hydrogen peroxide formation by the NADPH oxidase Nox4. J Biol Chem 286(15):13304–13313
21. Touyz RM, Montezano AC (2012) Vascular Nox4: a multifarious NADPH oxidase. Circ Res 110(9):1159–1161
22. Kaludercic N, Mialet-Perez J, Paolocci N et al (2014) Monoamine oxidases as sources of oxidants in the heart. J Mol Cell Cardiol 73:34–42
23. Hare JM (2004) Nitroso–redox balance in the cardiovascular system. N Engl J Med 351(20): 2112–2114
24. Lacza Z, Pankotai E, Busija DW (2009) Mitochondrial nitric oxide synthase: current concepts and controversies. Front Biosci (Landmark Ed) 14:4436–4443
25. Nediani C, Raimondi L, Borchi E et al (2011) Nitric oxide/reactive oxygen species generation and nitroso/redox imbalance in heart failure: from molecular mechanisms to therapeutic implications. Antioxid Redox Signal 14(2):289–331
26. Immenschuh S, Schroder H (2006) Heme oxygenase-1 and cardiovascular disease. Histol Histopathol 21(6):679–685
27. Griendling KK, Minieri CA, Ollerenshaw JD et al (1994) Angiotensin II stimulates NADH and NADPH oxidase activity in cultured vascular smooth muscle cells. Circ Res 74(6):1141–1148
28. Hiraoka W, Vazquez N, Nieves-Neira W et al (1998) Role of oxygen radicals generated by NADPH oxidase in apoptosis induced in human leukemia cells. J Clin Invest 102(11):1961–1968
29. Irani K, Xia Y, Zweier JL et al (1997) Mitogenic signaling mediated by oxidants in Ras-transformed fibroblasts. Science 275(5306):1649–1652
30. Sundaresan M, Yu ZX, Ferrans VJ et al (1995) Requirement for generation of H2O2 for platelet-derived growth factor signal transduction. Science 270(5234):296–299
31. Lijnen PJ, Piccart Y, Coenen T et al (2012) Angiotensin II-induced mitochondrial reactive oxygen species and peroxiredoxin-3 expression in cardiac fibroblasts. J Hypertens 30(10):1986–1991
32. Brennan M, Wu W, Fu X et al (2002) A tale of two controversies defining both the role of peroxidases in nitrotyrosine formation in vivo using eosinophil peroxidase and myeloperoxidase-deficient mice, and the nature of peroxidase-generated reactive nitrogen species. J Biol Chem 277(20):17415–17427
33. Zhang R, Brennan M, Shen Z et al (2002) Myeloperoxidase functions as a major enzymatic catalyst for initiation of lipid peroxidation at sites of inflammation. J Biol Chem 277(48):46116–46122
34. De Marchi U, Mancon M, Battaglia V et al (2004) Influence of reactive oxygen species production by monoamine oxidase activity on aluminum-induced mitochondrial permeability transition. Cell Mol Life Sci (CMLS) 61(19–20):2664–2671
35. Fearon IM, Faux SP (2009) Oxidative stress and cardiovascular disease: novel tools give (free) radical insight. J Mol Cell Cardiol 47(3):372–381
36. Carrasco Guerra HA, Palacios-Prü E, Dagert de Scorza C et al (1987) Clinical, histochemical, and ultrastructural correlation in septal endomyocardial biopsies from chronic chagasic patients: detection of early myocardial damage. Am Heart J 113(3):716–724
37. Wen X, Gupta S, Zago MP et al (2012) Defects of mtDNA replication impaired mitochondrial biogenesis during trypanosoma cruzi infection in human cardiomyocytes and chagasic patients: the role of Nrf1/2 and antioxidant response. J Am Heart Assoc 1(6):e003855
38. Wen JJ, Gupta S, Guan Z et al (2010) Phenyl-alpha-tert-butyl-nitrone and benzonidazole treatment controlled the mitochondrial oxidative stress and evolution of cardiomyopathy in chronic chagasic rats. J Am Coll Cardiol 55(22):2499–2508
39. Wen JJ, Garg NJ (2010) Mitochondrial complex III defects contribute to inefficient respiration and ATP synthesis in the myocardium of *T. cruzi* infected mice. Antioxid Redox Signal 12(1):27–37

40. Malle E, Marsche G, Arnhold J et al (2006) Modification of low-density lipoprotein by myeloperoxidase-derived oxidants and reagent hypochlorous acid. Biochim Biophys Acta (BBA)-Mol Cell Biol Lipids 1761(4):392–415
41. Goldstein S, Meyerstein D, Czapski G (1993) The fenton reagents. Free Radic Biol Med 15(4):435–445
42. Lourenco AP, Fontoura D, Henriques-Coelho T et al (2012) Current pathophysiological concepts and management of pulmonary hypertension. Int J Cardiol 155(3):350–361
43. Dhiman M, Wan X, Popov V et al (2013) MnSODtg mice control myocardial inflammatory and oxidative stress and remodeling responses elicited in chronic Chagas disease. J Am Heart Assoc 2(5):e000302
44. Halliwell B (1991) Reactive oxygen species in living systems: source, biochemistry, and role in human disease. Am J Med 91(3):S14–S22
45. Eiserich JP, Baldus S, Brennan M-L et al (2002) Myeloperoxidase, a leukocyte-derived vascular NO oxidase. Science 296(5577):2391–2394
46. Babior BM (1999) NADPH oxidase: an update. Blood 93(5):1464–1476
47. Candeias LP, Stratford MR, Wardman P (1994) Formation of hydroxyl radicals on reaction of hypochlorous acid with ferrocyanide, a model iron (II) complex. Free Radic Res 20(4):241–249
48. Bedard K, Krause K-H (2007) The NOX family of ROS-generating NADPH oxidases: physiology and pathophysiology. Physiol Rev 87(1):245–313
49. Weintraub NL (2002) Nox response to injury. Arterioscler Thromb Vasc Biol 22(1):4–5
50. BelAiba RS, Djordjevic T, Petry A et al (2007) NOX5 variants are functionally active in endothelial cells. Free Radic Biol Med 42(4):446–459
51. Carnesecchi S, Carpentier J-L, Foti M et al (2006) Insulin-induced vascular endothelial growth factor expression is mediated by the NADPH oxidase NOX3. Exp Cell Res 312(17):3413–3424
52. Schröder K, Zhang M, Benkhoff S et al (2012) Nox4 is a protective reactive oxygen species generating vascular NADPH oxidase. Circ Res 110(9):1217–1225
53. Paravicini TM, Touyz RM (2008) NADPH oxidases, reactive oxygen species, and hypertension clinical implications and therapeutic possibilities. Diabetes Care 31(Supplement 2):S170–S180
54. Cathcart MK (2004) Regulation of superoxide anion production by NADPH oxidase in monocytes/macrophages contributions to atherosclerosis. Arterioscler Thromb Vasc Biol 24(1):23–28
55. Liu J, Yang F, Yang XP et al (2003) NAD(P)H oxidase mediates angiotensin II-induced vascular macrophage infiltration and medial hypertrophy. Arterioscler Thromb Vasc Biol 23(5):776–782
56. Dhiman M, Garg N (2011) NADPH oxidase inhibition ameliorates *T. cruzi* induced myocarditis during Chagas disease. J Pathol 225(4):583–596
57. Berry CE, Hare JM (2004) Xanthine oxidoreductase and cardiovascular disease: molecular mechanisms and pathophysiological implications. J Physiol 555(3):589–606
58. Malik U, Hundley N, Romero G et al (2011) Febuxostat inhibition of endothelial-bound XO: implications for targeting vascular ROS production. Free Radic Biol Med 51(1):179–184
59. White C, Darley-Usmar V, Berrington W et al (1996) Circulating plasma xanthine oxidase contributes to vascular dysfunction in hypercholesterolemic rabbits. Proc Natl Acad Sci 93(16):8745–8749
60. Mervala EM, Cheng ZJ, Tikkanen I et al (2001) Endothelial dysfunction and xanthine oxidoreductase activity in rats with human renin and angiotensinogen genes. Hypertension 37:414–418
61. Afonso PV, Janka-Junttila M, Lee YJ et al (2012) LTB $_4$ is a signal-relay molecule during neutrophil chemotaxis. Dev Cell 22(5):1079–1091
62. Koshino T, Takano S, Houjo T et al (1998) Expression of 5-lipoxygenase and 5-lipoxygenase-activating protein mRNAs in the peripheral blood leukocytes of asthmatics. Biochem Biophys Res Commun 247(2):510–513
63. Poeckel D, Funk CD (2010) The 5-lipoxygenase/leukotriene pathway in preclinical models of cardiovascular disease. Cardiovasc Res 86(2):243–253
64. Yoshimura R, Inoue K, Kawahito Y et al (2004) Expression of 12-lipoxygenase in human renal cell carcinoma and growth prevention by its inhibitor. Int J Mol Med 13(1):41
65. Zhou YJ, Wang JH, Li L et al (2007) Expanding expression of the 5-lipoxygenase/leukotriene B_4 pathway in atherosclerotic lesions of diabetic patients promotes plaque instability. Biochem Biophys Res Commun 363(1):30–36
66. Reilly KB, Srinivasan S, Hatley ME et al (2004) 12/15-Lipoxygenase activity mediates inflammatory monocyte/endothelial interactions and atherosclerosis in vivo. J Biol Chem 279(10):9440–9450
67. McNally AK, Chisolm GM 3rd, Morel DW et al (1990) Activated human monocytes oxidize low-density lipoprotein by a lipoxygenase-dependent pathway. J Immunol 145(1):254–259
68. Yla-Herttuala S, Rosenfeld ME, Parthasarathy S et al (1990) Colocalization of 15-lipoxygenase mRNA and protein with epitopes of oxidized low density lipoprotein in macrophage-rich areas of atherosclerotic lesions. Proc Natl Acad Sci 87(18):6959–6963
69. Yla-Herttuala S, Rosenfeld ME, Parthasarathy S et al (1991) Gene expression in macrophage-rich human atherosclerotic lesions. 15-lipoxygenase and acetyl low density lipoprotein receptor messenger RNA colocalize with oxidation specific lipid-protein adducts. J Clin Invest 87(4):1146–1152
70. Wu ML, Ho YC, Lin CY et al (2011) Heme oxygenase-1 in inflammation and cardiovascular disease. Am J Cardiovasc Dis 1(2):150–158
71. Wang CY, Chau LY (2010) Heme oxygenase-1 in cardiovascular diseases: molecular mechanisms and clinical perspectives. Chang Gung Med J 33(1):13–24

72. Idriss NK, Blann AD, Lip GY (2008) Hemoxygenase-1 in cardiovascular disease. J Am Coll Cardiol 52(12):971–978
73. Searle J, Shih J, Muller R et al (2013) The role of myeloperoxidase (MPO) for prognostic evaluation in sensitive cardiac troponin I negative chest pain patients in the emergency department. Eur Heart J: Acute Cardiovasc Care 2(3):203–210
74. Hampton MB, Kettle AJ, Winterbourn CC (1998) Inside the neutrophil phagosome: oxidants, myeloperoxidase, and bacterial killing. Blood 92(9):3007–3017
75. Pendergraft WF III, Schmaier AH, Mahdi F (2007) Myeloperoxidase interacts with endothelial cell-surface cytokeratin 1 and modulates bradykinin production by the plasma kallikrein-kinin system. Am J Pathol 171(1):349–360
76. Astern JM (2007) Myeloperoxidase in vascular disease and autoimmunity. Dissertation, Pathology and Laboratory Medicine
77. Su HS, Nahrendorf M, Panizzi P et al (2012) Vasculitis: molecular imaging by targeting the inflammatory enzyme myeloperoxidase. Radiology 262(1):181–190
78. Eggers KM, Dellborg M, Johnston N et al (2010) Myeloperoxidase is not useful for the early assessment of patients with chest pain. Clin Biochem 43(3):240–245
79. El Kebir D, József L, Pan W et al (2008) Myeloperoxidase delays neutrophil apoptosis through CD11b/CD18 integrins and prolongs inflammation. Circ Res 103(4):352–359
80. Nicholls SJ, Hazen SL (2005) Myeloperoxidase and cardiovascular disease. Arterioscler Thromb Vasc Biol 25(6):1102–1111
81. Undurti A, Huang Y, Lupica JA et al (2009) Modification of high density lipoprotein by myeloperoxidase generates a pro-inflammatory particle. J Biol Chem 284(45):30825–30835
82. Holvoet P (2008) Relations between metabolic syndrome, oxidative stress and inflammation and cardiovascular disease. Verhandelingen-Koninklijke Academie voor Geneeskunde van België 70(3):193
83. Dhiman M, Estrada-Franco JG, Pando JM et al (2009) Increased myeloperoxidase activity and protein nitration are indicators of inflammation in patients with Chagas' disease. Clin Vaccine Immunol 16(5):660–666
84. Reuter S, Gupta SC, Chaturvedi MM et al (2010) Oxidative stress, inflammation, and cancer: how are they linked? Free Radic Biol Med 49(11):1603–1616
85. Santolini J (2011) The molecular mechanism of mammalian NO-synthases: a story of electrons and protons. J Inorg Biochem 105(2):127–141
86. Xie Q-W, Cho HJ, Calaycay J et al (1992) Cloning and characterization of inducible nitric oxide synthase from mouse macrophages. Science 256(5054):225–228
87. Tousoulis D, Kampoli AM, Tentolouris Nikolaos Papageorgiou C et al (2012) The role of nitric oxide on endothelial function. Curr Vasc Pharmacol 10(1):4–18
88. Knowles RG, Moncada S (1994) Nitric oxide synthases in mammals. Biochem J 298(2):249
89. Rafikov R, Fonseca FV, Kumar S et al (2011) eNOS activation and NO function: structural motifs responsible for the posttranslational control of endothelial nitric oxide synthase activity. J Endocrinol 210(3):271–284
90. Shaul PW (2002) Regulation of endothelial nitric oxide synthase: location, location, location. Annu Rev Physiol 64(1):749–774
91. Venugopal SK, Devaraj S, Yuhanna I et al (2002) Demonstration that C-reactive protein decreases eNOS expression and bioactivity in human aortic endothelial cells. Circulation 106(12):1439–1441
92. Galea E, Feinstein DL (1999) Regulation of the expression of the inflammatory nitric oxide synthase (NOS2) by cyclic AMP. FASEB J 13(15):2125–2137
93. Munoz-Bravo C, Gutierrez-Bedmar M, Gomez-Aracena J et al (2013) Iron: protector or risk factor for cardiovascular disease? Still controversial. Nutrients 5(7):2384–2404
94. Reddy BR, Wynne J, Kloner RA et al (1991) Pretreatment with the iron chelator desferrioxamine fails to provide sustained protection against myocardial ischaemia-reperfusion injury. Cardiovasc Res 25(9):711–718
95. DePalma R, Hayes V, Chow B et al (2010) Ferritin levels, inflammatory biomarkers, and mortality in peripheral arterial disease: a substudy of the Iron (Fe) and Atherosclerosis Study (FeAST) Trial. J Vasc Surg 51(6):1498–1503
96. Heitzer T, Schlinzig T, Krohn K et al (2001) Endothelial dysfunction, oxidative stress, and risk of cardiovascular events in patients with coronary artery disease. Circulation 104(22):2673–2678
97. Neunteufl T, Heher S, Katzenschlager R et al (2000) Late prognostic value of flow-mediated dilation in the brachial artery of patients with chest pain. Am J Cardiol 86(2):207–210
98. Epstein FH, Vane JR, Änggård EE et al (1990) Regulatory functions of the vascular endothelium. N Engl J Med 323(1):27–36
99. Vanhoutte PM, Boulanger CM, Mombouli JV (1995) Endothelium-derived relaxing factors and converting enzyme inhibition. Am J Cardiol 76(15):3E–12E
100. Cai H, Harrison DG (2000) Endothelial dysfunction in cardiovascular diseases: the role of oxidant stress. Circ Res 87(10):840–844
101. Heinecke J (2006) Lipoprotein oxidation in cardiovascular disease: chief culprit or innocent bystander? J Exp Med 203(4):813–816
102. Imhoff BR, Hansen JM (2009) Extracellular redox status regulates Nrf2 activation through mitochondrial reactive oxygen species. Biochem J 424(3):491–500

103. Díaz-Vélez CR, García-Castiñeiras S, Mendoza-Ramos E et al (1996) Increased malondialdehyde in peripheral blood of patients with congestive heart failure. Am Heart J 131(1):146–152
104. Sobotka PA, Brottman MD, Ze W et al (1993) Elevated breath pentane in heart failure reduced by free radical scavenger. Free Radic Biol Med 14(6):643–647
105. Weitz Z, Birnbaum A, Skosey J et al (1991) High breath pentane concentrations during acute myocardial infarction. Lancet 337(8747):933–935
106. McMurray J, Chopra M, Abdullah I et al (1993) Evidence of oxidative stress in chronic heart failure in humans. Eur Heart J 14(11):1493–1498
107. Harrison DG, Gongora MC (2009) Oxidative stress and hypertension. Med Clin N Am 93(3):621–635
108. Sugamura K, Keaney J Jr (2011) Reactive oxygen species in cardiovascular disease. Free Radic Biol Med 51(5):978–992
109. Pearson TA, Mensah GA, Alexander RW et al (2003) Markers of inflammation and cardiovascular disease application to clinical and public health practice: a statement for healthcare professionals from the centers for disease control and prevention and the American Heart Association. Circulation 107(3):499–511
110. Empana JP, Jouven X, Canouï-Poitrine F et al (2010) C-reactive protein, interleukin 6, fibrinogen and risk of sudden death in European middle-aged men: the PRIME study. Arterioscler Thromb Vasc Biol 30(10):2047–2052
111. Pepys MB, Hirschfield GM (2003) C-reactive protein: a critical update. J Clin Invest 111(12):1805–1812
112. Lincoff AM, Kereiakes DJ, Mascelli MA et al (2001) Abciximab suppresses the rise in levels of circulating inflammatory markers after percutaneous coronary revascularization. Circulation 104(2):163–167
113. Nichols TC, Fischer TH, Deliargyris EN et al (2001) Role of nuclear factor-kappa B (NF-κB) in inflammation, periodontitis, and atherogenesis. Ann Periodontol 6(1):20–29
114. Gupta S, Dhiman M, Wen JJ et al (2011) ROS signalling of inflammatory cytokines during trypanosoma cruzi infection. Adv Parasitol 76:153
115. Davì G, Guagnano MT, Ciabattoni G et al (2002) Platelet activation in obese women: role of inflammation and oxidant stress. JAMA 288(16):2008–2014
116. Virdis A, Schiffrin EL (2003) Vascular inflammation: a role in vascular disease in hypertension? Curr Opin Nephrol Hypertens 12(2):181–187
117. Singh K, Xiao L, Remondino A et al (2001) Adrenergic regulation of cardiac myocyte apoptosis. J Cell Physiol 189(3):257–265
118. Kuster GM, Pimentel DR, Adachi T et al (2005) α-Adrenergic receptor–stimulated hypertrophy in adult rat ventricular myocytes is mediated via thioredoxin-1–sensitive oxidative modification of thiols on Ras. Circulation 111(9):1192–1198
119. Nakagami H, Takemoto M, Liao JK (2003) NADPH oxidase-derived superoxide anion mediates angiotensin II-induced cardiac hypertrophy. J Mol Cell Cardiol 35(7):851–859
120. Satoh M, Shimoda Y, Akatsu T et al (2006) Elevated circulating levels of heat shock protein 70 are related to systemic inflammatory reaction through monocyte Toll signal in patients with heart failure after acute myocardial infarction. Eur J Heart Fail 8(8):810–815
121. Matsuzawa A, Ichijo H (2005) Stress-responsive protein kinases in redox-regulated apoptosis signaling. Antioxid Redox Signal 7(3–4):472–481
122. Aukrust P, Gullestad L, Ueland T et al (2005) Inflammatory and anti-inflammatory cytokines in chronic heart failure: potential therapeutic implications. Ann Med 37(2):74–85
123. Witowski J, Ksiazek K, Jorres A (2004) Interleukin-17: a mediator of inflammatory responses. Cell Mol Life Sci (CMLS) 61(5):567–579
124. Sen CK, Packer L (1996) Antioxidant and redox regulation of gene transcription. FASEB J 10(7):709–720
125. Dhawan S, Singh S, Aggarwal BB (1997) Induction of endothelial cell surface adhesion molecules by tumor necrosis factor is blocked by protein tyrosine phosphatase inhibitors: role of the nuclear transcription factor NF-kappa B. Eur J Immunol 27(9):2172–2179
126. Innamorato NG, Rojo AI, Garcia-Yague AJ et al (2008) The transcription factor Nrf2 is a therapeutic target against brain inflammation. J Immunol 181(1):680–689
127. Moriuchi H, Moriuchi M, Fauci AS (1997) Nuclear factor-kappa B potently up-regulates the promoter activity of RANTES, a chemokine that blocks HIV infection. J Immunol 158(7):3483–3491
128. Bjorkbacka H (2006) Multiple roles of Toll-like receptor signaling in atherosclerosis. Curr Opin Lipidol 17(5):527–533
129. Nordberg J, Arner ES (2001) Reactive oxygen species, antioxidants, and the mammalian thioredoxin system. Free Radic Biol Med 31(11):1287–1312
130. Li Y, Huang T-T, Carlson EJ et al (1995) Dilated cardiomyopathy and neonatal lethality in mutant mice lacking manganese superoxide dismutase. Nat Genet 11(4):376–381
131. Chen Z, Siu B, Ho YS et al (1998) Overexpression of MnSOD protects against myocardial ischemia/reperfusion injury in transgenic mice. J Mol Cell Cardiol 30(11):2281–2289
132. Tsutsui H, Kinugawa S, Matsushima S (2009) Mitochondrial oxidative stress and dysfunction in myocardial remodelling. Cardiovasc Res 81(3):449–456
133. Ku Y, Liu M, Ku C et al (2013) Relationship between vitamin D deficiency and cardiovascular disease. World J Cardiol 5(9):337

134. Ness A, Powles J (1997) Fruit and vegetables, and cardiovascular disease: a review. Int J Epidemiol 26(1):1–13
135. García-Andradea M, González-Laredoa R, Rocha-Guzmána N et al (2013) Mesquite leaves (Prosopis laevigata), a natural resource with antioxidant capacity and cardioprotection potential. Ind Crop Prod 44:336–342
136. Vijay T, Dhana Rajan M, Sarumathy K et al (2011) Phytochemical screening by GC-MS and cardio-protective activity of Pimpinella Tirupatiensis (Pt) on doxorubicin induced cardiotoxicity in albino rats. Int J Pharmacol Toxicol Sci 2:8–21
137. Lippi G, Franchini M, Favaloro E et al (2010) Moderate red wine consumption and cardiovascular disease risk: beyond the "French paradox". In: Seminars in thrombosis and hemostasis. Thieme Medical Publishers, New York
138. Wollin S, Jones P (2001) Alcohol, red wine and cardiovascular disease. J Nutr 131(5):1401–1404
139. Lin J, Tsai S (1999) Chemoprevention of cancer and cardiovascular disease by resveratrol. Proc Natl Sci Counc Repub China B 23(3):99–106
140. Baur J, Sinclair D (2006) Therapeutic potential of resveratrol: the in vivo evidence. Nat Rev Drug Discov 5(6):493–506
141. Hao H, He L (2004) Mechanisms of cardiovascular protection by resveratrol. J Med Food 7(3):290–298
142. Perez-Vizcaino F, Duarte J, Andriantsitohaina R (2006) Endothelial function and cardiovascular disease: effects of quercetin and wine polyphenols. Free Radic Res 40(10):1054–1065
143. Banerjee S, Maulik S (2002) Effect of garlic on cardiovascular disorders: a review. Nutr J 1(1):4
144. Vazquez-Prieto M, Gonzalez R, Renna N et al (2010) Aqueous garlic extracts prevent oxidative stress and vascular remodeling in an experimental model of metabolic syndrome. J Agric Food Chem 58(11):6630–6635
145. Elkayam A, Mirelman D, Peleg E et al (2001) The effects of allicin and enalapril in fructose-induced hyperinsulinemic hyperlipidemic hypertensive rats. Am J Hypertens 14(4):377–381
146. Rashid A, Khan H (1985) The mechanism of hypotensive effect of garlic extract. JPMA J Pak Med Assoc 35(12):357
147. Kim-Park S, Ku D (2000) Garlic elicits a nitric oxide-dependent relaxation and inhibits hypoxic pulmonary vasoconstriction in rats. Clin Exp Pharmacol Physiol 27(10):780–786
148. Sendl A, Elbl G, Steinke B et al (1992) Comparative pharmacological investigations of Allium ursinum and Allium sativum. Planta Med 58(01):1–7
149. Auer W, Eiber A, Hertkom E, Benheim H et al (1989) Hypertonie and hyperlipidamie: in leichterenauch Knoblauch. Der Allgemeinarzi 3:205–208
150. Silagy C, Neil H (1994) A meta-analysis of the effect of garlic on blood pressure. J Hypertens 12(4):463–468
151. Son E, Mo S, Rhee D et al (2006) Inhibition of ICAM-1 expression by garlic component, allicin, in gamma-irradiated human vascular endothelial cells via downregulation of the JNK signaling pathway. Int Immunopharmacol 6(12):1788–1795
152. McCullough ML, Peterson JJ, Patel R et al (2012) Flavonoid intake and cardiovascular disease mortality in a prospective cohort of US adults. Am J Clin Nutr 95:454–464
153. Bjelakovic G, Nikolova D, Gluud C (2013) Meta-regression analyses, meta-analyses, and trial sequential analyses of the effects of supplementation with beta-carotene, vitamin A, and vitamin E singly or in different combinations on all-cause mortality: do we have evidence for lack of harm? PLoS One 8(9):e74558
154. Arab L, Khan F, Lam H (2013) Tea consumption and cardiovascular disease risk. Am J Clin Nutr 98(6):1651S–1659S
155. Bohn S, Ward N, Hodgson J et al (2012) Effects of tea and coffee on cardiovascular disease risk. Food Funct 3(6):575–591
156. Rajesh M, Mukhopadhyay P, Batkai S et al (2010) Cannabidiol attenuates cardiac dysfunction, oxidative stress, fibrosis, and inflammatory and cell death signaling pathways in diabetic cardiomyopathy. J Am Coll Cardiol 56(25):2115–2125

Redox Biology of Aging: Focus on Novel Biomarkers

Kanti Bhooshan Pandey and Syed Ibrahim Rizvi

Abstract

Oxidative stress-mediated damage to functional macromolecules followed by their accumulation has been reported as a major factor behind aging and related consequences. Excessive generation of reactive oxygen species and/or depleted cellular defense impairs the balance between prooxidants and antioxidants causing shift in cellular redox state which leads to physiological dysfunctions making the body prone to external deleterious agents. In the present chapter, we have discussed the aging process and have focused on novel markers of oxidative damage that reflect the cellular redox status relevant to age-related studies incorporating the previously well-established markers of the aging process.

Keywords

Oxidative stress • Biomarkers • Aging • Redox balance

1 Introduction

Aging is characterized by deteriorative alterations in the post-reproductive phase cumulatively lowering the level of fitness and the ability to maintain redox state [1, 2]. Studies performed on large variety of aging model systems demonstrate that the main molecular characteristic of aging is macromolecular damage in the cells making them susceptible to various untoward changes [1, 3]. Thus, broadly aging may be defined as an inherent biological process that manifests within an organism at genetic, molecular, cellular, and systemic level [4].

To date, more than 300 theories are in existence to explain the mechanisms involved in the aging process; however, none of them exactly describes the cause of aging and associated attritions in functional capacity. Neither there is an explanation for the vast variations in the maximum life span of different species [5, 6]. Among all the theories, the "free radical theory of aging" proposed by Harman in 1956 is the most accepted which describes most of the physiological alterations responsible for cell senescence followed by decline in physiological fitness during aging [4, 5, 7]. According to this hypothesis, the endogenously generated oxygen

K.B. Pandey (✉) • S.I. Rizvi
Department of Biochemistry, University of Allahabad, Allahabad 211002, Uttar Pradesh, India
e-mail: kantibiochem@gmail.com

V. Rani and U.C.S. Yadav (eds.), *Free Radicals in Human Health and Disease*,
DOI 10.1007/978-81-322-2035-0_18, © Springer India 2015

free radicals are the agents of stochastic damage. In the context of aerobic biological systems, free radicals are frequently referred to as reactive oxygen species (ROS) since most of the free radicals in these systems contain oxygen in their structure [8].

ROS are produced inside the cell as a by-product during normal metabolic process including mitochondrial respiration and detoxification reactions. ROS are highly reactive and can cause damage to vital biomolecules of the cell. Normally ROS exist in the cell in balance with inherent antioxidants; however, a condition of oxidative stress occurs when this critical balance is disrupted due to excessive generation of ROS or weakened endogenous defense mechanisms or both [9]. Oxidative stress causes detrimental consequences and in the context of the aging resulting in the age-dependent accrual of macromolecular structural damage [4, 10].

ROS production and accumulation has been reported as a common denominator in many diseases and can lead to severe cellular damage leading to physiological dysfunction and cell death in virtually all aerobes [8]. When oxidative stress occurs, cells function to counteract the damaging effects of oxidants and to restore redox balance by resetting critical homeostatic parameters. Such cellular activity results in overexpression or silencing of genes encoding defensive enzymes, transcription factors, and structural proteins [11]. According to the free radical theory of aging, oxidative stress increases with increasing age; this condition leads to accumulation of oxidation products of lipids, proteins, nucleic acids, sugars, and sterols followed by disturbance in redox state of the cell/tissue, causing cellular dysfunction and making the body prone to external deleterious agents [2, 3] (Fig. 1).

There is mounting realization that the structural and functional damage-based hypothesis of aging process actually involves shift in cellular redox state since most of the antiaging interventions have reported to reconstruct the overall cellular redox state [2, 12, 13]. Many studies have reported that oxidative stress causes a prooxidizing shift in the thiol redox state and the resulting dysfunction of the redox-sensitive proteins [14, 15].

In this chapter, we have described the novel markers of oxidative stress-mediated damages that reflect the cellular redox status relevant to age-related studies. The chapter also incorporates the other well-established markers of aging process.

2 Age-Associated Changes in Redox State of Macromolecules

2.1 Glutathione System During Aging

Cellular redox state is collectively determined by the reduction potentials and reducing capacities of the redox duos, such as glutathione reduced (GSH)/glutathione oxidized (GSSG), nicotinamide adenine dinucleotide phosphate reduced (NADPH)/nicotinamide adenine dinucleotide phosphate oxidized ($NADP^+$), nicotinamide adenine dinucleotide reduced (NADH)/nicotinamide adenine dinucleotide oxidized (NAD^+), cysteine/cystine, etc.; nonetheless, the GSH/GSSG couple is regarded as the primary arbiter of the tissue redox state since it covers about fourfold higher concentration than the other redox couples [15]. Besides this, GSH/GSSG couple is metabolically linked to the other redox couples directly or indirectly through donations of reducing equivalents for the reduction of their oxidized forms. Another unique feature of GSH oxidation/reduction reactions is the involvement of two-electron transfers; other redox couples involve single electrons [2].

GSH is the major endogenous nonprotein antioxidant molecule in the cell that plays various vital roles including neutralization of ROS by direct reactions, providing reducing equivalents for the enzyme-mediated removal of H_2O_2 and

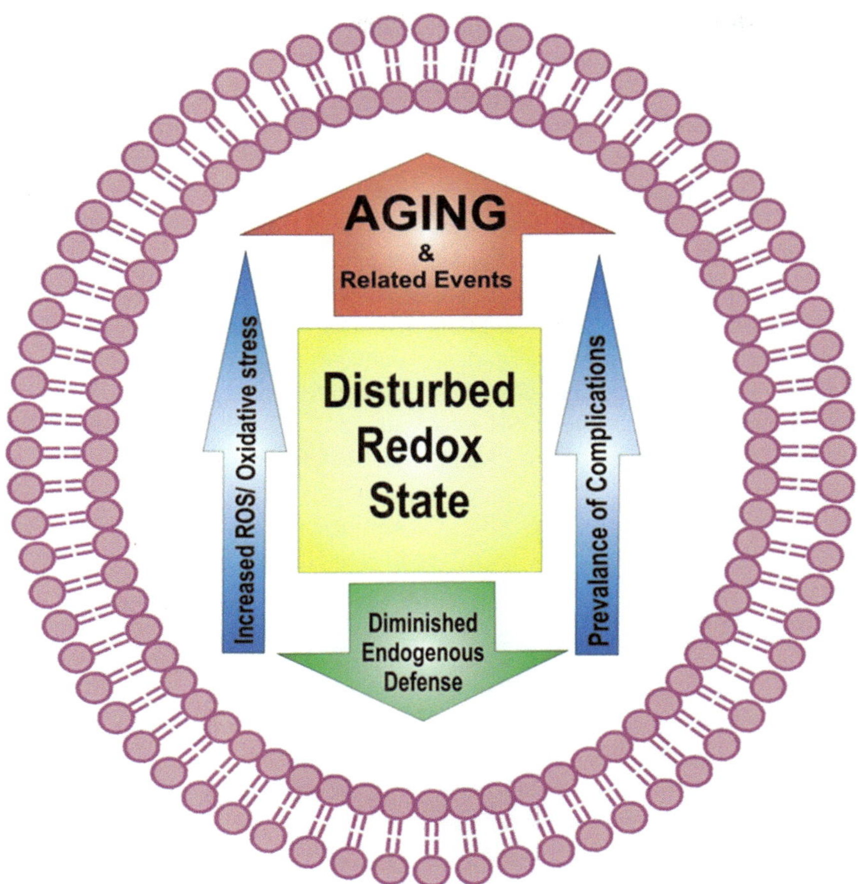

Fig. 1 Diagrammatic representation of relationship among diminished endogenous defense system, increased reactive oxygen species (ROS)/oxidative stress, prevalence of physiological complications, and aging. Excessive generation of ROS due to weak antioxidant systems or exposure to radiations, pesticides, xenobiotics, or malnutrition originates condition of oxidative stress. Oxidative stress is deleterious in many ways; it damages cellular biomolecules and thus disturbs cellular redox state followed by physiological complications which ultimately results in aging and related events

lipid peroxides, maintenance of protein thiol groups, and conjugation and excretion of xenobiotics [16]. The key functional component of GSH is the thiol group on the cysteinyl residue, which can act both as a reductant and as a nucleophile. In the presence of an oxidant, GSH is oxidized to glutathione disulfide (GSSG) (Reaction-1) which is later reconverted to GSH by the NADPH-dependent glutathione disulfide reductase (Reaction-2). The change in the amounts and ratios of GSH and GSSG represents the redox potential of the cell [15].

$$H_2O_2 + 2GSH \xrightarrow{\text{glutathioneperoxidase}} GSSG + 2H_2O \quad (1)$$

$$GSSG + NADPH + H^+ \xrightarrow{\text{glutathione disulfide reductase}} 2GSH + NADP^+ \quad (2)$$

An age-dependent depletion in GSH level and GSH:GSSG ratio has been reported in many clinical as well as laboratory studies [17–19]. Study on mice from 4 to 26 months of age revealed that GSH:GSSG ratios decreased in almost all tissues examined. Glutathione redox potential decreased from −11 to −25 mV in the tissue homogenates and −4.5 to −15 mV in the mitochondria during the same period [20]. The largest decline in the tissue redox potential was observed in the brain (−22 mV) and the eye (−25 mV). Age-dependent decline of glutathione redox potential in studies from insects to mammals establishes GSH as a potent marker of cellular redox status during aging [2].

2.2 Status of Proteins During Aging

Among various classes of ROS-induced structural modifications in biomolecules in the cell, protein oxidation is considered to be the most consequential functionally, since proteins act as transporters, enzymes, receptors, transcription factors, and cytoskeleton components [21]. Attack of ROS can lead to oxidation of amino acid residue side chains, formation of protein-protein cross linkages, and oxidation of protein backbone resulting in their fragmentation [22].

Since proteins are the much sophisticated biomolecule in the cell, even a slight alteration in their structure can result in impairment of various cell functions. It has been estimated that 30–50 % of cellular proteins are altered or dysfunctional in cells of older animals due to free radical damage [22, 23]. Many markers of protein oxidative damage are used by different laboratories including lipofuscin and branched-chain amino acids; however, not all of them mark age-dependent changes, and some lack reproducibility. We have only described here the potent biomarkers of oxidative damage to protein during aging.

2.2.1 Protein Carbonyls

Protein carbonyls are the most frequently measured biomarker of protein oxidation during aging and age-related pathologies [22–24]. Oxidation of proteins can lead to generation of protein carbonyls, which are most often formed on the amino acids lysine, arginine, proline, and threonine and by the fragmentation products of peptide bond cleavage reactions. In addition, protein carbonyls are also produced by the addition of carbonyl groups such as aldehydes and ketones, as side chains on these amino acids [21, 25].

Estimation of protein carbonyls gets superiority over the other markers to estimate oxidative damages in proteins due to the fact that protein carbonyls are chemically stable and can be stored at −80 °C for 3 months without changes in detectability [26]. The amount of protein carbonyls in tissue homogenates has been widely reported to increase with age in variety of species including humans [24, 27]. In addition, experimental regimens that are known to prolong life span, such as caloric restriction in rodents and reduction of metabolic rate in insects and other poikilotherms, have been reported to cause a decrease in the amounts of protein carbonyls in tissue homogenates, thus providing a direct evidence for consideration of protein carbonyls as a reliable marker of the aging process [3, 27]. Several studies have shown age-related increase in protein carbonyl level in healthy human subjects [28, 29].

2.2.2 Advanced Oxidation Protein Products

Advanced oxidation protein products (AOPPs) are defined as dityrosine containing cross-linked protein products and are considered as reliable marker to estimate the degree of protein oxidation [30]. The action of chloraminated oxidants, mainly hypochlorous acid and chloramines, is produced by myeloperoxidase in activated neutrophils which lead to the formation of AOPPs.

AOPPs are considered as the novel biomarker to estimate the extent of protein oxidation in the samples [30]. Biochemical characterization has revealed that AOPPs are carried by plasma proteins, especially albumin [31]. AOPPs can be formed in vitro by exposure of serum albumin to hypochlorous acid (HOCl). Thus, AOPPs might be formed during oxidative stress by reaction of plasma proteins with chlorinated oxidants and have been well known to mark the oxidant-mediated protein damage. Elevated levels of AOPPs during aging and age-induced disease have been reported in recent studies [29, 31–33]. In our study, carried out on 80 normal healthy subjects of both sexes between the ages of 18 and 85 years, we have observed an age-dependent increase in plasma AOPP level which correlated significantly with the total antioxidant capacity of the plasma [29]. Study performed by Cakatay and coworkers [33] on hepatic tissues of D-galactose-induced aged rats demonstrated greater susceptibility to oxidative protein damage as evidenced by elevated level of AOPPs. They concluded that this alteration in proteins may be operative in the many age-related liver diseases, which are pertinent to increased oxidative stress and altered redox homeostasis [33].

2.2.3 Advanced Glycation End Products

Advanced glycation end products (AGEs) are formed inside the cell by a nonenzymatic reaction of protein with carbohydrates [34]. AGEs are reported to be involved in the pathogenesis of many life-threatening diseases including diabetes, cancer, and many neuro disorders [34, 35]. Initially described by L. Maillard, it was known that AGEs are only derived from the reaction of carbonyl and amino compounds; however, recent findings have established that AGEs are heterogeneous group of compounds which are generated nonenzymatically by the reaction of reducing sugars and other α-carbonyl compounds with amino groups, not only present on proteins but also on lipids and nucleic acids [34].

In addition to in vivo formation, AGEs are also generated by heating of foods. The preparation of most meals is associated with heating, which accelerates the formation of glycation adducts [36]. Contribution of food-derived AGEs in disease pathophysiology is still under investigation; however, it has been reported that AGEs from food are able to cross the intestinal border or at least able to affect enterocytes/immune cells in the gut [36].

AGEs critically influence structural as well as functional properties of proteins. Many enzymes have been reported to reflect altered activities due to the presence of AGEs [37, 38]. The occurrence of AGEs cross-linking with other vital molecules of the cell results into compromised organ function including heart and kidney dysfunction [39]. Age-dependent accumulation of AGEs has been reported in many tissues including cartilage, peripheral fluids, and skin collagen [34, 35]. Increased deposition of AGEs in kidney is reported as one of the basic causes of diabetic nephropathy in aged individuals [34]. AGE-mediated tissue stiffness is a side effect of reduced protein flexibility associated with atherosclerosis during aging [40]. Elevated level of AGEs has been reported during aging in the lens [41]. Summarizing, AGEs are the major contributor of initiation and development of many age-related events.

2.3 Lipid Peroxidation During Aging

Peroxidation of lipids is the most extensively studied ROS-induced reactions in the body. ROS-induced lipid peroxidation occurs when a reactive hydrogen atom is extracted from the methylene group of an unsaturated fatty acid. Interestingly, once this process begins, lipid peroxidation spreads as an ROS-induced chain reaction until the levels of peroxidation are sufficiently high to result in the production of a non-radical molecule [42].

Cell membranes are highly influenced by lipid peroxidation because of high concentration of lipids in their structure. Lipids are the important constituent of the cell which function as steroid hormones, retinoic acids, and prostaglandins. Their peroxidation affects the cell in various ways [8, 42]. The first oxidation products of lipids are limited in their function as they are either volatile or highly reactive. Some of them can easily react to generate secondary oxidation products like malondialdehyde, 4-hydroxy-2-,3-trans-nonenal (HNE), isoprostanes, or oxysterols [43]. These secondary oxidation products influence gene expression and protein synthesis, and these can lead to further damage by cross-linking proteins. The secondary oxidation products of lipids can be both mutagenic and carcinogenic and play an important role in aging and disease progression. We have explained here only the most prominent and reproducible markers of lipid peroxidation relevant in the aging process.

2.3.1 Malondialdehyde

Malondialdehyde (MDA) is a well-established marker used to investigate the oxidative damage on lipids during many degenerative human diseases [3, 19, 44]. MDA can react with the free amino group of proteins, phospholipids, and nucleic acids leading to their structural modification that can induce dysfunction of the immune system. A high level of MDA can be detected in cell degradation after cell injury or disease [45, 46].

The cell membranes are prone to lipid peroxidation under oxidative stress that involves cleavage of polyunsaturated fatty acids at their double bonds leading to the formation of MDA. Studies on different model systems of aging and associated diseases have shown increased level of MDA in older population [24, 47]. Studies carried out on normal healthy young-, middle-, and old-aged subjects of both sexes in Indian and European population show a significant positive correlation between the erythrocyte MDA level and human age [18, 19]. Since the measurement of MDA is easy and fast to perform and the results are very reproducible, it is one of the most important biomarkers for the evaluation of the status of lipid peroxidation during aging.

2.3.2 Isoprostanes

Isoprostanes are other potent marker of lipid peroxidation [3]. Isoprostanes are prostaglandin-like substances produced in vivo by esterification of arachidonic acid. Isoprostanes have been measured in clinical trials and observational studies to determine the role of lipid peroxidation in aging and disease. It is noted that oxidative stress induces the production of prostaglandins in human cells [48] and that some prostaglandins can produce intracellular stress [49].

Among the many forms of the isoprostanes, the F2-isoprostanes are considered the best available biomarkers of lipid peroxidation and oxidative stress in vivo [50]. The use of F2-isoprostanes as biomarker of lipid peroxidation gets significance because F2-isoprostanes are detectable in liquid form in all body fluids and in their esterified form in biological tissues, indicating physiological levels of oxidative stress [50, 51]. Besides this, determination of F2-isoprostanes as an index of lipid peroxidation and extent of oxidative stress has many advantages over other potential biomarkers of oxidative stress including its chemical stability, a specific product of oxidation formed in vivo being unaffected by diet [51, 52]. Elevated level of F2-isoprostane has been reported in many age-related diseases including atherosclerosis, arthritis, diabetes, and renal failure [53–55].

2.3.3 4-Hydroxynonenal

Specific aldehyde product of lipid peroxidation, 4-hydroxynonenal (HNE), has recently been recognized as a reliable marker of lipid peroxidation during oxidative stress. HNE is a nine carbon amphiphilic product formed during lipid peroxidation when n-6-polyunseturated fatty acids such as arachidonic acid and linoleic acid are attacked by peroxidative free radicals [56]. Besides this, HNE is also reported to form enzymatically in microsomes [57].

The lipophilic property of HNE makes it moving inside the cell and also moving to other cells. During an encounter with proteins, HNE interact with their amino and thiol groups and form covalent bond with amino acids and thus modifying them. Many proteins have been showing impaired functions due to the modification with HNE including membrane-bound transporters, protein chaperons, and proteins of electron transport chain [56, 58].

Accumulating evidence suggests that HNE contributes a major role in pathogenic cellular changes that cause aging and other age-induced diseases such as diabetes and coronary diseases [56, 58, 59]. Age-dependent increase in HNE level has been reported in most of the studies in which HNE have been measured [60, 61]. In human as well as in animal models, elevated levels of HNE in atherosclerosis lesions have been reported [61]. Stroke is the major cause of morbidity and mortality in elderly. HNE is reported to cause dysfunction of cardiac myocytes by disruption of the actin cytoskeleton and deregulation of cellular calcium homeostasis [62]. Alzheimer's and Parkinson's diseases are the most common neuro disorders in aging population. HNE are believed to contribute to the dysfunction and death of neurons [59].

3 Age-Associated Damages in Nucleic Acids

3.1 DNA Damage During Aging

DNA damage represents a critical threat to cell function. If DNA damage is severe or its accumulation exceeds its elimination by DNA repair mechanisms, cellular senescence or apoptosis will occur, and this may contribute to the aging process [63].

ROS oxidize DNA bases and the ribose sugar ring leading to sites of base loss and strand breaks. It is estimated that the rate of damage to DNA by free radicals is about 1,000–1,000,000 hits per day in a single cell [64]. There are well-established reports that an increased baseline level of DNA oxidation is associated with several age-related diseases including cardiovascular diseases, diabetes, cancer, and many neurodegenerative and renal diseases [63, 65–67]. We have described and explained the most potent biomarkers of DNA damages that may be implicated in the aging process.

3.1.1 8-Oxo-7,8-dihydroguanine and 8-Oxo-7,8-dihydro-2′deoxyguanosine

Attack of ROS to DNA generates lesions in DNA that are highly mutagenic. Among the many lesions generated by ROS, 8-oxo-7,8-dihydroguanine (8-oxoGua) is considered the most prominent and degenerative [3]. 8-OxoGua is generated by oxidative damage of DNA at the C-8 position of guanine, with an estimated number of 100–500 8-oxoGua bases arising daily in the genome [68]. Interestingly, it is observed that if left unrepaired, this lesion can result in G-to-T transversion. Oxidation of DNA can also result in 8-oxo-7,8-dihydro-2'deoxyguanosine (8-oxodG) lesions, an alteration to the guanine nucleoside; however, it is reported that 8-oxoGua levels are approximately ten times higher than those of 8-oxodG [69]. Studies done at different laboratories in different countries have reported increased level of 8-oxoGua and 8-oxodG in aging populations [70].

3.1.2 Double-Stranded DNA Breaks

Double-stranded DNA breaks (DSBs) are other most toxic lesions to DNA which are generated by free radical attack. Though DSBs can be generated as a consequence of normal cellular processes like oxidative respiration, however, this rate increases manyfold in case of severe oxidative attack. DSBs are formed when both DNA strands encounter DNA damage, within 10–20 base pairs of each other, resulting in a break of the phosphodiester bond [3]. DSBs are so toxic that the presence of only 1–10 DSBs may induce p53-dependent G1 arrest followed by cell death [71]. Determination of DSBs as marker of DNA damage gets significance since DSBs cannot be repaired by the numerous template-directed repair systems, as both DNA strands are broken [3].

DSBs influence the cellular responses in many ways. It causes phosphorylation of histone H2AX. It is calculated that thousands of H2AX molecules adjacent to the break site become phosphorylated within minutes of the generation of a DSB, resulting in the formation of gamma-H2AX (g-H2AX) foci [72]. Detection of g-H2AX foci has been used as a biomarker for cancer and its development [73], and detection of foci at eroded telomere ends has been used as a marker of aging [74]. Recent studies claim positive relationship between levels of g-H2AX and age. Sedelnikova et al. in 2002 found that levels of endogenous g-H2AX foci increased with age [75]. In a recent report compiled by Schurman and coworkers using lymphocytes from individuals in the Baltimore Longitudinal Study of Aging, an increase in g-H2AX foci with age was documented [76].

4 RNA Damage During Aging

4.1 ROS-Mediated RNA Damage During Aging

Similar to DNA, RNA are also potentially targeted by ROS. There are several reports in humans which document that RNA are more severely attacked by free radicals and are more prone to persistent oxidative damage than DNA since RNA are mainly single stranded, leaving it easily accessible to ROS, and there is no identified active repair mechanism for oxidized RNA [77, 78]. Besides this, RNA are less protected by proteins compared to DNA, and cytoplasmic RNA are near to the mitochondria, the main source of ROS production [78].

Increased RNA oxidization products are reported in many age-associated consequences such as in dementia with Lewy bodies and Alzheimer's disease [79–81]. Since RNA oxidation is influenced by environment and not by genetics [82], measurement of RNA damage provides better understanding toward influence of oxidative stress during aging.

4.1.1 7,8-Dihydro-8-oxo-guanosine (8-OHG)

There are many RNA oxidation products that indicate the RNA damage occurs in the cell frequently; however, most of them lack the indication of the extent of damage and reproducibility. It was proposed that since RNA molecules are very similar to DNA, the oxidative lesions observed in DNA could also be observed on the corresponding bases in RNA [83], and these lesions may serve as marker of oxidative damage in RNA with increased accuracy.

In the past years, the oxidized product of guanosine, 7,8-dihydro-8-oxo-guanosine (8-OHG), has been actively investigated [84]. 8-OHG is the most examined oxidation products of the RNA due to its similarity to the 8-oxoG lesion in DNA and the ability to use many of the same methodologies verified on DNA substrates for the analysis of RNA oxidation products. Determination of 8-OHG in aging cell may mark the oxidative stress-mediated damage in RNA.

5 Plasma Membrane Redox System During Aging

Plasma membrane of the cell plays an important role in many cellular functions including nutritional transport, signal transduction, and regulation of cellular ion homeostasis [85]. Since the plasma membrane is composed of proteins and lipids and these biomolecules are susceptible to oxidative modifications, the plasma membrane is always under threat of oxidative damage that leads to impairments in its deformability, fluidity, and activities of membrane transporters which actively contribute in aging and related cellular consequences [86].

Recent studies suggest that each eukaryotic cell contains a group of oxide reductases, collectively known as plasma membrane redox system (PMRS), which play a very dynamic role in maintaining the redox state of cell [87, 88]. This PMRS transfers electrons from intracellular donors such as nicotinamide adenine dinucleotide reduced (NADH) and/or ascorbate (ASC) to extracellular acceptors. Although the exact physiological function through which PMRS maintains redox state is under investigation, proposed functions include maintenance of redox state of sulfhydryl residues in membrane proteins, neutralization of oxidative stressors outside the cells, stimulation of cell growth, recycling of vitamin E, reduction of lipid hydroperoxides, and reduction of ferric ion prior to iron uptake by a transferring-independent pathway [89, 90].

It has been investigated that the PMRS along with ascorbate free radical (AFR) reductase works to maintain the extracellular concentration of ASC by using an electron derived from intracellular ASC or other donors [90]. The concerted action of PMRS and AFR reductase seems very significant since ASC is the primary antioxidant present in the plasma, providing a first line of defense against stressors. It also serves as a cofactor in many enzyme reactions, including

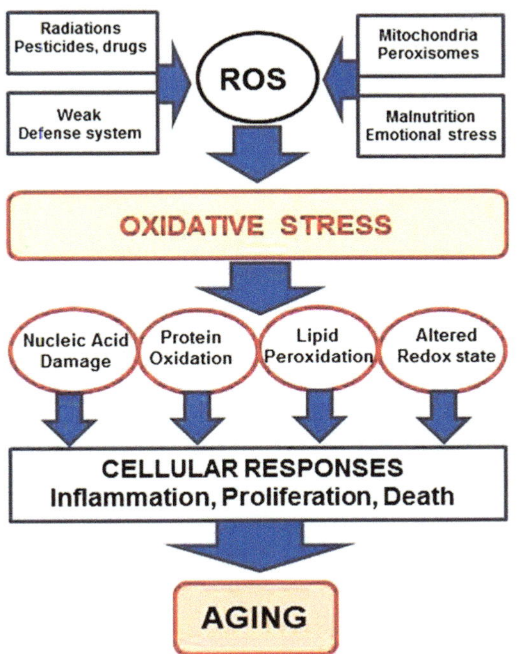

Fig. 2 Flow diagram of the oxidative stress-mediated aging process. *ROS* reactive oxygen species

those involved in the synthesis of catecholamines and peptide hormones [91]. Despite such crucial functions of ASC, humans are unable to synthesize ASC in the body due to lack of functional L-gulonolactone oxidase, the final enzyme of the ASC biosynthetic pathway in mammals [92]. In the presence of an oxidant, ASC is oxidized first to AFR and then to dehydroascorbate (DHA), which is unstable and undergoes irreversible hydrolysis to 2,3-diketo-L-gulonic acid, resulting in decreased level of the vitamin. Two molecules of AFR can react with each other to form one each of ASC and DHA.

It has been reported that activities of PMRS and AFR reductase increase with increase in age and that the activation of PMRS along with AFR reductase is a protective mechanism that operates to maintain the ASC level in the plasma, and thereby minimizing oxidative stress during aging [13, 90, 93]. Hyun and coworkers have reported that the enhancement of the PMRS is a mechanism by which caloric restriction may counteract mitochondrial dysfunction and oxidative stress in the brain during aging [88]. All these documentations emphasize that PMRS plays a very significant role in maintaining the redox state and may be considered as novel biomarker of aging. It may be summarized that the oxidative stress-mediated damage to the vital biomolecules followed by their impaired functions plays a major role in progression and development of aging and related pathological events (Fig. 2).

6 Conclusion

Oxidative stress-mediated shift in redox state plays a very crucial role in the aging process and other related consequences. Various biochemical parameters that reflect the balance between oxidant and prooxidant stage get directly affected and altered during aging. Many parameters mark the extent of oxidative damages; however, not all of them can be used as biomarker of aging process due to influence of various factors including sex and types of tissues incorporated in the study. In this context, there is a need for identification of novel biomarkers to represent age-related impairments. Parameters included in the chapter are reproducible and least affected by other factors and may be successfully used to assess aging-related changes in redox state.

Acknowledgement The work is supported by Council of Scientific and Industrial Research (CSIR), New Delhi, India, in the form of Research Associateship to KBP. SIR acknowledges the support of SERB-DST in the form of grant SR/SO/HS-147/2010.

References

1. Rattan SIS (2006) Theories of biological aging: genes, proteins and free radicals. Free Radic Res 40:10–12
2. Sohal RS, Orr WC (2012) The redox stress hypothesis of aging. Free Radic Biol Med 52:539–555
3. Jacob KD, Noren Hooten N et al (2013) Markers of oxidant stress that is clinically relevant in aging and age-related disease. Mech Ageing Dev 134:139–157
4. Harman D (2009) Origin and evolution of the free radical theory of aging: a brief personal history, 1954–2009. Biogerontology 10:773–781

5. Vina J, Borrás C, Miquel J (2007) Theories of ageing. IUBMB Life 59:249–254
6. Cefalu CA (2011) Theories and mechanisms of aging. Clin Geriatr Med 27:491–506
7. Harman D (1956) Ageing: a theory based on free radical and radiation chemistry. J Gerontol 11:298–300
8. Halliwell B, Gutteridge JMC (2007) Cellular responses to oxidative stress: adaptation, damage, repair, senescence and death. In: Free radicals in biology and medicine, 4th edn. Oxford University Press, New York
9. Sies H (1986) Biochemistry of oxidative stress. Angew Chem Int Ed 25:1058–1071
10. Sohal RS, Allen RG (1990) Oxidative stress as a causal factor in differentiation and aging: a unifying hypothesis. Exp Gerontol 25:499–522
11. Junqueira VB, Barros SB, Chan SS et al (2004) Aging and oxidative stress. Mol Asp Med 25:5–16
12. Rizvi SI, Jha R (2011) Strategies for the discovery of anti-aging compounds. Expert Opin Drug Discov 6:89–102
13. Pandey KB, Rizvi SI (2013) Resveratrol up-regulates the erythrocyte plasma membrane redox system and mitigates oxidation-induced alterations in erythrocytes during aging in humans. Rejuvenation Res 16:232–240
14. Droge W (2002) Free radicals in the physiological control of cell function. Physiol Rev 82:47–95
15. Schafer FQ, Buettne GR (2001) Redox environment of the cell as viewed through the redox state of the glutathione disulfide/glutathione couple. Free Radic Biol Med 30:1191–1212
16. Wu G, Fang YZ, Yang S et al (2004) Glutathione metabolism and its implications for health. J Nutr 134:489–492
17. Erden-Inal M, Sunal E, Kanbak G (2002) Age-related changes in the glutathione redox system. Cell Biochem Funct 20:61–66
18. Gil L, Siems W, Mazurek B et al (2006) Age associated analysis of oxidative stress parameters in human plasma and erythrocytes. Free Radic Res 40:495–505
19. Rizvi SI, Maurya PK (2007) Markers of oxidative stress in erythrocytes during aging in humans. Ann N Y Acad Sci 1100:373–382
20. Rebrin I, Kamzalov S, Sohal RS (2003) Effects of age and caloric restriction on glutathione redox state in mice. Free Radic Biol Med 35:626–635
21. Hawkins CL, Morgan PE, Davies MJ (2009) Quantification of protein modification by oxidants. Free Radic Biol Med 46:965–988
22. Beal MF (2002) Oxidatively modified proteins in aging and disease. Free Radic Biol Med 32:797–803
23. Levine RL, Stadtman ER (2001) Oxidative modification of proteins during aging. Exp Gerontol 36:1495–1502
24. Pandey KB, Rizvi SI (2010) Markers of oxidative stress in erythrocytes and plasma during aging in humans. Oxidative Med Cell Longev 3:2–12
25. Uchida K, Stadtman ER (1993) Covalent attachment of 4-hydroxynonenal to glyceraldehyde-3-phosphate dehydrogenase. A possible involvement of intra- and intermolecular cross-linking reaction. J Biol Chem 268:6388–6393
26. Griffiths HR (2000) Antioxidants and protein oxidation. Free Radic Res 33:47–58
27. Sohal RS (2002) Role of oxidative stress and protein oxidation in the aging process. Free Radic Biol Med 33:37–44
28. Voss P, Siems W (2006) Clinical oxidation parameters of aging. Free Radic Res 40:1339–1349
29. Pandey KB, Mehdi MM, Maurya PK et al (2010) Plasma protein oxidation and its correlation with antioxidant potential during human aging. Dis Markers 29:31–36
30. Witko-Sarsat V, Friedlander M, Capeillère-Blandin C et al (1996) Advanced oxidation protein products as a novel marker of oxidative stress in uremia. Kidney Int 9:1304–1313
31. Witko-Sarsat V, Friedlander M, Nguyen Khoa T et al (1998) Advanced oxidation protein products as novel mediators of inflammation and monocyte activation in chronic renal failure. J Immunol 161:2524–2532
32. Pandey KB, Mishra N, Rizvi SI (2010) Protein oxidation biomarkers in plasma of type 2 diabetic patients. Clin Biochem 43:508–511
33. Cakatay U, Aydın S, Atukeren P et al (2013) Increased protein oxidation and loss of protein-bound sialic acid in hepatic tissues of D-galactose induced aged rats. Curr Aging Sci 6:135–141
34. Nass N, Bartling B, Navarrete Santos A et al (2007) Advanced glycation end products, diabetes and ageing. Z Gerontol Geriatr 40:349–356
35. Höhn A, König J, Grune T (2013) Protein oxidation in aging and the removal of oxidized proteins. J Proteome 92:132–159
36. Somoza V (2007) The maillard reaction in food and medicine. Mol Nutr Food Res 51:381–382
37. Ahmed N, Thornalley PJ (2005) Peptide mapping of human serum albumin modified minimally by methylglyoxal in vitro and in vivo. Ann N Y Acad Sci 1043:260–266
38. Zeng J, Davies MJ (2005) Evidence for the formation of adducts and S- (carboxymethyl) cysteine on reaction of alpha-dicarbonyl compounds with thiol groups on amino acids, peptides, and proteins. Chem Res Toxicol 18:1232–1241
39. Simm A, Wagner J, Gursinsky T et al (2007) Advanced glycation endproducts: a biomarker for age as an outcome predictor after cardiac surgery? Exp Gerontol 42:668–675
40. Badenhorst D, Maseko M, Tsotetsi OJ et al (2003) Cross-linking influences the impact of quantitative changes in myocardial collagen on cardiac stiffness and remodelling in hypertension in rats. Cardiovasc Res 57:632–641
41. Viteri G, Carrard G, Birlouez-Aragon I et al (2004) Age-dependent protein modifications and declining proteasome activity in the human lens. Arch Biochem Biophys 427:197–203

42. Niki E (2009) Lipid peroxidation: physiological levels and dual biological effects. Free Radic Biol Med 47:469–484
43. Siems W, Grune T (2005) Lipid peroxidation measurements methodological approaches and clinical importance. In: Grune T (ed) Free radicals and diseases: gene expression, cellular metabolism and pathophysiology, vol 367, NATO science series. IOS Press, Oxford, pp 11–21
44. Pandey KB, Mishra N, Rizvi SI (2009) Myricetin may provide protection against oxidative stress in type 2 diabetic erythrocytes. Z Naturforsch C 64:626–630
45. Pandey KB, Rizvi SI (2009) Protective effect of resveratrol on formation of membrane protein carbonyls and lipid peroxidation in erythrocytes subjected to oxidative stress. Appl Physiol Nutr Metab 34:1093–1097
46. Lykkesfeldt J (2007) Malondialdehyde as biomarker of oxidative damage to lipids caused by smoking. Clin Chim Acta 380:50–58
47. Radak Z, Zhao Z, Goto S et al (2011) Age-associated neurodegeneration and oxidative damage to lipids, proteins and DNA. Mol Asp Med 32:305–315
48. Malek A, Sager R, Schneider H (2001) Effect of hypoxia, oxidative stress and lipopolysaccharides on the release of prostaglandins and cytokines from human term placental explants. Placenta 22:45–50
49. Kondo M, Oya-Ito T, Kumagai T et al (2001) Cyclopentenone prostaglandins as potential inducers of intracellular oxidative stress. J Biol Chem 276:12076–12083
50. Roberts LJ, Morrow JD (2000) Measurement of F (2)-isoprostanes as an index of oxidative stress in vivo. Free Radic Biol Med 28:505–513
51. Montuschi P, Barnes P, Roberts LJ 2nd (2007) Insights into oxidative stress: the isoprostanes. Curr Med Chem 14:703–717
52. Gopaul NK, Halliwell B, Anggrd EE (2000) Measurement of plasma F2-isoprostanes as an index of lipid peroxidation does not appear to be confounded by diet. Free Radic Res 33:115–127
53. Gopaul NK, Anggard EE, Mallet AI et al (1995) Plasma 8-epi-PGF2 alpha levels are elevated in individuals with noninsulin dependent diabetes mellitus. FEBS Lett 368:225–229
54. Davies SS, Roberts LJ II (2011) F2-isoprostanes as an indicator and risk factor for coronary heart disease. Free Radic Biol Med 50:559–566
55. Basu S, Whiteman M, Mattey DL et al (2001) Raised levels of F2-isoprostanes and prostaglandin F2a in different rheumatic diseases. Ann Rheum Dis 60:627–631
56. Mattson MP (2009) Roles of the lipid peroxidation product 4-hydroxynonenal in obesity, the metabolic syndrome, and associated vascular and neurodegenerative disorders. Exp Gerontol 44:625–633
57. Esterbauer H, Benedetti A, Lang J et al (1986) Studies on the mechanism of formation of 4-hydroxynonenal during microsomal lipid peroxidation. Biochim Biophys Acta 876:154–166
58. Poli G, Schaur RJ, Siems WG et al (2008) 4-hydroxynonenal: a membrane lipid oxidation product of medicinal interest. Med Res Rev 28:569–631
59. Hardas SS, Sultana R, Clark AM et al (2013) Oxidative modification of lipoic acid by HNE in Alzheimer disease brain. Redox Biol 1:80–85
60. Kinoshita M, Sakamoto T, Kashio A et al (2013) Age-related hearing loss in Mn-SOD heterozygous knockout mice. Oxidative Med Cell Longev 2013:325702
61. Shoeb M, Ansari NH, Srivastava SK et al (2013) 4-hydroxynonenal in the pathogenesis and progression of human diseases. Curr Med Chem 21:230–237
62. VanWinkle WB, Snuggs M, Miller JC et al (1994) Cytoskeletal alterations in cultured cardiomyocytes following exposure to the lipid peroxidation product, 4-hydroxynonenal. Cell Motil Cytoskeleton 28:119–134
63. Chen JH, Hales CN, Ozanne SE (2007) DNA damage, cellular senescence and organismal ageing: causal or correlative? Nucleic Acids Res 35:7417–7428
64. Ames BN, Shigenaga MK, Hagen TM (1993) Oxidants, antioxidants, and the degenerative diseases of aging. Proc Natl Acad Sci U S A 90:7915–7922
65. Collins AR, Gedik CM, Olmedilla B et al (1998) Oxidative DNA damage measured in human lymphocytes: large differences between sexes and between countries, and correlations with heart disease mortality rates. FASEB J 12:1397–1400
66. Malins DC, Johnson PM, Wheeler TM et al (2001) Age-related radical-induced DNA damage is linked to prostate cancer. Cancer Res 61:6025–6028
67. Morocz M, Kalman J, Juhasz A et al (2002) Elevated levels of oxidative DNA damage in lymphocytes from patients with Alzheimer's disease. Neurobiol Aging 23:47–53
68. Lindahl T (1993) Instability and decay of the primary structure of DNA. Nature 362:709–715
69. Andreoli R, Mutti A, Goldoni M et al (2011) Reference ranges of urinary biomarkers of oxidized guanine in (20-deoxy) ribonucleotides and nucleic acids. Free Radic Biol Med 50:254–261
70. Siomek A, Gackowski D, Rozalski R et al (2007) Higher leukocyte 8-oxo-7,8-dihydro-20-deoxyguanosine and lower plasma ascorbate in aging humans? Antioxid Redox Signal 9:143–150
71. Huang LC, Clarkin KC, Wahl GM (1996) Sensitivity and selectivity of the DNA damage sensor responsible for activating p53-dependent G1 arrest. Proc Natl Acad Sci U S A 93:4827–4832
72. Pilch DR, Sedelnikova OA, Redon C et al (2003) Characteristics of gamma-H2AX foci at DNA double-strand breaks sites. Biochem Cell Biol 81:123–129
73. Ivashkevich A, Redon CE, Nakamura AJ (2012) Use of the gamma-H2AX assay to monitor DNA damage and repair in translational cancer research. Cancer Lett 327:123–133
74. Nakamura AJ, Chiang YJ, Hathcock KS et al (2008) Both telomeric and non-telomeric DNA damage are

determinants of mammalian cellular senescence. Epigenetics Chromatin 1:6
75. Sedelnikova OA, Rogakou EP, Panyutin IG et al (2002) Quantitative detection of (125)IdU-induced DNA double-strand breaks with gamma-H2AX antibody. Radiat Res 158:486–492
76. Schurman SH, Dunn CA, Greaves R et al (2012) Age-related disease association of endogenous gamma-H2AX Foci in mononuclear cells derived from leukapheresis. PLoS One 7:e45728
77. Henriksen T, Hillestrom PR, Poulsen HE et al (2009) Automated method for the direct analysis of 8-oxo-guanosine and 8-oxo-20-deoxyguanosine in human urine using ultraperformance liquid chromatography and tandem mass spectrometry. Free Radic Biol Med 47:629–635
78. Hofer T, Badouard C, Bajak E et al (2005) Hydrogen peroxide causes greater oxidation in cellular RNA than in DNA. Biol Chem 386:333–337
79. Nunomura A, Chiba S, Kosaka K et al (2002) Neuronal RNA oxidation is a prominent feature of dementia with Lewy bodies. Neuro Rep 13:2035–2039
80. Isobe C, Abe T, Terayama Y (2009) Homocysteine may contribute to pathogenesis of RNA damage in brains with Alzheimer's disease. Neurodegener Dis 6:252–257
81. Martinet W, Knaapen MWM, De Meyer GRY et al (2002) Elevated levels of oxidative DNA damage and DNA repair enzymes in human atherosclerotic plaques. Circulation 106:927–932
82. Broedbaek K, Ribel-Madsen R, Henriksen T et al (2011) Genetic and environmental influences on oxidative damage assessed in elderly Danish twins. Free Radic Biol Med 50:1488–1491
83. Bellacosa A, Moss EG (2003) RNA repair: damage control. Curr Biol 13:482–484
84. Tanaka M, Han S, Kupfer PA et al (2011) An assay for RNA oxidation induced abasic sites using the aldehyde reactive probe. Free Radic Res 45:237–247
85. Wang X, Wu Z, Song G et al (1999) Effects of oxidative damage of membrane protein thiol group on erythrocyte membrane viscoelasticities. Clin Hemorheol Microcirc 21:137–146
86. Pandey KB, Rizvi SI (2011) Biomarkers of oxidative stress in red blood cells. Biomed Pap 155:131–136
87. Rizvi SI, Jha R, Maurya PK (2006) Erythrocyte plasma membrane redox system in human aging. Rejuvenation Res 9:470–474
88. Hyun DH, Emerson SS, Jo DG et al (2006) Calorie restriction upregulates the plasma membrane redox system in brain cells and suppresses oxidative stress during aging. Proc Natl Acad Sci U S A 103:19908–19912
89. VanDuijn MM, Van den Zee J, VanSteveninck J et al (1998) Ascorbate stimulates ferricyanide reduction in HL-60 cells through a mechanism distinct from the NADH-dependent plasma membrane reductase. J Biol Chem 273:13415–13420
90. Rizvi SI, Pandey KB, Jha R et al (2009) Ascorbate recycling by erythrocytes during aging in humans. Rejuvenation Res 12:3–6
91. Harrison FE, May JM (2009) Vitamin C function in the brain: vital role of the ascorbate transporter SVCT2. Free Radic Biol Med 46:719–730
92. Nishikimi M, Fukuyama R, Minoshima S et al (1994) Cloning and chromosomal mapping of the human nonfunctional gene for L-gulono-gamma-lactone oxidase, the enzyme for L-ascorbic acid biosynthesis missing in man. J Biol Chem 269:13685–13688
93. Pandey KB, Rizvi SI (2012) Upregulation of erythrocyte ascorbate free radical reductase by tea catechins: correlation with their antioxidant properties. Food Res Int 46:46–49

Part IV

Strategies to Ameliorate Oxidative Stress Induced Diseases

Antioxidant Supplements: Friend or Foe?

Saikat Sen and Raja Chakraborty

Abstract

The role of antioxidant compounds against harmful effect of free radical is acknowledged widely. In 1980s, antioxidants had emerged as a "miracle substance" and have come to public attention. Subsequently, the use of antioxidant supplements has increased significantly among the common people in the hope of preventing acute/chronic diseases. A number of both recent and previous investigations primarily epidemiological and observational studies have reported the beneficial effect of antioxidant supplements. Similar effect of antioxidant supplements was also confirmed by different randomized controlled trials (RCTs). However, the previous reports were not consistent, and potential clinical benefit derived from supplemental antioxidants for the general population is still under wide debate. Though to some extent, antioxidant supplementation did not seem to confer clinical benefit, it may increase the risk of particular disease. The reliance on antioxidant supplements for preventing disease is a concern. Although existing information does not allow a final and conclusive assessment of the relevance of regular intake of antioxidant supplementation for health, it does provide the basis for its rational consideration. This chapter addresses the specific aspects of antioxidant supplementation in health and disease in light of epidemiological evidences particularly RTCs, highlighting that foods rich in antioxidant compounds are generally safe than the regular intake of antioxidant supplements, exhibiting interesting health favorable effects.

S. Sen (✉) • R. Chakraborty
Department of Pharmacy, Assam Down Town University, Gandhinagar, Panikhaiti, Guwahati, Assam, 781 026, India
e-mail: saikat.pharm@rediffmail.com

V. Rani and U.C.S. Yadav (eds.), *Free Radicals in Human Health and Disease*,
DOI 10.1007/978-81-322-2035-0_19, © Springer India 2015

Keywords
Antioxidant supplement • Oxidative stress • Disease • Randomized controlled trials • Foods

1 Introduction

Oxidative stress, a condition associated with the production of excess of free radical beyond the capacity of the antioxidant cascade. Imbalance between oxidants and antioxidants leads to many physiological/biochemical changes and may contribute to different diseases. Reactive oxygen species (ROS) and reactive nitrogen species (RNS) are the by-products of cellular redox process and are responsible for both beneficial and harmful effects. The delicate balance between these two opposite effects is certainly a key feature of life [1, 2]. When the generation of ROS/RNS is increased, the risk of biological damage can also be increased, resulting in the development of many pathological conditions [3]. Pollution, cigarette smoking, chemicals, drugs, illness, stress, radiation, and unhealthy lifestyle can aggravate the generation of free radicals. Antioxidants scavenge ROS/RNS and suppress free radical generation, resulting in improved redox balance. Endogenous enzymatic or non-enzymatic antioxidant system acts along with exogenous antioxidants that function interactively and synergistically to neutralize the excess free radicals and protect the cells against oxidative stress [1, 3].

A healthy diet provides numerous vitamins, abundant micronutrients, and trace elements that are indispensable for good health. Several observational studies have found the positive relationship between higher intake of fruits and vegetables and reduced risk of chronic diseases. It is commonly established that fruits and vegetable which contain several vitamins and micronutrients can also act as antioxidant and free radical scavenging effect of those antioxidant substance may be beneficial for health. However, exactly which mechanism and which specific dietary constituents of fruits and vegetables might be helpful is not clear [4, 5]. Three decades back in the 1980s, antioxidant had emerged as "miracle substance," when scientists have recognized the role of free radical in the pathogenesis of different diseases and aging. Thus antioxidants have received considerable public attention, and intake of antioxidants either through healthy diet (fruits and vegetables) or as supplements became apparent in general people. Currently, there has been a strong general belief that intakes of antioxidant supplements beyond the daily intake of food/vegetable may have a role in the prevention of many diseases. In contrast to the hypothesis "antioxidant supplements offer only positive effect," several randomized controlled trials (RCTs) have been carried out. But many of these studies have failed to confirm any effect of antioxidant supplements on hard endpoints such as morbidity and mortality [4–6].

The previous overoptimistic attitude has obviously called for a more realistic and long-term assessment to find out the beneficial/harmful/no effect of antioxidant supplements. In this chapter, the necessity of antioxidant supplements to maintain optimal health and prevent chronic diseases is discussed based on the current evidences. However, causal inferences are hard to establish from observational investigations, but conclusive scientific understanding on the role of antioxidant supplements in health promotion and mechanisms underlying the interplay between antioxidant and health in relation to disease prevention are also briefly reviewed.

2 Antioxidants in Supplements and Their Role

Since 1990s, the field of antioxidant research is moving rapidly due to its potential as promising preventive agents. In this light, efforts have been made to confirm the effects of natural and synthetic antioxidants on the development and prevention of diseases. Fruits and vegetables are sources of numerous vitamins and micronutrients like vitamin C, vitamin E, β-carotene, flavonoid, and selenium, etc. [5–8]. There are thousands of different natural substances that can act as antioxidants. The most common are vitamin C, vitamin E, β-carotene, flavonoids, phenols, polyphenols, phytoestrogens, and several carotenoids, along with the minerals like selenium, zinc, manganese, glutathione, coenzyme Q10, lipoic acid, and many more. Along with natural antioxidants, several synthetic phenolic antioxidants (i.e., butylated hydroxyanisole, butylated hydroxytoluene, and propyl gallate) have also come to the attention of pharmaceutical and food processing industries as they effectively prevent the formation of free radicals, inhibit oxidation, and act as chelating agents [9]. These synthetic antioxidants are generally used in processed fruits and vegetables, soft drinks, margarine, and canned shellfish. But beneficial interactions of antioxidant substances with physical and chemical noxae are compared to those leading to adverse effects, for example, radiosensitization, increased toxicity of other chemicals, increased mutagen activity, and increased tumor yield from chemical carcinogens [1]. Natural compounds are generally preferred in comparison with synthetic one with the hope of safety and less toxicity. Since the mid-1990s, food industries have increased the sales of natural and organic foods as consumers have expressed fear about the safety of preservatives and additives in the food [9]. In vitro and in vivo experiments have shown that natural antioxidants confer the protection against free radical-mediated lipid peroxidation, DNA and protein oxidation, and oxidative stress-related mitochondrial dysfunction [1]. Table 1 discusses the general antioxidant mechanism of different dietary supplements.

While there is no doubt that the correct redox balance as well as optimum balance between endogenous and exogenous antioxidant capacity is crucial to life, the curative power of antioxidants has often been overestimated (Figs. 1 and 2) [9]. Some studies have shown that people with low intakes of antioxidant-rich foods were at greater risk for developing chronic diseases than other people who ate plenty of these foods. Two decades back, clinical trials began with testing the efficacy of single substances, like β-carotene and vitamin E, as weapons against heart disease, cancer, etc. [10]. Subsequently, clinical trials of different antioxidant supplements alone and in combination have started in a large scale. Even before the results of these trials, there was a huge hype describing the benefits of "antioxidant supplements," mostly as a business strategy. Unfortunately, the trials were mixed, but most have not observed the hopes for benefit. These mostly disappointing results haven't stopped the business and popularity of antioxidants. Indeed, antioxidant supplements represent a $500 million industry which is still increasing. Antioxidants are consumed by people in addition to their normal diet or added to the breakfast cereals, sports bars, energy drinks, and other processed foods without the advice of healthcare professionals, with the hope that they can promote the health and prevent the diseases [10, 11]. This section describes the findings of clinical or large-scale trials and meta-analysis of RCTs when the antioxidant supplements were used alone and in combination and possible explanations for the differences.

2.1 Vitamin C

Vitamin C is considered as a potent antioxidant, and small amount of vitamin C can protect the indispensable biomolecules from oxidative stress. Several disorders like high blood pressure, atherosclerosis, stroke, gallbladder disease, and different cancers have been associated with low level of vitamin C [6, 12, 13]. The optimum quantity of vitamin C intake, status of plasma level, and deficiency of vitamin C remain a matter of controversy. The recommended dietary allowances (RDA) of vitamin C for adults (>19 years) are 90 mg/day for men and 75 mg/day for women. The intake of 100 mg/day of ascorbic acid is found to be enough to saturate the body pools (neutrophils, leukocytes, and other tissues) in healthy individuals, which

Table 1 Different dietary antioxidants and their general mechanism

Antioxidants	General mechanism
Vitamin E	Quenches or neutralizes the excited triplet states of oxygen and different radicals like superoxide anion, hydroxyl, alkoxyl, peroxyl, hydroperoxyl, nitrogen dioxide, nitroxide, peroxynitrite radicals
	Beaks lipid peroxidation chain reactions
	Upregulates the antioxidant enzymes and regenerate several other antioxidants such as α-tocopheroxyl, urate, and β-carotene radical cation from their respective radical species
Vitamin C	Scavenges superoxide, hydroxyl radicals
	Neutralizes oxidants from stimulated neutrophils
	Mediates electron transfer to ascorbate-dependent peroxidases or regenerating membrane bound vitamin E. Thus, indirectly reduce lipid peroxidation in cell membranes
	Reducing power ability
	Recycles oxidized glutathione to reduce glutathione
Coenzyme Q_{10}	It inhibits lipid peroxidation
	Reduces mitochondrial oxidative stress
	Recycles vitamin E
Selenium	Scavenging active oxygen species as an essential constituent of glutathione peroxidase, thioredoxin reductase
Carotenoids [*i.e., β-carotene, lycopene, lutein, zeaxanthin*]	Quench singlet oxygen
	α-Carotene and β-carotene and β-cryptoxanthin can be converted to vitamin A
	Lutein and zeaxanthin can stabilize the membrane integrity and can competently act as secondary antioxidants
	Neutralize reactive radicals via electron transfer and generating carotenoid radical cations
	May interact with peroxyradicals to form a large resonance stabilized radical
	May scavenge free radicals through hydrogen atom mechanism transfer, resulting in alkyl radicals
	Lycopene can interact with superoxide anions and may produce carotenoid radical anion
	Carotenoids act as lipophilic antioxidants and prevent the oxidative damage on polyunsaturated fatty acids
	They can act as chain-breaking antioxidants by inhibiting the propagation of lipid peroxidation
	Can scavenge lipid peroxyl radicals
	Inactivation of lipoxygenase activity
Flavonoids [*i.e., quercetin, catechins, proanthocyanidins, etc.*] and nonflavonoid polyphenolics [*i.e., chlorogenic acid, secoisolariciresinol, resveratrol*]	Act by chelating prooxidant transition metal ions (Fe^{2+})
	Scavenge hydroxyl radical, superoxide anions, alkoxyl radicals, lipid peroxyl, and lipid alkoxyl radicals
	Act as singlet oxygen quenchers and scavenge of free radicals directly
	Inhibit lipid peroxidation processes
	Inhibiting the catalytic activity of several enzymes (i.e., xanthine oxidase, lipoxygenase, cyclooxygenase, and NADPH oxidase) eliciting ROS formation
	Interfere with inducible nitric oxide synthase activity
	Decrease the number of immobilized leukocytes during reperfusion

Exogenous Antioxidants
- used as supplements
Vitamins
Ex:
Vitamin C
Vitamin E
Minerals/
Trace elements
Ex:
Zinc
Selenium
Coenzyme Q10
Isolated
Phytochemicals
Ex:
Carotenoids: β-carotene, lycopene, lutein
Flavonols: quercetin
Isoflavones: genistein
Flavones: luteolin
Plant extracts
Ex:
Ginkgo biloba extract
Ginseng extract

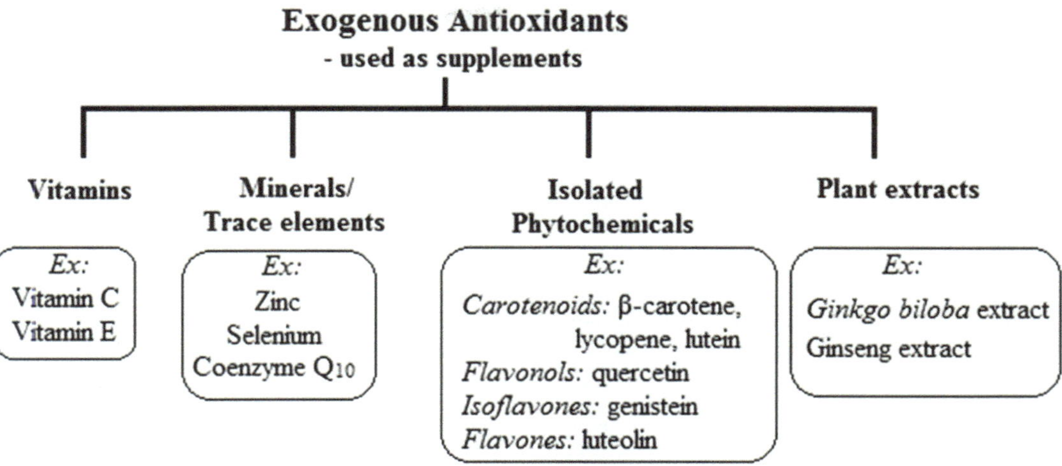

Fig. 1 Different endogenous antioxidants used as supplements

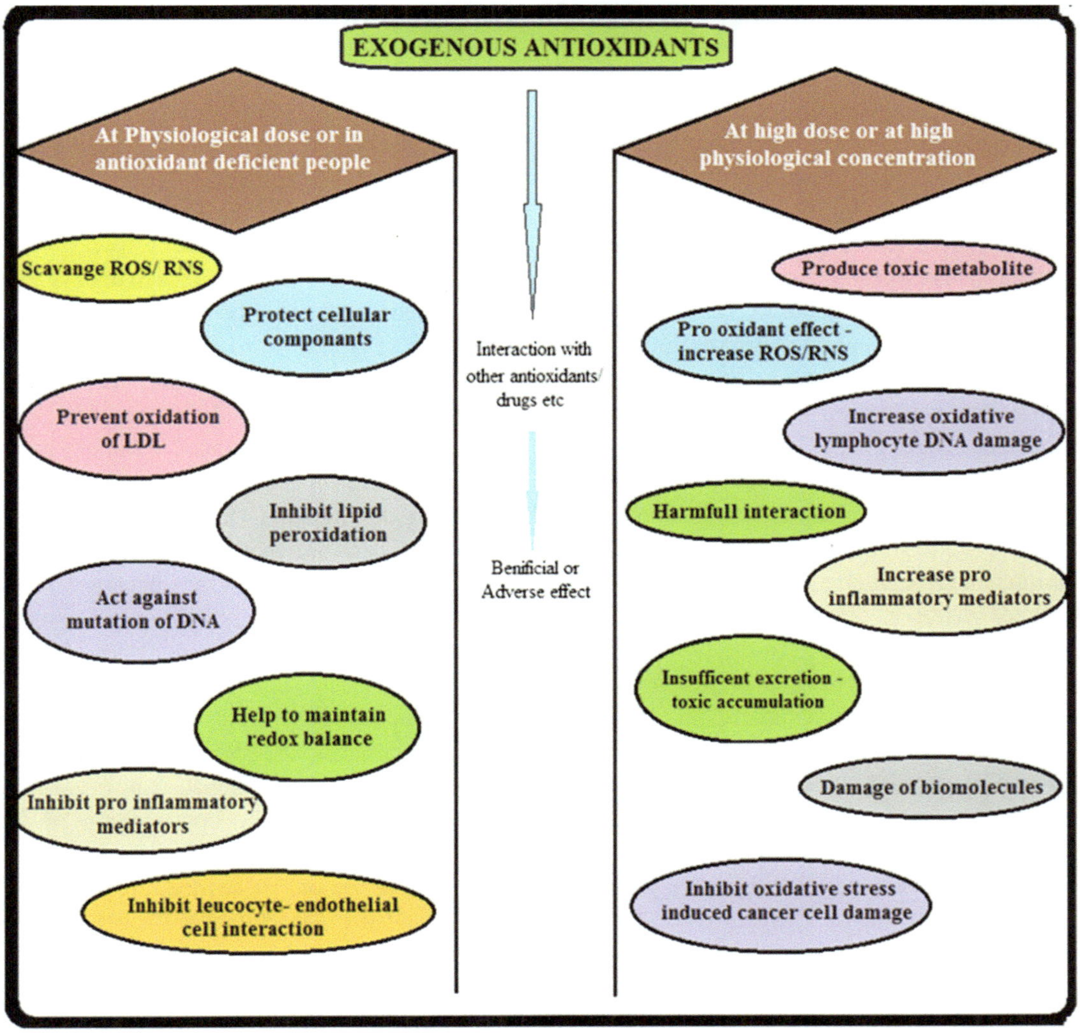

EXOGENOUS ANTIOXIDANTS
At Physiological dose or in antioxidant deficient people
Scavange ROS/ RNS
Protect cellular componants
Prevent oxidation of LDL
Inhibit lipid peroxidation
Act against mutation of DNA
Help to maintain redox balance
Inhibit pro inflammatory mediators
Inhibit leucocyte- endothelial cell interaction
Interaction with other antioxidants/ drugs etc
Benificial or Adverse effect
At high dose or at high physiological concentration
Produce toxic metabolite
Pro oxidant effect - increase ROS/RNS
Increase oxidative lymphocyte DNA damage
Harmfull interaction
Increase pro inflammatory mediators
Insufficent excretion - toxic accumulation
Damage of biomolecules
Inhibit oxidative stress induced cancer cell damage

Fig. 2 Double-edged effects of exogenous antioxidants

results optimal vitamin C level in plasma of 70 mmol/l [6, 14, 15]. It was recently reported that more than 20 % of Canadians do not get sufficient vitamin C from their diet and about one third of the population takes vitamin C in the form of supplements [16]; similar condition may persist in other countries also.

Several preclinical and preliminary studies have reported the beneficial effect of vitamin C in certain pathological condition like cold, wound, atherosclerosis, cancer, lung damage, and oxidative stress in CNS, sperm [13, 14]. Some investigations have suggested that the ingestion of ascorbic acid (1–2 g) effectively prevents/ameliorates common cold. But the beneficial effect of ascorbic acid supplement remains controversial, as most of RCTs did not find an overall significant effect of ascorbic acid supplement [14, 15]. A number of studies have demonstrated that large dose of vitamin C after onset of cold did not reduce the duration or severity of cold symptoms. Though some past trials have found that ascorbic acid reduced the severity and duration of cold symptoms during the period of infection or may avert it, this effect may link with the stimulation of immune response by ascorbic acid [14, 15]. A meta-analysis of available RCTs found that vitamin C supplementation significantly lowers serum uric acid, but authors have suggested for further study to establish the role of vitamin C in the treatment of gout [17].

The role of vitamin C in cardiovascular diseases (CVDs) is also a matter of concern. A number of animal studies have suggested that vitamin C confers beneficial effect against oxidative stress-related diseases. Several short-term studies on human also have supported the phenomena that vitamin C supplements improve the lipid profile and antioxidant capacity in patient with CVDs [12]. However, most of the long-term studies where myocardial infarction (MI), stroke, and CVD death were considered as ultimate outcome have failed to find any beneficial effect of vitamin C [12, 13]. The hypothesis regarding the protective effect of vitamin C against CVDs is supported by different mechanisms/researches like vitamin C protects oxidation of isolated LDL against oxidative stress, inhibits lipid oxidation, inhibits leukocyte–endothelial cell interactions induced by cigarette smoking, etc. [12, 18]. Some cross-sectional and longitudinal studies have reported that the risk of CVDs is inversely related to vitamin C intake and plasma vitamin C concentration [18]. But few observational studies have found the negative correlation between the risk of cardiovascular complications and dietary intake of vitamin C alone or in combination of other antioxidants, though most of the RCTs did not support this phenomenon [12, 18]. However, contradictory results have also been reported by different investigations carried out in different parts of the world. It was reported that high consumption of vitamin C reduced the risk of death from CVDs in Finnish and American women only, while another cohort study found beneficial effects of vitamin C on cardiovascular mortality in both sexes [12, 13, 18]. Majority of recent cohort studies did not find any impact of vitamin C supplementation in coronary heart disease mortality. However, a follow-up study of NHANES-I found that regular use of a vitamin C supplement decreased the standardized mortality ratio for cardiovascular mortality by 48 % and for all-cause mortality by 26 %. However, the results and conclusions of these studies are not consistent [19].

Vitamin C works against mutation of deoxyribonucleic acid (DNA), and this might be a key reason for the expected clinical value of vitamin C in the treatment of certain cancers. The consumption of vitamin C much higher than the RDA may decrease the risk or risk factors for certain types of cancer and may lengthen the life span of cancer patients. In several existing population-based studies, the incidence of cancer (like skin, cervical, and, possibly, breast cancer) may be prevented by the consumption of foods rich in vitamin C [6, 15]. The Department of Agriculture and the National Cancer Institute, USA, recommended the consumption of five fruits and vegetables daily, which may be useful in cancer prevention. These foods are rich in vitamin C, and if these advices are followed, daily vitamin C intake will be 210–280 mg, depending on food cofactors [15, 20]. Though this discussion indicated that vitamin C is essential in cancer prevention, fruits/foods with high level of vitamin C also contain many other beneficial nutrients and antioxidants. Therefore it is quite difficult to say that only vitamin C is responsible for cancer preventive effect. Nobel laureate Pauling and Cameron suggested that ascorbic acid in high dose may be beneficial in the treatment of cancer. Experimental animal studies and in vitro experiments have extrapolated the promising anticancer effect of ascorbic acid as it inhibited human mammary tumor growth in mice, decreased the incidence of kidney tumors in hamsters, inhibited the growth of a number of malignant and nonmalignant cell lines, exerted cytotoxic effect onto some human tumor cells (neuroblastoma, osteosarcoma, and retinoblastoma), etc. [14]. Unfortunately long-term RCTs with vitamin C alone or with other antioxidant supplements in cancer patients were discouraging. A recent study reported that the incidence of melanoma was higher only in women who received antioxidant supplement but not in men [21]. Thus, conclusive evidence or precise mechanism on the possible protective/beneficial effect of ascorbic acid supplementation is lacking.

Prooxidant versus antioxidant property, interaction with other dietary substances, and concentration of ascorbic acid are the key aspect to explain the possible harmful effects of vitamin C. In vitro studies have shown that low concentrations of ascorbic acid act as a prooxidant, but as an antioxidant at higher levels [22]. However, at elevated oral intake, vitamin C (e.g., 500 mg/day over 6 weeks) is found to produce prooxidant effects by increasing oxidative lymphocyte DNA damage in healthy human volunteers. In high concentration vitamin C can negatively impact the intactness of the gastrointestinal lining [23]. Several in vitro and in vivo animal experiments have demonstrated that high intake of iron along with ascorbic acid could promote in vivo lipid peroxidation of LDL and therefore could enhance the risk of atherosclerosis, though some literatures are also available suggesting the reverse effect [14]. Agus et al. have found that the tumor cells hold large amounts of vitamin C, and it is speculated that high amount of vitamin C in cancer cells may impede with chemotherapy or radiotherapy since these therapies induce cell death by oxidative mechanism. Thus, supplementation with ascorbic acid might make cancer treatment less effective [14, 24]. Several other characteristics upon the administration of vitamin C also need detail study. One of the key mechanisms of vitamin C in relation to its anticancer activity is the formation of H_2O_2. It was reported that the concentration of H_2O_2 induced by pharmacologic vitamin C is much more than those concentrations that regulate normal cellular processes. Whether transient changes in H_2O_2 level have long-term effects on human body is still unknown [25]. Recently, it has been reported that ascorbic acid causes decomposition of lipid hydroperoxide in the presence of transition metals to DNA-reactive bifunctional electrophiles, suggesting that ascorbic acid may enhance mutagenesis and risk of cancer [14]. The data of vitamin C inducing harmful effect on human are conflicting and inconsistent. However, in this regard more mechanistic and human in vivo studies are warranted.

2.2 Vitamin E

Vitamin E comprises a group of potent, lipid-soluble, chain-breaking antioxidants, which is considered first line of defense against lipid peroxidation. α-Tocopherol is the most abundant form of vitamin E in nature. The effect of vitamin E supplement on different human disease conditions like cancer and CVDs was investigated by several small and large-scale long-period studies. The most recent RDA for vitamin E is 15 mg [22 international units (IU) RRR or 33 IU all-rac] α-tocopherol [26]. Some studies have reported that vitamin E supplement was associated with a decreased risk of CVD mortality and nonfatal CVD events. In the all-female Nurses' Health Study, 37 % reduction in major coronary heart disease (CHDs) was observed when vitamin E supplement was used after 8-year follow-up, while in all-male Health Professionals Follow-Up Study (after 4 years), 25 % reduction was observed for incident coronary disease. But the use of vitamin E supplement with a minimum dose of

100 IU/day for not less than 2 years is required for these beneficial effects [19]. However, majority of RTCs have failed to prove the beneficial effect for vitamin E supplementation in CVD disease and mortality. Schurks et al. investigated the influence of vitamin E supplementation on total incident of ischemic and hemorrhagic stroke using a meta-analysis (nine trials, total 118,765 participants), and it was found that vitamin E supplementation results in a significant increase of the risk of hemorrhagic stroke by 22 %, but reduce the risk of ischemic stroke by 10 %. The effect of vitamin E supplementation on nonfatal myocardial infarction in patients with preexisting coronary artery disease was investigated, and it was found that supplementation decreased the disease by 3 %. Thus, vitamin E supplementation was not associated with reduction in the incidence of total cardiovascular or all-cause mortality. Researchers did not find any serious adverse effects associated with prophylactic use of vitamin E and have suggested that vitamin E supplementation could be helpful for secondary prevention of nonfatal myocardial infarction [27–29].

The relationship between cancer and vitamin E supplements is quite complex. Alkhenizan and Hafez have investigated 12 RCTs and correlated the effect of vitamin E supplements alone and with other supplements with total mortality, cancer mortality, total incidence of cancer, and incidence of lung, stomach, esophageal, pancreatic, prostate, breast, and thyroid cancers. They have proposed that vitamin E supplementation was not beneficial in the reduction of total mortality, cancer incidence, or cancer mortality, but may be helpful in the reduction of prostate cancer incidence. Some of the RCTs have reported opposite phenomenon. A Selenium and Vitamin E Cancer Prevention Trial which includes 35,533 men from 427 study sites in the United States, Canada, and Puerto Rico has found that dietary supplementation with vitamin E and selenium in combination did not produce any effect on prostate cancer risk, but the risk of prostate cancer among healthy men was significantly increased by vitamin E supplementation (400 IU/day) [30, 31].

There is evidence that high dose of vitamin E (>400 IU/day) supplements may increase all-cause mortality compared to placebo. Miller and colleagues documented 12,504 deaths from 19 RCTs, though few limitations restrict the wide acceptance of the meta-analysis [27]. Abner and colleagues also conducted a similar meta-analysis of 57 trials (sample sizes vary 28–39,876, duration varies 1–10.1 years, total 246,371 subjects and 29,295 all-cause deaths), but they did not find any relationship between dose and risk of mortality, and proposed that vitamin E supplementation up to dose 5,500 IU/day has no effect on all-cause mortality [32]. Berry et al. in their research (Bayesian meta-analysis) also have proposed similar phenomenon [33]. Recommendations for vitamin E supplementation are not consistent in different observational studies. Data from clinical trials are also limited, and many of them did not find consistent beneficial effect of vitamin E supplements in the prevention of chronic diseases and mortality.

Source and bioavailability of vitamin E along with the food habit during vitamin E supplementation is also important while discussing its effect. Thus, for a valid assessment of vitamin E supplements, it would be important to measure the circulating amount of vitamin E and to select people who have moderately low α-tocopherol and elevated oxidative stress status for inclusion in RCTs [26]. Recently, several non-antioxidant activities of vitamin E have attracted the attention. Vitamin E interferes with vitamin K and decreases the likelihood of clot formation. The consumption of vitamin E in excess may increase the risk of bleeding. Several clinical trials have found that vitamin E supplementation may reduce the risk of cardiovascular events and death from cardiovascular disease. Therefore, antithrombotic activity of vitamin E also may have important role in this regard along with the antioxidant potential of vitamin E [26]. Recently, a metabolite of a vitamin E, namely, 2,7,8-trimethyl-2-(β-carboxyethyl)-6-hydroxychroman, has gained much attention due to its strong natriuretic effect. This effect is an important determinant in hypertension, congestive heart failure, and cirrhosis [34]. Prooxidant effect of tocopherol is well established which is also responsible for the formation of free radicals by initiating the reduction of transition state metals. This prooxidant effect of tocopherol is responsible for worsening the fatal myocardial infarctions in a clinical study with vitamin E supplements [34]. Vitamin E generates quinones (as a metabolite) when it produces its antioxidant effect. These metabolites are known to generate oxygen radicals and oxidize cellular components. Quinones also produce Michael adducts with cellular thiols, depending upon the nature, the number, and the position of substituents on the quinine. The toxicity varies with the ability of these different quinines to form Michael adducts [34]. Tocopherols and several tocopherol esters restrain glutathione S-transferase P 1-1 (GSTP1-1) which is present in the skin. A laboratory experiment has confirmed that mice lacking GSTP1-1 have a greater risk for skin tumorigenesis. Vitamin E has also been reported to be a complete tumor promoter in mouse skin. Thus it is quite possible that "GSTP1-1 inhibiting effect" of vitamin E may be responsible for abovesaid carcinogenic activity. This urges to confirm the potential risk of the application of vitamin E, and another mechanism relates to it [34]. Thus diverse biological effects of vitamin E advocate a reevaluation including the dose, kinetics, metabolism, interaction of vitamin E, and the biological effects of the metabolites.

2.3 Selenium

Selenium (Se), an essential trace element, is required to retain optimal human health and provide protection against oxidative damage. Foods are foremost natural source of Se, but the beneficial and detrimental effect of Se depends on a narrow exposure window. Se is an important component of several antioxidant enzymes, such as glutathione peroxidase (GPx), thioredoxin reductase (TrxR), and iodothyronine deiodinases (IDD) [35]. Selenium deficiency is associated with an increased risk of cancer, infection, male infertility, and some neurologic conditions [36]. Epidemiological studies have suggested positive, negative, as well as no association between selenium supplementation and the risk of different types of cancer. Positive effect of selenium supplement on cancer was reported by different RCTs. Cancer preventive effect of Se supplementation alone was investigated by a meta-analysis of 9 RCTs in populations with a low baseline serum Se level and in high-risk populations for cancer [36]. A recently updated Cochrane review examined the association between Se supplementation and cancer risk and concluded that higher intake of Se might help to decrease the risk of bladder cancer, prostate cancer, and cancer mortality; but no relation was found between Se intake and risk of breast cancer. Most of the RCTs have reported that Se might help to prevent gastrointestinal and prostate cancers, but these results need to be confirmed in more appropriately designed RCTs. A large-scale RCT in 35,533 men (age ≥50 years) from different countries was discontinued after 5.5 years when no association between Se supplementation (200 μg/day) with or without vitamin E and prostate cancer risk was observed [36, 37].

The role of Se in cardiovascular disease has also yielded conflicting conclusions. Some observational studies have reported that poor Se concentration increases the risk of hypertension and coronary heart disease. However, these conclusions have not been supported by most of the observational studies. Most of these researches have suggested that there is no appreciable relation between the level of Se intake and risk of heart disease or cardiac death. Few of the investigations also have highlighted that the risk of CVDs increases with increase level of Se. Flores-Mateo et al. examined the association of Se supplements with CVDs/CHDs using 14 cohort studies, 11 case control studies, and 6 randomized trials. Association between Se concentrations and the risk of CHDs was reported usually in observational studies, though the validity of this association is uncertain. Some RCTs have found the beneficial effect of Se supplement in CVDs, but the findings are largely inconclusive. Thus authors have suggested that Se supplements should be avoided for cardiovascular disease prevention until the effect of Se supplementation is properly established [37, 38].

A randomized trial found that Se supplementation (200 μg/day) may increase risk for diabetes

[39]. Inverse relationship between serum Se level with thyroid volume, risk of goiter, and risk of thyroid tissue damage in people with mild iodine deficiency was also investigated, though RCTs of Se supplementation in patients with thyroid disease had mixed result. Se supplementation with other antioxidants was found to increase episodic memory, but selenium's independent role in this regard was unclear [39].

GPx (a selenoprotein) is considered as an important endogenous antioxidant found in human, which converts H_2O_2 into water, protecting the cell from oxidative damage, while another selenoprotein, thioredoxin reductase, is involved in the reduction of oxidized thioredoxin in a NADPH-dependent manner, serving a key role in regulating the cellular redox status [40]. The role of Se in respect to the risk of heart diseases and mortality is still uncertain and controversial. Se as antioxidant selenoenzymes and selenoproteins may be beneficial to decrease the production of oxidized LDL, and thioredoxin reductase was also found to play an important role in preventing oxidative stress and increasing NO bioavailability and, therefore, would reduce the incidence of heart diseases. However, non-oxidative mechanisms like downregulation of the LDL receptor have also been reported in animal studies [35]. Some researchers proposed that the recommended daily level of Se consumption in humans should be between 50 and 200 μg/day. But if the intake is increasing as little as 850–900 μg/day, it may lead to selenium toxicity [40]. Forms of selenium, concentrations, species, and its interaction with other dietary supplements are responsible for complexities observed in selenium toxicity. For example, in adult animals, inorganic selenium was found to be more toxic, and it was also reported that selenium toxicity was increased by dietary supplementation with methionine and vitamin E in rats [40]. Selenocysteine was found to induce apoptosis in human cancer cells by increasing ROS, and this can be considered as one of the important mechanisms of anticancer properties of Se. But several other mechanisms like anticancer effect mediated through p53, altering the sensitivity to chemotherapy and radiotherapy, enhancement of immunity, interactions of Se that affect protein synthesis and the cycle of cell division, and formation of anticancer selenium metabolites are also important in this regard [40, 41]. Thioredoxin reductases and three iodothyronine deiodinases are also responsible for controlling a variety of metabolic processes by regulating the activation and inactivation of thyroid hormones [40]. However,, the exact mechanism of selenium toxicity remains unclear, but there are many literatures which have found that in in vitro conditions, Se compounds may generate active oxygen species and this may be considered as a possible mechanism of selenium toxicity [42, 43]. Prooxidant effect particularly in the form of selenite has also been investigated. Se in its inorganic form reacts with tissue thiols (i.e., glutathione) to form selenotrisulphides, which reacts with other thiols to generate oxygen free radicals, such as superoxide anion. Organic diselenides are transformed into selenols in the presence of thiols to produce ROS. Methyl selenide formation is responsible for the formation of superoxide radicals that is also considered as important mechanism of selenium toxicosis. Besides these mechanisms selenium can have inhibitory effects on thiol proteins, for instance, those which have antioxidant effect [43]. Thus it is rational to believe that the use of Se supplements should be reevaluated in light of this growing knowledge.

2.4 Coenzyme Q10

Coenzyme Q10 (CoQ10) is a vitamin-like substance which is produced by the tissues in small quantities. CoQ10, also known as ubiquinone, primarily works as a vital intermediate of the electron transport system in the mitochondria. Adequate quantity of CoQ10 is essential for cellular respiration and ATP production. CoQ10 protects biomolecules from oxidative damage and is capable of recycling and regenerating other antioxidants such as vitamins E and C, thus conferring some benefits under certain circumstances when ingested [44, 45]. Recently, low levels of CoQ10 have been reported in several disease conditions, including cardiovascular diseases (cardiomyopathy, congestive heart failure, angina, acute myocardial infarction, arrhythmias, and mitral valve prolapse), neurological disorders (Parkinson's disease, Huntington's disease), cancers such as breast cancer, diabetes mellitus, male infertility, thyroid disorders, migraine, renal failure, periodontal disease, muscular dystrophy, acquired immunodeficiency syndrome (AIDS), and asthma [44], and thus CoQ10 supplementation can be helpful in these disease conditions.

A meta-analysis of different clinical trials (3 RCTs, 1 crossover study, and 8 open label studies) concluded that CoQ10 supplements have a beneficial effect on hypertensive patients, as CoQ10 decreases systolic blood pressure by up to 17 mm Hg and diastolic blood pressure by up to 10 mm Hg without significant side effects. The association of CoQ10 (150–300 mg/kg) with endothelial function was investigated, and it was found that CoQ10 supplement significantly improved the endothelial function. A recent meta-analysis of 13 RCTs concluded that there was a significant improvement in the ejection fraction in patients with congestive heart failure following CoQ10 supplementation (60–300 mg/day). Authors have suggested for further well-designed studies [46–48].

Influence of CoQ10 supplements in other disease conditions was also investigated. Coenzyme Q10 supplementation was found to improve gingival health, immune response in gum tissues, diseased gum conditions, migraine headache, and glycemic control in different human clinical trials. Coenzyme Q10 effects on sperm have also been reported in oligoasthenoteratozoospermia infertile men. Coenzyme Q10 supplementation for 3 months can attenuate oxidative stress in seminal plasma, which result in the improvement of semen parameters and antioxidant enzyme activity. Shults et al. have reported that CoQ10 was safe and well tolerated up to 1,200 mg/day and found that CoQ10 supplement slows down the progressive deterioration of functions in Parkinson's diseases [49–51]. But most of the RCTs have some limitations including the study duration, number of people, etc. Thus large, prolong RCTs are essential to reconfirm the clinical benefit of CoQ10 supplements.

CoQ10 was found to attenuate the oxidized low-density lipoprotein (oxLDL)-induced production of ROS and improve the antioxidant capacity. The attenuation of oxLDL-mediated downregulation of endothelial nitric oxide synthase (eNOS) and upregulation of inducible nitric oxide synthase (iNOS) by CoQ10 was also reported [52]. Current understandings are also suggesting that in addition to the antioxidant activity, CoQ10 also improves cardiac bioenergetics, endothelial function, and vasodilatory effect, exerts direct membrane-stabilizing activity, preserves myocardial Na^+-K^+ ATPase activity, stabilizes the integrity of Ca^{2+}-dependent slow channels, and rectifies mitochondrial "leak" of electrons during oxidative respiration [53]. Thus, it will not be right to say that only antioxidant mechanism of CoQ10 is responsible for its beneficial effect and increased quantity of CoQ10 than its normal level may also produce adverse effect.

Some investigations have reported that lycopene and CoQ10 supplementations increase the concentration of glutathione peroxidase. Another research has found that CoQ10 is responsible for increased levels of extracellular superoxide dismutase [54, 55]. The observed safety level of CoQ10 is 1200 mg/day/person. Overall, evidences from preclinical and clinical studies indicate that CoQ10 is highly safe for use as a dietary supplement. But at the same time higher level of CoQ10 in postmenopausal women may be connected with increased breast cancer risk. Some in vitro studies have found that CoQ10 can promote the survival of tumor cells treated with anticancer drugs. Theoretically, antioxidant effect of CoQ10 can diminish the effect of chemotherapeutic agents and radiotherapy as these therapies act by inducing oxidative stress in tumor cell. Some researchers have found that CoQ10 is contraindicated for pregnant or lactating women and those with diabetes and high blood pressure. It was also advised to avoid excessive exercise while taking CoQ10. CoQ10 may possibly interfere with warfarin and thus increase the risk of blood clotting or bleeding [56]. The clinical benefit with CoQ10 supplements is nothing short of dramatic, and only antioxidant activity is not responsible for its favorable effect. Thus more attention and awareness is essential before consumption of CoQ10 supplements without professional advice.

2.5 Flavonoids

Dietary flavonoids represent a wide range of polyphenolic compounds that are found in fruits, vegetables, grains, herbs, and beverages. Different types of flavonoids include flavonols (i.e., quercetin, kaempferol), flavones (i.e., apigenin, luteolin), flavanones (i.e., naringenin), flavan-3-ols (i.e., catechin, epicatechin), isoflavones (i.e., phytoestrogens), and anthocyanidins (i.e., cyanidin, delphinidin). Flavonoids exhibit a potent antioxidant activity and are responsible for several biological effects [57].

Effects of multi-flavonoid supplement on vascular and hemodynamic parameters in older prehypertensive patients were investigated, and it was found that flavonoid supplementation for 2 weeks in a double-blind, counterbalanced design results in a significant decrease in systolic blood pressure and mean arterial pressure in the treatment group with no changes in vascular parameters [58]. A meta-analysis has investigated the effect of food containing flavonoid on cardiovascular disease and its risk factors. Authors have observed that chocolate increases flow-mediated dilatation after acute or chronic intake and reduces systolic, diastolic blood pressure, while isolated soy protein significantly decreases the diastolic blood pressure and LDL cholesterol. Acute black tea intake was found to enhance systolic and diastolic blood pressure, while green tea was found effective in reducing LDL. Effects of several other flavonoid-containing products have already been investigated, but investigators failed to find their efficacy due to insufficient evidence [57]. Blostein-Fujii et al. investigated the influence of citrus flavonoid supplementation on the vulnerability of lipoprotein for oxidation in type II diabetic women using a short-term study and concluded that citrus flavonoid supplementation (about 1 g/day, 3 weeks) has no beneficial effect [59]. Neuroprotective properties, neuronal function enhancing and neurogenesis-stimulating effect of flavonoid have been reported which indicated that flavonoids may enhance cognitive function. Overall evidences are suggesting a positive association between flavonoid consumption and cognitive function. Most of the RCTs have reported the positive association, and few studies have also found the negative effect, but most of the studies had null findings [60]. Although general animal experiments, some epidemiological, longitudinal, and observational studies have suggested that flavonoid consumption may have beneficial effect on cardiac health, cognitive performance, and others, but a lot of researches and clinical trials are required to find the exact mechanism and to establish the positive effect of flavonoid supplements.

Interest in the possible health benefits of flavonoid supplements has increased owing to potent in vitro antioxidant and free radical scavenging activities of flavonoid. Vegetables, fruits, spices, tea, and cocoa are rich in natural sources of flavonoids. But in the recent past, dietary supplements of flavonoids have become increasingly popular as an alternative source. Flavonoids in vegetables and fruits contain a complex mixture of secondary plant metabolites and not only flavonoids per se. Thus, replacing these complex mixtures of secondary plant metabolites by single purified compounds as dietary supplements will not be wise [61]. If flavonoids are given as dietary supplements, the dose and interactions need to be taken into account. For example, the flavonol quercetin has been marketed by different industries as a dietary supplement with recommended daily doses of up to 1 g and more, whereas the daily consumption of quercetin from foods has been estimated to be 10–100 mg only. Higher dose of flavonoid may affect trace element, folate, and vitamin C status [61]. There is increasing evidence from human clinical studies that the absorption and bioavailability of specific flavonoids is greatly higher than originally believed [62]. However, the upper limit for plasma concentrations of polyphenols before the elaboration of adverse effects is unknown for many polyphenols [63]. Thus supplements containing such supraphysiological flavonoid levels may be responsible for harmful effects through prooxidative action. At high concentration under certain specific environment (high pH, presence of iron), phenolic antioxidants can start an auto-oxidation process and behave like prooxidants. Some studies have reported that phenolic acids in high concentration endorse carcinogenicity through H_2O_2-induced DNA damage and resultant genotoxicity. Thus, diet-derived flavonoids are generally considered as safe, based on their long history of use [63]. The overall beneficial effect of flavonoid supplements on health is uncertain, and intake of large quantities of them should not yet be encouraged.

2.6 Carotenoids

More than 600 carotenoids are available in nature, as complex with proteins or as crystalline carotenoid complexes. β-Carotene is the most abundant form of provitamin A, found in vegetables and fruits. β-Carotene effectively quenches singlet oxygen and reduces peroxyl free radical reactions [64]. It is one of the major antioxidant supplements that has been investigated widely to find its effect on health and diseases.

Druesne-Pecollo et al. after reviewing the results of nine RCTs have concluded that β-carotene supplementation did not produce any significant effect on the incidence of all cancers combined, pancreatic cancer, colorectal cancer, prostate cancer, breast cancer, melanoma, and nonmelanoma skin cancer. However,, β-carotene supplementation (20–30 mg/day) has been found to increase the incidence of lung and stomach cancers in smokers and asbestos workers significantly compared to placebo group. Another similar type of meta-analysis (six RCTs and total participants 40,544) has revealed that β-carotene supplements had no preventive effect on either cancer incidence or mortality, even though the risk of urothelial cancer, especially bladder cancer, had increased. A marginally increase in cancer risk with β-carotene supplements was observed among current smokers [65, 66]. Gallicchio et al. have reported that β-carotene supplementation did not have any influence on the risk of developing lung cancer. Thus,, several cohort studies have reported the inverse associations between different carotenoids and lung cancer, but most of the results are small and not statistically significant [67].

Investigators have also analyzed the effect of other carotenoids and reported inverse association with lung cancer risk, comparing the lowest with the highest intake except for lutein [pooled RR: 0.89 (0.79, 1.00), 0.80 (0.72, 0.89), 0.86 (0.77, 0.97), and 0.89 (0.79, 1.00) for α-carotene, β-cryptoxanthin, lycopene, and lutein–zeaxanthin, respectively] [67]. Protective effect of β-carotene supplementation on sunburn was reported, though the duration of supplementation had a significant influence on the effect size [68]. Positive protective effect of antioxidant supplements on health was demonstrated when β-carotene was given along with other antioxidants. However, most of the RCTs and clinical evidences did not support the fact that β-carotene supplementation has preventive effect on cancer; instead some have reported the negative effect of β-carotene supplementation on cancer risk.

Prooxidant effect of carotenoids in high concentrations or in case of tocopherol deficiencies has been investigated in in vitro and animal studies [64]. β-Carotene was found to scavenge nitrogen oxides in cigarette smoke, which results generation of β-apo-carotenals and other carotene oxidation products. If these products are not effectively neutralized by other antioxidants such as tocopherol and ascorbate, it may begin cell damage that could lead to neoplasm. On one hand, it is difficult to say that prooxidant effect of β-carotene is responsible for the enhancement of lung cancer in smokers taking the supplement. On the other hand, some studies have proposed that prooxidant effect of β-carotene may be responsible for its cytotoxic effect in tumor cells [64]. Environmental differences are also able to modify the effect of carotenoids. For example, carotenoids act as chain-breaking antioxidants at relatively low-oxygen concentration, but at higher oxygen level, carotenoid radical could react with oxygen to produce a carotenoid peroxyl radical which acts as a prooxidant and induces lipid peroxidation [69]. In this point, finding out the mechanism for beneficial/toxic effect of carotenoids will be the task of ongoing and future studies. At the same time to find the proper doses and combination, lengths of treatment, and effect of metabolites is more advantageous.

3 Antioxidant Supplements in Pathological Conditions

3.1 CVDs and Antioxidant Supplements

In vitro studies and in vivo animal studies have reported the beneficial effect of different antioxidant compounds against CVDs. This information was supported by several early in vivo observational studies, particularly for vitamin E in the prevention of coronary artery disease. On the basis of these studies, the consumption of antioxidant supplements by the general population increased substantially and became a multibillion dollar industry [70]. Dietary antioxidants like vitamins C and E, coenzyme Q10, and flavonoids are thought to play an important role in CVDs. But the use of supplements alone or in combination results varied clinical endpoint in different individuals. A meta-analysis of 50 RCTs which include 294,478 participants has not found any evidence which can support the use of vitamin and antioxidant supplements for the prevention of CVDs. But researchers have suggested that vitamin and antioxidant supplementation may be associated with a marginally increased risk of angina pectoris, while vitamin E supplementation is associated with a decreased risk of myocardial infarction (MI), but these effects are not consistence throughout the trials [71, 72]. Katsiki and Manes have reviewed 22 RCTs, to find the role of vitamins E, C, and A alone and in combination in the prevention of atherosclerosis. The authors did not find any effect of these supplements. It was also reported that vitamin E alone and in combination with other antioxidants does not confer any benefit in relation to the prevention of all-cause mortality or cardiovascular mortality. Similar phenomena have also been reported by several other meta-analyses [73]. A meta-analysis has investigated the effect of β-carotene and reported that β-carotene has no beneficial effect on all-cause mortality and cardiovascular mortality, but exerts little harmful effect on patients who were at high risk of lung cancer due to smoking [73]. Ye and Song in their meta-analysis proposed that vitamin E supplementation alone and in combination with vitamin C increases the risk of CHD risk, though vitamin C had no significant connection with CHD risk. No association was observed between β-carotene (1 mg/day) or vitamin C (30 mg/day) intake and the risk of CHD. Knekt et al. in their pooled analysis (nine cohort studies) have found that higher overall intakes of vitamin C for 10 years were associated with lower CHD rates [72–74]. The effect of ROS is tremendous in intracellular level. ROS acts as a cellular second messenger that induces hypertrophy, mitogenic effect, vasoconstriction, decreased NO availability, and stimulation of the production of multiple cytokines and chemokines. All these effects take place before the ROS is moved to the extracellular space and to the blood to induce oxidation of the LDL. The dietary antioxidant supplement targets the extracellular antioxidant rather than the intracellular mechanism of generating ROS. These observations to some extent may be helpful to explain the no effect antioxidant supplements in clinical outcome of patients with CAD [71]. Of course there is strong evidence which supports the fact that different effects (like scavenging ROS/RNS directly, inhibiting xanthine oxidase, inhibiting lipid peroxidation) of natural antioxidants could be beneficial in the prevention or treatment of CVDs/CHDs. But several other non-antioxidant mechanism also need to be considered, like quercetin also can act as reducing, vasodilatory, antiplatelet, and anti-atherogenic compound which is responsible for its vasoprotective effect [23]. Researches concerning nutritional regimens have shown that the consumption of large amounts of fruits and vegetables results decrease in the incidences of CVDs. Although the precise mechanisms for this beneficial effect are elusive, one of the possible explanations includes the consumption of antioxidants present in fruits and vegetables.

3.2 Antioxidant Supplements and Cancer

Experimental studies using animal models and in vitro cancer cell lines have found that antioxidants such as vitamin E, vitamin C, β-carotene, and Se could be useful to prevent different types of cancers. More than 200 epidemiologic studies published in this area have also indicated that low dietary intake of antioxidant-rich fruits and vegetables may increase the risk of cancer [75]. However, currently available evidence from different RCTs was insufficient to prove whether antioxidant supplements were beneficial toward the prevention of cancer.

Authors of a meta-analysis of 22 RCTs (total subjects 161,045) have reported that antioxidant supplements do not confer effect on overall primary and secondary prevention of cancer. They also found that the use of antioxidant supplements significantly increased the bladder cancer risk and cautioned that the effects of antioxidant supplements on human health, particularly in relation to cancer, should not be overemphasized. Bjelakovic et al. in their meta-analysis have investigated eight RCTs and found that antioxidant supplements (β-carotene; vitamins A, C, and E; and Se alone or in combination) may increase the development of colorectal adenoma in three low-bias risk trials (1.2, 0.99–1.4), but significantly reduced its development in five high-bias risk trials (0.59, 0.47–0.74). Thus authors concluded that antioxidant supplements may not confer any beneficial effect on primary or secondary prevention of colorectal adenoma and suggested for further research. Another meta-analysis (14 RCTs, $n=170525$) by author reviewed the effect of antioxidant supplements on gastrointestinal cancers and concluded that antioxidant supplementation does not produce any significant effect on esophageal, gastric, colorectal, pancreatic, and liver cancer incidences compared to placebo. But they have analyzed that Se may confer beneficial effect. Investigators have also reported that β-carotene alone or with vitamin A/vitamin E may increase mortality [75–77]. Similar phenomenon has also been reported by others where β-carotene at doses of 20–30 mg/day was associated with increased risk not only for lung cancer but also for gastric cancer, in smokers and asbestos workers [65]. The effect of antioxidant supplementation on chemotherapeutic efficacy is also doubtable and needs large, well-designed studies. Block et al. have reviewed several researches to find the effect of glutathione, melatonin, vitamin A, an antioxidant mixture, vitamin C, NAC, vitamin E, and ellagic acid, but failed to draw any conclusive evidence regarding the beneficial effect of antioxidant supplementation during chemotherapy. Several trails have reported that antioxidant supplementation increased survival times or increased tumor responses, or both. Recent researches and clinical trials indicated that high-dose antioxidant supplements (particularly β-carotene and vitamins A and E) may increase the risk of mortality in cancer, rather than reduce it [78, 79].

Some of the recent evidences strongly discourage the consumption of antioxidants during chemotherapy and radiotherapy. One of the probable mechanisms in this regard is that antioxidants may protect cancer cells against oxidative stress-induced apoptosis. Although some pilot studies in humans have reported that high-dose multiple dietary antioxidants along with radiotherapy or chemotherapy may increase the efficacy of cancer treatments, the mechanism remains unknown [80]. Prooxidant mechanism of several dietary antioxidants may also be responsible for the harmful effect as discussed in earlier section. Furthermore, certain categories of patients display own unique responses to endogenous ROS production and exogenous antioxidants. Several reports have indicated that people with low levels of ROS may become more susceptible to cancer by taking antioxidant supplements. The probable mechanism is that deficient ROS generation further suppressed in those subjects, thereby numbing the rate of protective apoptosis [80]. Evidences from different trials are indicating that antioxidant supplementation may have a favorable effect on cancer incidence only in healthy subjects who are not exposed to cancer risk and who have a particularly low baseline status, but not in healthy subjects with adequate antioxidant status. On the other hand, antioxidant supplements in high dose may have harmful effect in high-risk subjects without any clinical symptoms in whom the initial phase of cancer development has already started [81]. Thus a lot of high-quality RCTs and researches are still required in this regard.

3.3 Liver Diseases and Antioxidant Supplement Innervations

The association of oxidative stress and several liver diseases has been well reported. Accordingly, antioxidant supplements are viewed as a potential treatment strategy for various liver diseases. But the evidence supporting these suggestions is equivocal and needs further clarification.

Beneficial interaction of vitamin E supplementation and other putative free radical scavengers has been proposed for nonalcoholic steatohepatitis. This observation is based on initial observational studies and preclinical studies. Several RCTs have failed to find any positive effect of vitamin E in this regard. Lirussi et al. systematically examined the effects of antioxidant supplements on patients with nonalcoholic fatty liver disease or nonalcoholic steatohepatitis and proposed that antioxidant supplements may have variable effect on different enzyme levels. However, limited radiological and histological data warrant large prospective RCTs to find any definite conclusion [82, 83]. The consumption of antioxidant supplementation (vitamin E, 800 mg; vitamin C, 500 mg; and zinc, 40 mg) for 6 months is found beneficial for hepatitis C-infected patients. The intake of supplements provided an antioxidant protection, thus attenuating oxidation processes related to the disease. An RTC found that innervations with daily dose of ascorbic acid (500 mg), D-alpha-tocopherol (945 IU), and Se (200 μg) for 6 months had no effects on alanine aminotransferase, viral load, or oxidative markers [84, 85]. Bjelakovic et al. in their meta-analysis did not find any evidence which can support or refute antioxidant supplements (β-carotene; vitamins A, C, E; and Se) in patients with liver diseases like autoimmune liver diseases, viral hepatitis, alcoholic liver disease, and cirrhosis (any etiology). The authors have reported that antioxidant supplements significantly increased the activity of gamma glutamyl transpeptidase and suggested for more RCTs [86].

Intake of large amount of vitamin A beyond RDA is responsible for liver toxicity. When vitamin A is consumed in higher amount, the potential toxic form of Vitamin A is generates of retinol and retinoic acid in the absence of fat, both of which are known to be potentially toxic forms of vitamin A. β-Carotene is a precursor to vitamin A. Several reports also confirmed that supplements of beta-carotene are thought to accelerate the progression of alcoholic liver disease and may be responsible for other liver problems [87, 88]. High dose of zinc found to associate with liver problems. High-dose ingestion of zinc may be responsible for transient increase in the activities of liver enzyme [89]. Prooxidant effect of other antioxidant supplements also may be responsible for liver problem. Several advance researches are still required to establish the toxic/beneficial effect of different antioxidant supplements on hepatocyte.

3.4 Antioxidant Supplements in Diabetes Mellitus and Its Complications

Exogenous antioxidants can reimburse the lower plasma antioxidant levels which were observed in diabetic and in prediabetic individuals. Several investigators have suggested that dietary supplementation with antioxidants is associated with decreased risk of diabetes and its complications in animal and human. Antioxidants can also induce some changes that could be beneficial in reducing insulin resistance and protecting vascular endothelium. Few prospective epidemiological studies demonstrated that high serum vitamin E levels are connected with decreased risk of diabetes [90].

A cohort study on 4,304 people has found that the consumption of vitamin E (α-tocopherol, β-tocopherol, δ-tocopherol, and β-tocotrienol), carotenoids, and β-cryptoxanthin significantly reduced the risk of type 2 diabetes, though no association between intake of vitamin C and type 2 diabetes risk was observed. Akbar et al. in their systemic review and meta-analysis have found that dietary antioxidant supplementation did not affect on plasma glucose or insulin levels, suggesting that supplements do not have any role in the pathogenesis of insulin resistance. However, antioxidant supplementation results in a significant decrease in HbA1C levels, suggesting that antioxidant supplement may confer some benefit in the prevention of type 2 diabetes complications [90, 91].

The effect of antioxidant supplements against neuropathic ulcer of diabetic foot when administered with polarized light treatment was investigated and concluded that supporting therapy with CoQ10, α-lipoic acid, and vitamin E was effective to control the type 2 diabetes complications. Although the exact beneficial interaction of lipoic acid is unknown, in vitro studies have suggested that lipoic acid maintains the intracellular reduced glutathione level and blocks the activation of serine kinases that are related with insulin resistance [92, 93]. In another study, it was reported that antioxidant supplement (vitamin C 1,000 mg/day and vitamin E 400 IU/day) when given to 50 non-insulin-dependent diabetes patients for 10 days reduces the oxidative damage produced by nitric oxide and other free radicals, improving the ocular surface milieu. Garcia-Medina et al. in their 5-year follow-up study have found that oral antioxidant supplementation could be a beneficial adjunctive long-term therapy in the treatment of nonproliferative diabetic retinopathy [94, 95]. Several current researches have suggested that dietary α-tocopherol or other tocopherols, tocotrienols, carotenoids, flavonols, flavones, and vitamin C were not connected with the risk of type 2 diabetes. Kataja-Tuomola has also found that the intake of antioxidant supplement (α-tocopherol or β-carotene) did not prevent type 2 diabetes, macrovascular complications in diabetes, or mortality among diabetic subjects [96]. Interference with the chain reaction of lipid peroxidation, scavenging of the free radicals, and increase in the intracellular glutathione levels are considered as a key mechanism for the beneficial effect of vitamin E and vitamin C [93]. Natural β-carotene is found to affect on glutathione and part of its redox cycle enzymes by probable augmentation of its regeneration. Thus improving the antioxidant status may slow the rate of progress of vascular complications in diabetes mellitus [96]. The effect of zinc supplementation on diabetic individual was evaluated from different clinical trial evidences and concluded that zinc supplementation has beneficial effects on glycemic control and promotes healthy lipid parameters. Significant decrease in systolic and diastolic blood pressures after zinc supplementation was also observed in that study [97, 98]. Pancreatic exocrine cells and pancreatic ß-cells secrete zinc, which plays a key role as ionic signaling in large number of cells and tissues. Some researchers have confirmed that zinc also may be responsible for gastrointestinal hemorrhage and acute pancreatitis. Zinc excretion mainly is via the pancreas under normal circumstances. Prolonged intake of supplements may cause accumulation of zinc, thus resulting in impairment of the pancreatic function [89]. It was well established that exercise improves insulin action and reduces diabetes risk. In general, exercise causes generation of excessive free radicals, which are responsible for the activation of molecular defense systems against stress, and metabolizes carbohydrates more efficiently. Both of these effects are required to prevent diabetes and possibly other diseases. Therefore some scientists cautioned to choose appropriate antioxidant supplement in proper dose to the individuals at risk for diabetes or suffering from type 2 diabetes, especially if they exercise regularly to improve their health [99].

3.5 Effect of Antioxidant Supplements in the Prevention/Treatment of Other Disorders

Cataract is one of the foremost causes of blindness among the elderly. Epidemiological investigations have recommended that intake of foods containing antioxidant potential may be useful to prevent cataract, but the role of individual antioxidant micronutrient or antioxidant supplement on the cataract is still not investigated properly. Effects of several antioxidants like vitamin C, vitamin E, β-carotene, lycopene, lutein, zeaxanthin, α-carotene, β-cryptoxanthin, zinc, and Se were investigated in different RCTs. So far, clinical trials provided little evidence to beneficial effect of antioxidant supplements on cataract development, as most of the investigators did not find any beneficial effect of these supplements. A meta-analysis has reported that current evidences from RCTs do not support the use of antioxidant vitamin supplements (vitamin E, β-carotene) to prevent age-related macular degeneration (AMD). Moreover, people suffering from AMD, or early signs of the disease, may get some benefit from supplements (vitamin E, vitamin C, β-carotene, zinc), but potential adverse effect of high-dose antioxidant supplementation must be considered [100, 101]. It was also reported that the consumption of zinc supplements >50 mg/day may depress the immune response, while chronic exposure to selenium compounds may be responsible for several adverse health effects [102].

A systemic review was performed to evaluate the benefits of antioxidant treatment in schizophrenia and concluded that antioxidant supplementation may confer beneficial effect on symptomatic schizophrenia patients, especially those suffering from positive symptoms [103]. Effect of CoQ10 was reviewed from different RCTs and found that CoQ10 may provide some minor treatment benefits in Parkinson's disease. Allen et al. have reviewed 40 investigations and suggested that comparatively low dietary intakes of vitamins A and C result in significant increased odds of asthma and wheeze, but vitamin E intake does not appear to be associated with asthma status [104].

3.6 Effect of Antioxidant Supplements on Aging and Mortality

The hypothesis of free radical theory in relation of aging is based on the fact that ROS/RNS are responsible for the age-related damage at the cellular and tissue levels. Attention toward the consumption of antioxidant supplementation mainly in Western countries is still growing, but the supporting evidence is still scarce and equivocal. The discrepancy exists between observational data and the clinical trials, which could also be due to the variation between lifelong exposures to an antioxidant-rich diet versus a limited exposure to antioxidant supplements. Several epidemiological investigations have shown that antioxidant supplementation may decrease the risk of several clinical conditions; however these types of observations are generally not universal [102]. Current evidences do not support the blind use of antioxidant supplementation to prevent age-related pathophysiological modifications and clinical conditions. Several issues like efficacy, safety, underlying mechanism, and interactions are still needed to be addressed properly.

Several investigators have studied the effect of antioxidant supplements on mortality, and some of them are discussed here. Macpherson and colleagues in their meta-analysis (21 articles on RCT of antioxidant vitamin/mineral supplement and mortality) have proposed that multivitamin–multimineral has no effect on all-cause mortality (RR, 0.98; 95 % CI, 0.94, 1.02). However, authors have found a trend for a reduced risk of all-cause mortality across primary prevention trials (RR, 0.94; 95 % CI, 0.89, 1.00). Treatment with antioxidant multivitamin-multimineral also had no effect on mortality due to vascular causes (RR, 1.01; 95 % CI, 0.93, 1.09) or cancer (RR: 0.96; 95 % CI: 0.88, 1.04). Bjelakovic et al. have reviewed 68 RCTs which include a total of 232,606 participants (385 publications) to find the effect of β-carotene, vitamin A, vitamin C, vitamin E, and Se either alone or in combination with mortality. Authors have concluded that there was no significant effect of antioxidant supplements on mortality (RR, 1.02; 95 % CI, 0.98–1.06) when all low- and high-bias risk trials were pooled together. But overall analysis has suggested that β-carotene, vitamin A, and vitamin E supplement may increase mortality. Authors did not find any effect of vitamin C and Se on mortality and proposed for further study. The reexamined 66 RCTs used in previous study has reported that among the RCTs 24 had a positive outcome, 39 had a null outcome, and 3 had a negative outcome [102, 105–107]. We have examined the literature on describing the effect of different antioxidant supplements; some of them are described below and given in Table 2.

Table 2 Effect of antioxidant supplements on human health (compliance of some population-based/randomized trials)

Treatment	Outcome evaluated	Sample	Final outcome and references
Vit. E (400 IU every other day) and vit. C (500 mg/day) separately for long term [synthetic source]	Major cardiovascular events (nonfatal MI, stroke, and CVD death)	14,641 US male physicians (initial age ≥50 years)	No effect [108]
	Prostate and total cancer	14,641 US physicians, among them 1,307 have a history of prior cancer	No effect [109]
Vit. C (500 mg) + vit. E (400 mg) for 2 months daily	Lipid profiles and antioxidant capacity in CVD patients	40 CVD patients with age 57.7±10.6 years in Iran	Beneficial [110]
Vit. C (1 g) + vit. E (400 IU) daily for 11 weeks	Muscle performance, blood redox biomarkers, hemolysis after exercise	28 healthy men of the United Kingdom	No effect [111]
Vit. C (500 mg/day), vit. E (600 IU every other day), β-carotene (50 mg every other day) [long term] in different combinations	Secondary prevention of cardiovascular events	8,171 female aged ≥40 with self-reported history of CVD, or at least three cardiac risk factors	No effect [112]
	Primary prevention of type 2 diabetes		No effect [113]
	Cancer risk	7627 females (≥40 years) with a history of CVD or ≥3 CVD risk factors	No effect [114]
Se (200 μg/day) or vit. E (400 IU/day) or Se + vit. E [3 years]	Prostate cancer and prespecified secondary outcomes, including lung, colorectal, and overall primary cancer	35533 men from 427 participating sites in the United States, Canada, and Puerto Rico	No effect [115]
Se (200 μg/day) or vit. E (400 IU/day) or Se + vit. E [7–12 years]	Prostate cancer incidence	34887 men from the United States, Canada, and Puerto Rico	Negative for vit. E, no effect for other s[31]
Vit. E (600 IU) of natural source on alternate days [long term]	Risks of major cardiovascular event and total invasive cancer	39,876 healthy US women (age at least 45 years	No effect [116]
	Age-related macular degeneration		No effect [117]
Vit. C (120 mg), vit. E (30 mg), β-carotene (6 mg), Se (0.1 mg), zinc (20 mg) in combination [nearly 7 years]	Change in vascular structure and function	13,017 healthy men (aged 45–60) and women (aged 35–60) of France	No effect [118]
120 mg vit. C, 30 mg vit. E, 6 mg β-carotene, 100 mg Se, and 20 mg zinc in combination [7.5 years]	Risk of skin cancers	French adults (7,876 women and 5,141 men)	Negative in women, null for men [21]
Low, moderate, and high quantity of micronutrients (zinc and folate) and antioxidants (vit. C, vit. E, and β-carotene)	Semen quality	97 healthy nonsmoking male volunteers of California	Positive [119]

(continued)

Table 2 (continued)

Treatment	Outcome evaluated	Sample	Final outcome and references
Quercetin + vit. C (166 mg + 133 mg/capsule), α-lipoic acid (300 mg/capsule) [4 weeks with 2 weeks of washout period]	Blood level of inflammatory mediators and severity in rheumatoid arthritis patients	20 patients, Korea	No effect [120]
Low dose of lycopene, lutein, β-carotene, α-tocopherol, Se in different combinations for 12 weeks	Parameters related to skin structure	39 volunteers with healthy (normal skin of skin type 2)	Positive [121]
Mixture of 120 mg vit. C, 30 mg vit. E. 6 mg β-carotene, 100 μg Se and 20 mg zinc for 2 years	Pathogenesis of thrombosis and arteriosclerosis	186 presumably healthy volunteers of France	Positive [122]
Mixture of β-carotene (15 mg), lutein (15 mg), lycopene (15 mg) for 3 months	Biological markers of oxidative stress and LDL oxidizability	175 healthy male volunteers from European country	No effect [123]
Fruits and vegetables (nutrient value: 272 mg vit. C, 31 mg all-rac-α-tocopherol, 400 μg folic acid, for 90 days	In vivo lipid and protein oxidation (oxidative biomarker)	77 healthy men (39 nonsmokers and 38 smokers)	No effect [124]
Vit. C (100 mg/day), vit. E (100 mg/day), β-carotene (6 mg/day) and Se (50 μg/day) for 3 months	Chromosomal damage	86 (28 myocardial infarction survivors and 58 healthy people), from Bolivia	Beneficial [125]
Vit. E (75 mg), vit. C (650 mg), β-carotene (15 mg) for 6 months	Respiratory functions of workers exposed to high levels of ozone	47 street workers of Mexico City	Beneficial [126]
Vit. C (500 mg), vit. E (200 IU), co-enz Q10 (60 mg), and Se (100 mcg) for 6 months	Arterial compliance, humoral factors, and inflammatory markers	70 patients with at least two cardiovascular risk factors	Beneficial [127]
Vit. C (120 mg), vit. E (30 mg), β-carotene (6 mg), Se (100 μg), zinc (20 mg) (nearly 76 months)	Health-related quality	8,112 healthy French adults	No effect [128]
Vit. C (120 mg), vit. E (30 mg), β-carotene (6 mg), Se (100 μg), zinc (20 mg), follow-up time was 7.5 years	Metabolic syndrome	5,220 men and women, France	No effect [129]
	Fasting plasma glucose (FPG) and its association with dietary intakes or plasma antioxidants	3,146 people, France	No effect on FPG, Negative effect of dietary β-carotene and FPG [130]
β-carotene (50 mg on alternate days) [for 12 years]	Age-related maculopathy incidence	22,071 healthy male US physicians (age 40–84 year)	No effect [131]
	Nonmelanoma skin cancer		No effect [132]

(continued)

Table 2 (continued)

Treatment	Outcome evaluated	Sample	Final outcome and references
Vit. E (300 mg/day) for about 3 years	Biomarker of lipid peroxidation (cardiovascular risk factors)	144 participants with at least one major CVD risk factor (aged ≥50 years)	No effect [133]
Vit. E 500 IU [for 4 years]	Incidence or rate age-related macular degeneration	1,193 healthy volunteers (age 55–80 years)	No effect [134]
91 mg (136 IU) of d-α-tocopherol + 250 mg of slow release vitamin C (twice daily for 3 years)	Progression of carotid atherosclerosis	520 smoking and nonsmoking men and postmenopausal women (45–69 years) with serum cholesterol ≥5.0 mmol/L	Beneficial [135]
Natural source vitamin E (400 IU/day)	Cancer incidence and deaths, major cardiovascular events (myocardial infarction, stroke, and death), heart failure, angina, and revascularizations	9,541 patients at high risk for cardiovascular events (≥55 years)	Negative effect on heart failure, no effect on other events [136]
Se (200 μg/day) [follow-up of 7.7 years]	Incidence of type 2 diabetes	1,202 persons who did not have type 2 diabetes at baseline	Negative [39]
Lutein [20 mg/day for first 3 months and 10 mg/day for next 3 months]	Macular pigment optical density, visual acuity, and macular function	126 patients with age-related macular degeneration, Austria	Beneficial [137]
Quercetin (500 mg) + vit. C (250 mg), quercetin (500 mg), vit. C (250 mg), 8 weeks	Markers of inflammation and oxidative stress	60 men with systematic and regular exercise	Beneficial for combined group, other groups no effect [138]
Lutein esters (6 mg) + retinol (750 mg) + vit. C (250 mg) + vit. E (34 mg) + zinc (10 mg) + copper (0.5 mg) [9 months]	Contrast sensitivity in people with age-related macular disease	25 patients with age-related maculopathy and atrophic age-related macular degeneration	No effect [139]
Se (100 μg, 200 μg, or 300 μg/day) [min. 6 months]	Thyroid function	501 elderly UK volunteers	No effect [140]
Long-term Se (200 μg daily)	Preventing the recurrence of nonmelanoma skin cancer	1,312 residents of the Eastern United States (with a history of ≥2 basal cell or squamous cell carcinoma)	Beneficial in males with low plasma Se [141]
Lutein and zeaxanthin (0.5 mg + 0.02 mg/kg/day) [7th day of life until 40th week of postmenstrual age or until discharge]	Total antioxidant status	77 preterm infants (gestational age ≤34 weeks)	No effect [142]
Lutein/zeaxanthin (10 mg/2 mg), omega-3 long-chain polyunsaturated fatty acids (1 g) or combination [follow-up of 4.7 years]	Risk for cataract surgery	4,203 patients (age 50–85 years, at risk for progression to advanced age-related macular degeneration)	No effect [143]
Coenzyme Q10 (200 mg/day) for 3 months	Catalase, superoxide dismutase, F(2)-isoprostanes in seminal plasma	60 infertile men with idiopathic oligoasthenoteratozoospermia	Beneficial [50]

(continued)

Table 2 (continued)

Treatment	Outcome evaluated	Sample	Final outcome and references
α-Tocopherol (50 mg), β-carotene (20 mg), or combination [for 5–8 years]	Incidence of lung cancer	29,133 men aged 50–69 years who smoked five or more cigarettes daily	Negative effect for β-carotene group, no effect for α-tocopherol group [144]
Flavonoid extract (Colladeen, 320 mg oligomeric procyanidins) daily	Premenstrual fluid retention and leg health	30 subjects	Beneficial [145]
natural vitamin E (500 mg) for 6 weeks	Effect on bronchial hyperresponsiveness	Patient taking at least one dose of inhaled corticosteroid/day with a positive skin prick test to one of three common allergens, and bronchial hyperresponsiveness to methacholine	No effect [146]

4 Issues Associated with the use of Antioxidant Supplements

Randomized, placebo-controlled trials have offered little support that, consumption of antioxidant supplements provides substantial protection against CVDs, cancer, neurodegenerative diseases, or other chronic conditions. The results of the largest such trials have been mostly concluded with negative or null effect. Overall effects are really confusing and warrant further investigations. But some of the points in this regard are needed to be considered in explaining these variable results.

4.1 Who Needs Antioxidant Supplements

It was well established that ROS-induced oxidative stress is involved in pathogenesis of different diseases and exogenous antioxidants play a vital role in maintaining the delicate balance between oxidation and antioxidation in living systems [1, 147]. But who needs the antioxidant supplements? It was suggested that people consuming healthy diet rich in antioxidant components don't require antioxidant supplements. Antioxidants might help to avert diseases in individuals who are under increased oxidative stress even if they don't prevent them in other people.

4.2 Excessive Consumption of Antioxidant Supplements

High dose of supplements may be responsible for several physiological adverse effects. The consumption of high dose of Se beyond the recommended daily level may lead to selenium toxicity [40]. Higher dose of flavonoid was found to alter the trace element, folate, and vitamin C status [61]. In high concentration phenolic acids may exert carcinogenic, genotoxic effect through H_2O_2-induced DNA damage [63]. Epigallocatechin-3-gallate, a dietary antioxidant existing in green tea at its pharmacological dose (30 and 60 mg/kg), produced antianxiety effect in mice; however at 150 mg/kg, this tea polyphenol caused 100 % mortality probably due to its high hepatotoxic effect. Several investigations have suggested that green tea can be used as a healthy drink which has chemopreventive potential against cancer development, but if consumed very frequently (>1 l/day), it has been connected with increased frequency of esophageal cancer in some countries [147]. So before the consumption of supplements, people should know about the daily recommended level and whether he/she is suffering from the deficiency of that antioxidant supplements.

4.3 Pro-antioxidant Activity

"Prooxidant versus antioxidant property" of antioxidant supplements is one of the important determinants for the toxic effect of supplements. Several antioxidant supplements like vitamin C, vitamin E, β-carotene, phenolic antioxidants, etc., were found to possess prooxidant activity in higher concentration as discussed earlier. This prooxidant effect can enhance the generation of free radical, thus producing harmful effects. The prooxidant and possible harmful effects of an antioxidant depend on the concentration of antioxidant, redox potential, interaction with other antioxidants, existence of transition metals, and activity and level of endogenous antioxidants [148].

4.4 Interactions

Interaction with drugs, other supplements, or any other endogenous/exogenous substance should need active consideration. For example, vitamin E interferes with vitamin K which can impair clot formation [26], high intake of iron along with ascorbic acid may induce lipid peroxidation of LDL [14], certain flavonoids may cause flavonoid–drug interactions [149], and CoQ10 may possibly interfere with warfarin and thus increase the risk of blood clotting or bleeding [56]. Thus the effect of antioxidant supplements is also needed to be addressed.

4.5 Pharmacokinetic Variation

Absorption, distribution, metabolism, and excretion may differ from an individual to an individual. In the gastrointestinal tract, if nutrients and phytochemicals are available in soluble and bioaccessible form, they may produce antioxidant activity after being taken up by the epithelium. To be bioactive in other organs, other factors of bioavailability such as absorption from gut mucosa, transportation in place of action, metabolite produced by phase I and phase II, and excretion do play a role [147]. Excretion plays a vital role, for example, prolonged intake of supplements may cause accumulation of zinc in pancreas and thus results impairment of the pancreatic function [89]. Thus a reevaluation including the dose, metabolism, and kinetics is required.

4.6 Metabolite

Most of the investigations don't measure or consider the metabolite generates. Some of the metabolites if produced in excess quantity can produce toxic effect. For example, quinines are the metabolite of vitamin E known to generate oxygen radicals and oxidize cellular components [34]. Thus the estimation of the metabolites is also required to explain the effect of supplements.

4.7 Physiological Condition, Disease, and Therapy

CoQ10 may contraindicate for pregnant or lactating women, those with diabetes and high blood pressure [56]. It was also advised to avoid excessive exercise while taking CoQ10. Some studies have also suggested that the consumption of antioxidant supplements can interfere with chemotherapy/radiotherapy as these therapies kill the cancer cell by producing oxidative stress [4, 24]. At high pH and in the presence of iron, phenolic antioxidants when consumed in large dose can start an auto-oxidation process and behave like prooxidants [63]. Even though not yet well proved, the genetic background could also be responsible for the harmful effects of antioxidants [148].

Gender differences also may be responsible for variable effects of antioxidant supplements. Plasma levels of β-carotene and vitamin C were found more in women than in men after antioxidant supplement therapy (vitamin C, vitamin E, β-carotene, Se, and zinc) for 7.5 years [81]. Significant reduction in serum ferritin and hematocrit level was observed in especially women after competitive interaction between zinc and iron [89].

5 Perspectives and Future Directions

The consequences resulting through the use of antioxidant supplements are uncertain. Antioxidant supplements may play dual roles, performing as double-edged swords. The initial promising effects of antioxidant supplements against CVDs, cancer, and other oxidative stress-related diseases can be derived from observational studies, followed by generally "no effect" or "harmful effect" reported from a large number of RCTs. Although the role of oxidative stress in different acute and chronic diseases is largely established, the value of antioxidant supplement strategies is still controversial. This chapter has highlighted the misconception regarding "only beneficial effect of antioxidant supplements" and "antioxidant supplements are not associated with any adverse effect." Important point of criticism is the prospect to get experimental results "from the bench to the bedside." The risks which are associated with dietary supplements on the basis of observational studies have been documented. The use of antioxidant supplements in the treatment/prevention of human disease states has not been as successful as might have been predicted due to intrinsic pharmacokinetic or pharmacodynamic limitations. Implying health benefits of antioxidant supplementation in the general population is contrary to the evidence; furthermore, inappropriate use or surplus use of antioxidant supplements will put people at risk and will benefit only to the people involved in producing or selling of supplement products. In Table 3, we

Table 3 Some major limitation and few recommendations related to research on antioxidant supplements

Limitations	Recommendations
1. Majority of the RCTs are short term with less number of subjects	*Recommendations for future investigations*
2. When multicomponent supplements are used, the function of each component or synergistic effects of the components is not discussed properly	1. Perform RCTs with high methodological quality
3. When antioxidant supplement used for certain disease – no data available on other organ effect	2. Effect of genetic factor, environmental factor, and sociological factors should be investigated
4. "Concentration response relationship" not described thoroughly	3. More consideration should be given on RCTs rather than observational studies
5. Limited comparative reports on the beneficial effect of normal foods (mixture of antioxidants) vs. particular antioxidant supplement	4. Interaction of antioxidant supplement with current therapy or diet is also needed to be investigated
6. Effect of same antioxidant supplement on different group of populations (i.e., smokers vs. nonsmoker, Asian vs. African, poor vs. rich) was not investigated properly	5. "Supplement intake–plasma concentration–clinical endpoint relationship" should be investigated
	6. Changes in oxidative stress followed by the use of supplements should be monitored using appropriate, reliable, and sensitive biomarkers, and the actual benefits should be assessed
	Recommendations for people/health professionals
	1. People should be educated about the exact outcome of antioxidant supplement use
	2. In case of specific reason, specific antioxidant supplements should be taken but with the advice of health professionals
	3. General people should avoid regular intake of antioxidant supplements till the proper conclusion was reached
	4. In present condition, intake of foods rich in antioxidant compounds should be preferred than the supplements

pointed out some major drawbacks and few recommendations related to research on antioxidant supplements.

New strategies are needed to clarify the exact role of antioxidant supplements in light of different variables which may have impact on health effect of supplements. Although it is possible that some population groups may benefit from antioxidant supplementation, the evidence is so equivocal that it is inappropriate to make the sweeping recommendation for antioxidant supplementation. At the moment, currently available data are insufficient to recommend the routine use of antioxidant supplements. Factors responsible for unsatisfactory and discrepant results from observational, randomized, and clinical studies may include differences in the gender, age, body weight, duration of treatment, interaction, dosages, and dietary habits. Based on recent meta-analysis of RCT, including the current studies, it may be imperative to reevaluate the selection of antioxidant supplements. Thus indiscriminate widespread use of antioxidant supplements should be cautioned. Governments and regulating agencies also should strictly evaluate their efficacy and safety, before the marketing of antioxidant supplements as medicinal products. A balanced diet including fruits and vegetable is a complex combination of antioxidant as well as other potentially valuable micronutrients and macronutrients, which may, thus, work with several kinetics and dynamics. Well-defined long-term clinical trials are essential to assess the efficacy of the use of antioxidant supplements. Meanwhile, a well-balanced diet rich in fruits and vegetables is highly recommended until the remaining riddles in the antioxidant supplement puzzle are solved.

References

1. Sen S, Chakraborty R (2011) In: Silvana A, Hepel M (eds) Oxidative stress: diagnostics, prevention, and therapy, ACS symposium series 1083. American Chemical Society, Washington, DC, pp 1–37
2. Valko M, Leibfritz D, Moncol J et al (2007) Free radicals and antioxidants in normal physiological functions and human disease. Int J Biochem Cell Biol 39:44–84
3. Sen S, Chakraborty R, Sridhar C et al (2010) Free radicals, antioxidants, diseases and phytomedicines: current status and future prospect. Int J Pharma Sci Rev Res 3:91–100
4. Bjelakovic G, Gluud C (2007) Surviving antioxidant supplements. JNCI 99:742–743
5. Bjelakovic G, Nikolova D, Gluud LL et al (2012) Antioxidant supplements for prevention of mortality in healthy participants and patients with various diseases. Cochrane Database Syst Rev (2) Art no.: CD007176. doi:10.1002/14651858.CD00717
6. Lykkesfeldt J, Poulsen HE (2009) Is vitamin C supplementation beneficial? Lessons learned from randomized controlled trials. Br J Nutr 2009:1–9
7. Brambilla D, Mancuso C, Scuderi MR et al (2008) The role of antioxidant supplement in immune system, neoplastic, and neurodegenerative disorders: a point of view for an assessment of the risk/benefit profile. Nutr J 7:29
8. Lu J, Lina PH, Yaoa Q et al (2010) Chemical and molecular mechanisms of antioxidants: experimental approaches and model systems. J Cell Mol Med 14:840–860
9. Brewer MS (2011) Natural antioxidants: sources, compounds, mechanisms of action, and potential applications. Compr Rev Food Sci Food Saf 10:221–247
10. Anonymous (2013) Antioxidants: beyond the hype. Harvard School of Public Health. http://www.hsph.harvard.edu/nutritionsource/antioxidants. Accessed 14 May 2013
11. Anonymous (2013) Dietary supplements: what you need to know. Office of dietary supplements, National Institute of Health, US. http://ods.od.nih.gov/HealthInformation/DS_WhatYouNeedToKnow.aspx. Accessed 14 May 2013
12. Li Y, Schellhorn HE (2007) New developments and novel therapeutic perspectives for vitamin C. J Nutr 137:2171–2184
13. Iqbal K, Khan A, Khattak MMAK (2004) Biological significance of ascorbic acid (vitamin C) in human health – a review. Pak J Nutr 3:5–13
14. Naidu KA (2003) Vitamin C in human health and disease is still a mystery? An overview. Nutr J 2:1–10
15. Deruelle F, Baron B (2008) Vitamin C: is supplementation necessary for optimal health. J Altern Complemen Med 14:1291–1298
16. Garriguet D (2010) The effect of supplement use on vitamin C intake. Stat Can Health Rep 21(1):1–6, Catalogue no. 82-003-XPE
17. Juraschek SP, Miller ER, Gelber AC (2011) Effect of oral vitamin C supplementation on serum uric acid: a meta-analysis of randomized controlled trials. Arthritis Care Res 63:1295–1306
18. Padayatty SJ, Katz A, Wang Y et al (2003) Vitamin C as an antioxidant: evaluation of its role in disease prevention. J Am Coll Nutr 22:18–35
19. U.S. Preventive Services Task Force (2003) Routine vitamin supplementation to prevent cancer and

cardiovascular disease: recommendations and rationale. Ann Intern Med 139:51–55
20. Anonymous (2013) Vitamin C (Ascorbic acid). University of Maryland Medical Center. http://www.umm.edu/altmed/articles/vitamin-c-000339.htm. Accessed 14 May 2013
21. Hercberg S, Ezzedine K, Guinot C et al (2007) Antioxidant supplementation increases the risk of skin cancers in women but not in men. J Nutr 137:2098–2105
22. Buettner GR, Jurkiewicz BA (1996) Catalytic metals, ascorbate and free radicals: combinations to avoid. Rad Res 145:532–541
23. Bouayed J, Bohn T (2012) Nutrition, well-being and health. In: Bouayed J, Bohn T (eds) Dietary derived antioxidants: implications on health. InTech, Croatia, pp1–22.http://www.intechopen.com/books/nutrition-well-being-and-health/dietary-derived-antioxidants-implication-on-health
24. Agus DB, Vera JC, Golde DW (1999) Stromal cell oxidation: a mechanism by which tumors obtain vitamin C. Cancer Res 59:4555–4558
25. Levine M, Padayatty SJ, Espey MG (2011) Vitamin C: a concentration-function approach yields pharmacology and therapeutic discoveries. Adv Nutr 2:78–88
26. Traber MG, Frei B, Beckman JS (2008) Vitamin E revisited: do new data validate benefits for chronic disease prevention? Curr Opin Lipidol 19:30–38
27. Miller ER III, Pastor-Barriuso R, Dalal D et al (2005) Meta-analysis: high-dosage vitamin E supplementation may increase all-cause mortality. Ann Intern Med 142:37–46
28. Schurks M, Glynn RJ, Rist PM et al (2010) Effects of vitamin E on stroke subtypes: meta-analysis of randomised controlled trials. Br Med J 341:c5702
29. Alkhenizan AH, Al-Omran MA (2004) The role of vitamin E in the prevention of coronary events and stroke: meta-analysis of randomized controlled trials. Saudi Med J 25:1808–1814
30. Alkhenizan A, Hafez K (2007) The role of vitamin E in the prevention of cancer: a meta-analysis of randomized controlled trials. Ann Saudi Med 27:409–414
31. Klein EA, Thompson IM Jr, Tangen CM (2011) Vitamin E and the risk of prostate cancer. JAMA 306:1549–1556
32. Abner EL, Schmitt FA, Mendiondo MS et al (2011) Vitamin E and all-cause mortality: a meta-analysis. Curr Aging Sci 4:158–170
33. Berry D, Wathen JK, Newell M (2009) Bayesian model averaging in meta-analysis: vitamin E supplementation and mortality. Clin Trials 6:28–41
34. Bast A, Haenen GRMM (2002) The toxicity of antioxidants and their metabolites. Environ Toxicol Pharmacol 11:251–258
35. Tinggi U (2008) Selenium: its role as antioxidant in human health. Environ Health Prev Med 13:102–108
36. Lee E, Myung S, Jeon Y et al (2011) Effects of selenium supplements on cancer prevention: meta-analysis of randomized controlled trials. Nutr Cancer 63:1185–1195
37. The Office of Dietary Supplements (2013) Selenium. National Institute of Health U.S. http://ods.od.nih.gov/factsheets/Selenium-HealthProfessional/. Accessed 14 May 2013
38. Flores-Mateo G, Navas-Acien A, Pastor-Barriuso R et al (2006) Selenium and coronary heart disease: a meta-analysis. Am J Clin Nutr 84:762–773
39. Stranges S, Marshall JR, Natarajan R et al (2007) Effects of long-term selenium supplementation on the incidence of type 2 diabetes. Ann Intern Med 147:217–223
40. Morgan KL (2008) Genetic and pharmacologic analysis of the mechanisms of selenium toxicity in Caenorhabditis elegans. Ph.D. thesis, School of Medicine University of Pittsburgh
41. Spallholz JE (2001) Selenium and the prevention of cancer. Part II: mechanisms for the carcinostatic activity of Se compounds. The Bull of Selenium-Tellurium Dev Assoc :1–12
42. Seko Y, Imura N (1997) Active oxygen generation as a possible mechanism of selenium toxicity. Biomed Environ Sci 10:333–339
43. Mezes M, Balogh K (2009) Prooxidant mechanisms of selenium toxicity – a review. Acta Biologica Szeged 53:15–18
44. Anonymous (2007) Coenzyme Q10. Altern Med Rev 12:159–168
45. Anonymous (2013) HPRC dietary supplements classification system: Coenzyme Q10. Human Performance Resource Center. http://hprc-online.org/dietary-supplements/files/monograph-coq. Accessed 24 May 2013
46. Rosenfeldt FL, Haas SJ, Krum H et al (2007) Coenzyme Q10 in the treatment of hypertension: a meta-analysis of the clinical trials. J Hum Hypertens 21:297–306
47. Gao L, Mao Q, Cao J et al (2012) Effects of coenzyme Q10 on vascular endothelial function in humans: a meta-analysis of randomized controlled trials. Atherosclerosis 221:311–316
48. Fotino AD, Thompson-Paul AM, Bazzano LA (2013) Effect of coenzyme Q10 supplementation on heart failure: a meta-analysis. Am J Clin Nutr 97:268–275
49. Anonymous (2013) Coenzyme Q10. http://en.wikipedia.org/wiki/Coenzyme_Q10. Accessed 24 May 2013
50. Nadjarzadeh A, Shidfar F, Amirjannati N et al (2014) Effect of coenzyme Q10 supplementation on antioxidant enzymes activity and oxidative stress of seminal plasma: a double-blind randomised clinical trial. Andrologia 46(2):177–183
51. Shults CW, Oakes D, Kieburtz K et al (2002) Effects of coenzyme Q10 in early Parkinson diseases. Arch Neurol 59:1541–1550

52. Tsai K, Huang Y, Kao C, Yang D et al (2012) A novel mechanism of coenzyme Q10 protects against human endothelial cells from oxidative stress-induced injury by modulating NO-related pathways. J Nut Biochem 23:458–468
53. Kumar A, Kaur H, Devi P et al (2009) Role of coenzyme Q10 (CoQ10) in cardiac disease, hypertension and Meniere-like syndrome. Pharmacol Therap 124:259–268
54. Boutet M, Roland L, Thomas N, Bilodeau JF (2009) Specific systemic antioxidant response to preeclampsia in late pregnancy: the study of intracellular glutathione peroxidases in maternal and fetal blood. Am J Obstet Gynecol 200(5):530.e1–530.e7
55. Landmesser U, Merten R, Spiekermann S et al (2000) Vascular extracellular superoxide dismutase activity in patients with coronary artery disease: relation to endothelium-dependent vasodilation. Circulation 101:2264–2270
56. Anonymous (2013) Coenzyme Q10. RCT summary for professionals reliable cancer therapies. http://www.reliablecancertherapies.com/sites/default/files/documents/coenzymeq10_for_prof.pdf. Accessed 24 Oct 2013
57. Hooper L, Kroon PA, Rimm EB et al (2008) Flavonoids, flavonoid-rich foods, and cardiovascular risk: a meta-analysis of randomized controlled trials. Am J Clin Nutr 88:38–50
58. Curry CD (2012) The effects of a multi-flavonoid supplement on vascular and hemodynamic parameters in older pre-hypertensives. Dissertation, Appalachian State University
59. Blostein-Fujii A, DiSilvestro RA, Frid D et al (1999) Short term citrus flavonoid supplementation of type II diabetic women: no effect on lipoprotein oxidation tendencies. Free Radic Res 30:315–320
60. Macready AL, Kennedy OB, Ellis JA et al (2009) Flavonoids and cognitive function: a review of human randomized controlled trial studies and recommendations for future studies. Genes Nutr 4:227–242
61. Sarah Egert E, Rimbach G (2011) Which sources of flavonoids: complex diets or dietary supplements? Adv Nutr 2:8–14
62. Ross JA, Kasum CM (2002) Dietary flavonoids: bioavailability, metabolic effects, and safety. Ann Rev Nutr 22:19–34
63. Martin KR, Appel CL (2010) Polyphenols as dietary supplements: a double-edged sword. Nutr Diet Suppl 2:1–12
64. Patrick L (2000) Beta-carotene: the controversy continues. Altern Med Rev 5:530–545
65. Druesne-Pecollo N, Latino-Martel P, Norat T et al (2009) Beta-carotene supplementation and cancer risk: a systematic review and metaanalysis of randomized controlled trials. Int J Cancer 127:172–184
66. Jeon YJ, Myung SK, Lee EH et al (2011) Effects of beta carotene supplements on cancer prevention: metaanalysis of randomized controlled trials. Nutr Cancer 63:1196–1207
67. Gallicchio L, Boyd K, Matanoski G et al (2008) Carotenoids and the risk of developing lung cancer: a systematic review. Am J Clin Nutr 88:372–383
68. Kopcke W, Krutmann J (2008) Protection from sunburn with β-carotene – a meta-analysis. Photochem Photobiol 84:284–288
69. Omaye ST, Krinsky NI, Kagan VE et al (1997) β-carotene: friend or foe? Fundam Appl Toxicol 40:163–174
70. Tinkel J, Hassanain H, Khouri SJ (2012) Cardiovascular antioxidant therapy: a review of supplements, pharmacotherapies, and mechanisms. Cardiol Rev 20:77–83
71. Kris-Etherton PM, Lichtenstein AH, Howard BV et al (2004) Antioxidant vitamin supplements and cardiovascular disease. Circulation 110:637–641
72. Myung S, Ju W, Cho B et al (2013) Efficacy of vitamin and antioxidant supplements in prevention of cardiovascular disease: systematic review and meta-analysis of randomized controlled trials. BMJ 346:1–22
73. The Heart Foundation (2010) Antioxidants in food, drinks and supplements for cardiovascular health. http://www.heartfoundation.org.au/SiteCollectionDocuments/Antioxidants-Summary-Evidence.pdf. National Heart Foundation of Australia. Accessed 11 May 2013
74. Riccioni G, Bucciarelli T, Mancini B et al (2007) Antioxidant vitamin supplementation in cardiovascular diseases. Ann Clin Lab Sci 37:89–95
75. Myung SK, Kim Y, Ju W et al (2010) Effects of antioxidant supplements on cancer prevention: meta-analysis of randomized controlled trials. Ann Oncol 21:166–179
76. Bjelakovic G, Nagorni A, Nikolova D et al (2006) Meta-analysis: antioxidant supplements for primary and secondary prevention of colorectal adenoma. Aliment Pharmacol Ther 24:281–291
77. Bjelakovic G, Nikolova D, Simonetti RG et al (2004) Antioxidant supplements for prevention of gastrointestinal cancers: a systematic review and meta-analysis. Lancet 364:1219–1228
78. Block KI, Koch AC, Mead MN et al (2007) Impact of antioxidant supplementation on chemotherapeutic efficacy: a systematic review of the evidence from randomized controlled trials. Cancer Treat Rev 33(5):407–418
79. Bereznicki L (2009) Antioxidants and the prevention of cancer. Pharmacy 28:686–691
80. Robinson NG (2013) The good, bad, and uncertain: combining antioxidants and chemotherapy. http://csuvets.colostate.edu/pain/Articlespdf/ChemotherapyandAntioxidantsTheGoodBadandUncertain082106.pdf. Accessed 14 Dec 2013
81. Herrera E, Jimenez R, Aruoma OI et al (2009) Aspects of antioxidant foods and supplements in health and disease. Nutr Rev 67(1):S140–S144
82. Al-Busafi SA, Bhat M, Wong P et al (2012) Antioxidant therapy in nonalcoholic steatohepatitis. Hepat Res Treat 2012:1–8

83. Lirussi F, Azzalini L, Orando S et al (2007) Antioxidant supplements for non-alcoholic fatty liver disease and/or steatohepatitis. Cochrane Database Syst Rev 1, CD004996
84. Farias MS, Budni P, Ribeiro CM et al (2012) Antioxidant supplementation attenuates oxidative stress in chronic hepatitis C patients. Gastroenterol Hepatol 35:386–394
85. Groenbaek K, Friis H, Hansen M et al (2006) The effect of antioxidant supplementation on hepatitis C viral load, transaminases and oxidative status: a randomized trial among chronic hepatitis C virus-infected patients. Eur J Gastroenterol Hepatol 18:985–989
86. Bjelakovic G, Gluud LL, Nikolova D et al (2010) Meta-analysis: antioxidant supplements for liver diseases – the Cochrane Hepato-Biliary Group. Aliment Pharmacol Ther 32:356–367
87. Penniston KL, Tanumihardjo SA (2006) The acute and chronic toxic effects of vitamin A. Am J Clin Nutr 83:191–201
88. Anonyms (2013) Liver disease, general. http://www.med.nyu.edu/content?ChunkIID=38611. Accessed 4 Dec 2013
89. Nriagu JO (2011) Encyclopedia of environmental health. Elsevier Science, Amsterdam/London. http://store.elsevier.com/product.jsp?isbn=9780444522733
90. Ruhe RC, McDonald RB (2001) Use of antioxidant nutrients in the prevention and treatment of type 2 diabetes. J Am Coll Nutr 20:363S–369S
91. Akbar S, Bellary S, Griffiths HR (2011) Dietary antioxidant interventions in type 2 diabetes patients: a meta-analysis. Br J Diabetes Vasc Dis 11:62–68
92. Palacka P, Kucharska J, Murin J et al (2010) Complementary therapy in diabetic patients with chronic complications: a pilot study. Bratisl Lek Listy 111:205–211
93. Arulselvan P, Umamaheswari A, Fakurazi S (2012) Therapeutic approaches for diabetes with natural antioxidants. In: Capasso A (ed) Medicinal plants as antioxidant agents: understanding their mechanism of action and therapeutic efficacy. Research Signpost, Trivandrum, pp 237–266
94. Peponis V, Papathanasiou M, Kapranou A et al (2002) Protective role of oral antioxidant supplementation in ocular surface of diabetic patients. Br J Ophthalmol 86:1369–1373
95. Garcia-Medina JJ, Pinazo-Duran MD, Garcia-Medina M et al (2011) A 5-year follow-up of antioxidant supplementation in type 2 diabetic retinopathy. Eur J Ophthalmol 21:637–643
96. Kataja-Tuomola M (2011) Antioxidants, weight change and risk of type 2 diabetes. Dissertation, University of Helsinki
97. Levy Y, Zaltzberg H, Ben-Amotz A (1999) β-carotene affects antioxidant status in non-insulin-dependent diabetes mellitus. Pathophysiology 6:157–161
98. Jayawardena R, Ranasinghe P, Galappatthy P et al (2012) Effects of zinc supplementation on diabetes mellitus: a systematic review and meta-analysis. Diabetol Metab Synd 4:13
99. Anonyms (2013) Research examines adverse effects of antioxidants. Friedrich Schiller University in Jena. https://www.uni-jena.de/en/PM090511_Antioxidants.html. Accessed on 23 Dec 2013
100. Pastor-Valero M (2002) Antioxidant micronutrients and cataract: a review of epidemiological evidence. Gac Sanit 16(2):29–40
101. Evans J (2008) Antioxidant supplements to prevent or slow down the progression of AMD: a systematic review and meta-analysis. Eye 22:751–760
102. Fusco D, Colloca G, Monaco MRL (2007) Effects of antioxidant supplementation on the aging process. Clin Interv Aging 2:377–387
103. Heshmatollah A, de Jong MME, Drexhage HA (2012) Does antioxidant supplementation improve symptoms in schizophrenia patients a systematic review. Erasmus J Med 3:42–45
104. Allen S, Britton JR, Leonardi-Bee JA (2009) Association between antioxidant vitamins and asthma outcome measures: systematic review and meta-analysis. Thorax 64:610–619
105. Macpherson H, Pipingas A, Pase MP (2013) Multivitamin-multimineral supplementation and mortality: a meta-analysis of randomized controlled trials. Am J Clin Nutr 97:437–444
106. Bjelakovic G, Nikolova D, Gluud LL et al (2007) Mortality in randomized trials of antioxidant supplements for primary and secondary prevention systematic review and meta-analysis. JAMA 297:842–857
107. Biesalski HK, Grune T, Tinz J et al (2010) Reexamination of a meta-analysis of the effect of antioxidant supplementation on mortality and health in randomized trials. Nutrients 2:929–949
108. Sesso HD, Buring JE, Christen WG et al (2008) Vitamins E and C in the prevention of cardiovascular disease in men: the physicians' health study II randomized trial. JAMA 300:2123–2133
109. Gaziano JM, Glynn RJ, Christen WG et al (2009) Vitamins E and C in the prevention of prostate and total cancer in men: the physicians' health study II, a randomized controlled trial. JAMA 301:52–62
110. Karajibani M, Montazerifar F, Hashemi M et al (2011) Effects of antioxidant vitamin supplements on lipid profiles and antioxidant capacity in cardiovascular patients. Rawal Med J 36:1–11
111. Theodorou AA, Nikolaidis MG, Paschalis V et al (2011) No effect of antioxidant supplementation on muscle performance and blood redox status adaptations to eccentric training. Am J Clin Nutr 93:1373–1383
112. Cook NR, Albert CM, Gaziano JM et al (2007) A randomized factorial trial of vitamins C, E, and beta-carotene in the secondary prevention of cardiovascular events in women: results from the Women's Antioxidant Cardiovascular Study (WACS). Arch Intern Med 167:1610–1618
113. Song Y, Cook NR, Albert CM et al (2009) Effects of vitamins C and E and β-carotene on the risk of type 2 diabetes in women at high risk of cardiovascular

disease: a randomized controlled trial. Am J Clin Nutr 90:429–437
114. Lin J, Cook NR, Albert C et al (2009) Vitamins C and E and beta carotene supplementation and cancer risk: a randomized controlled trial. J Natl Cancer Inst 101:14–23
115. Lippman SM, Klein EA, Goodman PJ et al (2009) Effect of selenium and vitamin e on risk of prostate cancer and other cancers. JAMA 301:39–51
116. Lee I, Cook NR, Gaziano JM et al (2005) Vitamin E in the primary prevention of cardiovascular disease and cancer. JAMA 294:56–65
117. Christen WG, Glynn RJ, Chew EY et al (2010) Vitamin E and age-related macular degeneration in a randomized trial of women. Ophthalmology 117:1163–1168
118. Zureik M, Galan P, Bertrais S et al (2004) Effects of long-term daily low-dose supplementation with antioxidant vitamins and minerals on structure and function of large arteries. Arterioscler Thromb Vasc Biol 24:1485–1491
119. Eskenazi B, Kidd SA, Mark AR et al (2005) Antioxidant intake is associated with semen quality in healthy men. Hum Reprod 20:1006–1012
120. Bae S, Jung W, Lee E et al (2009) Effects of antioxidant supplements intervention on the level of plasma inflammatory molecules and disease severity of rheumatoid arthritis patients. J Am Coll Nutr 28:56–62
121. Heinrich U, Tronnier H, Stahl W et al (2006) Antioxidant supplements improve parameters related to skin structure in humans. Skin Pharmacol Physiol 19:224–231
122. Arnaud J, Bost M, Vitoux D et al (2007) Effect of low dose antioxidant vitamin and trace element supplementation on the urinary concentrations of thromboxane and prostacyclin metabolites. J Am Coll Nutr 26:405–411
123. Hininger IA, Meyer-Wenger A, Moser U et al (2001) No significant effects of lutein, lycopene or b-carotene supplementation on biological markers of oxidative stress and ldl oxidizability in healthy adult subjects. J Am Coll Nutr 20:232–238
124. Jacob RA, Aiello GM, Stephensen CB et al (2003) Moderate antioxidant supplementation has no effect on biomarkers of oxidant damage in healthy men with low fruit and vegetable intakes. J Nutr 133:740–743
125. Dusinska M, Kazimırova A, Barancokova M et al (2003) Nutritional supplementation with antioxidants decreases chromosomal damage in humans. Mutagenesis 18:371–376
126. Romieu I, Meneses F, Ramirez M et al (1998) antioxidant supplementation and respiratory functions among workers exposed to high levels of ozone. Am J Respir Crit Care Med 158:226–232
127. Shargorodsky M, Debby O, Matas Z et al (2010) Effect of long-term treatment with antioxidants (vitamin C, vitamin E, coenzyme Q10 and selenium) on arterial compliance, humoral factors and inflammatory markers in patients with multiple cardiovascular risk factors. Nutr Metab 7:55
128. Briancon S, Boini S, Bertrais S et al (2011) Long-term antioxidant supplementation has no effect on health-related quality of life: the randomized, double-blind, placebo-controlled, primary prevention SU.VI.MAX trial. Int J Epidemiol 40:1605–1616
129. Czernichow S, Vergnaud A, Galan P et al (2009) Effects of long-term antioxidant supplementation and association of serum antioxidant concentrations with risk of metabolic syndrome in adults. Am J Clin Nutr 90:329–335
130. Czernichow S, Couthouis A, Bertrais S et al (2006) Antioxidant supplementation does not affect fasting plasma glucose in the supplementation with antioxidant vitamins and minerals (SU.VI.MAX) study in France: association with dietary intake and plasma concentrations. Am J Clin Nutr 84:395–399
131. Christen WG, Manson JE, Glynn RJ et al (2007) Beta-carotene and age-related maculopathy in a randomized trial of U.S. physicians. Arch Ophthalmol 125:333–339
132. Frieling UM, Schaumberg DA, Kupper TS et al (2000) A randomized, 12-year primary-prevention trial of beta carotene supplementation for nonmelanoma skin cancer in the Physicians' health study. Arch Dermatol 136:179–184
133. Chiabrando C, Avanzini F, Rivalta C et al (2002) Long-term vitamin E supplementation fails to reduce lipid peroxidation in people at cardiovascular risk: analysis of underlying factors. Curr Control Trials Cardiovasc Med 3:5
134. Taylor HR, Tikellis G, Robman LD et al (2002) Vitamin E supplementation and macular degeneration: randomised controlled trial. BMJ 325:11–14
135. Salonen JT, Nen KN, Salonen R et al (2000) Antioxidant supplementation in atherosclerosis prevention (ASAP) study: a randomized trial of the effect of vitamins E and C on 3-year progression of carotid atherosclerosis. J Intern Med 248:377–386
136. The HOPE and HOPE-TOO Trial Investigators (2005) Effects of long-term vitamin E supplementation on cardiovascular events and cancer. JAMA 293:1338–1347
137. Weigert G, Kaya S, Pemp B et al (2011) Effects of lutein supplementation on macular pigment optical density and visual acuity in patients with age-related macular degeneration. Invest Ophthalmol Vis Sci 52:8174–8178
138. Askari G, Ghiasvand R, Feizi A et al (2012) The effect of quercetin supplementation on selected markers of inflammation and oxidative stress. J Res Med Sci 17:637–641
139. Bartlett HE, Eperjesi F (2007) Effect of lutein and antioxidant dietary supplementation on contrast sensitivity in age-related macular disease: a randomized controlled trial. Eur J Clin Nutr 61:1121–1127
140. Rayman MP, Thompson AJ, Bekaert B (2008) Randomized controlled trial of the effect of selenium supplementation on thyroid function in the elderly in the United Kingdom. Am J Clin Nutr 87:370–378

141. Duffield-Lillico AJ, Reid ME, Turnbull BW et al (2002) Baseline characteristics and the effect of selenium supplementation on cancer incidence in a randomized clinical trial: a summary report of the nutritional prevention of cancer trial. Cancer Epidemiol Biomarkers Prev 11:630–639
142. Costa S, Giannantonio C, Romagnoli C (2013) Effects of lutein supplementation on biological antioxidant status in preterm infants: a randomized clinical trial. J Matern Fetal Neonatal Med 26(13):1311–1315
143. The Age-Related Eye Disease Study 2 (AREDS2) Research Group (2013) Lutein/zeaxanthin for the treatment of age-related cataract. JAMA Ophthalmol 131(7):843–850
144. Albanes D, Heinonen OP, Taylor PR et al (1996) α-tocopherol and β-carotene supplements and lung cancer incidence in the alpha-tocopherol, beta-carotene cancer prevention study: effects of base-line characteristics and study compliance. J Nat Cancer Inst 88:1560–1570
145. Christie S, Walker AF, Hicks SM et al (2004) Flavonoid supplement improves leg health and reduces fluid retention in pre-menopausal women in a double-blind, placebo-controlled study. Phytomedicine 11:11–17
146. Pearson PJK, Lewis SA, Britton J et al (2004) Vitamin E supplements in asthma: a parallel group randomised placebo controlled trial. Thorax 59:652–656
147. Bouayed J, Bohn T (2010) Exogenous antioxidants—double-edged swords in cellular redox state. Oxid Med Cell Longev 3(4):228–237
148. Villanueva C, Kross RD (2012) Antioxidant-induced stress. Int J Mol Sci 13:2091–2109
149. Harwood M, Danielewska-Nikiel B, Borzelleca JF et al (2007) A critical review of the data related to the safety of quercetin and lack of evidence of in vivo toxicity, including lack of genotoxic/carcinogenic properties. Food Chem Toxicol 45:2179–2205

Linking Toll-Like Receptors Signaling to Oxidative Damage: Potential Role in Cancer Therapy

Kanchan Karki, Deepti Pande, Reena Negi, Seema Khanna, Ranjana S. Khanna, and Hari D. Khanna

Abstract

Inflammation either promotes host defense or damages organs. Endogenous molecules generated upon oxidative damage can activate TLRs (toll-like receptors) that ultimately alert the innate immune system of danger. Although role of TLRs in the regulation of tissue injury is well established, however, their role in carcinogenesis is still obscure. In this chapter the main emphasis is to open new roads concerning the role of TLRs in devising new opportunities for drug development in cancer through manipulating immune responses.

Keywords

Toll-like receptor (TLR) • MicroRNAs • SiRNAs • DAMP • Inflammation

1 Introduction

TLRs are a family of proteins involved in the recognition of pathogen-associated molecular motifs that activate both innate (nonspecific) and adaptive (specific) immune responses and contribute to the immune system's capacity to efficiently combat pathogens [1]. TLRs derive their name from their similarity to the protein coded by the toll gene identified in *Drosophila* in 1985 by Christiane Nüsslein-Volhard [2] as a maternally derived factor necessary for dorsal–ventral axis formation of the developing zygote [3]. In vertebrates, TLRs were initially found to be expressed in all lymphoid tissues, but it is now becoming apparent that TLRs are widely expressed throughout the body and most highly expressed in peripheral blood leukocytes, including monocytes, B cells, T cells, granulocytes, and DCs (dendritic cells) [4]. Several members of the TLR family have been identified: 11 in humans and 13 in mouse [5]. Each TLR contains extracellular domain with leucine-rich repeats (LLR) motif, and these

K. Karki • D. Pande • R. Negi • H.D. Khanna (✉)
Department of Biophysics, Banaras Hindu University, Varanasi 221005, Uttar Pradesh, India
e-mail: hdkhanna@yahoo.co.in

S. Khanna
Department of General Surgery, Institute of Medical Sciences, Banaras Hindu University, Varanasi 221005, India

R.S. Khanna
Department of Chemistry, Faculty of Science, Banaras Hindu University, Varanasi 221005, India

V. Rani and U.C.S. Yadav (eds.), *Free Radicals in Human Health and Disease*,
DOI 10.1007/978-81-322-2035-0_20, © Springer India 2015

LLR comprise a binding domain for recognition of their respective pathogens. The TLR receptors have many structural similarities both extracellular and intracellular, but they differ from each other in ligand specificities and expression patterns, and presumably they have some variability in the signaling pathways and target genes they can activate [6]. TLRs can sense a broad range of microbial pathogen signatures (pathogen-associated molecular patterns) including surface components and nucleic acids. Upon binding of microbial ligands, TLRs activate signaling pathways and stimulate cytokine production and other parts of the innate immune response. TLRs that mainly serve to detect bacterial LPS and lipoproteins are located on the cell surface (TLR1, TLR2, TLR4, TLR5, TLR6, and TLR11), whereas those that mainly recognize viral RNA and bacterial DNA are located in late endosome–lysosome compartments (TLR3, TLR7, TLR8, and TLR9). TLR11 is present in mice, but not in humans. TLR1, TLR2, TLR4, and TLR6 initiate signaling by heterodimerization. TLR2 forms heterodimers with TLR1 or TLR6 which recognize bacterial triacylated and diacylated lipoproteins, respectively [7].

With time, data began to emerge showing that reactive oxygen species (ROS) and nitric oxide (NO) can cause tissue damage and generate a wide range of molecules including heat shock proteins, high-mobility group box 1 (HMGB1) and uric acid crystals, collectively called as DAMPs (damage-associated molecular patterns) or (CDAMPs) cell death-associated molecular patterns [8, 9], can also trigger PRRs (pattern recognition receptors) such as TLRs, and may signal or propagate alarm, recruit professional cells to clear the offending trigger, or promote immune activation in various tumor cells, tissues, or tumor cell lines (Table 1). For these functional features, DAMPs are also called "alarmins" [33, 34]. Proteins, lipids, and nucleic acids have all been shown to act as DAMPs that can activate TLR signaling when presented in the appropriate context [35, 36].

TLRs activate antigen-presenting cells (APCs) such as DCs and stimulation of both T-cell- and B-cell-mediated immune responses. TLRs have also a crucial role in maintaining tissue homeostasis by regulating tissue repair and regeneration. Moreover, TLR signaling has also been shown to regulate apoptosis with the expression of anti-apoptotic proteins or inhibitors of apoptosis [37]. On the one hand, evidences from genetic, clinical, and basic science studies support that elevated expression of some TLRs has been reported in many tumor cells, tissues, or tumor cell lines [10, 11, 13, 15, 16, 20, 22–25, 27, 30, 32, 38–41]. On the other hand, TLR signaling in tumor cells was also shown to reduce the proliferative capacity of tumor cells [42]. This is done by means of the induction of signaling cascades. The result of this process leads to an inflammatory response and activates the adaptive immune system both in infectious and noninfectious disease scenarios [43]. That is the reason why TLR-dependent pathways are considered relevant pharmacological targets for the treatment of different pathological conditions.

In this chapter an attempt has been made to understand the connection between oxidative stress, TLR cell signaling, and carcinogenesis and highlight how our improved understanding of these connections may provide novel preventive, diagnostic, and therapeutic strategies to reduce the health burden of cancer.

2 Association of TLRs with "Danger Model"

Two theories have been proposed to explain the activation of the immune system: "Infectious-nonself" or "stranger" model proposed by Charles Janeway [44] and "danger" model proposed by Polly Matzinger [45]. Charles Janeway proposed that the immune cells might recognize pathogen-associated molecular patterns. Years later, this model found support in the discovery of TLRs, which act as pattern recognition receptors that detect "infectious-nonself" molecules conserved in pathogens. However, the model was not able to explain autoimmune diseases [44]. To explain this, Polly Matzinger proposed a "danger" model, in which the immune system might be activated by alarm or "danger" signals from injured tissues upon exposure to pathogens, toxins, mechanical

Table 1 Endogenous activators of several TLRs and their association with various tumors

DAMPs	Activation of TLR		
Biglycan (proteoglycan)	TLR 4 [10], TLR2 [9]	Gastric [11]	Analysis of biglycan mRNA and protein concentrations in gastric cancer tissues confirms that biglycan-positive patients were strongly associated with lymph
Fibrinogen (protein peptide)	TLR 4 [12]	Breast [13]	The MCF-7 human breast cancer epithelial cell line has the ability to synthesize, secrete, and deposit FBG into the extracellular matrix (ECM)
Heparin sulfate	TLR 4 [14]	Lung [15], colon, bladder, breast [16]	Heparan sulfate (HS) cleavage by heparanase (endo-beta-D glucuronidase) undergoes continuous remodeling during development of cancer
HMGB1	TLR2, TLR4 [17, 18], TLR9 [19]	Breast, lung, colorectal, pancreatic, melanoma [20, 21], leukemia [22]	Necrotic cell death, resulting in massive HMGB1 release. HMGB1 stimulates the expression of vascular endothelial growth factor and platelet-derived growth factor signaling and helps to sustain this proangiogenic gene expression
Hsps	TLR2, TLR4 [13, 23]	Breast [24], lung [25]	Small molecule Hsp90 inhibitors bind to the ATP-binding pocket, inhibit chaperone function, and could potentially result in cytostasis or cell death. Inhibition of Hsp90 function has also proven effective in killing cancer
Hyaluronic acid	TLR2, TLR4 [26]	Pancreatic acid [27]	MIA PaCa-2, a human pancreatic carcinoma cell line, secreted hyaluronidases abundantly and generated readily detectable levels of LMW-HAs ranging from approximately 10 to 40 mers. The tumor-derived HA oligosaccharides were able to enhance CD44 cleavage and tumor cell motility
A100A8/A9	TLR4 [18, 28]	Breast, lung, colorectal [28]	S100A8-, S100A9-, and S100A12-abundant cell types
Tenascin-C	TLR4 [29]	Brain, breast, lung, skin, prostate, kidney, bladder, lymphoma, ovarian [30]	Tenascin-C is an adhesion modulatory extracellular matrix molecule that is highly expressed in the microenvironment of most solid tumors
Versican	TLR2 [31]	Bone, lung [32]	Depending on the cancer type, versican is expressed by either the cancer cells themselves or by stromal cells surrounding the tumor

damage, etc. These factors drive TLR activation during the progression of diseases, collectively called PAMPs and DAMPs [45]. TLR ligands in this case can be either microbial (exogenous) or host derived (endogenous). This model has been supported by the identification of uric acid and HMGB1, host-derived ligands that activate an immune response to dying cells by way of TLR4 activation [46, 47]. Host-derived molecule, e.g., HMGB1, released after tissue damage could activate the innate immune response through TLRs [17], even in the absence of pathogens, or cooperate with "infectious-nonself" molecules [47]. Thus, these studies enable to explain the molecular basis of some inflammatory diseases, autoimmunity, atherosclerosis, carcinogenesis, etc.

3 Toll-Like Receptors Signaling Pathways

There is a dramatic difference between the TLR signaling in response to DAMPs released by oxidative stress and PAMPs released by infective stress [17]. HMGB1 protein, a DAMP, activates IKKα and IKKβ, but lipopolysaccharide (LPS), a PAMP, increases activity of only IKKβ in cultured neutrophils and macrophages [31]. MD2 mediates TLR4 recognition of PAMPs, but CD14 mediates TLR4 recognition of DAMPs released by necrotic cells [29]. TLR signaling pathway has best been characterized in immune cells. The toll-like receptor signaling is mediated by a number of distinct pathways [the NF-kB (nuclear factor of kappa light polypeptide gene enhancer in B cells) pathway, p38 pathways, CREB pathway, JNK (c-Jun N-terminal kinase) pathway, IRF3 (interferon regulatory factor 3) and IRF7 (interferon regulatory factor 7) pathways] [18, 48]. After binding of TLR ligands, TLRs dimerize and transmit signals throughout the cell through one or more of four adaptor proteins: myeloid differentiation primary response gene 88 (MyD88), toll/interleukin-1 receptor domain-containing adaptor inducing interferon-β (TRIF), toll/interleukin-1 receptor domain-containing adaptor protein (TIRAP), and TRIF-related adaptor molecule (TRAM), where MyD88 is part of the signaling cascade of all TLRs except for TLR3. TLR3 follows TRIF-dependent pathway exclusively, while TLR4 follows both of these pathways [49]. TLR ligand binding stimulates MyD88 facilitation of phosphorylation of interleukin-1 (IL-1) receptor-associated kinase (IRAK)-1 [48, 50, 51]. Phosphorylated IRAK1 recruits and activates tumor necrosis factor receptor (TNFR)-associated factor 6 (TRAF6). TRAF6 can then activate transforming growth factor (TGF)-β-activated kinase 1 (TAK1). TAK1 is a mitogen-activated protein kinase (MAPK). This kinase is able to phosphorylate major transcription factors NF-kB and JNK. Activation of these molecules leads to transcription of genes associated with TLR activation. TLR signaling mediated via TRIF interacts with TRAF6 which stimulates interferon regulatory factors 3 and 7 to activate interferon-β (IFN-β) and thereby leads to activation of signal transduction and transcription of genes associated with this pathway. Subsequent activation of these pathways promotes the expression of oxidant-generating inducible nitric oxide synthase (iNOS) and a wide variety of proinflammatory cytokines, chemokines, and their receptors, including tumor necrosis factor alpha (TNF-α), interleukins (ILs), and macrophage inflammatory proteins (MIPs) [51, 52]. These factors initiate the inflammatory response, increase vascular permeability, direct DC and macrophage migration from the periphery to the central lymphoid organs, and regulate various aspects of adaptive immunity development (Fig. 1) [54].

4 Oxidative Stress and Toll-Like Receptor Signaling as a Key Pathway of Tumorigenesis

It is now generally accepted that cancer can be exacerbated by inflammation induced by biological, chemical, and physical factors. Free radicals are directly involved in the oxidative destruction of macromolecules such as lipids, proteins, and nucleic acids [34]. In order to counteract intracellular damage by oxidative stress molecule such as reactive oxygen species (ROS) and nitric oxide (NO), cells have developed a so-called intracellular antioxidant system. Antioxidants which are backbone of this cellular defense system regulate oxidative reactions by inhibiting, delaying, or hampering the oxidation of the substances [4, 8].

Inflammation and cancer are intimately related [55]. It is well known that persistent inflammatory conditions can induce cancer formation because cytokines and chemokines play a crucial role promoting angiogenesis, cancer cell survival, chemoresistance, and therefore tumor progression. Inflammatory cells release proangiogenic growth factors, which facilitate neovascularization that ultimately enhance inflammatory cell recruitment and promote chronic inflammation [56]. On the one hand, the increase of ROS is beneficial for the development and for the successful outcome of inflammation, tissue repair, and signal transduc-

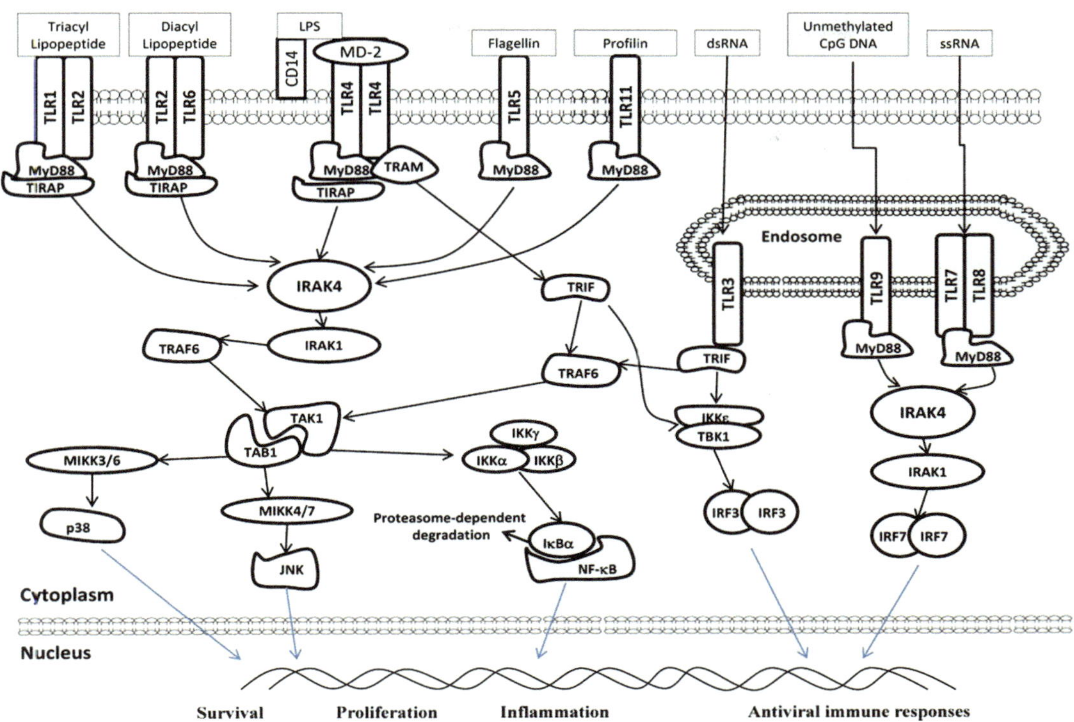

Fig. 1 Schematic representation of TLR signaling pathways [53]. TLRs localize to different subcellular compartments according to the molecular nature of the relevant ligands. TLR3, TLR7/8, and TLR9 are located at endolysosomal compartment and recognized dsRNA or ssRNA or unmethylated CpG DNA, respectively. In contrast, TLR2, TLR4, TLR5, and TLR11 are mainly displayed on plasma membrane and recognize indicated ligands. TLR2/1 heterodimers respond to triacyl lipopeptide, while TLR2/6 heterodimers recognize diacyl lipopeptide. CD14 and MD2 are accessory proteins required for LPS/TLR4 ligation. Following ligation with their respective ligands, TLRs recruit downstream adaptor molecules, such as MyD88 (all TLRs except TLR3) and TRIF (TLR3 and TLR4), to activate IRAK4, TRAF6, and IKKε/TBK1. These sequentially activate signaling pathway of NF-kB, p38, MAPK, and JNK, leading to a series of specific cellular responses related to cell survival, proliferation, and inflammation. IRF3 and IRF7 are also activated downstream of TLR3, TLR7, TLR8, and TLR9. These signaling mediate the production of type 1 interferon and antiviral immune responses. *MyD88* myeloid differentiation primary response protein 88, *IRAK* IL-1R-associated kinase, *TRAM* TRIF-related adaptor molecules, *TIRAP* TIR domain-containing adaptor protein, *TRAF* TNF receptor-associated factor, *TAB* TGF-β-activated kinase 1/MAP3K7-binding protein, *TAK* TGF-β-activated kinase, *IKK* inhibitor of kappa light polypeptide gene enhancer in B-cell kinase, *MAPK* mitogen-activated protein kinase

tion, leading to angiogenesis and cytokine production, and for the elimination of pathogenic agents by phagocytosis [35, 57–59]; however, on the other hand the consequences of inflammation persist with structural and functional alterations of tissues that are collectively referred to as chronic inflammatory processes [34]. In addition, these processes can contribute directly or indirectly to inactivate additional tumor suppressor genes within tumor cells or further increase expression of proto-oncogenes. In spite of its ideal outcome of inflammation, genetic instability due to persistent carcinoma cell oxidative stress therefore increases the malignant potential of the tumor [60]. The apoptotic cell death hinders inflammation and immune activation possibly because it inhibits the release of active DAMPs. Indeed, a DAMP, such as HMGB1, is preferentially released by necrotic rather than by apoptotic cells [61]. Moreover, in apoptotic cells, HMGB1 undergo oxidation-mediated inactivation due to cysteine residues which makes it unable to work as a DAMP [62]. As a further control of DAMP release, apoptotic cells are engulfed by scavenger cells before they reach a secondary necrotic state with delivery of intracellular content and possible trigger of an autoimmune reaction [63]. However, in the tumor setting, unscheduled necrotic death [64] results in a concomitant and unimpeded expression of DAMPs, such as HMGB1 and S100 protein family [65]. Studies have also examined that tumor cells may also contribute to maintain DAMPs bioactive.

Recent studies reported that TLRs are the cellular receptors that can sense these danger signals and represent a key molecular link between tissue injury, infection, and inflammation, e.g., a study reported that v-(2-carboxyethyl)pyrrole (CEP), end products of lipid oxidation generated during inflammation, and wound healing are recognized by toll-like receptor 2 (TLR2) [66]; versican, a breakdown product of oxidative stress, may activate fibroblasts and tumor-infiltrating myeloid cells through TLR2 and its coreceptors TLR6 and TLR1. These receptors and coreceptors are involved in the activation of the multiple types of chemokines and cytokines including TNF-α and IL-8 having tumorigenic properties and hence enhance tumor metastasis [65, 67, 68]. These findings suggest a new function of TLR2 as a sensor of oxidation-associated molecular patterns, and this mechanism can emerge as an important mechanism underlying numerous processes from tissue healing and remodeling to cancer progression.

Reactive oxygen species (ROS) produced by NADPH oxidase can function as defense (bactericidal) and signaling molecules related to innate immunity but, as previously mentioned when it occurs in excessive amount, promote a vicious cycle of various cellular responses and ultimately lead to cell death [69]. There are also some evidences associating TLR activation with membrane-associated and oxidant-generating enzyme complex NADPH oxidase in various inflammatory conditions. Interestingly, recent findings suggest that there may be a direct interaction between TLR4 and NADPH oxidase in mediating LPS-induced production of ROS [69]. Recent studies showed that reactive oxygen may also augment tumor cell migration, increasing the risk of invasion and metastasis in cancer cells [70, 71]. TLRs can activate NADPH oxidase, resulting in ROS production [69] and providing evidence that TLR and NADPH oxidase activities may be linked in various inflammatory conditions.

Besides that recently, naturally arising regulatory T cells (Tregs), originating from the thymus, have also been shown to express TLRs. Activation of TLRs in Tregs can increase or decrease their activity; it shows that the presence of CD4/CD25 Treg cells inhibits the generation of tumor antigen-specific cytotoxic responses. Removal of Treg cells with either anti-CD25 monoclonal antibodies (mAb) or low-dose cyclophosphamide enhanced tumor-specific cytotoxic responses. Using a cytokine-expressing cell-based vaccine results in significant antitumor effects against a transplantable tumor [50, 72]. HMGB1 directly enhances immune inhibitory functions of Tregs via receptor for advanced glycation end products (RAGE)-mediated mechanisms. HMGB1 effects on Treg may alter immune reactivity in the setting of chronic inflammatory states such as cancer [73]. Therefore, we may modulate inflammatory pathways for maintaining their pivotal role of immunological self-tolerance.

Mammalian microRNAs (miRNAs) are noncoding RNA oligonucleotides that have been highly conserved during evolution and have recently emerged as potent regulators of gene expression. MiRNAs regulate gene expression at posttranscriptional level and may function either as oncogenes or tumor suppressors. This link between miRNAs and TLR function has potential association with oxidative stress, inflammation, and cancer formation. MiR-155, related to inflammation and cancer, is highly expressed in B-cell lymphoma, breast cancers, lung cancers, and pancreatic adenocarcinomas [74]. Several miRNAs have been shown to be upregulated in response to TLR ligands, and many directly target components of the TLR signaling system [75]. A recent study showed that ligands for TLR2, TLR3, TLR4, and TLR9 could all induce the upregulation of miR-155 expression, through MyD88- and TRIF-dependent pathways. Mir-146 is also highly expressed in breast cancer [69, 70] and is a target of NF-kB signaling pathway.

Therefore, there is an essential need to reveal TLR signaling pathways for NF-kB depend regulation of unregulatory function of microRNAs.

5 Targeting TLRs in the Clinic

Conventional cancer treatments rely on four treatment modalities, namely, surgery, radiotherapy, cytotoxic chemotherapy, and hormone therapy. Most of these therapies are believed to directly attack and eliminate tumor cells. Now researchers are focusing on immunotherapy of cancer. Cancer immunotherapy has been the focus of intense research since the late nineteenth century when Dr. William Coley, pioneer of cancer immunotherapy, demonstrated that bacterial components can contribute to cancer regression by eliciting an antitumor immune response [76]. Immunotherapy of cancer has become a more promising approach in the past decade [77].

One strategy in clinical development for TLR therapeutics includes global blockade or enhancement of individual TLR function using neutralizing antibodies, soluble TLR extracellular domains (ECDs), natural agonist or antagonists, and small molecule inhibitors. Several studies reported that a number of TLR agonists or antagonists can be successfully used as an immunotherapeutic agent in many types of cancers. Unlike cytokines, induced by TLR function which are well validated in these diseases and are successfully being targeted in the clinic, TLR-targeted drugs are still in preclinical development. There are currently approximately twenty drugs in preclinical development, with a further dozen or so in clinical trials [78–83]. TLRs occur early in pathways, and so inhibiting them might be more potent than inhibiting their downstream cytokine targets [83]. TLR4 and MyD88 play an important role in antitumor responses following chemotherapy and irradiation. The study of Apetoh et al. revealed that TLR4-deficient mice have significantly larger tumors after doxorubicin and oxaliplatin treatment or irradiation than wild-type mice [47]. Cell death induction by chemotherapy or irradiation induces the release of HMGB1 to subsequently trigger TLR4 activation in DC, enhance antigen presentation, and promote cytotoxic T-cell responses. It has been shown that high doses of TLR agonists can lead to apoptosis and directly kill both tumor cells and ancillary cells of the tumor microenvironment, whereas low doses of TLR agonists promote cancer growth [42]. Activation of different TLRs might exhibit antitumor effects by increasing vascular permeability and by recruiting leukocytes, resulting in tumor lysis by natural killer cells and cytotoxic T cells [84]. This proves that TLRs act as a double-edged sword for cancer treatment.

Agonists of TLR9 (CpG ODN), TLR3 (poly I:C), and TLR4 (endotoxin analogues) have been used to increase the innate immune response and activate antigen-presenting cells of the host. The TLR2/4 agonist, OM-174, and *Mycobacterium bovis* bacillus Calmette–Guerin (BCG) are promising molecules against cancer metastasis. OM-174 induces TNF-α secretion, and BCG is effective against superficial bladder tumors [78]. OK-432, a TLR4 agonist, has been successfully used as an immunotherapeutic agent in many types of malignancies, including oral squamous cell carcinoma and head and neck cancer [79]. MPLA is also a TLR4 agonist that has been clinically tested as an adjuvant for cancer vaccines. Imiquimod, a synthetic agonist of TLR7, has been proven very effective as monotherapy for basal cell carcinoma. TLR3 activation inhibits nasopharyngeal carcinoma metastasis via downregulation of chemokine receptor CXCR4 [80]. Oral administration of polysaccharide krestin (PSK), a selective TLR2 agonist, in neu transgenic mice significantly inhibits breast cancer growth [81]. Peptides have also been used in combination with TLR agonists for effective treatment and prevention of spontaneous breast tumors [85].

Agonists of TLR3, poly I:C (polyinosinic acid–polycytidylic acid) and poly A:U (polyadenylic–polyuridylic acid) and analogues of dsRNA (double-stranded RNA) can lead to apoptosis in human cancer cells and have been used as an adjuvant to treat cancer patients. Poly A:U has been safely used for treating breast or gastric cancers as a monotherapy. Poly I:C could activate and mature human DC and induce apoptosis [86] and has been found to directly prolong CD4+ T cell survival via interactions with TLR3 in mice [87]. It was recently reported that activation of TLR3 by serial poly I:C treatment could strongly suppress the growth of prostate cancer in a mouse model of prostate tumors [88]. The combination of poly I:C, 5-fluorouracil, and IFN-a induced the highest level of apoptosis in colon carcinoma cells [89]. In addition, combination of poly I:C and ATO (arsenic trioxide) exerted a synergistic anticancer effect that provides the basis for the applicability of these agents for tumor therapy. Poly I:C could induce apoptosis in HCC (hepatocellular carcinoma cells) by the activation of mitochondrial death pathway that required ROS generation and TLR3 signaling activation [90]. These data demonstrated that poly I:C and poly A:U treatment alone or combined with activated TLR3 pathway synergistically induced apoptosis and improved the resultant antitumor activity.

Another aspect of study is excessive activation of TLR signaling pathways that may lead to unwarranted inflammation with hazardous outcomes, including septic shock or inflammatory diseases and carcinogenesis [10, 11, 13, 15, 16, 20, 22–25, 27, 30, 32, 38–41]; therefore, there is an essential need for clinical development of TLR antagonist-based therapeutic drugs. TLR4AsiRNA, TLR4BsiRNA, and TLR4CsiRNA were found to significantly inhibit TLR4 expression in MDA-MB-231 breast cell lines at both mRNA and protein levels. Knockdown of TLR4 gene in MDA-MB-231 resulted in a dramatic reduction of breast cancer cell viability and could inhibit proliferation and survival of breast cancer cells [56]. A novel anti-inflammatory aminosaccharide compound DFK1012 inhibits immune responses caused by TLR and NLR (NOD-like receptor) activation via proteolysis of proinflammatory cytokines induced by TLR and NLR stimulation in macrophages [91]. DFK1012 may represent a novel class of potential therapeutic agents aimed at the treatment of inflammatory disorders. Therefore, when different TLRs are used in combination with their agonist or antagonist or antigen isolated from tumor, it may increase the effect of vaccination and may evoke specific innate immunity against cancer [53].

6 Blockade of DAMP Activation

Suppressing DAMP activation of TLRs or block coreceptors or accessory molecules essential for DAMP activation offers a host of new potential targets for treating inflammatory diseases that may be viable alternatives to current approaches. Evidence that blockade of these mediators can ameliorate disease in human studies is beginning to emerge [92]. Antagonist of TLR cell signaling has been already mentioned above. However, current knowledge about the kinetics of expression or release of DAMPs and their turnover during disease progression is also very essential for development of immunotherapeutic drugs. Chen et al. [93] showed that CD24, a heat-stable antigen and a GPI-anchored protein, binds to DAMPs such as HMGB1, hsp70, and hsp90 and suppresses their activation of inflammatory signaling pathways. Dysfunction of this pathway might contribute to the etiology of autoimmune diseases and likewise may offer means to selectively inhibit DAMP activity [93, 94]. Various techniques targeting HMGB1 are subjected to trial. HMGB1 in the cancer initiates TLR4-dependent responses that contribute to neovascularization [95]. Technical breakthroughs in targeting HMGB1-TLR4 signaling cascade are urgently required for a potential therapeutic technique against cancer development, progression, and especially metastasis. It may constitute a novel therapeutic approach to angiogenesis-related diseases [96, 97].

7 Conclusions and Perspectives

These studies strongly suggest that TLR activation is essential for provoking the innate immune response and enhancing the adaptive immunity against pathogens, but TLR family members are also involved in carcinogenesis. Therefore, the intensity and duration of TLR responses must be tightly controlled. Recently, immunotherapy targeting TLRs against human malignancies is drawing attention. Using TLRs in combination with their agonist or antagonist, we can evoke specific innate immunity against cancer, but the picture of using either agonist or antagonist for TLRs is somewhat complicated in cancer therapy and cannot be decided immediately because of the scarcity of sufficient information. Hence, it is essential to improve our understanding of the mechanisms by which inflammatory cascades are activated as a result of oxidative damage. Important components of these connections are DAMPs released from damaged cells. DAMPs play various functions in tumors: They mediate endothelial cell activation, angiogenesis, and stem cell migration and contribute to recruit and activate innate immune cells to release cytokines that stimulate tumor growth and progression. Collectively they orchestrate a tumor-supporting microenvironment that is an indispensable participant in the neoplastic process.

Taking systematic review of the current literature into consideration, we can conclude that targeting these pathways can be one of the breakthroughs for diagnosis of specific cancer and for devising strategies to limit the detrimental consequences of the inflammatory response to tumorigenesis.

Acknowledgment This work is supported by Banaras Hindu University, Varanasi.

References

1. Kawai T, Akira S (2006) TLR signaling. Cell Death Differ 13:816–825
2. Anderson KV, Jürgens G, Nüsslein-Volhard C (1985) Establishment of dorsal-ventral polarity in the Drosophila embryo: genetic studies on the role of the Toll gene product. Cell 42:779–789
3. Hansson GK, Edfeldt K (2005) Toll to be paid at the gateway to the vessel wall. Arterioscler Thromb Vasc Biol 25:1085–1087
4. Gill R, Tsung A, Billiar T (2010) Linking oxidative stress to inflammation: toll-like receptors. Free Radic Biol Med 48:1121–1132
5. Uematsu AS, Takeuchi SO (2006) Pathogen recognition and innate immunity. Cell 124:783–801
6. Janeway C, Medzhitov R (2002) Innate immune recognition. Annu Rev Immunol 20:197–216
7. Akira S, Takeda K (2004) Toll-like receptor signaling. Nat Rev Immunol 4:499–511
8. Stanley MA, Pett MR, Coleman N (2007) HPV: from infection to cancer. Biochem Soc Trans 35:1456–1460
9. Galamb O, Sipos F, Spisák S (2009) Potential biomarkers of colorectal adenoma-dysplasia-carcinoma progression: mRNA expression profiling and in situ protein detection on TMAs reveal 15 sequentially upregulated and 2 downregulated genes. Cell Oncol 31:19–29
10. Muehleisen B, Jiang SB, Gladsjo JA et al (2012) Distinct innate immune gene expression profiles in non-melanoma skin cancer of immunocompetent and immunosuppressed patients. PLoS One 7:40754
11. Wang B, Li GX, Zhang SG et al (2011) Biglycan expression correlates with aggressiveness and poor prognosis of gastric cancer. Exp Biol Med 236:1247–1253
12. Smiley ST, King JA, Hancock WW (2001) Fibrinogen stimulates macrophage chemokine secretion through toll-like receptor 4. J Immunol 167:2887–2894
13. Rybarczyk BJ, Simpson-Haidaris PJ (2000) Fibrinogen assembly, secretion, and deposition into extracellular matrix by MCF-7 human breast carcinoma cells. Cancer Res 60:2033–2039
14. Vollmer J, Tluk S (2005) Immune stimulation mediated by autoantigen binding sites within small nuclear RNAs involves toll -like receptors 7 and 8. J Exp Med 202:1575–1585
15. Cohen E, Doweck I, Doweck I (2008) Heparanase is overexpressed in lung cancer and correlates inversely with patient survival. Cancer 113:1004–1011
16. McKenzie EA (2007) Heparanase: a target for drug discovery in cancer and inflammation. Br J Pharmacol 151:1–14
17. Park JS, Svetkauskaite D, He Q et al (2004) Involvement of toll-like receptors 2 and 4 in cellular activation by high mobility group box 1 protein. J Biol Chem 279:7370–7377
18. Vogl T, Tenbrock K, Ludwig S et al (2007) Mrp8 and Mrp14 are endogenous activators of toll-like receptor 4, promoting lethal, endotoxin-induced shock. Nat Med 13:1042–1049
19. Ivanov S, Dragoi AM, Wang X et al (2007) A novel role for HMGB1 in TLR9-mediated inflammatory responses to CpG-DNA. Blood 110:1970–1981
20. Beijnum JR, Nowak-Sliwinska P, Van den Boezem E et al (2012) Tumor angiogenesis is enforced by autocrine regulation of high-mobility group box 1. Oncogene 32:363–374
21. Vakkila J, Lotze MT (2004) Inflammation and necrosis promote tumor growth. Nat Rev Immunol 4:641–648
22. Yan Y, Xie M, Kang R et al (2012) HMGB1 is a therapeutic target for leukemia. Am J Blood Res 2:36–43
23. Xie W, Huang Y, Xie W et al (2010) Bacteria peptidoglycan promoted breast cancer cell invasiveness and adhesiveness by targeting toll-like receptor 2 in the cancer cells. PLoS One 5:e10850
24. Messaoudi S, Peyrat JF, Brion JD et al (2008) Recent advances in Hsp90 inhibitors as antitumor agents. Anti Cancer Agents Med Chem 8:761–782

25. Luo W, Rodina A, Chiosis G (2008) Heat shock protein 90: translation from cancer to Alzheimer's disease treatment? BMC Neurosci 9:S7
26. Schaeffler A, Gross P (2009) Fatty acid-induced induction of Toll-like receptor-4/nuclear factor-κB pathway in adipocytes links nutritional signalling with innate immunity. Immunology 126:233–245
27. Sugahara KN, Hirata T, Hayasaka H et al (2006) Tumor cells enhance their own CD44 cleavage and motility by generating hyaluronan fragments. J Biol Chem 281:5861–5868
28. Foell D, Wittkowski H, Vogl T et al (2007) S100 proteins expressed in phagocytes: a novel group of damage-associated molecular pattern molecules. J Leukoc Biol 81:28–37
29. Kim HM, Park BS, Kim JI et al (2007) Crystal structure of the TLR4–MD-2 complex with bound endotoxin antagonist eritoran. Cell 130:906–917
30. Orend G, Chiquet-Ehrismann R (2006) Tenascin-C induced signaling in cancer. Cancer Lett 244:143–163
31. Kim S, Takahashi H (2009) Carcinoma-produced factors activate myeloid cells through TLR2 to stimulate metastasis. Nature 457:102–106
32. Ricciardelli C (2009) The biological role and regulation of versican levels in cancer. Cancer Metastasis Rev 28:233–245
33. Kono H, Rock KL (2008) How dying cells alert the immune system to danger. Nat Rev Immunol 8:279–289
34. Toyokuni S, Okamoto K, Yodoi J et al (1995) Persistent oxidative stress in cancer. FEBS Lett 358:1–3
35. Fialkow L, Wang Y, Downey GP (2007) Reactive oxygen and nitrogen species as signaling molecules regulating neutrophil function. Free Radic Biol Med 42:153–164
36. Shin DM, Yang CS, Jo EU (2008) Mycobacterium tuberculosis lipoprotein-induced association of TLR2 with protein kinase C ζ in lipid rafts contributes to reactive oxygen species-dependent inflammatory signaling in macrophages. Cell Microbiol 10:1893–1905
37. Ioannou S, Voulgarelis M (2010) Toll-like receptors, tissue injury, and tumourigenesis. Mediat Inflamm 2010:581837
38. Kim WY, Lee JW, Choi JJ et al (2008) Increased expression of toll-like receptor 5 during progression of cervical neoplasia. Int J Gynecol Cancer 18:300–305
39. Lee JW, Choi JJ, Seo ES et al (2007) Increased toll-like receptor 9 expression in cervical neoplasia. Mol Carcinog 46:941–947
40. Koski GK, Czerniecki BJ (2005) Combining innate immunity with radiation therapy for cancer treatment. Clin Cancer Res 11:7–11
41. Kelly MG, Alvero AB, Chen R et al (2006) TLR-4 signaling promotes tumor growth and paclitaxel chemoresistance in ovarian cancer. Cancer Res 66: 3859–3868
42. O'Connell RM, Taganov KD, Boldin MP et al (2007) MicroRNA-155 is induced during the macrophage inflammatory response. Proc Natl Acad Sci U S A 104:1604–1609
43. Pasare C, Medzhitov R (2004) Toll-like receptors and acquired immunity. Semin Immunol 16:23–26
44. Janeway CA (1989) Approaching the asymptote? Evolution and revolution in immunology. Cold Spring Harb Symp Quant Biol 54:1–13
45. Matzinger P (2002) The danger model: a renewed sense of self. Science 296:301–305
46. Shi Y, Evans JE, Rock KL (2003) Molecular identification of a danger signal that alerts the immune system to dying cells. Nature 425:516–521
47. Apetoh L, Ghiringhelli F, Tesniere A et al (2007) Toll-like receptor 4-dependent contribution of the immune system to anticancer chemotherapy and radiotherapy. Nat Med 13:1050–1059
48. Tahara T, Arisawa T, Wang F et al (2007) Toll-like receptor 2 −196 to −174 del polymorphism influences the susceptibility of Japanese people to gastric cancer. Cancer Sci 98:1790–1794
49. Nieters A, Beckmann L, Deeg E et al (2006) Gene polymorphisms in toll-like receptors, interleukin-10, and interleukin-10 receptor alpha and lymphoma risk. Genes Immun 7:615–624
50. Woodgett JR, Ohashi PS (2005) GSK3: an in-toll-erant protein kinase? Nat Immunol 6:751–742
51. Re F, Strominger JL (2004) Heterogeneity of TLR-induced responses in dendritic cells: from innate to adaptive immunity. Immunobiology 209:191–198
52. McDermott EP, O'Neill LAJ (2002) Ras participates in the activation of p38 MAPK by interleukin-1 by associating with IRAK, IRAK2, TRAF6, and TAK-1. J Biol Chem 277:7808–7815
53. So EY, Ouchi T (2010) The application of toll like receptors for cancer therapy. Int J Biol Sci 6:675–681
54. Muzio N, Polentarutti N, Bosisio D et al (2000) Toll-like receptors: a growing family of immune receptors that are differentially expressed and regulated by different leukocytes. J Leukoc Biol 67:450–456
55. Coussens LM, Werb Z (2002) Inflammation and cancer. Nature 420:860–867
56. Karin M, Lawrence T, Nizet V (2006) Innate immunity gone awry: linking microbial infections to chronic inflammation and cancer. Cell 124:823–835
57. Alemán A, Schierloh P, Barrera SS et al (2004) Mycobacterium tuberculosis triggers apoptosis in peripheral neutrophils involving toll-like receptor 2 and p38 mitogen protein kinase in tuberculosis patients. Infect Immun 72:5150–5158
58. Yamamoto S, Shimizu S, Kiyonaka S et al (2007) TRPM2-mediated Ca2+ influx induces chemokine production in monocytes that aggravates inflammatory neutrophil infiltration. Nat Med 14:738–747
59. Blüml S, Rosc B, Lorincz A et al (2008) The oxidation state of phospholipids controls the oxidative burst in neutrophil granulocytes. J Immunol 181:4347–4353
60. Szatrowski TP, Nathan CF (1991) Production of large amounts of hydrogen peroxide by human tumor cells. Cancer Res 51:794–798
61. Piccinini AM, Midwood KS (2010) DAMPening inflammation by modulating TLR signalling. Mediat Inflamm 10:1–21

62. Kazama H, Ricci JE, Herndon JM et al (2008) Induction of immunological tolerance by apoptotic cells requires caspase-dependent oxidation of high-mobility group box-1 protein. Immunity 29:21–32
63. Huot J, Houle F, Marceau F et al (1997) Oxidative stress-induced actin reorganization mediated by the p38 mitogen-activated protein kinase/heat shock protein 27 pathway in vascular endothelial cells. Circ Res 80:383–392
64. Rust W, Kingsley K, Petnicki T et al (1999) Heat shock protein 27 plays two distinct roles in controlling human breast cancer cell migration on laminin-5. Mol Cell Biol Res Commun 1:196–202
65. Foell D, Wittkowski H, Roth J (2007) Mechanisms of disease: a 'DAMP' view of inflammatory arthritis. Nat Clin Pract Rheumatol 3:382–390
66. West XZ, Malinin NL, Merkulova AA et al (2010) Oxidative stress induces angiogenesis by activating TLR2 with novel endogenous ligands. Nature 467: 972–976
67. Satta N, Kruithof EKO, Reber G et al (2008) Induction of TLR2 expression by inflammatory stimuli is required for endothelial cell responses to lipopeptides. Mol Immunol 46:145–157
68. Curtin JF, Liu N, Candolfi M et al (2009) HMGB1 mediates endogenous TLR2 activation and brain tumor regression. PLoS Med 6(1):e10
69. Moldovan L, Irani K, Moldovan NI et al (1999) The actin cytoskeleton reorganization induced by Rac1 requires the production of superoxide. Antioxid Redox Signal 1:29–43
70. Wang X, Martindale JL, Liu Y et al (1998) The cellular response to oxidative stress: influences of mitogen-activated protein kinase signalling pathways on cell survival. Biochem J 15:291–300
71. Vulcano M, Dusi S, Lissandrini D et al (2004) Toll receptor-mediated regulation of NADPH oxidase in human dendritic cells. J Immunol 173:5749–5756
72. Ercolini AM, Ladle BH, Manning EA et al (2005) Recruitment of latent pools of high-avidity CD8+ T cells to the antitumor immune response. J Exp Med 20:1591–1602
73. Wild CA, Bergmann C, Fritz G et al (2012) HMGB1 conveys immunosuppressive characteristics on regulatory and conventional T cells. Int Immunol 24:485–494
74. Quinn SR, O'Neill LA (2011) A trio of microRNAs that control Toll-like receptor signaling. Int Immunol 23:421–425
75. Taganov KD, Boldin MP, Chang KJ et al (2006) NF-κB-dependent induction of microRNA miR-146, an inhibitor targeted to signaling proteins of innate immune responses. Proc Natl Acad Sci U S A 103: 12481–12486
76. Coley WB (1893) The treatment of malignant tumors by repeated inoculations of erysipelas: with a report of ten original cases. Am J Med Sci 10:487–511
77. Rosenberg SA, Yang JC, Restifo NP (2004) Cancer immunotherapy: moving beyond current vaccines. Nat Med 10:909–915
78. Garay RP, Viens P, Bauer J et al (2007) Cancer relapse under chemotherapy: why TLR2/4 receptor agonists can help. Eur J Pharmacol 563:1–17
79. Okamoto M, Oshikawa T, Tano T et al (2003) Involvement of Toll-like receptor 4 signaling in interferon-γ production and antitumor effect by streptococcal agent OK-432. J Natl Cancer Inst 95:316–326
80. Zhang Y, Sun R, Liu B et al (2009) TLR3 activation inhibits nasopharyngeal carcinoma metastasis via downregulation of chemokine receptor CXCR4. Cancer Biol Ther 9:1826–1830
81. Hailing L, Yang Y, Gad E et al (2010) Polysaccharide Krestin is a novel TLR2 agonist that mediates inhibition of tumor growth via stimulation of CD8 T cells and NK cells. Clin Cancer Res 17:67–76
82. Nava-Parada P, Forni G, Knutson KL et al (2007) Peptide vaccine given with a toll-like receptor agonist is effective for the treatment and prevention of spontaneous breast tumors. Cancer Res 67: 1326–1334
83. Bhattacharya D, Yusuf N (2012) Expression of toll-like receptors on breast tumors: taking a toll on tumor microenvironment. Int J Breast Cancer 716564
84. Muccioli M, Sprague L, Nandigam H et al (2012) Toll-like receptors as novel therapeutic targets for ovarian cancer. ISRN Oncol 2012:642141
85. Lacour J, Lacour F, Spira A et al (1980) Adjuvant treatment with polyadenylic-polyuridylic acid (Polya.Polyu) in operable breast cancer. Lancet 2:161–164
86. Dvorak HF (1986) Tumors: wounds that do not heal: similarities between tumor stroma generation and wound healing. N Engl J Med 315:1650–1659
87. Gelman AE, Zhang J, Choi Y et al (2004) Toll-like receptor ligands directly promote activated CD4+ T cell survival. J Immunol 172:6065–6073
88. Chin AI, Miyahira AK, Covarrubias A et al (2010) Toll-like receptor 3-mediated suppression of TRAMP prostate cancer shows the critical role of type I interferons in tumor immune surveillance. Cancer Res 70:2595–2603
89. Taura M, Fukuda R, Suico MA et al (2010) TLR3 induction by anticancer drugs potentiates poly I:C-induced tumor cell apoptosis. Cancer Sci 101:1610–1617
90. Shen P, Jiang T, Lu H et al (2011) Combination of Poly I:C and arsenic trioxide triggers apoptosis synergistically via activation of TLR3 and mitochondrial pathways in hepatocellular carcinoma cells. Cell Biol Int 35:803–810
91. Lee KH, Liu YJ, Biswas A et al (2011) A novel aminosaccharide compound blocks immune responses by toll-like receptors and nucleotide-binding domain, leucine-rich repeat proteins. J Biol Chem 286: 5727–5735
92. Lotze MT, Zeh HJ, Rubartelli A et al (2007) The grateful dead: damage-associated molecular pattern molecules and reduction/oxidation regulate immunity. Immunol Rev 200:60–81

93. Chen GY, Tang J, Zheng P et al (2009) CD24 and siglec-10 selectively repress tissue damage induced immune responses. Science 323:1722–1725
94. Liu Y, Chen GY, Zheng P (2009) CD24-Siglec G/10 discriminates danger- from pathogen-associated molecular patterns. Trends Immunol 30:557–561
95. Ohmori H, Luo Y, Kuniyasu H (2011) Non-histone nuclear factor HMGB1 as a therapeutic target in colorectal cancer. Expert Opin Ther Targets 15:183–193
96. Lin Q, Yang XP, Fang D et al (2011) High-mobility group box-1 mediates toll-like receptor 4 – dependent angiogenesis. Integr Physiol Exp Med 31: 1024–1032
97. Liu L, Li YH, Niu YB et al (2010) An apple oligogalactan prevents against inflammation and carcinogenesis by targeting LPS/TLR4/NF-κB pathway in a mouse model of colitis-associated colon cancer. Carcinogenesis 31:1822–1832

Management of Inflammation Using Cellular Redox Modifiers

Rahul Checker, Deepak Sharma, and Santosh K. Sandur

Abstract

The interactions between reactive oxygen species (ROS) and intracellular protein machinery have become increasingly important in the study of both normal and pathological processes. Further, several investigators have reported that cellular redox status plays an important role in the regulation of immune responses, especially in T cell signal transduction, gene expression, and functions. Glutathione, cysteine, and thioredoxin are three key regulators of cellular redox homeostasis, and the cellular redox balance is maintained by interconvertible oxidized and reduced form of these redox couples that exist in a dynamic equilibrium. Any change in this equilibrium can significantly modulate the crucial signaling pathways and thus alter cellular functions. Although it is well accepted that inflammatory and immune responses are highly susceptible to changes in redox status, contrasting reports showing differential effects of oxidative stress have resulted in incongruity over its role in immune activation and signaling. A growing body of evidence in the literature indicates that cellular oxidative stress and redox status are central to the signaling and expression of inflammatory genes involving NF-κB activation. Further, the area of oxidant and redox control of inflammatory processes and the immune system is relatively unexplored and a lot need to be accomplished. Addition or administration of exogenous/endogenous redox-active agents (which can either restore or perturb the existing redox status of the cells) is one of the means to unravel the link between oxidative stress and inflammation. This chapter focuses on our current understanding of the role of cellular redox status in inflammatory responses using biological response modifiers. It also emphasizes the need for future research on the impact of local changes in

R. Checker • D. Sharma • S.K. Sandur (✉)
Radiation Biology and Health Sciences Division,
Bio-Science Group, Bhabha Atomic Research Centre,
Trombay, Mumbai 400085, India
e-mail: sskumar@barc.gov.in

V. Rani and U.C.S. Yadav (eds.), *Free Radicals in Human Health and Disease*,
DOI 10.1007/978-81-322-2035-0_21, © Springer India 2015

the redox environment in different cellular compartments on gene regulation brought about by redox-sensitive transcription factors.

Keywords
Reactive oxygen species • Inflammation • Endogenous redox-active molecules • Exogenous redox-active molecules

1 Introduction

The term immunity is derived from the Latin word *Immunis* meaning "to make safe." The collection of cells, tissues, and molecules that mediate resistance to infections is called the immune system, and the coordinated reaction of these cells and molecules to infectious microbes is called the immune response. Inflammation (Latin, *inflammare*, to set on fire) is the body's protective reaction to injurious stimuli like infections, ionizing radiation, harmful chemicals, foreign particles, trauma, autoantibodies, and physical or thermal stress. Acute inflammation involves pathogen killing as well as tissue repair processes and helps to restore homeostasis at infected or damaged sites. Acute inflammatory response is normally well regulated so that it does not cause excessive damage to the host and is self-limiting due to the involvement of negative feedback mechanisms. However, inflammatory responses that fail to regulate themselves can become chronic and contribute to the development and progression of disease. Chronic inflammation is implicated in development of several degenerative diseases such as rheumatoid arthritis, atherosclerosis, congestive heart failure, Alzheimer's, and asthma. Hence, suppression of chronic inflammation may prevent occurrence of inflammation-associated disorders. The immunosuppressive regimens currently employed involve the use of multiple drugs such as corticosteroids, cyclosporin A, tacrolimus, sirolimus, azathioprine, etc. [1, 2]. Each of these drugs targets a discrete site in the lymphocyte activation cascade, but all of them exhibit distinct side effects on normal tissues which limits their long-term usage.

Reactive oxygen species (ROS) have been implicated as important regulators of T cell signal transduction, gene expression, and functions. They serve as signaling messengers to mediate various biological responses including cell proliferation, angiogenesis, innate immunity, programmed cell death, and senescence. Addition of antioxidants has been shown to inhibit mitogen-induced proliferation and production of IL-2, which is required for optimal T cell expansion [3, 4]. Also, studies carried out using alloantigen-stimulated cultures have suggested that antigen-mediated T cell activation requires ROS production [5]. In contrast to these studies, other reports show that exposure to exogenous oxidative stress suppresses mitogen or antigen-induced T cell activation. Such contrasting reports underscore the critical role of cellular redox balance in regulating immune responses. It may be argued that the level of exogenous oxidative stress is higher than endogenous levels of ROS. Alternatively, this exogenously induced oxidative stress may act in a coordinated fashion with endogenous ROS in eliciting a particular biological response. Further, determining the threshold of ROS for generation of a particular response beyond which it may become deleterious is critical under these conditions. It is well understood and documented that the redox balance of a cell is maintained by interconvertible oxidized and reduced forms of intracellular redox couples including GSH/GSSG, cysteine/cystine, and oxidized/reduced thioredoxin [6]. Under normal physiologic conditions, cells maintain redox balance through generation and elimination of reactive oxygen/nitrogen species. ROS include radical species such as superoxide ($O_2^{\cdot-}$) and hydroxyl radical (HO·), along with non-radical species such as hydrogen peroxide (H_2O_2). Reactive nitrogen species (RNS) include nitric oxide ($NO^{\cdot}$) and peroxynitrite ($ONOO^-$)

[7]. Both exogenous sources (like UV irradiation, X-rays, gamma rays, pollutants, and chemicals) and endogenous sources (like NAD(P)H oxidase (Nox), cytochrome c oxidase, nitric oxide synthase (NOS), xanthine oxidase, and mitochondrial electron transport chain) contribute to the formation of intracellular ROS. In order to maintain redox homeostasis, cells are equipped with enzymatic and nonenzymatic antioxidant systems that scavenge the ROS/RNS. Perturbation of these mechanisms can alter the delicate intracellular redox balance leading to oxidative stress.

Therapeutic targeting of the inflammatory response pathways is one of the thrust areas of current research and has significant clinical application in prevention of chronic inflammatory and autoimmune diseases, including atherosclerosis, type 2 diabetes, multiple sclerosis, and Alzheimer's disease. While several approaches including use of neutralizing antibodies against cytokines, chemokines, and their receptors and small molecule inhibitors of proinflammatory signaling proteins like IκB kinase (IKK), nuclear factor-κB (NF-κB), activator protein 1 (AP-1), mitogen-activated protein kinases (MAPK), and janus kinases (JAK) have been employed in the past, efforts are ongoing to develop complementary strategies to suppress inflammation. In this direction, ROS, which now have a well-recognized role as modifiers of innate and adaptive immunity, can be considered as potential targets for management of inflammation using cellular redox modifiers. Further, recent advances that highlight the role of cellular redox balance in regulating inflammatory signaling offer promise in this area of translational biomedical research (Fig. 1). Therefore, this chapter focuses on the role of endogenous and exogenous redox active molecules to act as anti-inflammatory agents. There are very few endogenous agents which have been reported to modify inflammatory responses. Among them we have selected agents that act via modulating the cellular redox balance. Although there are several exogenous molecules that have been explored to restrain inflammation, we preferred molecules that have been used in our laboratory or have been extensively studied for their anti-inflammatory activities (Tables 1 and 2).

2 Endogenous Redox-Active Molecules

2.1 Bilirubin

One of the key metabolic pathways in the mammalian system is the degradation of protoheme derived from heme-containing proteins such as hemoglobin and cytochrome P450 protein. During this process, heme oxygenase-1 (HO-1) converts heme to biliverdin, carbon monoxide, and iron, and biliverdin reductase reduces biliverdin to bilirubin (BR) at the cost of NADPH. Nanomolar concentrations of bilirubin have been shown to inhibit ROS generation in vitro [8]. Several researchers, including our laboratory, have demonstrated the powerful anti-inflammatory properties of BR in vitro and in vivo. The anti-inflammatory effects of bilirubin have been shown to be mediated via its ability to modulate the costimulatory pathway, suppress immune transcription factors, and downregulate expression of MHC class II, nitric oxide (NO), TNF-α, and iNOS [9, 10]. Reports show that clinically relevant concentrations of unconjugated bilirubin (UCB) induce apoptosis and necrosis in immune cells by depleting cellular GSH are critical in understanding the immunosuppression associated with hyperbilirubinemia [11]. Further, bilirubin also modulates the expressions of proinflammatory and proapoptotic genes and increases Foxp3(+) T regulatory (Treg) cells at the site of transplantation [12]. The ability of bilirubin in improving the survival and attenuating liver injury in response to lipopolysaccharide (LPS) infusion demonstrates its potent in vivo efficacy [13].

2.2 Hydrogen Sulfide

Hydrogen sulfide (H_2S) is a colorless, water-soluble, flammable gas that has the characteristic smell of rotten eggs. In mammalian systems, H_2S production is carried out by three endogenous enzymes, namely, cystathionine β-synthase (CBS), cystathionine γ-lyase (CGL or CSE), and 3-mercaptopyruvate sulfurtransferase (3MST).

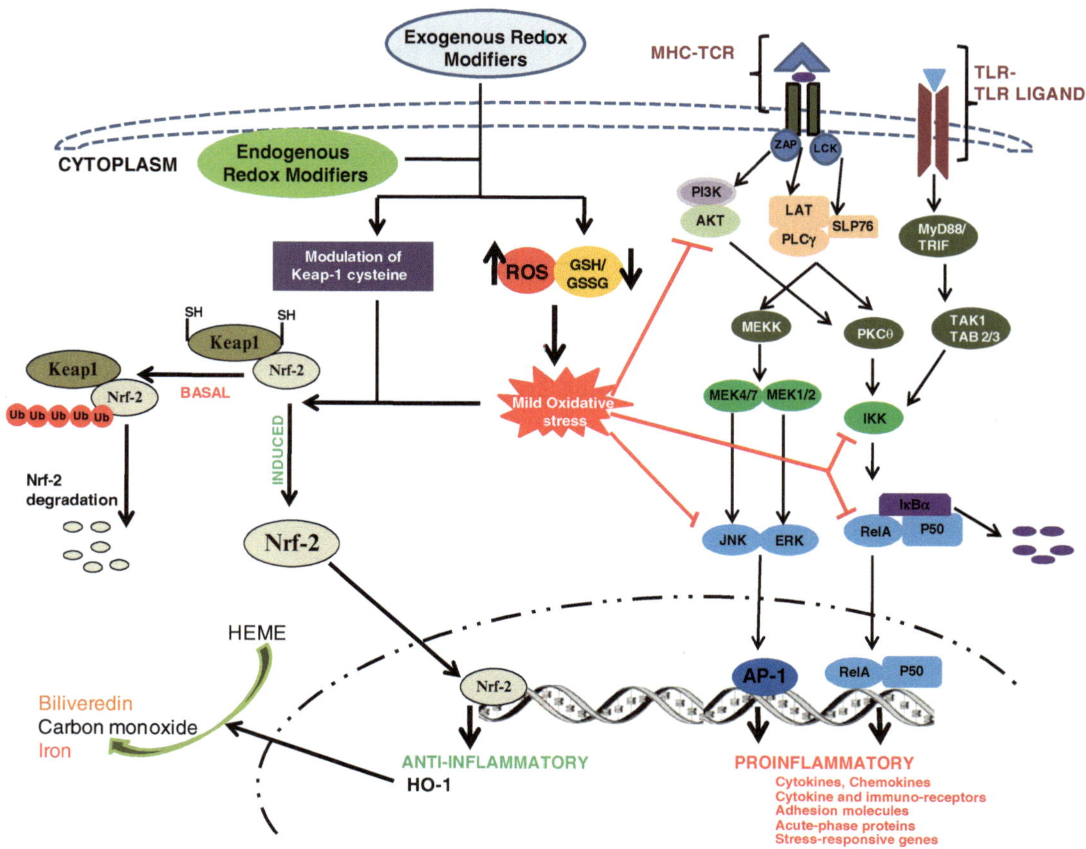

Fig. 1 Alterations in intracellular signaling pathways under oxidative stress conditions. *Red arrows* indicate inhibition

H_2S is produced in tissue homogenates at a rate in the range of 1–10 pmoles per second per mg protein, resulting in micromolar extracellular concentrations [14]. H_2S has also been reported to protect neurons from oxidative stress by increasing the production of the antioxidant glutathione. Literature suggested that H_2S reduces oxidative stress through two distinct mechanisms: it can act as a direct scavenger of ROS and also upregulate antioxidant defense through a nuclear-factor-E2-related factor-2 (Nrf-2)-dependent signaling pathway [15]. H_2S administration increases nuclear localization of Nrf-2 and expression of antioxidants (heme oxygenase-1 and thioredoxin 1), Bcl-2, and Bcl-xL and inactivates proapoptotic protein Bad [16]. The exact role of H_2S during inflammation is not very clear, and there is a great deal of discrepancy in the reports pertaining to the anti-inflammatory effects and the proinflammatory effects of endogenous and pharmacological H_2S. Studies on the effects of NaHS (sodium hydrosulfide) and a slow-releasing H_2S donor (GYY4137) showed that GYY4137 inhibited IL-1β, IL-6, TNF-α, NO, and PGE-2 and increased the synthesis of the anti-inflammatory cytokine IL-10. On the contrary, NaHS elicited a biphasic effect on proinflammatory mediators. It decreased the synthesis of IL-1β, IL-6, NO, PGE-2, and TNF-α at low concentrations but increased them at higher doses [17]. Importantly, lack of availability of enzyme-specific inhibitors and use of physiologically irrelevant H_2S donors have hampered significant and rapid progress in studying the biological activities of H_2S.

Table 1 Molecular targets of endogenous cellular redox modifiers

TARGET \ MOLECULE	Bilirubin	Hydrogen sulfide	Hydrgen peroxide	Glutathione	Nitric oxide	15d-PGJ2
IMMUNOREGULATORY SIGNALING PROTEINS						
AP-1	NR	NR	↑	↑	↑↓	↓
NF-κB	↓	↓	↓	↓	↑↓	↓
Nrf-2	↑	↑	↑	↑	↑	↑
JAK/STAT3	NR	↓	↑	↓	↓	↓
ERK	↑	↑	↑	↓	↓	↓
IKK	NR	↓	↓	↓	↓	↓
PI3K/AkT	↑	↑	↑	NR	↑	↓
FoxP3	↑	NR	↑	NR	↑	NE
NFAT	↑	NR	↑	NR	↑	↓
INFLAMMATORY MEDIATORS						
CRP	↓	↓	NR	NR	↓	NR
Pro-inflammatory cytokines (TNF-α, IL-6, IL-1β)	↓	↓	↓	↓	↓	↓
Anti-inflammatory cytokines (IL-10)	↓	↑	NR	↓	↓	↑
iNOS	↓	↓	↓	↓	↓	↓
Cox	↑	↓	NR	↑	↓	↓
ADHESION MOLECULES						
ICAM-1	↓	↓	↑	↓	↓	↓

↓ - Decreased activity; ↑ - Increased activity NR – No report; NE – No effect

2.3 Hydrogen Peroxide

Hydrogen peroxide (H_2O_2) was first isolated by Thenard in 1818, and it was quickly recognized that high concentrations of H_2O_2 result in cell injury by damaging key cellular molecules such as DNA and lipids [18]. Enzymes involved in increasing H_2O_2 levels include the family of NADPH oxidases [19] and other oxidases such as xanthine oxidase [20], 5-lipoxygenase, and superoxide dismutase [21]. Apart from regulating several cellular events, H_2O_2 is also involved in controlling transcription factors, cell proliferation, and apoptosis [22]. Hydrogen

Table 2 Molecular targets of exogenous cellular redox modifiers

MOLECULE / TARGET	Curcumin	Plumbagin	gamma-tocotrienol	Resveratrol	Epigallo-catechin Gallate	Chlorophyllin
IMMUNOREGULATORY SIGNALING PROTEINS						
AP-1	↓	↓	↓↑	↓	↓	↓
NF-κB	↓	↓	↓	↓	↓	↓
Nrf-2	↑	↑	↑	↑	↑	↑
JAK/STAT3	↓	↓	↓	↓	↓	↓
ERK	↓	↓	↓	↓	↓	↓
IKK	↓	↓	↓	↓	↓	NR
PI3K/AkT	↓	↓	↓	↓	↓	↓
FoxP3	↓↑	↑	NR	↓	↑	↑
NFAT	↓	↓	NR	↓	↓	NR
INFLAMMATORY MEDIATORS						
CRP	↓	NR	↓	↓	↓	NR
Pro-inflammatory cytokines (TNF-α, IL-6, IL-1β)	↓	↓	↓	↓	↓	↓
Anti-inflammatory cytokines (IL-10)	↑	NR	NE	↑	↑	NR
iNOS	↓	↓	↓	↓	↓	↓
Cox	↓	↓	↓	↓	↓	↓
ADHESION MOLECULES						
ICAM-1	↓	NR	↓	↓	↓	NR

↓ - Decreased activity; ↑ - Increased activity ↓↑ - Dual activity NR – No report; NE – No effect

peroxide alone and in conjunction with the amplification activity of myeloperoxidase (MPO) is responsible for bacterial killing [23]. The role of H_2O_2 in either promoting or inhibiting inflammation is still contentious. Hydrogen peroxide inhibits the response of human peripheral blood mononuclear cells (PBMCs) to phytohemagglutinin, concanavalin A, and also in mixed lymphocyte culture assays as measured in terms of T cell activation and proliferation, IL-2 secretion, and cytotoxic T cell generation [24]. Further, Oliveira-Marques et al. [25] proposed that H_2O_2 has a fine-tuning regulatory role, comprising both a proinflammatory control loop that increases pathogen removal and an anti-inflammatory control loop, which avoids an exacerbated harmful inflammatory response [25]. H_2O_2 at low extracellular concentrations (up to 25 μM) stimulates NF-κB nuclear translocation induced by TNF-α, whereas at higher doses, it inhibits NF-κB activation. Further, exposure of bone marrow or peritoneal neutrophils to H_2O_2 is associated with reduced nuclear translocation of NF-κB and decreased production of the NF-κB-dependent cytokines TNF-α and macrophage inhibitory protein-2 [26]. Recent reports demonstrate that H_2O_2 also has anti-inflammatory effects on neutrophil activation and inflammatory processes, such as hyperoxia-induced acute lung injury, in which activated neutrophils play a major role [27].

2.4 Glutathione

Glutathione (gamma-glutamyl-cysteinyl-glycine; GSH) is the most abundant low-molecular-weight thiol present in the cytoplasm. It is an important intracellular antioxidant and a redox-potential regulator that plays a vital role in drug detoxification and cellular protection against oxidative damage by eliminating free radicals, peroxides, and toxins. Upon oxidation, GSH forms GSH disulfide (GSSG) which can be reduced by a specific enzyme, glutathione reductase. The key role of GSH as an antioxidant is demonstrated by many studies showing that GSH depletion using chemicals like buthionine sulfoximine (BSO) (inhibitor of GSH synthesis) has an aggravation effect in many disease models. On the other hand, replenishing GSH levels with precursors of its synthesis such as N-acetylcysteine(NAC)or2-oxothiazolidine-4-carboxylic acid has protective effects. Antioxidants [28] and glutathione precursors [29, 30] have been shown to downregulate cytokine synthesis, immune activation, and downstream processes. Among several agents that are used to replenish or deplete GSH, NAC, and BSO are of particular importance as they exhibit antagonistic effects on a proinflammatory signal. NAC, an antioxidant and a GSH precursor [31], suppresses cytokine production [32] and ameliorates ROS-mediated injury [31]. In contrast, BSO, which depletes GSH by irreversibly inhibiting γ-glutamyl cysteine synthetase (γ-GCS), the rate-limiting enzyme in the biosynthesis of glutathione [33], enhances secretion of cytokines like TNF-α, IL-6, and IL-8 and induces intracellular accumulation of GSSG, but reduces the concentration of GSH [34, 35]. Besides inhibiting inflammation, GSH also acts as a regulatory mediator during an immune response usually in a direction beneficial to the host [36]. Apart from being a powerful antioxidant, the GSH/GSSG couple also acts as a redox buffer, regulating the redox state of "redox-sensitive proteins." This regulatory mechanism is based on the fact that many cysteine residues in proteins that are physiologically in their reduced state can in fact exist in various oxidation states. These include interchain disulfides, mixed disulfides with small molecular mass thiols (e.g., GSH, glutathionylation or with cysteine, cysteinylation), S-nitrosylation, and oxidation to sulfinic, sulfenic, or sulfonic acids. The ability of cells to reverse these oxidations (with the possible exception of sulfonic acids) makes these modifications likely mechanisms of protein regulation which is similar to other posttranslational modifications like protein phosphorylation. Protein glutathionylation can be either anti-inflammatory or proinflammatory depending on the protein that is undergoing S-thiolation. For instance, various proteins in the NF-κB pathway are regulated by glutathionylation. Glutathionylation of p65-NF-κB leads to its inactivation [37] and glutathionylation of p50-NF-κB inhibits its DNA binding [38] and thus glutathionylation would be anti-inflammatory. The ability of glutaredoxin (a thioltransferase that catalyzes specific and efficient deglutathionylation of proteins) to both decrease [39] and augment [40] production of inflammatory cytokines, depending on the experimental model, further confirms the contrasting role of glutathionylation.

2.5 Nitric Oxide

Nitric oxide (NO) is a gaseous signaling molecule that regulates various physiological and pathophysiological responses in the human body including circulation and blood pressure, platelet function, host defense, and neurotransmission. NO is synthesized from L-arginine in a reaction catalyzed by a family of nitric oxide synthase enzymes. Three different NOS isoforms have been characterized. The neuronal NOS (nNOS) is predominantly expressed in neurons in the brain and peripheral nervous system [41]. Endothelial NOS (eNOS) is mainly expressed in endothelial cells [42]. Both nNOS and eNOS are constitutively expressed and are inactive in resting cells. The third isoform of the NOS family is the inducible NOS (iNOS). iNOS is not expressed in most resting cells. Upon exposure to microbial products, such as LPS and dsRNA or proinflammatory cytokines, iNOS expression is induced which can constantly produce high levels of NO for prolonged periods [43]. NO has a biphasic action on oxidative killing of isolated cells: low concentrations protect against oxidative killing, while higher doses enhance killing, and these two effects occur by distinct mechanisms [44].

During an inflammatory response, NO is involved in several functions like regulation of signaling cascades and transcription factors, regulation of leukocyte rolling, migration, cytokine production, proliferation, and apoptosis [43, 45–47]. Specific iNOS inhibitors have been shown to be anti-inflammatory in various forms of experimentally induced inflammation, such as arthritis and colitis [47–49], and selective iNOS inhibitors are under development for treatment of inflammatory diseases. NO has been shown to regulate NF-κB-mediated gene transcription by S-nitrosylation of p50 subunit [50]. Mazzoni et al. [51] showed that myeloid suppressor cells secrete NO in response to signals from activated T cells, including IFN-γ and a contact-dependent stimulus, and inhibit T cell proliferation via NO-dependent inhibition of JAK1 and JAK3 [51]. Thus, NO, like many other inflammatory mediators, seems to have both pro- and anti-inflammatory effects depending on the concentration of NO and the physiological environment.

2.6 15-Deoxy-Δ12,14-Prostaglandin J2

The prostaglandins (PGs) are a family of structurally related molecules that are produced by cells in response to a variety of extrinsic stimuli and regulate cellular growth, differentiation, and homeostasis. The J series of cyclopentenone prostaglandins such as 15-deoxy-Δ12,14-PGJ2 (15d-PGJ2) are electrophilic lipid-signaling mediators derived from the nonenzymatic dehydration of PGD2, a major product of the cyclooxygenase pathway. Prostaglandins (PGs) are autacoids synthesized from 20 carbon-containing polyunsaturated fatty acids, principally arachidonic acid (AA) generated from membrane phospholipids and derived from dietary sources [52]. PGs contribute to inflammation, smooth muscle tone, hemostasis, thrombosis, parturition, and gastrointestinal secretion [53].

Prostaglandin J2 (PGJ_2) is a major cyclooxygenase product formed in a variety of tissues and cells and has marked effects on a number of biological processes, including platelet aggregation, relaxation of vascular and nonvascular smooth muscle, and nerve cell function. It also acts as an endogenous ligand for peroxisome proliferator-activated receptor γ (PPARγ) via covalent binding to the cysteine residue within the ligand-binding domain of the receptor, thus promoting adipocyte differentiation [54]. 15d-PGJ2 can also form covalent adducts with cysteine residues of other signaling proteins via Michael addition, leading to the activation of important redox-sensitive signaling pathways such as the antioxidant response pathway regulating the expression of phase II enzymes. 15d-PGJ2 induces multiple phase II enzymes and other antioxidant proteins via Nrf-2-dependent mechanism and also induces expression of anti-inflammatory and cytoprotective protein, HO-1, in various cell types, such as mouse macrophages [55], human mesangial cells [56], and rat pheochromocytoma (PC12) cells [57]. The anti-inflammatory action of 15d-PGJ2 is via suppression of IκB kinase activity leading to inhibition of IκBα degradation and reduction of p65 nuclear translocation [58].

3 Exogenous Redox-Active Molecules

3.1 Curcumin

Curcumin (CR), a yellow pigment from *Curcuma longa*, is a major component of turmeric and is commonly used as a spice and food-coloring material. The pleiotropic biological properties of curcumin have also been considered to be associated with its antioxidant property. Since free radical-mediated peroxidation of membrane lipids and oxidative damage of DNA and proteins are believed to be associated with a variety of chronic pathological complications such as cancer, atherosclerosis, neurodegenerative diseases, and aging, curcumin is being tried for protection against oxidative-stress-mediated pathological conditions. Curcumin has been shown to inhibit lipid peroxidation and effectively scavenge superoxide and peroxyl radicals [59]. The presence of the phenolic, β-diketone, as well as the methoxy groups contribute to the free radical-scavenging activity of curcumin. A few reports showing curcumin as prooxidant indicated that the prooxidant and antioxidant effects of curcumin are dependent on dose and the chemical environment (e.g., availability of free redox metal ions) [60]. The free radical-scavenging and antioxidant characteristics of curcumin are mediated through induction of the Nrf-2 signaling pathway [61]. Curcumin has also been shown to possess potent anti-inflammatory properties via inhibiting the production of proinflammatory cytokines (IL-8, MIP-1α, MCP-1, IL-1β, and TNF-α) [62] and suppression of important molecular targets involved in inflammation such as NF-κB, AP-1, and signal transducer and activator of transcription (STAT) proteins [63, 64]. Curcumin also suppresses the expression of the inflammatory mediators like COX-2, TNF-α, 5-LOX, IL-1β, IL-6, IL-8, MIP-1α, adhesion molecules, C-reactive protein (CRP), CXCR-4, and others [65–68]. The direct binding of curcumin to IkBα kinase was reported as the mechanism of its NF-κB-suppressing activity [63]. Thus curcumin could suppress inflammation by modulating the expression and activation of multiple proteins and transcription factors involved in inflammation. Further, the beneficial effects of oral curcumin on inflammatory diseases have been investigated in a number of clinical studies [69]. The modern age data on multiple cellular targets of curcumin substantiates grandma's recipe for control of acute and chronic inflammation.

3.2 Plumbagin

Plumbagin (PG), a naphthoquinone (5-hydroxy-2-methyl-1,4-naphthoquinone), is found in the plants of Plumbaginaceae, Droseraceae, Ancestrocladaceae, and Dioncophyllaceae families. The root of *Plumbago zeylanica* (also called Chitrak), a major source of plumbagin, has been used in traditional Indian medicine since 750 BC as an antiatherogenic, cardiotonic, hepatoprotective, and neuroprotective agent. Plumbagin has been shown to exert several therapeutic biological effects including anticancer, antiproliferative, chemopreventive, chemotherapeutic, and radiosensitizing properties in experimental animals as well as in tumor cells in vitro [70]. Recently, several investigators have shown its potent anti-inflammatory properties in vitro and in vivo. Plumbagin inhibits T cell proliferation, cell cycle progression, cytokine secretion, and expression of early and late activation markers CD69 and CD25 respectively via modulation of cellular redox and inhibition of IκBα degradation and NF-κB activation and prevents Graft Versus Host Disease-induced mortality in mice [71, 72]. Similar anti-inflammatory effects of plumbagin in human PBMNCs have also been reported in both allergic and non-allergic individuals [73]. In vivo also, plumbagin is effective as shown by prevention of encephalitogenic T cell responses, suppression of experimental autoimmune encephalomyelitis (EAE), and decreased T cell proliferation and IL-2 production in arthritic mice [74, 75]. Plumbagin has been used as a part of a multicomponent plant-based formulation which it was found to be effective in the management of chronic obstructive pulmonary disease without any side effects [76]. The strong anti-inflammatory effect of plumbagin also provides clues for its possible use as a cancer chemopreventive agent.

3.3 Gamma-Tocotrienol

Vitamin E is a well-known antioxidant and is an important nutrient in the human diet that is readily available in lipid-rich plant products. Tocopherols are predominant in olive, sunflower, corn, and soya beans oils, and tocotrienols are the major vitamin E components of palm oil, barley oil, and rice bran oil [77]. The tocopherols are saturated forms of vitamin E, whereas the tocotrienols are unsaturated and possess an isoprenoid side chain. The unsaturated side chain present on tocotrienols facilitates their better cellular uptake than tocopherols. This higher cellular uptake has been attributed to their superior biological activities as compared to tocopherols. Palm oil represents one of the most abundant natural sources of tocotrienols [78]. The distribution of vitamin E in palm oil is 30 % tocopherols and 70 % tocotrienols [79]. Like tocopherols, tocotrienols exhibit antioxidant activities, and most of their effects can be linked to its antioxidant function [80]. Studies have also reported that tocotrienols possess lipid-lowering, antiatherogenic, blood-pressure-lowering, antidiabetic, neuroprotective, and anti-inflammatory effects. Tocotrienols exhibit antioxidant effects by scavenging the chain-propagating peroxyl radical [81]. Several investigators have shown that γ-tocotrienols exhibit potent anti-inflammatory activity by inhibiting NF-κB activation [82, 83]. Further, γ-tocotrienol has been shown to suppress the expression of TNF-α [82], IL-1β [84], IL-6 [85], iNOS [86], and Cox-2 [82], all of which mediate inflammation. γ-Tocotrienol (GT) is more effective in suppressing T cell proliferation and cytokine production compared to α-tocotrienol (AT) when present continuously in the culture. Cellular uptake studies with tocotrienols showed higher accumulation of GT compared to AT which might be responsible for the superior biological activity of GT [87]. The data on tocotrienols clearly demonstrate that the dual responses in terms of anti-inflammatory or immunostimulatory actions depend on the duration for which cells are incubated with this molecule.

3.4 Resveratrol

Resveratrol (3,4,5-trihydroxystilbene; RT) is a phytoalexin, with established health potential due to its antioxidant, anticancer, and anti-inflammatory properties. It is found in variety of dietary sources including grapes, plums, peanuts, and wines, especially red wines and to a much lesser extent in white wines. Resveratrol research gained momentum following the paradoxical observation that a low incidence of cardiovascular diseases may coexist with a high-fat diet intake and moderate consumption of red wine [88], a phenomenon known as the French paradox. One of the important biological activities ascribed to resveratrol involves its antioxidant potential. The antioxidant activity of resveratrol is mediated via multiple mechanisms, including inhibiting production of reactive oxygen species by inflammatory cells [89], scavenging free radicals, and enhancing the levels of endogenous antioxidants by activation of Nrf-2 pathway [90]. Resveratrol is known to increase the levels of several antioxidant enzymes, including glutathione peroxidase, glutathione S-transferase, and glutathione reductase [91]. Similar to a variety of dietary antioxidants, resveratrol also possesses anti-inflammatory effects that are mediated through inhibition of transcription factors such as NF-κB and AP-1 [92]. Resveratrol also suppresses the expression of several important inflammatory mediators including TNF-α, IL-1β, IL-6, iNOS genes, and cyclooxygenases [93, 94].

3.5 Epigallocatechin Gallate

Green tea (*Camellia sinensis*) is an extremely popular beverage worldwide by virtue of its characteristic aroma, flavor, and multiple health benefits [95]. The major catechins in green tea are epigallocatechin-3-gallate (EGCG), epicatechin-3-gallate, epigallocatechin, and epicatechin. EGCG is the major catechin in green tea and accounts for 50–80 % representing 200–300 mg/brewed cup of green tea [96]. The presence of phenolic groups that are sensitive to oxidation and can generate quinone is responsible for the biological activities of EGCG [97]. EGCG has been shown to be anti-inflammatory and protective in several studies using animal models. EGCG supplementation inhibits splenocyte proliferation, T cell division, cell cycle progression, IL-2 receptor expression, and IL-2 accumulation, suggesting an impeded IL-2/IL-2 receptor signaling [98]. Mice fed with EGCG showed reduced ex vivo T cell division and cell cycle progression, and this effect was more pronounced in CD4+ than in CD8+ T cells. EGCG diet (1 %) produces more proinflammatory cytokines TNF-α, IL-6, IL-1β, and PGE-2 in their splenocytes and macrophages and less IL-4 in splenocytes [99]. Administration of EGCG to EAE (experimental autoimmune encephalomyelitis) mice reduces clinical symptoms, brain inflammation, and neuronal damage [100], and EGCG was also able to reduce the incidence of CIA (collagen-induced arthritis) together with a reduced expression of IFN-γ, TNF-α, and Cox-2 [101]. EGCG interferes with both T cell growth and effector function via blockade of the catalytic activities of the 20S/26S proteasome complex, resulting in intracellular accumulation of IκBα and subsequent inhibition of NF-κB activation [100].

3.6 Chlorophyllin

Chlorophyllin (CHL) is a water-soluble mixture of sodium-copper salts of green plant pigment chlorophyll, the ubiquitous photosynthetic green pigment present in the plants as well as in nutritional supplements such as extracts from *Spirulina* and *Chlorella vulgaris*. Chlorophyll has been credited with several beneficial properties, and CHL has been shown to be better than the parent compound as evident from antimutagenic studies. It has chemopreventive, antimutagenic, and anticarcinogenic properties. It has been distinctly demonstrated that CHL is a good scavenger of OH and deoxyribose peroxyl radicals [102]. ESR (Electron Spin Resonance) experiments also confirmed that CHL has potent antioxidant ability involving scavenging of various physiologically important ROS. Several studies showed the potent antioxidant effect of CHL in vitro and in vivo [103]. The immunomodulatory properties of CHL have also been reported, and it was shown to inhibit in vitro lymphocyte proliferation and activation-induced cell death. Chlorophyllin treatment of mice led to splenomegaly with an increase in the T cells, B cells, and macrophage numbers and also increased the number of peritoneal exudate cells (PEC). Increased phagocytic activity was seen in PEC obtained from CHL-treated mice, and CHL administration to mice immunized with sheep red blood cells (SRBC) augmented both humoral and cell-mediated immune responses [104]. Further, CHL treatment was also able to inhibit the homeostatic proliferation of CD4+ T cells in lymphopenic mice [105].

4 Conclusion

The concept that cellular redox status plays a central role as regulator of cellular functions and signaling pathways has gained significant recognition over the past several years. It is widely accepted that oxidative stress and inflammation go hand in hand, and antioxidants can be employed as anti-inflammatory mediators. The evidence supporting this concept is based largely on the finding that H_2O_2 activates the transcription factor NF-κB which has many inflammatory cytokines among its target genes, while thiol antioxidants inhibit its activation. Several studies have reported that thiol antioxidants, including GSH or NAC, inhibit cytokine production. However, considering the complex nature of biological systems and the intricate network of signaling pathways involved during an immune response, a unidirectional approach of implicating oxidative stress in disease pathogenesis is outdated. This is further supported by the fact that despite many clinical trials using a large number of molecules, the inhibitors of oxidative stress (both synthetic and plant-derived antioxidants) are confined to their use as dietary supplements or as alternative medicine and have failed to make the impact as far as their clinical use is concerned. Hence, the progression toward a concept of a more prominent role of cellular redox balance, as opposed to a one-way view, is desirable for development of novel strategies for management of inflammation and immune disorders.

5 Future Directions

A major gap in our understanding of redox signaling is the mechanisms by which these molecules transduce their cellular signals. The emerging evidence of reversible oxidation of protein cysteines as a means of redox regulation during conditions of low oxidative stress provides one of the more acceptable mechanisms for such signaling. It is now widely accepted that depending on the level of ROS, different redox-sensitive transcription factors are activated/inhibited and regulate distinct biological responses. The area of redox-based signaling has reached an exciting juncture, bolstered by the notion that oxidative processes can be precisely orchestrated to evoke diverse functional responses. As described in the previous sections, both endogenous and exogenous modulators of oxidative stress have proved to be potent modulators of immune responses. Many of the above-described molecules have been shown to attenuate inflammation and immune disorders via modulation of cellular redox status leading to oxidative modification of proteins of physiological and pathophysiological importance. However, the challenges ahead involve understanding of how oxidative stress activates/inactivates specific molecules accurately without affecting other redox-sensing molecules. It is also desirable to investigate the pleiotropic effects of redox modifiers other than the purpose for which they are employed. These pleiotropic effects may directly or indirectly influence (inhibit or exacerbate) the overall outcome of the study. The other important aspect of this upcoming area of research is the time of application of these redox modifiers, i.e., prior to inflammation or during the peak of inflammatory response or during the resolving phase of inflammation as this might again greatly influence the clinical outcome. Further, it is required to study the modulation of redox status by these molecules in vivo using different disease models and identify novel approaches to fine-tune the redox alterations induced by these molecules to achieve the desired outcome.

References

1. Allison AC (2000) Immunosuppressive drugs: the first 50 years and a glance forward. Immunopharmacology 47:63–83
2. Sherman LA, Chattopadhyay S (1993) The molecular basis of allorecognition. Annu Rev Immunol 11:385–402
3. Fidelus RK, Ginouves P, Lawrence D et al (1987) Modulation of intracellular glutathione concentrations alters lymphocyte activation and proliferation. Exp Cell Res 170:269–275
4. Novogrodsky A, Ravid A, Rubin AL et al (1982) Hydroxyl radical scavengers inhibit lymphocyte mitogenesis. Proc Natl Acad Sci U S A 79:1171–1174
5. Chaudhri G, Clark IA, Hunt NH et al (1986) Effect of anti-oxidants on primary alloantigen- induced T cell activation and proliferation. J Immunol 137:2646–2652
6. Schafer FQ, Buettner GR (2001) Redox environment of the cell as viewed through the redox state of the glutathione disulfide/glutathione couple. Free Radic Biol Med 30:1191–1212
7. Kamata H, Hirata H (1999) Redox regulation of cellular signalling. Cell Signal 11:1–14
8. Baranano DE, Rao M, Ferris CD et al (2002) Biliverdin reductase: a major physiologic cytoprotectant. Proc Natl Acad Sci U S A 99:16093–16098
9. Wang WW, Smith DL, Zucker SD (2004) Bilirubin inhibits iNOS expression and NO production in response to endotoxin in rats. Hepatology 40:424–433
10. Liu Y, Li P, Lu J et al (2008) Bilirubin possesses powerful immunomodulatory activity and suppresses experimental autoimmune encephalomyelitis. J Immunol 181:1887–1897
11. Khan NM, Poduval TB (2011) Immunomodulatory and immunotoxic effects of bilirubin: molecular mechanisms. J Leukoc Biol 90:997–1015
12. Rocuts F, Zhang X, Yan J et al (2010) Bilirubin promotes de novo generation of T regulatory cells. Cell Transplant 19:443–451
13. Lee SS, Gao W, Mazzola S et al (2007) Heme oxygenase-1, carbon monoxide, and bilirubin induce tolerance in recipients toward islet allografts by modulating T regulatory cells. FASEB J 21:3450–3457
14. Szabo C (2007) Hydrogen sulphide and its therapeutic potential. Nat Rev Drug Discov 6:917–935

15. Kimura Y, Kimura H (2004) Hydrogen sulfide protects neurons from oxidative stress. FASEB J 18:1165–1167
16. Calvert JW, Jha S, Gundewar S et al (2009) Hydrogen sulfide mediates cardioprotection through Nrf2 signaling. Circ Res 105:365–374
17. Whiteman M, Li L, Rose P et al (2010) The effect of hydrogen sulfide donors on lipopolysaccharide-induced formation of inflammatory mediators in macrophages. Antioxid Redox Signal 12:1147–1154
18. Plaine HL (1955) The effect of oxygen and hydrogen peroxide on the action of a specific gene and on tumor induction in Drosophila melanogaster. Genetics 40:268–280
19. Bedard K, Krause KH (2007) The NOX family of ROS-generating NADPH oxidases: physiology and pathophysiology. Physiol Rev 87:245–313
20. Pritsos CA (2000) Cellular distribution, metabolism and regulation of the xanthine oxidoreductase enzyme system. Chem Biol Interact 129:195–208
21. Demiryurek AT, Wadsworth RM (1999) Superoxide in the pulmonary circulation. Pharmacol Ther 84:355–365
22. Rojkind M, Dominguez-Rosales JA, Nieto N et al (2002) Role of hydrogen peroxide and oxidative stress in healing responses. Cell Mol Life Sci 59:1872–1891
23. Klebanoff SJ (2005) Myeloperoxidase: friend and foe. J Leukoc Biol 77:598–625
24. Freed BM, Rapoport R, Lempert N (1987) Inhibition of early events in the human T-lymphocyte response to mitogens and alloantigens by hydrogen peroxide. Arch Surg 122:99–104
25. De Oliveira-Marques V, Cyrne L, Marinho HS et al (2007) A quantitative study of NF-κB activation by H2O2: relevance in inflammation and synergy with TNF-α. J Immunol 178:3893–3902
26. Zmijewski JW, Zhao X, Xu Z et al (2007) Exposure to hydrogen peroxide diminishes NF-kappaB activation, IkappaB-alpha degradation, and proteasome activity in neutrophils. Am J Physiol Cell Physiol 293:C255–266
27. Zmijewski JW, Lorne E, Zhao X et al (2009) Antiinflammatory effects of hydrogen peroxide in neutrophil activation and acute lung injury. Am J Respir Crit Care Med 179:694–704
28. Reimund JM, Allison AC, Muller CD et al (1998) Antioxidants inhibit the in vitro production of inflammatory cytokines in Crohn's disease and ulcerative colitis. Eur J Clin Invest 28:145–150
29. Jeannin P, Delneste Y, Life P et al (1995) Interleukin-12 increases interleukin-4 production by established human Th0 and Th2-like T cell clones. Eur J Immunol 25:2247–2252
30. Pena LR, Hill DB, McClain CJ (1999) Treatment with glutathione precursor decreases cytokine activity. J Parenter Enter Nutr 23:1–6
31. Bernard GR (1991) N-acetylcysteine in experimental and clinical acute lung injury. Am J Med 91:54S–59S
32. Tsuji F, Miyake Y, Aono H et al (1999) Effects of bucillamine and N-acetyl-L-cysteine on cytokine production and collagen-induced arthritis (CIA). Clin Exp Immunol 115:26–31
33. Griffith OW, Meister A (1979) Potent and specific inhibition of glutathione synthesis by buthionine sulfoximine (S-n-butyl homocysteine sulfoximine). J Biol Chem 254:7558–7560
34. Gosset P, Wallaert B, Tonnel AB et al (1999) Thiol regulation of the production of TNF-α, IL-6 and IL-8 by human alveolar macrophages. Eur Respir J 14:98–105
35. Haddad JJ, Land SC (2000) The differential regulation of apoptosis factors in the alveolar epithelium is redox sensitive and requires NF- κB (Rel A) – selective targeting. Biochem Biophys Res Commun 271:257–267
36. Ghezzi P (2011) Role of glutathione in immunity and inflammation in the lung. Int J Gen Med 4:105–113
37. Qanungo S, Starke DW, Pai HV et al (2007) Glutathione supplementation potentiates hypoxic apoptosis by S-glutathionylation of p65-NFκB. J Biol Chem 282:18427–18436
38. Pineda-Molina E, Klatt P, Vazquez J et al (2001) Glutathionylation of the p50 subunit of NF-kappaB: a mechanism for redox-induced inhibition of DNA binding. Biochemistry 40:14134–14142
39. Chung S, Sundar IK, Yao H et al (2010) Glutaredoxin 1 regulates cigarette smoke-mediated lung inflammation through differential modulation of I{kappa} B kinases in mice: impact on histone acetylation. Lung Cell Mol Physiol 299:L192–203
40. Shelton MD, Distler AM, Kern TS et al (2009) Glutaredoxin regulates autocrine and paracrine proinflammatory responses in retinal glial (müller) cells. J Biol Chem 284:4760–4766
41. Boissel JP, Schwarz PM, Forstermann U (1998) Neuronal-type NO synthase: transcript diversity and expressional regulation. Nitric Oxide 2:337–349
42. Shaul PW (2002) Regulation of endothelial nitric oxide synthase: location, location, location. Annu Rev Physiol 64:749–774
43. Bogdan C (2001) Nitric oxide and the immune response. Nat Immunol 2:907–916
44. Joshi MS, Ponthier JL, Lancaster JR (1999) Cellular antioxidant and pro-oxidant actions of nitric oxide - reactive nitrogen species, reactive oxygen species, transition metal ions and the vascular system. Free Radic Biol Med 27:1357–1366
45. Clancy RM, Amin AR, Abramson SB (1998) The role of nitric oxide in inflammation and immunity. Arthritis Rheum 41:1141–1151
46. Rawlingson A (2003) Nitric oxide, inflammation and acute burn injury. Burns 29:631–640
47. Cross RK, Wilson KT (2003) Nitric oxide in inflammatory bowel disease. Inflamm Bowel Dis 9:179–189
48. Vallance P, Leiper J (2002) Blocking NO synthesis: how, where and why? Nat Rev Drug Discov 1:939–950

49. Moncada S, Higgs EA (1995) Molecular mechanisms and therapeutic strategies related to nitric oxide. FASEB J 9:1319–1330
50. DelaTorre A, Schroeder RA, Punzalan C et al (1999) End toxin-mediated S-nitrosylation of p50 alters NF-kappa B-dependent gene transcription in ANA-1 murine macrophages. J Immunol 162:4101–4108
51. Mazzoni A, Bronte V, Visintin A et al (2002) Myeloid suppressor lines inhibit T cell responses by an NO-dependent mechanism. J Immunol 168:689–695
52. Smith WL (1989) The eicosanoids and their biochemical mechanisms of action. Biochem J 259:315–324
53. Tsuboi K, Sugimoto Y, Ichikawa A (2002) Cardiovascular pharmacology: endothelial control: endothelial control. Prostaglandins Other Lipid Mediat, 1st edn. Elsevier, UK
54. Kliewer SA, Lenhard JM, Willson TM et al (1995) A prostaglandin J2 metabolite binds peroxisome proliferator-activated receptor gamma and promotes adipocyte differentiation. Cell 83:813–819
55. Lee TS, Tsai HL, Chau LY (2003) Heme oxygenase-1 protects gastric mucosal cells against non-steroidal anti-inflammatory drugs. J Biol Chem 278:19325–19330
56. Zhang X, Lu L, Dixon C et al (2004) Stress protein activation by the cyclopentenone prostaglandin 15-deoxy-Δ12,14-prostaglandin J2 in human mesangial cells. Kidney Int 65:798–810
57. Kim JW, Li MH, Jang JH et al (2008) 15-Deoxy-Δ12,14-prostaglandin J2 rescues PC12 cells from H_2O_2-induced apoptosis through Nrf2-mediated upregulation of heme oxygenase-1: potential roles of Akt and ERK1/2. Biochem Pharmacol 76:1577–1589
58. Giri S, Rattan R, Singh AK et al (2004) 5-Aminoimidazole-4-carboxamide-1-β-D-ribofuranoside inhibits cancer cell proliferation in vitro and in vivo via AMP-activated protein kinase. J Immunol 173:5196–5208
59. Priyadarsini KI, Maity DK, Naik GH et al (2003) Role of phenolic O-H and methylene hydrogen on the free radical reactions and antioxidant activity of curcumin. Free Radic Biol Med 35:475–484
60. Banerjee A, Kunwar A, Mishra B et al (2008) Concentration dependent antioxidant/pro-oxidant activity of curcumin: studies from AAPH induced hemolysis of RBCs. Chem Biol Interact 174:134–139
61. Balstad TR, Carlsen H, Myhrstad MC et al (2011) Coffee, broccoli and spices are strong inducers of electrophile response element-dependent transcription in vitro and in vivo – studies in electrophile response element transgenic mice. Mol Nutr Food Res 55:185–197
62. Abe Y, Hashimoto S, Horie T (1999) Curcumin inhibition of inflammatory cytokine production by human peripheral blood monocytes and alveolar macrophages. Pharmacol Res 39:41–47
63. Aggarwal S, Ichikawa H, Takada Y et al (2006) Curcumin (diferuloylmethane) down-regulates expression of cell proliferation and antiapoptotic and metastatic gene products through suppression of IkappaBalpha kinase and Akt activation. Mol Pharmacol 69:195–206
64. Bharti AC, Donato N, Aggarwal BB (2003) Curcumin (diferuloylmethane) inhibits constitutive and IL-6-inducible. Stat 3 phosphorylation in human multiple myeloma cells. J Immunol 171:3863–3871
65. Skommer J, Wlodkowic D, Pelkonen J (2007) Gene-expression profiling during curcumin-induced apoptosis reveals downregulation of CXCR4. Exp Hematol 35:84–95
66. Shakibaei M, Schulze-Tanzil G, John T et al (2005) Curcumin protects human chondrocytes from IL-l1beta-induced inhibition of collagen type II and beta1-integrin expression and activation of caspase-3: an immunomorphological study. Ann Anat 187:487–497
67. Shishodia S, Amin HM, Lai R et al (2005) Curcumin (diferuloylmethane) inhibits constitutive NF-kappaB activation, induces G1/S arrest, suppresses proliferation, and induces apoptosis in mantle cell lymphoma. Biochem Pharmacol 70:700–713
68. Li L, Aggarwal BB, Shishodia S et al (2004) Nuclear factor-κB and IκB are constitutively active in human pancreatic cells and their down-regulation by curcumin (Diferuloylmethane) is associated with the suppression of proliferation and the induction of apoptosis. Cancer 101:2351–2362
69. Basnet P, Skalko-Basnet N (2011) Curcumin: an anti-inflammatory molecule from a curry spice on the path to cancer treatment. Molecules 16:4567–4598
70. Padhye S, Dandawate P, Yusufi M et al (2012) Perspectives on medicinal properties of plumbagin and its analogs. Med Res Rev 32:1131–1158
71. Checker R, Sharma D, Sandur SK et al (2009) Anti-inflammatory effects of plumbagin are mediated by inhibition of NF-kappaB activation in lymphocytes. Inter Immunopharmacol 9:949–958
72. Sharma D, Kumar SS, Checker R et al (2009) Spatial distribution, kinetics, signaling and cytokine production during homeostasis driven proliferation of CD4+ T cells. Mol Immunol 46:2403–2412
73. Kohli V, Sharma D, Sandur SK et al (2011) Immune responses to novel allergens and modulation of inflammation by vitamin K3 analogue: a ROS dependent mechanism. Inter Immunopharmacol 11:233–243
74. Jia Y, Jing J, Bai Y et al (2011) Amelioration of experimental autoimmune encephalomyelitis by plumbagin through down-regulation of JAK-STAT and NF-< kappa >B signaling pathways. PLoS One 6:e27006
75. Poosarla A, NR D, Athota RR et al (2011) Modulation of T cell proliferation and cytokine response by Plumbagin, extracted from Plumbago zeylanica in collagen induced arthritis. BMC Complement Alternat Med 11:114
76. Murali PM, Rajasekaran S, Paramesh P et al (2006) Plant-based formulation in the management of chronic obstructive pulmonary disease: a randomized double-blind study. Respir Med 100:39–45

77. Theriault A, Chao JT, Wang Q et al (1999) Tocotrienol: a review of its therapeutic potential. Clin Biochem 32:309–319
78. Elson CE (1992) Tropical oils: nutritional and scientific issues. Crit Rev Food Sci Nutr 31:79–102
79. Sundram K, Sambanthamurthi R, Tan YA (2003) Palm fruit chemistry and nutrition. Asia Pac J Clin Nutr 12:355–362
80. Colombo ML (2010) An update on vitamin E, tocopherol and tocotrienol perspectives. Molecules 15:2103–2113
81. Packer L, Weber SU, Rimbach G (2001) Molecular aspects of alpha-tocotrienol antioxidant action and cell signalling. J Nutr Biochem 131:369–373
82. Ahn KS, Sethi G, Krishnan K et al (2007) Gamma-tocotrienol inhibits nuclear factor-kappaB signaling pathway through inhibition of receptor-interacting protein and TAK1 leading to suppression of anti-apoptotic gene products and potentiation of apoptosis. J Biol Chem 282:809–820
83. Kaileh M, Sen R (2010) Role of NF-κB in the anti-inflammatory effects of tocotrienols. J Am Coll Nutr 29:334–339
84. Norazlina M, Lee PL, Lukman HI et al (2007) Effects of vitamin E supplementation on bone metabolism in nicotine-treated rats. Singapore Med J 48:195–199
85. Ahmad NS, Khalid BA, Luke DA et al (2005) Tocotrienol offers better protection than tocopherol from free radical-induced damage of rat bone. Clin Exp Pharmacol Physiol 32:761–770
86. Wu SJ, Liu PL, Ng LT (2008) Tocotrienol-rich fraction of palm oil exhibits anti-inflammatory property by suppressing the expression of inflammatory mediators in human monocytic cells. Mol Nutr Food Res 52:921–929
87. Wilankar C, Sharma D, Checker R et al (2011) Role of immunoregulatory transcription factors in differential immunomodulatory effects of tocotrienols. Free Radic Biol Med 51:129–143
88. Renaud S, De Lorgeril M (1992) Wine, alcohol, platelets, and the French paradox for coronary heart disease. Lancet 339:1523–1526
89. Martinez J, Moreno JJ (2000) Effect of resveratrol, a natural polyphenolic compound, on reactive oxygen species and prostaglandin production. Biochem Pharmacol 59:865–870
90. Kode A, Rajendrasozhan S, Caito S et al (2008) Resveratrol induces glutathione synthesis by activation of Nrf2 and protects against cigarette smoke-mediated oxidative stress in human lung epithelial cells. Am J Physiol 294:478–488
91. Yen GC, Duh PD, Lin CW (2003) Effects of resveratrol and 4-hexylresorcinol on hydrogen peroxide-induced oxidative DNA damage in human lymphocytes. Free Radic Res 37:509–514
92. Manna SK, Mukhopadhyay A, Aggarwal BB (2000) Resveratrol suppresses TNF-induced activation of nuclear transcription factors NF-kappa B, activator protein-1, and apoptosis: potential role of reactive oxygen intermediates and lipid peroxidation. J Immunol 164:6509–6519
93. Qureshi AA, Guan XQ, Reis JC et al (2012) Inhibition of nitric oxide and inflammatory cytokines in LPS-stimulated murine macrophages by resveratrol, a potent proteasome inhibitor. Lipids Health Dis 11:76
94. Khanduja KL, Bhardwaj A, Kaushik G (2004) Resveratrol inhibits N-nitrosodiethylamine-induced ornithine decarboxylase and cyclooxygenase in mice. J Nutr Sci Vitaminol 50:61–65
95. Siddiqui IA, Asim M, Hafeez BB et al (2011) Green tea polyphenol EGCG blunts androgen receptor function in prostate cancer. FASEB J 25:1198–1207
96. Khan N, Afaq F, Saleem M et al (2006) Targeting multiple signaling pathways by green tea polyphenol (−)-epigallocatechin-3-gallate. Cancer Res 66:2500–2505
97. Lambert JD, Elias RJ (2010) The antioxidant and pro-oxidant activities of green tea polyphenols: a role in cancer prevention. Arch Biochem Biophys 501: 65–72
98. Wu D, Guo Z, Ren Z et al (2009) Green tea EGCG suppresses T cell proliferation through impairment of IL-2/IL-2 receptor signaling. Free Radic Biol Med 47:636–643
99. Pae M, Ren Z, Meydani M et al (2012) Dietary supplementation with high dose of epigallocatechin-3-gallate promotes inflammatory response in mice. J Nutr Biochem 23:526–531
100. Aktas O, Prozorovski T, Smorodchenko A et al (2004) Green tea epigallocatechin-3-gallate mediates T cellular NF-kappa B inhibition and exerts neuroprotection in autoimmune encephalomyelitis. J Immunol 173:5794–5800
101. Haqqi TM, Anthony DD, Gupta S et al (1999) Prevention of collagen-induced arthritis in mice by a polyphenolic fraction from green tea. Proc Natl Acad Sci U S A 96:4524–4529
102. Kumar SS, Chaubey RC, Devasagayam TP et al (1999) Inhibition of radiation-induced DNA damage in plasmid pBR322 by chlorophyllin and possible mechanism(s) of action. Mutat Res 425: 71–79
103. Kumar SS, Shankar B, Sainis KB (2004) Effect of chlorophyllin against oxidative stress in splenic lymphocytes in vitro and in vivo. Biochim Biophys Acta 1672:100–111
104. Sharma D, Kumar SS, Sainis KB (2007) Antiapoptotic and immunomodulatory effects of chlorophyllin. Mol Immunol 44:347–359
105. Sharma D, Kumar SS, Raghu R et al (2007) Differential modulation of mitogen driven proliferation and homeostasis driven proliferation of T cells by rapamycin, Ly294002 and chlorophyllin. Mol Immunol 44:2831–2840

Oxidative Stress, Antioxidant Status, and Redox Signaling in Carcinogenesis

Deepti Pande, Reena Negi, Kanchan Karki, Seema Khanna, U.S. Dwivedi, Ranjana S. Khanna, and Hari D. Khanna

Abstract

An increase in ROS production or a decrease in ROS-scavenging capacity due to exogenous stimuli or endogenous metabolic alterations can disrupt redox homeostasis, leading to an overall increase of intracellular ROS levels or oxidative stress. Oxidative stress promotes damage to the cell structure including proteins, lipids, membranes, and DNA, strongly implicating such damage in the etiology of cancer. The extent to which oxidative damage contributes to the process of carcinogenesis is not yet clear. Besides the role of oxidants in the induction of mutations, it is apparent that ROS mediate cell-signaling pathways that are involved in cell growth regulatory pathways and are thus instrumental in the process of carcinogenesis. The activation of transcription factors including MAP-kinase/AP-1 and NF-κB pathways has a direct effect on cell proliferation and apoptosis. Thus damage, mutations, and altered gene expression are all key players in the process of carcinogenesis. The involvement of oxidants appears to be the common denominator to all these events supporting the idea that tumor microenvironment plays an important role in cancer initiation and progression. An altered pro-oxidant–antioxidant balance may lead to an increased oxidative damage and consequently play an important role in the carcinogenesis. Significantly raised NF-κB activation, VEGF level, and cell proliferation index in carcinoma indicate definite correlation of oxidative stress and carcinogenesis. Biomarkers of oxidative stress have the potential to help establish pathogenic stage and risk for disease.

D. Pande • R. Negi • K. Karki • H.D. Khanna (✉)
Department of Biophysics, Banaras Hindu University, Varanasi 221005, Uttar Pradesh, India
e-mail: hdkhanna@yahoo.co.in

S. Khanna
Department of General Surgery, Institute of Medical Sciences, Banaras Hindu University, Varanasi 221005, India

U.S. Dwivedi
Department of Urology, Institute of Medical Sciences, Banaras Hindu University, Varanasi 221005, India

R.S. Khanna
Department of Chemistry, Faculty of Science, Banaras Hindu University, Varanasi 221005, India

V. Rani and U.C.S. Yadav (eds.), *Free Radicals in Human Health and Disease*, DOI 10.1007/978-81-322-2035-0_22, © Springer India 2015

Keywords
Oxidative damage • Lipids • Proteins • Catalase • Superoxide dismutase • Glutathione peroxidase

1 Introduction

Oxygen is an important element that plays significant role in many processes of the human body including cellular metabolism and intercellular and intracellular signaling and acts as a key component for an effective immune system and response [1]. Although beneficial, it is accepted that oxygen, through ROS generation, can react with DNA, proteins, and other cellular components resulting in their damage. ROS can induce direct cellular injury, which may trigger a cascade of radical reactions enhancing secondary ROS generation. Cells attempt to neutralize these ROS cascades with antioxidants. There is a critical balance between oxidants and antioxidant defenses [2]. Oxidative stress-induced cell injury has been reported to have a role in the pathogenesis of a number of diseases, including cancer, atherosclerosis, diabetes mellitus, epilepsy, radiation damage, cellular aging, reperfusion damage, inflammatory diseases, and parkinsonism [3]. Considerable evidence has emerged in recent years implicating ROS as having an important role in the initiation of cellular injury which can lead to cancer development. Alterations in oxidant–antioxidant profile are known to occur in cancer.

1.1 Reactive Oxygen Species (ROS)

Reactive oxygen species (ROS) are derived from the metabolism of molecular oxygen [4]. ROS include superoxide anion radicals (O_2^-), singlet oxygen (O_2), hydrogen peroxide (H_2O_2), and highly reactive hydroxyl radical (OH^-). The deleterious effects of oxygen are said to result from its metabolic reduction to these highly reactive and toxic species [5].

ROS normally exist in all aerobic cells in balance with biochemical antioxidants. Oxidative stress occurs when this critical balance is disrupted because of excess production of ROS, antioxidant depletion, or both. To counteract the oxidant effects and to restore redox balance, cells must reset important homeostatic parameters. ROS are not always harmful metabolic by products, when tightly regulated ROS can act as intracellular signaling molecules [6, 7]. When a free radical reacts with a non-radical molecule, the target molecule is converted to a radical, which may further react with another molecule. The primary formation of most of the ROS is the reduction of molecular oxygen with the formation $O_2^{-\bullet}$ [8, 9]. Superoxide radicals may also serve as precursors for other reactive oxygen species.

Superoxide undergoes a dismutation to form H_2O_2 spontaneously or enzymatically (I). Superoxide can also react with nitric oxide ($NO^\bullet$) to form peroxynitrite ($ONOO^-$) [10].

$$\text{(I)} \quad O_2^{-\bullet} + O_2^{-\bullet} + 2H^+ \rightarrow H_2O_2 + O_2$$

Hydrogen peroxide is more stable than $O_2^{-\bullet}$. It can diffuse through the plasma membrane and if not scavenged locally by catalase or glutathione peroxidase, can promote radical reactions far from its origin. The most reactive and potentially harmful radical has been considered to be $OH^\bullet$ because the lifetime of $OH^\bullet$ is extremely short; it can thus be expected to react at or close to its site of formation. Most of the toxicity of free radicals in vivo is thought to arise from reactions catalyzed by metal ions, such as Cu(II) and Fe(III) [11]. The hydroxyl radical is generated from H_2O_2 through the Fenton reaction catalyzed by the transition metals iron or copper (II):

$$(II) \quad Fe^{2+} + H_2O_2 \rightarrow Fe^{3+} + OH^{\bullet} + OH^{\bullet}$$

Or from $O_2^{-\bullet}$ and H_2O_2 through the Haber–Weiss reaction catalyzed by iron or copper (III)

$$(III) \quad O_2^{-\bullet} + H_2O_2 \underset{Catalyst}{\overset{Metal}{\rightarrow}} O_2 + OH^{\bullet} + OH^{\bullet}$$

In living cells, the major sources of endogenous ROS are hydrogen peroxide and superoxide anion, which are generated as by-products of cellular metabolism such as mitochondrial respiration [12]. Alternatively, hydrogen peroxide may be converted into water by the enzyme catalase or glutathione peroxidase. Variability or inductive changes in the expression of these enzymes can significantly influence cellular redox potential. ROS can cause tissue damage by reacting with lipids in cellular membranes, nucleotide in DNA [13], sulfhydryl groups in proteins [14], and cross-linking/fragmentation of ribonucleoproteins [15]. The relatively unreactive superoxide anion radical is converted by superoxide dismutase (SOD) into H_2O_2 [16].

ROS generation results from exposure to numerous exogenous agents and events including pollution, sunlight, X-rays, smoking and alcohol [17] ionizing radiation, cytokines, growth factors, chemotherapeutic drugs, environmental toxins, hyperthermia, and macrophages during the inflammatory response [18, 19].

Typically, low concentration of ROS is essential for normal physiological functions like gene expression, cellular growth, and defense against infection. Sometimes they also act as the stimulating agents for biochemical processes within the cell [20, 21]. Cells communicate with each other and respond to extracellular stimuli through biological mechanisms called cell signaling or signal transduction [22]. Signal transduction is a process enabling information to be transmitted from the outside of a cell to various functional elements inside the cell. Signal transduction is triggered by extracellular signals such as hormones, growth factors, cytokines, and neurotransmitters [23]. Signals sent to the transcription machinery responsible for expression of certain genes are normally transmitted to the cell nucleus by a class of proteins called transcription factors. By binding to specific DNA sequences, these factors regulate the activity of RNA polymerase II. These signal transduction processes can induce various biological activities, such as muscle contraction, gene expression, cell growth, and nerve transmission.

While ROS are predominantly implicated in causing cell damage, they also play a major physiological role in several aspects of intracellular signaling and regulation [21]. It is a well-known feature that cells are capable of generating endogenously and constitutively ROS which are utilized in the induction and maintenance of signal transduction pathways involved in cell growth and differentiation.

Most cell types have been shown to elicit a small oxidative burst generating low concentrations of ROS when they are stimulated by cytokines, growth factors, and hormones, e.g., interleukin-1b (IL-1b), interleukin-6 (IL-6), interleukin-3 (IL-3), tumor necrosis factor-α (TNF-α), angiotensin II (ANGII), platelet-derived growth factor (PDGF), nerve growth factor (NGF), transforming growth factor-b1 (TGF-b1), granulocyte–macrophage colony-stimulating factor (GM-CSF), and fibroblast growth factor (FGF-2) [23]. ROS exert their effects through the reversible oxidation of active sites in transcription factors such as nuclear factor-kappa B (NF-κB) and activator protein-1 (AP-1) leading to gene expression and cell growth [24]. ROS can also cause indirect induction of transcription factors by activating signal transduction pathways [25]. ROS also appear to serve as secondary messengers in many pathways. This led to the assumption that the initiation and/or proper functioning of several signal transduction pathways relies on the action of ROS as signaling molecules which may act on different levels in the signal transduction cascade. ROS can thus play a very important physiological role as secondary messengers [26, 27].

1.2 Oxidative Stress

Oxidative stress is defined as an imbalance between pro-oxidants (free radical species) and the body's scavenging ability (antioxidants) [28].

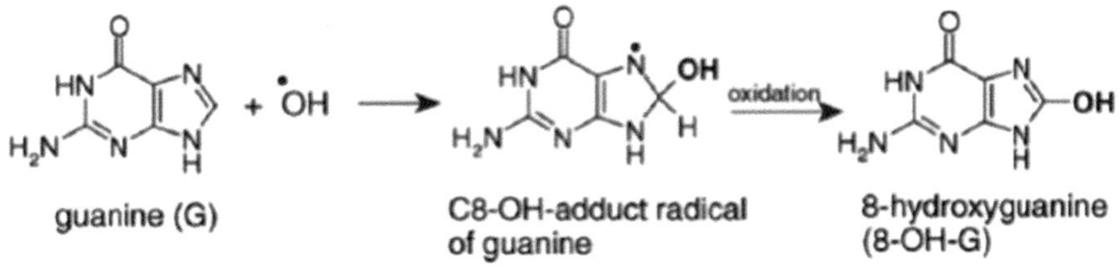

Fig. 1 Formation of 8-OHdG

It may be due to either increased production of reactive oxygen species (ROS) or decreased levels of antioxidants (enzymatic and nonenzymatic) or both [29]. Most of these oxygen-derived species are produced at a low level by normal aerobic metabolism, and the damage they cause to cells is constantly repaired. However, under the severe levels of oxidative stress that cause necrosis, the damage causes ATP depletion and prevents controlled apoptotic death [30]. Oxidative stress could be considered as a large increase (becoming less negative) in the cellular reduction potential, or a large decrease in the reducing capacity of the cellular redox couples, such as glutathione [31]. Many biochemical compounds, namely, nucleic acid, amino acid, protein, lipid, lipoprotein, carbohydrate, and macromolecules of collagen tissue, can be damaged irreversibly or reversibly by free radicals [32].

1.3 Oxidative Damage to DNA

Oxidative DNA damage is a major source of the mutation load in living organisms, with more than 100 oxidative DNA adducts having been identified [33–36]. The estimated frequency of oxidative DNA damage in human cells is 104 lesions/cell/day [37]. Being highly reactive, the hydroxyl radical is the predominant ROS that targets DNA [33]. Hydrogen peroxide, a precursor to hydroxyl radical, is less reactive and more readily diffusible and thus more likely to be involved in the formation of oxidized bases through Fenton and Haber–Weiss reactions [38, 39]. ROS-induced DNA damage can result in single- or double-strand breakage, base modifications, deoxyribose modification, and DNA cross-linking. Cell death, DNA mutation, replication errors, and genomic instability can occur if the oxidative DNA damage is not repaired prior to DNA replication [40–43]. The most extensively studied and most abundant oxidative DNA lesion produced is 8-hydroxydeoxy guanosine (8-OHdG), which is mutagenic in bacterial and mammalian cells [44].

Numerous studies have demonstrated that 8-OHdG levels are elevated in various human cancers [45–48] and in animal models of tumors [49, 50]. 8-OHdG in its stable syn conformation can pair with both cytosine and adenine. If the A:G mismatch is not repaired, a G:C to T:A transversion will occur, commonly found in mutated oncogenes and tumor suppressor genes [44, 51, 52]. During DNA replication, ROS can also react with dGTP in the nucleotide pool to form 8-OHdG. Therefore, in addition to the G:C to T:A caused by 8-OHdG in the DNA template, it is postulated that during DNA replication, 8-OHdG in the nucleotide pool can be incorporated into DNA opposite dC or dA on the template strand, resulting in A:T to C:G transversions [36, 44] (Fig. 1).

Based on this evidence, 8-OHdG has been widely used as a biomarker of oxidative DNA damage, and measurement of 8-OHdG level is applied to evaluate the load of oxidative stress [53, 54]. In addition, RNS, produced during the process of chronic inflammation, can cause nitrative DNA damage to form 8-nitroguanine. The formation of 8-nitroguanine has been observed in various human samples, and experimental evidence has suggested that 8-nitroguanine is a mutagenic DNA lesion, which preferentially leads to G:T transversions [55]. The assessment of oxidative DNA damage products in various biological matrices, such as serum and/or urinary

8-OHdG or 8-nitroguanine, could be important to understanding the role of oxidative stress in cancer development and disease intervention.

1.4 Oxidative Damage to Lipids

Lipid peroxidation is a form of oxidative damage in cell membranes determined as free radicals reacting with polyunsaturated fatty acids [56]. Polyunsaturated fatty acids (PUFAs) are abundant in cellular membranes and also in low-density lipoproteins (LDL). They comprise mainly n-3 and n-6 PUFAs and are especially sensitive to free radicals [57, 58].

The interaction of ROS and lipids consists of three different steps: initiation, propagation, and termination. In the initiation phase, conjugated dienes are formed as hydrogen atom is abstracted from a lipid methylene group. Conjugated dienes absorb ultraviolet light at 230–235 nm, and their measurement is regarded as an accurate and repeatable marker of lipid peroxidation [59]. The molecular oxygen reacts with carbon-centered free radicals, and thus lipid hydroperoxides (LOOH) are formed [60]. LOOH may cause alterations in membrane structure and function which, however, may be reversible [56]. LOOH are formed earlier in the pathway leading to the formation of malondialdehyde (MDA) and are the source of highly reactive aldehydes [61]. The aldehydes, for example, MDA and 4-hydroxynonenal (HNE), have the potential to modify DNA and proteins. They are also capable of inducing apoptosis or necrosis in various cells [56]. Measuring MDA levels in the plasma or serum provides a suitable in vivo index of lipid peroxidation [62]. TBA (thiobarbituric acid) reacts with numerous chemical species (including nucleic acids, amino acids, proteins, phospholipids, and aldehydes) when heated under acidic conditions and produces a typical pink chromophore which can be measured by UV or fluorescence detection [63]. Numerous in vivo studies have supported the view that lipid peroxidation has a key role in carcinogenesis [64]. Also the production of ROS by circulating inflammatory cells in tumor tissues might promote lipid peroxidation, which also additionally gives rise to active oxygen species [65].

1.5 Oxidative Damage to Proteins

Proteins are major targets for ROS/RNS because of their high overall abundance in biological systems; they are primarily responsible for most functional processes within cells. It has been estimated that proteins can scavenge the majority (50–75 %) of reactive species (ROS/RNS) generated [66]. Exposure of proteins to ROS/RNS may alter every level of protein structure from primary to quaternary, causing major physical changes in protein structure. Oxidative damage to proteins is induced either directly by ROS/RNS or indirectly by reaction of secondary by-products of oxidative stress and can occur via different mechanisms, leading to peptide backbone cleavage, cross-linking, and/or modification of the side chain of virtually every amino acid [66–68]. Most protein damage is irreparable, and oxidative changes of protein structure can have a wide range of downstream functional consequences, such as inhibition of enzymatic and binding activities, increased susceptibility to aggregation and proteolysis, increased or decreased uptake by cells, and altered immunogenicity [69–73]. Individual proteins may display different susceptibilities to oxidative attack, linked to the variable proportions and distributions of sulfhydryl groups, Fe–S clusters, reduced heme moieties, copper prosthetic groups, sequence motifs, and residues exposed on the molecular surface. Oxidative damage to proteins may be important in vivo not only in its own right (affecting, e.g., the functions of receptors, enzymes, and transport proteins) but also because it can contribute to secondary damage to other biomolecules, for example, inactivation of DNA repair enzymes and loss of fidelity of damaged DNA polymerases in replicating DNA. The major fate of oxidized proteins is catabolism by proteasomal and lysosomal pathways, but some functionally inactive proteins appear to be poorly degraded, form protein aggregates, and accumulate in separate compartments within cells or the extracellular environment [67, 69]. The accumulation of such damaged material increases during the normal aging process may contribute to a range of human pathologies and is able to act as an inhibitor of the proteasome [69, 74, 75]. Because of this decreased capacity for removal of oxidized proteins, the accumulation of misfolded and damaged proteins is accelerated. The vicious circle of decreased proteolysis and accumulation of increasing amounts of oxidatively damaged proteins continues until the protein aggregates cause metabolic dysfunctions or the initiation of apoptotic or necrotic events. Protein carbonyls may be generated by the oxidation of several amino acid side chains (e.g., in Lys, Arg, Pro, and Thr) by the formation of Michael adducts between Lys, His, and Cys residues and unsaturated aldehydes, forming ALEs and by glycation/glycoxidation of Lys amino groups, forming advanced glycation end products [67, 68, 72, 73, 76]. The formation of carbonyl compounds is the most general and widely used marker of severe protein oxidation both in vitro and in vivo, with several assays developed for the quantification of these species [76]. The chemical stability of protein carbonyls makes them suitable targets for laboratory measurement and is also useful for their storage [73]. As a marker of oxidative damage to proteins, carbonyls have been shown to accumulate during aging, ischemia–reperfusion injury, chronic inflammation, cystic fibrosis, and many of age-related diseases in a variety of organisms [77, 78].

1.6 Antioxidant System

According to the commonly accepted definition by Halliwell and Gutteridge [11], an antioxidant is "any substance that delays, prevents or removes oxidative damage to a target molecule". Antioxidants are molecules or compounds that act as free radical scavengers. The effect of reactive oxygen and nitrogen species is balanced by the antioxidant action of nonenzymatic antioxidants, as well as by antioxidant enzymes. Under normal conditions, there is a balance between both the activities and the intracellular levels of these antioxidants. This balance is essential for the survival of organisms and their health. Such antioxidant defense is extremely important as they represent the direct removal of free radicals (pro-oxidants), thus providing maximal protection for biological sites [2, 79].

Antioxidants may be synthesized in the body or obtained from the diet [68]. The most efficient enzymatic antioxidants involve superoxide dismutase, catalase, and glutathione peroxidase [80]. Nonenzymatic antioxidants involve vitamin C, vitamin E, carotenoids, thiol antioxidants (glutathione, thioredoxin, and lipoic acid), natural flavonoids, and other compounds [81]. Some antioxidants act in a hydrophilic environment, others in a hydrophobic environment, and some act in both environments of the cell. Certain antioxidants are able to regenerate other antioxidants and thus restore their original function. The redox cycles of vitamins E and C form such an antioxidant network [82]. The capacity to regenerate one antioxidant by another is driven by the redox potentials of the (reduction/oxidation) couple. There is a link between increased levels of ROS and disturbed activities of enzymatic and nonenzymatic antioxidants in tumor cells.

2 Enzymatic Antioxidant

2.1 Superoxide Dismutase (SOD)

Superoxide dismutase (SOD) is one of the most effective intracellular enzymatic antioxidants. It catalyzes the dismutation of $O_2^{\bullet-}$to O_2 and to the less-reactive species H_2O_2. SOD destroys $O_2^{\bullet-}$ with remarkably high reaction rates, by successive oxidation and reduction of the transition metal ion [80].

Superoxide dismutase exists in several isoforms, differing in the nature of the active metal center and amino acid constituency, as well as their number of subunits, cofactors, and other features. In humans there are three forms of SOD: cytosolic Cu–Zn-SOD, mitochondrial Mn-SOD, and extracellular SOD (EC-SOD) [83].

Cu–Zn–SOD specifically catalyzes the dismutation of the superoxide anion to oxygen and water [80]. Mitochondrial Mn-SOD cycles from Mn(III) to Mn(II) and back to Mn(III) during the two-step dismutation of superoxide. Mn-SOD is one of the most effective antioxidant enzymes that has antitumor activity [84]. In line with this is the fact that the concentrations of transition metal ions have been found to be significantly reduced in some tumors. For certain tumor cells the activity of total SOD (Cu–Zn-SOD and Mn-SOD) has also been found to be reduced [83]. Extracellular superoxide dismutase (EC-SOD) is a secretory, tetrameric, copper- and zinc-containing glycoprotein, with a high affinity for certain glycosaminoglycans such as heparin and heparin sulfate [80]. Its regulation in mammalian tissues occurs primarily in a manner coordinated by cytokines, rather than as a response of individual cells to oxidants.

2.2 Glutathione Peroxidase (GPx)

Glutathione peroxidase [80] is a selenium-containing tetrameric glycoprotein, that is, a molecule with four selenocysteine amino acid residues. As the integrity of cellular and subcellular membranes depends heavily on glutathione peroxidase, the antioxidative protective system of glutathione peroxidase, itself, depends heavily on the presence of selenium. All GPx enzymes are known to add two electrons to reduce peroxides by forming selenoles (Se–OH). The antioxidant properties of these seleno-enzymes allow them to eliminate peroxides as potential substrates for the Fenton reaction. GPx acts in conjunction with the tripeptide glutathione (GSH), which is present in cells in high (micromolar) concentrations. The substrate for the catalytic reaction of GPx is H_2O_2 or organic peroxide ROOH. GPx decomposes peroxides to water (or alcohol) while simultaneously oxidizing GSH:

$$2GSH + H_2O_2 - G \rightarrow Px\,GSSG + 2H_2O$$

$$2GSH + ROOH - G \rightarrow PxGSSG + ROH + H_2O$$

Significantly, GPx competes with catalase for H_2O_2 as a substrate and is the major source of protection against low levels of oxidative stress.

2.3 Catalase

Catalase is an enzyme present in the cells of plants, animals, and aerobic (oxygen-requiring) bacteria [80]. Catalase is located in a cell organelle called the peroxisome. The enzyme very efficiently promotes the conversion of hydrogen peroxide to water and molecular oxygen. It is a tetrameric 240 KD antioxidant enzyme; each monomer contains a hem as prosthetic group at the catalytic center [85]. Catalase can oxidize different toxins, such as formaldehyde, formic acid, and alcohols. In doing so, it uses hydrogen peroxide according to the following reaction [86]:

$$H_2O_2 + H_2R \quad \rightarrow \quad 2H_2O + R$$

The significantly decreased capacity of a variety of tumors for detoxifying hydrogen peroxide is linked to a decreased level of catalase. The Michaelis–Menten constant (Km) of catalase is higher than the Km of glutathione peroxidase (GPx), which suggests that catalase scavenges H_2O_2 efficiently at high H_2O_2 concentrations [87].

3 Nonenzymatic Antioxidant

3.1 Vitamin E (α-Tocopherol)

Vitamin E is the major lipid-soluble antioxidant present in lipid membranes and human plasma lipoproteins. It exists in 8 different isoforms, of which alpha-tocopherol is biologically the most important. Vitamin E is a strong inhibitor of apoptosis and a stabilizer of biological membranes and is known to act on all steps of membrane oxidative damage [64, 88]. Alpha-tocopherol functions in vivo as a strong protector against lipid peroxidation and also blocks nitrosamine formation [89, 90]. Erhola et al. [91] reported that it is the final main antioxidant consumed after exposure to oxidants. Even though alpha-tocopherol is the major or even the only lipid-soluble chain-breaking antioxidant in human plasma, it is responsible for only 2–3 % of the total peroxyl radical trapping capacity (TRAP) of plasma [92, 93]. Depending on the circumstances, alpha-tocopherols can either initiate or inhibit apoptosis [94]. In the presence of iron, alpha-tocopherol may exert pro-oxidant effects, this, however, not being likely to occur in the human body since most transition metals are bound to proteins [95]. Other functions of vitamin E include inhibition of cell adhesion, proliferation and protein kinase activity, as well as enhancement of immunity and modulation of gene expression.

3.2 Vitamin C

Vitamin C, ascorbic acid, is a water-soluble chain-breaking antioxidant, known to be the most effective aqueous phase antioxidant in human plasma [95]. It plays a vital role in various biological processes [96]. Ascorbic acid also inhibits lipid peroxidation, oxidation of low-density lipoproteins, and protein oxidation [95, 97]. The concentrations of ascorbic acid are much lower in plasma than in human tissues. Studies suggest that ascorbic acid is the first plasma antioxidant to be consumed after exposure to oxidants. Ascorbic acid is easily oxidized to dehydroascorbic acid, which is known to have effects similar to those of ascorbic acid in vivo. Ascorbic acid can react directly with superoxide, hydroxyl radicals, and singlet oxygen. Vitamins C and E can also spare glutathione and prevent its oxidation [96]. Vitamin C is important in recycling the tocopherol radical of vitamin E to an active reduced state [98]. Of note, ascorbic acid may display pro-oxidant effects in the presence of free transition metal catalysts.

3.3 Vitamin A

Vitamin A plays an important role in cellular function, development, and maintenance of normal visual acuity [99]. In addition, vitamin A is known as an important natural antioxidant [100]. Vitamin A is a fat-soluble vitamin present in many lipid substances. β-carotene, present in membranes, is converted into vitamin A when the body needs it. Although the mechanism of its in vivo action is unclear, β-carotene is suggested to deactivate ROS (in particular singlet oxygen and lipid radicals) and to reduce lipid peroxidation [101, 102]. Although less important than vitamin E inside the system, β-carotene and vitamin A act with vitamin C and vitamin E in order to protect cells against ROS [103].

3.4 Trace Elements

Through ROS-mediated reactions, metals cause DNA damage, lipid peroxidation, and protein modification. Metals also cause activation of nuclear transcription factors, activation of various signaling proteins, cell cycle arrest, and apoptosis. Metal-induced oxidative stress explains some, but not all of the carcinogenic effects of metals. Metals affect growth receptors, tyrosine and serine/threonine kinases, and nuclear transcription factors by ROS-dependent and ROS-independent mechanisms. Previous studies of the effects of metals and metal-induced ROS on signaling and carcinogenesis have focused on a single biological pathway, enzymes, or genes.

Zinc antioxidant function may be related to several factors. First, zinc is an essential component of Cu–Zn-SOD, one of the cell's first lines of defense against reactive oxygen species (ROS), which functions to remove the superoxide anion (O^{2-}). The second potential mechanism for zinc antioxidant effects is the antagonism of redox-active transition metals, such as iron or copper, and the prevention of oxidation of sulfhydryl groups within proteins. In chemically defined systems, zinc can prevent hydroxyl radical formation by transition metals. Most biological molecules cannot be damaged at a significant rate by direct reactions with oxygen, superoxide, or hydrogen peroxide. However, they can be oxidized by hydroxyl radicals (OH^-). Although the precise mechanism by which zinc acts as an antioxidant is unclear, compromised zinc status clearly has a significant impact on the antioxidant capacity of the cell. Increased oxidative stress is associated with low cellular zinc. Zinc status may affect redox-sensitive signals and ultimately may alter signal pathways involved in stress response, DNA repair, chromosome breaks, and apoptosis [104–108]. Cancer patients have low zinc status as compared to healthy controls [109].

Magnesium (Mg) is the second most prevalent intracellular cation and is involved in the metabolic activity of the cell. Because copper is an essential component of several endogenous antioxidant enzymes, and that free radicals have been proposed to play a role in the process of carcinogenesis, the effects of dietary copper levels on the development of cancer have been investigated [110].

The weight of evidence from in vitro and in vivo assays indicates that copper is not genotoxic [111]. However, in vitro studies have shown that cancer cells in a high copper environment find it easy to proliferate into tumor [112, 113]. Therefore, it has been proposed that copper-lowering drug may stabilize advanced cancer. Similarly to iron, copper is a well-known pro-oxidant and may participate in metal-catalyzed peroxidation of lipids.

Selenium appears to function as an antimutagenic agent, preventing the malignant transformation of normal cells. These protective effects of Se seem to be primarily associated with its presence in the glutathione peroxidases (GSH-Pxs) and thioredoxin reductase, which are known to protect DNA and other cellular components from oxidative damage [114, 115]. Generally, selenoenzymes are known to play roles in the control of cell division, oxygen metabolism, detoxification process, induction of apoptosis in cancer cells, and functioning of the immune system. Other modes of actions involve inactivation of oncogenes. Additionally, in vitro studies have shown that selenium compounds are able to inhibit oxidative stress-induced DNA damage, and carcinogen-induced covalent DNA adducts formation.

Further in vitro studies have shown that selenium is able to induce apoptosis and inhibits cell growth in transformed cells [116]. This implies that selenium compounds are able to induce the p53 gene. Selenium is involved in signal transduction via activation of MAPKs and transcription factors such as AP- 1, NF-κB which influences gene expression and cell growth [117]. Selenium has been shown to regulate p53. In addition, it has been suggested that the activity of p53 is selenium dependent and redox influenced. The molecular basis for selenium in inducing apoptosis in cancer cells occurs via mediation of cell-signaling targets.

Premalignant human breast cancer cells incubated with methylselenic acid exhibited growth inhibition and induction of apoptosis. The effect of selenium on molecular processes involving cell signaling and apoptosis occurs mainly via a redox-dependent mechanism. Selenium and GSH-Px levels have been found to be decreased in patients with carcinoma of the uterine cervix. Two clinical intervention trials on the protective effect of selenium against cancer performed in the USA have shown that selenium intake did not reduce the risk of skin cancer; however, significant reductions in the secondary end points in lung, colon, and prostate cancer incidence were observed.

From a compilation of biochemical, animal, and human data, links have been proposed between increased levels of iron in the body and an enhanced risk of a variety of diseases including vascular disease, cancer, and certain neurological conditions [118, 119]. Iron-mediated formation of ROS leading to DNA and lipid damage appears to result from an exaggeration of the normal function of iron, which is to transport oxygen to tissues. Iron excess is believed to generate oxidative stress, understood as an increase in the steady-state concentration of reactive oxygen and nitrogen species. The "free" or "catalytic" form of iron mediates the production of reactive oxygen species via the Fenton and Haber–Weiss reactions, thereby inducing oxidative stress [120]. The toxicity of superoxide anion and hydrogen peroxide arises from their iron-dependent conversion into the extremely reactive hydroxyl radical (OH) (Haber–Weiss reaction) that causes severe damage to membranes, proteins, and DNA. Free iron is essential in initiating and sustaining the chain reaction of lipid peroxidation [121]. Iron-induced free radical damage to DNA appears to be important for the development of cancer, and cancer cells are known to grow rapidly in response to iron [122].

4 ROS and Carcinogenesis

Oxidative stress plays a role in various clinical conditions such as malignant diseases, diabetes mellitus, atherosclerosis, chronic inflammation, viral infection, and ischemia–reperfusion injury [3]. Carcinogenesis is associated with various epigenetic mechanisms, which can alter intra- and intercellular communications and gene expression, affecting cell proliferation, differentiation, and apoptosis. In addition to classical epigenetic events such as DNA methylation and histone acetylation, the major mechanisms include changes in the concentrations of signal molecules (hormones, growth factors, fatty acids, etc.); modulation of cell receptors; drug-, hormone-, and fatty acid-metabolizing enzymes; and interference with intracellular signal transduction pathways [123].

Damage to DNA by ROS has been widely accepted as a major cause of cancer [124]. In patients with diseases associated with a risk of cancer indicates an increased rate of oxidative DNA damage [15, 125]. ROS can damage DNA, and division of the cells with unpaired or misrepaired damage leads to mutations [126, 127]. Indeed these species can act at several steps, and in multistage carcinogenesis, it is now assumed that ROS are involved both in the initiation and progression of cancer [128]. ROS can cause oxidative DNA and protein damage, damage to tumor suppressor genes [129, 130], and oxidative stress has been shown to induce malignant transformation of cells in culture [131].

While ROS are predominantly implicated in causing cell damage, they also play a major physiological role in several aspects of intracellular signaling and regulation [132]. The abnormal behavior of neoplastic cells can often be traced to an alteration in cell signaling mechanisms, such as receptor or cytoplasmic tyrosine kinases, altered levels of specific growth factors, intracellular processes for conveying membrane signals to the nucleus, portions of the transcription apparatus, and genes involved in the cell cycle and the regulation of DNA replication (Fig. 2). It has been clearly demonstrated that ROS interfere with the expression of a number of genes and signal transduction pathways [23]. Because ROS are oxidants by nature, they influence the redox status and may, according to their concentration, cause either a positive response (cell proliferation) or a negative cell response (growth arrest or cell death). As already mentioned above, while high concentrations of ROS cause cell death or even necrosis, the effects of ROS on cell proliferation occurred exclusively at low or transient concentrations of radicals. Low concentrations of superoxide radical and hydrogen peroxide in fact stimulate proliferation and enhanced survival in a wide variety of cell types. ROS can thus play a very important physiological role as secondary messengers [26].

The "two-faced" character of ROS is substantiated by growing body of evidence that ROS within cells act as secondary messengers in intracellular signaling cascades, which induce and maintain the oncogenic phenotype of cancer cells; however, ROS can also induce cellular senescence and apoptosis and can therefore function as antitumorigenic species. Oxidative stress induces a cellular redox imbalance which has been found to be present in various cancer cells compared with normal cells; the redox imbalance thus may be related to oncogenic stimulation. DNA mutation is a critical step in carcinogenesis, and elevated levels of oxidative DNA lesions (8-OHdG) have been noted in various tumors. It appears that the DNA damage is predominantly linked with the initiation process. On the other hand, iron, copper, chromium, cobalt, vanadium, cadmium, arsenic, or nickel can mediate the formation of free radicals (e.g., Fenton chemistry), evoking DNA damage (both mitochondrial and nuclear) and damage to lipids and proteins [43]. Therefore, ROS play a key role during both growth factor and receptor signaling, and these second messengers are recognized to have a synergistic function for anchorage-dependent growth signaling. Many findings support a role of ROS as key second messengers, through redox regulation of several downstream targets. Deregulation of these redox pathways may help to transform the cells to elude the native apoptotic response and to activate cancer cell growth [133].

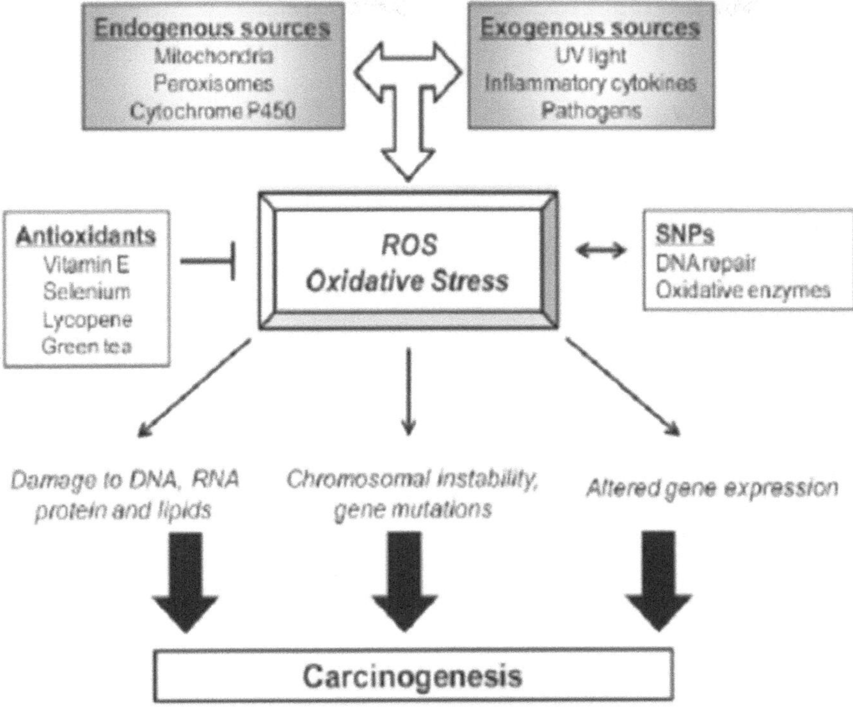

Fig. 2 Reactive oxygen species and their role in the development of human cancer

ROS also activate activating protein-1 (AP-1) and nuclear factor-kappa B (NF-κB) signal transduction pathways which in turn lead to the transcription of genes involved in cell growth regulatory pathways [43]. NF-κB pathway plays important roles in the control of cell proliferation, differentiation, apoptosis, stress response, cell signaling transduction, and other physiological processes. P53 is a gene whose disruption is associated with more than half of all human cancers [134]. The p53 protein guards a cell cycle checkpoint, and inactivation of p53 allows uncontrolled cell division.

Tumor promotion can be attributed to cellular proliferation, differentiation, and apoptosis, which could be secondary to alteration in the activity of tumor suppressor gene and oncogene products. The cumulative production of ROS through either endogenous or exogenous insults is common for many types of cancer cell that are linked with altered redox regulation of cellular signaling pathways. NF-κB is a nuclear transcription factor that was first identified by Sen and Baltimore in 1986. It is ubiquitously expressed and participates in a wide range of biological processes involved in cell survival, differentiation, inflammation, and growth [135]. This dimeric transcription factor is composed of different members of the Rel family, consisting of p50 (NF-κB1), p52 (NF-κB2), c-Rel, v-Rel, Rel A (p65), and Rel B [136]. Normally, NF-κB dimers are sequestered in the cytoplasm in an inactive state through binding to inhibitory IκB proteins. Activation of NF-κB occurs in response to a wide spectrum of extracellular stimuli that promote the dissociation of IκBs by sequential phosphorylation and proteolytic degradation, a process that depends on the IκB kinase (IKK) complex. Degradation of IκB proteins allows the entry of NF-κB into the nucleus where it binds κB-regulatory elements [137]. All NF-κB proteins are capable of homodimerization or heterodimerization with the other NF-κB proteins with the exception of RelB, which can only form heterodimers. Though most NF-κB dimer combinations result in the regulation of similar sets of

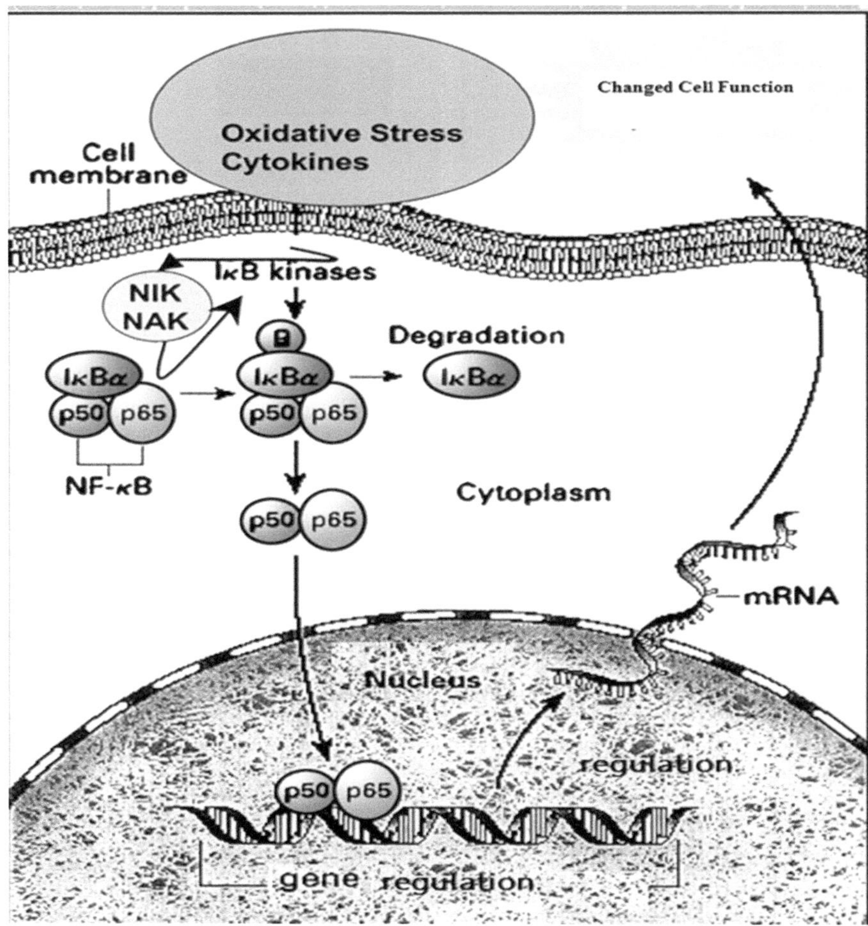

Fig. 3 Schematic diagram representing activation of NF-κB

genes, the ability to interact in various homo- and heterodimer configurations contributes to their ability to bind with varying affinities to κB sites in distinct DNA sequences, and they thus regulate unique, as well as overlapping, gene sets (Fig. 3).

The transcription of NF-κB-dependent genes influences the levels of ROS in the cell, and in turn, the levels of NF-κB activity are also regulated by the levels of ROS, but the exact molecular mechanism involved in this regulation is not known. NF-κB is a direct target for oxidation, which can affect its ability to bind to DNA [138]. Depending on the context, ROS can both activate and inhibit NF-κB signaling. A high degree of complexity characterizes ROS interactions with NF-κB pathways owing to the capability for ROS to act in many ways and at numerous places simultaneously. In addition, NF-κB activation has been linked to the carcinogenesis process through promotion of angiogenesis and tumor cell invasion and metastasis [135].

5 Stress Signals that Regulate p53 into the Route to Carcinogenesis

ROS, the principal mediators of oxidative stress, induce responses such as apoptosis or permanent growth arrest/senescence in normal cells. The p53 protein is a zinc-binding protein containing several reactive cysteines, and its key biochemical property sequence-specific DNA binding is dependent upon redox regulation in vitro [139]. The p53 plays an important role in protecting cells from tumorigenesis. It is categorized as a tumor suppressor because of its ability to halt the cell cycle or initiate apoptosis if cells are damaged beyond repair. As a tumor suppressor, p53 is essential for preventing inappropriate cell proliferation and maintaining genome integrity following genotoxic stress [140, 141]. p53 is activated and emerges as a pivotal regulatory protein which triggers diverse biological responses, both at the level of a single cell as well as in the whole organism [78, 139–141].

Mutational inactivation of p53 has been found to be involved in 50 % of human cancers, which indicates the importance of p53 in human carcinogenesis [142]. p53 is activated in response to a variety of stimuli, such as UV radiation, hypoxia, and nucleotide deprivation. Redox modification at a posttranslational level often occurs by reduction or oxidation of disulfide bonds. Because of this, p53 is considered one of the oxidative stress response transcription factors [143]. p53 activation involves an increase in overall p53 protein level as well as qualitative changes in the protein through extensive posttranslational modification, thus resulting in activation of p53-targeted genes [144]. The primary role of p53 in tumor suppression can be attributed to its ability to act as a sequence-specific transcription factor which regulates expression of different cellular genes to modulate various cellular processes [145], although protein–protein interactions may also play a role. In response to various types of stress, p53 is accumulated in the nucleus and binds to specific sites in the regulatory regions of p53-responsive genes and then strongly promotes the transcription of such genes [140]. The p53 downstream targets are differentially activated depending on the cell type, extent of the damage which has influenced p53 activation, and various other as yet unidentified parameters [146]. There is clear evidence on the role of p53 as a trans-activator or trans-repressor of genes involved in the production and control of ROS. Therefore, this redox sensitivity may be one of the biochemical mechanisms by which p53 acts as a sensor of multiple forms of stress [147]. Reactive species can increase p53 activity either indirectly by producing DNA damage or directly by promoting p53 phosphorylation. Angiogenesis and the development of metastasis are intrinsically connected. The cancer cell invasion is a critical point for cancer metastasis. The tumor microenvironment, which is highly heterogeneous in terms of nutrient supply, pH, and oxygenation, plays a major role in the ability of tumor cells to resist stress by altering gene expression and cellular functions of cancer cells [148, 149]. The presence of inflammatory cells, myofibroblasts, and endothelial cells in the tumor microenvironment supports not only the growth of tumor cells but also its ability to resist stress and therefore facilitates metastasis [148]. In this context, the growth factors and cytokines that facilitate communication between the tumor cells and the stromal compartment are particularly important. The process results from an imbalance between positive and negative angiogenic regulators released by both tumor cells and host cells [150]. Angiogenesis essentially involves endothelial cells and is regulated by several stimuli (hypoxia, growth factors, cytokines, and oxidative stress), which contribute to the modulation of vascular endothelial growth factor (VEGF), angiopoietin (ANG), EGF, fibroblast growth factor (FGF), transforming growth factor (TGF), platelet-derived growth factor (PDGF), and their receptors [151].

Among these angiogenic factors, VEGF is a potent angiogenic peptide [152, 153]. It stimulates angiogenesis and increases vascular permeability. It is a soluble dimeric 34–46 kDa protein which has been implicated in endothelial cell proliferation and migration [154, 155]. VEGF expression has been demonstrated in human malignancies including breast, lung, colorectal, bladder, ovarian, renal, and stomach cancer [156–158]. VEGF has been found to be upregulated by conditions associated with the generation of free radicals and reactive oxygen intermediates. Free radicals have been suggested as a possible cause of endothelial damage in a variety of situations including pathogenesis of atherosclerosis, reperfusion injury, endotoxic shock, cancer, aging, and several other conditions, including inflammatory diseases [159–162]. Oxidative stress is a potent factor in vascular cell proliferation and is considered to be involved in pathophysiology of cancers [160].

ROS may promote cell survival and proliferation, thus contributing to cancer development [163]. ROS can coordinate a variety of redox-sensitive transcription factors, such as NF-κB, Nrf2, HIF, and p53 [164]. A moderate increase of ROS often leads to NF-κB activation, with subsequent induction of genes encoding for proteins that inhibit apoptosis, including Bcl-xL, cellular inhibitors of apoptosis (cIAPs), and TNFR-associated factors (TRAF) [165]. NF-κB promotes cell survival mainly through the regulation of several oxidative stress-inducible genes, like those encoding for heme oxygenase-1, catalase, GPx, SOD, and TRX [166]. Under basal or low oxidative stress, p53 exerts antioxidant effects, thereby protecting cells from the accumulation of DNA damage and favoring repair mechanisms and thus allowing cell survival [167]. If the cell lacks p53, damaged DNA continues to replicate and is transmitted to the daughter cells. Apoptosis and cell cycle arrest may be reduced when p53 is mutated, allowing tumor development and progression. Increased generation of reactive oxygen species (ROS) and an altered redox status have long been observed in cancer cells, and recent studies suggest that this biochemical property of cancer cells can be exploited for therapeutic benefits. Cancer cells in advanced-stage tumors frequently exhibit multiple genetic alterations and high oxidative stress.

References

1. Sitkovsky M, Lukashev D (2005) Regulation of immune cells by local tissue oxygen tension: HIF-1α and adenosine receptors. Nat Rev Immunol 5:712–721
2. Sies H (1991) In: Sies H (ed) Oxidative stress, introduction in oxidative stress, oxidants and antioxidants. Academic Press, London
3. Sohal R (2002) Role of oxidative stress and protein oxidation in the aging process. Free Radic Biol Med 33:4433–4437
4. Halliwell B, Gutteridge J (1999) Free radicals in biology and medicine. Oxford University Press, New York
5. Buechter DD (1988) Free radicals and oxygen toxicity. Pharm Res 5:253–260
6. Scandalios JG (2002) The rise of ROS. Trends Biochem Sci 27:483–486
7. Klein JA, Ackerman SL (2003) Oxidative stress, cell cycle, and neurodegeneration. J Clin Invest 111:785–793
8. Fridovich I (1983) Superoxide radical: an endogenous toxicant. Annu Rev Pharmacol Toxicol 23:239–257
9. Halliwell B, Gutteridge JMC (1989) Free radicals in biology and medicine, 2nd edn. Clarendon Press, Oxford
10. Beckman JS, Chen J, Ischiropoulos H et al (1994) Oxidative chemistry of peroxynitrite. Methods Enzymol 233:229–240
11. Halliwell B, Gutteridge JMC (2007) Free radicals in biology and medicine, 4th edn. Oxford University Press, Oxford
12. Nohl H, Koziov AV, Gille L et al (2003) Cell respiration and formation of reactive oxygen species: facts and artifacts. Biochem Soc Trans 31:1308–1311
13. Ahsan H, Ali A, Ail R (2003) Oxygen free radicals and systemic autoimmunity. Clin Exp Immunol 31:398–404
14. Knight JA (1995) Diseases related to oxygen derived free radicals. Ann Clin Lab Sci 25:111–121
15. Waris R, Turkson J, Hassanein T et al (2005) Virus constitutively activates STAT-3 via oxidative stress: role of STAT-3 in HCV replication. J Virol 79:1569–1580
16. Aruoma OI, Halliwell B, Dizdaroglu M (1989) Iron ion-dependent modification of bases in DNA by the superoxide radical-generating system hypoxanthine/xanthine oxidase. J Biol Chem 264:13024–13028
17. Gotz ME, Kunig G, Reiderer Y (1994) Oxidative stress: free radical production in neuronal degeneration. Pharmacol Ther 63:37–122
18. Finkel T, Holbrook NJ (2000) Oxidants, oxidative stress and the biology of ageing. Nature 408:239–247
19. Costa V, Moradas-Ferreira P (2001) Oxidative stress and signal transduction in Saccharomyces cerevisiae: insights into ageing, apoptosis and diseases. Mol Asp Med 22:217–246
20. Remacle J, Raes M, Toussaint O et al (1995) Low levels of reactive oxygen species as modulators of cell function. Mutat Res 316:103–122
21. Droge W (2002) Free radicals in the physiological control of cell function. Physiol Rev 82:47–95
22. Poli G, Leonarduzzi G, Biasi F et al (2004) Oxidative stress and cell signaling. Curr Med Chem 11:1163–1182
23. Thannickal VJ, Fanburg BL (2000) Reactive oxygen species in cell signaling. Am J Physiol Lung Cell Mol Physiol 279:L1005–L1028
24. Schreck R, Baeuerl PA (1991) A role for oxygen radicals as second messengers. Trends Cell Biol 11:39–42
25. Finkel T (1998) Oxygen radicals and signaling. Curr Opin Cell Biol 10:248–253
26. Lowenstein CJ, Dinerman JL, Snyder SH (1994) Nitricoxide- a physiological messenger. Ann Intern Med 120:227–237

27. Storz P (2005) Reactive oxygen species in tumor progression. Front Biosci 10:1881–1896
28. Agarwal A, Gupta S, Sharma RK (2005) Role of oxidative stress in female reproduction. Reprod Biol Endocrinol 3:28
29. Irashad M, Chaudhuri PS (2002) Oxidant- antioxidant system: role and significance in human body. Ind J Exp Biol 40:1233–1239
30. Beal MF (2005) Mitochondria take centre stage in aging and neurodegeneration. Ann of Neurol 58:495–505
31. Schafer F, Buettner G (2001) Redox environment of the cell as viewed through the redox state of the glutathione disulfide/glutathione couple. Free Radical Bio Med 30:191–212
32. Halliwell B (1994) Free radicals, antioxidants, and human disease: curiosity, cause, or consequence? Lancet 344:721–724
33. Lu AL, Li X, Gu Y et al (2001) Repair of oxidative DNA damage: mechanisms and functions. Cell Biochem Biophys 35:141–170
34. Von Sonntag C (1987) New aspects in the free-radical chemistry of pyrimidine nucleobases. Free Radic Res Commun 2:217–224
35. Dizdaroglu M (1992) Oxidative damage to DNA in mammalian chromatin. Mutat Res 275:331–342
36. Demple B, Harrison L (1994) Repair of oxidative damage to DNA: enzymology and biology. Annu Rev Biochem 63:915–948
37. Fraga CG, Shigenaga MK, Park JW et al (1990) Ames BN Oxidative damage to DNA during aging: 8-hydroxy-2′-deoxyguanosine in rat organ DNA and urine. Proc Natl Acad Sci U S A 87:4533–4537
38. Guyton KZ, Kensler TW (1990) Oxidative mechanisms in carcinogenesis. Br Med Bull 49:523–544
39. Barber DA, Harris SR (1994) Oxygen free radicals and antioxidants: a review. Am Pharm NS 34:26–35
40. Marnett LJ (2000) Oxyradicals and DNA damage. Carcinogenesis 21:361–370
41. Cooke JP (2003) NO and angiogenesis. Atheroscler Suppl 4:53–60
42. Klaunig JE, Kamendulis LM (2004) The role of oxidative stress in carcinogenesis. Annu Rev Pharmacol Toxicol 44:239–267
43. Valko M, Rhodes CJ, Moncol J et al (2006) Free radicals, metals and antioxidants in oxidative stress-induced cancer. Chem Biol Interact 160:1–40
44. Cheng KC, Cahill DS, Kasai H et al (1992) 8-Hydroxyguanine, an abundant form of oxidative DNA damage, causes G-T and A-C substitutions. J Biol Chem 267:166–172
45. Miyake H, Hara I, Kamidono S et al (2004) Oxidative DNA damage in patients with prostate cancer and its response to treatment. J Urol 171:1533–1536
46. Weiss JM, Goode EL, Ladiges WC et al (2005) Polymorphic variation in hOGG1 and risk of cancer: a review of the functional and epidemiologic literature. Mol Carcinog 42:127–141
47. Diakowska D, Lewandowski A, Kopec W et al (2007) Oxidative DNA damage and total antioxidant status in serum of patients with esophageal squamous cell carcinoma. Hepatogastroenterology 54:1701–1704
48. Tanaka H, Fujita N, Sugimoto R et al (2008) Hepatic oxidative DNA damage is associated with increased risk for hepatocellular carcinoma in chronic hepatitis C. Br J Cancer 98:580–586
49. Gottschling BC, Maronpot RR, Hailey JR et al (2001) The role of oxidative stress in indium phosphide-induced lung carcinogenesis in rats. Toxicol Sci 64:28–40
50. Muguruma M, Unami A, Kanki M et al (2007) Possible involvement of oxidative stress in piperonyl butoxide induced hepatocarcinogenesis in rats. Toxicology 236:61–75
51. Shibutani S, Takeshita M, Grollman AP (1991) Insertion of specific bases during DNA synthesis past the oxidation-damaged base 8-oxodG. Nature 349:431–434
52. Grollman AP, Moriya M (1993) Mutagenesis by 8-oxoguanine: an enemy within. Trends Genet 9:246–249
53. Gao CM, Takezaki T, Wu JZ et al (2004) Polymorphisms in thymidylate synthase and methylenetetrahydrofolate reductase genes and the susceptibility to esophageal and stomach cancer with smoking. Asian Pac J Cancer Prev 5:133–138
54. Hwang ES, Bowen PE (2007) DNA damage, a biomarker of carcinogenesis: its measurement and modulation by diet and environment. Crit Rev Food Sci Nutr 47:27–50
55. Kawanishi S, Hiraku Y (2006) Oxidative and nitrative DNA damage as biomarker for carcinogenesis with special reference to inflammation. Antioxid Redox Signal 8:1047–1058
56. Lopaczynski W, Zeisel SH (2001) Antioxidants, programmed cell death, and cancer. Nutr Res 21:295–307
57. Feurgard C, Bayle D, Gue'zingar F et al (1998) Effects of ionizing radiation (neutrons/gamma rays) on plasma lipids and lipoproteins in rats. Radiat Res 150:43–51
58. De Zwart LL, Meerman JHN, Commandeur JNM et al (1999) Biomarkers of free radical damage applications in experimental animals and in humans. Free Radic Biol Med 26:202–226
59. Clarkson PM, Thompson HS (2000) Antioxidants: what role do they play in physical activity and health? Am J Clin Nutr 72:637S–646S
60. Zieba M, Suwalski M, Kwiatkowska S et al (2000) Comparison of hydrogen peroxide generation and the content of lipid peroxidation products in lung cancer tissue and pulmonary parenchyma. Respir Med 94:800–805
61. Urso ML, Clarkson PM (2003) Oxidative stress, exercise, and antioxidant supplementation. Toxicology 189:41–54
62. Morabito F, Cristani M, Saija A et al (2004) Lipid peroxidation and protein oxidation in patients affected by Hodgkin's lymphoma. Mediat Inflamm 13:381–383

63. Seljeskog E, Hervig T, Mansoor MA (2006) A novel HPLC method for the measurement of thiobarbituric acid reactive substances (TBARS). A comparison with a commercially available kit. Clin Biochem 39:947–954
64. Gago-Dominguez M, Castelao JE (2006) Lipid peroxidation and renal cell carcinoma: further supportive evidence and new mechanistic insights. Free Radic Biol Med 40:721–733
65. Seven A, Civelek S, Inci E et al (1999) Evaluation of oxidative stress parameters in blood of patients with laryngeal carcinoma. Clin Biochem 32:369–373
66. Davies MJ, Fu S, Wang H, Dean RT (1999) Stable markers of oxidant damage to proteins and their application in study of human disease. Free Radic Biol Med 27:1151–1161
67. Dean RT, Fu S, Stocker R et al (1997) Biochemistry and pathology of radical-mediated protein oxidation. Biochem J 324:1–18
68. Davies MJ (2005) The oxidative environment and protein damage. Biochim Biophys Acta 1703:93–109
69. Grune T, Merker K, Sandig G et al (2003) Selective degradation of oxidatively modified protein substrates by the proteasome. Biochem Biophys Res Commun 305:709–718
70. Ischiropoulos H (2003) Biological selectivity and functional aspects of protein tyrosine nitration. Biochem Biophys Res Commun 305:776–783
71. Requena JR, Levine RL, Stadtman ER (2003) Recent advances in the analysis of oxidized proteins. Amino Acids 25:221–226
72. Stadtman ER, Levine RL (2003) Free radical-mediated oxidation of free amino acids and amino acid residues in proteins. Amino Acids 25:207–218
73. Stadtman ER, Moskovitz J, Levine RL (2003) Oxidation of methionine residues of proteins: biological consequences. Antiox Redox Signal 5:577–582
74. Keck S, Nitsch R, Grune T et al (2003) Proteasome inhibition by paired helical filament-t in brains of patients with Alzheimer's disease. J Neurochem 85:115–122
75. Sitte N, Huber M, Grune T et al (2000) Proteasome inhibition by lipofuscin/ceroid during postmitotic ageing of fibroblasts. FASEB J 14:1490–1498
76. Dalle-Donne I, Giustarini D, Colombo R et al (2003) Protein carbonylation in human diseases. Trends Mol Med 9:169–176
77. Kettle AJ, Chan T, Osberg I et al (2004) Myeloperoxidase and protein oxidation in the airways of young children with cystic fibrosis. Am J Respir Crit Care Med 170:1317–1323
78. Levine AJ (1997) P53, the cellular gatekeeper for growth and division. Cell 88:323–331
79. Vertuani S, Angusti A, Manfredini S (2004) The antioxidants and proantioxidants network: an overview. Curr Pharm Design 10:1677–1694
80. Mates JM, Perez-Gomez C, De Castro IN (1999) Antioxidant enzymes and human diseases. Clin Biochem 32:595–603
81. McCall MR, Frei B (1999) Can antioxidant vitamins materially reduce oxidative damage in humans? Free Rad Biol Med 26:1034–1053
82. Sies H, Stahl W, Sevanian A (2005) Nutritional, dietary and postprandial oxidative stress. J Nutr 135:969–972
83. Landis GN, Tower J (2005) Superoxide dismutase evolution and life span regulation. Mech Ageing Dev 126:365–379
84. Behrend L, Henderson G, Zwacka RM (2003) Reactive oxygen species in oncogenic transformation. Biochem Soc Trans 31:1441–1444
85. Vainshtein G (1998) Three-dimensional structure of the enzyme catalase. Nature 293:411–412
86. Fita M, Albert G (1985) The active center of catalase. J Mol Biol 185:21–37
87. Jones DP, Eklow L, Thor H et al (1981) Metabolism of hydrogen peroxide in isolated hepatocytes: relative contributions of catalase and glutathione peroxidase in decomposition of endogenously generated H_2O_2. Arch Biochem Biophys 210:505–516
88. Kolanjiappan K, Manoharan S, Kayalvizhi M (2002) Measurement of erythrocyte lipids, lipid peroxidation, antioxidants and osmotic fragility in cervical cancer patients. Clin Chim Acta 326:143–149
89. Slater TF (1984) Free-radical mechanisms in tissue injury. Biochem J 222:1–15
90. Niki E, Yamamoto Y, Komuro E et al (1991) Membrane damage due to lipid oxidation. Am J Clin Nutr 53:201S–205S
91. Erhola M, Kellokumpu-Lehtinen P, Metsa-Ketela T et al (1996) Effects of anthracyclin-based chemotherapy on total plasma antioxidant capacity in small cell lung cancer patients. Free Radic Biol Med 21:383–390
92. Burton GW, Joyce A, Ingold KU (1983) Is vitamin E the only lipid-soluble, chain-breaking antioxidant in human blood plasma and erythrocyte membranes? Arch Biochem Biophys 221:281–290
93. Uotila JT, Kirkkola AL, Rorarius M et al (1994) The total peroxyl radical trapping ability of plasma and cerebrospinal fluid in normal and preeclamptic parturients. Free Radic Biol Med 16:581–590
94. Bartsch H, Nair U, Risch A et al (2000) Genetic polymorphism of CYP genes, alone or in combination, as a risk modifier of tobacco-related cancers. Cancer Epidemiol Biomarkers Prev 9:3–28
95. Frei B, England L, Ames BN (1989) Ascorbate is an outstanding antioxidant in human blood plasma. Proc Natl Acad Sci U S A 86:6377–6381
96. Duarte TL, Lunec J (2005) Review: when is an antioxidant not an antioxidant? A review of novel actions and reactions of vitamin C. Free Radic Res 39:671–686
97. Frei B (1991) Ascorbic acid protects lipids in human plasma and low-density lipoprotein against oxidative damage. Am J Clin Nutr 54:1113–1118
98. Niki E (1987) Interaction of ascorbate and alpha-tocopherol. Ann NY Acad Sci 498:186–199

99. Smith J, Steinemann TL (2000) Vitamin A deficiency and the eye. Int Ophthalmol Clin 40:83–91
100. Livrea MA, Tesoriere L (1998) Antioxidant activity of vitamin A within lipid environment. Subcell Biochem 30:113–143
101. Powers SK, Lennon SL (1999) Analysis of cellular responses to free radicals: focus on exercise and skeletal muscle. Proc Nutr Soc Med 58:1025–1033
102. May JM, Qu ZC, Whitesell RR et al (1996) Ascorbate recycling in human erythrocytes: role of GSH in reducing dehydroascorbate. Free Radic Biol Med 20:543–551
103. Groussard C, Rannou-Bekono F, Machefer G et al (2003) Changes in blood lipid peroxidation markers and antioxidants after a single sprint anaerobic exercise. Eur J Appl Physiol 89:14–20
104. Ho E, Ames BN (2002) Low intracellular zinc induces oxidative DNA damage, disrupts p53, NF-κB and AP-1 binding and affects DNA repair in a rat glioma cell line. Proc Natl Acad Sci U S A 99:16770–16775
105. Olin KL, Shigenaga MK, Ames BN et al (1993) Maternal dietary zinc influences DNA strand break and 8-hydroxy-2'-deoxyguanosine levels in infant rhesus monkey liver. Proc Soc Exp Biol Med 203:461–466
106. Taylor CG, Bettger WJ, Bray TM (1988) Effect of dietary zinc or copper deficiency on the primary free radical defense system in rats. J Nutr 118:613–621
107. Oteiza PI, Clegg MS, Zago MP et al (2000) Zinc deficiency induces oxidative stress and AP-1 activation in 3T3 cells. Free Radic Biol Med 28:1091–1099
108. Golub MS, Gershwin ME, Hurley LS et al (1985) Studies of marginal zinc deprivation in rhesus monkeys: infant behavior. Am J Clin Nutr 42:1229–1239
109. Federico A, Iodice P, Federico P et al (2000) Effects of selenium and zinc supplementation on nutritional status in patients with cancer of digestive tract. Eur J Clin Nutr 55:293–297
110. Daniel KG, Harbach RH, Guida WC et al (2004) Copper storage diseases: Menkes, Wilson's, and cancer. Front Biosci 9:2652–2662
111. Olivares M, Pizarro F, Speisky H et al (1998) Copper in infant nutrition: safety of World Health Organization provisional guideline value for copper content of drinking water. J Pediatr Gastroenterol Nutr 26:251–257
112. Coates RJ, Weiss NS, Daling JR et al (1989) Cancer risk in relation to serum copper levels. Cancer Res 49:4353–4356
113. Wu TJ, Sempos CT, Freudenheim JL et al (2004) Serum iron, copper and zinc concentrations and risk of cancer mortality in US adults. Ann Epidemiol 14:195–201
114. Trueba GP, Sanchez GM, Giuliani A (2004) Oxygen free radical and antioxidant defense mechanism in cancer. Front Biosci 9:2029–2044
115. Schrauzer GN (2000) Anticarcinogenic effects of selenium. Cell Mol Life Sci 57:1864–1873
116. Sinha R, Said TK, Medina D (1996) Organic and inorganic selenium compounds inhibit mouse mammary cell growth in vitro by different cellular pathways. Cancer Lett 107:277–284
117. Makropoulos V, Bruning T, Schulze-Osthoff K (1996) Selenium mediated inhibition of transcription factor NF-kappa B and HIV-1 LTR promoter activity. Arch Toxicol 70:277–283
118. Berg D, Gerlach M, Youdim MBH (2001) Brain iron pathways and their relevance to Parkinson's disease. J Neurochem 79:225–236
119. Siah CW, Trinder D, Olynyk JK (2005) Iron overload. Clin Chim Acta 358:24–36
120. Toyokuni S (1996) Iron-induced carcinogenesis: the role of redox regulation. Free Rad Biol Med 20:553–566
121. McCord JM (1998) Iron, free radicals, and oxidative injury. Sem Hematol 35:5–126
122. Sahu SC (1992) Dietary iron and cancer: a review. Environ Carcinog Ecotoxicol Revs 10:205–237
123. Hofmanová J, Machala M, Kozubík A (2004) Epigenetic mechanisms of the carcinogenic effects of xenobiotics and in vitro methods of their detection. Folia Biol (Praha) 46:165–173
124. Ames BN (1983) Dietary carcinogens and anticarcinogens: oxygen radicals and degenerative diseases. Science 221:1256–1264
125. Takeuchi T, Morimoto K (1993) Increased formation of 8-hydroxydeoxyguanosine, an oxidative DNA damage, in lymphoblasts from Fanconi's, anemia patients due to possible catalase deficiency. Carcinogenesis 14:1115–1120
126. Higinbotham KG, Rice JM, Diwan BA et al (1992) GGT to GTT transversions in codon 12 of the K-ras oncogene in rat renal sarcomas induced with nickel subsulfide or nickel subsulfide/iron are consistent with oxidative damage to DNA. Cancer Res 52:4747–4751
127. Lunec J, Holloway KA, Cooke MS et al (2002) Urinary 8-oxo-2-deoxyguanosine: redox regulation of DNA repair in vivo? Free Rad Biol Med 33:875–885
128. Moller P, Wallin H (1998) Adduct formation, mutagenesis and nucleotide excision repair of DNA damage produce by reactive oxygen species and lipid peroxidation product. Mutat Res 410:271–290
129. Wei H (1992) Activation of oncogenes and/or inactivation of antioncogenes by reactive oxygen species. Med Hypotheses 39:267–270
130. Bohr VA, Dianov GL (1999) Oxidative DNA damage processing in nuclear and mitochondrial DNA. Biochimie 81:155–160
131. Weitzman SA, Gordon LI (1990) Inflammation and cancer: role of phagocyte generated oxidants in carcinogenesis. Blood 76:655–663
132. Palmer HJ, Paulson KE (1997) Reactive oxygen species and antioxidants in signal transduction and gene expression. Nutr Rev 55:353–361
133. Chiarugi P, Fiaschi T (2007) Redox signaling in anchorage-dependent cell growth. Cell Signal 19:672–682
134. Sun Y, Oberley LW (1996) Redox regulation of transcriptional activators. Free Rad Biol Med 21:335–348

135. Sethi G, Sung B, Aggarwal BB (2008) Nuclear factor-kappa B activation: from bench to bedside. Exp Biol Med 233:21–31
136. Baeuerle PA, Baltimore D (1996) NF-kappa B: ten years after. Cell 87:13–20
137. Hacker H, Karin M (2006) Regulation and function of IKK and IKK-related kinases. Sci STKE 357:13
138. Pantano C, Reynaert NL, van der Vliet A et al (2006) Redox-sensitive kinases of the nuclear factor-κB signaling pathway. Antioxid Redox Signal 8:1791–1806
139. Ventura A, Kirsch DG, Mclaughlin ME et al (2007) Restoration of p53 function leads to tumour regression in vivo. Nature 445:661–665
140. Kern SE, Kinzler KW, Bruskin A et al (1991) Identification of p53 as a sequence-specific DNA binding protein. Science 252:1708–1711
141. Vogelstein B, Lane D, Levine AJ (2000) Surfing the p53 network. Nature 408:307–310
142. Harris CC, Hollstein M (1992) p53 tumor suppressor gene. Princ Pract Oncol 6:1–12
143. Han ES, Muller FL, Pérez VI et al (2008) The in vivo gene expression signature of oxidative stress. Physiol Genomics 34:112–126
144. Fritsche M, Haessler C, Brandner G (1993) Induction of nuclear accumulation of the tumor-suppressor protein p53 by DNA-damaging agents. Oncogene 8:307–318
145. Farmer G, Bargonetti J, Zhu H et al (1992) Wild-type p53 activates transcription in vitro. Nature 358:83–86
146. Oren M (2003) Decision making by p53: life, death and cancer. Cell Death Differ 10:431–442
147. Hainaut P, Mann K (2001) Zinc binding and redox control of p53 structure and function. Antioxid Redox Signal 3:611–623
148. Gatenby RA, Gillies RJ (2008) A microenvironmental model of carcinogenesis. Nat Rev 8:56–61
149. Smallbone K, Gatenby RA, Gillies RJ (2007) Metabolic changes during carcinogenesis: potential impact on invasiveness. J Theor Biol 244:703–713
150. Rak J, Filmus J, Finkenzeller G et al (1995) Oncogenes as inducers of tumor angiogenesis. Cancer Metastasis Rev 14:263–277
151. Carmeliet P, Jain RK (2000) Angiogenesis in cancer and other diseases. Nature 407:249–257
152. Bergers G, Benjamin LE (2003) Tumorigenesis and the angiogenic switch. Nat Rev Cancer 3:401–410
153. Pepper MS, Ferrara N, Orci L et al (1992) Potent synergism between vascular endothelial growth factor and basic fibroblast growth factor in the induction of angiogenesis in vitro. Biochem Biophys Res Commun 189:824–831
154. Connolly DT, Heuvelman DM, Nelson R et al (1989) Tumor vascular permeability factor stimulates endothelial cell growth and angiogenesis. J Clin Invest 84:1470–1478
155. Nicosia RF (1998) What is the role of vascular endothelial growth factor-related molecules in tumor angiogenesis? Am J Pathol 153:11–16
156. Gasparini G, Toi M, Gion M et al (1997) Prognostic significance of vascular endothelial growth factor protein in node-negative breast carcinoma. J Nat Cancer Inst 89:137–149
157. Cox G, Jones JL, Walker RA et al (2000) Angiogenesis and non-small cell lung cancer. Lung Cancer 27:81–100
158. O'Byrne KJ, Dobbs N, Propper D et al (1999) Vascular endothelial growth factor platelet counts and prognosis in renal cancer. Lancet 353:1494–1495
159. Yang W, de Bono DP (1997) A new role for vascular endothelial growth factor and fibroblast growth factors: increasing endothelial resistance to oxidative stress. FEBS Lett 403:139–142
160. Wink DA, Vodovotz Y, Laval J et al (1998) The multifaceted roles of nitric oxide in cancer. Carcinogens 19:711–721
161. Das UN (2002) A radical approach to cancer. Med Sci Monit 8:79–92
162. Bartoli GM, Galeotti T (1979) Growth- related lipid peroxidation in tumour microsomal membranes, and mitochondria. Biochim Biophys Acta 574:537–541
163. Burdon RH (1995) Superoxide and hydrogen peroxide in relation to mammalian cell proliferation. Free Radic Biol Med 18:775–794
164. Trachootham D, Lu W, Ogasawara MA et al (2008) Redox regulation of cell survival. Antioxid Redox Signal 10:1343–1374
165. Karin M, Lin A (2002) NF-κB at the crossroads of life and death. Nat Immunol 3:221–227
166. Ishii T, Itoh K, Takahashi S et al (2000) Transcription factor Nrf2 coordinately regulates a group of oxidative stress-inducible genes in macrophages. J Biol Chem 275:16023–16029
167. Sablina AA, Budanov AV, Ilyinskaya GV et al (2005) The antioxidant functions of the p53 tumor suppressor. Nat Med 11:1306–1313

Interplay Among Bacterial Resistance, Biofilm Formation and Oxidative Stress for Nosocomial Infections

Reema Gabrani, Garima Sharma, Shweta Dang, and Sanjay Gupta

Abstract

Nosocomial infections are leading threat as 5–10 % of hospitalised patients ensue to approximately 90,000 deaths per year. Implanted medical devices and procedures accrue to higher rates of infection and add to considerable socio-economic burden. The problem gets compounded by the increased risk of biofilm formation on indwelling medical devices. Bacteria within biofilm are much more resistant to antibiotic treatment as compared to planktonic cultures. Surface adhesion molecules keep the bacteria tethered to the surface and molecular changes within bacteria and its complex structure contribute towards development of resistance. Reactive oxygen species (ROS) are the last product of various metabolic pathways of bacterial cells which help the bacteria in the development of biofilm and antibiotic resistance. Certain bactericidal drugs have shown bacterial killing by internal production of ROS. Bacteria mediate SOS repair response to ROS which can introduce mutations in their genome leading to development of resistance. Thus, ROS plays an important role in the generation of resistance in bacterial biofilm towards antibiotics.

Keywords

Antibiotics • *P. aeruginosa* • Quorum sensing • Reactive oxygen species • *S. epidermidis* • Therapeutics

1 Introduction

Nosocomial infection or hospital-acquired infection (HAI) refers to the infection occurring in patients during hospital visit favoured by its environment. Nosocomial infections are mainly caused by bacterial, fungal and viral pathogens enhanced

R. Gabrani (✉) • G. Sharma • S. Dang • S. Gupta
Department of Biotechnology, Jaypee Institute of Information Technology, A-10, Sector-62, Noida 201307, Uttar Pradesh, India
e-mail: reema.gabrani@jiit.ac.in

V. Rani and U.C.S. Yadav (eds.), *Free Radicals in Human Health and Disease*,
DOI 10.1007/978-81-322-2035-0_23,

by the use of instrumentation or devices for incubation, delivery of therapeutic agents or drainage of body fluids during patient care as supportive measures. These nosocomial infections occur among 7–12 % of the hospitalised patients globally with more than 1.4 million people suffering from the infectious complications acquired in the hospital [1].

Bacterial nosocomial infections can be deadly at times especially in immunocompromised patients. Advancement in medical sciences has led to increased use of invasive devices that have also resulted in high rates of infection [2]. According to International nosocomial infection control consortium (2010) both gram-positive (*Staphylococcus* spp.) and gram-negative bacteria (*Pseudomonas*, *Klebsiella* spp. *Escherichia coli*) could be deadly and acquire antibiotic resistance. Gram-positive bacteria are the most frequent reasons for hospital-acquired infection, and currently, coagulase-negative staphylococci (CoNS) especially *Staphylococcus epidermidis* has been shown to contribute significantly to nosocomial bacteraemia. Staphylococcus species cause a wide variety of lung, bone, heart and bloodstream infections and are frequently resistant to antibiotics. Beta-haemolytic streptococci form a part of mouth, skin, intestine and upper respiratory tract and thus in cases of infection can cause complete haemolysis of red blood cells. The gram-negative bacteria Enterobacteriaceae (e.g. *Escherichia coli*, *Proteus*, *Klebsiella*, *Enterobacter*, *Serratia marcescens*) may colonise sites when the host defences are compromised (catheter insertion, bladder catheter, cannula insertion) and cause serious infections [3–5].

Rampant usage of antibiotics has resulted in drug-resistant strains of bacteria which are the leading causes of nosocomial infection. The majority (over 70 %) of bacterial pathogens that cause fatal infections are likely to be resistant to at least one of the drugs commonly used in the treatment for bacterial infections [6, 7]. Moreover, it has been reported that certain bacteria thrive in our system as biofilm and become increasingly resistant to antibiotic treatment. It has been commonly observed that in medical device-related infections, bacteria adhere to the surface of devices conditioned by biological fluids and develop into biofilm. Biofilm are much more challenging to treat as they show increased resistance to bacteria. It has been postulated that oxidative stress results in the production of reactive oxygen species (ROS) and reactive nitrogen intermediates (RNI) which changes the atmospheric conditions for biofilm. Drug-induced oxidative stress can lead to SOS response in bacterial cells which can contribute towards resistance to antibiotic. Thus, oxidative stress can mediate the bacterial pathogenesis by affecting antibiotic resistance and biofilm formation.

2 Biofilm

Biofilm is formed when the bacterial community attaches to the surface of adherent and forms a polymeric matrix. Bacterial biofilm formation and maintenance is highly dependent on the extrapolymeric substances secreted by these structured cells. These substances differ with respect to bacteria and alter the surrounding localised environment. The matrices formed by different bacteria are very diverse in nature. These differences can result due to small changes in the composition of extracellular matrix.

Different stages in biofilm formation have been characterised and certain common features across the bacterial species are initiation, accumulation, maturation and biofilm detachment and dispersal (Fig. 1). Biofilm initiation involves the direct attachment of cells on abiotic surface or interface by electrostatic and hydrophobic interaction and also cell-cell interaction. The second stage is accumulation of bacterial cells. Once bacterial cell gets adhered to the solid surface, it starts further accumulating through intercellular aggregation and finally results in a multilayered biofilm. During biofilm formation, fluid-filled channels are formed and the biofilm gets a specific 3D structure. Extracellular polysaccharide, protein and recently studied extracellular DNA constitute the extracellular polymeric substance matrix. Once a mature biofilm gets established, individual cells or cell aggregates get released. Many factors like mechanical force, nutrient depletion, waste products and pH changes are responsible for disassembly of bacterial biofilm.

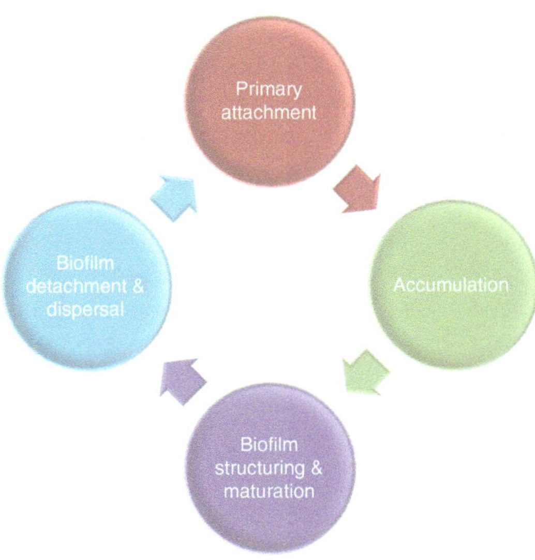

Fig. 1 Various steps involved in the biofilm development

Biofilm cluster detachment in turn results in systemic spread of biofilm fragments [8–10]. The cells within the biofilm are more resistant to antibiotics than planktonic cells. The reasons attributed to this resistance could be low growth rate of cells, expression of certain resistant genes and inability of antibiotics to penetrate the biofilm. Persister cells which survive the antimicrobial treatment have been shown to specifically contribute towards recurrent infections attributed to biofilm. Biofilm formation is controlled by several genes and factors in bacterial cells. Here we have discussed biofilm formation and genetic structure of two model organisms, *Pseudomonas aeruginosa* for gram-negative bacteria and *S. epidermidis* for gram-positive bacteria.

2.1 *P. aeruginosa* Biofilm Formation

P. aeruginosa biofilm formation is mainly influenced by small signals known as quorum sensors which are produced and secreted by bacteria. Quorum-sensing (QS) signals which include N-acylhomoserine lactones (AHL) and the 4-quinolones act as auto-inducers for biofilm differentiation. Various quinolones have been studied till date, out of which Pseudomonas quinolone signal (2-heptyl-3-hydroxy-4-quinolone) is the most studied. Quorum sensing is regulated by two main pathways, las and rhl. The las system includes LasR and LasI genes and rhl system encompasses RhlR and Rhl1 genes. LasR and RhlR genes encode transcription activator proteins and LasI and Rhl1, auto-inducer synthase genes, help in the synthesis of N-(3-oxododecanoyl) homoserine lactone (3OC12-HSL) and N-butyryl-homoserine lactone, respectively. The Las signalling system regulates the expression of extracellular toxins which help in *P. aeruginosa* virulence. Both las and rhl system help in the maturation and differentiation of biofilm [11].

The matrix rich in secreted extracellular polymeric substances (EPS) consists of Pel, Psl and Alginate polysaccharides, proteins, lipids and extracellular DNA (eDNA). *P. aeruginosa* exopolysaccharides, alginate and Psl are encoded by 12 gene operon each from PA3540 to PA3551 (on PAO1 genome) and PA2231 to PA2242, respectively. Pel is encoded by a seven gene operon from PA3058 to PA3064. The polysaccharides formed by three separate gene clusters provide mesh-like matrix which connects the biofilm cells and helps in pellicle formation and gives distinct wrinkled appearance. Another component, eDNA, plays an important role in the initial stages of biofilm formation by holding the cells together. The cells growing in these "slime matrix" are resistant to antibiotics (e.g. aminoglycosides, ß-lactam antibiotics, fluoroquinolones) [12].

Recently, it has been shown that phenazine, electrochemically active compounds, controlled by QS system, regulates the production of pyocyanin which interacts with molecular oxygen. It results in production of ROS which can lead to cellular injury and release of DNA. Subsequent studies reported eDNA in biofilms formed by *P. aeruginosa* results from lysis of small population of cells. Lysis and release of eDNA are gulated by two different pathways. One pathway is linked to quorum sensing and is responsible for relarge amount of eDNA release and second pathway is quorum-sensing independent and releases basal level of eDNA. It was experimentally observed that biofilms formed by QS mutants with decreased eDNA levels were more susceptible to SDS treatment indicating that eDNA plays a crucial role in stabilising the biofilm matrix.

In summary, the process of *P. aeruginosa* biofilm development involves the attachment of planktonic cells to a solid surface which forms microcolonies. QS-regulated rhamnolipid production aids in microcolony formation. Cells migrate and spread over substratum, resulting into a flat, uniform mat; the microcolonies later grow forming stalk and mushroom-like structure. Rhamnolipids are responsible for maintaining open channels and mushroom cap formation. EPS matrix is produced aided by eDNA release and Pel polysaccharide production which are under QS control. Cells disperse from biofilm with the help of rhamnolipid during various stages of biofilm maturation and can resume planktonic mode of growth [13–19].

2.2 *S. epidermidis* Biofilm Formation

The initial attachment of the planktonic cells to the surface is controlled by various physiochemical features like hydrophobic interactions and van der Waals forces. The staphylococcal surface proteins, SSP-1 and SSP-2, and the major autolysin, AtlE protein help in attachment to the surface apart from other identified proteins like the fibrinogen-binding Fbe and SdrG, SdrF and SdrH cell-surface-associated proteins. Another nonprotein component teichoic acid binds to fibronectin and mediates binding of *S. epidermidis* to the surface. The cells start piling on each other and extracellular molecules especially polysaccharide intercellular adhesin (PIA) and the biofilm associated proteins – Aap (accumulation-associated protein) and Bhp (Bap homologue protein) play a crucial role in the development phase of biofilm. PIA is involved in the haemagglutinating activity and is encoded by enzymes synthesised by intercellular adhesion (ica) operon. The *ica* operon has one regulatory gene *icaR* and four structural proteins are encoded by *ica*ADBC.

The mature form of biofilm has mushroom-like structure and fluid-filled channels that help in delivery of nutrients and oxygen to all cells in the biofilm and also enable the removal of metabolic waste. The mature form has gelatinous matrix, glycocalyx, which aids the stability and robustness of biofilm. This slimy layer plays an important role in imparting resistance to antibiotics. Finally, the bacterial cells can detach from biofilm and disseminate and colonise in distant areas. Biofilm formation is facilitated by the luxSQS system and the accessory gene regulator (agr) system. QS system helps the biofilm to adapt under the altered condition of growth and nutrient requirement. It regulates gene expression and coordinates the functioning of well-organised bacterial community [27].

These sessile communities make cells more resistant to conventional antibiotics. There are many hypotheses that relate the biofilm resistance nature to conventional therapies. One states that antibiotics are unable to penetrate bacterial biofilm; the second indicates towards the nutritional limitation of biofilms so they can survive in starvation condition and the third mechanism is the development of protective biofilm phenotype [8, 28]. The various genes required for biofilm formation for selected bacteria are listed in Table 1.

3 Bacterial Resistance

Antibiotics have been used very effectively to combat the infections caused by bacteria. Major antibiotic mechanisms are listed in Table 2. The development of resistance towards antibiotics has been a major hurdle in treatment of infections caused by these bacteria. Many factors responsible for the bacterial resistance against the antibiotics are discussed.

3.1 Antibiotic Inactivation

Bacterial cells are able to produce several enzymes which are hydrolytic in nature. The classical hydrolytic amidases are the β-lactamases that cleave the β-lactam ring of the penicillin and cephalosporin antibiotics. These enzymes act on the functional site of antibiotic and make it nonfunctional. Many gram-negative and gram-positive bacteria produce such enzymes, and more than 200 different types of β-lactamase

Table 1 Key genetic players in biofilm formation for selected organisms

S. No.	Organism	Genes for biofilm formation	Functions	Reference
1	*E. coli*	*omp*, *rcs*, *slp*	Initial steps of *E. coli* biofilm formation on abiotic surfaces	[20]
2	*S. mutans*	*comC*, *comD*, *comE* and *comX*	Initiation of biofilm formation	[21]
3	*Campylobacter jejuni*	*cdtB*, *cst-II*, *ggt* and *virB11*	Adherence, invasion, colonisation, molecular mimicry and cytotoxin production	[22]
4	*P. aeruginosa*	*alg*, *pel*, *psl*	Extracellular matrix production for biofilm formation	[23]
5	*S. epidermidis*	*icaA*, *icaD*, *icaB*, *icaC*, *aap*, *bhp*	For production of full slime synthesis, biofilm accumulation	[24]
6	*Enterococcus faecalis*	*esp*, *fsr*, *gelE*, *sprE*	Genes are closely related to the biofilm formation	[25]
7	*Listeria monocytogenes*	*sigB*, *lbrA*	Biofilm synthesis	[26]

Table 2 Various mode of action of conventional antibiotics

S. No.	Antibiotic action	Examples	Mode of action	Reference
1	Inhibition of cell wall synthesis	Penicillins Cephalosporins Telavancin	Mainly inhibit peptidoglycan biosynthesis required for cell wall synthesis	[29, 30]
2	Inhibition of protein synthesis	Macrolides Chloramphenicol Glycylglycines	Bind with the 30S or 50S subunit of ribosome to inhibit protein synthesis	[29, 31]
3	Interference with nucleic acid synthesis	Rifampicin Quinolones	Block the essential enzymes like DNA-directed RNA polymerase, type II topoisomerases DNA gyrase and topoisomerase IV for DNA synthesis	[29]
4	Inhibition of a metabolic pathway	Sulfonamides Trimethoprim	Inhibit the main metabolic activities like folate synthesis	[29]
5	Disorganisation of the cell membrane	Polymyxins Daptomycin	Increase bacterial membrane permeability by efflux of potassium from the bacterial cell	[32, 33]

have been identified. β-lactamases are classified into four groups on the basis of functional characteristics, including preferred antibiotic substrate. Extended-spectrum β-lactamases (ESBLs) mediate resistance to all penicillins, third-generation cephalosporins (e.g. ceftazidime, cefotaxime, ceftriaxone) and aztreonam, but not cephamycins (cefoxitin and cefotetan) and carbapenems. ESBLs are very diverse: more than 180 different ESBLs have been identified. They are most commonly detected in *Escherichia coli*, *Klebsiella pneumonia* and *Proteus mirabilis*, but have also been found in other *Enterobacteriaceae*. Other hydrolytic enzyme examples include esterases that have been linked to macrolide antibiotic resistance and ring-opening epoxidases causing resistance to fosfomycin [34–37].

3.2 Antibiotic Inactivation by Group Transfer

The most diverse family of resistant enzymes is the group of transferases. These enzymes inactivate antibiotics (aminoglycosides, chloramphenicol, streptogramin, macrolides or rifampicin) by chemical substitution (adenylyl, phosphoryl or acetyl groups are added to the periphery of the antibiotic molecule). The modified antibiotics are affected in their binding to a target. Chemical strategies include *O*-acetylation and *N*-acetylation, *O*-phosphorylation, *O*-nucleotidylation, *O*-ribosylation, *O*-glycosylation and thiol transfer. These all covalent modification strategies require a co-substrate for their activity (ATP, acetyl-CoA, NAD+, UDP-glucose or glutathione), and consequently these processes are restricted to the cytoplasm [38–40].

3.3 Antibiotic Inactivation by Redox Process

The oxidation or reduction of antibiotics has been frequently exploited by pathogenic bacteria. Few examples include the oxidation of tetracycline antibiotics by the TetXenzyme; *Streptomyces virginiae*, producer of the type A streptogramin antibiotic virginiamycin M1, protects itself from its own antibiotic by reducing a critical ketone group to an alcohol [37, 38].

3.4 Target Modification

Target modification is the second major cause of antibiotic resistance. Due to alteration in the binding site (target), the antibiotics are unable to bind and hence result in resistance. *MecA* gene-encoded Penicillin- binding protein 2a (PBP2a) transpeptidase in *S. aureus* is responsible for developing methicillin resistance (methicillin-resistant *S. aureus*, MRSA). Penicillin-binding proteins catalyse the transglycosylase and transpeptidase activities necessary for cell wall synthesis. In the presence of β-lactam antibiotic, the transpeptidase activity is lost, whereas PBP2A of MRSA is resistant to β-lactam acylation and can catalyse the transpeptidation necessary for cell wall synthesis [39–41].

3.5 Peptidoglycan Structure Alteration

Peptidoglycan is the one of the essential components of cell wall of bacterial cell which remains a major target for many of antibiotics. The presence of mutations in the penicillin-binding domain of penicillin-binding proteins (PBPs) results in decreased affinity to β-lactam antibiotics. Vancomycin (glycopeptide) inhibits cell wall synthesis of gram-positive bacteria by binding C-terminal acyl-D-alanyl-D-alanine (acyl-D-Ala-D-Ala)-containing residues in peptidoglycan precursor. Resistance is achieved by altering the target site by changing the D-Ala-D-Ala to D-alanyl-D-lactate (D-Ala-D-Lac) or D-alanyl-D-serine (D-Ala-D-Ser) at the C-terminus, which inhibits the binding of vancomycin, and thus the resistance is developed [42–44].

3.6 Protein Synthesis Interference

Mutations in 23S rRNA close to the sites of methylation have also been associated with resistance to the macrolide group of antibiotics in a range of organisms. In addition to multiple mutations in the 23S rRNA, alterations in the L4 and L22 proteins of the 50S subunit have been reported in macrolide-resistant *S. pneumonia*. Mutations in the 16S rRNA gene confer resistance to the aminoglycosides [45–47].

3.7 DNA Synthesis Interference

Some antibiotics like fluoroquinolones interact with the DNA gyrase and topoisomerase IV enzymes and prevent DNA replication and transcription. Resistance is conferred by mutations in specific regions of the structural genes that sufficiently alter these enzymes preventing the binding of antibiotics. The most common mutations in this region cause resistance through decreased drug affinity for the altered gyrase–DNA complex [48, 49].

3.8 Efflux Pumps and Outer Membrane (OM) Permeability

The efflux pumps are the membrane proteins which play a role in exporting out the proteins and maintenance of intracellular concentrations of the cell. Mutation in MexAB-OprM efflux pump of *P. aeruginosa* provides resistance against β-lactams antibiotics. Gram-positive as well as gram-negative bacteria can have either single-drug and/or multiple-drug efflux pumps [50–57].

4 ROS in Antibiotic Resistance and Biofilm

Biofilm is affected by several factors like temperature, pH and medium composition. Genetic variation in bacterial cells thriving in biofilm can help them to adapt to different environmental conditions. It is now being reported that oxidative stress influences the diversity in biofilm formation. Oxidative stress induced inside biofilm results in generation of ROS, including superoxide (O_2•–), hydrogen peroxide (H_2O_2), hydroxyl radical (OH•), peroxyl radical (ROO•) and singlet oxygen ($1O_2$). These oxidants could be produced due to aerobic bacterial respiration though limited in biofilm but sufficient to produce oxidants. The other reasons could be reduced production of antioxidants, alteration of electron transport chain or induction by metabolites such as phenazine.

Bacteria are also subjected to oxidative stress response inside the phagocytic cell as a part of innate immunity to combat the infection. ROS such as H_2O_2 and hydroxyl radical help in containing the infection. Bacteria generally have comprehensive defence system encoded by genes like superoxide dismutases (sod) and catalases (kat) that can scavenge reactive oxygen, stress regulators like perR and sigB and recA encoding DNA repair genes [58].

In *S. epidermidis*, oxidation-sensing regulator, AbfR (aggregation and biofilm formation regulator) has been identified. Oxidative stress such as H_2O_2 mediates the up-regulation of *abfR* gene which not only controls response to oxidative stimuli but also helps in bacterial aggregation and biofilm formation. OxyR, PerR and OhrR are three well-studied sensing transcriptional regulators for peroxide. Out of these three, OxyR and PerR respond to H_2O_2, whereas OhrR senses organic peroxide. OhrR is a member of MarR family proteins which is a widely distributed protein family of bacteria and archaea. MarR family proteins MgrA, SarZ and SarA of some gram-positive bacteria *Staphylococcus aureus* have also been reported to utilise cysteine oxidation to sense the oxidative stress which regulates antibiotic resistance and virulence [59].

A new oxidative stress-sensing and response *ospR* (oxidative stress response and pigment production Regulator) gene has been identified in *P. aeruginosa*. *OspR* has been shown to sense oxidative stress and regulate β-lactam resistance and genes involved in quorum sensing and tyrosine metabolism. A highly conserved Cys residue, Cys-24, has been shown to be responsible for sensing oxidative stress [60].

Bactericidal antibiotics like quinolones, aminoglycosides and β-lactams induce production of hydroxyl radicals which help in the killing of both gram-negative and gram-positive bacteria. The bacteria compensate by increasing the expression of TCA-cycle genes which lead to enhanced NADH consumption followed by burst in superoxide production. ROS has the capability to damage nucleic acids, proteins and cell membrane (Fig. 2). The breaks in the double-stranded DNA activate DNA repair by recombinatorial DNA repair genes. SOS activates the expression of error-prone polymerases which may carry out the repair via nucleotide excision, base excision or recombination pathways. Error-prone repair results in hypermutability which helps the cells to adapt to altered conditions and generates diversity. This generation of diversity aids in survival of antibiotic-resistant strains and poses fresh challenges in combating the infection [61].

5 Therapeutics

Due to rampant use of antibiotics, bacteria develop resistance towards antibiotics and nosocomial infection further gets complicated by the formation of biofilm for which treatment is increasingly becoming difficult. So, researchers are focusing towards the identification and discovery of novel therapeutics for treating nosocomial infection. About 350 genes mainly play a

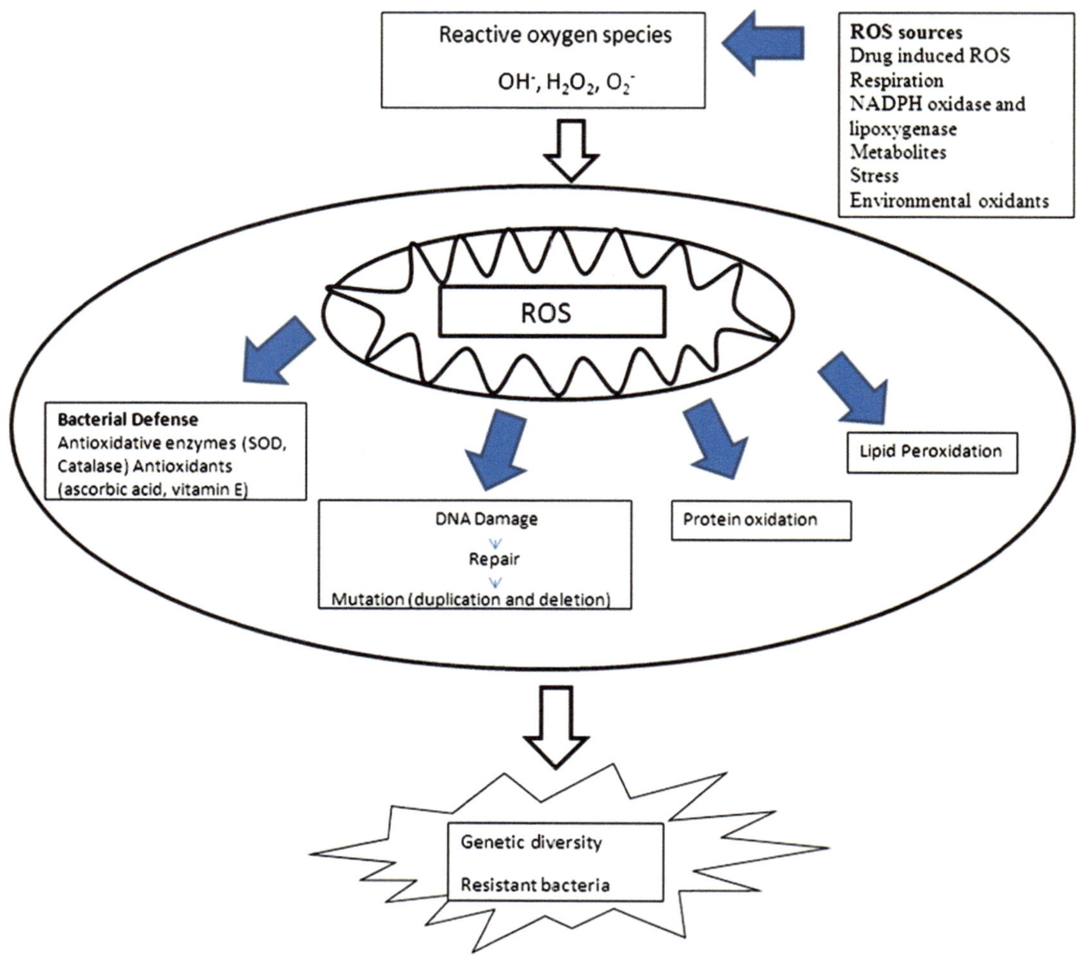

Fig. 2 Role of ROS in biofilm formation

role in quorum-sensing system, designing of agonist against these QS system helps in the disruption of biofilm at a very nascent stage. Both natural and synthetic compounds are being explored to treat bacterial infections. Various medicinal plants and their active compounds like Baicalin, isothiocyanate rich plant named Horseradish and ajoene from garlic have recently been reported as QS inhibitors [62, 63]. Various molecules involved in biofilm formation are potential targets for generation of monoclonal antibodies (mAb). MAb have been developed against accumulation-associated protein (AAP) expressed on S. *epidermidis* cell wall which helps in the adhesion and accumulation of bacterial cells as biofilm [64]. MAb have also been identified against different epitopes of Psl which is required for the formation and maintenance of biofilm in *P. aeruginosa* [65] and screened for their efficiency. Enzymes like DNase I and DNase 1L2 exhibit potential antibiofilm activity against eDNA which helps in biofilm formation [66]. Bacteriophages have the ability to infect bacterial cells and PAK-P3 and P3-CHA bacteriophages from Myoviridae family have been reported for degradation of *P. aeruginosa* polysaccharides [67]. Photodynamic therapy (PDT), a non-antibiotic therapy using photosensitising compounds which gets activated by low power laser, has been shown to be effective against gram-positive and gram-negative resistant bacteria [68]. Nanoparticles because of controlled and sustained release and effective killing of microorganism are being explored as novel delivery system against biofilm and ROS system. Small antimicrobial peptides, Peptoids 1 and 1-C13, LL-37, other cathelicidins of bovine origin (BMAP-27, BMAP-28) and the artificial peptide P19 (9/B), have also exhibited potential antibiofilm activity against *P. aeruginosa* [69].

6 Conclusion

Nosocomial infections are the most common infections during hospital visit. The most common types of nosocomial infections are surgical wound infections, respiratory infections, genitourinary infections, as well as gastrointestinal infections. These infections are often caused by breaches of infection control practices and procedures, unclean and non-sterile environmental surfaces and/or ill employees. Causative microorganisms usually attach to surfaces of abiotic substances and produce extracellular polysaccharides, resulting in the formation of a biofilm. Biofilms are becoming a serious threat for public health because of the increase of their resilient nature to treatment and development of resistance and alteration in genetic structure which incurs the biofilm more resistant to treatment. Though the first antibiotic was discovered in the 1940s, but till date the infectious disease due to developing resistance nature of bacteria is a surmountable challenge. The frequent as well as overuse of antibiotics leads to antibiotic resistance among bacteria and to problems in treatment. Increasingly, oxidative stress has also been shown to play an important role in biofilm formation. Oxidative stress leads to the generation of various reactive oxygen species (ROS), including superoxide (O_2•–), hydrogen peroxide (H_2O_2), hydroxyl radical (OH•), peroxyl radical (ROO•) and singlet oxygen ($1O_2$) which can cause DNA damage. Repair machinery can introduce the mutations in DNA which further alters the genetic makeup of cell and can result in resistant bacteria. The identification of genes and proteins involved in the process can help to develop the new therapeutic strategies to combat the infection, especially due to resistant and biofilm forming bacteria.

References

1. Kramer A, Schwebke I, Kampf G (2006) How long do nosocomial pathogens persist on inanimate surfaces? A systematic review. BMC Infect Dis 6:130. doi:10.1186/1471-2334-6-130
2. Inweregbu JD, Pittard A (2005) Nosocomial infections. Continuing education in Anaesthesia. Crit Care & Pain 5:14–17
3. Hsueh P-R, Chen W-H, Luh K-T (2006) Relationships between antimicrobial use and antimicrobial resistance in Gram-negative bacteria causing nosocomial infections from 1991–2003 at a university hospital in Taiwan. Int J Antimicrob Agents 26:463–472
4. Robert C, Welliver MD, McLaughlin S (1984) Unique epidemiology of nosocomial infection in a children's hospital. Am J Dis Child 138(2):131–135
5. Robert A, Weinstein RG et al (2005) Overview of nosocomial infections caused by gram-negative bacilli. Clin Infect Dis 41(6):848–854
6. Harold C (1992) The crisis in antibiotic resistance. Science 257:1064–1073
7. Philip S, Stewart J, Costerton JW (2001) Antibiotic resistance of bacteria in biofilms. Lancet 358:135–138
8. Merle E, Olson, Ceri H, Douglas WM (2002) Biofilm bacteria: formation and comparative susceptibility to antibiotics. Can J Vet Res 66(2):86–92
9. Beloin C, Ghigo JM (2005) Finding gene-expression patterns in bacterial biofilms. Trends Microbiol 13(1):16–19
10. Pratt LA, Roberto K (1999) Genetic analyses of bacterial biofilm formation. Curr Opin Microbiol 2:598–603
11. Sharma G, Rao S, Bansal A et al (2014) *Pseudomonas aeruginosa* biofilm: potential therapeutic targets. Biologicals 42:1–7
12. Govan JR, Deretic V (1996) Microbial pathogenesis in cystic fibrosis: mucoid *Pseudomonas aeruginosa* and Burkolderia cepacia. Microbiol Rev 60:539–574
13. Musken M, Di Fiore S, Dotsch A et al (2006) Genetic determinants of *Pseudomonas aeruginosa* biofilm establishment. Microbiology 156:431–441
14. McKnight SL, Iglewski BH, Pesci EC (2000) The Pseudomonas quinolone signal regulates rhl quorum sensing in *Pseudomonas aeruginosa*. J Bacteriol 182:2702–2708
15. Cao H, Krishnan G, Goumnerov B et al (2000) A quorum sensing-associated virulence gene of *Pseudomonas aeruginosa* encodes a LysR-like transcription regulator with a unique self-regulatory mechanism. J Sci Proc Natl Acad Sci U S A 98:14613–14618
16. Diggle SP, Cornelis P, Williams P et al (2006) 4-Quinolone signalling in *Pseudomonas aeruginosa*: old molecules, new perspectives. J Sci Med Microbiol 296:83–91
17. Pearson JP, Gray KM, Passador L et al (1994) Structure of autoinducer required for expression of *Pseudomonas aeruginosa* virulence genes. Proc Natl Acad Sci U S A 91:197–201

18. Gambello MJ, Kaye S, Iglewski BH (1993) LasR of *Pseudomonas aeruginosa* is a transcriptional activator of alkaline protease gene (apr) and an enhancer of exotoxin A expression. Infect Immun 61:1180–1184
19. Friedman L, Kolter R (2004) Genes involved in matrix formation in *Pseudomonas aeruginosa* PA14 biofilm. J Sci Mol Microbiol 51:675–690
20. Prigent-Combaret C, Vidal O, Dorel C et al (1999) Abiotic surface sensing and biofilm-dependent regulation of gene expression in *Escherichia coli*. J Bacteriol 181(19):5993–6002
21. Li YH, Tang N, Aspiras MB, Lau PC et al (2002) A quorum-sensing signaling system essential for genetic competence in *Streptococcus mutans* is involved in biofilm formation. J Bacteriol 184:2699–2708
22. González-Hein G, Huaracán B, García P et al (2014) Prevalence of virulence genes in strains of *Campylobacter jejuni* isolated from human, bovine and broiler. Microbiology 44:1223–1229
23. Cole SJ, Records AR, Orr MW, Linden SB, Lee VT (2014) Catheter-associated urinary tract infection by *Pseudomonas aeruginosa* is mediated by exopolysaccharide independent biofilms. Infect Immun 82(5):2048–2058
24. Zhou S, Chao X, Fei M, Dai Y, Liu B (2013) Analysis of *S. epidermidis* icaA and icaD genes by polymerase chain reaction and slime production: a case control study. BMC Infect Dis 13:242
25. Diani M, Esiyok OG, Ariafar MG et al (2014) The interactions between esp, fsr, gelE genes and biofilm formation and pfge analysis of clinical *Enterococcus faecium* strains. Afr J Microbiol Res 8:129–137
26. Wassinger A, Zhang L, Tracy E et al (2013) Role of a GntR-family response regulator LbrA in *Listeria monocytogenes* biofilm formation. PLoS One 8:e70448
27. McCann T, Brendan FG, Sean PG (2008) *Staphylococcus epidermidis* device-related infections: pathogenesis and clinical management. Maureen 60:1551–1571
28. Lucas RH, David AD, Michael JM et al (2005) Aminoglycoside antibiotics induce bacterial biofilm formation. Nature 436:1171–1175
29. Strohl WR (1997) Biotechnology of antibiotics. Marcel Dekker Inc, New York
30. Benton B, Breukink E, Visscher et al (2007) Telavancin inhibits peptidoglycan biosynthesis through preferential targeting of transglycosylation: evidence for a multivalent interaction between telavancin and lipid II. Int J Antimicrob Agents 29:S51–S52
31. Leach KL, Swaney SM, Colca JR et al (2007) The site of action of oxazolidinone antibiotics in living bacteria and in human mitochondria. Mol Cell 26:393–402
32. Tenover FC (2006) Mechanisms of antimicrobial resistance in bacteria. Am J Med 119:3–10
33. Straus SK, Hancock RWE (2006) Mode of action of the new antibiotic for Gram-positive pathogens daptomycin: comparison with cationic antimicrobial peptides and lipopeptides. Biochim Biophys Acta 1758:1215–1223
34. Kotra LP, Mobashery S (1999) Mechanistic and clinical aspects of b-lactam antibiotics and b-lactamases. Arch Immunol Ther Exp (Warsaw) 47:211–216
35. Poole K (2004) Resistance to b-lactam antibiotics. Cell Mol Life Sci 61:2200–2223
36. Bush K, Jacoby GA, Medeiros AA (1995) A functional classification scheme for b-lactamases and its correlation with molecular structure. Antimicrob Agents Chemother 39:1211–1233
37. Bonnet R (2004) Growing group of extended-spectrum beta-lactamases, The CTX-M enzymes. Antimicrob Agents Chemother 48:1–14
38. Yazawa K, Mikami Y, Maeda A et al (1994) Phosphorylative inactivation of rifampicin by *Nocardia otitidiscaviarum*. J Antimicrob Chemother 33:1127–1135
39. Nakamura A, Miyakozawa I, Nakazawa K et al (2000) Detection and characterization of a macrolide 2′-phosphotransferase from *Pseudomonas aeruginosa* clinical isolate. Antimicrob Agents Chemother 44:3241–3242
40. Matsuoka M, Sasaki T (2004) Inactivation of macrolides by producers and pathogens. Curr Drug Targets Infect Disord 4:217–240
41. Yang W, Moore IF, Koteva KP et al (2004) TetX is a flavin-dependent monooxygenase conferring resistance to tetracycline antibiotics. J Biol Chem 279:52346–52352
42. Nagai K, Davies TA, Jacobs MR et al (2002) Effects of amino acid alterations in penicillin-binding proteins (PBPs) 1a, 2b, and 2x on PBP affinities of penicillin, ampicillin, amoxicillin, cefditoren, cefuroxime, cefprozil, and cefaclor in 18 clinical isolates of penicillin-susceptible, -intermediate, and -resistant pneumococci. Antimicrob Agents Chemother 46:1273–1280
43. Kosowska K, Jacobs MR, Bajaksouzian S et al (2004) Alterations of penicillin-binding proteins 1A, 2X, and 2B in *Streptococcus pneumoniae* isolates for which amoxicillin MICs are higher than penicillin MICs. Antimicrob Agents Chemother 48:4020–4022
44. Dowson CG, Coffey TJ, Spratt BG (1994) Origin and molecular epidemiology of penicillin-binding-protein-mediated resistance to b-lactam antibiotic. Trends Microbiol 2:361–366
45. Weisblum B (1998) Macrolide resistance. Drug Resist Updat 1:29–41
46. Spigaglia P, Mastrantonio P (2002) Analysis of macrolide-lincosamide- streptogramin B (MLSB) resistance determinant in strains of *Clostridium difficile*. Microb Drug Resist 8:45–53
47. Ackermann G, Degner A, Cohen SH et al (2003) Prevalence and association of macrolide-lincosamide-streptogramin B (MLSB) resistance with resistance to moxifloxacin in *Clostridium difficile*. J Antimicrob Chemother 51:599–603
48. Khodursky AB, Zechiedrich EL, Cozzarelli NR (1995) Topoisomerase IV is a target of quinolones in *Escherichia coli*. Proc Natl Acad Sci U S A 92:11801–11805
49. Ince D, Zhang X, Silver LC et al (2002) Dual targeting of DNA gyrase and topoisomerase IV: target interactions

of garenoxacin (BMS-284756, T-3811ME), a new desfluoroquinolone. Antimicrob Agents Chemother 46:3370–3380

50. Nikaido H, Zgurskaya HI (1999) Antibiotic efflux mechanisms. Curr Opin Infect Dis 12:529–536
51. Webber MA, Piddock LJ (2003) The importance of efflux pump in bacterial antibiotic resistance. J Antimicrob Chemother 51:9–11
52. Van Veen HW, Konings WN (1997) Drug efflux proteins in multidrug resistant bacteria. Biol Chem 378:769–777
53. Poole K (2001) Multidrug efflux pumps and antimicrobial resistance in *P. aeruginosa* and related organisms. J Mol Microbiol Biotechnol 3:225–264
54. Gotoh N, Tsujimoto H, Poole K et al (1995) The outer membrane protein OprM of *Pseudomonas aeruginosa* is encoded by oprK of the mexA-mexB-oprK multidrug resistance operon. Antimicrob Agents Chemother 39:2567–2569
55. Köhler T, Epp SF, Curty LK et al (1999) Characterization of MexT, the regulator of the MexE-MexF-OprN multidrug efflux system of *Pseudomonas aeruginosa*. J Bacteriol 181:6300–6305
56. Langton KP, Henderson PJF, Herber RB (2005) Antibiotic resistance: multidrug efflux proteins, a common transport mechanism? Nat Prod Rep 22:439–451
57. Putman M, Van HW, Konings WN (2000) Molecular properties of bacterial multidrug transporters. Microbiol Mol Biol Rev 64:672–693
58. Suo Y, Huang Y, Liu Y et al (2012) The expression of superoxide dismutase (SOD) and a putative ABC transporter permease is inversely correlated during biofilm formation in *Listeria monocytogenes* 4bG. PLoS One. doi:10.1371/journal.pone.0048467
59. Liu X, Sun X, Wu Y et al (2013) Oxidation-sensing regulator AbfR regulates oxidative stress responses, bacterial aggregation, and biofilm formation in *Staphylococcus epidermidis*. J Biol Chem 288(6): 3739–3752
60. Lefu L, Thomas S, Barbara IK et al (2010) *Pseudomonas aeruginosa* OspR is an oxidative stress sensing regulator that affects pigment production, antibiotic resistance and dissemination during infection. Mol Microbiol 75:76–91
61. Boles BB, Pradeep K (2008) Endogenous oxidative stress produces diversity and adaptability in biofilm communities. Proc Natl Acad Sci U S A 105:12503–12508
62. Zeng Z, Qian L, Cao L et al (2008) Virtual screening for novel quorum sensing inhibitors to eradicate biofilm formation of Pseudomonas aeruginosa. Appl Microbiol Biotechnol 79:119e26
63. Jakobsen TH, Van Gennip M, Phipps RK, Alhede M et al (2012) Ajoene, a sulfur-rich molecule from garlic inhibits genes controlled by quorum sensing. Antimicrob Agents Chemother 56:2314e25
64. Sun D, Accavitti MA, Bryers JD (2005) Inhibition of biofilm formation by monoclonal antibodies against *Staphylococcus epidermidis* RP62A accumulation-associated protein. Clin Diagn Lab Immunol 12:93–100
65. Digiandomenico A, Warrener P, Hamilton M et al (2012) Identification of broadly protective human antibodies to Pseudomonas aeruginosa exopolysaccharide Psl by phenotypic screening. J Sci Exp Med 209:1273e87
66. Eckhart L, Fischer H, Barken KB et al (2007) DNase1L2 suppresses biofilm formation by *Pseudomonas aeruginosa* and *Staphylococcus aureus*. J Sci Br J Dermatol 156:1342e5
67. Alemayehu D, Casey PG, McAuliffe O et al (2012) Bacteriophages φMR299-2 and φNH-4 can eliminate *Pseudomonas aeruginosa* in the murine lung and on cystic fibrosis lung airway cells. mBio 3:e00029
68. Biel MA, Sievert C, Usacheva M et al (2011) Reduction of endotracheal tube biofilms using antimicrobial photodynamic therapy. J Sci Lasers Surg Med 4:586e90
69. Dean SN, Bishop BN, Van Hoek ML (2011) Susceptibility of *Pseudomonas aeruginosa* biofilm to alpha-helical peptides: D-enantiomer of LL-37. J Sci Front Microbiol 2:122e8

Antioxidants and Other Potent Strategies to Reduce Oxidative Stress in Semen

Amrit Kaur Bansal

Abstract

Oxidative stress (OS) has been considered a major contributory factor to the male infertility. It is the result of imbalance between the reactive oxygen species (ROS) and antioxidants in the body which can lead to sperm damage, deformity, and eventually male infertility. Although high concentrations of the ROS cause sperm pathology (ATP depletion) leading to insufficient axonemal phosphorylation, lipid peroxidation, and loss of motility and viability, but many evidences demonstrate that low and controlled concentrations of these ROS play an important role in sperm physiological processes such as capacitation, acrosome reaction, and signaling processes to ensure fertilization. ROS are also generated during cryopreservation of spermatozoa for AI practices. To reduce the oxidative stress, there are certain compounds and reactions which dispose, scavenge, and suppress the formation of ROS or oppose their actions are called antioxidants. Currently, many antioxidants are under investigation. The supplementation of a cryopreservation extender with antioxidant has been shown to provide a cryoprotective effect on mammalian sperm quality. This chapter explains the impacts of oxidative stress and reactive oxygen species on spermatozoa functions, causes of ROS generation, and antioxidative strategies to reduce this OS. This study also suggests that antioxidant supplementation could be of clinical importance in prolonging the spermatozoal storage for assisted reproductive techniques (ARTs) like artificial insemination (AI), in vitro fertilization (IVF), and intrauterine insemination (IUI) purposes.

Keywords

Spermatozoa • ROS • Oxidative stress • Antioxidants • Cryopreservation

A.K. Bansal (✉)
Department of Veterinary Gynaecology and Obstetrics, Guru Angad Dev Veterinary and Animal Sciences University, Ludhiana 141004, India
e-mail: bansal2amrit@yahoo.co.in

V. Rani and U.C.S. Yadav (eds.), *Free Radicals in Human Health and Disease*,
DOI 10.1007/978-81-322-2035-0_24, © Springer India 2015

1 Introduction

The importance of antioxidants in the preservation of sperm functions under in vitro conditions is indicated by a large number of studies [1]. Antioxidants are the main defense mechanisms against the oxidative stress, induced by free radicals [2]. A variety of biological and chemical antioxidants that attack reactive oxygen species (ROS) and inhibit lipid peroxidation (LPO) are presently under investigation. In semen, antioxidants are present in seminal plasma, as well as in sperm cells. Seminal plasma contains three important enzymatic antioxidants such as superoxide dismutase (SOD), catalase (CAT), and glutathione peroxidase/reductase (GPX/GRD) system and a ray of nonenzymatic antioxidants such as ascorbate, urate, vitamin E, vitamin C, pyruvate, glutathione, albumin, taurine, and hypotaurine (Fig. 1) [2].

Researchers studying the markers of male fertility/infertility focused their attention on the oxidative stress in mammalian spermatozoa. Most of the recent data published on the ROS and antioxidant therapies are limited to human spermatozoa [3–7]. This chapter emphasizes the impacts of oxidative stress and antioxidants on various functions of sperm and their utility in reproduction. It is also suggested that antioxidant supplementation could be of clinical importance in prolonging the spermatozoal storage for artificial insemination (AI), in vitro fertilization (IVF), and intrauterine insemination (IUI) purposes. Further studies are required to understand the antioxidant strategies or the mechanisms, whereby ROS and endogenous antioxidants produced in sperm cells influence the reproductive processes and thereby, promoting the fertility of crossbred cattle bull spermatozoa.

1.1 Free Radicals

Free radicals are major types of ROS. They induce cellular damages when they pass an unpaired electron to the nearby cellular structures, resulting in oxidation of cell membrane lipids, amino acids, and nucleic acids [8, 9]. Free radicals are also known as a necessary evil for intracellular signaling involved in the normal process of cell proliferation, differentiation, and migration [10–12]. In the reproductive tract, free radicals also play a dual role and can modulate various reproductive functions [13]. Excess of free radicals generation frequently involves an error in spermiogenesis,

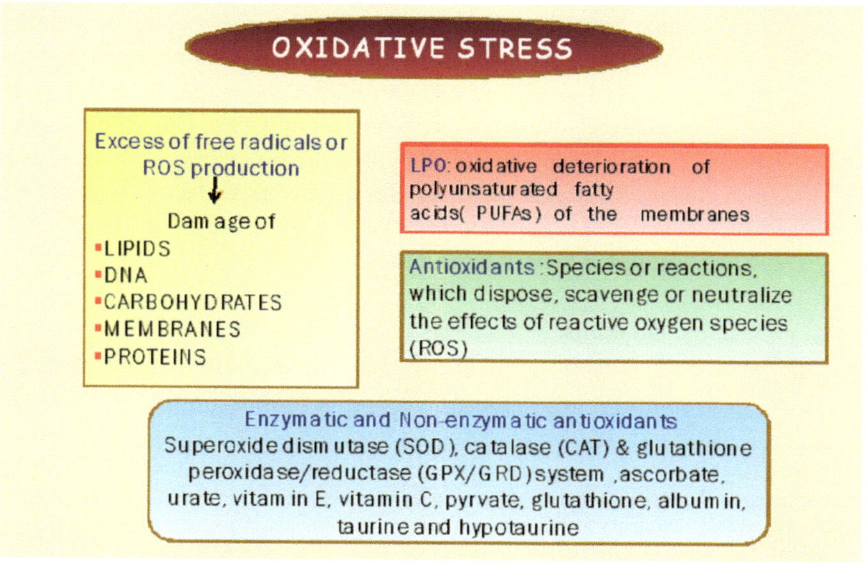

Fig. 1 Causes of oxidative stress and various antioxidants to reduce its level

resulting in the release of spermatozoa from the germinal epithelium, exhibiting abnormally high levels of cytoplasmic retention [14].

1.2 Reactive Oxygen Species (ROS)

Reactive oxygen species such as hydrogen peroxide (H_2O_2), the superoxide anion (O_2^-), the hydroxyl radical ($^{\bullet}OH$), the hypochlorite radical ($^{\bullet}OHCl$), etc., have been shown to influence the gamete functions and embryo development. ROS are formed as necessary by-product during normal enzymatic reactions of inter- and intracellular signaling pathways [15].

Although controlled generation of ROS may have physiological functions such as signaling molecules (second messengers) in many different cell types, however, their uncontrolled production is considered an important factor in aging, diet, and health and in the etiology of pathologic conditions, such as myocardial infarction, cataract, or rheumatoid arthritis [15].

1.2.1 Classification of ROS

ROS represent a broad category of molecules that indicate the collection of radicals (hydroxyl ion, superoxide, nitric oxide, peroxyl, etc.), non-radicals (ozone, single oxygen, lipid peroxides, hydrogen peroxide), and oxygen derivatives [16]. Recently reactive nitrogen species (RNS) such as nitrous oxide, peroxynitrite, nitroxyl ion, and free nitrogen radicals are also considered a subclass of ROS [17, 18]. In mammalian sperm, nitric oxide (NO) inhibits both motility and sperm competence for zona binding [19].

1.2.2 Causes of ROS Generation in Semen

There are many causes of ROS production in semen, but important ones are the following:

- In male, a hypothetical NADH oxidase at the level of sperm membrane and low sperm diphorase (mitochondrial NADH-dependent oxidoreductase) are the two main ROS-producing systems [20].
- In bovine semen, ROS are generated primarily by dead spermatozoa via an aromatic amino acid oxidase catalyzed reaction [21].
- Under normal physiological conditions, seminal plasma and normal spermatozoa do not produce ROS, but morphologically abnormal spermatozoa can produce [15].
- Leukocytes and immature spermatozoa are also the main sources of ROS [29]. Leukocytes particularly neutrophils and macrophages have been associated with excessive ROS production, and they ultimately cause sperm dysfunction [9, 22–28].
- Absence of endogenous defense mechanisms and exposure of gametes and embryos to various manipulation techniques are also the causes of ROS generation under in vitro conditions [13].

1.2.3 Beneficial and Adverse Effects of ROS in Reproduction

Mammalian spermatozoal membranes are rich in polyunsaturated fatty acids (PUFAs) and thus are sensitive to oxygen-induced damage mediated by ROS. When ROS attack the sperm membranes, it results in decreased sperm motility, rapid loss of intracellular ATP leading to axonemal damage, decreased sperm viability, and increased mid-piece sperm morphological defects with deleterious effects on sperm capacitation and acrosome reaction (Fig. 2) [15, 29].

ROS cause adverse effects on the sperm plasma membrane, DNA, and physiological processes, thereby, affecting the quality of spermatozoa. In some instances, ROS beneficially affect few marine species such as sea urchin. In case of sea urchin, hydrogen peroxide (H_2O_2) plays a pivotal role in the formation of fertilization envelope, which prevents polyspermy and protects the embryo from oxidative insult from its environment [15].

The assumption that ROS can influence male fertility has received substantial scientific support [30]. In humans, the amount of ROS have been used in semen specimens prepared for in vitro fertilization (IVF) to assess the degree of damage to spermatozoa. With the great understanding of

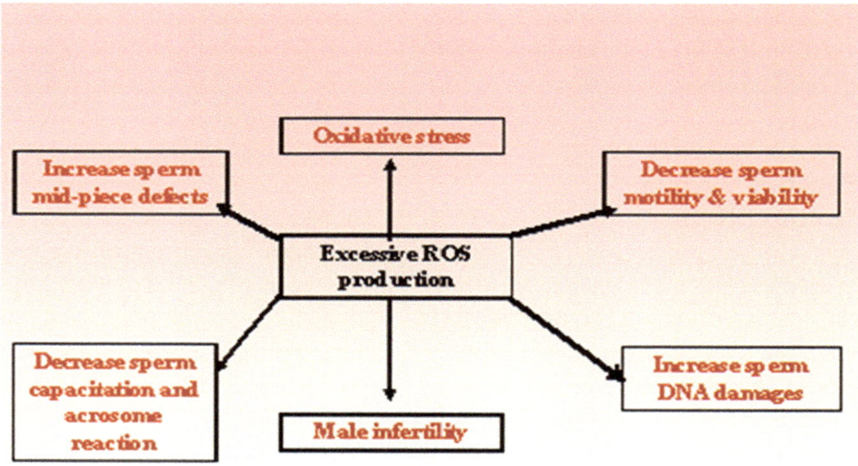

Fig. 2 Effects of excessive ROS production

the critical role played by the ROS in fertilization process, researchers have attempted to associate and, thus, predict IVF success rate using ROS level [2].

The relationship between ROS and sperm motion parameters is controversial [15]. It is well established that excess ROS cause peroxidative damage of sperm membrane. As sperm membrane lipids are very susceptible to peroxidation, therefore, their peroxidation alters the architecture of lipid matrix, changes viscosity, stimulates phospholipase A_2 (PLA_2) activity, etc. [31]. In human sperm, the peroxidative damage appears to be initiated by an excessive generation of ROS and results in loss of membrane fluidity and integrity so that the spermatozoa are no longer able to exhibit the biological response [32].

Excess ROS formation is positively correlated with the abnormal sperm concentration, morphology, and motility. In humans, ROS formation decreases when motility is greater than 60 %, suggesting that in infertile men, a sperm suspension with a high concentration of immotile spermatozoa has a greater probability of production of ROS than a highly motile sperm suspension [15]. Role of ROS has been implicated in over a hundred of diseases such as arthritis, connective tissue disorder, carcinogenesis, aging, toxin exposure, physical injury, infection, and acquired immunodeficiency syndrome, and method for counteracting ROS impacts on reproductive tissues with antioxidants is still in its infancy.

In addition, in humans, ROS may also affect the sperm axoneme, inhibit mitochondrial functions, and affect the synthesis of DNA, RNA, and proteins. The principal cytotoxic reactive oxygen intermediates in ROS associated damage is probably hydrogen peroxide (H_2O_2), generated by intracellular dismutation of superoxide anion ($^{\bullet}O_2^-$) [33]. ROS are important mediators of sperm functions. These may have been shown to influence gamete functions and embryo development. Evidences have been presented to support the contention that ROS and especially superoxide anion ($^{\bullet}O_2^-$) are required for late stage of embryo development such as two germ cell layers and egg cylinder [34].

1.3 Lipid Peroxidation and Its Effect on Semen

1.3.1 Lipid Peroxidation (LPO) in Sperm Membranes

The mechanism of ROS-induced damage to spermatozoa includes an oxidative attack on the sperm membrane lipids leading to initiation of lipid peroxidation (LPO) cascade [15].

Mammalian spermatozoa are known to be susceptible to loss of motility in the presence of exogenous oxidant, as a consequence of LPO [35]. The susceptibility of ruminant spermatozoa to oxidative attack is due to abundance of PUFAs in sperm plasma membrane. The presence of PUFAs is responsible for membrane fluidity and flexibility, which helps the sperm to engage in membrane fusion events associated with the fertilization. Unfortunately, the presence of double bonds in these molecules makes them vulnerable to free radicals attack and the initiation of LPO cascade. This results in subsequent loss of membrane and morphological integrity, impaired cell functions, and induction of sperm apoptosis [36].

Concerning with the chemistry of LPO in spermatozoa, it implies that once this process has been initiated, its propagation is impeded, leading to accumulation of lipid peroxides (LOO·) in the sperm plasma membrane [15]. Supplementation of transition metal ions such as Fe^{2+} to the sperm suspension results in a sudden acceleration of LPO and loss of sperm functions such as motility and viability [37]. The key intermediates in spontaneous LPO are the lipid hydroperoxides (LOOH·)

1.3.2 Lipid Peroxidation: Detrimental Effects on Sperm Functions

Lipid peroxides are spontaneously generated in the sperm plasma membrane and are released by the action of phospholipase A_2. They are capable of inducing DNA damage and decrease in fertility during storage of semen [38]. The peroxides are generally associated with decreased sperm functions and viability but also have a significant enhancing effect on the ability of spermatozoa to bind with homologous and heterologous zona pellucida [39].

2 Oxidative Stress in Semen Health

2.1 Oxidative Stress and Its Impacts on Semen

One of the most important factors contributing to poor quality semen has been reported to be oxidative stress [36]. Oxidative stress is a condition associated with an increased rate of cellular damage induced by oxygen and oxygen-derived oxidants commonly known as ROS [40]. Uncontrolled production of ROS that exceeds the antioxidant capacity of the semen leads to oxidative stress (OS), which is harmful to spermatozoa [41]. When the balance between the production of ROS and availability of antioxidants in semen gets disturbed, it results in oxidative stress. Continuous interaction of the animal physiological system with the free radicals leads to cumulative damage of proteins, lipids, DNA, carbohydrates, and membranes [29].

The presence of high concentration of long chain polyunsaturated fatty acids (PUFAs) within the lipid structure of sperm cells requires efficient antioxidant systems to defend against peroxidative damages. The protective antioxidant system within the spermatozoa is primarily of cytoplasmic origin. The axosome and associated dense fibers of the middle piece in sperm are covered by mitochondria that generate energy from intracellular stores of ATP. These are responsible for sperm motility. It is hypothesized that additives/antioxidants displayed cryoprotective effects on the functional integrity of the axosome and mitochondria, thereby improving the post-thawed motility.

Oxidative stress plays a major role in etiology of defective sperm functions via mechanisms involving the induction of peroxides damage to the plasma membrane [42]. Increase in oxidative stress may inhibit/alter the action of antioxidant enzymes, which culminates in increased LPO, decreased sperm motility, and viability functions and ultimately leads to infertility (Fig. 2) [29]. Oxidative stress is also known to affect the integrity of sperm genome inducing high frequency of single and double DNA breaks [38]. The oxidative stress induced by white blood cells has a damaging effect on PUFAs of the sperm phospholipids resulting in decreased membrane fluidity [14].

2.2 Cryopreservation/Freezing Thawing: Oxidative Stress

The cryopreservation of spermatozoa has allowed specific opportunities for the conservation of genetic resources through sperm banks, the guarantee of constant commercial supply of semen, and collaboration in breed improvement programs by means of the artificial insemination (AI) techniques [43]. Semen cryopreservation is an important procedure which allows specific advantages to livestock industry [44]. The low success rates of the cryopreserved semen, as compared to natural breeding, due to sublethal damage that is not completely understood, withhold its wider acceptability in the field. The basic studies on semen preservation and cryopreservation understand the male fertility factors for effective fertilization.

Freezing/thawing of sperm sample is routinely performed in cattle breeding industries in order to perform artificial insemination. These procedures are known to produce ROS in sperm samples. During cryopreservation, semen is exposed to cold shock and atmospheric oxygen, which in turn increases the susceptibility to lipid peroxidation due to higher production of ROS [45]. As the sperm plasma membrane is one of the key structures affected by cryopreservation [46, 47], sperm cryopreservation and thawing is associated with increased ROS production and decreased antioxidant level. Both freezing and thawing cause tremendous alterations in cell water volume. Spermatozoa discard most of their cytoplasm during the terminal stages of differentiation and lack the significant cytoplasmic component containing antioxidants that counteract the damaging effect of ROS and LPO [43]. Due to this, spermatozoa are susceptible to LPO during cryopreservation and thawing [43], which confers considerable mechanical stress on the cell membrane [47]. It has been noted in humans that ROS level has a positive correlation with the extent of apoptotic sperms [48]. Despite recent morphological advances, cryopreservation exerts detrimental effects on spermatozoa that lead to significant decrease in sperm viability and motility and ultimately in decreased cryopreserved sperm rates (CSR). The fertility potential of cryopreserved mammalian spermatozoa is lower than that of fresh sperm [49]. Frozen-thawed ram sperm has shown serious cryopreservation damages and thus a highly reduced fertilizing capacity [50]. Long-term (freezing) and short-term (liquid) storage of sperm may lead to membrane deterioration due to membrane phase transition occurring in the regions of highly specialized sperm plasma membrane [44]. Antioxidant capacity of semen may, however, be insufficient in preventing LPO during the freezing-thawing processes. The improvement of cryopreservation technique requires depth knowledge of the gamete physiology and the biochemical processes occurring during semen collection, processing, and freezing and thawing. Cryopreservation induces extensive biophysical and biochemical changes in the membrane of spermatozoa that ultimately decrease the fertility potential of the cells [51]. Procedure of cryopreservation increases premature capacitation of spermatozoa [52]. These alterations may not affect only the motility but also reduce the life span, ability to interact with the female reproductive tract, and fertility potentials of the spermatozoa. Freezing and thawing processes also lead to the generation of reactive oxygen species (ROS) [45]. Excessive production of ROS during cryopreservation has been associated with the reduced post-thaw motility, viability, membrane integrity, antioxidant status, fertility, and sperm functions. The post-thaw motility of the cryopreserved buffalo semen is poor and the success rate of IVF with buffalo sperm is only 10–20 % as compared to cattle which is 30–35 % [50].

Damage due to oxidative stress may be reduced or bypassed by the inclusion of antioxidants prior to freezing processes [45]. Tris-based extenders are frequently used for semen cryopreservation in domestic animals [53]. In recent years, cryoprotectants like taurine and trehalose were supplemented to the freezing extender of bull [54], boar [55], ram [43, 45], goat [56], and dog [57] spermatozoa to improve the semen characteristics after cryopreservation. The sulfonic amino acid, taurine, acts as an antioxidant and can traverse the sperm plasma membrane and inhibits lipid peroxidation and protects the cells against the accumulation of ROS [54]. Trehalose, a nonreducing disaccharide, has a protective role against osmotic effect and forms specific interactions with membrane phospholipids, rendering hypertonic media, causing cellular osmotic dehydration before freezing, and hence decreasing the amount of cell injury by ice crystallization.

3 Strategies to Reduce Oxidative Stress in Semen

3.1 Antioxidants

Spermatozoa are protected by various antioxidants and antioxidant enzymes in the seminal plasma or in spermatozoa itself to prevent oxidative damages (Fig. 3) [58]. An antioxidant that reduces oxidative stress and improves sperm motility could be useful in the management of male infertility [59]. Antioxidants are the agents, which break the oxidative chain reaction, thereby, reduce the oxidative stress [8, 60]. Vitamin E (antioxidant) may directly quench the free radicals such as peroxyl and alkoxyl (ROO·) generated during ferrous ascorbate-induced LPO; thus, it is suggested a major chain-breaking antioxidant [61]. Antioxidants, in general, are the compounds and reactions, which dispose, scavenge, and suppress the formation of ROS or oppose their actions. Mn^{2+} enhances sperm motility, viability, capacitation, and acrosome reaction by decreasing the oxidative stress [62, 63]. Thiol groups also play an important role in detoxification and antioxidation of ROS, besides maintaining the intracellular redox status. These groups serve as defense mechanisms of sperm cells to fight against oxidative stress [62]. A variety of biological and chemical antioxidants that attack ROS and LPO are presently under investigation [64]. Diluter supplemented with antioxidant or combination of antioxidants prior to the cryopreservation process may be recommended to facilitate the enhancement of sperm cryopreservation technique for the goat breeding industry [36]. Supplementation of antioxidants to the media during IVF procedures may enhance sperm quality, normal pronuclear formation, and embryo development to the blastocyst stage [65].

3.1.1 Enzymatic Antioxidants

Enzymatic antioxidants are also known as natural antioxidants; they neutralize the effect of ROS and thus prevent the damages of the cellular structures. Enzymatic antioxidants are composed of superoxide dismutase (SOD), catalase, glutathione peroxidase (GPx), and glutathione reductase (GR) [66]. SOD spontaneously dismutase

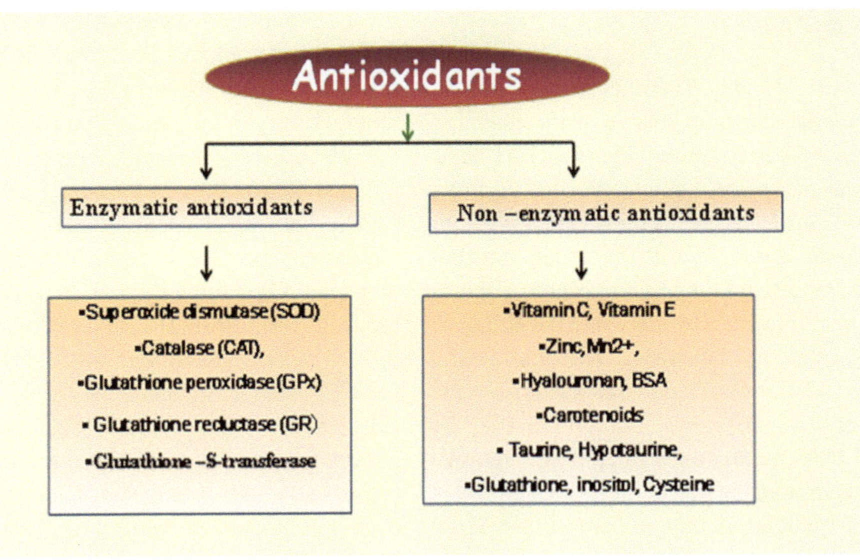

Fig. 3 Schematic representation of enzymatic and nonenzymatic antioxidants

superoxide anion ($O^-_2{}^\bullet$) to form O_2 and H_2O_2, while catalase converts H_2O_2 to O_2 and H_2O as shown in following equation [64].

$$2\left(O^-_2{}^\bullet\right) + 2H^+ \overset{SOD}{\rightarrow} H_2O_2 + O_2$$

$$H_2O_2 \overset{Catalase}{\rightarrow} H_2O + \tfrac{1}{2}O_2$$

SOD protects spermatozoa against spontaneous O_2 toxicity and LPO [64]. SOD and catalase also remove $O^-_2{}^\bullet$ generated by NADPH oxidase in neutrophils and play a major role in decreasing LPO and protecting spermatozoa against oxidative damage [64]. Catalase is mainly found in ram and cattle spermatozoa, and its potential role is to control oxidative stress caused by H_2O_2,, and thus, it prevents aging process in sperm [43].

Glutathione peroxidase (GSHPx), a selenium-containing antioxidant enzyme with glutathione as the electron donor, removes peroxyl (ROO·) radicals from various peroxides including H_2O_2 and results in conversion of glutathione reduced (GSH) to glutathione peroxide(GSSG) in sperm. On the other hand, glutathione reductase (GR) regenerates reduced GSH from GSSG as shown in the following equation [29]:

$$2GSH + H_2O_2 \overset{GSHPx}{\rightarrow} GSSG + 2H_2O$$

$$GSSG + NADPH + H^+ \overset{GR}{\rightarrow} 2GSH + NADP^+$$

GSH peroxidase and GSH reductase may directly act as antioxidant enzymes involved in the inhibition of sperm lipid peroxidation (Fig. 4). GSH has a likely role in sperm nucleus decondensation. Thus, in view of the great number of mitochondria in spermatozoa, these antioxidant mechanisms are important in the maintenance of sperm motility, rate of hyperactivation, and the ability of sperm to undergo acrosome reaction during sperm preparation techniques especially in the absence of seminal plasma. A high GSH/GSSG ratio will help spermatozoa to combat oxidative stress. It seems that the role of these GSH enzymes and their associated mechanisms are related to infertility in men and is an important area for further investigation [29].

3.1.2 Nonenzymatic Antioxidants

Nonenzymatic antioxidants are also known as synthetic antioxidants or dietary supplements. The body's complex antioxidant system is influenced by dietary intake of antioxidants and vitamins and minerals such as vitamin C, vitamin E, zinc, taurine, hypotaurine, and glutathione (Fig. 3) [2].

3.1.2.1 Glutathione

It is a molecule found at mM level in a number of cells and is able to react with many ROS directly [45]. GSH is also a cofactor for GSHPx that catalyzes the reduction of toxic H_2O_2 and other hydroperoxides, protecting the mammalian cells from oxidative stress [45]. Glutamine (5 mM) has been provided a cryoprotective effect by improving post-thaw motility, membrane integrity, and catalase enzyme activity in ram semen [44].

Glutathione in the reduced form is a tripeptide (GSH, γ glutamyl cysteinyl glycine) and is widely distributed in living organisms and often presents in high concentration in plants and animal cells [67, 68]. It participates in numerous metabolic processes including the respiration of germinating seedlings and transport of amino acids [68]. Glutathione (GSH) is the major non-protein sulfhydryl compound in mammalian cells that is known to have numerous biological functions [69]. It plays a prominent role in detoxification and antioxidation of exogenous and endogenous compounds, as well as maintaining the intracellular redox status. Glutathione is a natural reservoir of reducing power, which can be quickly used by the cells as a defense against oxidative stress [69].

The sulfhydryl groups (-SH) of glutathione confers its protective action against oxidative damages. Glutathione exists in two forms: the reduced form (GSH) and the oxidized form (GSSG) [69]. Glutathione is widely distributed thiol in animal organism, not only in somatic cells but also in the gametes [69]. It is an essential cofactor of number of enzymes that serve to preserve the -SH group of protein in the reduced state by means of disulfide interchange [68]. Thiol groups in both spermatozoa and seminal plasma are almost protein associated [68].

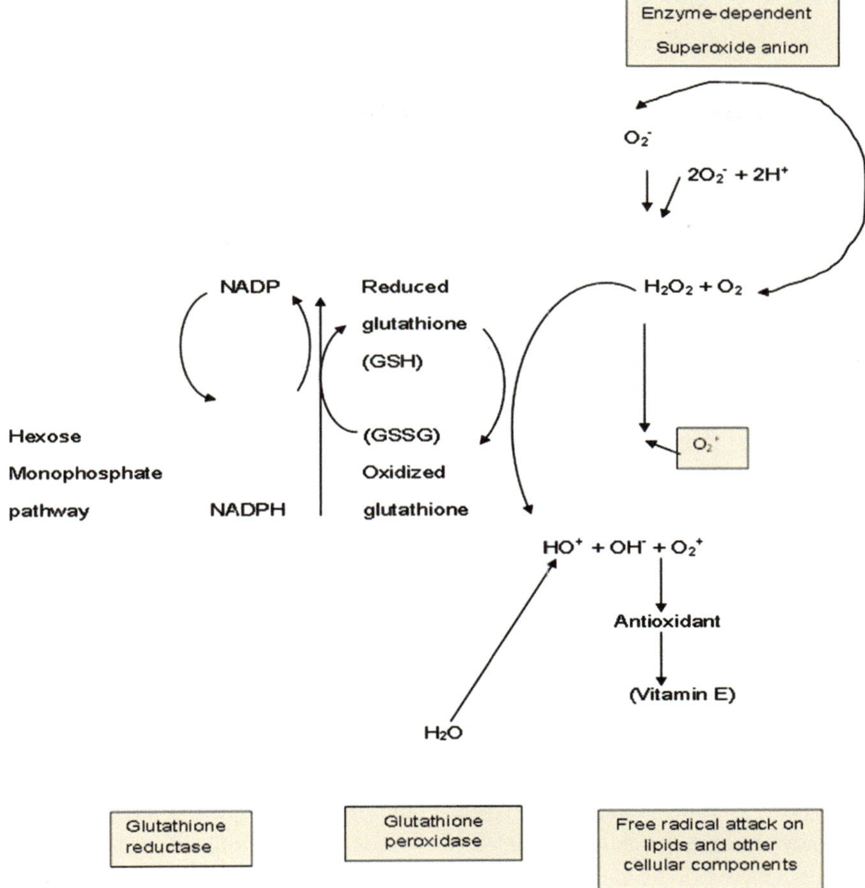

Fig. 4 Glutathione cycle and other antioxidants in protecting sperm against lipid peroxidation

Evidences have shown that membrane sulfhydryl groups play regulatory role in membrane modulation. It is, therefore, suggested that these sulfhydryl groups play a very important role in sperm motility and metabolism. These groups used as infertility assessment in unexplained male infertility and can be targeted for contraceptive research [70].

The basic function of GSH in mammalian semen is related to its interaction with other systems, as a preventive mechanism against ROS. This scavenging function of GSH helps the spermatozoa to counteract many deleterious effects such as lipid peroxidation of plasmalemma, loss of motility, leakage of intracellular enzymes, and damage of chromatin [69].

3.1.2.2 Inositol

Supplementation of **inositol** to the extender can improve the motility of frozen-thawed bull sperm [36]. Inositol has cryoprotective and antioxidative properties, resulting in higher antioxidant activity, acrosome integrity, and intact morphological rates [36].

3.1.2.3 Cysteine

It is a low molecular weight amino acid containing thiol (-SH); it is a precursor of intracellular glutathione (GSH) [50]. It has been shown to penetrate the cell membrane easily, enhancing the intracellular GSH biosynthesis both in vivo and in vitro and protecting the membrane lipids and proteins due to indirect radical scavenging properties [25]. It is also thought that GSH synthesis under in vitro conditions may be impaired, because of deficiency of cysteine in the media, due to its high instability and autoxidation to cysteine [50]. Cysteine has cytoprotective effect on the functional integrity of axosome and mitochondria improving post-thawed sperm motility [50]. Cysteine has been shown to prevent the loss in motility of frozen-thawed bull, ram, and goat semen. It has also improved the viability, the chromatin structure, and membrane integrity of boar sperm during liquid preservation [50] and has enhanced porcine oocytes maturation and fertilization under in vitro conditions [25].

3.1.2.4 Trehalose or Taurine

It is a sulfonic amino acid and acts as nonenzymatic scavenger that plays an important role in the protection of spermatozoa against ROS [52]. Trehalose performs better cryoprotective role by improving post-thaw fertilizing ability in ram, bull, and mouse sperm [43]. Recent study has demonstrated that, in ram, trehalose shows its antioxidant property when semen is incubated at 37 °C for 3 h [43]. Tuarine displayed antioxidative properties by elevating catalase level in close association with superoxide dismutase concentration in ram, rabbit, and bull spermatozoa [52].

3.1.2.5 Hyaluronan

Hyaluronan is an essential component of the extracellular matrix and non-sulfated glycosaminoglycan and is involved in important physiological functions such as motility, capacitation of spermatozoa, and preserve post-thaw spermatozoa viability and in vitro membrane stability [43]. Hyaluronan improves sperm motility, viability, and membrane integrity after freezing and thawing procedures and decrease polyspermy with declining motility in humans and boars [43].

3.1.2.6 Bovine Serum Albumin (BSA)

BSA is known to eliminate free radicals generated by oxidative stress and the protection of membrane integrity of sperm cells from heat shock during freezing and thawing of canine semen [50]. Albumin used in sperm washing procedure is likely to serve as antioxidant by providing thiol groups required for "chain-breaking" antioxidant activity.

3.1.2.7 Carotenoids

Carotenoids such as beta carotene and lycopene are also important components of antioxidant defense. Beta carotene protects the plasma membrane against LPO in rat [50].

3.1.2.8 Vitamin E

Vitamin E appears to be the first line of defense against the peroxidation of PUFAs contained in the cellular and subcellular membrane phospholipids, because of its lipid solubility [71]. The phospholipids of mitochondria, endoplasmic reticulum, and plasma membrane possess high affinity for α-tocopherol, and the vitamin E appears to be concentrated at these sites.

It protects sperm membrane against oxidative damage (Fig. 4) [72]. It is a well-documented antioxidant and has been shown to inhibit free radical-induced damage to sensitive cell membrane [73]. The antioxidant action of α-tocopherol is effective at higher oxygen concentration, and thus, it is not surprising that it tends to be concentrated in those lipid structures that are exposed to the higher O_2 partial pressure. Vitamin E is the major chain-breaking antioxidant in membrane by directly neutralizing superoxide anion ($^{\bullet}O_2^-$), hydrogen peroxide (H_2O_2), and hydroxyl radical ($OH^{\bullet}$) [15]. α-Tocopherol is classified as a chain-breaking antioxidant because of its ability to break the lipoperoxidative chain reaction through its interaction with lipid peroxyl and alkoxyl radicals [74]. It is a powerful antioxidant and has been shown to afford some protection to mammalian cells from oxidative attack generated both under in vivo and in vitro conditions [74]. Damage caused by iron-catalyzed peroxidation can be prevented by including α-tocopherol (vitamin E) to the medium [15]. It breaks the free radical chain reaction by forming a relatively stable radical tocopheroxy at a concentration of 10 mM [15].

3.1.2.9 Trace Metal Ions

The antioxidant and/or pro-oxidant potential of three trace metal ions, namely, aluminum (Al), manganese (Mn), and selenium (Se), has been studied [75]. The effects of Al and Mn have been found to be anion dependent, and manganese proved to be the trace metal ion of choice [75]. It is well known that Mn^{2+} is a potent inhibitor of in vitro LPO in variety of systems [76]. However, the inhibitory mechanisms of Mn^{2+} on LPO have not been fully elucidated [76]. The action of Mn^{2+} at high concentration is not due to scavenging of lipid radicals; but rather Mn^{2+} inhibited the LPO by counteracting LOO$^{\bullet}$ (lipid peroxyl radicals), which are water soluble, resulting in inhibition of LPO [76]. Probably, Mn^{2+} ions compete with iron (Fe^{2+}) at anionic oxygen of phosphate groups in phospholipids to inhibit LPO [76]. It is reported that 1 μM Mn^{2+} significantly inhibited MDA production in LPO of brain homogenates, which may contain a trace amount of endogenous iron [77].

4 Role of Antioxidants in Sperm Capacitation and Acrosome Reaction

The following model for the role of vitamin E in sperm capacitation and acrosome reaction is suggested, based on the study by Bansal [78] and the relevant literature [24, 79–82]. ROS initiates a cascade of events to promote capacitation and acrosome reaction. These events can be inhibited by the antioxidative nature of vitamin E. By quenching the superoxide anion and H_2O_2, vitamin E inhibits the generation of ROS which lead to inactivation of membrane-bound enzyme adenylate cyclase (AC). Thus, vitamin E interferes with the first step of the ROS mediating capacitation and acrosome processes, thereby, inactivating the whole signal transduction cascade. On inactivation of adenylate cyclase, cAMP generation from the ATP gets slowed down, thus decreasing the level of cAMP. But high concentrations of cAMP and H_2O_2 are required for phoshorylating tyrosine receptors and activating the enzyme phospholipase C (PLC). This enzyme further cleaves the peroxidized chain to unesterified fatty acids which are removed later to promote the fusion events. Increase in level of cAMP is also required to stimulate Na^{2+}/Ca^{2+} or Ca^{2+} channel receptors or openings, which lead to increases in intracellular calcium (Ca_i^{2+}) level by enhancing Ca^{2+} uptake by the sperm. Ca_i^{2+} enters the sperm cell and activates the enzyme phospholipase C (PLC). This enzyme catalyzes the cleavage of protein phosphatidylinositol bisphosphate (PIP_2) into two products diacyl glycerol (DAG) and inositol triphosphate (IP_3). DAG activates protein kinase C (membrane-bound enzyme), which in turn catalyzes the removal of unesterified fatty acids from the sperm membrane, thus promoting the acrosome reaction. Therefore, it is suggested that vitamin E inhibits or decreases the rate of acrosome reaction through its antioxidative property. It is concluded that vitamin E has negative effect on the acrosome reaction, although it improves other functions of sperm such as motility, viability, and membrane integrity.

Following model of signal transduction for the role of Mn^{2+} in enhancing % hyperactivity is suggested based on the study by Bansal [77] and relevant literature [83–86]. Messenger systems such as adenylate cyclase, guanylate cyclase, and calamodulin are highly affected by intracellular salts, such as Mn^{2+}, Mg^{2+}, and Ca^{2+}. Out of these, Mn^{2+} is a well-known potent stimulator of adenylate cyclase activity, which in turn enhances the level of cAMP. Increase in the level of cAMP with the Mn^{2+} supplementation phosphorylates many proteins that are involved in the movements or flagellar beating of spermatozoa. Increase in the level of cAMP also stimulates the Ca^{2+} uptake by the cell, thus, increasing the level of intracellular calcium (Ca_i^{2+}). At a higher level, Ca_i^{2+} increases the membrane integrity and viability. These properties of the sperm are required for its optimal functioning under normal and oxidative stress conditions. Elevation of Ca_i^{2+} in flagellum of spermatozoa drives hyperactivation in rat [87] and bull [88] spermatozoa. It is thought that the intake of Ca^{2+} is slow during capacitation but rapid at the time of acrosome reaction. Guraya et al. suggested that adequate level of Ca^{2+} is required in sperm, once the capacitation gets completed [89]. Larsen [90] found that high concentration of Ca_i^{2+} is related to the cell death. Based on the present and above studies, it is suggested that supplementation of Mn^{2+} to the bull spermatozoa permits adequate rise in Ca_i^{2+} level without decreasing their viability. Further, Mn^{2+} has beneficial effects on sperm survival and % hyperactivity during capacitation and acrosome reaction.

Sidhu et al. [91] found that production of ROS during storage of semen at 4 °C or incubation at 37 °C for capacitation caused deleterious effects on sperm. Bansal [92] postulated that Mn^{2+} play an important role in improving the quality of cattle bull semen by its scavenging property i.e., reduction in production of ROS during storage and capacitation of spermatozoa. She also suggested that Mn^{2+} may be used as a potent antioxidant/additive to sperm samples to be used for assisted reproductive techniques (ARTs).

Since long it has been known that supplementation of culture media with antioxidants such as ROS scavengers, disulfide reducing, or divalent chelators prolongs the motility of reactivated bull spermatozoa after freezing and thawing [93]. It has been suggested that antioxidant therapy appears to be efficient not only in vitro but also in vivo [93]. Numerous antioxidants have proven beneficial in protecting damaging effects of ROS on sperm movement and against oxidative damages [47].

5 Conclusions

Evaluation of OS and the use of antioxidants are not routine in clinical practice. The immediate need is to simplify and validate the evaluation of ROS and OS status so that it can be performed routinely without the use of sophisticated equipments. Also, it is important to establish reference values for ROS above which antioxidants could be used for male infertility treatment. The dose and duration of these antioxidants should also be determined and standardized. This study suggests that antioxidant supplementation could be of clinical importance in prolonging the spermatozoal storage for artificial insemination (AI), in vitro fertilization (IVF), and intrauterine insemination (IUI) purposes. Further studies are required to understand the antioxidant strategies or the mechanisms whereby ROS and endogenous antioxidants produced in sperm cells influence the reproductive processes and thereby, promoting the fertility of crossbred cattle bull spermatozoa.

References

1. Aitken RJ, Keith TJ, Robertson SA (2012) Reactive oxygen species and review sperm function – in sickness and in health. J Androl 33:1096–1106
2. Agarwal A, Gupta S, Sharma RK (2005) Role of oxidative stress in female reproduction. Reprod Biol Endocrinol 3:28
3. Aitken RJ, Clarkson JS (1988) Generation of reactive oxygen species, lipid peroxidation and human sperm function. Biol Reprod 40:183–197
4. Aitken RJ, Gordon E, Harkiss D et al (1998) Relative impact of oxidative stress on the functional competence and genomic integrity of human spermatozoa. Biol Reprod 59:1037–1046
5. Alvarez JG, Storey BT (1984) Lipid peroxidation and the reaction of superoxide and hydrogen peroxide in mouse spermatozoa. Biol Reprod 30:833–841
6. Dandekar SP, Nadkarni GD, Kulkarni VS et al (2002) Lipid peroxidation and antioxidant enzymes in male infertility. J Postgrad Med 48:186–189
7. Bansal AK (2000) Role of Mn^{2+} and Cd^{2+} in the human ejaculated spermatozoa under oxidative stress – lipid peroxidation, thiol redox ratio and motility. M.Sc. (Hons) thesis, Panjab University, Chandigarh
8. Miller JK, Slebodzinska EB (1993) Oxidative stress, antioxidants and animal function. J Dairy Sci 76:2812–2823
9. Ochsendorf FR (1999) Infections in the male genital tract and reactive oxygen species. Hum Reprod Update 5:399–420
10. Agarwal A, Nallela KP, Allamaneni SS et al (2004) Role of antioxidants in treatment of male infertility: an overview of the literature. Reprod Biomed Online 8:616–627
11. Rhee SG (2006) Cell signaling H_2O_2, a necessary evil for cell signaling. Science 312:1882–1883
12. Ford WC (2001) Reactive oxygen species and sperm. Hum Fertil (Camb) 4:77–78
13. Agarwal A, Makker K, Sharma R (2008) Clinical relevance of oxidative stress in male factor infertility an update. Am J Reprod Immunol 59:2–11
14. Sanocka D, Kurpisz M (2004) Reactive oxygen species and sperm cells. Reprod Biol Endocrinol 2:12–26
15. Sharma RK, Agarwal A (1996) Role of reactive oxygen species in male infertility. Urology 48:835–850
16. Agarwal A, Prabakaran SA (2005) Mechanism, measurement and prevention of oxidative stress in female reproductive physiology. Indian J Exp Biol 43:963–974
17. Sikka SC (2001) Relative impact of oxidative stress on male reproductive function. Curr Med Chem 8:851–862

18. Darley- Usmar V, Wiseman H, Halliwell B (1995) Nitric oxide and oxygen radicals. A question of balance. FEBS Lett 369:131
19. Kenkel S, Rolf C, Nieschlag E (2001) Occupational risk for male infertility: an analysis of patients attending a tertiary referral centre. Int J Androl 24:318–326
20. Gavella M, Lipovac V (1992) NADH dependent oxidoreductase (diaphorase) activity and isozyme pattern of sperm in infertile men. Arch Androl 28:135–141
21. Sariozkan S, Bucak MN, Tuncer PB et al (2009) The influence of cysteine and taurine on microscopic – oxidative stress parameters and fertilizing ability of bull semen following cryopreservation. Cryobiology 58:134–138
22. Garrido N, Meseguer M, Simon C et al (2004) Proxidative and antioxidative imbalance in human semen and its relation with male infertility. Asian J Androl 6:59–65
23. Aitken RJ, Baker HW (1995) Seminal leukocytes: passengers, terrorist or good Samaritans? Hum Reprod 10:1736–1739
24. Shalika S, Dugan K, Smith RD et al (1996) The effect of positive semen bacterial and ureaplasma culture on in vitro fertilization success. Hum Reprod 11:2789–2792
25. Aitken RJ, Fisher HM, Fulton N et al (1997) Reactive oxygen species generation by human spermatozoa is induced by exogenous NADPH and inhibited by the flavoprotein inhibitors diphenylene iodonium and quinacrine. Mol Reprod Dev 47:468–482
26. Hendin BN, Kolettis PN, Sharma RK et al (1999) Varicocele is associated with elevated spermatozoa reactive oxygen species production and diminished seminal plasma antioxidant capacity. J Urol 161:1831–1834
27. Pasqualotto FF, Sharma RK, Polts JM et al (2000) Seminal oxidative stress in patients with chronic prostates. Urology 55:881–885
28. Sharma RK, Posqualotto AE, Nelson DR et al (2001) Relationship between seminal white blood cell counts and oxidative stress in men treated at an infertility clinic. J Androl 22:575–583
29. Saleh RA, Agarwal A, Kandirali E et al (2002) Leukocytospermia is associated with increased reactive oxygen species production by human spermatozoa. Fertil Steril 78:1215–1224
30. Sikka SC (1996) Oxidative stress and role of antioxidants in normal and abnormal sperm functions. Front Biosci 1:e78–e86
31. Agarwal A, Allamaneni SSR, Nallella KP et al (2005) Correlation of reactive oxygen species levels with the fertilization rate after *in vitro* fertilization: a qualified meta-analysis. Fertil Steril 84:228–231
32. Chatterjee S, Gagnon C (2001) Production of reactive oxygen species by spermatozoa undergoing cooling, freezing and thawing. Mol Reprod Dev 59:451–458
33. O'Flaherty C, Beconi M, Beorlegui N (1997) Effect of natural antioxidants, superoxide dismutase and hydrogen peroxide on capacitation of frozen – thawed bull spermatozoa. Andrology 29:269–275
34. Irvine DS (1996) Glutathione as a treatment for male infertility. Rev Reprod 1:6–12
35. Kodama H, Kuribayashi Y, Gagnon C (1996) Effect of sperm lipid peroxidation on fertilization. J Androl 17:151–157
36. Rao B, Soufir JC, Martin M et al (1989) Lipid peroxidation in human spermatozoa as related to mid piece abnormalities and motility. Gamete Res 24:127–134
37. Bucak MN, Sariozkan S, Tuncer PB et al (2010) The effect of antioxidants on post-thawed Angora goat (Capra hircus ancryensis) sperm parameters, lipid peroxidation and antioxidant activities. Small Rumin Res 89:24–30
38. Bansal AK, Bilaspuri GS (2008) Effect of ferrous sulphate and ascorbic acid on sperm motility, viability and lipid peroxidation of crossbred cattle bull spermatozoa. Animal 2:100–104
39. Twigg J, Fulton N, Gomez E et al (1998) Analysis of the impact of intracellular reactive oxygen species generation on the structural and functional integrity of human spermatozoa: lipid peroxidation, DNA fragmentation and effectiveness of antioxidants. Hum Reprod 13:1429–1436
40. Aitken RJ, Clarkson JS, Fishel S (1989) Generation of reactive oxygen species, lipid peroxidation and human sperm function. Biol Reprod 40:183–197
41. Sikka SC, Rajasekaran M, Hellstrom WJ (1995) Role of oxidative stress and antioxidants in male infertility. J Androl 16:464–468
42. Desai NR, Mahfouz R, Sharma R et al (2010) Reactive oxygen species levels are independent of sperm concentration, motility and abstinence in a normal healthy, proven fertile man: a longitudinal study. Fertil Steril 94:1541–1543
43. Bansal AK, Bilaspuri GS (2010) Impacts of oxidative stress and antioxidants on semen functions. Vet Med Int 2011:1–7
44. Bucak MN, Atessahin A, Varisli O et al (2007) The influence of trehalose, taurine, cysteamine and hyaluronan on ram semen. Microscopic and oxidative stress parameters after freeze-thawing process. Theriogenology 67:1060–1067
45. Bucak MN, Tuncer PB, Sariozkan S et al (2009) Comparison of the effects of glutamine and an amino acid solution on post-thawed ram sperm parameters, lipid peroxidation and anti-oxidant activities. Small Rumin Res 81:13–17
46. Bucak MN, Atessahin, Yuce A (2008) Effect of antioxidant and oxidative stress parameters on ram semen after the freeze – thawing process. Small Rumin Res 75:128–134
47. Agarwal A, Nallella KP, Allamaneni SSR et al (2004) Role of antioxidants in treatment of male infertility: an overview of the literature. Reprod Biol Med 8:616–627
48. Yousef MI, Abdallah GA, Kamel KI (2003) Effect of ascorbic acid and vitamin E supplementation on semen quality and biochemical parameters of male rabbits. Anim Reprod Sci 76:99–111
49. Cheema RS, Bansal AK, Bilaspuri GS (2009) Manganese provides antioxidant protection for sperm cryopreservation that may offer new consideration for clinical fertility. Oxidative Med Cell Longev 2:147–154

50. Said TM, Grunewald S, Paasch U et al (2005) Effects of magnetic activated cell sorting on sperm motility and cryopreserved rates. Fertil Steril 83:1442–1446
51. Uysal O, Bucak MN (2007) Effects of oxidized glutathione, bovine serum albumin, cysteine and lycopene on the quality of frozen thawed ram semen. Acta Vet Brno 76:383–390
52. Chatterjee S, Lamirande E, Gagnon C (2001) Cryopreservation alters membrane sulfhydryl status of bull spermatozoa: protection by oxidized glutathione. Mol Reprod Dev 60:498–506
53. Reddy NSS, Mohanarao GJ, Atreja SK (2010) Effects of adding taurine and trehalose to a tris – based egg yolk extender on buffalo (Bubalua bubalis) sperm quality following cryopreservation. Anim Reprod Sci 119:183–190
54. Purdy PH (2006) A review on goat sperm cryopreservation. Small Rumin Res 6:215–225
55. Chen Y, Foote RH, Brockett CC (1993) Effect of sucrose, trehalose, hypotaurine, taurine and blood serum on survival of frozen bull sperm. Cryobiology 30:423–431
56. Funahashi H, Sano T (2005) Selected antioxidants improve the function of extended boar semen stored at 10°C. Theriogenology 63:1605–1616
57. Atessahin A, Bucak MN, Tuncer PB et al (2008) Effects of antioxidant additives on microscopic and oxidative parameters of Angara goat semen following the freeze-thawing process. Small Rumin Res 77:38–44
58. Martins-Bersa A, Rocha A, Mayenco-Aguirre A (2009) Effects of taurine supplementation and ionophore concentrations on post-thaw acrosome reaction of dog spermatozoa. Theriogenelogy 71:248–253
59. Kim JG, Parthasarathy S (1998) Oxidation and spermatozoa. Semen Reprod Endocrinol 16:235–239
60. Bansal AK, Bilaspuri GS (2008) Effect of manganese on bovine sperm motility, viability, and lipid peroxidation *in vitro*. Anim Reprod CBRA 5:90–96
61. Kumar H, Mahmood S (2001) The use of fast acting antioxidants for the reduction of cow placental retention and subsequent endometritis. Ind J Anim Sci 71:650–653
62. Bansal AK, Bilaspuri GS (2009) Antioxidant effect of vitamin E on motility, viability and lipid peroxidation in cattle spermatozoa under oxidative stress. Anim Sci Paper Rep 27:5–14
63. Bilaspuri GS, Bansal AK (2008) Mn^{2+} A potent antioxidant and stimulator of sperm capacitation and acrosome reaction in crossbred cattle bulls. Arch Anim Breed 51:149–158
64. Bansal AK, Bilaspuri GS (2008) Oxidative stress alters membrane sulfhydryl status, lipid and phospholipid contents of crossbred cattle bull spermatozoa. Anim Reprod Sci 104:398–404
65. Du Plessis SS, Makker K, Desai NR et al (2008) Impact of oxidative stress on IVF. Expert Rev Obstet Gynecol 3:539–554
66. Goncalves FS, Barretto LSS, Arruda RP et al (2010) Effect of antioxidants during bovine *in vitro* fertilization procedures on spermatozoa and embryo development. Reprod Domest Anim 45:129–135
67. De Lamirande E, Gagnon C (1995) Impact of reactive oxygen species on spermatozoa: a balancing act between beneficial and detrimental effects. Hum Reprod 10(1):15–21
68. Saleman JJ, Kufra FS, Matt A et al (2005) Role of glutathione in reproductive tract secretions on mouse pre-implantation embryo development. Biol Reprod 73:308–314
69. Li TK (1975) The glutathione and thiol content of mammalian spermatozoa and seminal plasma. Biol Reprod 12:641–646
70. Luberda Z (2005) The role of glutathione in mammalian gametes. Reprod Biol 5:5–17
71. Nivsarkar M, Cherian B, Patel S (1998) A regulatory role of sulfhydryl groups in modulation of sperm membrane conformation by heavy metals: Sulfhydryl groups as markers for infertility assessment. Biochem Biophys Res Commun 247:716–718
72. Horton HR, Moran LH, Ochs RS et al (2002) Principles of biochemistry, 3rd edn. Prentice Hall, Upper Saddle River, pp 221–222
73. Verma A, Kanwar KC (1999) Effect of vitamin E on human sperm motility and lipid peroxidation *in vitro*. Asian J Androl 1:151–154
74. Sinclair S (2000) Male infertility: nutritional and environmental consideration. Altern Med Rev 5:28–38
75. Donnelly ET, McClure N, Lewis SEM (1999) Antioxidant supplementation *in vitro* does not improve human sperm motility. Fertil Steril 72:484–495
76. Anand RK, Kanwar U (2001) Role of same trace metal ions in placental membrane lipid peroxidation. Biol Trace Elem Res 82:61–75
77. Tampo Y, Yonaha M (1992) Antioxidant mechanism of Mn (II) in phospholipids peroxidation. Free Radic Biol Med 13:115–120
78. Shukla GS, Chandra VC (1981) Manganese toxicity: lipid peroxidation in rat brain. Acta Pharmacol Toxicol 48:95–100
79. Bansal (2006) Effects of antioxidants on crossbred cattle bull spermatozoa under oxidative stress. Ph.D. thesis, Punjab Agricultural University, Ludhiana
80. O'Flaherty C, Beorlegui NB, Beconi MT (1999) Reactive oxygen species requirements for bovine sperm capacitation and acrosome reaction. Theriogenology 52:289–301
81. O'Flaherty C, de Lamirande E, Gagnon C (2005) Reactive oxygen species and protein kinases modulate the level of phosphor-MEK like proteins during human sperm capacitation. Biol Reprod 73:94–105
82. de Lamirande E, Jiang H, Zini A et al (1997) Reactive oxygen species and sperm physiology. Rev Reprod 2:48–54
83. Visconti PE, Kopf GS (1998) Regulation of protein-phosphorylation during sperm capacitation. Biol Reprod 59:1–6
84. Lapointe S, Ahmad I, Buhr MM et al (1996) Modulation of post-thaw motility, survival, calcium uptake and fertility of bovine sperm by magnesium and manganese. J Dairy Sci 79:2163–2169

85. Garbers DL, Kopf GS (1980) The regulation of spermatozoa by calcium and cyclic nucleotides. In: Greengard P, Roinson GA (eds) Advances in cyclic nucleotides research. Raven, New York, pp 251–305
86. Tash JS, Means AR (1983) Cyclic adenosine 3′-5′ monophosphate, calcium and protein phosphorylation in flagellar motility. Biol Reprod 28:75–104
87. Lindemann CB, Gotz JS (1988) Calcium regulation of flagellar curvature and swimming pattern in triton X-100 extracted rat sperm. Cell Motil Cytoskeleton 10:420–431
88. Lindemann CB, Kanous KS, Gardner TK (1991) The interrelationship of calcium and cAMP mediated effects on reactivated mammalian sperm models in comparative spermatology: 20 years after edited by Bacetti B. Ed. Raven, New York, p 491
89. Guraya SS (1999) Cellular and molecular biology of capacitation and acrosome reaction in spermatozoa. Int Rev Cytol 199:1–66
90. Larsen CJ (1994) The BCL_2 gene is a prototype of a gene family that controls programmed cell death (apoptosis). Ann Genet 37:121–134
91. Sidhu KS, Sundhey R, Guraya SS (1984) Stimulation of capacitation and the acrosome reaction in ejaculated buffalo (*Bubalus bubalis*) sperm and the effect of a sperm motility factor. Int J Androl 7:324–333
92. Bansal AK (2013) Manganese: a potent antioxidant in semen. Iran J Appl Anim Sci 3:217–221
93. Tarin JJ, Brines J, Cono A (1998) Is antioxidant therapy a promising strategy to improve human reproduction? Antioxidant may protect against infertility. Hum Reprod 13:1415–1424

Nano-encapsulation of a Natural Polyphenol, Green Tea Catechins: Way to Preserve Its Antioxidative Potential

Shweta Dang, Sonal Gupta, Rakhi Bansal, Javed Ali, and Reema Gabrani

Abstract

Emanating from extensive research carried out all over the globe, green tea has been acknowledged for plethora of pharmacological activities like anti-inflammatory, antimicrobial, antitumor, antiaging, and many more. This wide array of health effects have been attributed to green tea catechins (GTCs). These GTCs correspond to the class of antioxidants which scavenge the precarious free radicals in the body and, thus, prevent the progression of various diseases. However, poor bioavailability, short half-life, stability issues, and short shelf life hamper its use as a therapeutic agent. To address these limitations, various encapsulation techniques have been explored by the scientists. The encapsulation techniques employed for green tea and its catechins range from microparticles, microcapsules, nanoparticles, and lipid nanocapsules to self assembly approaches like liposomes, micelles, and microemulsions. Encapsulation not only helps in enhancing the aqueous solubility and stability of the GTCs but also provides for controlled and sustained release thus protecting their biological/pharmacological activity in systemic circulation. At the same time, sustained release is likely to enhance bioavailability and thus can reduce the likelihood of repeated use. The aim of this chapter is to discuss the role of green tea as an antioxidant and various nano-encapsulation strategies to enhance the efficacy of related products. It also provides an insight into some of the commercial and patented green tea-based products.

Keywords

Green tea • Nanoemulsions • Nanoparticles • Drug delivery • Bioavailability • Antioxidant

S. Dang (✉) • S. Gupta • R. Bansal • R. Gabrani
Department of Biotechnology, Jaypee Institute of Information Technology, A-10, Sector-62, Noida 201307, Uttar Pradesh, India
e-mail: shweta.dang@jiit.ac.in

J. Ali
Faculty of Pharmacy, Jamia Hamdard, Hamdard Nagar, New Delhi 110062, India

V. Rani and U.C.S. Yadav (eds.), *Free Radicals in Human Health and Disease*,
DOI 10.1007/978-81-322-2035-0_25, © Springer India 2015

1 Introduction

Free radical is an oxygen-containing molecule that has one or more unpaired electrons, making it highly reactive with other molecules. High level of free radicals can cause indiscriminate damage to biological molecules such as DNA, protein, or lipid, leading to loss of biological functions and even cell death. The imbalance between the production of free radicals and the ability of body to detoxify their harmful effects leads to a condition known as oxidative stress [1]. Oxidative stress is now believed to be one of the major causes of harmful diseases like cancer, neurodegenerative diseases such as Parkinson's disease and Alzheimer's disease, chronic fatigue syndrome, fragile X syndrome, heart and blood vessel disorders, atherosclerosis, heart failure, heart attack, and inflammatory diseases [2–6]. Superoxide anion ($O_2^{\bullet-}$), hydroxyl ion ($OH^{\bullet}$), alkoxy ($RO^{\bullet}$) and peroxy radicals ($ROO^{\bullet}$), and peroxynitrite ($ONOO^{\bullet-}$) are considered to be the most common free radicals responsible for the various pathophysiological conditions [7]. The human body produces endogenous molecules both *enzymatic* and *nonenzymatic* to combat the effects of increase in free radicals. The enzymes superoxide dismutase (SOD) catalase and glutathione peroxidase (GP_X) have a transition metal at their core, which is capable of taking on different valences as they transfer electrons during the detoxification process [8]. Nonenzymatic defenses including ascorbate (vitamin C), α-tocopherol (vitamin E), bilirubin, etc. act as scavengers of free radicals directly or prevent the production of free radicals through sequestration of redox-active metals like iron and copper [9].

A variety of plants have shown activity against free radicals due to the presence of polyphenols and other antioxidants present in them. These plant polyphenols are comprised of flavonoids, cinnamic acid derivatives, curcumin, caffeine, catechins, gallic acid derivatives, anthocyanins, and tannins [10].

GTCs have been well reported for their antioxidative potential. Numerous studies on green tea have highlighted its role in the prevention or treatment of various diseases [11]. Green tea has proved to be highly effective as anticarcinogenic [12], antimicrobial [13–15], antiviral [16–18], anti-inflammatory, antidiabetic, and hypocholesterolemic agent [19] in the few recent years. Studies have reported green tea to be effective in preventing atherosclerosis and cardiovascular diseases and modulate cholesterol metabolism [20–22]. These pharmacological properties of green tea are ascribed to the presence of major polyphenol components known as catechins [23, 24]. There are four types of catechins, namely, epicatechin (EC), epicatechin gallate (ECG), epigallocatechin (EGC), and epigallocatechin gallate (EGCG). Out of these EGCG is the most abundant and potent catechin [25] and has proved to be as effective as synthetic antioxidants with less harmful effects [26].

Non-phenolic secondary metabolites such as allicin, carotenoids, thiols, jasmonic acid, eicosapentaenoic acid, ascopyrones, retinal, and melatonin have also shown excellent antioxidant activity [27, 28]. A list of various plants, their antioxidants, and the parts from which these antioxidants are derived is shown in Table 1.

The challenges faced by the effective delivery of GTCs and other natural compounds can be addressed by different nano-encapsulation approaches. Nanocarriers are delivery systems in the nanometer size range (50–500 nm) containing encapsulated, dispersed, adsorbed, or conjugated drugs. Nanoscale drug delivery systems have the ability to improve the pharmacokinetics and increase biodistribution of therapeutic agents to target organs and have the desirable advantage of improving solubility of hydrophobic compounds in aqueous medium to render them suitable for administration [51].

2 Green Tea as a Natural Antioxidant

Green tea (*Camellia sinensis*) originated thousands of years back in China and is the second most consumed beverage in the world. Green tea or the non-fermented type is obtained from the young leaves and terminal apical buds by drying

Table 1 Selected plants and plant based products exhibiting antioxidant activity

Plant name	Plant part used	Antioxidant	Pharmacological action	References
Camellia sinensis L. (green tea)	Leaves	Polyphenols and flavonoids	Antimicrobial, anticancer, anti-inflammatory, antidiabetic, etc.	[25]
Punica granatum L. (anar)	Peels	Tannins and phenolic compounds	Protection against carbon tetrachloride induced hepatotoxicity	[29–31]
Cinnamomum cassia L. (dalchini)	Bark	Cinnamaldehyde, cinnamon, and other cinnamic acid derivatives	Glucose intolerance, diabetes, antimicrobial activity, anticancer, quench hydroxyl radical, and hydrogen peroxide	[32, 33]
Vitis vinifera L. (grape seed)	Seeds	Catechin, epicatechin, dimeric, trimeric, and tetrameric proanthocyanidins	Hepatic fibrosis, ischemia–reperfusion injury, cancer, reduces free radical production	[34–36]
Ginkgo biloba L. (maidenhair tree)	Leaves	Flavonoids glycoside (myricetin and quercetin) and phenolic compounds	Inhalation of leaves decoction is helpful in bronchial asthma, cerebral insufficiency, Alzheimer's disease, and schizophrenia	[37–39]
Phyllanthus emblica L. (amla)	Fruits	Vitamin C, emblicannin A and B	Anti-stress, immunomodulatory effects, inflammation, cancer, age-related renal disease, and diabetes	[40]
Curcuma longa L. (turmeric, haldi)	Roots, leaves	Curcumin, eugenol, camphene	Blood purifier, antifungal, antibacterial, anticancer, anti-inflammatory	[41, 42]
Daucus carota L. (carrot)	Root	Carotenoids, flavonoids, glycosides	Piles, jaundice, leprosy, urinary problems, and bronchitis	[43]
Foeniculum vulgare L. (fennel, saunf)	Fruit	Anethole, limonene, anisaldehyde, volatile oil	Carminative, throat infections, bronchitis, purgative, diuretic, and kidney disorders	[44]
Glycyrrhiza glabra L. (mulethi)	Roots	Glycyrrhizin, flavonoids, liquiritin, isoliquiritin, 2-methyl isoflavones	Bronchitis, asthma, diuretic, ulcers	[45]
Psoralea corylifolia L. (babchi)	Seed	Fixed oils, essential oils, bakuchiol	Aphrodisiac, scabies, protect mitochondria against oxidative stress	[46]
Ocimum sanctum L. (tulsi)	Leaves	Volatile oils, eugenol, thymol	Carminative, protect against throat infections, expectorant, and other stomach and throat disorders	[47]
Withania somnifera L. (ashwagandha)	Leaves, seeds	Steroidal compounds like withanolide, withanine	Increase immunity, hepatoprotective	[48, 49]
Solanum nigrum L. (nightshade)	Leaves	Phenolic compounds and flavonoids	Hepatoprotective, diuretic, laxative, gonorrhea	[50]

and steaming the fresh leaves, whereas the traditional black tea also includes postharvest fermentation step in preparation [52]. The number of free hydroxyl groups, the presence of ortho-hydroxylation on B-ring of flavonoid molecules, a C2-C3 double bond in C-ring, and the presence of 3-hydroxyl groups are reported as the main conditions of antiradical and antioxidant properties [26].

2.1 Chemical Composition of Green Tea

Green tea has a complex chemical composition, containing proteins (15–20 % dry weight) and amino acids (1–4 % dry weight) such as theanine, glutamic acid, tryptophan, glycine, serine, aspartic acid, tyrosine, valine, leucine, threonine, arginine, and lysine; carbohydrates (5–7 % dry weight)

Table 2 Chemical contents of green tea

Constituent	Content (% dry weight)
Proteins	15–20
Amino acids	1–4
Carbohydrates	5–7
Mineral and trace elements	5
Lipids	7
Polyphenols	30
Fibers	26
Pigments	2

such as cellulose, pectins, glucose, fructose, and sucrose; minerals and trace elements (5 % dry weight) such as calcium, magnesium, chromium, manganese, iron, copper, zinc, molybdenum, selenium, sodium, phosphorus, cobalt, strontium, nickel, potassium, fluorine, and aluminum; and trace amounts of lipids (linoleic and α-linolenic acids), sterols (stigmasterol), vitamins (B, C, E), xanthene bases (caffeine, theophylline), pigments (chlorophyll, carotenoids), and volatile compounds (aldehydes, alcohols, esters, lactones, hydrocarbons) [53].

Green tea contains polyphenols (30 % of the dry weight), which include flavonols, flavondiols, flavonoids, and phenolic acids. Most of the green tea polyphenols (GTPs) are flavonols, commonly known as catechins which are found in greater amounts in green tea than in black or Oolong tea [54]. Table 2 provides the chemical constituents of green tea.

2.2 Antioxidative Potential of Green Tea: Central Functioning Factor

As depicted in Fig. 1, green tea acts as a multifaceted agent that is responsible for various mechanisms involved in free radical formation and their detoxification inside the human body.

- *Free radical scavengers*: GTCs and polyphenols act as scavengers of physiologically harmful free radical such as superoxide anion ($O_2^{\bullet -}$), hydroxyl ion ($OH^{\bullet}$), alkoxy ($RO^{\bullet}$) and peroxy radicals ($ROO^{\bullet}$), and peroxynitrite ($ONOO^{\bullet -}$). The antioxidant activity is attributed to the presence of free hydroxyl groups in the structure of catechins [55, 56].
- *Inhibition of transcription factors*: Green tea inhibits the activation of redox-sensitive transcription factors such as activation protein-1 (AP-1) and nuclear factor-κB (NF-κB) [57].
- *Chelators*: Transition metals such as iron and copper catalyze the free radical formation; green tea has the ability to chelate these metal ions and thus inhibit the formation of free radicals [58].
- *Inhibition of prooxidant enzymes*: Peroxynitrite ($ONOO^{\bullet -}$) and other NO derivatives are capable of damaging DNA and proteins. Green tea polyphenols effectively inhibit the formation of peroxynitrite ($ONOO^{\bullet -}$) radical [59].

2.3 Green Tea in Lab and Market: Current Scenario

Though a large number of studies have proved the benefits of green tea, a vast difference was observed in the efficiency exhibited by free form of green tea in vitro as compared to in vivo results [60]. The less efficiency of green tea in vivo was attributed to the chemically unstable catechins [61]. Catechins have shown to readily undergo oxidation in solution, causing the loss of hydrogen atoms, formation of a semiquinone radical intermediate, and quinone oxidized products [62]. pH and oxygen concentration was found to be the most crucial factor influencing the stability of catechins, and it has been shown that the rate of oxidation increases as the pH increases [63]. Su and associates showed that the half-life of a mixture of catechins reduced from 24 h at pH 5.0 to approximately 1 h at pH 7.4 [64]. The short half-life was also reported to be due to rapid systemic clearance resulting in poor oral bioavailability of <2–5 % [65, 66]. Catechins were reported to show apparent permeability (P_{app}) of $0.8–3.5 \times 10^{-7}$ cm/s when permeability studies were done using Caco-2 [67] which are exceptionally low considering the classic poorly permeable compound mannitol, which exhibits P_{app} values in the range of $6–10 \times 10^{-6}$ cm/s [68]. Moreover, environmental conditions like exposure to high

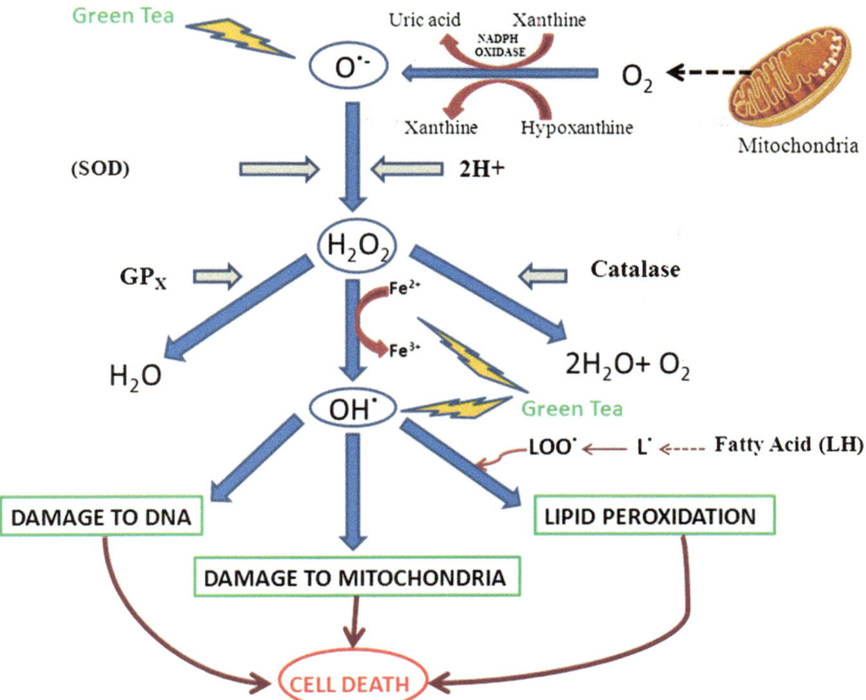

Fig. 1 Mechanism of free radical generation and their detoxification by endogenous enzymes superoxide dismutase (*SOD*), catalase, and glutathione peroxidase (GP_X). Fatty acid (*LH*) is converted to fatty acid free radicals ($L^{\cdot}$) causing lipid peroxidation. Green tea acts as a scavenger for superoxide free radical, hydrogen peroxide, and hydroxyl ion

humidity and light also resulted in oxidation/degradation of green tea and its active constituents [69].

Thus, both crude green tea extract and catechins have been encapsulated in various formulations which can be administered in the body via several routes like oral and transdermal for targeting different disease conditions [70]. These formulations are used to improve the stability and bioavailability of green tea and are either applied locally or consumed orally for systemic action. They address the problems associated with conventional consumption of green tea.

Various formulations composed of green tea extract or its constituents are under patent protection and already available commercially. Tables 3 and 4 summarize some of the patents and commercial products, respectively.

Though these products have shown marked response commercially, the efficacies are hindered largely due to poor bioavailability. To overcome these limitations, nano-encapsulation strategies have been explored.

3 Advanced Dosage Forms for Green Tea Extract and Its Catechins

Hydrogen peroxide generation ability of GTCs is a central component with respect to its role in combating various diseases. GTCs play an important role in scavenging free radicals, and while doing so they themselves might get oxidized. Therefore, nano-encapsulation of these catechins could sustain their antioxidative potential and functional stability, as described in the following section. Furthermore, sustained release of GTCs from a nanoformulation would lead to a lower dose requirement due to decreased plasma fluctuations and, therefore, could be a valuable tool

Table 3 Some of the patents related to green tea and its catechins

Formulations	Summary of invention	Patent no.	Inventors [71–88]
Nanoparticles	Stabilized, biocompatible gold nanoparticles encapsulating ECGC having high affinity toward cancer cells	US0129618	Katti et al. (2013)
Capsule	Composition for weight reduction comprising capsaicin, green tea extract, L-tyrosine, and caffeine	US7867526	Astrup and Toubro (2011)
Capsule	Soft gel capsules containing polymethoxylated flavones and palm oil tocotrienols for maintaining cardiovascular health	US7887852	Udell (2011)
Tablet/suspension	Composition comprising an effective amount of epicatechin for treating hypertension	US7875651	Romanczyk and Schmitz (2011)
Tablet/capsule	Composition for treating prostate cancer, comprising therapeutically effective amounts of supercritical extracts	US7744934	Newmark et al. (2010)
Tablet/capsule	Composition for modulating cytokines to regulate an inflammatory or immunomodulatory response	US7758903	Randolph and Roh-Schimdt (2010)
Gel	Improved topical medicaments useful in the treatment of conditions that are ameliorated by increased cell metabolism, circulation, and nerve function	US7704522	Morgan (2010)
Gel	Composition and method for eliminating foreign bodies from a host	US7807190	Kingsley (2010)
Dye	Composition for dyeing keratin fibers, comprising of pure plant dye and active metal in a mineral acting as a moderating agent	US7749286	Greaves and Greaves (2010)
Food bar	Nutritional supplement to enhance learning, academic, and behavioral functioning	US7771756	Schlesser (2010)
Emulsions	Bioactive complex composition having enhanced oxidative stability, emulsion stability, mineral-rich transparent beverages, and a wide range of functional health benefits	US7780873	Mora-gutierrez and Gurin (2010)
Resin	Chinese herb extract for treating dementia and preparation method	US7824714	Wu et al. (2010)
Powders/tablets	Composition for the treatment of obesity, comprising a catechol-rich extract of green tea	US6830765	Rombi (2004)
Liquid	Combination of nondigestible oligosaccharides and the green tea catechin, EGCG for the restoration and the maintenance of colon health	WO2004000045	Simmons and Dong (2003)
Emulsion	Use of a content of catechins or a content of green tea extract in cosmetic preparations for tanning the skin	US6399046	Schonrock and Max (2002)
Tablet/powder	Methods and compositions of treating cancer or solid tumors comprising the administration of a therapeutically effective amount of catechins	US6428818	Morre et al. (2002)
Effervescent tablet	Natural product formulation containing a concentrated green tea extract which increases the formulations absorption rate and bioavailability	US6299925	Xiong et al. (2001)
Powder	Process for preparing green tea extracts having the color and flavor suitable for incorporation into non-tea beverage matrixes	US5427806	Ekanayake et al. (1995)

Table 4 Some of the commercially available green tea formulations and their applications

Product name	Formulation	Company	Green tea volume	Indications (web reference: [128–144])
Balance Point for men	Capsule	Progressive Health Nutraceuticals (Colorado, USA)	150 mg	Multivitamin and natural rejuvenator made for men
Balance Point for women	Capsule	Progressive Health Nutraceuticals (Colorado, USA)	150 mg	Potent nutritional health protection for women
Hydroxycut Hardcore	Capsule	MuscleTech (USA)	–	Increases the fat-burning neurotransmitter norepinephrine
Methyl Ripped	Capsule	NxLabs Inc. (Wilmington, USA)	–	Fat burner while maintaining muscles
Men's SuperVites	Capsule	Heaven & Earths	10 mg	Broad spectrum nutritional support for men
Diet Rx	Capsule	Diet Rx (USA)	50 %	Natural appetite control
Mega-T	Capsule	Mega-T Ltd. (London, UK)	600 mg	Dietary supplement stimulating metabolism
TeaCare	Capsule	TeaCare Ltd. (UK)	150 mg	Maintains a healthy heart condition and also supports weight management
SNS Green Tea	Capsule	Serious Nutrition Solutions (New York, USA)	500 mg	Burn fat, suppresses appetite, enhances immune function, and maximizes performance
LifePak Nano	Capsule	Nu Skin (USA)	45 mg	Nutritional antiaging program
Slenderite	Pill	Natrol (Chatsworth, USA)	–	Slimming pills
Green Fusion Fat Burner	Soft gels	Sports Research	444 mg	Controls appetite and boosts the metabolism
Polyphenon E	Ointment	Mitsui Norin Co., Ltd. (Japan)	15 %	For treating genital warts
Veregen	Ointment	Veregen (New Jersey, USA)	15 %	Treatment of warts on the outside of the genitals and around the outside of the anus
Optivegetol Green Tea	Extract	Gattefosse (Berkshire, UK)	–	Antiaging/anti-wrinkle product and for sun and after-sun lines
GreenTea	Extract	Himalaya Healthcare	–	Weight reduction, lowers blood pressure and strengthens the walls of blood vessels, promotes gastric secretion and aids digestion
Green Tea Herbasol® Extract PG PF	Extract	Lipoid Kosmetik AG (Switzerland)	–	UV protectant, antioxidant, free radical scavenging

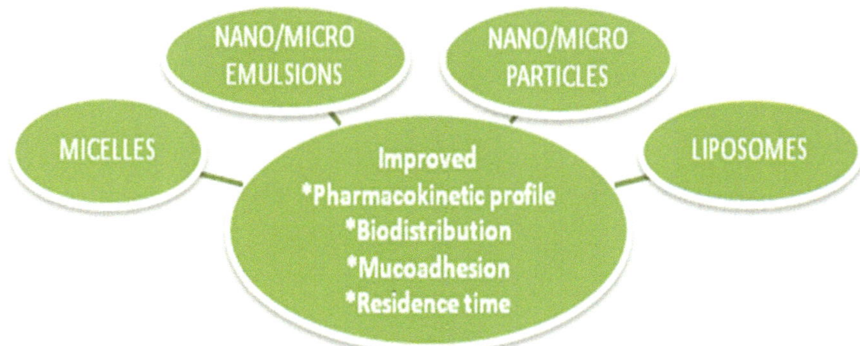

Fig. 2 Strategies to encapsulate green tea and its catechins

to limit the toxicity associated with a high dose of active natural products. This section summarizes all such formulation techniques which have been used to encapsulate green tea or isolated compounds, discussing their advantages (Fig. 2) and future developments and trends.

3.1 Nanoparticles

Novel drug delivery approach could be based on its encapsulation in nanoparticles. This would protect drugs against chemical and enzymatic degradation and potentially enhance the selective uptake of these particles [89]. Consequently, many researchers successfully attempted to encapsulate green tea and its catechins to enhance their therapeutic activity.

Siddiqui and his associates used polylactic acid–polyethylene glycol nanoparticles for the encapsulation of EGCG. In vitro and in vivo (performed in athymic nude mice) studies suggested that nano-encapsulation of chemopreventive agents like EGCG helps in retaining its biological activity due to increased half-life. Moreover, use of biodegradable polymers for nano-encapsulation of EGCG limited its toxicity [90].

In order to improve its pharmacokinetic profile, Italia and his associates modified double-emulsion method for the encapsulation of EGCG into poly(lactic-co-glycolic acid) polymer for oral consumption. The particles were of approximately 130 nm size with encapsulation efficiency of 70 %. In vivo studies were performed on a rat model of cyclosporine A-induced chronic nephrotoxicity. Results indicated that nanoparticulate formulation of EGCG administered orally or intraperitoneally was equally efficacious at one third dose and thus enhanced efficacy of EGCG after nano-encapsulation [91].

Chitosan exhibits various favorable characteristics like chemical inertness, thermal stability, biocompatibility, biodegradability, and non-toxicity [92].

Due to these features, chitosan is abundantly used as an effective polymer for nano-encapsulation purposes and also serves for sustained delivery of encapsulated agents when applied topically [93]. Therefore, Bing and his co-workers investigated the polyanion-initiated gelation process using chitosan–tripolyphosphate nanoparticles for the delivery of tea catechins through epithelial/mucosal lining. Chitosan was dissolved in acetic acid solution with sonication to form the homogeneous solution. The aqueous solution of tea catechins was added into chitosan solution for an appropriate time. The addition of tripolyphosphate to chitosan solution initiated ionic gelation, thereby resulting into spontaneous formation of tea catechin encapsulated into polymeric nanoparticles. It was observed that higher chitosan concentration led to larger nanoparticles, higher surface charge, and lower membrane permeability. This resulted in reduced burst effect and provided a means of controlled release [94].

Reducing agents are known to stabilize catechins by getting preferentially oxidized in their place [95]. In order to protect the catechin and EGCG encapsulated in chitosan–tripolyphosphate nanoparticles from degradation in potassium hydrogen phosphate buffer (pH 7.4), Dube and his associates compared the effectiveness of reducing agents like ascorbic acid, dithiothreitol, and tris(2-carboxyethyl)phosphine. The remaining levels of catechin and EGCG after 24 h incubation in the presence of reducing agents were high. Encapsulated catechin was degraded to 50 % of its initial level in 24 h while nonencapsulated catechin took only 8 h. Thus encapsulated catechin took thrice the time to degrade as compared to the nonencapsulated catechin. Similarly, encapsulated and nonencapsulated EGCG were degraded to 50 % of their initial levels in 40 and 10 min, respectively. These results demonstrate that ascorbic acid and tris(2-carboxyethyl)phosphine in combination provide the highest protection to catechins, and encapsulation in polymeric nanoparticles possibly plays an important role in enhancing the stabilization of catechins during oral administration [96].

Many edible products such as citrus juice and ascorbic acid have been formulated into tea beverages to increase the post digestion recovery of EC, ECG, EGC, and EGCG [97].

Poly(g-glutamic acid) (g-PGA) is an anionic, nontoxic, and edible polypeptide that has been used in foods, medicine, and cosmetics [98]. Tang and associates encapsulated tea catechins in chitosan: g-PGA (1.5:1.0) nanoparticles. Nanoparticles were prepared by dissolving tea catechins in deionized water. Tea catechin solutions were mixed with the aqueous g-PGA to obtain tea catechin/g-PGA mixtures. The tea catechin/g-PGA mixtures were added into an aqueous chitosan under magnetic stirring at room temperature. The self-assembled nanoparticles were formed by electrostatic interaction between g-PGA and chitosan, as depicted through FT-IR analysis. The nanoparticles of mean particle size of 134.5–147.8 nm were obtained with the loading efficiency of 13.8–23.5 %. Paracellular permeability of encapsulated tea catechins was predicted by measuring the TEER of Caco-2 cell monolayer. The permeability was found to increase to 24 % at pH 6.6 suggesting them to the suitable carriers in the transmucosal delivery of tea catechins in the area of the intestinal lumen [99].

Use of biocompatible and biodegradable polymers helps in protection of food ingredients against oxidation and other enzymatic reactions [100]. Chen and group prepared self-assembled tea catechin/gelatin nanoparticles for oral delivery. Aqueous gelatin was added into aqueous catechin in different concentrations under magnetic stirrer for approximately 10 min at room temperature. The self-assembled nanoparticles were collected by centrifugation, freeze-dried, and stored at room temperature. The mean particle size was less than 200 nm, and at 1:1 ratio of catechins and gelatin, more than 96 % entrapment was achieved. FT-IR spectral analysis indicated that phenomenon of self-assembly of nanoparticles occurs due to hydrogen bonding between aliphatic and aromatic hydroxyl groups of gelatin and catechins, respectively. Free radical scavenging assays showed that tea catechins could be protected by the nanoparticles and antioxidant activity of tea catechins was retained even after 3 weeks of storage. The results suggested that the more the content of catechin/gelatin ratio in the nanoparticles, the higher was the antioxidant activity. The tea catechin/gelatin nanoparticles exhibited around 40 % inhibition of degradation of gelatin by trypsin. The stability is attributed to specific interaction between gelatin and polyphenolic catechins due to which the active site recognized by trypsin is protected [101].

Gomes and his associates introduced a novel system that consisted of maltodextrin and gum arabic nanoparticles coated with egg yolk L-A-phosphatidylcholine to encapsulate and protect EGCG. Polysaccharide core was prepared by homogenization and spray dried. This polysaccharide core was found to be responsible for resistance against mechanical stress. The lipid coating was applied by the lipid film hydration method that resulted in increased size of particles as observed by dynamic light scattering. Results showed that the particle assemblies exhibited high retention efficiency of EGCG, even at physiological pH, hence opening the possibility of their use for intact delivery and controlled release of tea catechins [102].

In another study, Ma and group prepared solid lipid nanoparticles based on the phase behaviors of hot microemulsions. The pseudoternary phase diagram was constructed for the system of glyceryl monostearate (oil phase), mixed surfactants (polyoxyethylene stearate and poloxamer 188), and water at 60 °C until the mixture becomes transparent. Tea polyphenol solution was added gradually to the oil phase under mild vortexing to obtain thermodynamically stable water-in-oil microemulsion. Finally, the microemulsion was dispersed in cold aqueous polyoxyethylene stearate solution (0.8 %) and ultrasonicated for 2 h to obtain solid lipid nanoparticles of tea polyphenols. The mean particle size was less than 150 nm with polydispersity index of 0.5. A sustained transdermal penetration with the permeation coefficient approximately 3 μg.cm^2/h was achieved [103].

Smith and group investigated the ability of nanolipidic formulation of EGCG in improving its oral bioavailability for the treatment of neurodegenerative diseases including Alzheimer's disease and HIV-associated dementia. In vitro studies performed on SweAPP N2a (neuronal cells) showed that nanolipidic EGCG particles improved α-secretase enhancing ability up to 91 %. Oral bioavailability of nanolipidic EGCG particles in mouse models in vivo was found to be more than two fold as compared to free EGCG [104].

Manea and her associates developed a new method of encapsulating green tea extract in nanostructured lipid carriers (NLC) to check its antimicrobial against *Escherichia coli* K 12-MG1655 and antioxidant activity. They were obtained by loading green tea extract in three types of vegetable oils—grape seed oil (GSO), St. John's wort oil (HPO), and sea buckthorn oil (SBO)—using high-shear homogenization method (HSH). NLC made by GSO exhibited least Z-avg of 125.6 nm as compared to other two vegetable oils whose particle size increased as the concentration of green tea extract increased. SBO showed the least antibacterial activity whereas HPO showed the highest antibacterial activity (SBO < GSO < HPO). The antioxidant activity of NLCs–green tea extract was enhanced in comparison with that of pure green tea extract. NLC produced by Tween20 had less Z-avg than Tween80 [105].

The antitumor effects of green tea polyphenols were studied by Liang and group by encapsulating tea polyphenol (TP) extract (93 % pure) in chitosan nanoparticles (CNPs) using two types of chitosan biomaterials: *N-carboxymethyl chitosan* (Mv = 61 kDa, degree of deacetylation 83 %) and *chitosan hydrochloride* (Mv = 90 kDa, degree of deacetylation 85 %). The nanoparticles were produced by dissolving TP (10.94 mg) in chitosan hydrochloride solution (1.19 mg/mL) and then adding the carboxymethyl chitosan solution (3.63 mg/ml) dropwise into the chitosan hydrochloride and TP mixed solution under stirring at room temperature for 30 min. Results showed the drug content and encapsulation rate of TP-CNPs to be 16 % and 83 %, respectively. The cytotoxic effects of TP-CNPs were evaluated through MTT assays on HepG2 cells which showed that the TP-CNPs were effective in inhibiting the proliferation of cells and the inhibition increased with increasing concentration of TP-CNPs. The fluorescence microscopic studies showed the TP-CNPs to be causing less degree of cell apoptosis than TPs which was attributed to the slow release of TP in CNPs; however, the TEM images of HepG2 cells showed microvilli disappearance, margination, intracytoplasmic vacuoles, and mitochondrial swelling with the formation of apoptotic bodies which are the features of apoptosis, inferring TP-CNPs permeation into the cells, releasing TP and causing apoptosis. The cell apoptosis was detected by annexin V/PI double stain assay; TP-CNPs can inhibit the proliferation of HepG2, through necrosis and apoptosis induction in cancer cells as compared to TP, the CNPs which showed only tumor cell necrosis [106].

3.2 Micelles

Distinctive properties of a micelle lie in its unique structural composition. It forms aggregate in aqueous solution with hydrophilic head regions in contact with surrounding solvent, sequestering the hydrophobic tail regions in the micelle center in which sparingly soluble drugs can be encapsulated. Micelles can be used for targeted drug delivery due to their ability to form conjugates with a variety of surface ligands that confer specificity and signal them toward the target site. Moreover, corona part of micelles provides an impermeable shell to external proteins and enzymes and hence prevents hydrolysis and enzymatic degradation during transport to the target [107].

Researchers from the Institute of Bioengineering and Nanotechnology (Singapore) used chemically modified EGCG oligomers to serve as carrier for flavonoids, drugs, and proteins. They developed core-shell micellar nanocomplexes using self-assembly system in a gentle aqueous solution at room temperature. EGCG oligomers formed the inner core of micellar system and an outer shell of polyethylene glycol conferred stability, prolonged circulation half-life in the bloodstream, and controlled biodistribution. These micelles are targeted toward cancer cells due to selective expression of EGCG receptors on cancer cells (Web Reference: [143]).

EGCG is highly sensitive to oxidation thus limits its usage as the preventive medicine. Shpigelmann and his associates used thermally modified β-lactoglobulin (protein) to form co-assembled nanoparticles for the intact delivery of EGCG. Optimal nano-entrapment was obtained when EGCG was added to preheated (~80 °C) β-lactoglobulin solution during cooling and vortexing. Particle size analysis confirmed the size of thermally induced protein-EGCG co-assemblies to be smaller than 50 nm. A 33-fold lower initial degradation rate and a 3.2-fold slower degradation over 8 days were observed for nano-entrapped EGCG as compared to free EGCG. These results demonstrated that nano-encapsulation of EGCG conferred protection against oxidative degradation [108].

3.3 Liposomes

Pharmaceutical formulations in the form of liposomes can be used to deliver the intact drug to a targeted site. Liposomes allow spatial and temporal distribution of encapsulated compound and reduced cytotoxicity thereby increasing efficiency of the treatment [109]. They are amphipathic in nature and have been extensively used for entrapping both hydrophilic and hydrophobic compounds [110].

Fang and group developed liposomal formulations encapsulating tea catechins for transdermal delivery. This study demonstrated that the use of anionic surfactants such as deoxycholic acid and dicetyl phosphate in the presence of ethanol increased the catechin permeation from the liposomes by five- to seven fold as compared to the nonencapsulated EGCG. Among all the catechins, EGCG showed the highest encapsulation rate and in vivo skin deposition level in liposomes carried out in nude mouse model. Results suggested that the transdermal delivery of catechins can be enhanced by their encapsulation in liposomes [111].

Calcium pectinate gel beads are nontoxic, mechanically strong, biocompatible, and acid stable and, thus, have been widely used to encapsulate drugs. For the sustained delivery of catechins orally, Lee and his associates used external ionotropic gelation method for the preparation of calcium pectinate gel beads entrapping catechin-loaded liposomes. Lipid mixture of L-alpha-phosphatidylcholine and cholesterol was mixed with organic solvent and evaporated in a rotary evaporator to obtain a lipid film. The film was mixed with an aqueous catechin solution by rotary shaking to encapsulate catechins into liposomes. This suspension of liposomes was added to the pectin. This mixture was added dropwise to 100 ml of calcium chloride solution with gentle agitation to form the beads. Beads were washed with deionized water and then dried at 37 °C for 12 h. In vitro release of catechin-loaded liposomes entrapped within the beads was determined in simulated gastric fluid and simulated intestinal fluids. Liposome loaded with catechins showed low amount of release as compared to free catechins. However, when catechin-loaded liposomes were given hydroxypropyl methyl cellulose coating, maximum encapsulation was achieved and gave the best control over release initially in an hour, reaching a plateau phase, and after that a slow steady phase was observed in simulated intestinal fluid for up to 3 h [112].

Elastic liposomes are composed of undeformable vesicles and have been investigated to combine the advantages of two types of carrier into a single system, by obtaining drug-in-lipid-in-elastic liposomal formulation. Huang and co-workers prepared the liposomes containing soy phosphatidylcholine, cholesterol, and Tween 80 in the presence of ethanol by a thin-film method and subsequent sonication and extrusion. More than 80 % of the catechin was entrapped in the aqueous core of liposomes produced with Tween 80. Liposomes entrapping catechin remained stable in the presence of gastrointestinal fluid and in simulated intestinal fluid. The liposomes showed sustained release of catechin compared to that of an aqueous solution. In this study of brain distribution, these liposomes showed 2.9- and 2.7-fold higher catechin accumulation in the cerebral cortex and hippocampus as compared to the aqueous solution. These results indicated that oral bioavailability and brain regional distribution of catechin could be improved by utilizing elastic liposomes [113].

A novel formulation known as "phenolipids" was introduced by Ramadan, who incorporated phenolic compounds into phospholipids, such as soy lecithin, and provided a potential new application for encapsulated phenolics in the food and pharmaceutical industries [114].

Rashidinejad and associates similarly encapsulated GTCs and EGCG in soy lecithin liposomes. They used a hard fat cheese as a vehicle for inclusion of liposomes and to increase the antioxidant activity. Liposomes were prepared by simple method of high-shear homogenization. The catechins and EGCG were dissolved in 0.25 M acetate buffer, in which soy lecithin was dispersed with magnetic stirring followed by high-shear homogenization. The control groups were also subjected to same process conditions, one containing only catechins and EGCG with acetate buffer and second containing only soy lecithin in acetate buffer. Gel-filtered liposomes were obtained using Sephadex column and stored at 4 °C. The dynamic light scattering results showed the particle size to increase in the loaded liposomes from 133 to 169.7 nm in case of catechin and upto 173 nm in case of EGCG. The encapsulation efficiency >70 % with encapsulation yield of ~80 % was obtained. The zeta potential of the liposomes were studied which revealed that the polyphenols were located inside the liposomal structure rather than coating the surfaces. The loaded liposomes containing either polyphenol showed better stability compared to the empty liposomes suggesting that lecithin liposomes to be a promising encapsulating technique to protect and deliver tea catechins to the gut [115].

3.4 Microparticles

Microparticles have a much larger surface-to-volume ratio than at the macroscale and therefore can be used as effective drug delivery systems [116]. They have several advantages including sustained and controlled drug release and active as well as passive targeting of drugs by easy manipulation of particle size and surface characteristics [117].

To overcome the limitations of other drug dosages (low entrapment efficiency and requirement of high temperature), Wisuitiprot and group developed novel polymer-based microparticles using water-in-silicone emulsion method with green tea extract. Chitosan microparticles of diameter less than 5 μm were obtained by mixing chitosan with tripolyphosphate solution. Entrapment efficiency of chitosan microparticles was found to be improved at acidic pH of the tripolyphosphate solution thus resulted in slow release of catechins. At neutral pH, the release of catechins depended on their molecular stabilities, and increased degradation was observed with temperature. These results suggested that the degradation of tea catechins loaded in polymeric microparticles was less than that of free catechins [118].

Lee and co-workers investigated the optimal conditions for the preparation of calcium pectinate microparticles (reinforced with liposome and hydroxypropyl methylcellulose). They optimized various parameters including calcium chloride concentration (5.2 %), hydroxypropyl methylcellulose concentration (0.08 %), and hardening time (12.63 min) for the sustained-release of catechins. These microparticles released about 50 % of the entrapped catechin into simulated gastric fluid after 24 h, whereas complete catechin release was observed in simulated intestinal fluid after 8 h. Release studies performed in rat plasma confirmed that antioxidative effect of encapsulated catechin was maintained effectively as compared to free catechin. This study indicated that catechins encapsulated into microparticles can be used as sustained drug delivery carriers [119].

Fu and his associates prepared EGCG microparticles by using low-temperature spray-drying method. The diameter of EGCG microparticles obtained was approximately 30 μm, and that of lactose-added EGCG was found to be 40 μm. EC_{50} (half maximal effective concentration) value of spray-dried EGCG microparticles obtained at different temperatures was found to be approximately same to that of untreated EGCG. Using the drying temperatures ranging between 70 °C and 130 °C showed a little effect on antioxidant activity of EGCG since NMR spectrum studies indicated that EGCG did not undergo chemical structural change during encapsulation process. These results confirmed that low-temperature drying method is favorable for the maintenance of optimized chemopreventive activity of EGCG [120].

Sosa and group coprecipitated green tea polyphenols with a biodegradable polymer, polycaprolactone by a semicontinuous supercritical antisolvent process to improve the oral bioavailability. For the coprecipitation experiments, the above mentioned polymer was added to the extract in different concentration ratios and dissolved in an ultrasonic bath to obtain a solution. Afterward, the solution (approx. 350 ml) was fed to the precipitator until the liquid pump stops and pure CO_2 was fed during a period long enough to ensure the complete removal of organic solvent from the precipitator. After the decompression, a sample of the particles retained in the frit and vessel wall was collected. HPLC results showed that almost 90 % of polyphenols were retained in the coprecipitates. Drug release profile showed that 30 % of polyphenols were relatively rapidly released by diffusion, while the majority of them were encapsulated within the crystalline domains of polymer matrix. These results showed that to increase the drug release, polymers of low molecular weight and lower crystallinity should be tested [121].

Kafirin particles are the protein microspheres used for targeted and controlled drug delivery. Taylor and his associates used kafirin microparticles to encapsulate catechins. Catechin was dissolved separately in 70 % aqueous acetone. Polyphenol solution was mixed with freeze-dried kafirin microparticles by stirring. Samples were dried overnight at 25 °C. The resultant material was ground to a fine powder using a mortar and pestle to obtain catechin encapsulated into kafirin microparticles. Results indicated that release of antioxidant activity from kafirin microparticles encapsulating catechin was approximately 70 %. These results suggested a possibility of cross-linking between polyphenols and proteins (kafirin), hence protecting against enzymatic degradation under simulated gastric conditions. Kafirin microparticles have potential for encapsulating phenolic antioxidants to confer delayed and controlled release of the antioxidant activity over a prolonged period of time [122].

3.5 Microcapsules

Microcapsules are designed to release its contents when broken by pressure, dissolved, or melted. These microcapsules are stable against chemical attack when administered through various routes [123]. Microcapsules with dense or porous membranes can be produced by various techniques such as interfacial polymerization, in situ polymerization, and interfacial precipitation.

Shutava and his associates prepared polyphenol microcapsules carrying EGCG and gelatin by using the layer-by-layer assembly method. Films were assembled by sequential dipping of quartz slides into aqueous solutions of gelatin and EGCG for 15 min per layer with three intermediate washings in deionized water. $MnCO_3$ microcores were dispersed in deionized water using a bath sonicator; water was changed twice using centrifugation to precipitate the microparticles. EGCG in the layer-by-layer assemblies retained its antioxidant activity. Results suggested that the EGCG content in the polyphenol film material was as high as 30 % and these microcapsules protect encapsulated polyphenols against reactive oxygen species thereby prolonging their life and antioxidative efficiency [124].

Ethanol extracts of green tea (core material) were encapsulated as microcapsules using beta-cyclodextrin (wall material) by Liang and co-workers. The formulation was optimized by RSM using SAS software. Combined effect of four independent variables—core material/wall material ratio, cyclodextrin content, stirring time, and stirring temperature—were studied using a Box–Behnken design. The responses measured were encapsulation percentage and effects of green tea on bone mineral density (BMD); bone mineral content (BMC) was studied in 60 Kunming mice for 30 days. Ca, P, Zn, and Cu contents in blood and bone were measured by atomic absorption spectrophotometry. Results showed that the encapsulation yield of tea extract increased up to 96 % with the increase in core material/wall material ratio and beta-cyclodextrin content. A further increase in both of them caused the decline in the encapsulation yield. The animal studies showed that when cyclodextrin encapsulated tea extract was fed to young mice of 2 months, there was no change in BMD and BMC; however, there was a marked increase in both BMD and BMC in mice of 15 months. These indicated that beta-cyclodextrin encapsulation tea extract could markedly improve bone quality in aged animals [125].

3.6 Emulsion System

An emulsion presents a unique delivery system due to its ability to encapsulate both hydrophilic and hydrophobic compounds, induces rapid absorption, and also acts as a penetration enhancer [126].

In order to develop biocompatible carriers for oral delivery, Ru et al. prepared an oil-in-water emulsion of EGCG by high-pressure homogenization method. For the further stabilization of this emulsion system, ι-carrageenan and β-lactoglobulin were used. EGCG was first dissolved in polyethylene glycol–water (1:2) solution and then diluted with water. This mixture was added dropwise into 5 % canola oil (composed of 0.6 % ι-carrageenan and 0.4 % β-lactoglobulin) under high-speed homogenization. Finally, the emulsion was homogenized at ~15,000 psi pressure for three cycles to form stable and homogeneous oil-in-water emulsion. Cell proliferation inhibition studies showed a significant difference in the efficiency of EGCG in free form and emulsion system. About 50 μg/ml of free EGCG was able to inhibit cell proliferation, while the submicrometer emulsion of EGCG showed stronger anticancer effect at concentrations of 25 and 50 μg/ml. This result suggested that emulsion system enhanced the bioactivity of EGCG and further the pressure and the number of cycles applied during high-pressure homogenization influence the structure, size, and viscosity of emulsion system [127].

With an aim of improving the bioavailability of green tea extract, Kim and group prepared nano-emulsified green tea extract and investigated its antioxidant and hypolipidemic effects. Nanoemulsions were prepared by heating mixture A (cholesterol 2.5 %, phytosterol 2.5 %, Cetech-3 2.0 %, Cetech-5 2.0 %) and mixture B (cetyl phosphate 0.4 %, glycerine 7.0 %, water 92.6 %), mixing and homogenizing at 1,500 rpm for 5 min, followed by cooling at 45 °C. 10 % green tea extract and 15 % medium-chain fatty acid triglyceride (MCT) were added to the mixture and sprayed through a high-pressure microemulsifier at 13,500 psi. Mean particle size of 300 nm was obtained. In vivo analysis was carried out on the two identical groups of 57BL/6 mice. First group was fed on green tea extract and the other on nano-emulsified green tea extract for 4 weeks. Total and low-density lipoprotein cholesterol concentrations were significantly reduced after nano-emulsified green tea extract treatment in comparison with green tea extract, while antioxidative effect of green tea extract was found to be comparable with that of nanoemulsion. Protein expression of low-density lipoprotein receptor was significantly increased in the livers of nano-emulsified green tea extract-treated group as compared to the green tea extract-treated group. The results of in vivo studies suggested that the expression of 3-hydroxy-3-methylglutaryl coenzyme A reductase (a key enzyme involved in cholesterol biosynthesis) was significantly downregulated by nano-emulsified green tea extract. The expression of mRNA of sterol regulatory element-binding protein 2 (regulates the expression of HMG-CoA reductase gene) was reduced more in nano-emulsified green tea extract-treated group as compared to the green tea extract-treated group [19].

4 Conclusion

Oxidative stress is the major cause of a number of pathophysiological diseases in human body. Catechins present in green tea which are an important class of flavonoids due to their antioxidant property and other benefits are able to fight against therapeutic conditions including heart disease, hypercholesterolemia, dead brain cells, cancer, and microbial infections. Therefore, an efficient delivery system to stabilize and prolong the shelf life of green tea and its active components is needed. This chapter has summarized various nanocarrier-based delivery systems for encapsulation of green tea and its catechins, and results have demonstrated enhanced stability and bioavailability through oral and transepidermal route. The results of various studies prove primarily through in vitro results and cell line work that encapsulation of green tea and its active catechins enhances the antioxidant activity, stability against enzymatic degradation, and solubility issues. Higher encapsulation efficiencies were achieved via encapsulation in nanoparticles as compared to liposomes, micelles, or other systems. Nanoparticles using biodegradable polymers have been explored exhaustively for encapsulating catechins via various modes of preparation and show promising results in terms of enhancing the encapsulation, efficacy, and retention at the site of action. Some studies have even shown the possibility of targeting nano-encapsulated catechins to the regions of the brain for the treatment of central nervous system-related disorders. But more advanced in vivo work needs to be done to substantiate the results of the in vitro work in the future.

References

1. Gutteridge JM, Halliwell B (1992) Free radicals in biology and medicine. Oxford Science Publications, Oxford
2. Jenner P (2003) Oxidative stress in Parkinson's disease. Ann Neurol 53:S26–S36
3. Lyras L, Cairns NJ, Jenner A et al (1997) An assessment of oxidative damage to proteins, lipids, and DNA in brain from patients with Alzheimer's disease. J Neurochem 68:2061–2069
4. Dhalla NS, Temsah RM, Netticadan T (2000) Role of oxidative stress in cardiovascular diseases. J Hypertens 18:655–673
5. Kerr S, Brosnan MJ, McIntyre M et al (1999) Superoxide anion production is increased in a model of genetic hypertension: role of the endothelium. Hypertension 33:1353–1358
6. Kukreja RC, Hess ML (1992) The oxygen free-radical system: from equations through membrane–protein interactions to cardiovascular injury and protection. Cardiovasc Res 26:641–655
7. Cadenas E, Davies KJ (2000) Mitochondrial free radical generation, oxidative stress, and aging. Free Rad Biol Med 29:222–230
8. Mates JM (2000) Effects of antioxidant enzymes in the molecular control of reactive oxygen species toxicology. Toxicology 153:83–104
9. Kohen R, Nyska A (2002) Oxidation of biological systems: oxidative stress phenomena, antioxidants, redox reactions, and methods for their quantification. Toxicol Pathol 30:620–650
10. Bors W, Heller W, Michel C et al (1990) Flavonoids as antioxidants: determination of radical-scavenging efficiencies. Methods Enzymol 186:343–355
11. Pastore RL, Fratellone P (2006) Potential health benefits of green tea (Camellia sinensis): a narrative review. J Sci Healing 2:531–539
12. Gonzalez de Mejia E, Ramirez-Mares MV, Puangpraphant S (2009) Bioactive components of tea: cancer, inflammation and behavior. Brain Behav Immun 23:721–731
13. Sakanaka S, Aizawa M, Kim M et al (1996) Inhibitory effects of green tea polyphenols on growth and cellular adherence of an oral bacterium, Porphyromonas gingivalis. Biosci Biotech Biochem 60:745–749
14. Sharma A, Gupta S, Sarethy IP et al (2012) Green tea extract: possible mechanism and antibacterial activity on skin pathogens. Food Chem 135:672–675
15. Stoicov C, Saffari R, Houghton JM (2009) Green tea inhibits Helicobacter growth *in vivo* and *in vitro*. Int J Antimicrob Agents 33:473–478
16. Yamaguchi K, Honda M, Ikigai et al (2002) Inhibitory effects of (−)-epigallocatechin gallate on the life cycle of human immunodeficiency virus type 1 (HIV-1). Antivir Res 53:19–34
17. Weber JM, Ruzindana-Umunyana A, Imbeault L et al (2003) Inhibition of adenovirus infection and adenain by green tea catechins. Antivir Res 58:167–173
18. Song JM, Lee KH, Seong BL (2005) Antiviral effect of catechins in green tea on influenza virus. Antivir Res 68:66–74
19. Kim YJ, Houng SJ, Kim JH et al (2012) Nanoemulsified green tea extract shows improved hypocholesterolemic effects in C57BL/6 mice. J Nutr Biochem 23:186–191
20. Bursill CA, Roach PD (2006) Modulation of cholesterol metabolism by the green tea polyphenol

(-)-epigallocatechin gallate in cultured human liver (HepG2) cells. J Agric Food Chem 54:1621–1626
21. Miura Y, Chiba T, Tomita I et al (2001) Tea catechins prevent the development of atherosclerosis in apoprotein E-deficient mice. J Nutr 131:27–32
22. Andersen LF, Jacobs DR, Carlsen MH et al (2006) Consumption of coffee is associated with reduced risk of death attributed to inflammatory and cardiovascular diseases in the Iowa Women's Health Study. Am J Clin Nutr 83:1039–1046
23. Tang S, Sheehan D, Buckley DJ et al (2001) Antioxidant activity of added tea catechins on lipid oxidation of raw minced red meat, poultry and fish muscle. Int J Food Sci Technol 36:685–692
24. Atoui AK, Mansouri A, Boskou G et al (2005) Tea and herbal infusions: their antioxidant activity and phenolic profile. Food Chem 89:27–36
25. Cabrera C, Artacho R, Gimenez R (2006) Beneficial effects of green tea- a review. J Am Coll Nutr 25:79–99
26. Cao G, Sofic E, Prior RL (1996) Antioxidant capacity of green tea and common vegetables. J Agr Food Chem 44:3426–3431
27. Bendich A, Olson JA (1989) Biological actions of carotenoids. FASEB J 3:1927–1932
28. Goldman A (1995) Melatonin, a review. Brit J Clin Pharma 19:258–260
29. Guo CJ, Yang JJ, Wei JY et al (2003) Antioxidant activities of peel, pulp and seed fractions of common fruits as determined by FRAP assay. Nutr Res 23:1719–1726
30. Singh RP, Murthy KN, Jayaprakasha GK (2002) Studies on the antioxidant activity of pomegranate peel and seed extracts using *in vitro* models. J Agric Food Chem 50:81–86
31. Murthy KN, Jayaprakasha GK, Singh RP (2002) Studies on antioxidant activity of pomegranate peel extract using *in vivo* models. J Agric Food Chem 50:4791–4795
32. Anderson RA, Broadhurst CL, Polansky MM et al (2004) Isolation and characterization of polyphenol type-A polymers from cinnamon with insulin-like biological activity. J Agric Food Chem 52:65–70
33. Muricia MA, Egea I, Romozaro F et al (2004) Antioxidant evaluation in dessert spices compared with common food additives. Influence of irradiation procedure. J Agric Food Chem 52:1872–1881
34. Teresa EB, Yolanda GF, Rivas-Gonzalo JC et al (1992) Characterization of procyanidins of *Vitis vinifera* variety Tinta del Paris grape seeds. J Agric Food Chem 40:1794–1799
35. Şehirli O, Ozel Y, Dulundu E et al (2008) Grape seed extract treatment reduces hepatic ischemia reperfusion injury in rats. Phytother Res 22:43–48
36. Shao ZH, Becker LB, Vanden Hoek TL et al (2003) Grape seed proanthocyanidin extract attenuates oxidant injury in cardiomyocytes. Pharmacol Res 47:463–469
37. Kleijnen J, Knipschild P (1992) Ginkgo biloba. Lancet 340:1136–1139
38. Pietri S, Maurelli E, Drieu K et al (1997) Cardioprotective and antioxidant effects of the terpenoid constituents of *Ginkgo biloba* extract (EGb761). J Mol Cell Cardiol 29:733–742
39. Pincemail J, Dupuis M, Nasr C et al (1989) Superoxide anion scavenging effect and superoxide dismutase activity of *Ginkgo biloba* extract. Experientia 45:708–712
40. Katiyar CK, Brindavanam NB, Tiwari P et al (1997) Immunomodulation. Narosa Publishing House, New Delhi
41. Oshiro M, Kuroyanagi M, Uneo A (1990) Structures of sesquisterpenes from *Curcuma longa*. Phytochemistry 29:2201–2205
42. Sawasha SG, Yamagar VT, Gondkar PP et al (2003) Chemical control of leaf spot disease of turmeric (*Curcuma longa*). Pestology 27:23–24
43. Bishayee Sarkar A, Chaterjee M (1995) Hepatoprotective activity of (*Daucus carota* L.) against carbon tetrachloride intoxication in mouse liver. J Ethnopharmacol 47:69–74
44. Ruberto G, Bratta MT, Deans S et al (2000) Antioxidant and antimicrobial activity of *Foeniculum vulgare* L. and *Crithmum maritimum* essential oils. Plants Med 66:687–693
45. Mortez – Semnani K, Saeedi M, Shahnavaz B (2003) Comparison of antioxidant activity of extract from roots of liquorice (glycyrrhiza glabra l.) to commercial antioxidants in 2% hydroquinone cream. J Cosmet Sci 54:551–558
46. Haraguchi H, Inoue J, Tamura Y, Mizutani K (2000) Inhibition of mitochondrial lipid peroxidation by bakuchiol, a monoterpene from *Psoralea corylifolia* L. Planta Med 66:569–571
47. Devi PU, Ganasoundari A (1999) Modulation of glutathione and antioxidant enzymes by ocimum sanctum and role in protection against radiation injury. Indian J Exp Boil 37:262–268
48. Vande Velde V, Lavie D (1982) A α- 16 withanolide in *Withania somnifera* L. as possible precursor for α- side chain. Phytochemistry 21:731–733
49. Sudhir S, Bhudhiraja RD, Miglani GP et al (1986) Pharmacological studies on leaves of *Withania somnifera* L. Planta Med 52:61–63
50. Sultana S, Parwaiz S, Iqbal M et al (1995) Crude extracts of hepatoprotective plants, *Solanum nigrum* and *Cichorium intybus* inhibit free radical mediated DNA damage. J Ethnopharmocol 45:189–192
51. Gupta S, Gabrani R, Ali J et al (2011) Exploring novel approaches to vaginal drug delivery. Recent Pat Drug Deliv Formul 5:82–94
52. Yukihiko H (2001) Green tea: health benefits and applications. Marcel Dekker Inc., New York
53. Chacko SM, Thambi PT, Kuttan R et al (2010) Beneficial effects of green tea: a literature review. Chin Med 5:13
54. Vinson JA (2000) Black and green tea and heart disease: a review. Biofactors 13:127–132
55. Scott BC, Butler J, Halliwell B et al (1993) Evaluation of the antioxidant actions of ferulic acid and catechins. Free Radic Res Commun 19:241–253

56. Rice-Evans CA, Miller NJ, Paganga G (1996) Structure-antioxidant activity relationships of flavonoids and phenolic acids. Free Radic Biol Med 20:933–956
57. Yang CS, Maliakal P, Meng X (2002) Inhibition of carcinogenesis by tea. Annu Rev Pharmacol Toxicol 42:25–54
58. Brown JE, Khodr H, Hider RC et al (1998) Structural dependence of flavonoid interactions with Cu^{2-} ions: implications for their antioxidant properties. Biochem J 330:1173–1178
59. Surh YJ, Chun KS, Cha HH et al (2001) Molecular mechanisms underlying chemopreventive activities of anti-inflammatory phytochemicals: down-regulation of COX-2 and iNOS through suppression of NF-kappa B activation. Mutat Res 480–481:243–268
60. Baba S, Osakabe N, Natsume M, Muto Y et al (2001) In vivo comparison of the bioavailability of (+)-catechin, (-)-epicatechin and their mixture in orally administered rats. J Nutr 131:2885–2891
61. Mochizuki M, Yamazaki S, Kano K et al (2002) Kinetic analysis and mechanistic aspects of auto-oxidation of catechins. Biochim Biophys Acta 1568:35–44
62. Janeiro P, Brett AMO (2004) Catechin electrochemical oxidation mechanisms. Anal Chim Acta 518:109–115
63. Rahman K (2007) Studies on free radicals, antioxidants, and co-factors. Clin Interv Aging 2:219–236
64. Su YN, Leung LK, Huang Y et al (2003) Stability of tea theaflavins and catechins. Food Chem 83:189–195
65. Catterall F, King LJ, Clifford MN et al (2003) Bioavailability of dietary dose of 3H-labelled tea antioxidants (+)-catechin and (-)-epicatechin in rats. Xenobiotica 33:743–753
66. Cai Y, Anavy ND, Chow HSS (2002) Contribution of presystemic hepatic extraction to the low oral bioavailability of green tea catechins in rats. Drug Metab Dispos 30:1246–1249
67. Kadowaki M, Sugihara N, Tagashira T et al (2008) Presence or absence of a gallate moiety on catechins affects their cellular transport. J Pharm Pharmacol 60:1189–1195
68. Yang H, Finaly R, Teitelbaum DH (2003) Alteration in epithelial permeability and ion transport in a mouse model of total parenteral nutrition. Crit Care Med 31:1118–1125
69. Windlansky ME, Hamburg M, Anter E et al (2007) Acute EGCG supplementation reverses endothelial dysfunction in patients with coronary artery disease. J Am Coll Nutr 26:95–102
70. Batchelder RJ, Calder RJ, Thomas CP et al (2004) *in vitro* transdermal delivery of the major catechins and caffeine from extract of *camellia sinensis*. Int J Pharma 283:45–51
71. Astrup AV, Tourbo S (2011) Composition for weight reduction comprising capsaicin, green tea extract, L-tyrosine and caffeine. US Patent 7,867,526, 11 Jan 2011
72. Udell RG (2011) Soft gel capsules containing polymethoxylated flavones and palm oil tocotrienols. US Patent 7,887,852, 15 Feb 2011
73. Romanczyk Jr LJ, Schmitz HH (2011) Epicatechin for hypertension treatment. US Patent 7,875,651, 25 Jan 2011
74. Newmark T, Schulick P, Katz A (2010) Methods for modulating eicosanoid metabolism. US Patent 7,744,934, 29 June 2010
75. Randolph RK, Roh-Schimdt H (2010) Cytokine modulators and related methods of use. US Patent 7,758,903, 20 July 2010
76. Morgan C (2010) Topical medicament. US Patent 7,704,522, 27 Apr 2010
77. Kingsley JD (2010) Methods for eradicating lice and fleas from a host. US Patent 7,807,190, 5 Oct 2010
78. Greaves E, Greaves JT (2010) Composition for dyeing keratin fibers and a method of dyeing hair using same. US Patent 7,749,286, 6 July 2010
79. Schlesser JL (2010) Nutritional supplement to enhance learning, academic, and behavioral functioning. US Patent 7,771,756, 10 Aug 2010
80. Mora-gutierrez A, Gurin MH (2010) Bioactive complexes compositions and methods of use thereof. US Patent 7,780,873, 24 Aug 2010
81. Wu RY, Lai TH, Chyan YJ et al (2010) Chinese herb extract for treating dementia and preparation method thereof. US Patent 7,824,714, 2 Nov 2010
82. Rombi M (2004) Green tea extract for treating obesity. US Patent 6,830,765, 14 Dec 2004
83. Simmons DL, Dong C (2003) Liquid compositions comprising non-digestible oligosaccharides and green tea catechins, method and uses thereof. WIPO Patent WO2004000045, 31 Dec 2003
84. Schonrock U, Max H (2002) Use of a content of catechins or a content of green tea extract in cosmetic preparations for tanning the skin. US Patent 6,399,046, 4 June 2002
85. Morre DM, Morre JD, Cooper R et al (2002) Tea catechin formulations and processes for making same. US Patent 6,428,818, 6 Aug 2002
86. Xiong W, Quan D, Patel DC (2001) Effervescent green tea extract formulation. US Patent 6,299,925, 9 Oct 2001
87. Reed JC, Pellecchia M (2010) Methods and compounds useful to induce apoptosis in cancer cells. US Patent 7,812,058, 12 Oct 2010
88. Ekanayake A, Kirksey ST, Pultinas Jr EP (1995) Process for making a stable green tea extract and product. US Patent 5,427,806, 27 June 1995
89. Des Rieux A, Ragnarsson EG, Gullberg E et al (2005) Transport of nanoparticles across an in vitro model of the human intestinal follicle associated epithelium. Eur J Pharm Sci 25:455–465
90. Siddiqui IA, Adhami VM, Bharali DJ et al (2009) Introducing nanochemoprevention as a novel approach for cancer control: proof of principle with green tea polyphenol epigallocatichin-3-gallate. Cancer Res 69:1712–1716
91. Italia JL, Datta P, Ankola DD et al (2008) Nanoparticles enhance *per oral* bioavailability of poorly available molecules: epigallocatechin gallate nanoparticles ameliorates cyclosporine induced

nephrotoxicity in rats at three times lower dose than oral solution. J Biomed Nanotechnol 4:304–312
92. Dutta PK, Tripathi S, Mehrotra GK et al (2009) Perspectives for chitosan based antimicrobial films in food applications. Food Chem 114:1173–1182
93. Mi FL, Wu FB, Shyu SS et al (2002) Control of wound infections using bilayer chitosan wound dressing with sustainable antibiotic delivery. J Biomed Mater Res 59:438–449
94. Hu B, Pan C, Sun Y et al (2008) Optimization of fabrication parameters to produce chitosan-tripolyphosphate nanoparticles for delivery of tea catechins. J Agric Food Chem 56:7451–7458
95. Chen ZY, Zhu QY, Wong YF et al (1998) Stabilizing effect of ascorbic acid on green tea catechins. J Agric Food Chem 46:2512–2516
96. Dube A, Ng K, Nicolazzo JA et al (2010) Effective use of reducing agents and nanoparticle encapsulation in stabilizing catechins in alkaline solution. Food Chem 122:662–667
97. Green RJ, Murphy AS, Schulz B et al (2007) Common tea formulations modulate in vitro digestive recovery of green tea Catechins. Mol Nutr Food Res 51:1152–1162
98. Ogawa Y, Yamaguchi F, Yuasa K et al (2007) Efficient production of g- polyglutamic acid by Bacillus subtilis (natto) in jar fermenters. Biosci Biotech Biochem 61:1684–1687
99. Tang D, Yu SH, Ho YC et al (2013) Characterization of tea catechins-loaded nanoparticles prepared from chitosan and an edible polypeptide. Food Hydrocolloid 30:33–41
100. Hamidi M, Azadi A, Rafiei P (2008) Hydrogel nanoparticles in drug delivery. Adv Drug Deliv Rev 60:1638–1649
101. Chen YC, Yu SH, Tsai GJ et al (2010) Novel technology for the preparation of self- assembled catechin/gelatine nanoparticles and their characterization. J Agric Food Chem 58:6728–6734
102. Gomes JFPS, Rocha S, do Carmo Pereira M et al (2010) Lipid/particle assemblies based on maltodextrin-gum arabic core as bio-carriers. Colloids Surf B: Biointerfaces 76:449–455
103. Ma QH, Xia Q, Lu YY et al (2007) Preparation of tea polyphenols-loaded solid lipid nanoparticles based on the phase behaviors of hot microemulsions. Solid State Phenom 121–123:705–708
104. Smith A, Giunta B, Bickford PC et al (2010) Nanolipidic particles improve the bioavailability of epigallocatechin-3-gallate (EGCG) for the treatment of Alzheimer's disease. Int J Pharm 389:207–212
105. Manea AM, Vasile BS, Meghea A (2013) Antioxidant and antimicrobial activities of green tea extract loaded into nanostructured lipid carriers. C R Chimie 17(4):331–341
106. Liang J, Li F, Fang Y et al (2014) Cytotoxicity and apoptotic effects of tea polyphenol-loaded chitosan nanoparticles on human hepatoma HepG2 cells. Mater Sci Eng 36:7–13
107. Torchilin V (2007) Micellar nanocarriers: pharmaceutical perspective. Pharm Res 24:1–16
108. Shpigelmann A, Israeli G, Livney YD (2010) Thermally-induced protein–polyphenol co-assemblies: beta lactoglobulin-based nanocomplexes as protective nanovehicles for EGCG. Food Hydrocoll 24:735–743
109. Gomez-Hens A, Fernandez-Romero JM (2006) Analytical methods for the control of liposomal delivery systems. Trend Anal Chem 2:167–178
110. Vedha Hari BN, Chitra KP, Bhimavarapu R et al (2010) Novel technologies: a weapon against tuberculosis. Indian J Pharmacol 42:338–344
111. Fang JY, Hwang TL, Huang YL et al (2006) Enhancement of the transdermal delivery of catechins by liposomes incorporating anionic surfactants and ethanol. Int J Pharm 310:131–138
112. Lee JS, Chung D, Lee HG (2008) Preparation and characterization of calcium pectinate gel beads entrapping catechin-loaded liposomes. Int J Biol Macromol 42:178–184
113. Huang YB, Tsai MJ, Wu PC et al (2011) Elastic liposomes as carriers for oral delivery and the brain distribution of (+)-catechin. J Drug Target 19:708–718
114. Ramadan MF (2012) Antioxidant characteristics of phenolipids (quercetin-enriched lecithin) in lipid 482 matrices. Ind Crop Prod 36:363–369
115. Rashidinejad A, John Birch E, Sun-Waterhouse D et al (2014) Delivery of green tea catechin and epigallocatechin gallate in liposomes incorporated into low-fat hard cheese. Food Chem 156:176–183
116. Siepmann J, Siepmann F (2006) Microparticles used as drug delivery systems. Progr Colloid Polymer Sci 133:15–21
117. Padalkar AN, Shahi SR, Thube MW (2011) Microparticles: an approach for betterment of drug delivery system. Int J Pharm Res Dev 1:99–115
118. Wisuitiprot W, Somsiri A, Ingkaninan K et al (2011) A novel technique for chitosan microparticle preparation using a water/silicone emulsion: green tea model. Int J Cosmet Sci 33:351–358
119. Lee JS, Kim HW, Chung D et al (2009) Catechin-loaded calcium pectinate microparticles reinforced with liposome and hydroxypropylmethylcellulose: optimization and *in vivo* antioxidant activity. Food Hydrocoll 23:2226–2233
120. Fu N, Zhou Z, Jones TB et al (2011) Production of monodisperse epigallocatechin gallate (EGCG) microparticles by spray drying for high antioxidant activity retention. Int J Pharm 413:155–166
121. Sosa MV, Rodriguez-Rojo S, Mattea F et al (2010) Green tea encapsulation by means of high pressure antisolvent coprecipitation. J Supercritical Fluids 56:304–311
122. Taylor J, Taylor JRN, Belton PS et al (2009) Kafirin microparticle encapsulation of catechin and sorghum condensed tannins. J Agric Food Chem 57:7523–7528
123. Venkatasan P, Manavalan R, Valliappan K (2009) Microencapsulation: a vital technique in novel drug delivery system. J Pharm Sci Res 1:26–35
124. Shutava TG, Balkundi SS, Lvov YM (2009) (-)-Epigallocatechin gallate/gelatin layer-by-layer assembled films and microcapsules. J Colloid Interface Sci 330:276–283

125. Haidong L, Fang Y, Zhihonga T et al (2011) Study on preparation of cyclodextrin encapsulation tea extract. Int J Biol Macromol 49:561–566
126. Gupta S, Sahni JK, Ali J et al (2012) Development and characterization of green tea loaded microemulsion for vaginal infections. Adv Mat Lett 3:493–497
127. Ru Q, Yu H, Huang Q (2010) Encapsulation of epigallocatechin-3-gallate (EGCG) using oil-in-water (O/W) submicrometer emulsions stabilized by ι-carrageenan and β-lactoglobulin. J Agric Food Chem 58:10373–10381

Web References

128. Progressive health's balance point for men – product review. Available at: http://www.supplementnews.org/Progressive_Health/Balance_Point_for_Men. Accessed 25 June 2013
129. Progressive health's balance point for women – product review. Available at: http://www.supplementnews.org/Progressive_Health/Balance_Point_for_Women. Accessed 25 June 2013
130. Muscletech hydroxycut hardcore X- 120 caps. Available at: http://www.indiasupplement.com/muscletech-hydroxycut-hardcore-x-120-caps. Accessed 25 June 2013
131. Methyl ripped review. Available at: http://www.dietpillrating.com/methyl-ripped/. Accessed 26 June 2013
132. Heaven & Earth's women's super vites – product review. Available at: http://www.supplementnews.org/Heaven_&_Earth/Womens_Super_Vites. Accessed 26 June 2013
133. Diet-Rx pill natural appetite control without prescription. Available at: https://physicianformulas.com/store/scripts/prodview.asp?idproduct=319. Accessed 26 June 2013
134. Mega-T green tea dietary supplement (Caplets 120) – 60 days supply. Available at: http://www.aragonproducts.com/theproducts.cfm?master=6754. Accessed 28 June 2013
135. Teavigo by DSM and Teacare's collaboration continues to set new standards. Available at: http://www.dsm.com/en_US/html/dnp/news_items/23012008_Teavigo_and_Teacare.htm. Accessed 28 June 2013
136. Serious nutrition solutions. Available at: http://forum.bodybuilding.com/showthread.php?t=126764293&page=1. Accessed 28 June 2013
137. Pharmanex product information sheet. Available at: http://premarketsouthafrica.com/downloads/lifepak-nano.pdf. Accessed 29 June 2013.
138. Natrol slenderite- supplementary facts. Available at: http://www.vitacost.com/Natrol-Slenderite. Accessed 29 June 2013
139. Green tea fusion fat burner. Available at: http://www.netnutri.com/browse.cfm/4,3275.html. Accessed 29 June 2013
140. FDA approvals: Ziana, Kadian, Polyphenon E. Available at: http://www.medscape.com/viewarticle/548709. Accessed 29 June 2013.
141. About Veregen (Sinecatechins) ointment. Available at: http://www.veregen.com/veregenrx/pdver_web_default.html Accessed 30 June 2013
142. Gattefosse Optivegetol-Products. Available at: http://www.alfa-chemicals.co.uk/Divisions/PersonalCare/PersonalCare-Products/PersonalCare-ProductGroupDetails.aspx?p=324&Product%20Group=Gattefoss%C3%A9+Optivegetol%E2%84%A2 Accessed 30 June 2013.
143. Himalaya green tea. Available at: http://himalayawellness.com/products/personalcare/greentea.htm. Accessed 30 June 2013
144. Green Tea Herbasol. Available at: http://www.lipoid-kosmetik.com. Accessed 30 June 2013

Antioxidative Peptides Derived from Food Proteins

Dominic Agyei, Michael K. Danquah, Indira P. Sarethy, and Sharadwata Pan

Abstract

The search for natural antioxidants is an ongoing endeavour as an aid to combat the harmful effects of free radicals. Research advances in the past few decades have shown that, by controlled enzymatic hydrolysis, natural antioxidants can be produced from food proteins. In this chapter, the role of certain antioxidative peptides derived from food proteins is discussed in relation to their prospect in the prevention of oxidative stress. The molecular diversity of these food peptides is described together with their pharmacological effects and mechanisms of action in relation to antioxidation. The production of these peptides and the elucidation of their antioxidative peptides are also presented. Owing to their therapeutic potential, antioxidative peptides derived from food proteins can be incorporated as ingredients in functional foods, nutraceuticals and pharmaceuticals, where their biological activities may inhibit product oxidation or assist in the control and prevention of diseases induced by free radicals. However, further insightful research is needed to overcome certain scientific challenges and thereby increase and promote consumer acceptance of these natural antioxidants.

Keywords

Oxidative stress • Food proteins • Bioactive antioxidative peptides • Bioprocessing applications • Human health

D. Agyei
Department of Chemical Engineering, Monash University, Clayton, VIC 3800, Australia

M.K. Danquah
Department of Chemical Engineering, Monash University, Clayton, VIC 3800, Australia

Department of Chemical and Petroleum Engineering, Curtin University of Technology, Miri, Sarawak 98009, Malaysia

I.P. Sarethy (✉)
Department of Biotechnology, Jaypee Institute of Information Technology, Noida, Uttar Pradesh 201307, India
e-mail: indirap.sarethy@jiit.ac.in

S. Pan
Department of Chemical Engineering, Indian Institute of Technology Bombay, Mumbai, Powai 400076, India

V. Rani and U.C.S. Yadav (eds.), *Free Radicals in Human Health and Disease*, DOI 10.1007/978-81-322-2035-0_26, © Springer India 2015

1 Introduction: Oxidative Stress and Effects on Human Health

Free radicals (atoms with an unpaired electron) play an important role in biology. They are mainly derived from oxygen (as reactive oxygen species/ROS) and nitrogen (as reactive nitrogen species/RNS). The production of free radicals can be due to normal endogenous metabolic processes and pathophysiological states or via exposure to external physicochemical conditions (such as pollutants, smoke, xenobiotics and radiation) [1]. Free radicals are necessary for the body because they play crucial roles in cellular redox signalling and homeostasis. Thus, the body monitors and inhibits the production of high levels of free radicals by using a plethora of mechanisms for minimising free radical-induced damages and/or for repairing damages that occur. Endogenous enzymes (such as superoxide dismutase, catalase, glutathione peroxidase and glutathione reductase) and antioxidants from dietary sources (vitamin A, vitamin C and vitamin E) play a key role in the defence mechanism against free radicals [2]. The failure of these mechanisms to counteract the harmful effects of free radicals (ROS) results in tissue injury and is termed oxidative stress [1–3]. Free radical chemistry in the body is of much concern because the reaction of a radical with a non-radical (such as lipid, protein, carbohydrate and nucleic acid) results in a free radical chain reaction and the formation of new radicals which in turn can react with new macromolecules [3, 4]. Some of the harmful effects of free radical action include lipid peroxidation; protein carbonylation, oxidation and cross-linking; DNA base conversion and strand breaks; and protein/DNA cross-linking [4]. All these disrupt normal cellular metabolic and signalling mechanisms. Thus, oxidative stress has been implicated in the process of ageing and also in the aetiology of over a hundred major diseases such as neurodegenerative diseases (Lou Gehrig's disease, Parkinson's disease, Alzheimer's disease, Huntington's disease), cardiovascular diseases (myocardial infarction, strokes), sickle cell disease and chronic fatigue syndrome [1, 4, 5].

2 Significance of Antioxidants

2.1 Role of Antioxidants

Given the potential for tissue damage by free radicals, antioxidants are used to prevent and/or improve different diseased states. Endogenous antioxidant defences of the body can be classified into enzymatic and nonenzymatic types [2]. They can either be in the intracellular or extracellular medium or include enzymes such as superoxide dismutase (SOD), catalase (CAT), glutathione peroxidases (GPx), glutathione reductase (Gred), peroxiredoxins, etc. In connection with this, there are some chemical elements which have no antioxidant action themselves and are instead required for the activity of some antioxidant enzymes. These are called antioxidant nutrients and include selenium, copper, zinc, iron and manganese. The nonenzymatic antioxidants include glutathione (GSH), melatonin, uric acid, ubiquinol (coenzyme Q), α-tocopherol and tocotrienols (vitamin E), ascorbic acid (vitamin C), lipoic acid and carotenes (including retinol, vitamin A and β-carotene) among others [6]. Generally, the action of antioxidants is at the levels of prevention (where formation of ROS is stopped by enzymatic antioxidants), interception (via radical scavenging) and repair (via the action of repair enzymes) [1, 7].

2.2 Synthetic and Natural Antioxidants

Nonenzymatic antioxidants are obtained from supplements or antioxidant-containing foods and have been known to help reduce oxidative damage and reduce the risk in development of chronic diseases such as cancer and cardiovascular diseases [2, 6–8].

Synthetic and natural food antioxidants are also used as additives to protect food and medicine quality by preventing oxidative deterioration such as lipid oxidation. Examples of these include butylated hydroxytoluene (BHT), butylated hydroxyanisole (BHA), tert-butylhydroquinone (TBHQ), ethylene diamine tetraacetic acid (EDTA) and gallates such as propyl gallate, dodecyl gallate and octyl gallate. However, in recent years, preference for natural antioxidants is escalating as most synthetic antioxidants are disadvantaged with high toxicity and carcinogenic or mutagenic effects [2, 7]. For example, although BHT and BHA have a history of use as in food industry, cosmetics and therapeutic industry, they have recently been reported to be dangerous for human health, especially when used at high concentration. BHA and BHT are highly volatile and unstable at elevated temperatures [7]. As such, it is feared that their degradation products might exert toxic effects [9]. Moreover, legislation requirements are being tightened regarding the use of synthetic food additives because of recent toxicological studies and the increasing interest among consumers for 'natural' approaches to disease prevention or management. Natural antioxidants are of low cost and highly compatible with dietary intake and exert no harmful effects inside the human body. These make them promising alternatives to the use of synthetic antioxidants in food, cosmetic and therapeutic products [7]. The use of dietary natural antioxidants has been practised for centuries and reflects beliefs and practices of the age old saying of Hippocrates 'let food be thy medicine and thy medicine food'.

2.3 Bioactive Peptides in Disease Control

Bioactive peptides and hydrolysates derived from food protein sources have shown promising functions in modulating several endogenous physiological processes in living body systems. The biological activities exhibited by these products are due to the presence of special peptides in forms and quantities that trigger beneficial health effects in vitro and in vivo. These peptides trigger physiological responses such as antihypertensive, immunomodulatory, anticancer, antimicrobial and antioxidant activities [10–18]. Aside phytochemicals, several food-derived proteins and protein hydrolysates have been shown to exhibit various antioxidant activities [19, 20], and recent literature has identified several antioxidative food protein hydrolysates of plant, marine and animal origin which are utilised in health, pharmaceutical and food products as well as in the control of oxidative stress.

3 Antioxidative Peptides: Functionalities and Mechanism

Phenolic groups in plant-derived antioxidants allow the 'deactivation' of oxidant species, and this is the subject of intense research [21]. Similarly, due to their three-dimensional structure, proteins and peptides are able to perform certain physiological roles such as antioxidative activities in vitro and in vivo. Over the past few decades, research has unravelled a number of antioxidative peptides from a wide range of food proteins (see Table 1). Some examples of these are presented below.

3.1 Sources of Antioxidative Peptides

3.1.1 Plant Protein Derived

The classic source of nonenzymatic antioxidants is plant-derived polyphenols obtained from a wide array of plant-based foods and medicinal plants. However, in recent years, some plant protein hydrolysates have been shown to exhibit potent antioxidative properties.

Sweet Potato (*Ipomoea batatas*): Mu and Sun [35] have recently identified potent antioxidative preparations from an alcalase hydrolysate of sweet potato. Fractions of molecular weight below 3 kDa showed protective effects against DNA damage via hydroxyl radical-scavenging activity (50 % inhibitory concentration, IC_{50}= 1.74 mg mL^{-1}) and Fe^{2+}-chelating ability (IC_{50} of 1.54 mg mL^{-1}).

Table 1 Antioxidative properties of food-derived proteins and peptides

Type of food	Protein source of peptide or hydrolysate	Protein source of peptide or hydrolysate	Purification method	Antioxidative peptides identified	Antioxidant capacity measurement (in vitro)	Ref.
Plant	Soy protein	Alcalase	Ultrafiltration and consecutive chromatography (gel filtration, reversed phase)	Fractionated soy protein hydrolysates	LAPS, TBARS and peroxide value measurements	[22]
	Zein (corn protein)	Alcalase	Gel filtration, ultrafiltration and reversed-phase HPLC	YA LMCH	ABTS+, DPPH● and superoxide anion scavenging effects	[23]
	Rice endosperm protein	Alcalase, chymotrypsin, neutrase, papain and flavorase	Consecutive chromatography (ion exchange, gel filtration, reversed phase)	FRDEHKKKHDRGDEF	Superoxide anion, DPPH● and ●OH radical-scavenging activities, LAPS	[24]
Meat	Porcine myofibrillar protein	Papain, actinase E	Ion exchange and reversed-phase chromatography	DSGVT, IEAEGE, DAQEKLE, EELDNALN, VPSIDDQEELM	LAPS, DPPH● radical-scavenging capacity, metal-ion chelation	[25]
Chicken egg	Egg white	Pepsin	Unpurified	Egg white hydrolysates	ORAC, MDA	[26]
Milk	Whey proteins	Alcalase	Unpurified	Whey protein hydrolysates	DPPH● radical-scavenging capacity, Cu^{2+} ion chelation, TBARS reducing powers	[27]
	Casein from yak milk	Trypsin, pepsin, flavorase, flavorzyme and papain	Unpurified	Yak casein hydrolysates	DPPH●, superoxide ($O_2^{\bullet -}$) anion and hydrogen peroxide scavenging capacity	[28]

Fish	Blue mussel (*Mytilus edulis*)	Fermentation	Consecutive chromatography (ion exchange, gel filtration and reversed phase)	FGHPY	LAPS and •OH radical scavenging (ESR)	[29]
	Horse mackerel (*Megalaspsis cordyla*)	Pepsin, trypsin and α-chymotrypsin	Consecutive chromatography (ion exchange and gel filtration)	NHRYDR	DPPH•, •OH and peroxyl radical scavenging (ESR)	[30]
Mycoproteins / fungi	*Ganoderma lucidum*	Fermentation	ultrafiltration and gel-based ligand chromatography	Unidentified	LAPS measurements, •OH, superoxide anion radical scavenging, iron chelation activity	[31]
Algae	Marine algae (*Chlorella ellipsoidea*)	Papain, trypsin, pepsin and α-chymotrypsin	Consecutive chromatography (gel filtration and reversed-phase HPLC)	LNGDVW	DPPH•, •OH and peroxyl radical scavenging (ESR)	[32]
	Microalgae (*Pavlova lutheri*)	Yeast proteolytic enzymes	Consecutive chromatography (ion exchange and reversed phase)	MPGPLSPL	DPPH•, •OH and superoxide radical scavenging (ESR)	[33]
	Microalgae (*Chlorella vulgaris*)	Pepsin	$(NH_4)_2SO_4$ precipitation, and consecutive chromatography (gel filtration and ion exchange)	VECYGPNRPQF	DPPH•, •OH and superoxide, peroxyl and ABTS radical scavenging	[34]

Abbreviations: *ABTS*$^{\bullet+}$ 2,2′-azinobis(3-ethylbenzothiazoline-6-sulfonic acid), *LAPS* linoleic acid peroxidation system, *DPPH* 2,2-diphenyl-1-picrylhydrazyl, *ESR* electron spin resonance, *TBARS* thiobarbituric acid reactive substances, *ORAC* oxygen radical absorbance capacity, *MDA* malon-dialdehyde assay

These antioxidative properties were attributed to the presence of amino acids such as histidine, methionine, cysteine, tyrosine and phenylalanine, as well as other hydrophobic amino acids, all of which have antioxidant properties.

Aside the biological application, the production of antioxidative peptides from sweet potato is also an important environmental remediation strategy. This is because commercial production of sweet potato starch generates huge quantities of waste water rich in protein, sugars and minerals. These nutrients increase the economic cost of disposal of the waste water due to the high biochemical oxygen demand (BOD) and chemical oxygen demand (COD). Therefore, the recovery of bioactive sweet potato hydrolysates from starch waste water also assists in lowering pollution and reduces the wastage of protein resources [35].

Soybean (*Glycine max*): Soybean is one of the most protein-rich cultivated plants in the world. It has been known to possess multiple health-promoting functions due to the presence of several nutrients including bioactive proteins and peptides [36, 37]. Park and others [22] have reported a potent antioxidant hydrolysate fraction obtained by alcalase treatment of defatted soy proteins. Low molecular mass hydrolysate fractions (NMWCO <3 kDa) showed strong antioxidant properties when assayed with the thiobarbituric acid reactive substances assay. The potent antioxidant peptide was rich in the hydrophobic amino acids (phenylalanine, alanine and proline). In another study, Roblet and others [38] observed that ultrafiltration of soy protein hydrolysates obtained from a double enzymatic action of pepsin and pancreatin yielded peptides with antioxidant capacity (as assayed by oxygen radical absorbance capacity assay). Soy protein hydrolysates with enhanced antioxidant activity therefore have potential for use as a functional ingredient in food and medicinal products. Other well-studied plant protein sources of antioxidative peptides include rice endosperm proteins, yam tuber proteins, defatted peanut kernels, chickpea proteins, corn proteins (zein), alfalfa leaf proteins, sunflower proteins and wheat gluten [20, 39].

3.1.2 Animal Protein Derived

3.1.2.1 Marine Animal Foods

The biodiversity of the sea and marine life is huge and enormous. The marine environment is home to a wide range of organisms that contain natural compounds with medicinal properties.

Oyster: Recently, very potent antioxidative peptides have been isolated from alcalase hydrolysates of proteins from the Akoya pearl oyster (*Pinctada fucata*). When the peptides were formulated into cosmetic cream and applied topically, the preparation slowed down the process of lipid peroxidation (as evaluated by TBARS assay) [40]. In the same study, female Kunming mice were used as models for studying stimulated photo-ageing and response of skin elasticity, water content, wrinkling and histology of mice skin to UVB irradiation. *P. fucata* antioxidant peptides in the cosmetic cream were able to slow down the progress of wrinkle development and UVB irradiation-induced decrease in skin elasticity [40]. Such preparations could therefore be useful ingredients for non-invasive topical skin treatments to retard or reverse the effects of photo-ageing on human skin.

Fish: In another study, pepsin, trypsin and α-chymotrypsin, in an in vitro simulated gastro-intestinal digestion, were used to hydrolyse the skins of two marine fishes – horse mackerel (*Megalaspis cordyla*) and croaker (*Otolithes ruber*). After purification, two peptides (NHRYDR and GNRGFACRHA) were obtained, and both peptides exhibited high Fe^{2+} chelation comparable to EDTA, as well as anti-peroxidation of polyunsaturated fatty acid (PUFA) that was greater in activity than that of the natural antioxidant α-tocopherol [30]. Horse mackerel and croaker skin therefore have a potential as nutraceutical and bioactive foods wherein, upon consumption, they may be expected to protect against oxidative damage in living systems.

It has been estimated that the fish processing industry produces over 60 % by-product waste in the forms of skins, head, viscera, trimmings, liver, frames, bones and roes [41]. These by-products are rich in proteins and can therefore be processed into high-market-value products such as bioactive peptides. Chalamaiah and others [41] have reviewed the antioxidative properties of fish protein hydrolysates prepared from different parts of fish. A great variety of other marine fish protein hydrolysates have also been screened for their antioxidant properties [41–43], and the possible applications of these biomolecules are limitless.

3.1.2.2 Milk and Dairy Products

Milk and dairy proteins constitute the most widely studied and most abundant source of bioactive peptides, and these peptides can be encrypted in the caseins or whey proteins [10].

Casein: Peptides derived from a peptic digest of casein have been isolated and characterised for their free radical-scavenging activities [44]. These peptides were purified by a range of chromatographic methods (ion exchange, gel filtration and reversed-phase HPLC) yielding peptides with strong scavenging affinity for superoxide, hydroxyl and DPPH radicals. The peptides identified included EL, YFYPEL, FYPEL, YPEL and PEL. This demonstrates that a Glu-Leu sequence is important for the activity [44].

Whey: Whey proteins are important dietary source of branched-chain amino acids which have been shown to stimulate protein synthesis [45]. Additionally, from the results of preclinical studies in rodents, whey proteins have been shown to possess several health functions such as anti-inflammatory or anticancer and antihypertensive properties [46].

Antioxidant properties of whey protein hydrolysates and peptides have also been demonstrated. In their studies, Peña-Ramos and others [47] were able to purify peptides with lipid anti-peroxidation activities by reacting whey protein isolates with the commercial enzymes Alcalase, Protamex and Flavourzyme. The peptide fractions were purified by size exclusion chromatography and showed that higher-molecular-weight fractions (>45 kDa) possessed a higher TBARS inhibition effect than lower-molecular-weight fractions and hydrolysate mixtures. Amino acid composition assay revealed that the prevalence of histidine and hydrophobic amino acids was responsible for the antioxidative activities [47]. In another study, a range of different antioxidative activities were displayed by Alcalase-treated fractions of whey protein isolates. Compared with non-hydrolysed whey protein isolates (WPI), the hydrolysates showed greater radical-scavenging ability, Cu^{2+}-chelating effects and improved reducing power. When compared with the standard antioxidants BHA and sodium ascorbate, the WPI hydrolysates were superior in sequestering Cu^{2+} [27]. Antioxidant peptides have also been isolated from several commercial fermented milk products [48]. These studies show that enzyme-hydrolysed whey proteins have the potential as effective antioxidants that can act as a hydrogen donor, metal-ion chelator and radical stabiliser to inhibit lipid oxidation in foods. Whey protein wastes are an environmental nuisance in the dairy industry [49]. The conversion of whey proteins to useful natural antioxidants could therefore be a viable valorisation tool for making value out of whey waste, thereby providing a sustainable solution to the problem of whey protein disposal faced by the dairy industry.

A number of antioxidative peptides have also been identified in meat and egg proteins. These have not been discussed here in detail but are captured in Table 1.

3.1.3 Mycoproteins Derived

Edible Mushrooms: A significantly huge number of mushroom species have been used over the years as food and additives of nutraceutical and pharmaceutical preparations [6]. The antioxidant properties of most edible mushroom species are based on several compounds including proteins and peptides [2].

Sun et al. [31] isolated a potent peptide from fermented medical mushroom (*Ganoderma lucidum*). The major antioxidant fraction was a peptide designated as *G. lucidum* peptide (GLP) and exhibited a number of antioxidative properties. In soybean oil system, GLP showed higher antioxidant activity than BHT. Furthermore, GLP possessed hydroxyl radical-scavenging property (IC50 value of 27.1 μg/ml) and superoxide anion radical quenching abilities in a dose-dependent manner (IC50 value of 25 μg/ml). GLP also displayed substantial antioxidant activity in the rat liver tissue homogenates and mitochondrial membrane peroxidation systems. Auto-haemolysis of rat red blood cells was blocked by GLP in a dose-dependent manner. These demonstrate that the antioxidative properties of *G. lucidum* are mainly due to peptides [31].

3.1.4 Algae Derived

In terms of nutrition, algae are an unconventional source of dietary proteins, oils and antioxidants. However, they act as a useful source of these nutrients particularly to individuals with dietary restrictions such as vegetarians and patients with fish allergies [50].

Spirulina: In recent years, the microalga *Spirulina* is gaining rapid attention as a major health food among researchers and consumers. This is largely as a result of its high protein content (60–70 % by dry weight). *Spirulina* has been shown to possess hypolipidemic, antioxidant and anti-inflammatory bioactivities in a number of preclinical and clinical studies [51], and these activities have been due to the presence of specific bioactive protein and peptide sequences [52]. A number of antioxidative peptides have also been obtained from algae species such as *Chlorella vulgaris* [53], *Chlorella ellipsoidea* [32] and *Pavlova lutheri* [33].

3.2 Effects of Amino Acid on Antioxidative Properties

It has been observed that the biological function of antioxidative peptides is based on the amino acid composition and sequence. In fact, the levels, compositions and properties of free amino acids (FAAs) and peptides have been correlated to the antioxidant activities of protein hydrolysates [54]. Certain structural characteristics such the nature, charge, basicity, aromaticity and hydrophobicity of key amino acids in the primary sequence influence the three-dimensional (3D) conformation of a peptide and therefore affect the biological function [10, 55]. A strong structure-function relationship can be used to explain the antioxidative properties of most amino acids. Met and Cys possess a thiol group each, Tyr and Phe possess phenolic groups, while His has an imidazole ring which has been shown to be responsible for certain antioxidant properties involving chelating and lipid radical-trapping abilities [56]. Summarised in Table 2 are key roles played by some amino acids in peptides with antioxidative properties.

Table 2 Effects of amino acid on antioxidative properties

Amino acid type	Examples and effect
Electrically charged amino acids	H: imidazole group of histidine increases metal-ion chelation, scavenging of hydroxyl radical and quenching of active oxygen radicals
Hydrophobic amino acids	W: has scavenging ability
	V, L, G: when present at about 55 % of the sequence, they increase solubility of peptides in lipid phase. Peptides are therefore able to inhibit lipid peroxidation at the water-lipid interface
Sulphur-containing amino acids	C: acts as precursors for the synthesis of glutathione (γ-L-glutamyl-L-cysteinylglycine)
Acid amino acids	E, D: change redox cycling capacity and improve metal-ion binding
Aromatic amino acids	F, Y: aromatic rings enables the chelation of pro-oxidant metal ions

Adapted from [22, 56, 57]

3.3 Perspectives in the Production of Antioxidative Peptides

As discussed earlier, bioactive peptides are obtainable from protein biomolecules which in themselves do not trigger any physiological response but are only made ‘bioactive’ after enzymatic hydrolysis. Direct wet production of bioactive peptides can be due to gastrointestinal digestion, in vitro enzymatic digestion and whole cell fermentation of food proteins [10] (see Fig. 1).

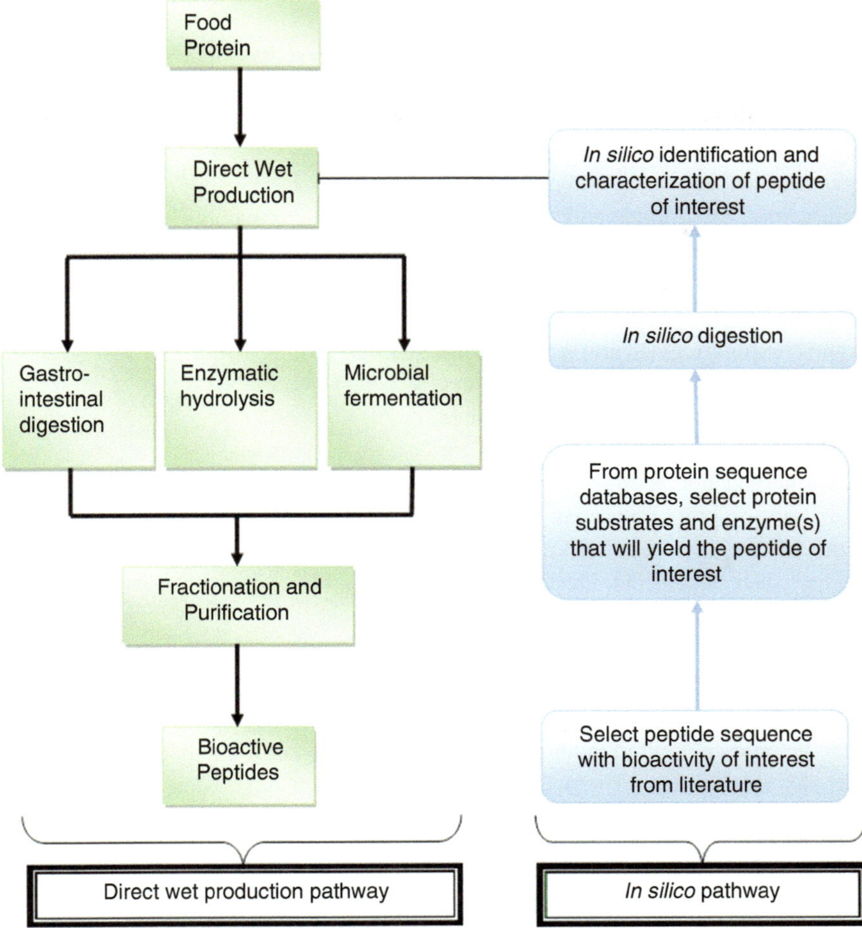

Fig. 1 Production of bioactive peptides

A growing body of evidence suggests that bioactive peptides are produced naturally from dietary proteins during gastrointestinal transit of the protein. However, the demerit of this production pathway is that the generation of bioactive peptides is uncontrolled. Also, peptides generated may be insufficient to trigger the necessary physiological response in adult humans [58].

The production of bioactive peptides by the action of enzymes on whole proteins is the most popular and widely studied approach. Enzymes employed include pepsin, trypsin, chymotrypsin, pancreatin, alcalase and thermolysin. Enzymes from bacterial and fungal sources have also been utilised in the production of bioactive peptides from various proteins [11, 46]. The production of bioactive peptides by exploiting the proteolytic properties of microbial proteases has been argued as cost-effective, economic and scalable [10, 16].

Microbial fermentation is also a widely used pathway for the production of bioactive peptides. This is because several microbial species, particularly the dairy starters, are highly proteolytic and therefore can generate peptides during the fermentation of proteins in dairy products such as cheese and yogurt. The proteolytic system of several lactic acid bacteria has been studied [59, 60], and this presents a plethora of live proteolytic microorganisms and their proteolytic enzymes which can be used in peptide production from food proteins [10].

In recent years, a number of in silico and 'omic' techniques have been employed in the production and studies of bioactive food protein peptides. Some of these 'omic' tools that are of primary interest to the study and applications of bioactive peptides derived from food proteins are nutrigenomics, food proteomics and food peptidomics. Advancement in 'omic' techniques has been served greatly by the amalgamation of high-throughput mass-spectrometry-based methodologies [61] and computational achievements in the development of new algorithms to analyse huge and bulky data in a quick and effective manner. These computational peptidomics and in silico techniques allow bioactive peptides of interest to be predicted from food proteins of known amino acid sequence by using enzymes of known catalytic specificity that will yield the peptide of interest (Fig. 1). Another area of advancement is the use of models such as the quantitative structure activity relationship (QSAR). QSAR works on the principle that the biological activity of a molecule can be predicted, often mathematically, based on its physicochemical or structural properties [62–65]. In silico, prediction and QSAR have been used to identify bioactive peptides from several common food proteins [66, 67].

4 Antioxidant Capacity Evaluation

In many ways, the estimation of antioxidant capacity of biomolecules is based on the biochemical pathways or mechanism of action exhibited by antioxidants. It can be observed from Table 1 that a large number of different methods are used to evaluate antioxidant capacity. This diversity makes it difficult to compare the antioxidant properties of one system with another. Also, in complex systems like foods or cells, no single assay is able to accurately reflect the mechanism of action of all radicals and antioxidants [57].

In most studies, an antioxidant assay could be one of two approaches. The first involves measuring the extent to which the preformed free radical is scavenged. The second approach involves estimating the presence of antioxidant against the generation of the radical [68]. Several techniques and assays have been used to estimate antioxidant properties (Table 3), but it must be borne in mind that specific methods have not yet been developed to determine the antioxidative capacity of peptides [20]. A number of other methods have also been discussed [69].

Table 3 Some antioxidant estimation assays

Assay approach	Antioxidant estimation assays
Free radical scavenging	Oxygen radical absorbance capacity (ORAC) assay
	Hydroxyl radical-scavenging activity (HRSA) assay
	Superoxide radical-scavenging activity (SRSA) assay
	2,2-Diphenyl-1-picrylhydrazyl (DPPH) radical-scavenging activity assay
	2,2′-Azinobis (3-ethylbenzothiazoline-6-sulfonic acid) (ABTS)
Inhibition of free radical generation	Fe^{2+} chelating (FC) activity
	Copper-chelating (CC) activity assay
	Lipid peroxidation inhibition activity (LPIA) assay
	Ferric reducing/antioxidant power (FRAP) assay
	Thiobarbituric acid reactive substances (TBARS) assay

5 Potential Applications of Antioxidative Peptides

Broadly, since antioxidative peptides are of 'natural' origin, they can be used as functional ingredients in food, nutraceutical, pharmaceutical and cosmeceutical products, where they will help to control deterioration caused by oxidation and also control oxidative stress in living body systems.

5.1 Antioxidative Peptides in Food Industry

The potential for antioxidative peptides to contribute to a healthier nutrition via ingesting has been widely discussed in the scientific community and is the subject of interest in the context of health-promoting functional foods [46]. There is an increasing belief in the benefits of using natural approaches for disease control and prevention. The result is a rapid growth in commercial markets for the sale of dietary supplements and functional foods. This is evidenced by the fact that the marketing of food products with health claims is a well-established industry in most developed countries. There is a booming market for antihypertensive and antihypercholesterolaemic peptides through products like BioZate (USA), Festivo (Finland), PeptoPro (Netherlands) and other products classified as Food for Specified Health Uses (FOSHU, Japan). However, no such strides have been made in the investment and marketing of antioxidative peptides [57]. The absence of large-scale production and purification technologies for bioactive peptides is another limiting factor to the commercialisation of bioactive peptides [18]. To circumvent this, whole protein hydrolysates with unpurified bioactive peptide mixtures are used to fortify functional foods. However, for the purposes of pharmaceutical applications, purified peptides must be used [57].

5.2 Antioxidative Peptides as Pharmaceutical and Dermatological Products

Proteins and peptides are native in living body systems where they are responsible for a host of physiological functions. The diverse physiological roles of food peptides make them suitable candidates for the development of therapeutic agents. In purified forms, food-derived peptides can be used as active pharmaceutical ingredients (APIs) in therapeutic products [16]. The rationale for the use of bioactive peptides in therapeutic and dermatological products hinges on the fact that these peptides are very diverse in nature and have a wide spectrum of therapeutic action. They have high absorption profile and biological activities, have high specificities, are low in toxicity and also have 'generally regarded as safe' (GRAS) status [70].

6 Some Research Challenges

It has been argued that the use of natural antioxidative peptides should be encouraged due to their toxicological safety (compared with synthetic compounds) and the better intestinal absorption profile. However, before this can be achieved, a number of research-related bottlenecks must be removed.

6.1 Allergenicity

One major concern associated with the use of antioxidants from food proteins is the potential for allergenic reactions – especially when the peptides are obtained from dairy, soy, nuts and eggs. All the constituents that are responsible for both pollen and food allergies are proteinaceous in nature. Also, several native proteins (such as cow milk) act as precursors for both bioactive peptides and allergenic peptides – giving both toxic and non-toxic effects. This may account for the reason why cow milk constitutes the leading cause of food allergies in infants and young children under 3 years [71, 72]. It has also been shown that some protein hydrolyses can also be considered as allergens because they may retain part of the allergenicity of the native protein [73]. It is therefore important that food-derived bioactive peptides be subjected to comprehensive safety assessment in order to identify any potential cytotoxic and allergenic properties.

6.2 Safety and Undesirable Organoleptic Properties of Peptide-Fortified Foods

Another issue of concern is the potential of peptides to impart undesirable organoleptic and functional properties such as bitterness and changes in product textures and colour. Aside the production of bioactive peptides, enzymatic treatment of most food proteins also results in the undesirable outcome of bitter peptides [74]. Peptides could also alter the texture (i.e. viscosity or gelation) and colour (due to Maillard reaction) of the final product [19]. These may be undesirable in functional food products that may need to be fortified with bioactive peptides. Further, certain products of Maillard reaction have been shown to be potentially carcinogenic [75]. Other factors that have impact on the biological efficacy of antioxidative peptides and therefore need to be studied and monitored include activity at low concentration, compatibility with food and nutraceutical matrices, bioavailability, fate of the peptides during gastrointestinal transit as well as permeability through cellular membranes [57].

7 Conclusion and Future Projections

Free radical damage contributes to the aetiology and aggravation of many chronic health conditions and ageing. Antioxidants have been used to control free radical formation and propagation, thereby controlling free radical-induced tissue damage. Although synthetic antioxidants have a long history of use in food and other systems, they are being phased out due to reports about their effect on human health. On the other hand, some peptides have been isolated from several food proteins and shown to have potent antioxidative activities. These peptides derived from food proteins serve as natural, effective and non-toxic alternatives to synthetic antioxidants. The antioxidative properties of food peptides have been well established in terms of their radical-scavenging properties, oxidant ion chelation properties and free radical production inhibition. These peptides therefore have the prospects of being incorporated as ingredients in functional foods, nutraceuticals and pharmaceuticals where their biological activities may assist in the control and prevention of diseases caused by oxidative stress and free radical damage. However, further insightful research on the safety, undesirable side reactions, stability and allergenicity is needed to firmly establish their therapeutic potency and justify their future use as ingredients in food and pharmaceutical products.

References

1. Devasagayam TP, Tilak JC, Boloor KK et al (2004) Free radicals and antioxidants in human health: current status and future prospects. J Assoc Physicians India 52:794–804
2. Stajic M, Vukojevic J, Knezevic A et al (2013) Antioxidant protective effects of mushroom metabolites. Curr Top Med Chem 13:2660–2676
3. Betteridge DJ (2000) What is oxidative stress? Metabolism 49:3–8
4. Ramalingam M, Kim S-J (2012) Reactive oxygen/nitrogen species and their functional correlations in neurodegenerative diseases. J Neural Transm 119:891–910
5. Patel VP, Chu CT (2011) Nuclear transport, oxidative stress, and neurodegeneration. Int J Clin Exp 4:215–229
6. Ferreira IC, Barros L, Abreu RM (2009) Antioxidants in wild mushrooms. Curr Med Chem 16:1543–1560
7. Lobo V, Patil A, Phatak A, Chandra N (2010) Free radicals, antioxidants and functional foods: impact on human health. Pharmacogn Rev 4:118–126
8. Papas AM (1999) Diet and antioxidant status. Food Chem Toxicol 37:999–1007
9. Hirose M, Takesada Y, Tanaka H et al (1998) Carcinogenicity of antioxidants BHA, caffeic acid, sesamol, 4-methoxyphenol and catechol at low doses, either alone or in combination, and modulation of their effects in a rat medium-term multi-organ carcinogenesis model. Carcinogenesis 19:207–212
10. Korhonen H, Pihlanto A (2006) Bioactive peptides: production and functionality. Int Dairy J 16:945–960
11. Korhonen H (2009) Milk-derived bioactive peptides: from science to applications. J Funct Foods 1:177–187
12. Gibbs B (2004) Production and characterization of bioactive peptides from soy hydrolysate and soy-fermented food. Food Res Int 37:123–131
13. Hartmann R, Meisel H (2007) Food-derived peptides with biological activity: from research to food applications. Curr Opin Biotechnol 18:163–169
14. Yang R, Zhang Z, Pei X et al (2009) Immunomodulatory effects of marine oligopeptide preparation from Chum Salmon (*Oncorhynchus keta*) in mice. Food Chem 113:464–470
15. Möller NP, Scholz-Ahrens KE, Roos N et al (2008) Bioactive peptides and proteins from foods: indication for health effects. Eur J Nutr 47:171–182
16. Agyei D, Danquah MK (2011) Industrial-scale manufacturing of pharmaceutical-grade bioactive peptides. Biotechnol Adv 29:272–277
17. Danquah MK, Agyei D (2012) Pharmaceutical applications of bioactive peptides. OA Biotechnol 1(2):5
18. Agyei D, Danquah MK (2012) Rethinking food-derived bioactive peptides for antimicrobial and immunomodulatory activities. Trends Food Sci Technol 23:62–69

19. Elias RJ, Kellerby SS, Decker EA (2008) Antioxidant activity of proteins and peptides. Crit Rev Food Sci Nutr 48:430–441
20. Samaranayaka AGP, Li-Chan ECY (2011) Food-derived peptidic antioxidants: a review of their production, assessment, and potential applications. J Funct Foods 3:229–254
21. Scalbert A, Johnson IT, Saltmarsh M (2005) Polyphenols: antioxidants and beyond. Am J Clin Nutr 81:215S–217S
22. Park SY, Lee J-S, Baek H-H et al (2010) Purification and characterization of antioxidant peptides from soy protein hydrolysate. J Food Biochem 34:120–132
23. Tang X, He Z, Dai Y et al (2009) Peptide fractionation and free radical scavenging activity of zein hydrolysate. J Agric Food Chem 58:587–593
24. Zhang J, Zhang H, Wang L et al (2010) Isolation and identification of antioxidative peptides from rice endosperm protein enzymatic hydrolysate by consecutive chromatography and MALDI-TOF/TOF MS/MS. Food Chem 119:226–234
25. Saiga A, Tanabe S, Nishimura T (2003) Antioxidant activity of peptides obtained from porcine myofibrillar proteins by protease treatment. J Agric Food Chem 51:3661–3667
26. Manso M, Miguel M, Even J et al (2008) Effect of the long-term intake of an egg white hydrolysate on the oxidative status and blood lipid profile of spontaneously hypertensive rats. Food Chem 109:361–367
27. Peng X, Kong B, Xia X et al (2010) Reducing and radical-scavenging activities of whey protein hydrolysates prepared with Alcalase. Int Dairy J 20:360–365
28. Mao X-Y, Cheng X, Wang X et al (2011) Free-radical-scavenging and anti-inflammatory effect of yak milk casein before and after enzymatic hydrolysis. Food Chem 126:484–490
29. Jung W-K, Rajapakse N, Kim S-K (2005) Antioxidative activity of a low molecular weight peptide derived from the sauce of fermented blue mussel, *Mytilus edulis*. Eur Food Res Technol 220:535–539
30. Sampath Kumar NS, Nazeer RA, Jaiganesh R (2012) Purification and identification of antioxidant peptides from the skin protein hydrolysate of two marine fishes, horse mackerel (*Megalaspis cordyla*) and croaker (*Otolithes ruber*). Amino Acids 42:1641–1649
31. Sun J, He H, Xie BJ (2004) Novel antioxidant peptides from fermented mushroom *Ganoderma lucidum*. J Agric Food Chem 52:6646–6652
32. Ko S-C, Kim D, Jeon Y-J (2012) Protective effect of a novel antioxidative peptide purified from a marine *Chlorella ellipsoidea* protein against free radical-induced oxidative stress. Food Chem Toxicol 50:2294–2302
33. Ryu B, Kang K-H, Ngo D-H et al (2012) Statistical optimization of microalgae Pavlova lutheri cultivation conditions and its fermentation conditions by yeast, *Candida rugopelliculosa*. Bioresour Technol 107:307–313
34. Sheih IC, Fang TJ, Wu T-K et al (2009) Anticancer and antioxidant activities of the peptide fraction from algae protein waste. J Agric Food Chem 58:1202–1207
35. Mu MZT-H, Sun M-J (2012) Sweet potato protein hydrolysates: antioxidant activity and protective effects on oxidative DNA damage. Int J Food Sci Technol 47:2304–2310
36. Huang W-Y, Davidge ST, Wu J (2012) Bioactive natural constituents from food sources – potential use in hypertension prevention and treatment. Crit Rev Food Sci Nutr 53:615–630
37. Saavedra L, Hebert EM, Minahk C et al (2013) An overview of "omic" analytical methods applied in bioactive peptide studies. Food Res Int 54:925–934
38. Roblet C, Amiot J, Lavigne C et al (2012) Screening of in vitro bioactivities of a soy protein hydrolysate separated by hollow fiber and spiral-wound ultrafiltration membranes. Food Res Int 46:237–249
39. Sarmadi BH, Ismail A (2010) Antioxidative peptides from food proteins: a review. Peptides 31:1949–1956
40. Wu Y, Tian Q, Li L et al (2013) Inhibitory effect of antioxidant peptides derived from *Pinctada fucata* protein on ultraviolet-induced photoaging in mice. J Funct Food 5:527–538
41. Chalamaiah M, Kumar BD, Hemalatha R et al (2012) Fish protein hydrolysates: proximate composition, amino acid composition, antioxidant activities and applications: a review. Food Chem 135:3020–3038
42. Ngo DH, Ryu B, Kim SK (2014) Active peptides from skate (*Okamejei kenojei*) skin gelatin diminish angiotensin-I converting enzyme activity and intracellular free radical-mediated oxidation. Food Chem 143:246–255
43. Klompong V, Benjakul S, Kantachote D et al (2007) Antioxidative activity and functional properties of protein hydrolysate of yellow stripe trevally (*Selaroides leptolepis*) as influenced by the degree of hydrolysis and enzyme type. Food Chem 102:1317–1327
44. Suetsuna K, Ukeda H, Ochi H (2000) Isolation and characterization of free radical scavenging activities peptides derived from casein. J Nutr Biochem 11:128–131
45. Kimball SR, Jefferson LS (2006) Signaling pathways and molecular mechanisms through which branched-chain amino acids mediate translational control of protein synthesis. J Nutr 136:227S–231S
46. Korhonen HJ, Marnila P (2013) In: Park YW, Haenlein FW (eds) Milk bioactive proteins and peptides. Wiley, West Sussex, pp 148–171
47. Peña-Ramos EA, Xiong YL, Arteaga GE (2004) Fractionation and characterisation for antioxidant activity of hydrolysed whey protein. J Agric Food Chem 84:1908–1918
48. Hernández-Ledesma B, Miralles B, Amigo L et al (2005) Identification of antioxidant and ACE-inhibitory peptides in fermented milk. J Agric Food Chem 85:1041–1048
49. Smithers GW (2008) Whey and whey proteins-'from gutter-to-gold'. Int Dairy J 18:695–704
50. Kovač DJ, Simeunović JB, Babić OB et al (2013) Algae in food and feed. Food Feed Res 20:21–32
51. Deng R, Chow T-J (2010) Hypolipidemic, antioxidant, and antiinflammatory activities of microalgae *Spirulina*. Cardiovasc Ther 28:e33–e45

52. Guan XY, Zhang WJ, Zhang XW et al (2009) A potent anti-oxidant property: fluorescent recombinant α-phycocyanin of *Spirulina*. J Appl Microbiol 106:1093–1100
53. Sheih IC, Wu T-K, Fang TJ (2009) Antioxidant properties of a new antioxidative peptide from algae protein waste hydrolysate in different oxidation systems. Bioresour Technol 100:3419–3425
54. Wu H-C, Chen H-M, Shiau C-Y (2003) Free amino acids and peptides as related to antioxidant properties in protein hydrolysates of mackerel (*Scomber austriasicus*). Food Res Int 36:949–957
55. Hancock REW, Sahl H-G (2006) Antimicrobial and host-defense peptides as new anti-infective therapeutic strategies. Nat Biotechnol 24:1551–1557
56. Je J-Y, Park P-J, Kim S-K (2005) Antioxidant activity of a peptide isolated from Alaska pollack (*Theragra chalcogramma*) frame protein hydrolysate. Food Res Int 38:45–50
57. Freitas AC, Andrade JC, Silva FM et al (2013) Antioxidative peptides: trends and perspectives for future research. Curr Med Chem 20:4575–4594
58. Gauthier SF, Pouliot Y, Saint-Sauveur D (2006) Immunomodulatory peptides obtained by the enzymatic hydrolysis of whey proteins. Int Dairy J 16:1315–1323
59. Agyei D, Potumarthi R, Danquah MK (2013) In: Kim S-K (ed) Production of *Lactobacilli* proteinases for the manufacture of bioactive peptides: part I – upstream processes. Wiley, Sussex, pp 207–229
60. Agyei D, Potumarthi R, Danquah MK (2013) In: Kim S-K (ed) Production of *Lactobacilli* proteinases for the manufacture of bioactive peptides: part II – downstream processes. Wiley, Sussex, pp 231–251
61. del Carmen Mena M, Albar JP (2013) Next generation instruments and methods for proteomics. In: Cifuentes A (ed) Foodomics: advanced mass spectrometry in modern food science and nutrition. Wiley, Hoboken, pp 15–67
62. Nakai S, Li-Chan E (1993) Recent advances in structure and function of food proteins: QSAR approach. Crit Rev Food Sci Nutr 33:477–499
63. Pripp AH, Isaksson T, Stepaniak L et al (2005) Quantitative structure activity relationship modelling of peptides and proteins as a tool in food science. Trends Food Sci Technol 16:484–494
64. Hansch C, Leo A, Hoekman DH (1995) Exploring QSAR: fundamentals and applications in chemistry and biology. American Chemical Society, Washington, DC
65. Carrasco-Castilla J, Hernández-Álvarez A, Jiménez-Martínez C et al (2012) Use of proteomics and peptidomics methods in food bioactive peptide science and engineering. Food Eng Rev 4: 224–243
66. Majumder K, Wu J (2010) A new approach for identification of novel antihypertensive peptides from egg proteins by QSAR and bioinformatics. Food Res Int 43:1371–1378
67. Gu Y, Majumder K, Wu J (2011) QSAR-aided *in silico* approach in evaluation of food proteins as precursors of ACE inhibitory peptides. Food Res Int 44:2465–2474
68. Re R, Pellegrini N, Proteggente A et al (1999) Antioxidant activity applying an improved ABTS radical cation decolorization assay. Free Radic Biol Med 26:1231–1237
69. Karadag A, Ozcelik B, Saner S (2009) Review of methods to determine antioxidant capacities. Food Anal Method 2:41–60
70. Zompra AA, Galanis AS, Werbitzky O et al (2009) Manufacturing peptides as active pharmaceutical ingredients. Futur Med Chem 1:361–377
71. Ludman S, Shah N, Fox AT (2013) Managing cows' milk allergy in children. BMJ (Clin Res Ed) 347:f5424
72. Koletzko S, Niggemann B, Arato A et al (2012) Diagnostic approach and management of cow's-milk protein allergy in infants and children: ESPGHAN GI Committee practical guidelines. J Pediatr Gastroenterol Nutr 55:221–229
73. Hartmann R, Wal JM, Bernard H et al (2007) Cytotoxic and allergenic potential of bioactive proteins and peptides. Curr Pharm Des 13:897–920
74. Kim H-O, Li-Chan ECY (2006) Quantitative structure – activity relationship study of bitter peptides. J Agric Food Chem 54:10102–10111
75. Friedman M (2005) In: Friedman M, Mottram D (eds) Biological effects of maillard browning products that may affect acrylamide safety in food. Springer, New York, pp 135–156